Plant Biochemistry

Third edition

Plant Biochemistry

Hans-Walter Heldt
in cooperation with Fiona Heldt

An update and translation of the German third edition

AMSTERDAM • BOSTON • HEIDELBERG • LONDON
NEW YORK • OXFORD • PARIS • SAN DIEGO
SAN FRANCISCO • SINGAPORE • SYDNEY • TOKYO

Acquisition Editor David Cella
Project Manager Justin Palmeiro
Editorial Coordinator Kelly Sonnack
Marketing Manager Linda Beattie
Cover Design Cate Barr
Composition SNP Best-set Typesetter Ltd., Hong Kong
Printer Courier

Elsevier Academic Press
200 Wheeler Road, Burlington, MA 01803, USA
525 B Street, Suite 1900, San Diego, California 92101-4495, USA

84 Theobald's Road, London WC1X 8RR, UK

This book is printed on acid-free paper. ∞

Library of Congress Cataloging-in-Publication Data
Application Submitted

British Library Cataloguing in Publication Data
A catalogue record for this book is available from the British Library

ISBN: 0-12-088391-0

For all information on all Elsevier Academic Press Publication
visit our Web site at www.books.elsevier.com

Printed in the United States of America
04 05 06 07 08 09 9 8 7 6 5 4 3 2 1

Dedicated to my teacher Martin Klingenberg

Preface

This textbook is written for students and is the product of more than three decades of my teaching experience. It intends to give a broad but concise overview of the various aspects of plant biochemistry, including molecular biology. I have attached importance to an easily understood description of the principles of metabolism but also have restricted the content in such a way that a student is not distracted by unnecessary details. In view of the importance of plant biotechnology, industrial applications of plant biochemistry have been pointed out wherever appropriate. Thus, special attention has been given to the generation and utilization of transgenic plants.

Since there are many excellent textbooks on general biochemistry, I have deliberately omitted dealing with elements such as the structure and function of amino acids, carbohydrates, and nucleotides; the function of nucleic acids as carriers of genetic information; and the structure and function of proteins and the basis of enzyme catalysis. I have dealt with topics of general biochemistry only when it seemed necessary for enhancing understanding of the problem in hand. Thus, this book is, in the end, a compromise between a general textbook and a specialized textbook.

This book is a translation of the third German edition. Compared to the previous English edition, it has been fully revised to take into account the progress in the field. In particular, the chapter about phytohormones has undergone extensive changes. As a year has passed since the third German edition was completed, the text has been thoroughly updated, taking into account the rapid development of the discipline in 2003.

I am very grateful to the many colleagues who, through discussions and the reprints they have sent me on their special fields, have given me the information that made it possible for me to write the various editions of this book. It was especially helpful for me that the colleagues in the following list went to great trouble to read critically one or more chapters, to draw my attention to mistakes, and to suggest improvements for which I am particularly grateful. My special thanks go to my coworker and wife, Fiona Heldt. Without her intensive support, it would not have been possible for me to write this book or to prepare the English version.

I have tried to eradicate as many mistakes as possible, probably not with complete success. I am therefore grateful for any suggestions and comments.

Hans-Walter Heldt
Göttingen, January 2004

Thanks to the following colleagues for their suggestions after looking through one or more chapters of the various editions of this book:

Prof. Jan Anderson, Canberra, Australia
Prof. John Andrews, Canberra, Australia
Prof. Tom ap Rees, Cambridge, Great Britain
Prof. Kozi Asada, Uji, Kyoto, Japan
Dr. Tony Ashton, Canberra, Australia
Prof. Murray Badger, Canberra, Australia
Prof. Peter Böger, Konstanz, Germany
Dr. Sieglinde Borchert, Göttingen, Germany
Prof. Peter Brandt, Berlin, Germany
Prof. Axel Brennicke, Ulm, Germany
Prof. David Day, Perth, Australia
Prof. Karl-Josef Dietz, Bielefeld, Germany
Dr. Wolfgang Dröge-Laser, Göttingen, Germany
Prof. Gerry Edwards, Pullman, Washington, USA
Prof. Walter Eschrich, Göttingen, Germany
Prof. Ivo Feußner, Göttingen, Germany
Prof. Ulf-Ingo Flügge, Cologne, Germany
Prof. Margrit Frentzen, Aachen, Germany
Prof. Wolf Frommer, Tübingen, Germany
Dr. R.T. Furbank, Canberra, Australia
Prof. Christine Gatz, Göttingen, Germany
Prof. Curtis V. Givan, Durham NH, USA
Prof. Gowindjee, Urbana, Ill. USA
Prof. Jan Eiler Graebe, Göttingen, Germany
Prof. Peter Gräber, Stuttgart, Germany
Prof. Dietrich Gradmann, Göttingen, Germany
Prof. Erwin Grill, Munich, Germany
Dr. Bernhard Grimm, Gatersleben, Germany
Prof. Wolfgang Haehnel, Freiburg, Germany
Dr. Iris Hanning, Munich, Germany
Dr. Marshall D. Hatch, Canberra, Australia
Prof. Ulrich Heber, Würzburg, Germany

Prof. Dieter Heineke, Göttingen, Germany
Dr. Klaus-Peter Heise, Göttingen, Germany
Prof. E. Heinz, Hamburg, Germany
Dr. Frank Hellwig, Göttingen, Germany
Dr. Gieselbert Hinz, Göttingen, Germany
Prof. Steven C. Huber, Raleigh, USA
Dr. Graham Hudson, Canberra, Australia
Dr. Colin Jenkins, Canberra, Australia
Prof. Wolfgang Junge, Osnabrück, Germany
Prof. Werner Kaiser, Würzburg, Germany
Dr. Steven King, Canberra, Australia
Prof. Martin Klingenberg, Munich, Germany
Prof. Gotthard H. Krause, Düsseldorf, Germany
Dr. Silke Krömer, Osnabrück, Germany
Prof. Werner Kühlbrandt, Heidelberg, Germany
Dr. Toni Kutchan, Munich, Germany
Prof. Hartmut Lichtenthaler, Karlsruhe, Germany
Dr. Gertrud Lohaus, Göttingen, Germany
Prof. Ulrich Lüttge, Darmstadt, Germany
Dr. John Lunn, Canberra, Australia
Prof. Enrico Martinoia, Neuchatel, Switzerland
Prof. Hartmut Michel, Frankfurt, Germany
Prof. K. Müntz, Gatersleben, Germany
Prof. Walter Neupert, Munich, Germany
Prof. Lutz Nover, Frankfurt, Germany
Prof. C. Barry Osmond, Canberra, Australia
Dr. Katharina Pawlowski, Göttingen, Germany
Prof. Birgit Piechulla, Rostock, Germany
Prof. Andrea Polle, Göttingen Germany
Prof. Klaus Raschke, Göttingen, Germany
Dr. Günter Retzlaff, Limburger Hof, Germany
Dr. Sigrun Reumann, Göttingen, Germany
Dr. Gerhard Ritte Potsdam, Germany
Prof. Gerhard Röbbelen, Göttingen, Germany
Prof. David Robinson, Heidelberg, Germany
Prof. Eberhard Schäfer, Freiburg, Germany
Prof. Dierk Scheel, Halle, Germany
Prof. Hugo Scheer, Munich, Germany
Prof. Renate Scheibe, Osnabrück, Germany
Prof. Ahlert Schmidt, Hannover, Germany
Dr. Karin Schott, Göttingen, Germany
Dr. Ulrich Schreiber, Würzburg, Germany

Dr. Danja Schünemann, Aachen, Germany
Prof. Gernot Schultz, Hannover, Germany
Prof. Jens D. Schwenn, Bochum, Germany
Prof. Jürgen Soll, Kiel, Germany
Prof. Martin Steup, Potsdam, Germany
Prof. Heinrich Strotmann, Düsseldorf, Germany
Prof. Gerhard Thiel, Darmstadt, Germany
Prof. Rudolf Tischner, Göttingen, Germany
Prof. Achim Trebst, Bochum, Germany
Prof. Hanns Weiss, Düsseldorf, Germany
Prof. Dietrich Werner, Marburg, Germany
Prof. Peter Westhoff, Düsseldorf, Germany

Contents

Introduction

Plant biochemistry examines the molecular mechanisms of plant life. One of the main topics is photosynthesis, which in higher plants takes place mainly in the leaves. Photosynthesis utilizes the energy of the sun to synthesize carbohydrates and amino acids from water, carbon dioxide, nitrate, and sulfate. Via the vascular system, a major part of these products is transported from the leaves through the stem into other regions of the plant, where they are required, for example, to build up the roots and supply them with energy. Hence, the leaves have been given the name *source*, and the roots the name *sink*. The reservoirs in seeds are also an important group of the sink tissues, and, depending on the species, act as a store for many agricultural products such as carbohydrates, proteins, and fat.

In contrast to animals, plants have a very large surface, often with very thin leaves in order to keep the diffusion pathway for CO_2 as short as possible and to catch as much light as possible. In the finely branched root hairs, the plant has an efficient system for extracting water and inorganic nutrients from the soil. This large surface, however, exposes plants to all the changes in their environment. They must be able to withstand extreme conditions such as drought, heat, cold, or even frost, as well as an excess of radiated light energy. Day after day the leaves must contend with the change between photosynthetic metabolism during the day and oxidative metabolism during the night. Plants encounter these extreme changes in external conditions with an astonishingly flexible metabolism in which a variety of regulatory processes take part. Since plants cannot run away from their enemies, they have developed a whole arsenal of defense substances to protect themselves from being eaten.

Plant agricultural production is the basis for human nutrition. Plant gene technology, which can be regarded as a section of plant biochemistry, makes a contribution to combat the impending global food shortage resulting from the enormous growth of the world population. The use of environmentally compatible herbicides and protection against viral or fungal infestation by means of gene technology is of great economic importance. Plant biochemistry also is instrumental in breeding productive varieties of crop plants.

Plants are the source of important industrial raw material such as fat and starch, but they also are the basis for the production of pharmaceuticals. It is to be expected that in the future gene technology will lead to the extensive use of plants as a means of producing sustainable raw material for industrial purposes.

The aim of this short list is to show that plant biochemistry is not only an important field of basic science explaining the molecular function of a plant, but also an applied science that, now at a revolutionary phase of its development, is in a position to contribute to the solution of important economic problems.

To reach this goal, it is necessary that sectors of plant biochemistry such as bioenergetics, the biochemistry of intermediary metabolism and the secondary plant compounds, as well as molecular biology and other sections of plant sciences, such as plant physiology and the cell biology of plants, cooperate closely with one another. Only the integration of the results and methods of working of the different sectors of plant sciences can help us to understand how a plant functions and to put this knowledge to economic use. This book describes how this may be achieved.

Since there are already many good general textbooks on biochemistry, the elements of general biochemistry are not considered here, and it is presumed that the reader will obtain the knowledge of general biochemistry from other textbooks.

1

A leaf cell consists of several metabolic compartments

In higher plants photosynthesis occurs mainly in the **mesophyll**, the chloroplast-rich tissue of leaves. Figure 1.1 shows an electron micrograph of a mesophyll cell and Figure 1.2 shows a diagram of the cell structure. The cell contents are surrounded by a **plasma membrane** called the plasmalemma and

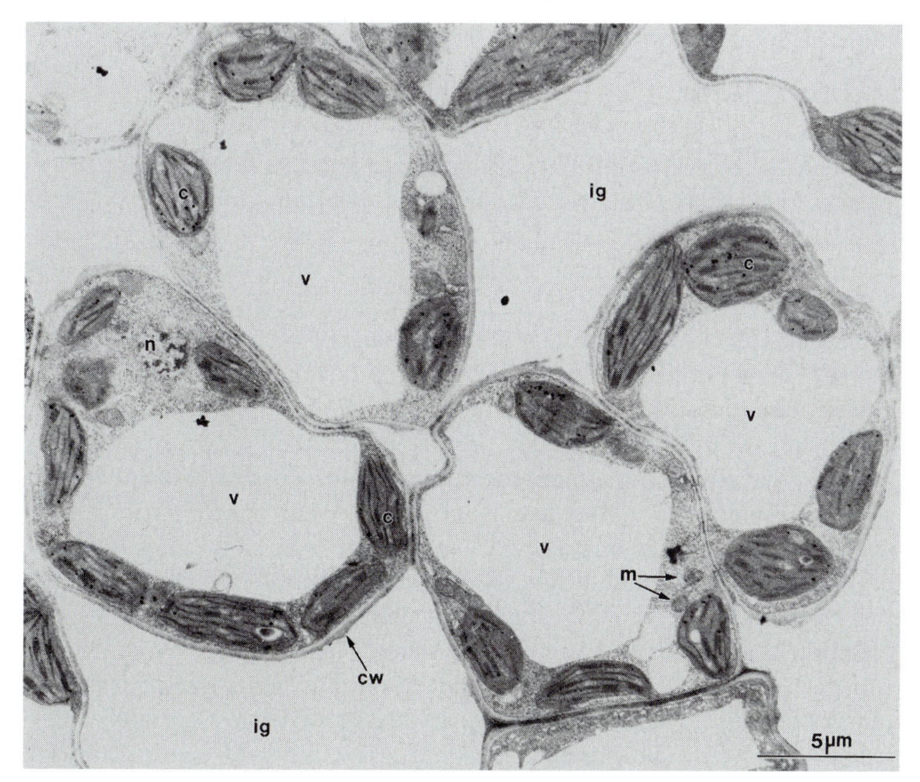

Figure 1.1 Electron micrograph of mesophyll tissue from tobacco. In most cells the large central vacuole is to be seen (v). Between the cells are the intercellular gas spaces (ig), which are somewhat enlarged by the fixation process. c: chloroplast, cw: cell wall, n: nucleus, m: mitochondrion. (By D. G. Robinson, Heidelberg.)

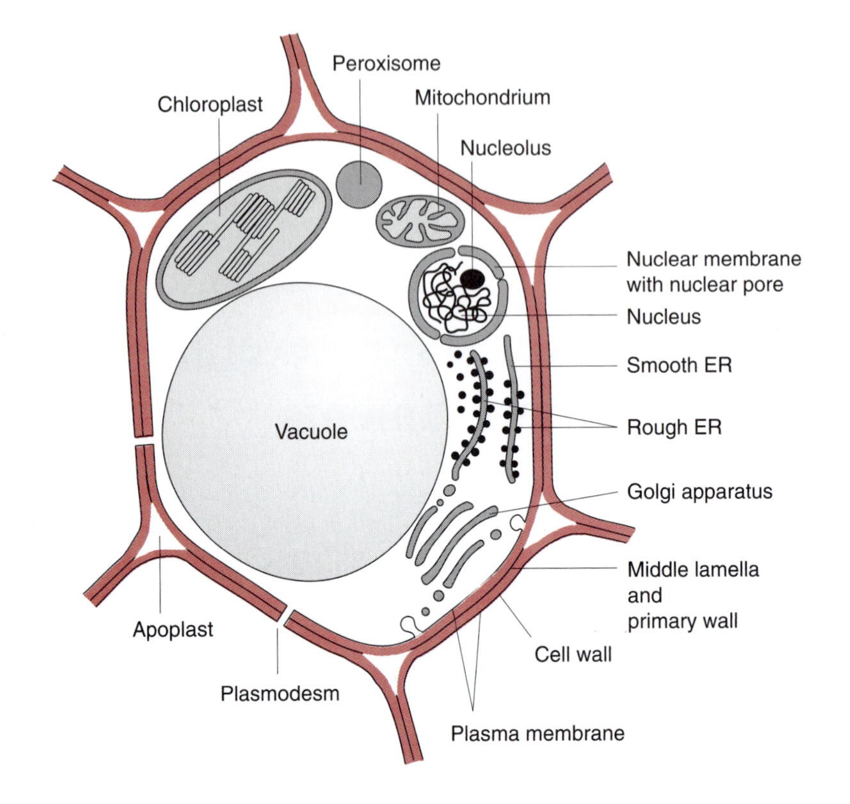

Figure 1.2 Diagram of a mesophyll cell.

are enclosed by a **cell wall**. The cell contains organelles, each with its own characteristic shape, which divide the cell into various compartments (subcellular compartments). Each compartment has specialized metabolic functions, which will be discussed in detail in the following chapters (see also Table 1.1). The largest organelle, the vacuole, usually fills about 80% of the total cell volume. Chloroplasts represent the next largest compartment, and the rest of the cell volume is filled with mitochondria, peroxisomes, the nucleus, the endoplasmic reticulum, the Golgi bodies, and, outside these organelles, the cell plasma, called **cytosol**. In addition, there are oil bodies derived from the endoplasmic reticulum. These oil bodies, which occur in seeds and some other tissues (e.g., root nodules), are storage organelles for triglycerides (see Chapter 15).

The **nucleus** is surrounded by the **nuclear envelope**, which consists of the two membranes of the endoplasmic reticulum. The space between the two membranes is known as the **perinuclear space**. The nuclear envelope is interrupted by **nuclear pores** with a diameter of about 50 nm. The nucleus contains chromatin, consisting of DNA double strands that are stabilized by being bound to basic proteins (histones). The genes of the nucleus are

Table 1.1: Subcellular compartments in a mesophyll cell* and some of their functions

	Percent of the total cell volume	Functions (incomplete)
Vacuole	79	Maintenance of cell turgor, store and waste depository
Chloroplasts	16	Photosynthesis, synthesis of starch and lipids
Cytosol	3	General metabolic compartment, synthesis of sucrose
Mitochondria	0.5	Cell respiration
Nucleus	0.3	Contains the genome of the cell. Reaction site of replication and
Peroxisomes		Reaction site for processes in which toxic transcription intermediates are formed
Endoplasmic Reticulum		Storage of Ca^{++} ions, participation in the export of proteins from the cell and in the transport of proteins into the vacuole
Oil bodies (Oleosomes)		Storage of triacylglycerols
Golgi bodies		Processing and sorting of proteins destined for export from the cells or transport into the vacuole

* Mesophyll cells of spinach; data by Winter, Robinson, and Heldt, 1994.

collectively referred to as the **nuclear genome**. Within the nucleus, usually off-center, lies the nucleolus, where ribosomal subunits are formed. These ribosomal subunits and the messenger RNA formed by transcription of the DNA in the nucleus migrate through the nuclear pores to the ribosomes in the cytosol, the site of protein biosynthesis. The synthesized proteins are distributed between the different cell compartments according to their final destination.

The cell contains in its interior the **cytoskeleton**, which is a three-dimensional network of fiber proteins. Important elements of the cytoskeleton are the **microtubuli** and the **microfilaments**, both macromolecules formed by the aggregation of soluble (globular) proteins. Microtubuli are tubular structures composed of **α** and **β tubuline** momomers. The microtubuli are connected to a large number of different motor proteins that transport bound organelles along the microtubuli at the expense of ATP. Microfilaments are chains of polymerized **actin** that interact with **myosin** to achieve movement. Actin and myosin are the main constituents of the animal muscle. The cytoskeleton has many important cellular functions. It is involved in the spatial organization of the organelles within the cell, enables thermal stability, plays an important role in cell division, and has a function in cell-to-cell communication.

1.1 The cell wall gives the plant cell mechanical stability

The difference between plant cells and animal cells is that plant cells have a cell wall. This wall limits the volume of the plant cell. Water taken up into the cell by osmosis presses the plasma membrane against the inside of the cell wall, thus giving the cell mechanical stability.

The cell wall consists mainly of carbohydrates and proteins

The cell wall of a higher plant is made up of about 90% carbohydrates and 10% proteins. The main carbohydrate constituent is **cellulose**. Cellulose is an unbranched polymer consisting of D-glucose molecules, which are connected to each other by glycosidic (β1·4) linkages (Fig. 1.3A). Each glucose unit is rotated by 180° from its neighbor, so that very long straight chains can be formed with a chain length of 2,000 to 25,000 glucose residues. About 36 cellulose chains are associated by interchain hydrogen bonds to a crystalline lattice structure known as a **microfibril**. These crystalline regions are impermeable to water. The microfibrils have an unusually high tensile strength, are very resistant to chemical and biological degradations, and are in fact so stable that they are very difficult to hydrolyze. However, many bacteria and fungi have cellulose-hydrolyzing enzymes (cellulases). These bacteria can be found in the digestive tract of some animals (e.g., ruminants), thus enabling them to digest grass and straw. It is interesting to note that cellulose is the most abundant organic substance on earth, representing about half of the total organically bound carbon.

Hemicelluloses are also important constituents of the cell wall. They are defined as those polysaccharides that can be extracted by alkaline solutions. The name is derived from an initial belief, which later turned out to be incorrect, that hemicelluloses are precursors of cellulose. Hemicelluloses consist of a variety of polysaccharides that contain, in addition to D-glucose, other

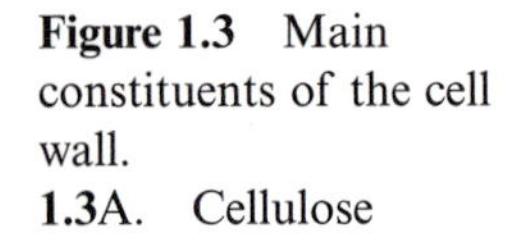

Figure 1.3 Main constituents of the cell wall.
1.3A. Cellulose

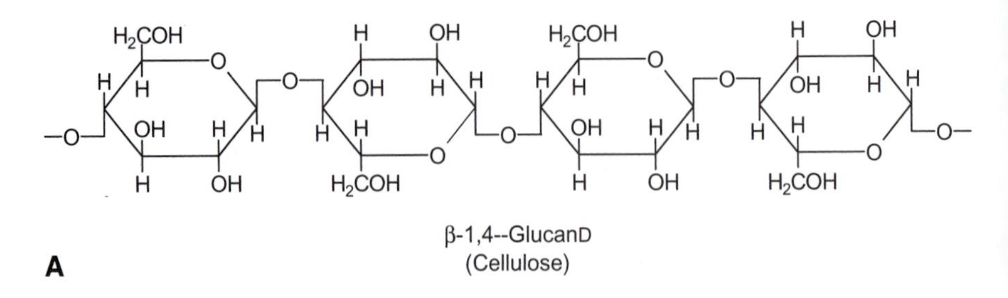

carbohydrates such as the hexoses D-mannose, D-galactose, D-fucose, and the pentoses D-xylose and L-arabinose. Figure 1.3B shows xyloglucan as an example of a hemicellulose. The basic structure is a β-1,4-glucan chain to which xylose residues are bound via (α1·6) glycosidic linkages, which in part are linked to D-galactose and D-fucose. In addition to this, L-arabinose residues are linked to the 2-OH group of the glucose.

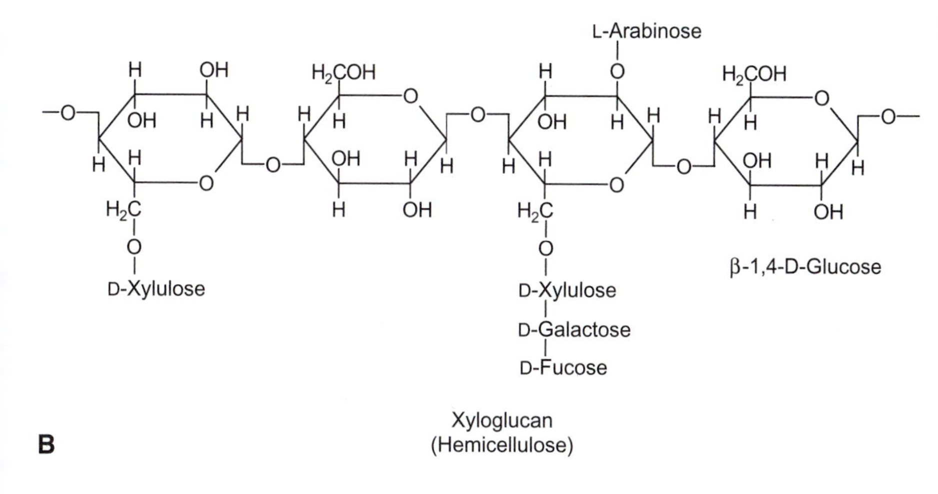

1.3B. A hemicellulose

Another major constituent of the cell wall is **pectin**, a mixture of polymers from sugar acids, such as D-galacturonic acid, which are connected by (α1·4) glycosidic links (Fig. 1.3C). Some of the carboxyl groups are esterified by methyl groups. The free carboxyl groups of adjacent chains are linked by Ca^{++} and Mg^{++} ions (Fig. 1.4). When Mg^{++} and Ca^{++} ions are absent, pectin is a soluble compound. The Ca^{++}/Mg^{++} salt of pectin forms an amorphous, deformable gel that is able to swell. The food industry makes use of this property of pectin when preparing jellies and jams.

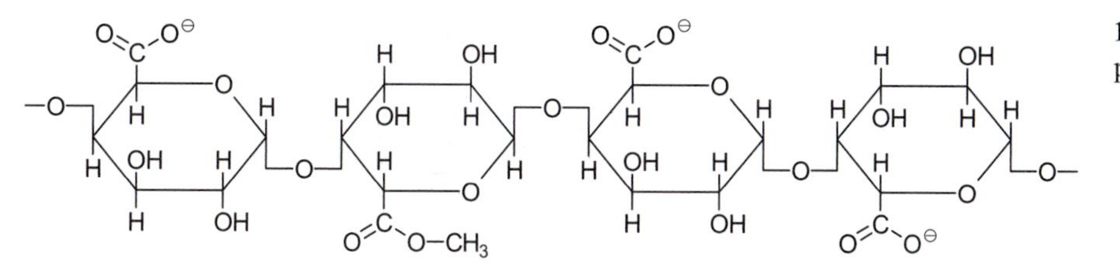

1.3C. Constituent of pectin

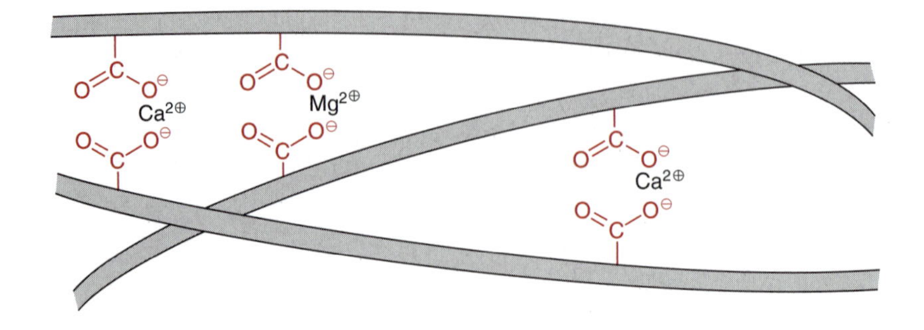

Figure 1.4 Ca^{++} and Mg^{++} ions mediate electrostatic interactions between pectin strands.

The structural proteins of the cell wall are connected by glycosidic linkages to the branched polysaccharide chains and belong to the class of proteins known as **glycoproteins**. The carbohydrate portion of these glycoproteins varies from 50% to over 90%. Cell walls also contain waxes (Chapter 15), cutin, and suberin (Chapter 18).

For a plant cell to grow, the very rigid cell wall has to be loosened in a precisely controlled way. This is facilitated by the protein **expansin**, which occurs in growing tissues of all flowering plants. It probably functions by breaking hydrogen bonds between cellulose microfibrils and cross-linking polysaccharides.

In a monocot plant, the **primary wall** (i.e., the wall initially formed after the growth of the cell) consists of 20% to 30% cellulose, 25% hemicellulose, 30% pectin and, 5% to 10% glycoprotein. It is permeable to water. Pectin makes the wall elastic and, together with the glycoproteins and the hemicellulose, forms the matrix in which the cellulose microfibrils are embedded. When the cell has reached its final size and shape, another layer, the **secondary wall**, which consists mainly of cellulose, is added to the primary wall. The microfibrils in the secondary wall are arranged in a **layered structure** like plywood (Fig. 1.5).

The incorporation of **lignin** in the secondary wall causes the lignification of plant parts and the corresponding cells die, leaving the dead cells with only a supporting function (e.g., for forming the branches and twigs of trees or the stems of herbaceous plants). Section 18.3 describes in detail how lignin is formed by the polymerization of the **phenylpropane derivatives** cumaryl alcohol, coniferyl alcohol, and sinapyl alcohol, resulting in a very solid structure. Dry wood consists of about 30% lignin, 40% cellulose, and 30% hemicellulose. After cellulose, lignin is the most abundant natural substance on earth.

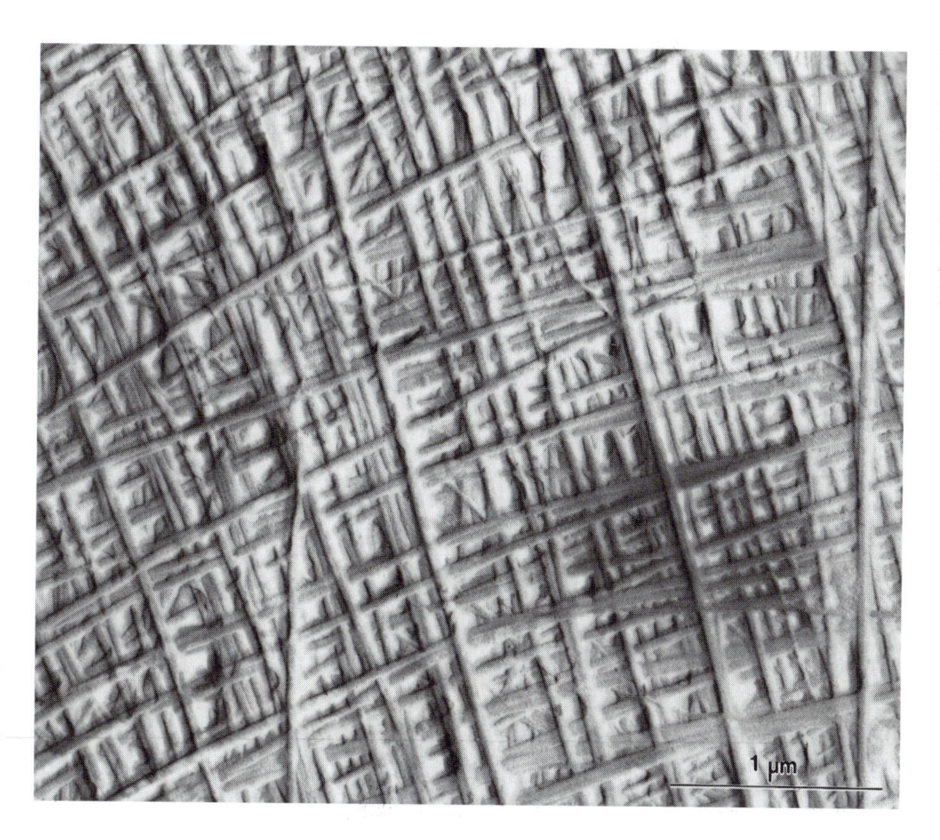

Figure 1.5 Cell wall of the green alga *Oocystis solitaria*. The cellulose microfibrils are arranged in a layer pattern, in which parallel layers are arranged one above the other. Freeze etching. (By D. G. Robinson, Heidelberg.)

Plasmodesmata connect neighboring cells

Neighboring cells are normally connected by **plasmodesmata** thrusting through the cell walls. The plasmodesmata allow mostly the passage of molecules up to a molecular mass of about 800 to 900 Dalton. They are permeable to the various intermediates of metabolism such as soluble sugars, amino acids, and free nucleotides. A single plant cell may contain from 1,000 to more than 10,000 plasmodesmata. These plasmodesmata connect many plant cells to form a single large metabolic compartment where the metabolites in the cytosol can move between the various cells by diffusion. This continuous compartment formed by different plant cells (Fig. 1.6) is called the **symplast**. In contrast, the spaces between cells, which are often continuous, are termed the extracellular space or the **apoplast** (Fig. 1.2).

Figure 1.7 shows a diagram of a plasmodesm. The tubelike opening through the cell wall is lined by the plasma membrane, which is continuous between the neighboring cells. In the interior of this tube there is another tubelike membrane structure, which is part of the endoplasmatic reticulum

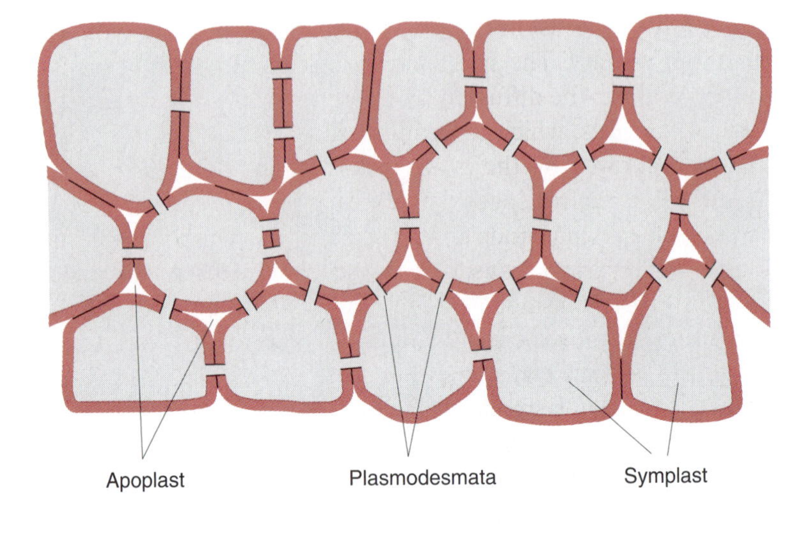

Figure 1.6 Plasmodesmata connect neighboring cells to form a symplast. The extracellular spaces between the cell walls form the apoplast. Schematic representation. Each of the connections shown actually consists of very many neighboring plasmodesmata.

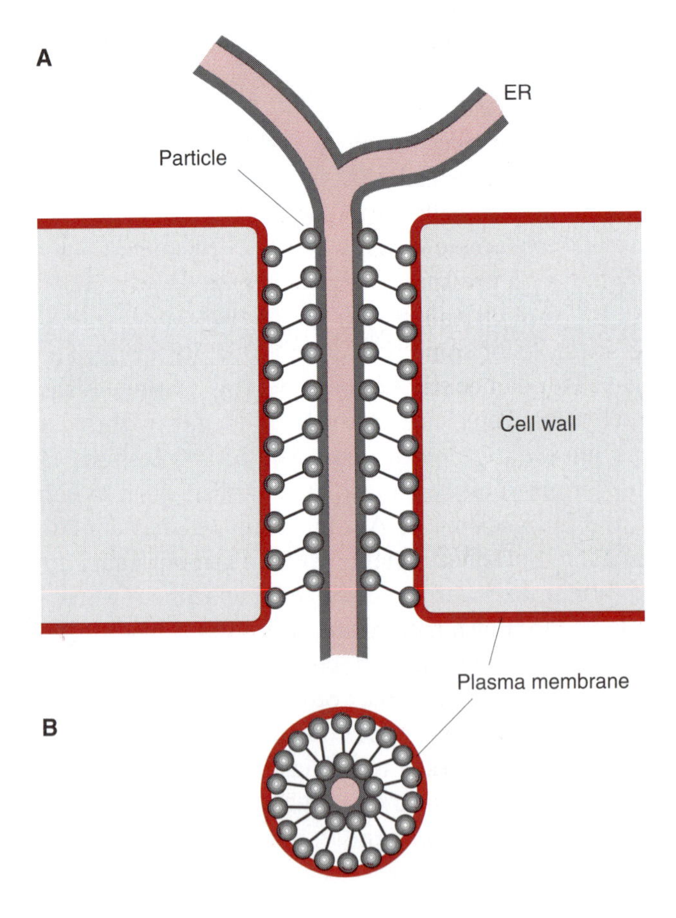

Figure 1.7 Diagram of a plasmodesm. The plasma membrane of the neighboring cells is connected by a tubelike membrane invagination. Inside this tube is a continuation of the endoplasmic reticulum. Embedded in the membrane of the ER and the plasma membrane are protein particles that are connected to each other. The spaces between the particles form the diffusion path of the plasmodesm. It is controversial whether a diffusion between the neighboring cells also takes place via the ER lumen.
A. cross-sectional view of the membrane
B. vertical view

(ER) of the adjacent cells. In this way the ER system of the entire symplast represents a continuity. The space between the plasma membrane and the ER membrane forms the diffusion pathway between the cytosol of adjacent cells. Protein particles, which are connected to each other, are attached to the outer tube formed by the plasma membrane and the ER membrane. It is assumed that the free space between these protein particles determines the aperture of the plasmodesm. A number of plant viruses, including the *Tobacco mosaic virus*, cause the synthesis of **virus movement proteins**, which can alter the plasmodesmata to such an extent that viral nucleic acids bound to the movement protein can slip through. Thus, after infecting a single cell, a virus can spread over the entire symplast. In the widening process of the plasmodesmata by virus movement proteins, the **cytoskeleton** appears to be involved. There are indications that this represents a general transport process of which the viruses take advantage. It is presumed that the cell's own movement proteins, upon the consumption of ATP, facilitate the transfer of macromolecules, such as RNA and proteins, from one cell to the next via the plasmodesmata. In this way, for example, transcription factors might be distributed as signals in a regulated mode via the symplast.

The plant cell wall can be lysed by cellulose and pectin hydrolyzing enzymes obtained from microorganisms. When leaf pieces are incubated with these enzymes, plant cells that have lost their surrounding cell wall can be obtained. These naked cells are called **protoplasts**. Protoplasts, however, are stable only in an **isotonic medium** in which the osmotic pressure corresponds to the osmotic pressure of the cell fluid. In pure water the protoplasts, as they have no cell wall, swell so much that they burst. In appropriate media, the protoplasts of some plants are viable, they can be propagated in cell culture, and they can be stimulated to form a cell wall and even to regenerate a whole new plant.

1.2 Vacuoles have multiple functions

The vacuole is enclosed by a membrane, called a **tonoplast**. The number and size of the vacuoles in different plant cells vary greatly. Young cells contain a larger number of smaller vacuoles but, taken as a whole, occupy only a minor part of the cell volume. When cells mature, the individual vacuoles amalgamate to form a **central vacuole** (Figs. 1.1 and 1.2). The increased volume of the mature cell is due primarily to the enlargement of the vacuole. In cells of storage or epidermal tissues, the vacuole often takes up almost the entire cellular space.

An important function of the vacuole is to maintain **cell turgor**. For this purpose, salts, mainly from inorganic and organic acids, are accumulated in the vacuole. The accumulation of these osmotically active substances draws water into the vacuole, which, in turn, causes the tonoplast to press the protoplasm of the cell against the surrounding cell wall. Plant turgor is responsible for the rigidity of nonwoody plant parts. The plant wilts when the turgor decreases due to lack of water.

Vacuoles have an important function in **recycling** those cellular constituents that are defective or no longer required. Vacuoles contain hydrolytic enzymes for degrading various macromolecules such as proteins, nucleic acids, and many polysaccharides. Structures, such as mitochondria, can be transferred by endocytosis to the vacuole and are digested there. For this reason one speaks of **lytic vacuoles**. The resulting degradation products, such as amino acids and carbohydrates are made available to the cell. This is especially important during **senescence** (see section 19.5) when prior to abscission, part of the constituents of the leaves are mobilized (e.g., to form seeds).

Last, but not least, vacuoles also function as **waste deposits**. With the exception of gaseous substances, leaves are unable to rid themselves of waste products or xenobiotics such as herbicides. These are ultimately deposited in the vacuole (Chapter 12).

In addition, vacuoles also have a **storage function**. Many plants use the vacuole to store reserves of nitrate and phosphate. Some plants store malic acid temporarily in the vacuoles in a diurnal cycle (see section 8.5). Vacuoles of storage tissues contain carbohydrates (section 13.3) and storage proteins (Chapter 14). Many plant cells contain different types of vacuoles (e.g., lytic vacuoles and protein storage vacuoles beside each other).

The storage function of vacuoles plays a role when utilizing plants as natural protein factories. It is now possible by genetic engineering to express economically important proteins (e.g., antibodies) in plants, where the vacuole storage system functions as a cellular storage compartment for accumulating these proteins in high amounts. Since normal techniques could be used for the cultivation and harvest of the plants, this method has the advantage that large amounts of proteins can be produced at low costs.

1.3 Plastids have evolved from cyanobacteria

Plastids are cell organelles which occur only in plant cells. They multiply by division and in most cases are inherited **maternally**. This means that all the plastids in a plant usually have descended from the **proplastids** in the egg

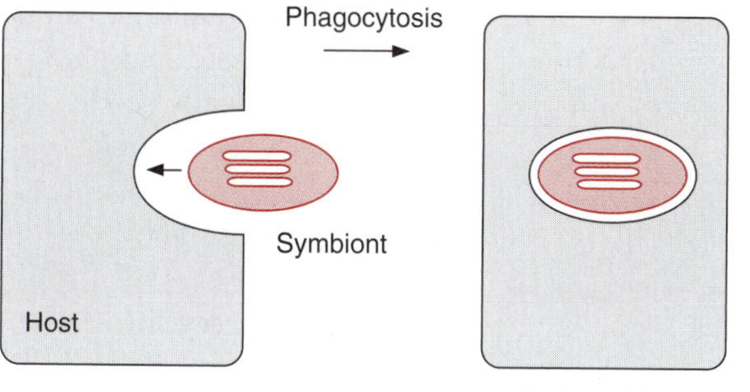

Figure 1.8 A cyanobacterium forms a symbiosis with a host cell.

cell. During cell differentiation, the proplastids can differentiate into green **chloroplasts**, colored **chromoplasts**, and colorless **leucoplasts**. Plastids possess their own circular chromosome as well as enzymes for gene duplication, gene expression, and protein synthesis. The plastid genome (**plastome**) has properties similar to that of the prokaryotic genome, as for instance in the cyanobacteria, but encodes only a minor part of the plastid proteins; the majority of these proteins are encoded in the nucleus and subsequently are transported into the plastids. The proteins encoded by the plastome comprise part of the proteins of photosynthetic electron transport and of ATP synthesis.

As early as 1883 the botanist Andreas Schimper postulated that plastids are evolutionary descendants of intracellular symbionts, thus founding the basis for the **endosymbiont hypothesis**. According to this hypothesis, the plastids descend from cyanobacteria, which were taken up by phagocytosis into a host cell (Fig. 1.8) and lived there in a symbiotic relationship. Through time these endosymbionts lost the ability to live independently because a large portion of the genetic information of the plastid genome was transferred to the nucleus. Comparative DNA sequence analyses of proteins from chloroplasts and from early forms of cyanobacteria allow the conclusion to be drawn that all chloroplasts of the plant kingdom derive from one symbiotic event. Therefore it is justified to speak of an **endosymbiotic theory**.

Proplastids (Fig. 1.9A) are very small organelles (diameter 1–1.5 µm). They are undifferentiated plastids found in the meristematic cells of the shoot and the root. They, like all other plastids, are enclosed by two membranes forming an envelope. According to the endosymbiont theory, the inner envelope membrane derives from the plasma membrane of the protochlorophyte and the outer envelope membrane from plasma membrane of the host cell.

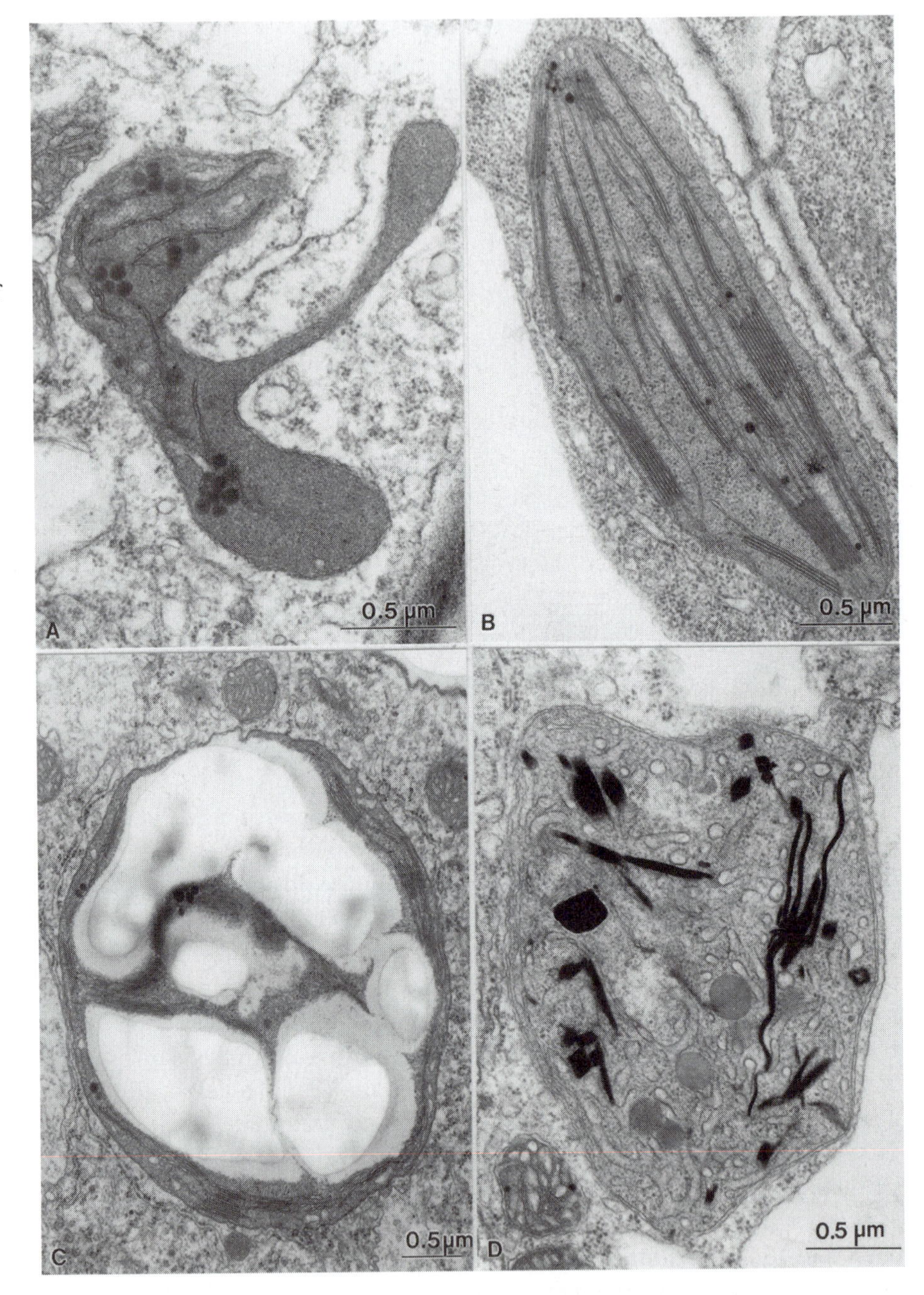

Figure 1.9 Plastids occur in various differentiated forms. A. Proplastid from young primary leaves of *Cucurbita pepo* (courgette); B. Chloroplast from mesophyll cell of tobacco leaf fixed at the end of the dark period; C. Leucoplast: amyloplast from the root of *Cestrum auranticum*; D. Chromoplast from petals also of *C. auranticum*. (By D. G. Robinson, Heidelberg.)

Chloroplasts (Fig. 1.9B) are formed by differentiation of the proplastids (Fig. 1.10). A mature mesophyll cell contains about 50 chloroplasts. By definition chloroplasts contain chlorophyll. However, they are not always green. In red and brown algae, other pigments mask the green color of the chloro-

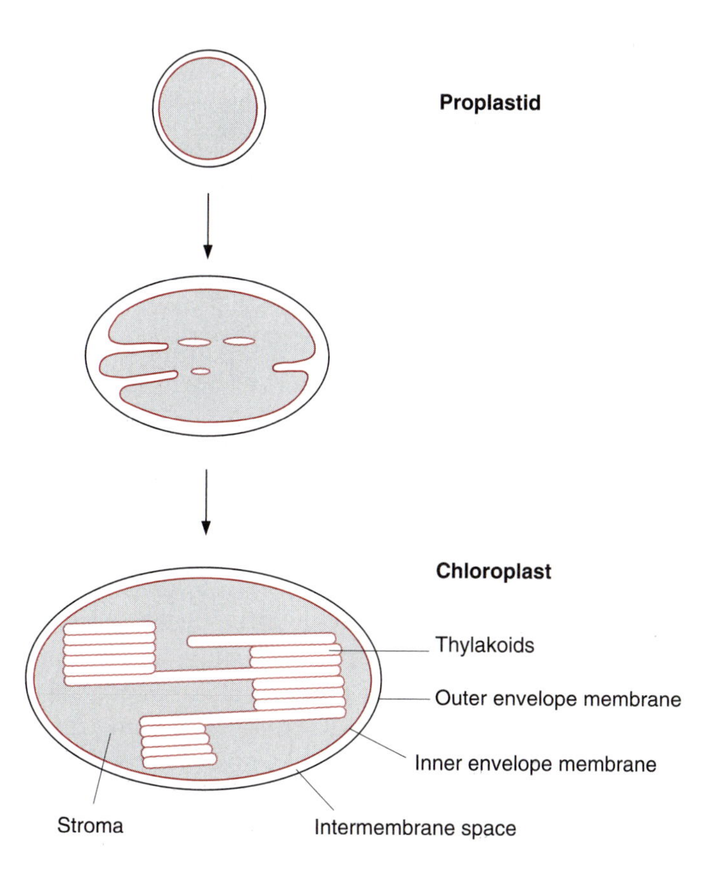

Figure 1.10 Scheme of the differentiation of a proplastid to a chloroplast.

phyll. Chloroplasts are lens-shaped and can adjust their position within the cell to receive an optimal amount of light. In higher plants their length is 3 to 10 μm. The two **envelope membranes** enclose the **stroma**. The stroma contains a system of membranes arranged as flattened sacks (Fig. 1.11), which were given the name **thylakoids** (in Greek, sac-like) by Wilhelm Menke in 1960. During differentiation of the chloroplasts, the inner envelope membrane invaginates to form thylakoids, which are subsequently sealed off. In this way a large membrane area is provided as the site for the photosynthesis apparatus (Chapter 3). The thylakoids are connected to each other by tubelike structures, forming a continuous compartment of the thylakoid space. Many of the thylakoid membranes are squeezed very closely together; they are said to be stacked. These stacks can be seen by light microscopy as small particles within the chloroplasts and have been named **grana**.

There are three different compartments in chloroplasts: the **intermembrane space** between the outer and inner envelope membrane; the **stroma**

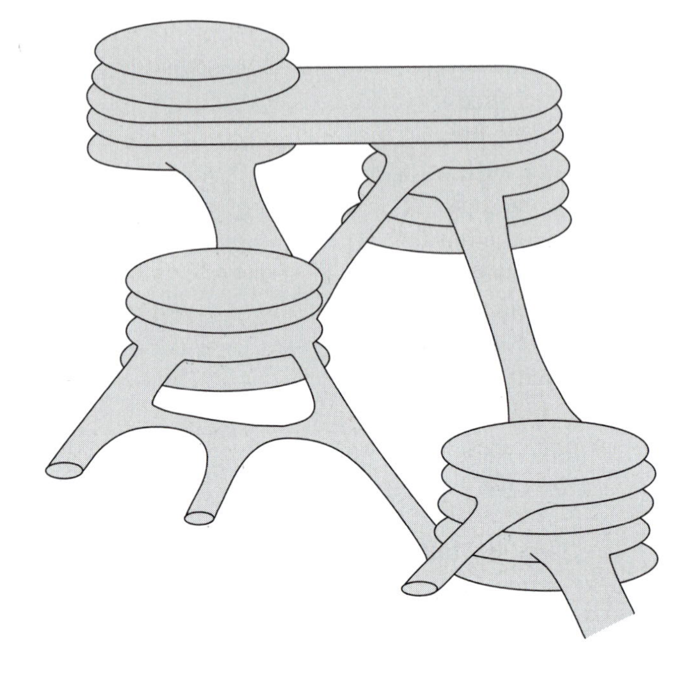

Figure 1.11 The grana stacks of the thylakoid membranes are connected by tubes, forming a continuous thylakoid space (thylakoid lumen). (After Weier and Stocking, 1963.)

space between the inner envelope membrane and the thylakoid membrane; and the **thylakoid lumen**, which is the space within the thylakoid membranes. The **inner envelope membrane** is a permeability barrier for metabolites and nucleotides, which can pass through only with the aid of specific translocators (section 1.9). In contrast, the **outer envelope membrane** is permeable to metabolites and nucleotides (but not to macromolecules such as proteins or nucleic acids). This permeability is due to the presence of specific membrane proteins called **porins**, which form pores permeable to substances with a molecular mass below 10,000 Dalton (section 1.11). Thus, the inner envelope membrane is the actual boundary membrane of the metabolic compartment of the chloroplasts so that the chloroplast stroma can be regarded as the "protoplasm" of the plastids. In comparison, the thylakoid lumen represents an external space that functions primarily as a compartment for partitioning protons to form a proton gradient (Chapter 3).

The stroma of chloroplasts contains **starch grains**. This starch serves mainly as a diurnal carbohydrate store, the starch formed during the day being a reserve for the following night (section 9.1). Therefore at the end of the day the starch grains in the chloroplasts are usually very large and, during the following night, become very small again. The formation of starch in plants always takes place in plastids.

Often structures that are not surrounded by a membrane are found inside the stroma. They are known as **plastoglobuli** and contain, among other substances, lipids, and plastoquinone. A particularly high amount of plastoglobuli is found in the plastids of senescent leaves, containing degraded products of the thylakoid membrane. About 10 to 100 identical plastid genomes are localized in a special region of the stroma known as the **nucleoide**. The ribosomes present in the chloroplasts are either free in the stroma or bound to the surface of the thylakoid membranes.

In leaves grown in the dark (etiolated plants), the plastids have a yellowish color and are termed **etioplasts**. These etioplasts contain some, but not all, of the chloroplast proteins. They are devoid of chlorophyll but contain instead membrane precursors, termed prolaminar bodies, which probably consist of lipids. The etioplasts are regarded as an intermediate stage of chloroplast development.

Leucoplasts (Fig. 1.9C) are a group of plastids that include many differentiated colorless organelles with very different functions (e.g., the **amyloplasts**), which act as a store for starch in non-green tissues such as root, tubers, or seeds (Chapter 9). Leucoplasts are also the site of lipid biosynthesis in non-green tissues. Lipid synthesis in plants is generally located in plastids. The reduction of nitrite to ammonia, a partial step of nitrate assimilation (Chapter 10), is also always located in plastids. In those cases in which nitrate assimilation takes place in the roots, leucoplasts are the site of nitrite reduction.

Chromoplasts (Fig. 1.9D) are plastids that, due to their high carotenoid content (Fig. 2.9), are colored red, orange, or yellow. They are the same size as chloroplasts but have no known metabolic function. Their main function may be to house the pigments of some flowers and fruit (e.g., the red color of tomatoes).

1.4 Mitochondria also result from endosymbionts

Mitochondria are the site of cellular respiration where substrates are oxidized for generating ATP (Chapter 5). Mitochondria, like plastids, multiply by division and are maternally inherited. They also have their own genome (consisting in plants of a large circular DNA strand and often several small circular DNA strands) and their own machinery for gene duplication, gene expression, and protein synthesis. The mitochondrial genome encodes only a small number of the mitochondrial proteins (Table 20.6); most of them are encoded in the nucleus. Mitochondria are of **endosymbiontic origin**. Phylogenetic experiments based on the comparison of DNA sequences led to the conclusion that all mitochondria derive from a single event in which a precursor **proteobacterium** entered an endosymbiosis with an anaerobic bacterium, probably an **archaebacterium**.

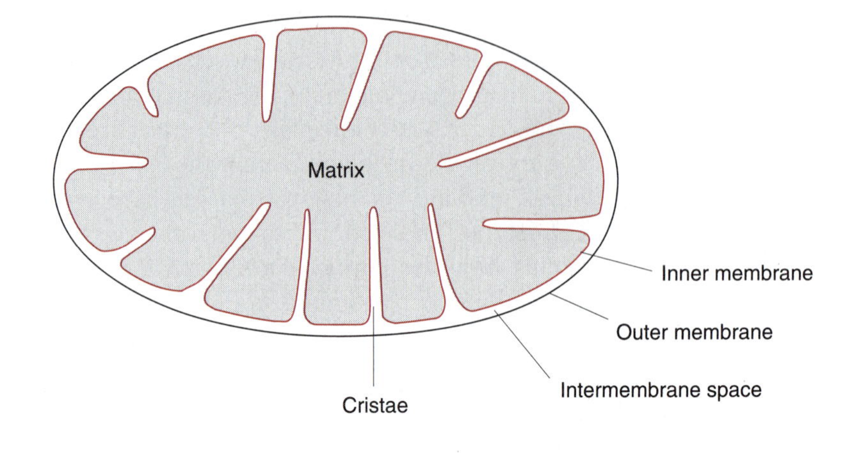

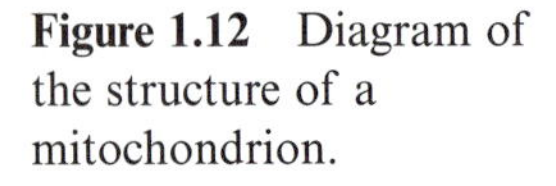

Figure 1.12 Diagram of the structure of a mitochondrion.

The endosymbiontic origin (Fig. 1.8) explains why the mitochondria are enclosed by two membranes (Fig. 1.12). Similar to chloroplasts, the **mitochondrial outer membrane** contains **porins** (section 1.11) that render this membrane permeable to molecules below a mass of 4,000 to 6,000 Dalton, such as metabolites and free nucleotides. The permeability barrier for these substances and the site of specific translocators (section 5.8) is the **mitochondrial inner membrane**. Therefore the **intermembrane space** between the inner and the outer membrane has to be considered as an external compartment. The "protoplasm" of the mitochondria, surrounded by the inner membrane, is called the **mitochondrial matrix**. The mitochondrial inner membrane contains the proteins of the respiratory chain (section 5.5). In order to enlarge the surface area of the inner membrane, it is invaginated in folds (**cristae mitochondriales**) or **tubuli** (Fig. 1.13) into the matrix. The structure of these membrane invaginations corresponds to the structure of the thylakoids, the only difference being that in the mitochondria these invaginations are not separated as a distinct compartment from the inner membrane. Also, the mitochondrial inner membrane is the site for formation of a **proton gradient**. Therefore the mitochondrial intermembrane space and the chloroplastic thylakoid lumen correspond to each other in functional terms.

1.5 Peroxisomes are the site of reactions in which toxic intermediates are formed

Peroxisomes, also termed microbodies, are small, spherical organelles with a diameter of 0.5 to 1.5 µm (Fig. 1.14) that, in contrast to plastids and

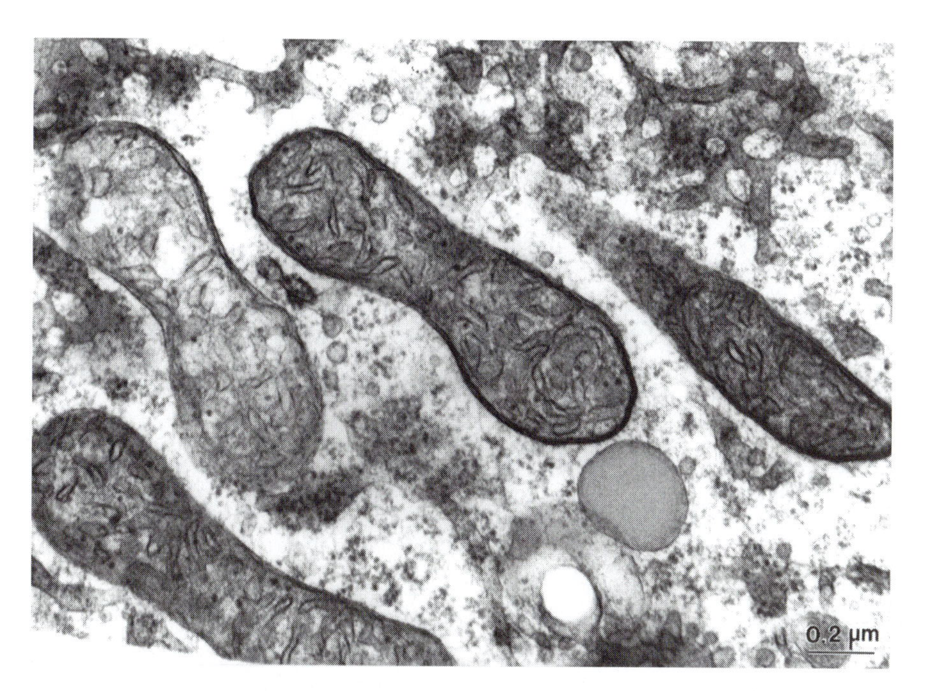

Figure 1.13 In mitochondria invaginations of the inner membrane result in an enlargement of the membrane surface. The figure shows mitochondria in a barley aleurone cell. (By D. G. Robinson, Heidelberg.)

mitochondria, are enclosed by only a single membrane. This membrane also contains porins. The peroxisomal matrix represents a specialized compartment for reactions in which toxic intermediates are formed. Thus peroxisomes contain enzymes catalyzing the oxidation of substances accompanied by the formation of $\mathbf{H_2O_2}$, and also contain **catalase**, which immediately degrades this H_2O_2 (section 7.4). Peroxisomes are a common constituent of eukaryotic cells. In plants there are two important differentiated forms: the **leaf peroxisomes** (Fig. 1.14A), which participate in photorespiration (Chapter 7); and the **glyoxysomes** (Fig. 1.14B), which are present in seeds containing oils (triacylglycerols) and play a role in the conversion of triacylglycerols to carbohydrates (section 15.6). They contain all the enzymes for fatty acid β-oxidation. The origin of peroxisomes is a matter of dispute. Some results indicate that peroxisomes are synthesized *de novo* from invaginations of the membranes of the endoplasmic reticulum, and other results indicate that peroxisomes are formed by division of preexisting peroxisomes, like plastids and mitochondria, with the one difference being that they have no genetic apparatus. A comparison of protein sequences has shown that peroxisomes from plants, fungi, and animals have a common ancestor. Whether this was also an endosymbiont, as in the case of mitochondria and plastids, but one that lost its genome, is still not clear.

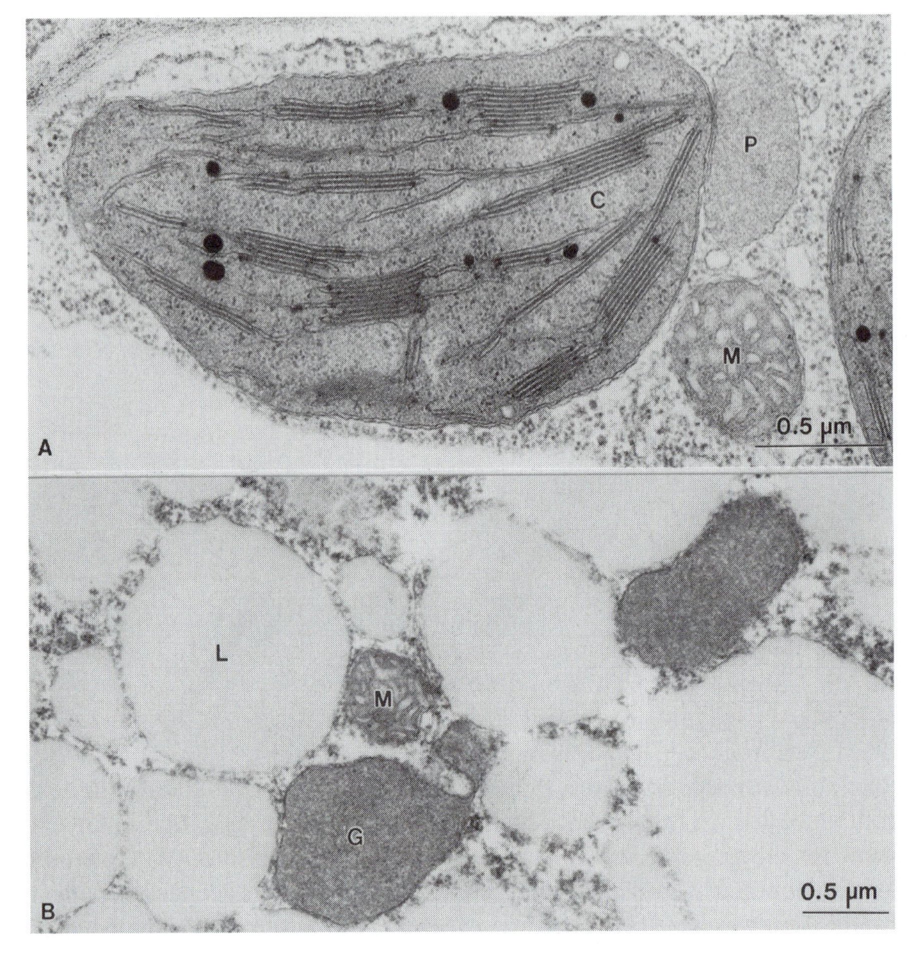

Figure 1.14 Peroxisomes. A. Peroxisomes from the mesophyll cells of tobacco. The proximity of peroxisome (P), mitochondrion (M), and chloroplast (C) reflects the rapid metabolite exchange between these organelles in the course of photorespiration (discussed in Chapter 7). B. Glyoxysomes from germinating cotyledons of *Cucurbita pepo* (courgette). The lipid degradation described in section 15.6 and the accompanying gluconeogenesis require a close contact between lipid droplets (L), glyoxysome (G), and mitochondrion (M). (By D. G. Robinson, Heidelberg.)

1.6 The endoplasmic reticulum and Golgi apparatus form a network for the distribution of biosynthesis products

In an electron micrograph, the **endoplasmic reticulum (ER)** appears as a labyrinth traversing the cell (Fig. 1.15). Two structural types of ER can be differentiated: the rough and the smooth forms. The **rough ER** consists of flattened sacs that are sometimes arranged in loose stacks of which the outer side of the membranes is occupied by **ribosomes**. The **smooth ER** consists primarily of branched tubes without ribosomes. Despite these morphological differences, the rough ER and the smooth ER are constituents of a continuous membrane system.

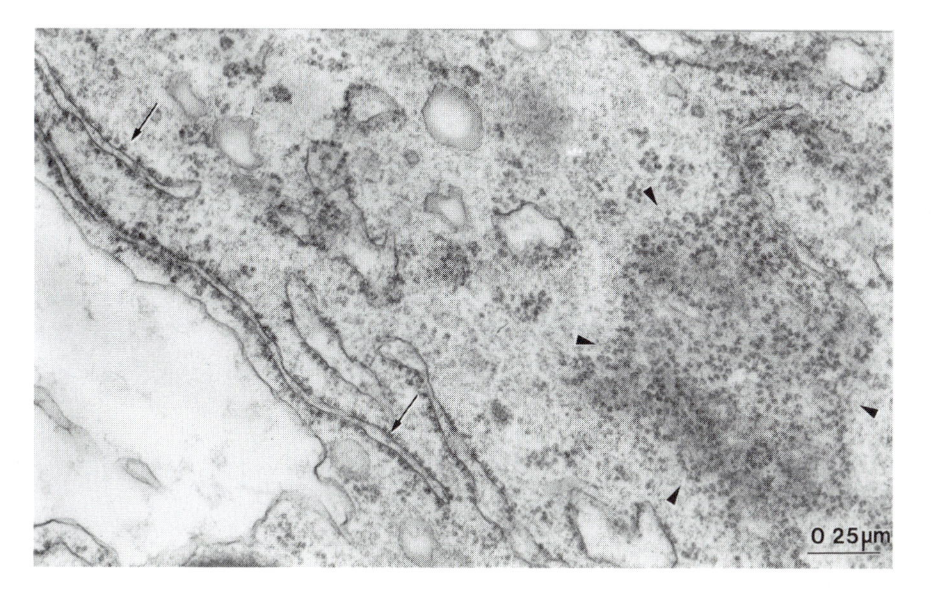

Figure 1.15 Rough endoplasmic reticulum, cross section (arrows) and tangential sections (arrowheads). The ribosomes temporarily attached to the membrane occur as polysome complexes (ribosome + mRNA). Section from the cell of a maturing pea cotyledon. (By D. G. Robinson, Heidelberg.)

The presence of ribosomes on the outer surface of the ER is temporary. Ribosomes are attached to the ER membrane only when the protein that they form is destined for the ER itself, for the vacuoles, or for export from the cell. These proteins contain an amino acid sequence (**signal sequence**) that causes the peptide chain during its synthesis to enter the lumen of the ER (section 14.5). A snapshot of the ribosome complement of the ER would show only those ribosomes that at the moment of fixation of the tissue are involved in the synthesis of proteins destined for import into the ER lumen. Membranes of the ER are also the site of membrane lipid synthesis, where the necessary fatty acids are provided by the plastids.

In seeds and other tissues, **oil bodies** (also called **oleosomes**) are present, which are derived from the ER membrane. The oil bodies store triglycerides and are of great economic importance since they are the storage site of oil plants, such as rape or olives. The oil bodies are enclosed by a half biomembrane only, of which the hydrophobic fatty acid residues of the membrane lipids project into the oil and the hydrophilic heads project into the cytosol (section 15.2).

In addition, the ER is a suitable storage site for the production of proteins in plants by genetic engineering. It is possible to provide those proteins with a signal sequence and the amino terminal ER-retention signal KDEL (Lys Asp Glu Leu). The ER of leaves is capable of accumulating large amounts of such extraneous proteins (up to 2.5 to 5% of the total leaf protein). It may be noted that the function of the ER is not affected when such large amounts of extraneous proteins are accumulated.

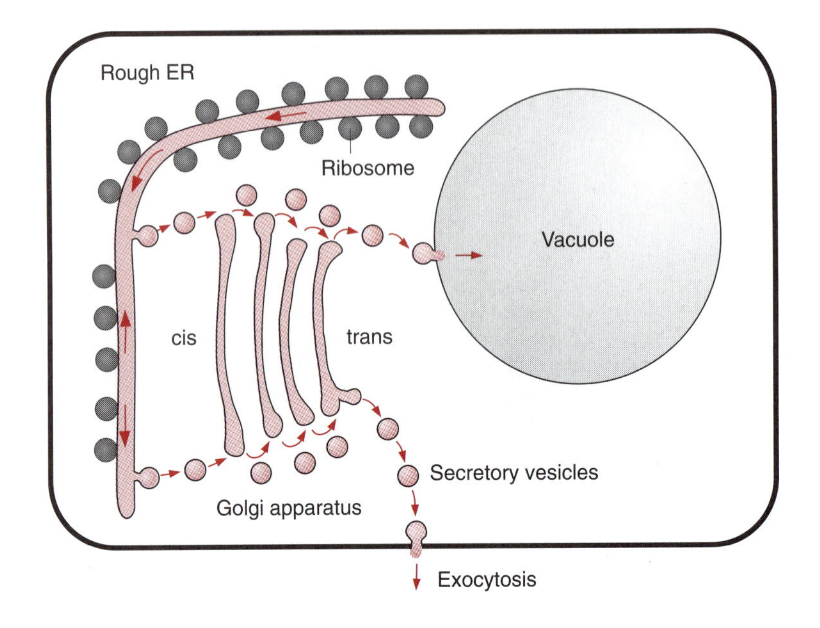

Figure 1.16 Scheme of the interplay between the endoplasmic reticulum and the Golgi apparatus in the transfer of proteins from the ER to the vacuoles and in the secretion of proteins from the cell.

In the ER lumen, proteins are often modified by N-glycosylation (attachment of hexose chains to amino acid residues; see section 17.7). Proteins channeled into the ER lumen are transferred to the *cis* side of the **Golgi apparatus** by membrane vesicles budding off from the ER (Fig. 1.16). These vesicles are covered on the outside by **coat proteins** consisting of six to seven different subunits. The Golgi apparatus, discovered in 1898 by the Italian Camillo Golgi, using a light microscope, consists of up to 20 curved discs arranged in parallel, the so-called **Golgi cisternae** or dictyosomes, which are surrounded by smooth membranes (not occupied by ribosomes) (Fig. 1.17). At both sides of the discs, vesicles of various size can be seen to bud off. The Golgi apparatus consists of the *cis* compartment, the middle compartment, and the *trans* compartment. During transport through the Golgi apparatus, proteins are often modified by O-glycosylation (attachment of hexose chains to serine and threonine residues).

Two mechanisms for transporting proteins through the Golgi apparatus are under discussion: (1) According to the **vesicle shuttle** model (Fig. 1.16), the proteins pass through the different cisternae by enbudding and vesicle transfer. Each cisterna has its fixed position. (2) According to the **cisternae progression** model, cisternae are constantly being newly formed by vesicle fusion at the *cis* side, and they then decompose to vesicles at the *trans* side. Present results show that both systems probably function in parallel.

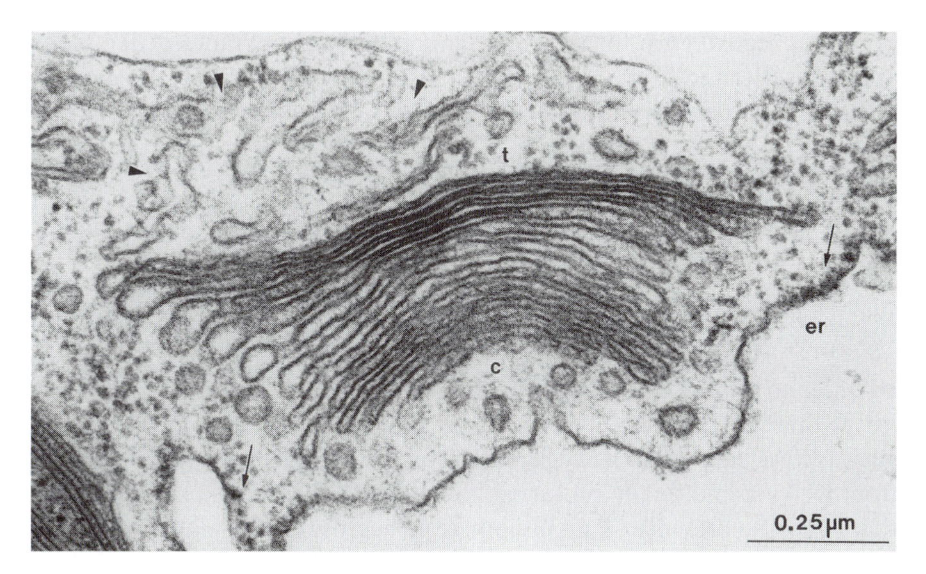

Figure 1.17 Golgi apparatus (dictyosome) in the green alga *Chlamydomonas reinhardii*. C = *cis* side, t = *trans* side. Arrowheads point to the *trans* Golgi network. The swollen endoplasmatic reticulum (ER) is typical for this cell. On the ER, ribosomes can be recognized, except in the area where vesicles bud off. (By D. G. Robinson, Heidelberg.)

In the Golgi apparatus, proteins are selected either to be removed from the cell by exocytosis (secretion) or to be transferred to lytic vacuoles or to storage vacuoles (section 1.2). Signal sequences of proteins act as sorting signals to direct proteins into the vacuolar compartment The proteins destined for the lytic vacuoles are transferred in **clathrin-coated vesicles**. Clathrin is a protein consisting of two different subunits (α-UE 180,000 Dalton, β-UE 35,000 to 40,000 Dalton). Both 3α- and 3β-subunits form a complex with three arms (Triskelion), which polymerizes to a hexagonal latticed structure surrounding the vesicle (Fig. 1.18). The transport into the storage vacuoles proceeds via other vesicles without clathrin. Secretion proteins, containing only the signal sequence for entry into the ER, reach the plasma membrane via **secretion vesicles** without a protein coat and are secreted by **exocytosis**.

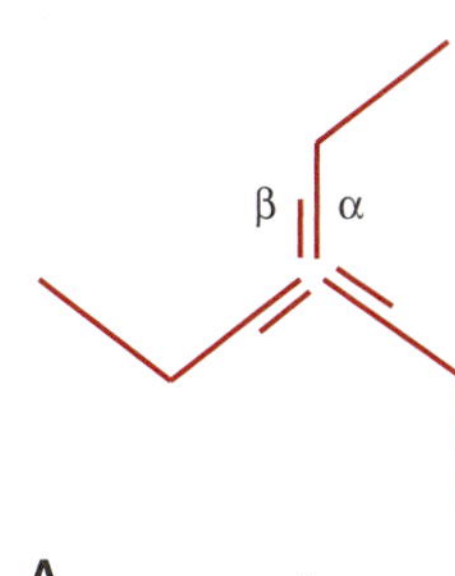

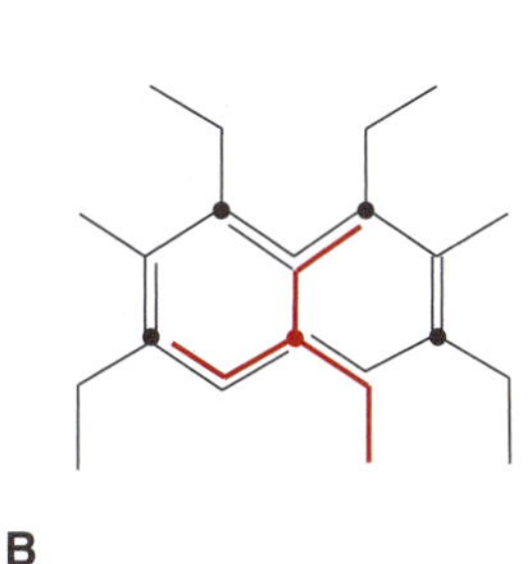

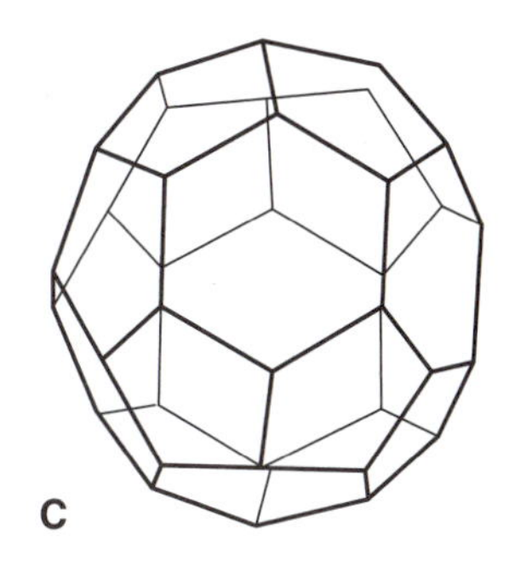

Figure 1.18 Model of the structure of clathrin-coated vesicles.. (A) 3α and 3β subunits of clathrin form a complex with three arms. (B) From this a hexagonal and pentagonal lattice (the latter not shown here) is formed by polymerization and this forms (C) the coat. (From Kleinig and Sitte.)

The collective term for the ER membrane, the membranes of the Golgi apparatus (derived from the ER), the transfer vesicles, and the nuclear envelope is the **endomembrane system**.

1.7 Functionally intact cell organelles can be isolated from plant cells

In order to isolate cell organelles, the cell has to be disrupted to such an extent that its organelles are released into the isolation medium. This forms what is known as a **cell homogenate**. In order to prevent the free organelles from swelling and finally rupturing, the isolation medium must be isotonic, that is, by the presence of an **osmotic** (e.g., sucrose), an osmotic pressure is generated in the medium, which corresponds to the osmotic pressure of the aqueous phase within the organelle. Media containing 0.3 mol/L sucrose or sorbitol usually are used for cell homogenization.

Figure 1.19 shows the protocol for the isolation of chloroplasts as an example. Small leaf pieces are homogenized by cutting them up within seconds using blades rotating at high speed, such as in a food mixer. It is important that the homogenization time is short; otherwise the cell organelles released into the isolation medium would also be destroyed. However, such homogenization works only with leaves with soft cell walls, such as spinach. In the case of leaves with more rigid cell walls (e.g., cereal plants), protoplasts are first prepared from leaf pieces as described in section 1.1. These protoplasts are then ruptured by forcing the protoplast suspension through a net with a mesh smaller than the size of the protoplasts.

The desired organelles can be separated and purified from the rest of the cell homogenate by differential or density gradient centrifugation. In the case of **differential centrifugation**, the homogenate is suspended in a medium with a density much lower than that of the cell organelles. In the gravitational field of the centrifuge, the sedimentation velocity of the particles depends primarily on the **particle size** (the large particles sediment faster than the small particles). As shown in Figure 1.19, taking the isolation of chloroplasts as an example, relatively pure organelle preparations can be obtained within a short time by a sequence of centrifugation steps at increasing speeds.

In the case of **density gradient centrifugation** (Fig. 1.20), the organelles are separated according to their **density**. Media of differing densities are assembled in a centrifuge tube so that the density increases from top to bottom. To prevent altering the osmolarity of the medium, heavy macro-

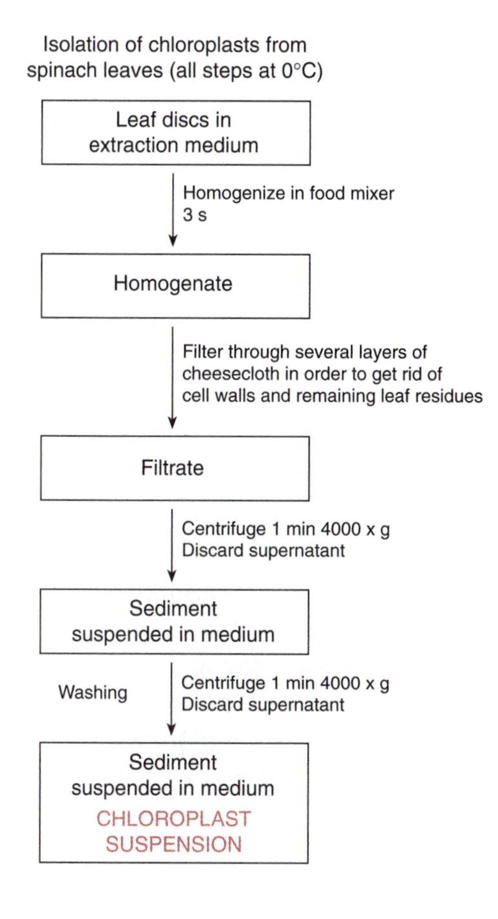
Isolation of chloroplasts from spinach leaves (all steps at 0°C)
Leaf discs in extraction medium
Homogenize in food mixer 3 s
Homogenate
Filter through several layers of cheesecloth in order to get rid of cell walls and remaining leaf residues
Filtrate
Centrifuge 1 min 4000 x g Discard supernatant
Sediment suspended in medium
Washing
Centrifuge 1 min 4000 x g Discard supernatant
Sediment suspended in medium
CHLOROPLAST SUSPENSION

Figure 1.19 Protocol for the isolation of functionally intact chloroplasts.

molecules [e.g., Percoll (silica gel)], are used to achieve a high density. The cell homogenate is layered on the density gradient prepared in the centrifuge tube and centrifuged until all the particles of the homogenate have reached their zone of equal density in the gradient. As this density gradient centrifugation requires high centrifugation speed and long running times, it is often used as the final purification step after preliminary separation by differential centrifugation.

By using these techniques it is possible to obtain functionally intact chloroplasts, mitochondria, peroxisomes, and vacuoles of high purity in order to study their metabolic properties in the test tube.

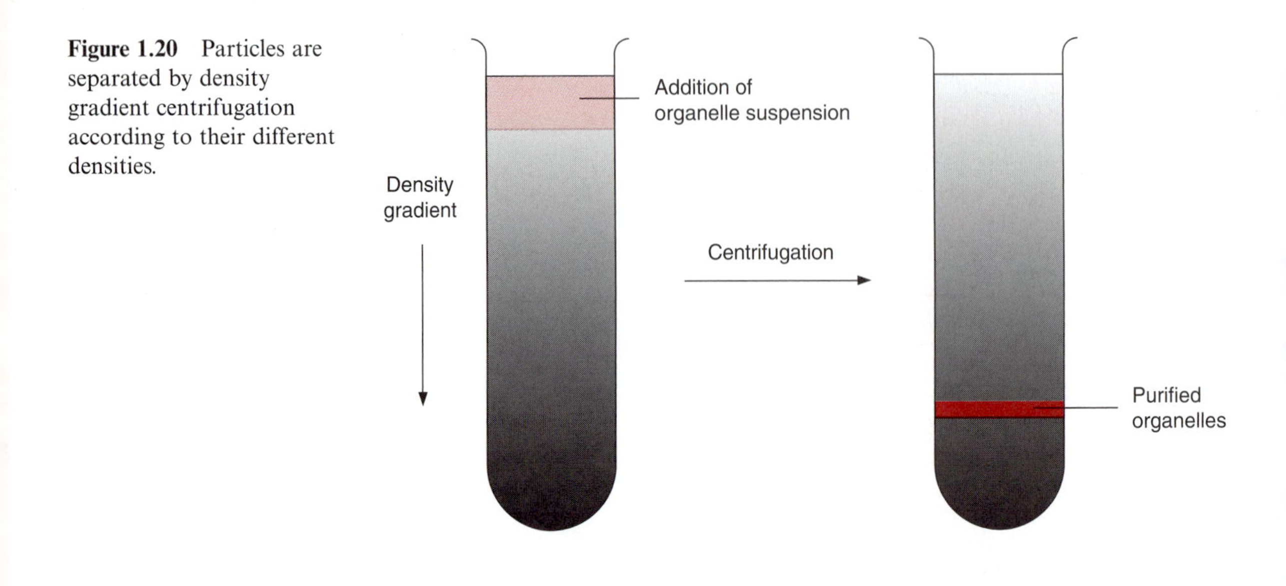

Figure 1.20 Particles are separated by density gradient centrifugation according to their different densities.

1.8 Various transport processes facilitate the exchange of metabolites between different compartments

Each of the cell organelles mentioned in the preceding section has a specific function in cell metabolism. The interplay of the metabolic processes in the various compartments requires a transfer of substances across the membranes of these cell organelles as well as between the different cells. This transfer of material takes place in various ways: by specific translocators, channels, pores, via vesicle transport, and in a few cases (e.g., CO_2 or O_2) by nonspecific diffusion through membranes. The vesicle transport and the function of the plasmodesmata have already been described.

Figure 1.21 describes various types of transport processes according to formal criteria. When a molecule moves across a membrane independent of the transport of other molecules, the process is called **uniport**, and when counter-exchange of molecules is involved, it is called **antiport**. The mandatory simultaneous transport of two substances in the same direction is called **symport**. A transport via uniport, antiport, or symport, in which a charge is also moved simultaneously, is termed **electrogenic transport**. A vectorial transport, which is coupled to a chemical or photochemical reaction, is named **active** or **primary active transport**. Examples of active transport are the transport of protons driven by the electron transfer of the photosynthetic electron transport chain (Chapter 3) or the respiratory chain (Chapter

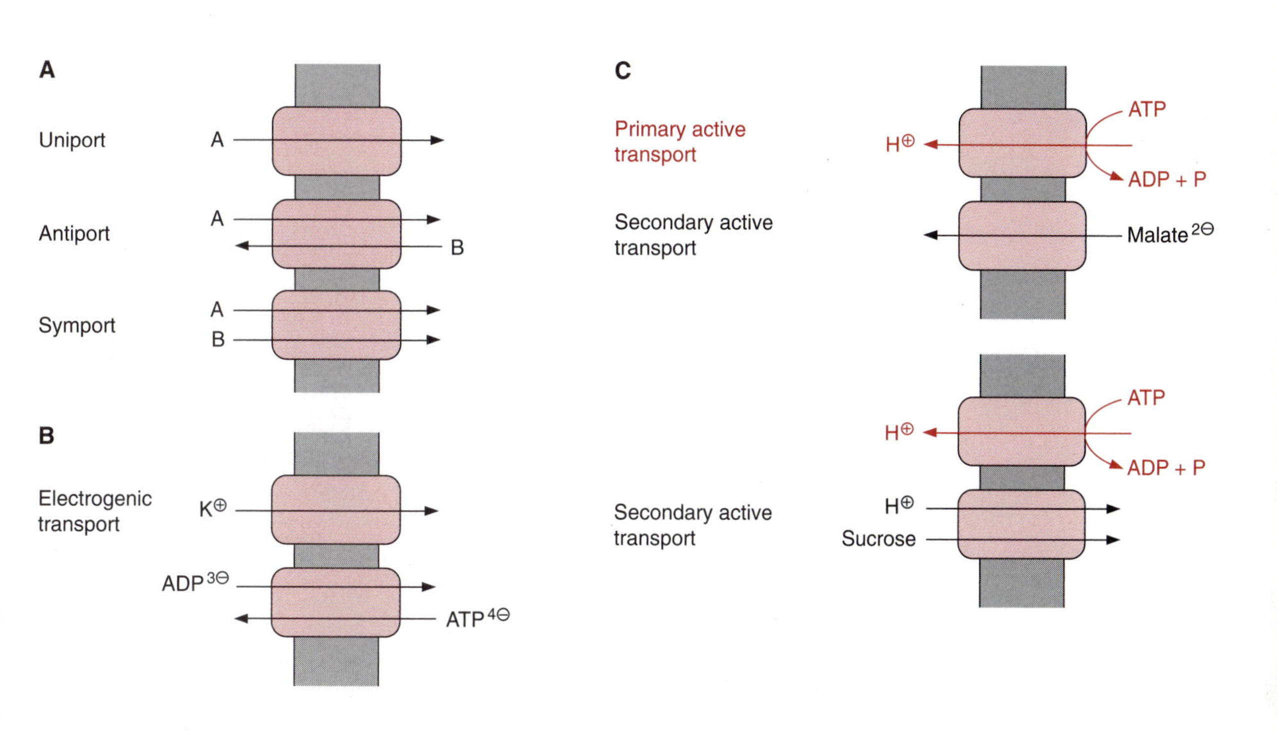

Figure 1.21 Classification of membrane transport processes.

5) or by the consumption of ATP (Fig. 1.21C). Such proton transport is electrogenic; the transfer of a positive charge results in the formation of a membrane potential. Another example of primary active transport is the ATP-dependent transport of glutathione conjugates into vacuoles (see section 12.2).

In a **secondary active transport**, the only driving force is an electrochemical potential across the membrane. In the case of an electrogenic uniport, the membrane potential can be the driving force by which a substrate is transported across the membrane against the concentration gradient. An example of this is the accumulation of malate in the vacuole (Fig. 1.21C; see also Chapter 8). Another example of secondary active transport is the transport of sucrose via an H^+-sucrose symport in which a proton gradient, formed by primary active transport, drives the accumulation of sucrose (Fig. 1.21C). This transport plays an important role in loading sieve tubes with sucrose (Chapter 13).

1.9 Translocators catalyze the specific transport of substrates and products of metabolism

Specialized membrane proteins catalyze a specific transport across membranes. In the past these proteins were called carriers, as it was assumed that after binding the substrate at one side of the membrane, they would diffuse through the membrane to release the substrate on the other side. We now know that this simple picture does not apply. Instead, transport can be visualized as a process by which a molecule moves through a specific pore. The proteins catalyzing such a transport are termed **translocators**. The triose phosphate-phosphate translocator of chloroplasts will be used as an example to describe the structure and function of such a translocator. This translocator enables the export of photoassimilates from the chloroplasts by catalyzing a counter-exchange of phosphate with triose phosphate (dihydroxyacetone phosphate or glyceraldehyde-3-phosphate) (Fig. 9.12). Quantitatively it is the most abundant transport protein in plants.

Silicone layer filtering centrifugation is a very useful tool (Fig. 1.22) for measuring the uptake of substrates into chloroplasts or other cell organelles. To start measurement of transport, the corresponding substrate is added to a suspension of isolated chloroplasts and is terminated by separating the chloroplasts from the surrounding medium by centrifugation through a silicone layer. The amount of substrate taken up into the separated chloroplasts is then quantitatively analyzed.

A hyperbolic curve is observed (Fig. 1.23) when this method is used to measure the uptake of phosphate into chloroplasts at various external concentrations of phosphate. At very low phosphate concentrations the rate of uptake rises proportionally to the external concentration, whereas at a higher phosphate concentration the curve levels off until a **maximal velocity** is reached (V_{max}). These are the same characteristics as seen in enzyme catalysis. During enzyme catalysis the substrate (S) is first bound to the enzyme (E). The product (P) formed on the enzyme surface is then released:

$$E + S \longrightarrow ES \xrightarrow{\text{Catalysis}} EP \longrightarrow E + P$$

The transport by a specific translocator can be depicted in a similar way:

$$S + T \longrightarrow ST \xrightarrow{\text{Transport}} TS \longrightarrow T + S$$

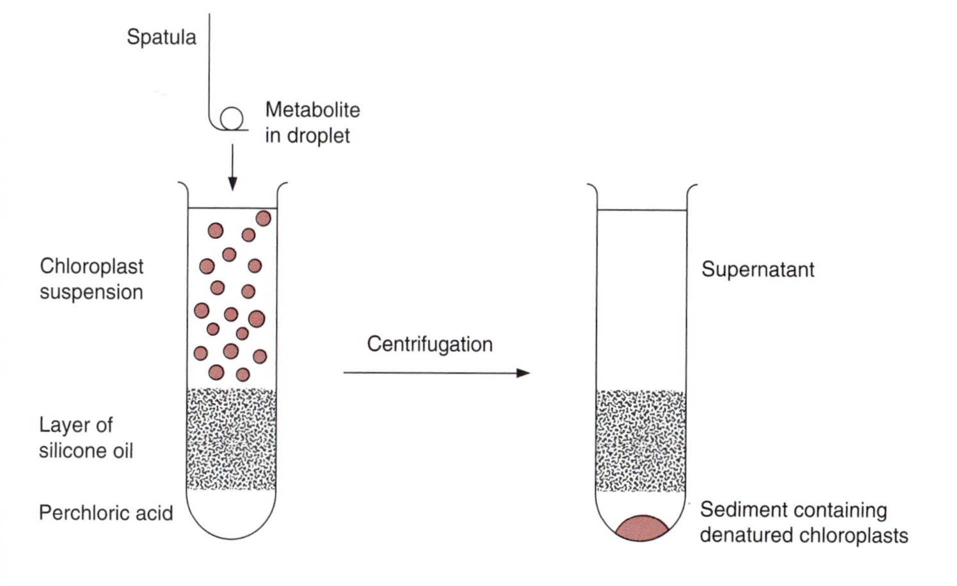

Figure 1.22 Silicone oil filtering centrifugation: measurement of the uptake of substances into isolated chloroplasts. For the measurement, the bottom of a centrifuge tube contains perchloric acid on which silicone oil is layered. The substance to be transported is added to the chloroplast suspension above the silicone layer using a small spatula. To simplify detection, metabolites labeled with radio isotopes (e.g., ^{32}P or ^{14}C) are usually used. The uptake of metabolites into the chloroplasts is terminated by centrifugation in a rapidly accelerating centrifuge. Upon centrifugation the chloroplasts migrate within a few seconds through the silicone layer into the perchloric phase, where they are denatured. That portion of the metabolite which has not been taken up remains in the supernatant. The amount of metabolite that has been taken up into the chloroplasts is determined by measurement of the radioactivity in the sedimented fraction. The amount of metabolite carried nonspecifically through the silicone layer, either by adhering to the outer surface of the plastid or present in the space between the inner and the outer envelope membranes, can be evaluated in a control experiment in which a substance is added (e.g., sucrose) that is known not to permeate the inner envelope membrane.

The substrate is bound to a specific binding site of the translocator protein (T), transported through the membrane, and then released from the translocator. The maximal velocity V_{max} corresponds to a state in which all the binding sites of the translocators are saturated with substrate. As is the case for enzymes, the K_m for a translocator corresponds to the substrate concentration at which transport occurs at half maximal velocity. Also in analogy to enzyme catalysis, the translocators usually show high **specificity** for the transported substrates. For instance, the chloroplast triose phosphate-phosphate translocator of C_3 plants (see section 9.1) transports

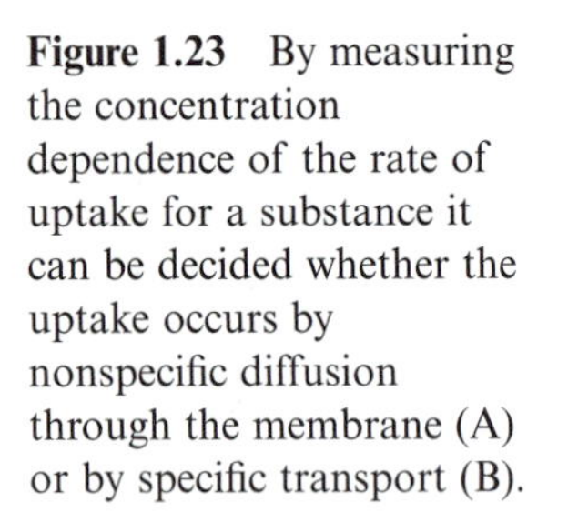
Figure 1.23 By measuring the concentration dependence of the rate of uptake for a substance it can be decided whether the uptake occurs by nonspecific diffusion through the membrane (A) or by specific transport (B).

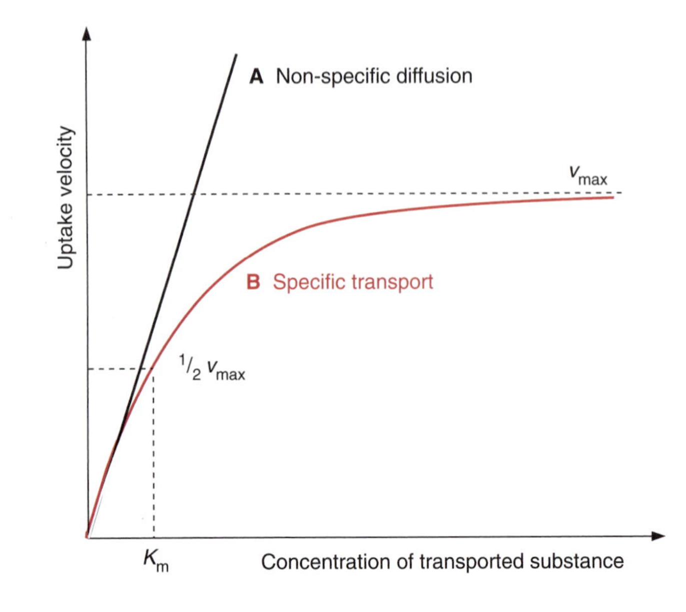

orthophosphate, dihydroxyacetone phosphate, glyceraldehyde-3-phosphate, and 3-phosphoglycerate, but not 2-phosphoglycerate. The various substrates compete for the binding site. Therefore, one substrate such as phosphate will be a **competitive inhibitor** for the transport of another substrate such as 3-phosphoglycerate. The triose phosphate-phosphate translocator of chloroplasts is an antiport, so that for each molecule transported inward (e.g., phosphate), another molecule (e.g., dihydroxyacetone phosphate) must be transported out of the chloroplasts.

Figure 1.24 shows a scheme of the transport process. The triose phosphate-phosphate translocator from chloroplasts, like translocators from mitochondria and other compartments, consists of two subunits that form a **gated pore**. Both subunits form a common substrate binding site, which is accessible either from the inside or from the outside, depending on the conformation of the translocator protein. The first step of the transport process is to bind a substrate (A) to the substrate binding site accessible from the outside. A **conformational change** then occurs and the substrate is finally released to the inner side. Another substrate (B) can now bind to the free binding site and thus be transported to the outside. An obligatory counter-exchange may be due to the fact that the shift of the binding site from one side of the membrane to the other can occur only when the binding site is occupied by a substrate. This mode of antiport is called a **ping-pong mechanism** (Fig. 1.24A). In the case of the chloroplast triose phosphate-

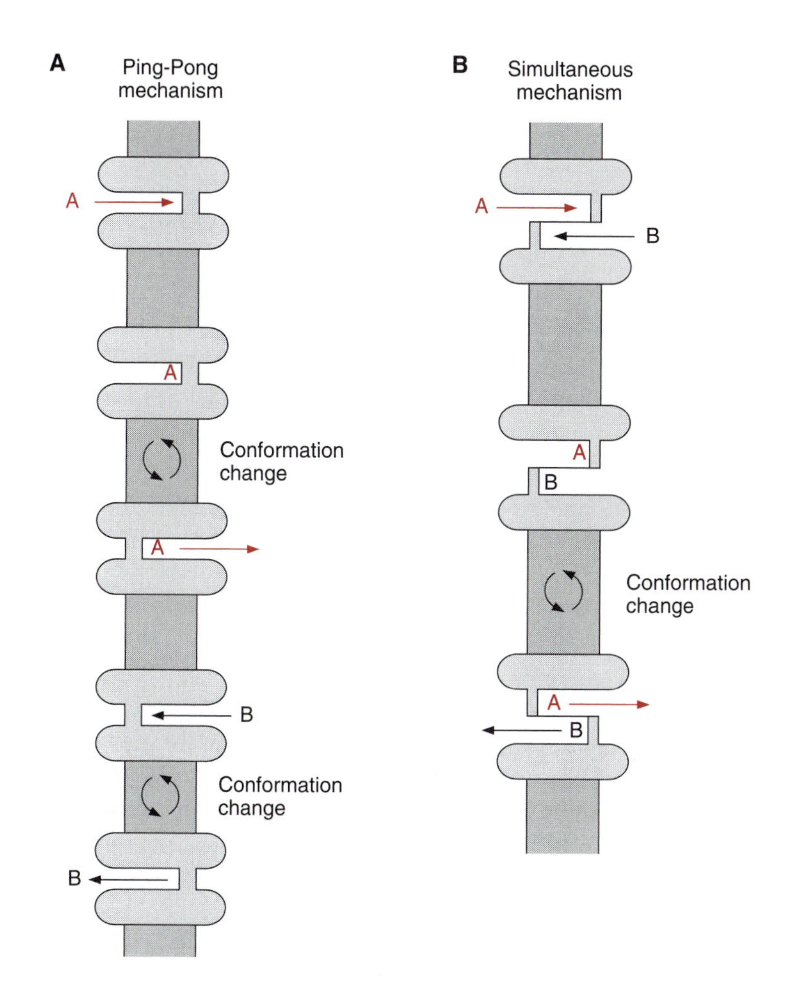

Figure 1.24 Antiport. Diagram of two possibilities for the counter-exchange of two substrate molecules (A, B). 1) Ping-pong mechanism: A translocator molecule catalyzes the transport of A and B sequentially. 2) Simultaneous mechanism: A and B are transported simultaneously by two translocator molecules tightly coupled to each other. See text.

phosphate translocator, the counter-exchange occurs according to the ping-pong mechanism. However, in many cases a counter-exchange proceeds by a **simultaneous mechanism** (Fig. 1.24B), in which two translocators are linked to each other in such a way that a change of conformation can occur only if both binding sites are either occupied or unoccupied. The known mitochondrial translocators operate by a simultaneous mechanism.

Translocators have a common basic structure

Translocators are **integral membrane proteins**. They traverse the lipid bilayer of the membrane in the form of **α-helices**. In such transmembrane α-helices the side chains have to be hydrophobic; therefore they contain hydrophobic amino acids such as alanine, valine, leucine, isoleucine, or phenylalanine.

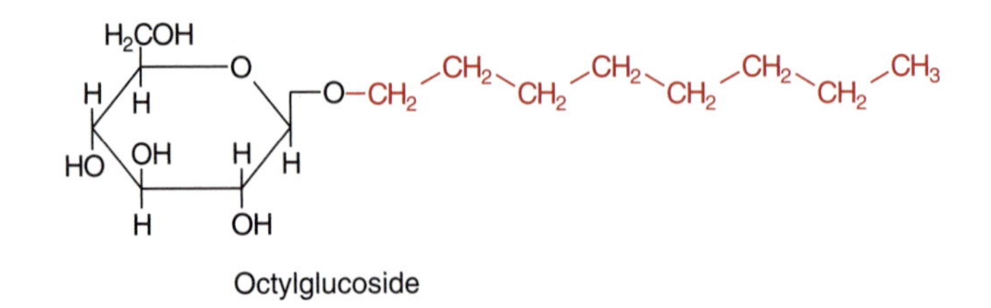

Figure 1.25
Octylglucoside, a glycoside composed from α-D-glucose and octyl alcohol, is a mild nonionic detergent that allows membrane proteins to be solubilized from the membranes without being denatured.

Membrane proteins are not soluble in water due to their high hydrophobicity. Mild nonionic detergents, such as octylglucoside (Fig. 1.26), can be used to dissolve these proteins from the membrane. The hydrophobic carbon chain of the detergent binds to the hydrophobic protein and, due to the glucose residues, the micelle thus formed is water-soluble. If the detergent is then removed, the membrane proteins aggregate to a sticky mass that cannot be solubilized again.

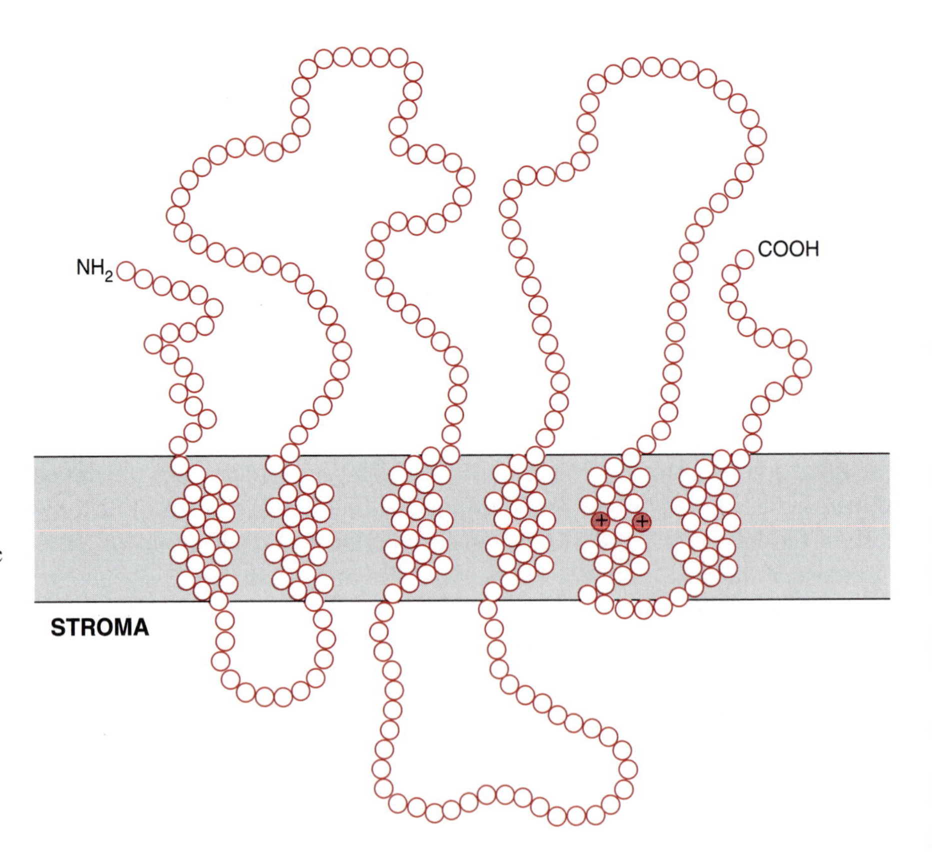

Figure 1.26 The triose phosphate-phosphate translocator from spinach forms six transmembrane helices. Each circle represents one amino acid. The likely positions of the transmembrane helices were evaluated from the hydrophobicity of the single amino acid residues. The amino acids, marked with red, containing a positive charge in helix 5, represent an arginine and a lysine. These amino acids probably provide the binding sites for the anionic substrates of the triose phosphate-phosphate translocator. (Data from Flügge et al., 1989.)

A number of translocator proteins have been isolated and purified using this method. This has allowed the analysis of their amino acid sequences. It can be predicted from the hydrophobicity of the amino acids in a peptide sequence which sections of the chain are likely to form transmembrane helices (hydropathy analysis). This procedure has led to the prediction that the 324 amino acids of the subunit of the chloroplast triose phosphate-phosphate translocator (Fig. 1.25) fold to form six transmembrane helices. Each contains about 20 amino acids and thus is large enough to span the envelope membrane with its cross-sectional distance of about 6 nm. In its functional form the chloroplast triose phosphate-phosphate translocator consists of two identical subunits. A comparison of sequences indicated that almost all the translocators from bacteria, plants, and animals known so far consist either of a dimer where each monomer has six transmembrane helices or of a single monomer with 12 transmembrane helices. Recently the three-dimensional structure of the mitochondrial ATP/ADP translocator has been analyzed by X-ray crystallography. (For the method, see section 3.3.) These studies demonstrated that the monomer of the translocator protein consists of six transmembrane helices in a barrel-like structure forming the translocation pore.

Aquaporins make cell membranes permeable for water

A comparison of the osmotic behavior of different cells and cell organelles shows that their membranes vary largely in their water permeability. The water permeability of a pure lipid bilayer is relatively low. Peter Agre from the Johns Hopkins University in Baltimore detected in kidney and blood cells proteins forming **membrane channels for water**, which he termed **aquaporins**. In 2003 he was awarded the Nobel Prize in Chemistry for this important discovery. It turned out that these aquaporins also occur in plants (e.g., in plasma membranes and membranes of the vacuole). Notably both types of membranes play a major role in the hydrodynamic response of a plant cell. A plant contains many aquaporin isoforms. Thus in the model plant *Arabidopsis thaliana* (section 20.1) about 30 different genes of the aquaporin family have been found; these are specifically expressed in the various plant organs. By regulation of gene expression and the regulation of the permeability of aquaporins present in the membranes, of which the mechanism is not yet resolved, the water permeability of the various plant cells is adapted to the environmental conditions.

X-ray structure analysis by electron cryomicroscopy showed that the subunits of the aquaporins each have six transmembrane helices (see section 3.3). In the membranes the subunits of the aquaporins are present as tetramers, of which each monomer apparently forms a channel, transporting 10^9 to 10^{11} water molecules per second. **The water channel** consists of a very narrow primarily **hydrophobic pore** with binding sites for only four H_2O molecules. These binding sites act as a **selection filter** for a specific water transport. It can be deduced from the structure that, for energetic reasons, these water channels are relatively impermeable for protons. It was observed recently that an aquaporin from the plasma membrane of tobacco also transports CO_2. This finding suggests that aquaporins may play a role as CO_2 transporters in plants.

The now commonly used term *aquaporin* is rather unfortunate, as aquaporins have an entirely different structure from the porins described in section 1.11. While the aquaporins, as well as the translocators and ion channels, as will be described in the following sections, are formed of transmembrane helices, the porins consist of β sheets.

1.10 Ion channels have a very high transport capacity

The chloroplast triose P-P translocator mentioned previously has a turnover number of 80 s^{-1} at 25°C, which means that it transports 80 substrate molecules per second. The turnover numbers of other translocators are in the range of 10 to 1,000 s^{-1}. Membranes also contain proteins which form **ion channels** that transport various ions at least three orders of magnitude faster than translocators (10^6–10^8 ions per second). They differ from the translocators in having a pore open to both sides at the same time. The flux of ions through the ion channel is so large that it is possible to determine the transport capacity of a single channel from the measurement of electrical conductivity.

The procedure for such single channel measurements, called the **patch clamp technique**, was developed by two German scientists, Erwin Neher and Bert Sakmann, who were awarded the Nobel Prize in Medicine or Physiology in 1991 for this research. The setup for this measurement (Fig. 1.27) consists of a glass pipette that contains an electrode filled with an electrolyte fluid. The very thin tip of this pipette (diameter about 1 μm) is sealed tightly by a membrane patch. The number of ions transported through this patch per unit time can be determined by measuring the electrical current [usually expressed as conductivity in Siemens (S)]. Figure 1.28 shows an example of the measurement of the single channel currents with the plasma membrane of broad bean guard cells. The recording of the change in current shows that the channel opens for various lengths of time and then closes again.

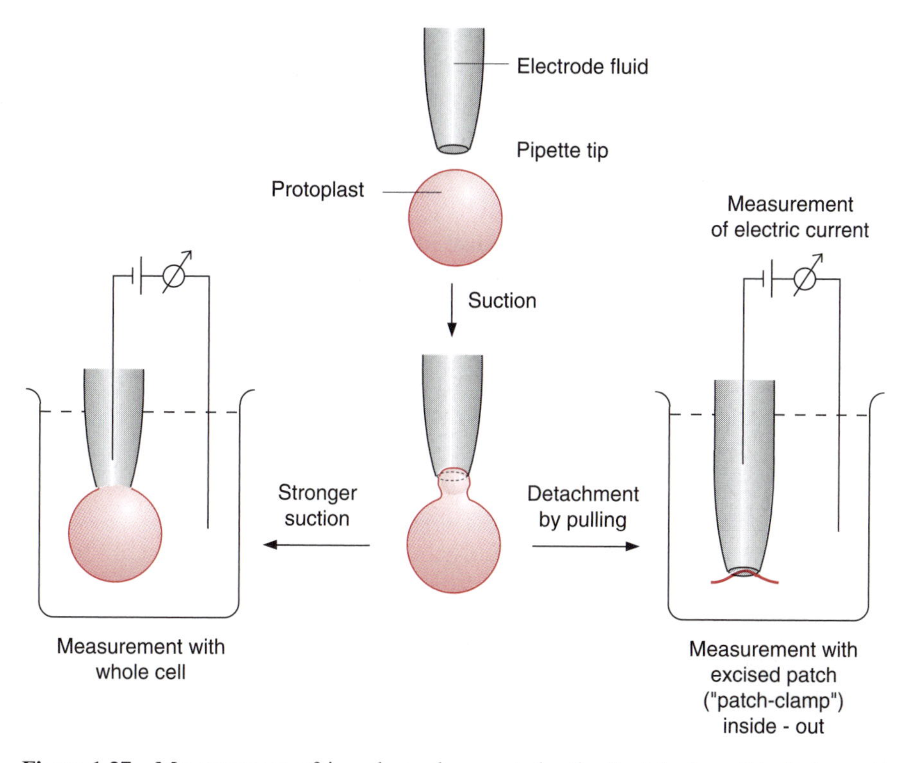

Figure 1.27 Measurement of ion channel currents by the "patch clamp" technique. A glass pipette with a diameter of about 1 µm at the tip, containing an electrode and electrode fluid, is brought into contact with the membrane of a protoplast or a cell organelle (e.g., vacuole). By applying slight suction, the opening of the pipette tip is sealed by the membrane. By applying stronger suction, the membrane surface over the pipette opening breaks, and the electrode within the pipette is now in direct electrical connection with the space inside the cell. In this way the channel currents can be measured for all the channels present in the membrane (measurement with whole cell). Alternatively, by slight pulling, the pipette tip can be removed from the protoplast or the vacuole with the membrane patch, which is sealed to the tip, being torn off from the rest of the cell. In this way the currents are measured only for those channels that are present in the membrane patch. A voltage is applied for measurement of the channel current and the current is measured after amplification.

This principle of **stochastic switching** between a **nonconductive state** and a defined **conductive state** is a typical property of ion channels. In the open state various channels have different conductivities, which can range between a few pS and several hundred pS. Moreover, various channels have characteristic mean open and close times, which can, depending on the channel, last from a few milliseconds to seconds. The transport capacity of the

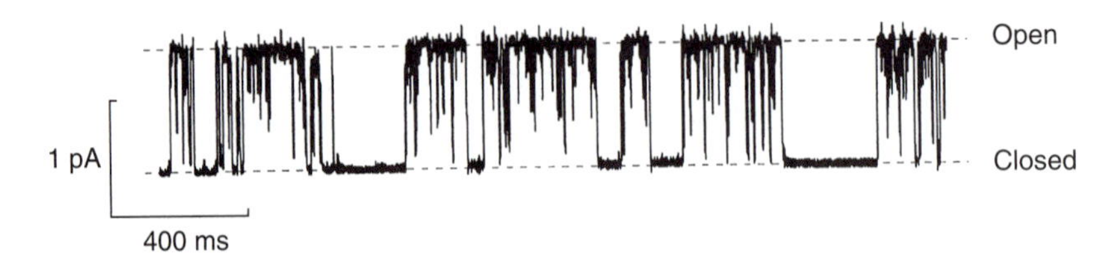

Figure 1.28 Measurement of single channel currents of the K^+ outward channel in a patch (Fig. 1.27) of the plasma membrane of guard cells from *Vicia faba*. (Outer medium 50 mM K^+, cytoplasmic side 200 mM K^+, voltage +35 mV.) (Data from Prof. G. Thiel, Darmstadt.)

channel per unit time therefore depends on the conductivity of the opened channel as well as on the mean duration of the open state.

Many ion channels have been characterized that are more or less specific for certain ions. Plants contain highly selective cation channels for H^+, K^+, and Ca^{++} and also selective anion channels for Cl^- and dicarboxylates, such as malate. Plant membranes, in contrast to those of animals, seem to possess no specific channels for Na^+ ions. The opening of many ion channels is regulated by the electric **membrane potential**. This means that membranes have a very important function in the electrical regulation of ion fluxes. Thus, in guard cells (section 8.1) the hyperpolarization of the plasma membrane (> −100 mV) opens a channel that allows potassium ions to flow into the cell (**K^+ inward channel**), whereas depolarization opens another channel by which potassium ions can leave the cell (**K^+ outward channel**). In addition, the opening of many ion channels is controlled by ligands such as Ca^{++} ions, protons, or by phosphorylation of the channel protein. This enables regulation of the channel activity by metabolic processes and by messenger substances (Chapter 19).

Up to now the amino acid sequences of many channel proteins have been determined. It emerged that certain channels (e.g., those for $K^{\pm}$ ions in bacteria, animals, and plants), are very similar. Roderick MacKinnon and coworkers from the Rockefeller University in New York resolved the three-dimensional structure of the K^+ channel for the bacterium *Streptomyces lividans* using X-ray structure analysis (section 3.3). These pioneering results, for which Roderick MacKinnon was awarded the Nobel Prize in Chemistry in 2003, have made it possible to recognize for the first time the molecular function of an ion channel. It has long been known that the channel protein is built from two identical subunits, each of which has two transmembrane helices connected by a **sequence of about 30 amino acids (loop)** (Fig. 1.29A).

It was already known that this loop is responsible for the ion selectivity of the channel. Structure analysis showed that a K^+ channel is built of four of these subunits (Fig. 1.29B, C). One helix of each subunit (the inner one) lines the channel while the other (outer) helix is directed toward the lipid membrane. The pore's interior consists of a channel filled with water, which is separated from the outside by a selection filter. This filter is formed from the loops of the four subunits mentioned previously. It has such a small pore that the K^+ ions first have to **strip off their hydrate coat** before they can pass through. In order to compensate for the large amount of energy required to dehydrate the K^+ ions, the pore is lined with a circular array of oxygen atoms that act as a **"water substitute"** and form a complex with the K^+ ions. The pore is also negatively charged to bind the cations. The K^+ ions entering the pore push those already bound in the pore through to the other side of the filter. Na^+ ions are too small to be complexed in the pore's selection filter; they are unable to strip off their hydrate coat and therefore their passage through the filter is blocked. This explains the K^+ channel's ion selectivity. This function of the aforementioned loops between transmembrane helices as a selectivity filter appears to be a common characteristic of K^+ ion channels in microorganisms and the animal and plant kingdom. Studies with K^+ channels from plants revealed that a decisive factor in the K^+ selectivity is the presence of glycine residues in the four loops of the selectivity filter. The recent elucidation of the three-dimensional structure of chloride channels from the bacteria *Salmonella* and *E. coli* (belonging to a large family of anion channels from prokaryotic and eukaryotic organisms), yielded analogous results. Also these channels, formed from transmembrane α-helices, contain in their interior loops, which function as selectivity filters by electrostatic interaction with and coordinating binding of the chloride anion. It appears now that the features the K^+ channel of *Streptomyces* shown above reveal a general principle of the mechanism by which a specific ion channel functions.

There are similarities between the basic structure of ion channels and translocators. It was shown that translocators, such as the chloroplast triose phosphate-phosphate translocator, can be converted by reaction with chemical agents into a channel, open to both sides at the same time, with an ion conductivity similar to the ion channels discussed previously. The functional differences between translocators and ion channels—the translocation pore of translocators is accessible only from one side at a time and transport involves a change of conformation, whereas in open ion channels the aqueous pore is open to both sides—is due to differences in the filling of the pore by peptide chains, providing the translocator pore with a gate.

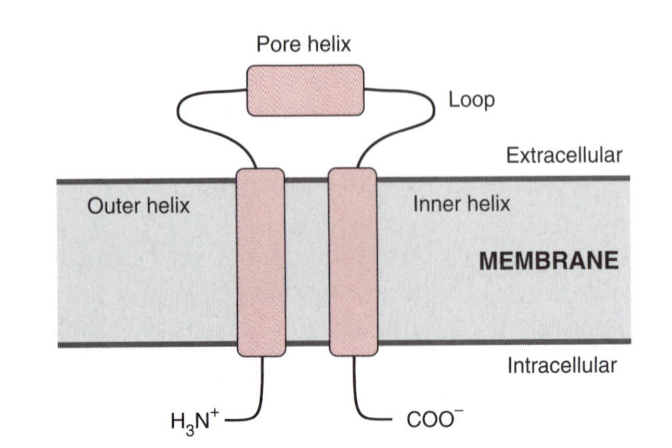
Pore helix
Loop
Extracellular
Outer helix
Inner helix
MEMBRANE
Intracellular
H_3N^+
COO^-

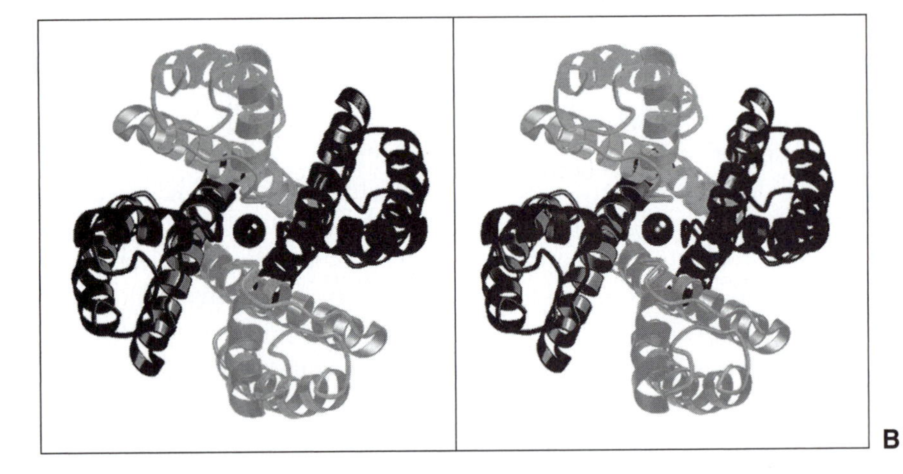
B

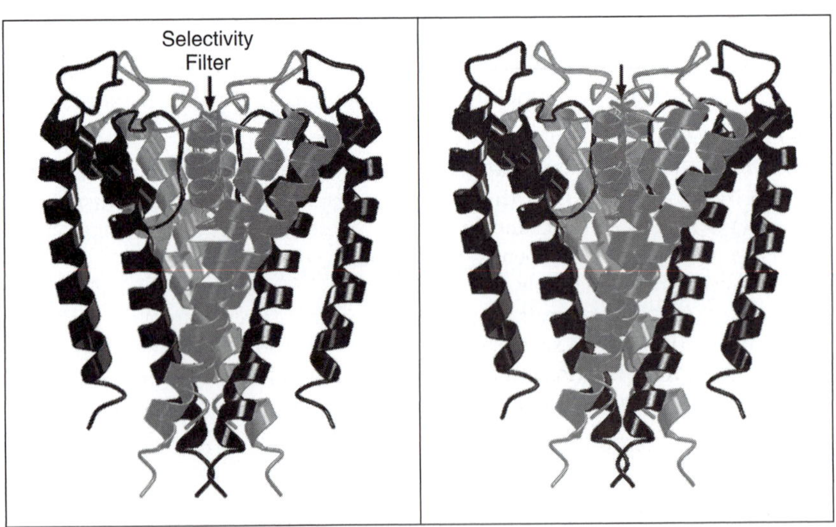
Selectivity Filter
C

1.11 Porins consist of β-sheet structures

As already mentioned, the outer membranes of chloroplasts and mitochondria appear to be unspecifically permeable to metabolites such as, for instance, nucleotides and sugar phosphates. This relatively unspecific permeability is due to pore-forming proteins named **porins**.

The size of the aperture of the pore formed by a porin can be determined by incorporating porins into an artificial lipid membrane that separates two chambers filled with an electrolyte (Fig. 1.30). Membrane proteins, which have been solubilized in a detergent, are added to one of the two chambers. Because of their hydrophobicity, the porin molecules incorporate, one after another, into the artificial lipid membrane, each time forming a new channel, which can be seen in the stepwise increase in conductivity. As each stepwise increase in conductivity corresponds to the conductivity of a single pore, it is possible to evaluate the size of the aperture of the pore from the conductivity of the electrolyte fluid. Thus the size of the aperture for the porin pore of mitochondria has been estimated to be 1.7 nm and that of chloroplasts about 3 nm.

Porins have been first identified in the outer membrane of gram-negative bacteria, such as *Escherichia coli.* In the meantime, several types of porins, differing in their properties, have been characterized. **General porins** form unspecific diffusion pores, consisting of a channel containing water, allowing the diffusion of substrate molecules. These porins consist of subunits with a molecular mass of about 30 kDa. Porins in the membrane often occur as **trimers**, in which each of the three subunits forms a pore. Porins differ

Figure 1.29 Structural model of the K^+ channel from *Streptomyces lividans* A. Diagram of the amino acid sequence of a channel protein monomer. The protein forms two transmembrane helices that are connected by a loop. There is still another helix within this loop, which, however, does not protrude through the membrane. B. Stereo pair of a view of the K^+ channel from the extracellular side of the membrane. The channel is formed by four subunits (marked black and red alternately), from which one transmembrane helix always lines the channel (inner helix). The ball symbolizes a K^+ ion. C. Stereo pair of a side view of the K^+ ion channel. The eight transmembrane helices form a spherical channel, which is connected to the wide opening by a selection filter. (Results of X-ray structure analysis by Doyle et al., 1998, with kind permission.) How to look properly at a stereo picture. Sit at a window and look into the distance. Push the picture quickly in front of your eyes without changing the focus. At first you will see three pictures unclearly. Focus your eyes so that the middle picture is the same size as those at either side of it. Now focus sharply on the middle picture. Suddenly you will see a very plastic picture of the spherical arrangement of the molecules.

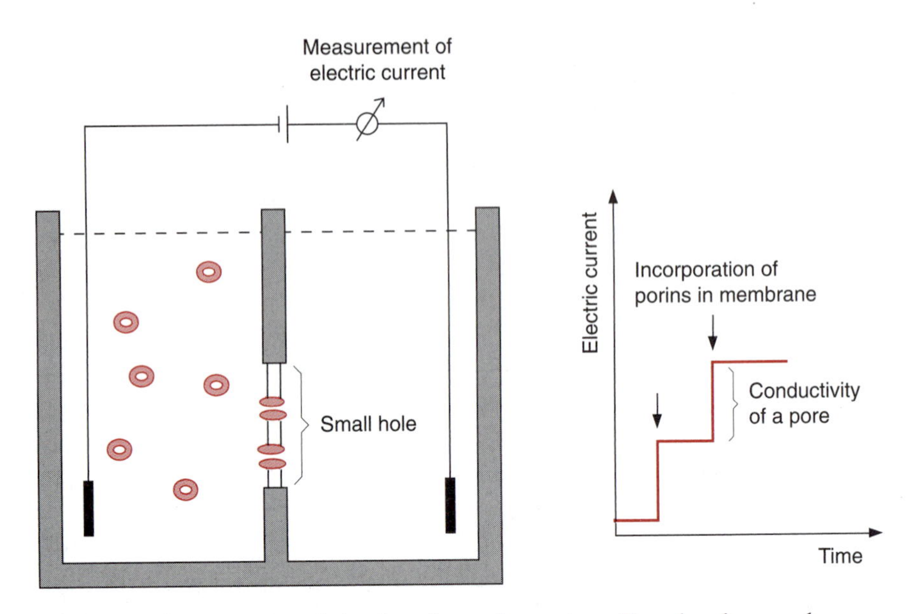

Figure 1.30 Measurement of the size of a porin aperture. Two chambers, each provided with an electrode and filled with electrolyte fluid, are separated from each other by a divider containing a small hole. A small drop containing a membrane lipid is brushed across this hole. The solvent is taken up into the aqueous phase and the remaining lipid forms a double layer, an artificial membrane. Upon the addition of a porin, which has been solubilized from a membrane, spontaneous incorporation of the single porin molecule into the artificial membrane occurs. The aqueous channel through the lipid membrane formed with each incorporation of a porin protein results in a stepwise increase of current activity.

distinctly from the translocator proteins discussed in the preceding in that they have no exclusively hydrophobic regions in their amino acid sequence, a requirement for forming transmembrane α-helices. Analysis of the three-dimensional structure of a bacterial porin by X-ray structure analysis (section 3.3) revealed that the walls of the pore are formed by **β-sheet** structures (Fig. 1.31). Altogether, 16 β-sheets, each consisting of about 13 amino acids and connected to each other by hydrogen bonds, form a pore (Fig. 1.32A). This structure resembles a **barrel** in which the β-sheets represent the barrel staves. Hydrophilic and hydrophobic amino acids alternate in the amino acid sequences of the β-sheets. One side of the β-sheet, occupied by hydrophobic residues, is directed toward the lipid membrane phase. The other side, with the hydrophilic residues, is directed toward the aqueous phase inside the pore (Fig. 1.32B). Compared with the ion channel proteins, the porins have an economical structure in the sense that a much larger channel is formed by one porin molecule than by a channel protein with a two times higher molecular mass. Another type of porin forms **selective**

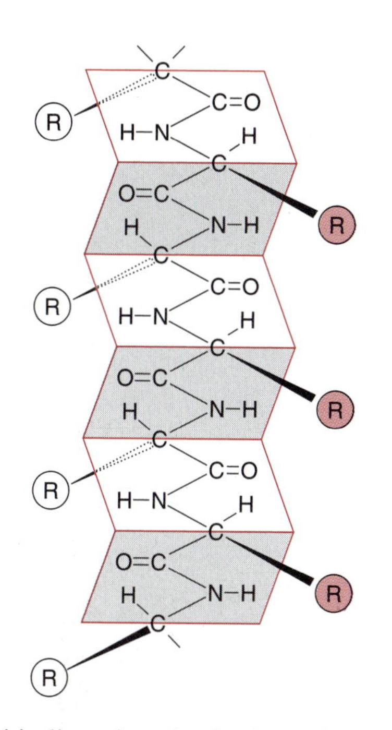

Figure 1.31 With β-sheet conformation the amino acid residues of a peptide chain are arranged alternately in front of and behind the surface of the sheet.

pores, which contain binding sites for ionic and nonionic substrates (e.g., carbohydrates). In *E. coli* a maltodextrin-binding porin was found to consist of 16 β-sheets, with loops in between, which protrude into the aqueous channel of the pore and contain the corresponding substrate binding sites.

The **mitochondrial porin** resembles in its structure the bacterial general porin. It also consists of 16 β-sheets. The measurement of porin activity in artificial lipid bilayer membranes (Fig. 1.30) revealed that the open pore had a slight anion selectivity. Applying a voltage of 30 mV closes the pore to a large extent and renders it cation-specific. For this reason the mitochondrial porin has been named voltage-dependent anion selective channel (**VDAC**). The physiological function of this voltage-dependent regulation of the pore opening remains to be elucidated.

In **chloroplasts** the outer envelope membrane was found to contain a porin with a molecular mass of 24 kDa (outer envelope protein, **OEP24**), forming an unspecific diffusion pore. OEP24 resembles in its function the mitochondrial VDAC, although there is no sequence homology between them. OEP24, in its open state, allows the diffusion of various metabolites.

Moreover, the outer envelope membrane of chloroplasts contains another porin (**OEP21**) forming an **anion selective channel**. OEP21 especially enables the diffusion of phosphorylated metabolites such as dihydroxyacetone phosphate and 3-phosphoglycerate. The opening of this pore is regulated by the binding of substrates. It is presently being investigated as to whether and to what extent the flux of metabolites across the outer enve-

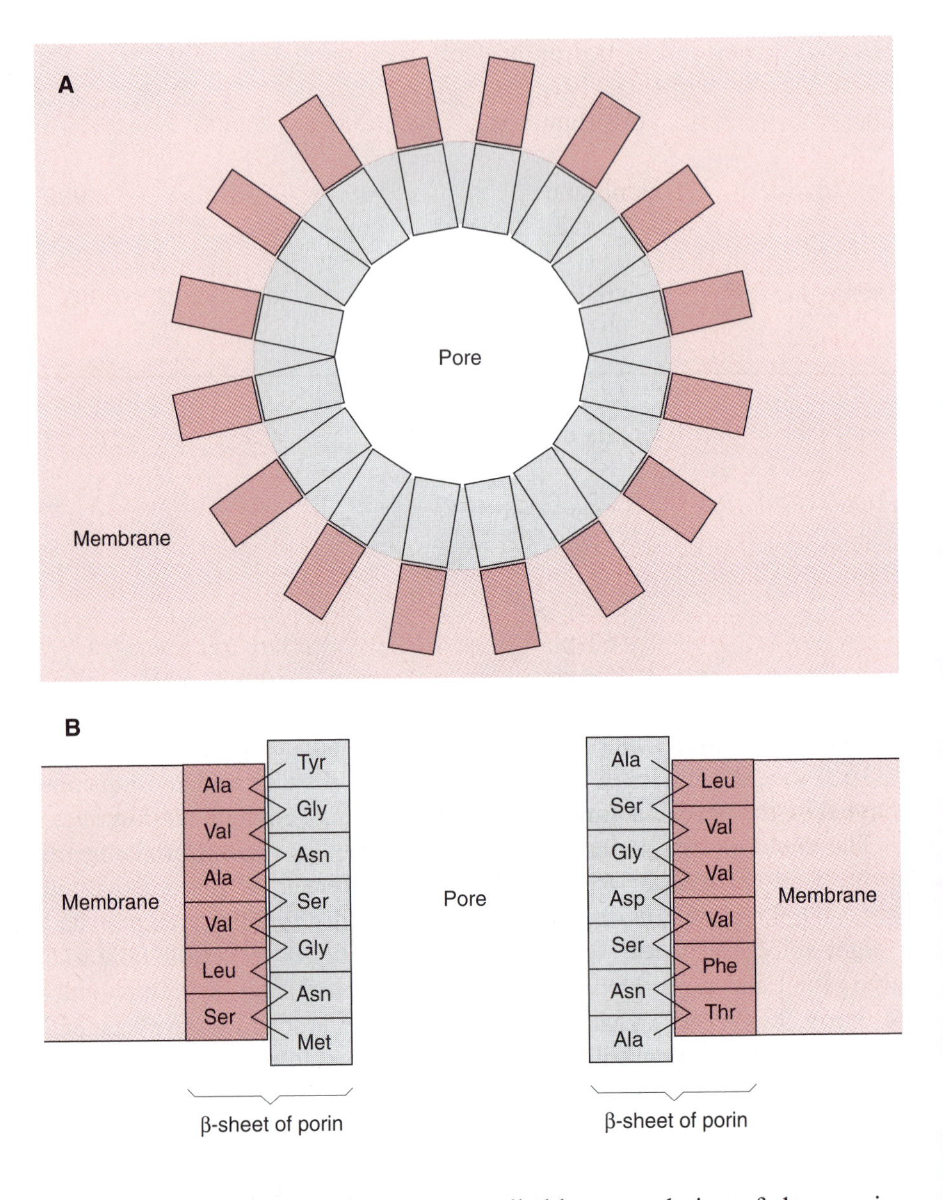

Figure 1.32 Diagram of the structure of a membrane pore formed by a porin. Figure A shows the view from above and figure B shows a cross section through the membrane. Sixteen β-sheet sequences of the porin molecules, each 13 amino acids long, form the pore. The amino acid residues directed toward the membrane side of the pore have hydrophobic character; those directed to the aqueous pore are hydrophilic.

lope membrane of chloroplasts is controlled by a regulation of the opening of OEP24 and OEP21. Another pore-forming protein is located in the membrane of peroxisomes, as will be discussed in Section 7.4.

Further reading

Aaziz, R., Dinant, S., Epel, B. L. Plasmodesmata and plant cytoskeleton. Trends Plant Sci 6, 326–330 (2001).

Benz, R. Permeation of hydrophilic solutes through mitochondrial porins. Biochim Biophys Acta 1197, 167–196 (1994).

Blatt, M. R. Plant potassium channels double up. Trends Plant Sci 2, 244–246 (1997).

Cavalier-Smith, T. Membrane heredity and early chloroplast evolution. Trends Plant Sci 5, 174–182 (2000).

Choi, D., Lee, Yi., Cho, H.-T., Kende, H. Regulation of expansin gene expression affects growth and development in transgenic rice plants. Plant Cell 15, 1386–1398 (2003).

Doyle, D. A., Cabrai, J. M., Pfuetzner, R. A., Kuo, A., Gulbis, J. M., Cohen, S. L., Chait, B. T., MacKinnon, R. The structure of the potassium channel: Molecular basis of K^+ conduction and selectivity. Science 280, 69–77 (1998).

Dutzler, R., Campbell, E. B., Cadene, M., Chait, B. T., MacKinnon, R. X-ray structure of a CIC chloride channel at 3.0 Å reveals the molecular basis of anion selectivity. Nature 415, 287–294 (2002).

Flügge, U.-I., Fischer, K., Gross, A., Sebald, W., Lottspeich, F., Eckerskorn, C. The triose phosphate-3-phosphoglycerate-phosphate translocator from spinach chloroplasts: Nucleotide sequence of a full-length cDNA clone and import of the in vitro synthesized precursor protein into chloroplasts. EMBO J 8, 39–46 (1989).

Flügge, U.-I. Transport in and out of plastids: Does the outer envelope membrane control the flow? Trends Plant Sci 5, 135–137 (2000).

Flügge, U.-I. Phosphate translocators in plastids. Annu Rev Plant Physiol Plant Mol Biol 50, 27–45 (1999).

Frandsen, G. I., Mundy, J., Tzen, J. T. C. Oil bodies and their associated proteins, oleosins and caleosin. Physiologia Plantarum 112, 301–307 (2001).

Gunning, D. E. S., Steer, M. W. Ultrastructure and the biology of plant cells. Edward Arnold, London (1975).

Jackson, D. Opening up the communication channels: Recent insights into plasmodesmal function. Curr Opin Plant Biol 3, 394–399 (2000).

Kjellbom, P., Larsson, C., Johansson, I., Karlsson, M., Johanson, U. Aquaporins and water homeostasis in plants. Trends Plant Sci 4, 308–314 (1999).

Krämer, R. Functional principles of solute transport systems: Concepts and perspectives. Biochim Biophys Acta 1185, 1–34 (1994).

Kuo, A., Gulbis, J. M., Antcliff, J. F., Rahman, T., Lowe, E. D., Zimmer, J., Cuthbertson, J., Ashcroft, F. M., Ezaki, T., Doyle, D. A. Crystal structureof the potassium channel KirBac1.1 in the closed state. Science 300, 1922–1926 (2003).

Li, X., Franceschi, V. R., Okita, T. W. Segregation of storage protein mRNAs on the rough endoplasmatic reticulum membranes of rice endosperm cells. Cell 72, 869–879 (2001).

Lucas, W. J., Wolf, S. Connections between virus movement, macromolecular signaling and assimilate allocation. Curr Opin Plant Biol 2, 192–197 (1999).

Maathius, F. J. M., Ichida, A. M., Sanders, D., Schroeder, J. I. Roles of higher plant K^+ channels. Plant Physiol 114, 1141–1149 (1997).

Maeser, P., Hosoo, Y., Goshima, S., Horie, T., Eckelman, B., Yamada, K., Yoshida, K., Bakker, E. P., Shinmyo, A., Oiki, S., Schroeder J. I., Uozumi N. Glycine residues in potassium channel-like selectivity filters determine potassium selectivity in four-loop-per-subunit HKT transporters from plants. Proc Natl Acad Sci USA 99, 6428–6433 (2002).

Moreira, D., Le Guyader, H., Philippe, H. The origin of red algae and the evolution of chloroplasts. Nature 405, 69–72 (2000).

Murata, K., Mitsuoka, K., Hirai, T., Walz, T., Agre, P., Heymann, J. B., Engel, A., Fujiyoshi, Y. Structural determinants of water permeation through aquaporin-1. Nature 407, 599–605, (2000).

Murphy, D. J., Herandez-Pinzon, I., Patel, K. Role of lipid body proteins in seeds and other tissues. J Plant Physiol 158, 471–478 (2001).

Nebenführ, A., Staehelin, L. A. Mobile factories: Golgi dynamics in plant cells. Trends Plant Sci 6, 160–167 (2001).

Overall, R. L., Blackman, L. M. A model on the macromolecular structure of plasmodesmata. Trends Plant Sci 307–311 (1996).

Pebay-Peyroula, E., Dahout-Gonzalez, C., Kahn, R., Trezeguet, V., Lauquin, G.-J.-M., Brandolin, G. Structure of mitochondrial ATP/ADP carrier in complex with carboxyatractyloside. Nature 426, 39–44 (2003),

Pennell, R. Cell walls: Structures and signals. Curr Opin Plant Biol 1, 504–510 (1998).

Pohlmeyer, K., Soll, J., Grimm, R., Hill, K., Wagner, R. A high-conductance solute channel in the chloroplastic outer envelope from pea. Plant Cell 10, 1207–1216 (1998).

Reiter, W. D. The molecular analysis of cell wall components. Trends Plant Sci 3, 27–32 (1998).

Robinson, D. G. Plant membranes. John Wiley and Sons, New York (1985).

Robinson, D. G. ed. The Golgi apparatus and the plant secretory pathway. Blackwell, Oxford, UK (2003).

Santoni, V., Gerbeau, P., Javot, H., Maurel, C. The high diversity of aquaporins reveals novel facets of plant membrane functions. Curr Opin Plant Biol 3, 476–481 (2000).

Sul, H., Han, B.-G., Lee, J. K., Wallan, P., Jap, B. K. Structural basis of water-specific transport through theAQP1 water channel. Nature 414, 872–878 (2001).

Tanner, W., Carpari, T. Membrane transport carriers. Annu Rev Plant Physiol Plant Mol Biol 47, 595–626 (1996).

Uehlein, N., Lovisolo, C., Siefritz, F., Kaldenhoff, R. The tobacco aquaporin NtAQP1 is a membrane CO_2 pore with physiological functions. Nature 425, 734–737 (2003).

Very, A.-A., Sentenac, H. Molecular mechanism and regulation of K^+ transport in higher plants. Annu Rev Plant Biol 54, 575–602 (2002).

Wandelt, C. I., Kahn, M. R. I., Craig, S., Schroeder, H. E., Spencer, D. E., Higgins, T. J. V. Vicilin with carboxy-terminal KDEL is retained in the endoplasmatic reticulum and accumulates to high levels in leaves of transgenic plants. Plant J 2, 181–192 (1992).

Ward, J. M. Patch-clamping and other molecular approaches for the study of plasma membrane transporters demystified. Plant Physiol 114, 1151–1159 (1997).

Winter, H., Robinson, D. G., Heldt, H. W. Subcellular volumes and metabolite concentrations in spinach leaves. Planta 193, 530–535 (1994).

Zhou, Y., Morais-Cabral, J. H., Kaufmann, A., MacKinnon, R. Chemistry of ion coordination and hydration revealed by a K^+ channel-Fab complex at 2.0Å resolution. Nature 414, 4348 (2001).

2

The use of energy from sunlight by photosynthesis is the basis of life on earth

Plants and cyanobacteria capture the light of the sun and utilize its energy to synthesize organic compounds from inorganic substances such as CO_2, nitrate, and sulfate to make their cellular material; they are photoautotrophic. In photosynthesis photon energy splits water into oxygen and hydrogen, the latter bound as NADPH. This process, termed the *light reaction*, takes place in the photosynthetic reaction centers embedded in membranes. It involves the transport of electrons, which is coupled to the synthesis of ATP. NADPH and ATP are consumed in a so-called dark reaction to synthesize carbohydrates from CO_2 (Fig. 2.1). The photosynthesis of plants and cyanobacteria created the biomass on earth, including the deposits of fossil fuels and atmospheric oxygen. Animals are dependent on the supply of carbohydrates and other organic substances as food; they are heterotrophic. They generate the energy required for their life processes by oxidizing the biomass, which has first been produced by plants. When oxygen is consumed, CO_2 is formed. It follows that the light energy captured by plants is also the source of energy for the life processes of animals.

2.1 How did photosynthesis start?

Measurements of the distribution of radioisotopes led to the conclusion that the earth was formed about 4.6 billion years ago. The earliest indicators of life on earth are fossils of bacteria-like structures, estimated to be 3.5 billion years old. There was no oxygen in the atmosphere when life on earth

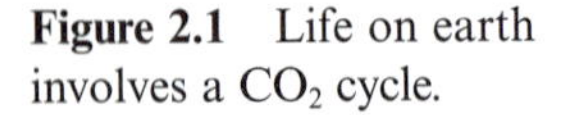
Figure 2.1 Life on earth involves a CO_2 cycle.

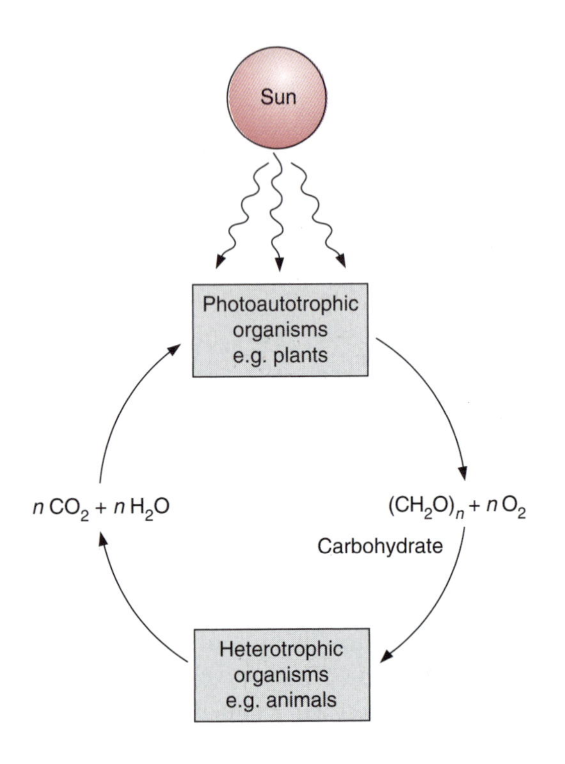

commenced. This is concluded from the fact that in very early sediment rocks iron is present as Fe^{2+}. Mineral iron is oxidized to Fe^{3+} in the presence of oxygen. According to our present knowledge, the earth's atmosphere initially contained components such as carbon dioxide, molecular hydrogen, methane, ammonia, prussic acid, and water.

In 1922 the Russian scientist Alexander Oparin presented the interesting hypothesis that organic compounds were formed spontaneously in the early atmosphere by the input of energy [e.g., in the form of ultraviolet radiation (there was no protective ozone layer), electrical discharges (lightning), or volcanic heat]. It was further postulated that these organic compounds accumulated in ancient seas and became the constituents of early forms of life. In 1953 the American scientists Stanley Miller and Harold Urey substantiated this hypothesis by simulating the postulated **prebiotic synthesis** of organic substances. They exposed a gaseous mixture of components present in the early atmosphere, consisting of H_2O, CH_4, NH_3 and H_2 to electrical discharges for about a week at 80°C. Amino acids (such as glycine and alanine) and other carboxylic acids (such as formic, acidic, lactic, and succinic acid) were found in the condensate of this experiment. Other investigators added substances such as CO_2, HCN, and formaldehyde to the gaseous mixture, and these experiments showed that many components of living cells (e.g., carbohydrates, fatty acids, tetrapyrroles, and the nucleobases adenine, guanine, cytosine, and uracil) were formed spontaneously by exposing a postulated early atmosphere to electric or thermal energy.

There is a hypothesis that the organic substances formed by the abiotic processes mentioned previously, accumulated in the ancient seas, lakes, and pools over a long period of time prior to the emergence of life on earth. There was no oxygen to oxidize the substances that had accumulated and no bacteria or other organisms to degrade them. Oparin had already speculated that a **"primordial" soup** was formed in this way, providing the building material for the origin of life. Since oxygen was not yet present, the first organisms must have been **anaerobes**.

It is widely assumed now that early organisms on this planet generated the energy for their subsistence by **chemolithotrophic metabolism**, for example, by the reaction:

$$FeS + H_2S \longrightarrow FeS_2 + H_2 \quad (\Delta G^{o\prime} = -42\,kJ/mol).$$

It seems likely that already at a very early stage of evolution the catalysis of this reaction was coupled to the generation of a proton motive force (section 4.1) across the cellular membrane, yielding the energy for the synthesis of ATP by a primitive ATP synthase (section 4.3). **Archaebacteria**, which are able to live anaerobically under extreme environmental conditions (e.g., near hot springs in the deep sea), and which are regarded as the closest relatives of the earliest organisms on earth, are able to produce ATP via the preceding reaction. It was probably a breakthrough for the propagation of life on earth when organisms evolved that were able to utilize the energy of the sun as a source for biomolecule synthesis, which occurred at a very early stage in evolution. The now widely distributed **purple bacteria** and **green sulfur bacteria** may be regarded as relics from an early period in the evolution of photosynthesis.

Prior to the description of photosynthesis in Chapter 3, the present chapter will discuss how plants capture sunlight and how the light energy is conducted into the photosynthesis apparatus.

2.2 Pigments capture energy from sunlight

The energy content of light depends on its wavelength

In Berlin at the beginning of the twentieth century Max Planck and Albert Einstein, two Nobel Prize winners, carried out the epoch-making studies proving that light has a dual nature. It can be regarded as an electromagnetic wave as well as an emission of particles, which are termed **light quanta** or **photons**.

The energy of the photon is proportional to its frequency v:

$$E = h \cdot v = h \cdot \frac{c}{\lambda} \tag{2.1}$$

where h is the Planck constant ($6.6 \cdot 10^{-34}$ J s) and c the velocity of the light ($3 \cdot 10^8$ m s^{-1}). λ is the wavelength of light.

The mole (abbreviated to mol) is used as a chemical measure for the amount of molecules and the amount of photons corresponding to $6 \cdot 10^{23}$ molecules or photons (Avogadro number N_A). The energy of one mol photons amounts to:

$$E == h \cdot \frac{c}{\lambda} \cdot N_A \tag{2.2}$$

In order to utilize the energy of a photon in a thermodynamic sense, this energy must be at least as high as the Gibbs free energy of the photochemical reaction involved. [In fact much energy is lost during energy conversion (see section 3.4), with the consequence that the energy of the photon must be higher than the Gibbs free energy of the corresponding reaction.] We can equate the Gibbs free energy ΔG with the energy of the absorbed light:

$$\Delta G = E = h \cdot \frac{c}{\lambda} \cdot N_A \tag{2.3}$$

The introduction of numerical values of the constants h, c, and N_A yields:

$$\Delta G = 6.6 \cdot 10^{-34} \cdot (\mathrm{J} \cdot \mathrm{s}) \cdot \frac{3 \cdot 10^8 (\mathrm{m})}{(\mathrm{s})} \cdot \frac{1}{\lambda(\mathrm{m})} \cdot \frac{6 \cdot 10^{23}}{(\mathrm{mol})} \tag{2.4}$$

$$\Delta G = \frac{119{,}000}{\lambda\,(\mathrm{nm})} \quad [\mathrm{kJ\,/\,mol\ photons}] \tag{2.5}$$

It is often useful to state the electrical potential (ΔE) of the irradiation instead of energy when comparing photosynthetic reactions with redox reactions, which will be discussed in Chapter 3:

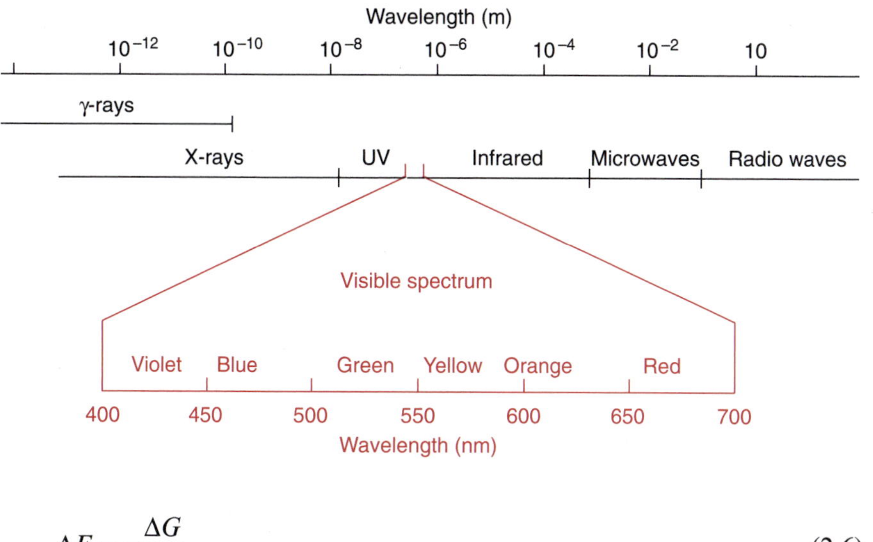

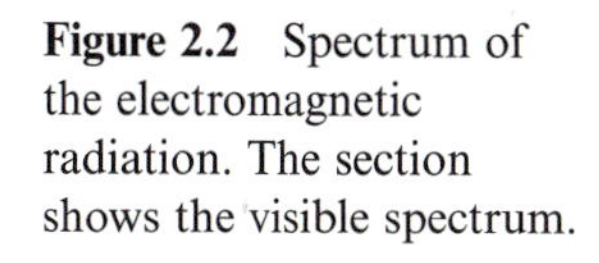
Figure 2.2 Spectrum of the electromagnetic radiation. The section shows the visible spectrum.

$$\Delta E = -\frac{\Delta G}{F} \tag{2.6}$$

where F = number of charges per mol = 96,480 Amp · s · mol^{-1}. The introduction of this value yields:

$$\Delta E = -\frac{N_A \cdot h \cdot c}{\mathrm{F} \cdot \lambda(\mathrm{nm})} = \frac{1231}{\lambda(\mathrm{nm})} \quad [\mathrm{Volt}] \tag{2.7}$$

The human eye perceives only the small range between about 400 and 700 nm of the broad spectrum of electromagnetic waves (Fig. 2.2). The light in this range, where the intensity of solar radiation is especially high, is utilized in plant photosynthesis. Bacterial photosynthesis, however, is able to utilize light in the infrared range.

According to equation 2.3 the energy of irradiated light is inversely proportional to the wavelength. Table 2.1 shows the light energy per mol photons for light of different colors. Consequently, violet light has an energy of about 300 kJ/mol photons. Dark red light, with the highest wavelength (700 nm) that can still be utilized by plant photosynthesis, contains 170 kJ/mol photons. This is only about half the energy content of violet light.

Chlorophyll is the main photosynthetic pigment

In photosynthesis of a green plant, light is collected primarily by **chlorophylls**, pigments that absorb light at a wavelength below 480 nm and between 550 and 700 nm (Fig. 2.3). When white sunlight falls on a chlorophyll layer, the green light with a wavelength between 480 and 550 nm is not absorbed, but is reflected. This is why plant chlorophylls and whole leaves are green.

Table 2.1: The energy content and the electrochemical potential difference of photons of different wavelengths

Wavelengths (nm)	Light color	Energy content kJ/mol Photons	ΔE e Volt
700	Red	170	1.76
650	Bright red	183	1.90
600	Yellow	199	2.06
500	Blue green	238	2.47
440	Blue	271	2.80
400	Violet	298	3.09

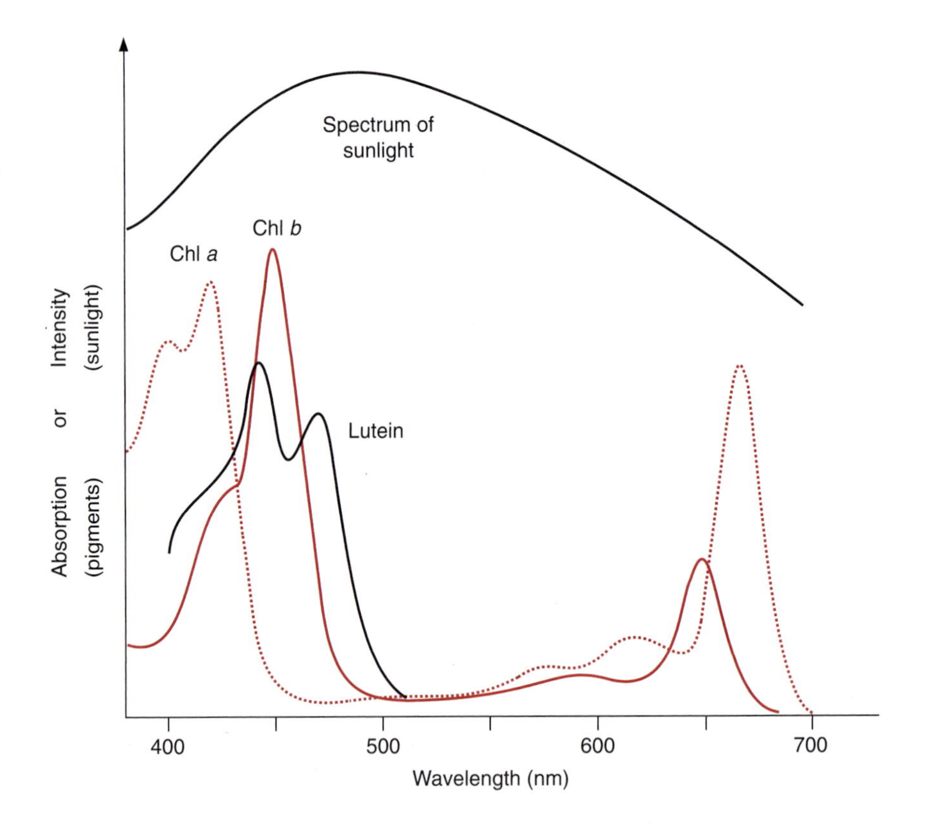

Figure 2.3 Absorption spectrum of chlorophyll-*a* (chl-*a*) and chlorophyll-*b* (chl-*b*) and of the xanthophyll lutein dissolved in acetone. The intensity of the sun's radiation at different wavelengths is given as a comparison.

Experiments carried out between 1905 and 1913 in Zurich and Berlin by Richard Willstätter and his collaborators led to the discovery of the structural formula of the green leaf pigment chlorophyll, a milestone in the history of chemistry. This discovery made such an impact that Richard Willstätter was awarded the Nobel Prize in Chemistry as early as 1915. There are different classes of chlorophylls. Figure 2.4 shows the structural formula

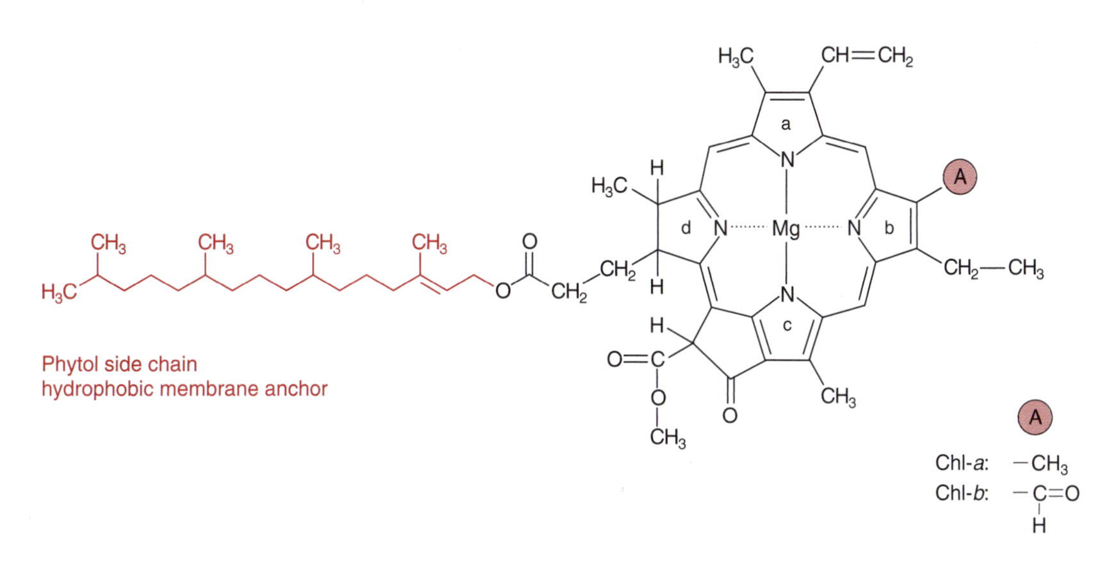

Figure 2.4 Structural formula of chlorophyll-*a*. In chlorophyll-*b* the methyl group in ring b is replaced by a formyl group (red). The phytol side chain gives chlorophyll a lipid character.

of chlorophyll-*a* and chlorophyll-*b* (**chl-*a*, chl-*b***). The basic structure is a ring made of four pyrroles, a **tetrapyrrole**, which is also named **porphyrin**. **Mg^{++}** is present in the center of the ring as the central atom. Mg^{++} is covalently bound with two N-atoms and coordinately bound to the other two atoms of the tetrapyrrole ring. A cyclopentanone is attached to ring c. At ring d a propionic acid group forms an ester with the alcohol **phytol**. Phytol consists of a long branched hydrocarbon chain with one C-C double bond. It is derived from an isoprenoid, formed from four isoprene units (section 17.7). This long hydrophobic hydrocarbon tail renders the chlorophyll highly soluble in lipids and therefore promotes its presence in the membrane phase. Chlorophyll always occurs bound to proteins. In ring b, chl-*b* contains a formyl residue instead of the methyl residue in chl-*a*. This small difference has a large influence on light absorption. Figure 2.3 shows that the absorption spectra of chl-*a* and chl-*b* differ markedly.

In plants, the ratio chl-*a* to chl-*b* is about three to one. Only chl-*a* is a constituent of the photosynthetic reaction centers (sections 3.6 and 3.8) and therefore it can be regarded as the central photosynthesis pigment. In a wide range of the visible spectrum, however, chl-*a* does not absorb light. This nonabsorbing region is named the **"green window."** The absorption gap is narrowed by the light absorption of chl-*b*, with its first maximum at a higher wavelength than chl-*a* and the second maximum at a lower wavelength. As shown in section 2.4, the light energy absorbed by chl-*b* can be transferred very efficiently to chl-*a*. In this way chl-*b* enhances the plant's efficiency for utilizing sunlight energy.

The structure of chlorophylls has remained remarkably constant during the course of evolution. Purple bacteria, probably formed more than 3 billion years ago, contain as photosynthetic pigment a bacteriochlorophyll-

a, which differs from the chlorophyll-*a* shown in Figure 2.4 only by the alteration of one side chain and by the lack of one double bond. This, however, influences light absorption; both absorption maxima are shifted outward and the nonabsorbing spectral region in the middle is broadened. This shift allows purple bacteria to utilize light in the infrared region.

The tetrapyrrole ring not only is a constituent of chlorophyll but also has attained a variety of other functions during evolution. It is involved in methane formation by bacteria with Ni as the central atom. With Co it forms **cobalamin** (vitamin B_{12}), which participates as a cofactor in reactions in which hydrogen and organic groups change their position. With Fe^{++} instead of Mg^{++} as the central atom, the tetrapyrrole ring forms the basic structure of **hemes** (Fig. 3.24), which, on the one hand, as cytochromes, function as redox carriers in electron transport processes (sections 3.7 and 5.5) and, on the other hand, as myoglobin or hemoglobin, store or transport oxygen in aerobic organisms. The tetrapyrrole ring in animal hemoglobin differs only slightly from the tetrapyrrole ring of chl-*a* (Fig. 2.4).

It seems remarkable that a substance that attained a certain function during evolution is being utilized after only minor changes for completely different functions. The reason for this functional variability of substances such as chlorophyll or heme is that their reactivity is governed to a great extent by the proteins to which they are bound.

Chlorophyll molecules are bound to chlorophyll-binding proteins. In a complex with proteins the absorption spectrum of the bound chlorophyll may differ considerably from the absorption spectrum of the free chlorophyll. The same applies for other light-absorbing substances, such as carotenoids, xanthophylls, and phycobilins, which also occur bound to proteins. These substances will be discussed in the following sections. In this text, free absorbing substances are called **chromophore** (Greek, carrier of color) and the chromophore-protein complexes are called **pigments**. Pigments are often named after the wavelength of their absorption maximum. Chlorophyll-a_{700} means a pigment of chl-*a* with an absorption maximum of 700 nm. Another common designation is P_{700}; this name leaves the nature of the chromophore open.

2.3 Light absorption excites the chlorophyll molecule

What happens when a chromophore absorbs a photon? When a photon with a certain wavelength hits a chromophore molecule that absorbs light of this wavelength, the energy of the photon excites electrons to a higher energy

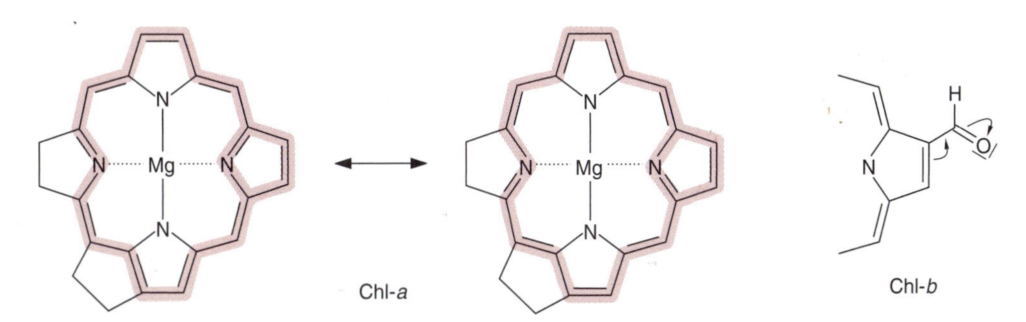

Figure 2.5 Resonance structures of chlorophyll-*a*. In the region marked red, the double bonds are not localized; the π electrons are distributed over the entire conjugated system. The formyl residue of chlorophyll-*b* attracts electrons and thus affects the π electrons of the conjugated system.

level. This occurs as an "all or nothing" process. According to the principle of energy conservation expressed by the first law of thermodynamics, the energy of the chromophore is increased by the energy of the photon, which results in an **excited state** of the chromophore molecule. The energy is absorbed only in discrete quanta, resulting in discrete excitation states. The energy required to excite a chromophore molecule depends on the chromophore structure. A general property of chromophores is that they contain many **conjugated double bonds**, 10 in the case of the tetrapyrrole ring of chl-*a*. These double bonds are delocalized. Figure 2.5 shows two possible resonance forms.

After absorption of energy, an electron of the conjugated system is elevated to a higher orbit. This excitation state is termed a **singlet**. Figure 2.6 shows a scheme of the excitation process. As a rule, the higher the number of double bonds in the conjugated system, the lower the amount of energy required to produce a first singlet state. For the excitation of chlorophyll, dark red light is sufficient, whereas butadiene, with only two conjugated double bonds, requires energy-rich ultraviolet light for excitation. The light absorption of the conjugated system of the tetrapyrrole ring is influenced by the side chains. Thus, the differences in the absorption maxima of chl-*a* and chl-*b* mentioned previously can be explained by an electron attracting effect of the carbonyl side chain in ring b of chl-*b* (Fig. 2.5).

The spectra of chl-*a* and chl-*b* (Fig. 2.3) each have two main absorption maxima, showing that each chlorophyll has two main excitation states. In addition, chlorophylls have minor absorption maxima, which for the sake of simplicity will not be discussed here. The two main excitation states of chlorophyll are known as the first and second singlet (Fig. 2.6). The absorption maxima in the spectra are relatively broad. At a higher resolution the spectra can be shown to consist of many separate absorption lines. This fine

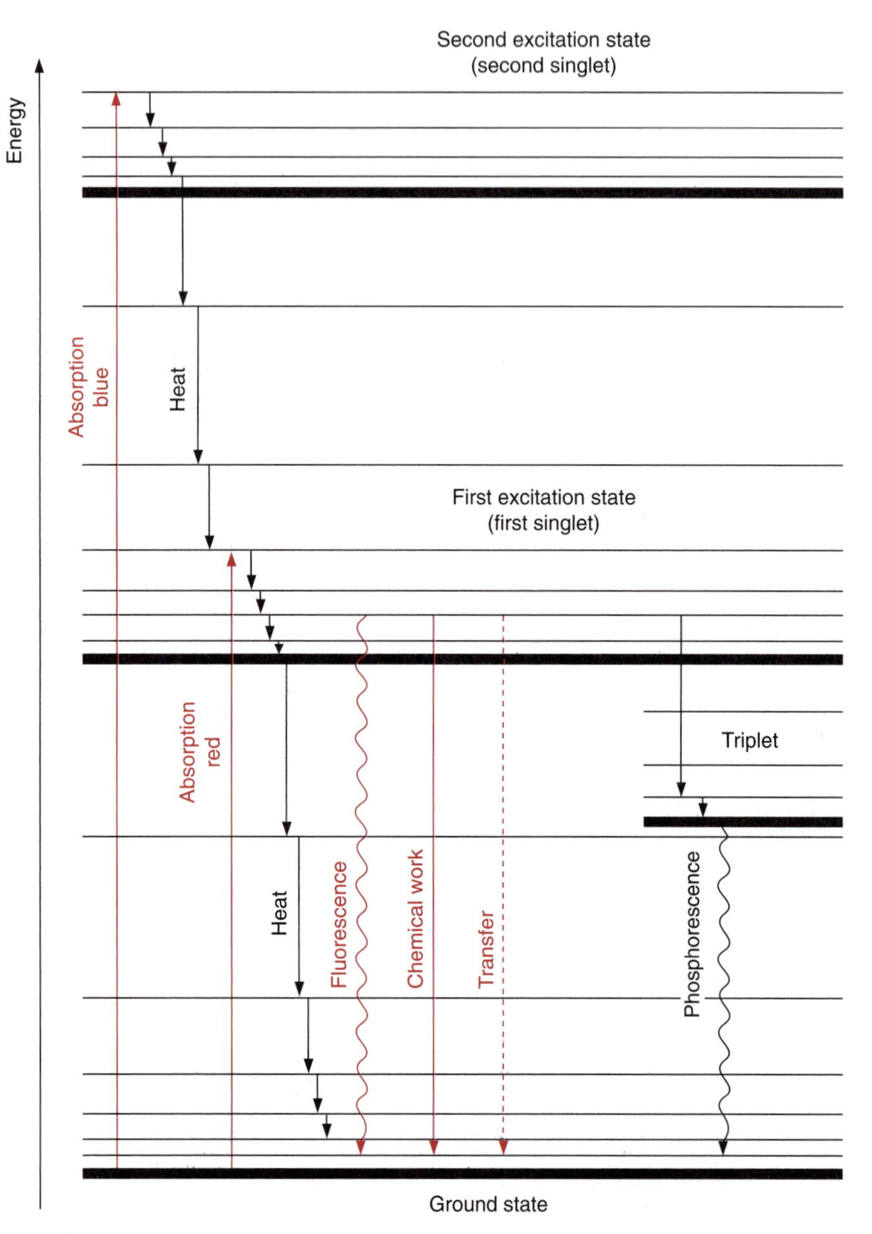

Figure 2.6 Scheme of the excitation states of chlorophyll-*a* and its return to the ground state, during which the released excitation energy is converted to photochemical work, fluorescent or phosphorescent light, or dissipated into heat. This simplified scheme shows only the excitation states of the two main absorbing maxima of the chlorophylls. The second excitation state shown here is in reality the third singlet.

structure of the absorption spectra is due to chlorophyll molecules in the ground and in the singlet states being in various states of **rotation** and **vibration**. In the energy diagram scheme of Figure 2.6 the various rotation and vibration energy levels are drawn as fine lines and the corresponding ground states as solid lines.

The energy levels of the various rotation and vibration states of the ground state overlap with the lowest energy levels of the **first singlet**. Analogously, the energy levels of the first and the second singlet also overlap. If a chlorophyll molecule absorbs light in the region of its absorption maximum in blue light, one of its electrons is elevated to the **second singlet** state. This second singlet state with a half-life of only 10^{-12} s is too unstable to use its energy for chemical work. The excited molecules lose energy in form of heat by rotations and vibrations until the first singlet state is reached. This first singlet state can also be attained by absorption of a photon of red light, which contains less energy. The first singlet state is much more stable than the second one; its half-life amounts to $4 \cdot 10^{-9}$ s.

The return of the chlorophyll molecule from the first singlet state to the ground state can proceed in different ways

1. The most important path for conversion of the energy released when the first singlet state returns to the ground state is its utilization for **chemical work**. The chlorophyll molecule transfers the excited electron from the first singlet state to an electron acceptor and a positively charged chlorophyll radical $chl^{+\cdot}$ remains. This is possible since the excited electron is bound less strongly to the chromophore molecule than in the ground state. Section 3.5 describes in detail how the electron can be transferred back from the acceptor to the $chl^{+\cdot}$ radical via an electron transport chain, by which the chlorophyll molecule returns to the ground state and the free energy derived from this process is conserved for chemical work. As an alternative, the electron deficit in the $chl^{+\cdot}$ radical may be replenished by another electron donor [e.g., water (section 3.6)].

2. The excited chlorophyll can return to the ground state by releasing excitation energy in the form of light; this light emittance is named **fluorescence**. Due to vibrations and rotations, part of the excitation energy is usually lost beforehand in the form of heat, with the result that the fluorescence light has less energy (longer in wavelength) than the energy of the excitation light, which was required for attaining the first singlet state (Fig. 2.7).

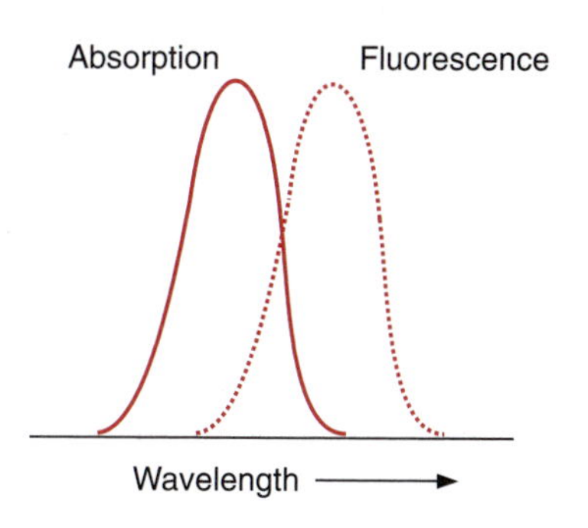

Figure 2.7 Fluorescent light generally has a longer wavelength than excitation light.

3. It is also possible that the return from the first singlet to the ground state proceeds in a stepwise fashion via the various levels of vibration and rotation energy, by which the energy difference is completely converted to heat.

4. By releasing part of the excitation energy in the form of heat, the chlorophyll molecule can attain an excited state of lower energy, called the first **triplet state**. This triplet state cannot be reached directly from the ground state by excitation. In the triplet state the spin of the excited electrons has been reversed. As the probability of a spin reversal is low, the triplet state does not occur frequently, but in the case of a very high excitation, part of the chlorophyll can reach this state. By emitting so-called **phosphorescent light**, the molecule can return from the triplet state to the ground state. Phosphorescent light is again lower in energy than the light required to attain the first singlet state. The return from the triplet state to the ground state requires a reversal of the **electron spin**. As this is rather improbable, the triplet state, in comparison to the first singlet state, has a relatively long life (half-life 10^{-4} to 10^{-2} s.). The triplet state of the chlorophyll has no function in photosynthesis *per se*. In its triplet state, however, the chlorophyll can excite oxygen to a singlet state, whereby the oxygen becomes very reactive with a damaging effect on cell constituents. Section 3.10 describes how the plant manages to protect itself from the **harmful singlet oxygen**.

5. The return to the ground state can be coupled with the excitation of a neighboring chromophore molecule. This transfer is important for the function of the antennae and will be described in the following section.

2.4 An antenna is required to capture light

In order to excite a photosynthetic reaction center, a photon with a defined energy content has to react with a chlorophyll molecule in the reaction center. The probability is very slight that a photon not only has the proper energy, but also hits the pigment exactly at the site of the chlorophyll molecule. Therefore efficient photosynthesis is possible only when the energy of photons of various wavelengths is captured over a certain surface by a so-called **antenna** (Fig. 2.8). Similarly, radio and television sets could not work without an antenna.

The antennae of plants consist of a large number of protein-bound chlorophyll molecules that absorb photons and transfer their energy to the reaction center. Only a few thousandths of the chlorophyll molecules in the leaf are constituents of the actual reaction centers; the remainder are contained in the antennae. Observations made as early as 1932 by Robert Emerson and William Arnold in the United States indicated that the large

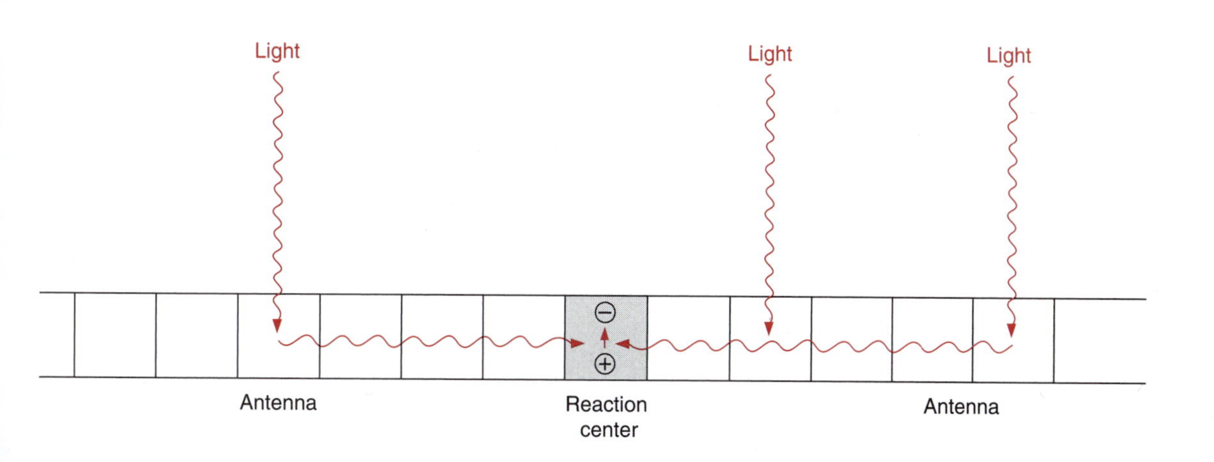

Figure 2.8 Photons are collected by an antenna and their energy is transferred to the reaction center. In this scheme the squares represent chlorophyll molecules. The excitons conducted to the reaction center cause a charge separation (see section 3.4).

majority of chlorophyll molecules are not part of the reaction centers. The two researchers illuminated a suspension of the green alga *chlorella* with light pulses of 10 μs duration, interrupted by dark intervals of 20 ms. The evolution of oxygen was used as a measure for photosynthesis. The light pulses were made so short that chlorophyll could undergo only one photosynthetic excitation cycle and a high light intensity was chosen in order to achieve maximum oxygen evolution. Apparently the photosynthetic apparatus was thus saturated with photons. Analysis of the chlorophyll content of the algae suspension showed that only one molecule of O_2 was formed per 2,400 chlorophyll molecules under saturating conditions.

In the following years Emerson refined these experiments and was able to show that when the pulses had a very low light intensity, the amount of oxygen formed increased proportionally with the light intensity. From this it was calculated that the release of one molecule of oxygen had a minimum quantum requirement of about **eight photons**. These results settled a long scientific dispute with Otto Warburg, who had concluded from his experiments that only four photons are required for the evolution of one molecule of O_2.

The results of Emerson and Arnold led to the conclusion that when the quantum requirement is evaluated at eight photons per molecule of O_2 formed, and, as was recognized later, two reaction centers require four photons each upon the formation of O_2, about **300 chlorophyll molecules** are associated with one reaction center. These are constituents of the **antennae**.

The antennae contain additional accessory pigments to utilize those photons where the wavelength corresponds to the green window between the absorption maxima of the chlorophylls. In higher plants these pigments are carotenoids, mainly **xanthophylls**, including lutein and the related violaxan-

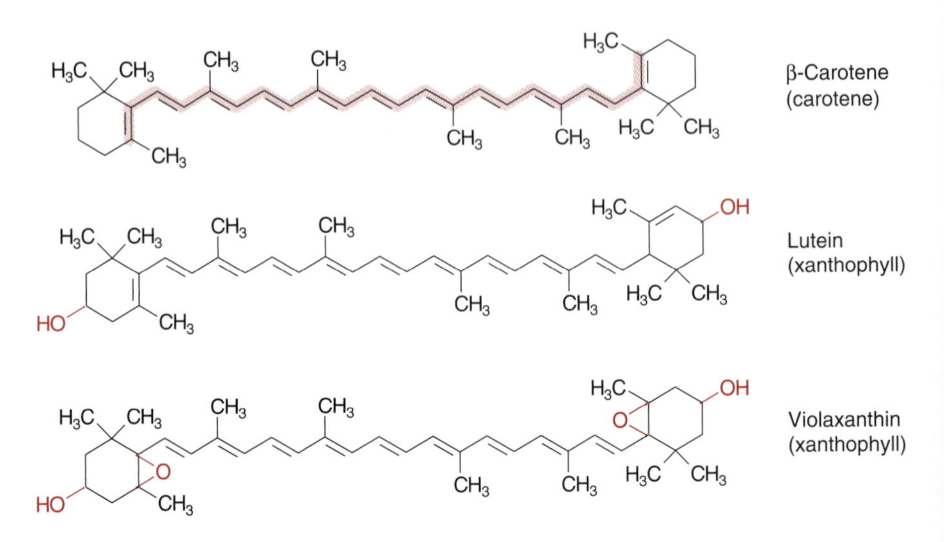

Figure 2.9 Structural formula of a carotene (β-carotene) and of two xanthophylls (lutein and violaxanthin). Due to the conjugated isoprenoid chain, these molecules absorb light and also have lipid character.

thin as well as **carotenes** with β-carotene as the main substance (Fig. 2.9). Moreover, an important function of these carotenoids in the antennae is to prevent the formation of the harmful triplet state of the chlorophylls previously mentioned (see section 3.10). Important constituents of the antennae in cyanobacteria are **phycobilins**, which will be discussed at the end of this section.

How is the excitation energy of the photons, which have been captured in the antennae, transferred to the reaction centers?

The possibility could be excluded that in the transfer of the energy in the antennae, electrons are transported from chromophore to chromophore in a sequence of redox processes, as in the electron transport chains of photosynthesis or of mitochondrial respiration (to be discussed later). Such an electron transport would have considerable activation energy. This is not the case, however, since a flux of excitation energy can be measured in the antennae at temperatures as low as 1° K. At these low temperatures light absorption and fluorescence still occur, whereas chemical processes catalyzed by enzymes are completely frozen. This makes it probable that the energy transfer in the antennae proceeds according to a mechanism that is related to those of light absorption and fluorescence.

When chromophores are positioned very close to each other, it is possible that the quantum energy of an irradiated photon is transferred from one chromophore to the next. Just as one quantum of light energy is named a photon, one quantum of excitation energy transferred from one molecule to the next is termed an **exciton**. A prerequisite for the transfer of excitons is

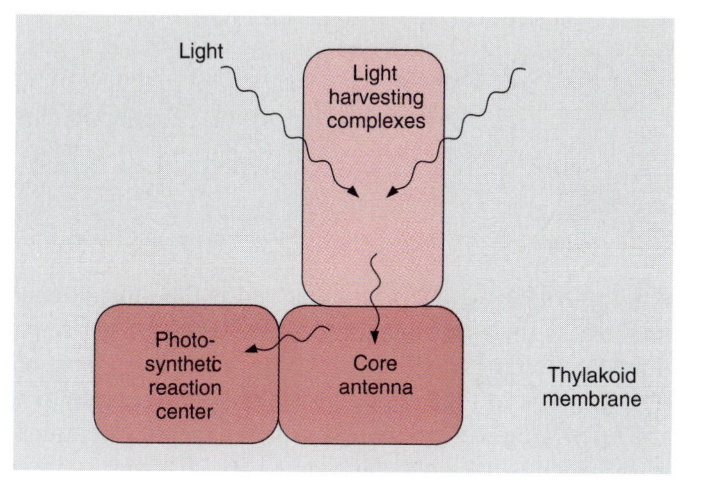

Figure 2.10 Basic scheme of an antenna.

that the chromophores involved are positioned in a specific way. This is arranged by proteins. Therefore the chromophores of the antennae always occur as protein complexes.

The antennae of plants consist of an inner part and an outer part (Fig. 2.10). The outer part, formed by the **light harvesting complexes (LHC**s), collects the light. The inner part of the antenna, consisting of the **core complexes**, is an integral constituent of the reaction centers; it also collects light and conducts the excitons collected in the outer part of the antenna into the photosynthetic reaction centers.

The LHCs are formed by polypeptides, which bind chl-*a*, chl-*b*, xanthophylls, and carotenes. These proteins, termed **LHC polypeptides**, are encoded in the nucleus. A plant contains many different LHC polypeptides. In a tomato, for instance, at least 19 different genes for LHC polypeptides have been found, which are very similar to each other. They are homologous, as they have all evolved from a common ancestral form. These LHC polypeptides form a multigene family.

Plants contain two reaction centers, which are arranged in sequence: a reaction center of photosystem II (**PS II**), which has an absorption maximum at 680 nm, and that of photosystem I (**PS I**) with an absorption maximum at 700 nm. The function of these reaction centers will be described in sections 3.6 and 3.8. Both photosystems have different LHCs.

The function of an antenna can be illustrated using the antenna of photosystem II as an example

The antennae of the PS II reaction center contain primarily four LHCs termed LHC-II*a–d*. The main component is **LHC-II*b***; it represents 67% of

Table 2.2: Composition of the LHC-II*b*-Monomer

Peptide:	232 Amino acids
Lipids:	1 Phosphatidylglycerol, 1 Digalactosyldiacylglycerol
Chromophores:	7 Chl-*a*, 5 Chl-b, 2 Lutein

(Kühlbrandt, 1994).

the total chlorophyll of the PS II antenna and is the most abundant membrane protein of the thylakoid membrane, thus lending itself to particularly thorough investigation. LHC-II*b* occurs in the membrane, most probably as a trimer. The monomer (Table 2.2) consists of a polypeptide to which two molecules of lutein (Fig. 2.9) are bound. The polypeptide contains one threonine residue, which can be phosphorylated by ATP via a protein kinase. Phosphorylation regulates the activity of LHC-II (section 3.10).

There has been a breakthrough in establishing the three-dimensional structure of LHC-II*b* by electron cryomicroscopy at 4 K of crystalline layers of LHC-II*b*-trimers (Fig. 2.11). The LHC-II*b*-peptide forms three transmembrane helices. The two lutein molecules span the membrane crosswise. The chl-*b*-molecules, where the absorption maximum in the red spectral region lies at a shorter wavelength than that of chl-*a*, are positioned in the outer region of the complexes. Only one of the chl-*a*-molecules is positioned in the outer regions; the others are all present in the center. Figure 2.12 shows a vertical projection of the probable arrangement of the monomers to form a trimer. The chl-*a* positioned in the outer region mediates the transfer of energy to the neighboring trimers or to the reaction center. The other LHC-II-peptides (*a*, *c*, *d*) are very similar to the LHC-II*b* peptide; they differ in only 5% of their amino acid sequence and it can be assumed that their structure is also very similar. The chl-*a*/chl-*b* ratio is much higher in LHC-II*a* and LHC-II*c* than in LHC-II*b*. Most likely LHC-II*a* and LHC-II*c* are positioned between LHC-II*b* and the reaction center.

Figure 2.13 shows a hypothetical scheme of the probable array of the PS II antenna. The outer complexes, consisting of LHC-II*b*, are present at the periphery of the antenna. The excitons captured by chl-*b* in LHC-II*b* are transferred to chl-*a* in the center of the LHC-II*b* monomers and are then transferred further by chl-*a*-contacts between the trimers to the inner antennae complexes. The inner complexes are connected by small chlorophyll-containing subunits to the core complex. This consists of the antennae proteins CP 43 and CP 47, which are closely attached to the reaction center (Fig. 3.22), and each contains about 15 chl-*a* molecules. Since the absorption maximum of chl-*b* is at a lower wavelength than that of chl-*a*, the transfer of excitons from chl-*b* to chl-*a* is accompanied by loss of energy as heat.

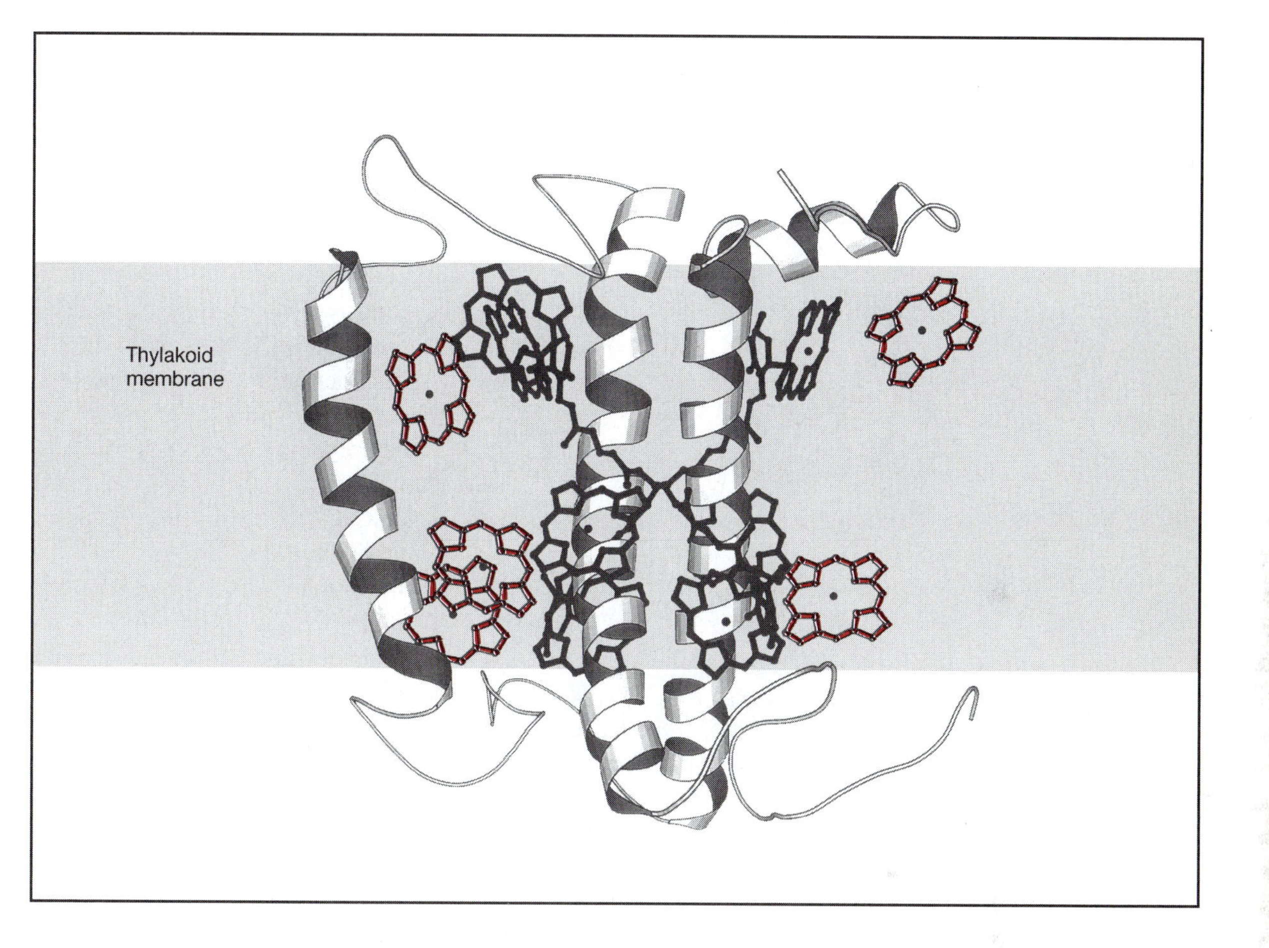

Figure 2.11 Sterical arrangement of the LHC-II*b* monomer in the thylakoid membrane, viewed from the side. Three α-helices of the protein span the membrane. Chlorophyll-*a* (black) and chlorophyll-*b* (red) are oriented almost perpendicularly to the membrane surface. Two lutein molecules (black) in the center of the complex act as an internal cross brace. (By courtesy of W. Kühlbrandt, Heidelberg.)

This promotes the flux of excitons from the periphery to the reaction center. The connection between the outer LHCs (LHC-II*b*) and the PS II can be interrupted by phosphorylation. In this way the actual size of the antenna can be adjusted to the intensity of illumination (section 3.10).

Photosystem I contains fewer LHCs than photosystem II (section 3.8) since its core antenna is larger than in PS II. The LHCs of PS I are similar to those of PS II. Sequence analysis shows that LHC-I and LHC-II stem from a common archetype. It has been suggested that in the phosphorylated state LHC-II*b* can also function as an antenna of PS I (see section 3.10).

The mechanism of the movement of excitons in the antenna is not yet fully understood. The exciton may be delocalized by being distributed over a whole group of chromophore molecules. On the other hand, the exciton may be present initially in a certain chromophore molecule and subsequently transferred to a more distant chromophore. This process of exciton transfer has been termed the **Förster mechanism**. The transfer of excitons between

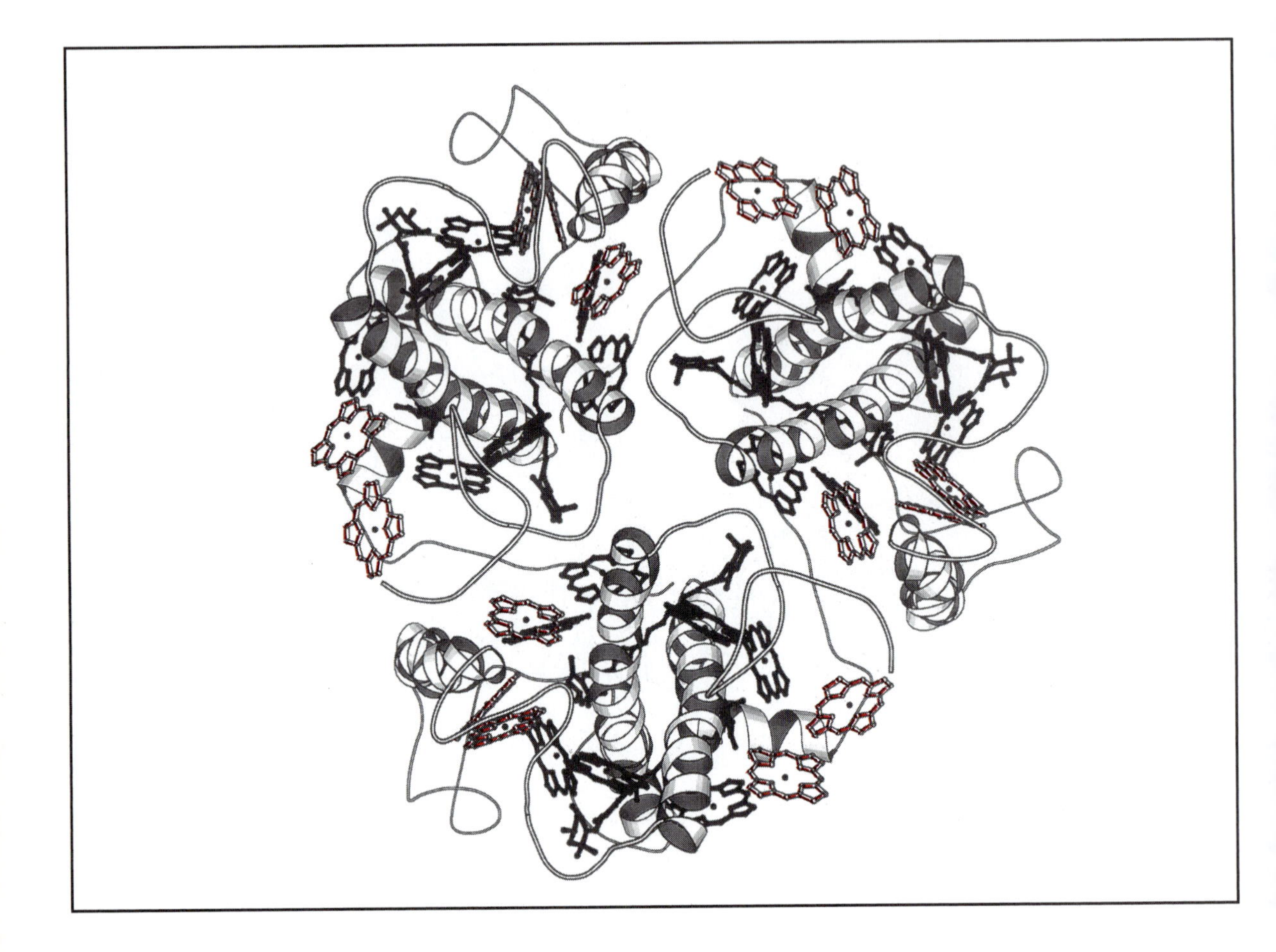

Figure 2.12 The LHC-II-trimer viewed from above from the stroma side. Within each monomer the central pair of helices form a left-handed supercoil, which is surrounded by chlorophyll molecules. The chl-*b* molecules (red) are positioned at the side of the monomers. (By courtesy of W. Kühlbrandt, Heidelberg.)

closely neighboring chlorophyll molecules within an LHC probably proceeds via **delocalized electrons** and the transfer between the LHCs and the reaction center via the Förster mechanism. Absorption measurements with ultrafast laser technique have shown that the exciton transfer between two chlorophyll molecules proceeds within 0.1 picoseconds (10^{-13} s). Thus the velocity of the exciton transfer in the antennae is much faster than the charge separation in the reaction center ($\approx$3.5 picoseconds) discussed in section 3.4. The reaction center functions as an **energy trap** for excitons present in the antenna.

Phycobilisomes enable cyanobacteria and red algae to carry out photosynthesis even in dim light

Cyanobacteria and red algae possess antennae structures that can collect light of very low intensity. These antennae are arranged as particles on top

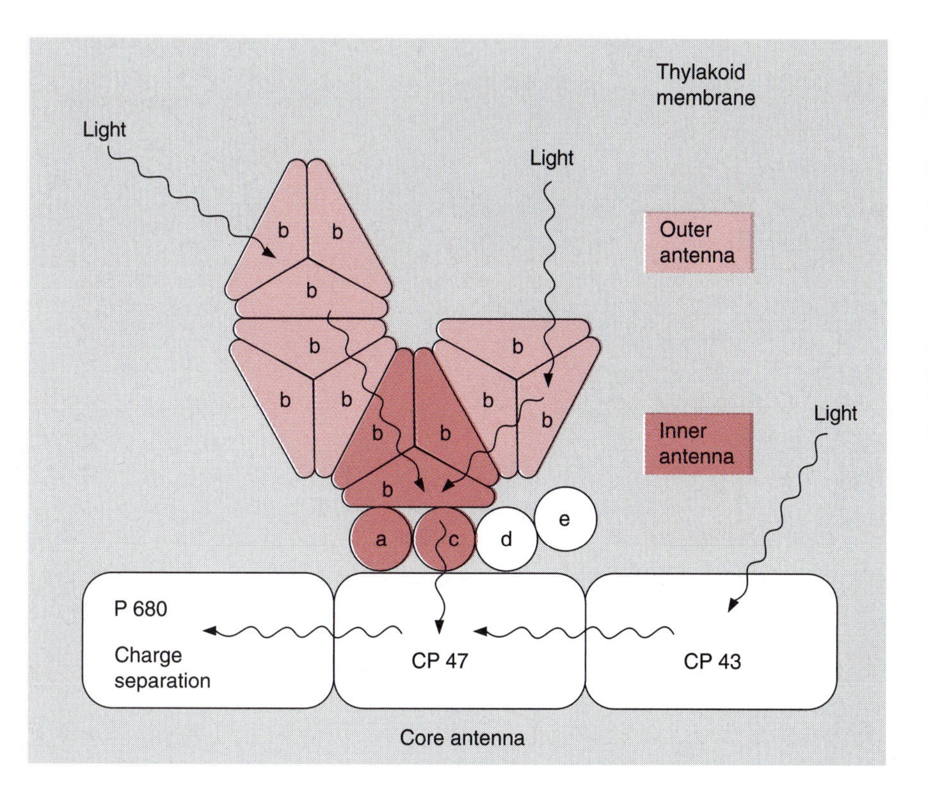

Figure 2.13 Scheme of the arrangement of the light harvesting complexes in the antenna of photosystem II in a plant viewed from above (after Thornber); a means LHC-II*a* and so on. The inner antenna complexes are linked by LHC-II*a* and LHC-II*c* monomers to the core complex. The function of the LHC-II*d* and LHC-II*e* monomers is not entirely known.

of the membrane near the reaction centers of photosystem II (Fig. 2.14). These particles, termed **phycobilisomes**, consist of proteins (**phycobiliproteins**), which are covalently linked with phycobilins. **Phycobilins** are open-chained tetrapyrroles and therefore are structurally related to the chlorophylls. Open-chained tetrapyrroles are also contained in bile, which explains the name *-bilin*. The phycobilins are linked to the protein by a thioether bond between an SH-group of the protein and the vinyl side chain of the phycobilin. The protein **phycoerythrin** is linked to the chromophore **phycoerythrobilin**, and the proteins **phycocyanin** and **allophycocyanin** to the chromophore **phycocyanobilin** (Fig. 2.15). The basic structure in the phycobiliproteins consists of a heterodimer, (α, β). Each of these subunits contains one to four phycobilins as a chromophore. Three of these heterodimers aggregate to a trimer $(\alpha, \beta)_3$ and thus form the actual building block of a phycobilisome. The so-called linker polypeptides function as "mortar" between the building blocks.

Figure 2.14 shows the structure of a phycobilisome. The phycobilisome is attached to the membrane by anchor proteins. Three aggregates of four to five $(\alpha, \beta)_3$ units form the core. This core contains the pigment allophycocyanin (AP) to which cylindrical rodlike structures are attached, each with

Figure 2.14 Scheme of a side view of the structure of a phycobilisome. The units shown consist of three α- and three β-subunits each, (After Bryanth.)

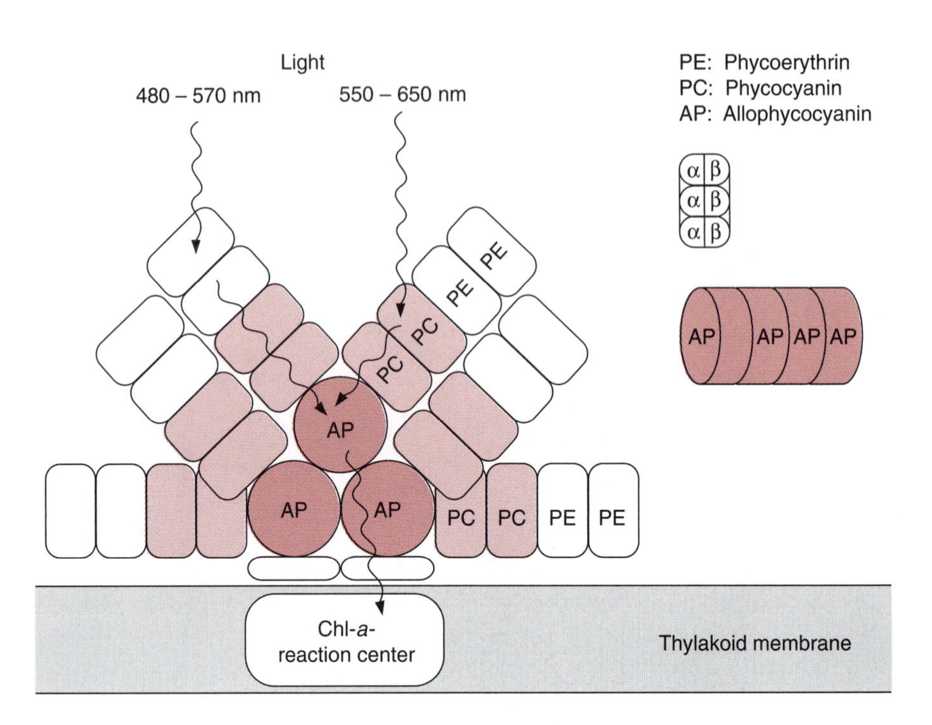

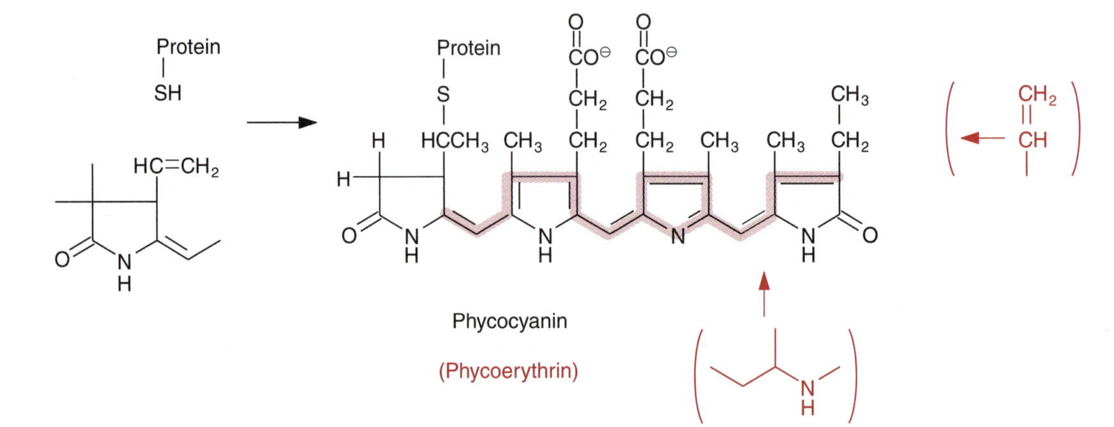

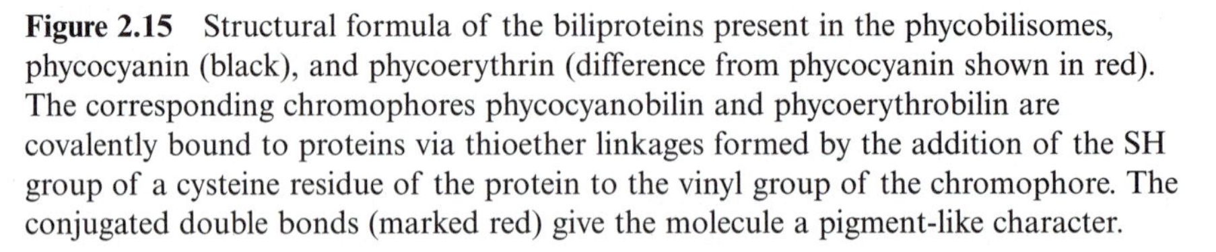

Figure 2.15 Structural formula of the biliproteins present in the phycobilisomes, phycocyanin (black), and phycoerythrin (difference from phycocyanin shown in red). The corresponding chromophores phycocyanobilin and phycoerythrobilin are covalently bound to proteins via thioether linkages formed by the addition of the SH group of a cysteine residue of the protein to the vinyl group of the chromophore. The conjugated double bonds (marked red) give the molecule a pigment-like character.

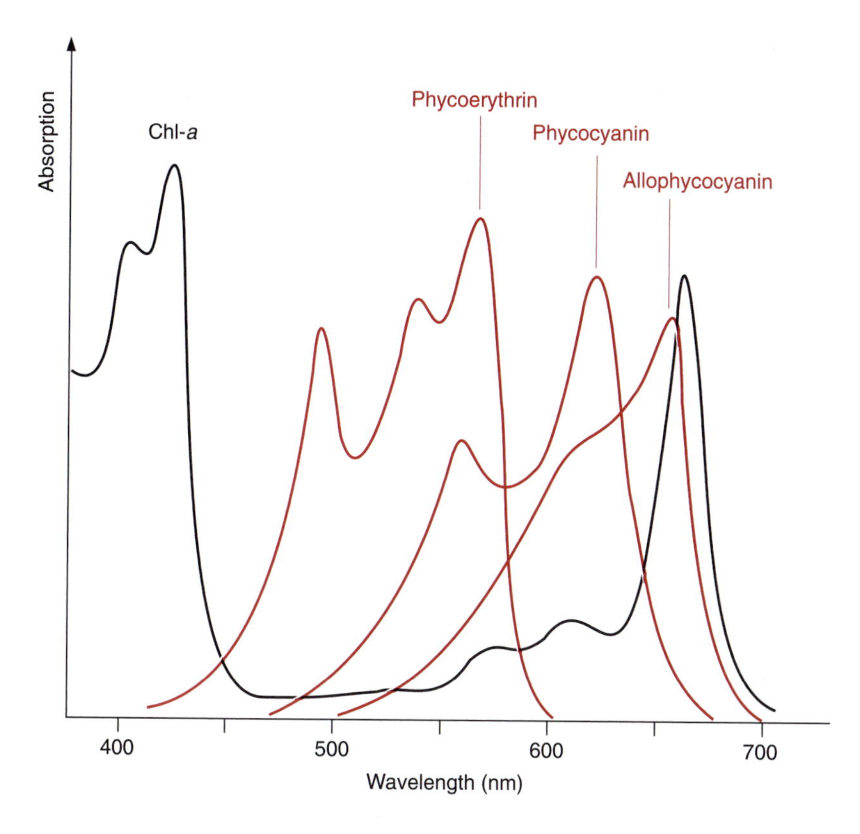

Figure 2.16 Absorption spectra of the phycobiliproteins phycoerythrin, phycocyanin, and allophycocyanin and, for the sake of comparison, also of chlorophyll-*a*.

four to six building blocks. The inner units contain mainly phycocyanine (PC) and the outer ones phycoerythrin (PE). The function of this structural organization is illustrated by the absorption spectra of the various biliproteins shown in Figure 2.16. The light of shorter wavelength is absorbed in the periphery of the rods by phycoerythrin and the light of longer wavelength in the inner regions of the rods by phycocyanin. The core transfers the excitons to the reaction center. A spatial distribution between the short wavelength absorbing pigments at the periphery and the long wavelength absorbing pigments in the center was shown in the preceding section for the PS II antennae of higher plants.

Due to the phycobiliproteins, phycobilisomes are able to absorb green light very efficiently (Fig. 2.16), thus allowing cyanobacteria and red algae to survive in deep water. At these depths, due to the green window of photosynthesis (Fig. 2.3), only green light is available, as the light of the other wavelengths is absorbed by green algae living in the upper regions of the water. The algae in the deeper regions are obliged to invest a large portion of their cellular matter in phycobilisomes in order to carry out photosynthesis at this very low light intensity. Biliproteins can amount to 40% of the total cellular protein of the algae. These organisms undertake an extraordinary expenditure to collect enough light for survival.

Further reading

Bada, J. L., Lazcano, A. Prebiotic soup—revisiting the Miller experiment. Science 300, 74–746 (2003).

Cerullo, G., Polli, D., Lanzani, G., De Silverstri, S., Hashimoto, H., Cogdell, R. J. Photosynthetic light harvesting by carotenoids: Detection of an intermediate excited state. Science 298, 2395–2398 (2002).

Emerson, R. The quantum yield of photosynthesis. Annu Rev Plant Physiol 9, 1–24 (1958).

Glazer, A. N. Photosynthetic accessory proteins with bilin prosthetic groups. The Biochemistry of Plants, Vol. 8 (eds. M. D. Hatch, N. K. Boardman), pp 51–96. Academic Press, New York (1981).

Govindjee, Knox, R. S., Amesz, J. (eds.) Photosynthetic unit: Antenna and reaction centers. Photosynth Res 48, 1–319 (1996).

Kühlbrandt, W. Structure and function of the plant light harvesting complex, LHC-II. CurrBiol 4, 519–528 (1994).

Nilsson, A., Stys, D., Drakenberg, T., Spangfort, M. D., Forsén, S., Allen, J. F. Phosphorylation controls the three-dimensional structure of plant light harvesting complex II. J Biol Chem 272, 18350–18357 (1997).

Ruban, A. V., Wentworth, M., Yakushevska, A. E., Andersson, J., Lee, P. J., Keegstra W., Dekker, J. P., Boekema, E. J., Jansson, S., Horton, P. Plants lacking the main light-harvesting complex retain photosystem II macro-organization. Nature 42, 648–652 (2003).

Scheer, H., Chlorophylls. CRC. Boca Raton, Fla. (1991).

Turconi, S., Weber, N., Schweitzer, G., Strotmann, H., Holzwarth, A. R. Energy charge separation kinetics in photosystem I. 2. Picosecond fluorescence study of various PSI particles and light-harvesting complex isolated from higher plants. Biochim Biophys Acta, 1187, 324–334 (1994).

Vogelmann, T. C., Nishio, J. N., Smith, W. K. Leaves and light capture light propagation and gradients of carbon fixation in leaves. Trends Plant Sci 1, 65–70 (1996).

Xiong, J., Fischer, W. M., Inoue, K., Nakahara, M., Bauer, C. E. Molecular evidence for the early evolution of photosynthesis. Science 289, 1724–1730 (2000).

3

Photosynthesis is an electron transport process

The previous chapter described how photons are captured by an antenna and conducted as excitons to the reaction centers. This chapter deals with the function of these reaction centers and describes how photon energy is converted to chemical energy to be utilized by the cell. As mentioned in Chapter 2, plant photosynthesis probably evolved from bacterial photosynthesis, so that the basic mechanisms of the photosynthetic reactions are alike in bacteria and plants. Bacteria have proved to be very suitable objects for studying the principles of photosynthesis since their reaction centers are more simply structured than those of plants and they are more easily isolated. For this reason, first bacterial photosynthesis and then plant photosynthesis will be described.

3.1 The photosynthetic machinery is constructed from modules

The photosynthetic machinery of bacteria is constructed from defined complexes, which also appear as components of the photosynthetic machinery in plants. As will be described in Chapter 5, some of these complexes are also components of mitochondrial electron transport. These complexes can be thought of as modules that developed at an early stage of evolution and have been combined in various ways for different purposes. For easier understanding, the functions of these modules in photosynthesis will be treated first as **black boxes** and a detailed description of their structure and function will be given later.

Purple bacteria have only one reaction center (Fig. 3.1). In this reaction center the energy of the absorbed photon excites an electron, which means it is raised to a negative redox state. The excited electron is transferred back to the ground state by an electron transport chain, called the **cytochrome-*b*/*c*$_1$ complex,** and the released energy is transformed to a chemical form, which is then used for the synthesis of biomass (proteins, carbohydrates). Generation of energy is based on coupling the electron transport with the transport of protons across the membrane. In this way the energy of the excited electron is conserved in the form of an electrochemical H^+-potential across the membrane. The photosynthetic reaction centers and the main components of the electron transport chain are always located in a membrane.

Via ATP-synthase the energy of the H^+-potential is used to synthesize ATP from ADP and phosphate. Since the excited electrons in purple bacteria return to the ground state of the reaction center, this electron transport is called **cyclic electron transport**. In purple bacteria the proton gradient is also used to reduce NAD via an additional electron transport chain named the **NADH dehydrogenase complex** (Fig. 3.1). By consuming the energy of the H^+-potential, electrons are transferred from a reduced substance (e.g., organic acids or hydrogen sulfide) to NAD. The ATP and NADH formed by bacterial photosynthesis are used for the synthesis of organic matter; especially important is the synthesis of carbohydrates from CO_2 via the Calvin cycle (see Chapter 6).

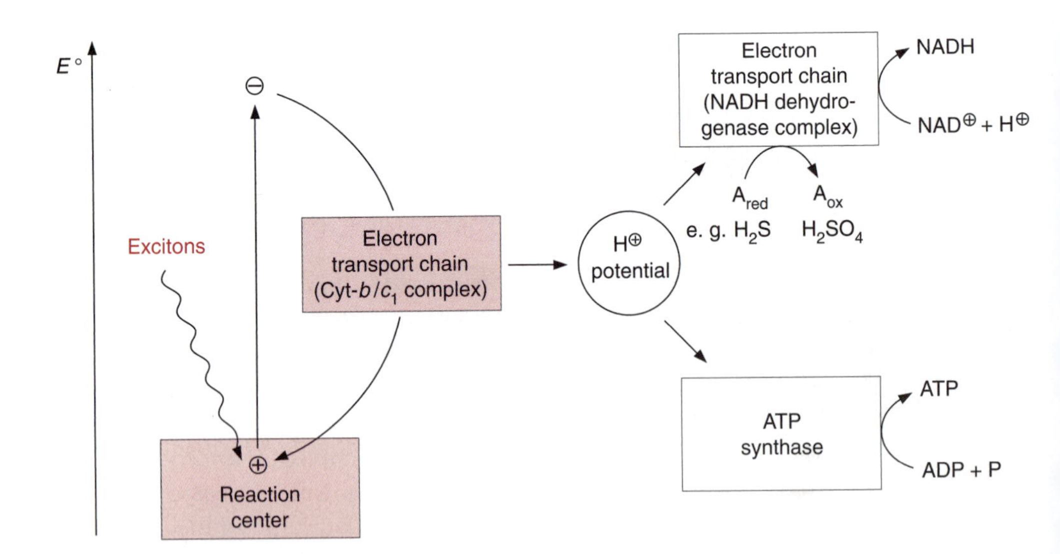

Figure 3.1 Scheme of the photosynthetic apparatus of purple bacteria. The energy of a captured exciton in the reaction center elevates an electron to a negative redox state. The electron is transferred to the ground state via an electron transport chain including the cytochrome-*b*/*c*$_1$ complex and cytochrome-*c*. Free energy of this process is conserved by formation of a proton potential which is used partly for synthesis of ATP and partly to enable an electron flow for the formation of NADH from electron donors such as H_2S.

The reaction center of **green sulfur bacteria** (Fig. 3.2) is homologous to that of purple bacteria, indicating that they have both evolved from a common precursor. ATP is also formed in green sulfur bacteria by cyclic electron transport. The electron transport chain (cytochrome-b/c_1 complex) and the ATP-synthase involved here are very similar to those in purple bacteria. However, in contrast to purple bacteria, green sulfur bacteria are able to form NADH by a **noncyclic electron transport process**. In this case, the excited electrons are transferred to the **ferredoxin-NAD-reductase complex,** which reduces NAD to NADH. Since the excited electrons in this noncyclic pathway do not return to the ground state, an electron deficit remains in the reaction center and is replenished by electron donors such as H_2S, ultimately being oxidized to sulfate.

Cyanobacteria and plants use water as an electron donor in photosynthesis (Fig. 3.3). As oxygen is liberated, this process is called **oxygenic photosynthesis**. Two photosystems designated II and I are arranged here in tandem. The machinery of oxygenic photosynthesis is made up of modules that have already been described in bacterial photosynthesis. The structure of the reaction center of photosystem II corresponds to that of the reaction center of purple bacteria, and that of photosystem I corresponds to the reaction center of green sulfur bacteria. The enzymes ATP-synthase and

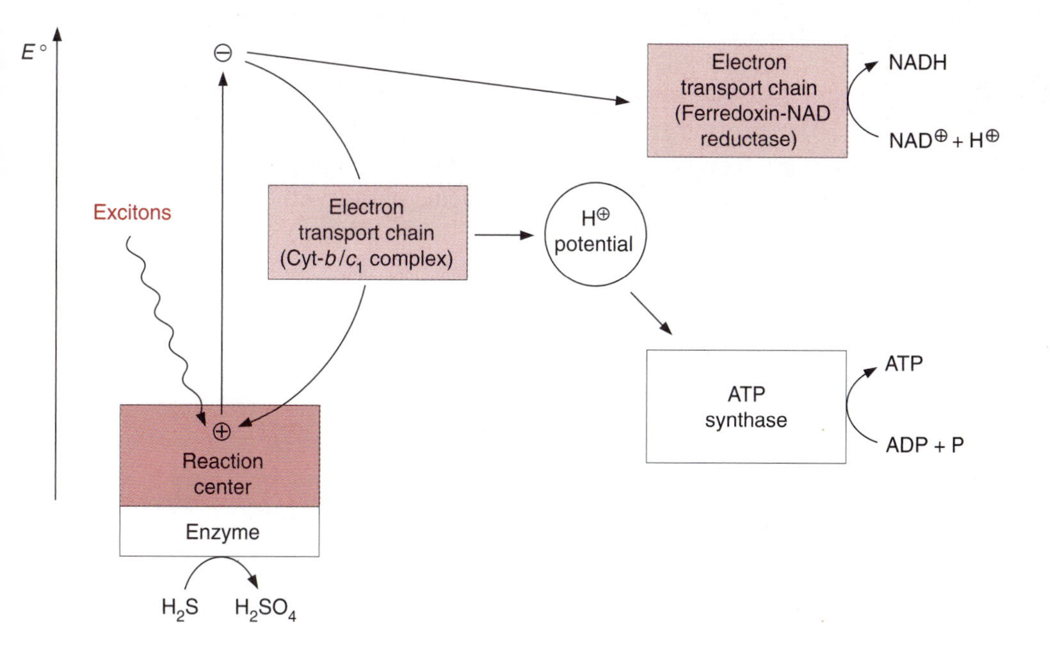

Figure 3.2 Scheme of the photosynthetic apparatus in green sulfur bacteria. In contrast to the scheme in Figure 3.1, part of the electrons elevated to a negative redox state are transferred via an electron transport chain (ferredoxin-NAD reductase) to NAD, yielding NADH. The electron deficit arising in the reaction center is compensated for by electron donors such as H_2S.

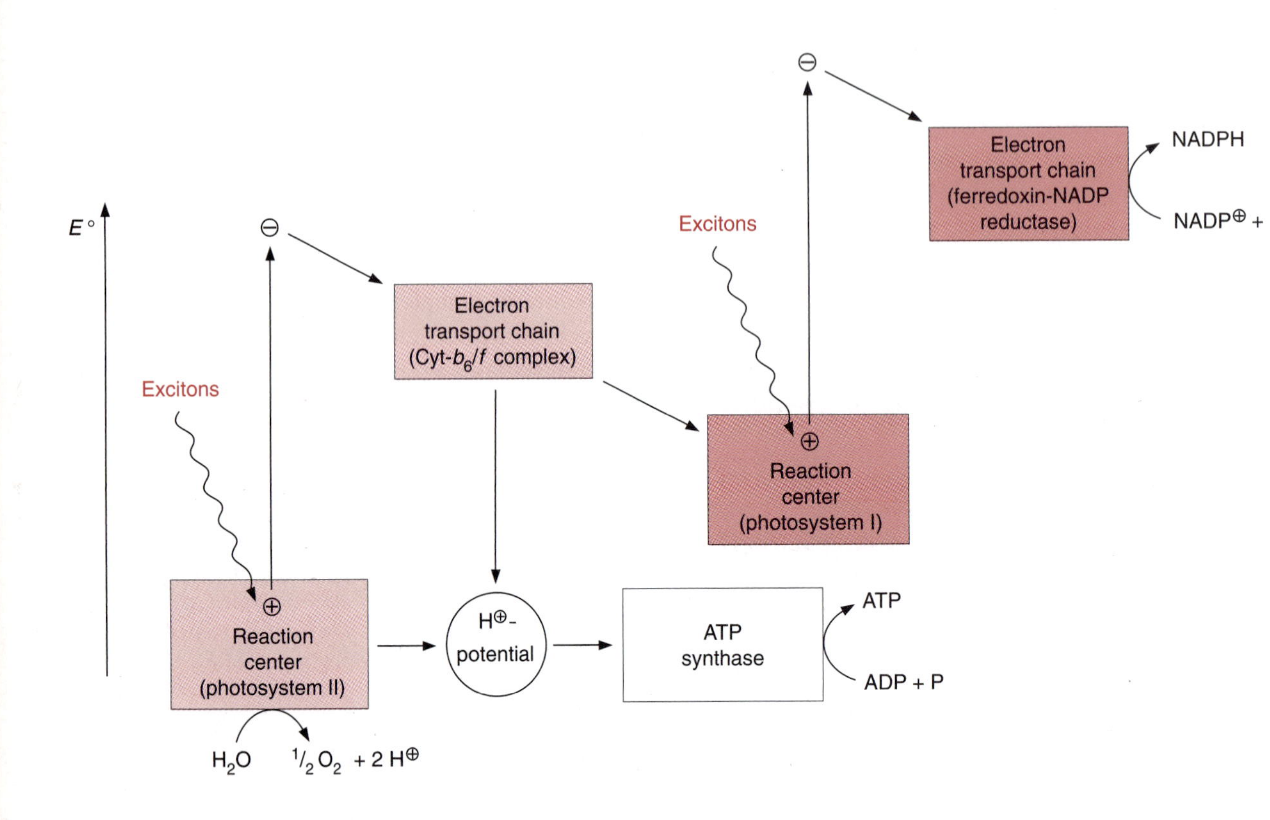

Figure 3.3 In the photosynthetic apparatus of cyanobacteria and plants, two reaction centers corresponding in their function to the photosynthetic reaction centers of purple bacteria and green sulfur bacteria (shown in Figs. 3.1 and 3.2) are arranged in sequence.

ferredoxin-NADP-reductase are very similar to those of photosynthetic bacteria. The electron transport chain of the **cytochrome-b_6/f complex** has the same basic structure as the cytochrome-b/c_1 complex in bacteria.

Four excitons are required in oxygenic photosynthesis to split one molecule of water:

$$H_2O + NADP^+ + 4\ \text{excitons} \longrightarrow 1/2\ O_2 + NADPH + H^+$$

In this noncyclic electron transport, electrons are transferred to NADP and protons are transported across the membrane to generate the proton gradient that drives the synthesis of ATP. Thus, for each mol of NADPH formed by oxygenic photosynthesis, about 1.5 molecules of ATP are generated simultaneously (section 4.4). Most of this ATP and NADPH is used for assimilating CO_2 and nitrate to give carbohydrates and amino acids. Oxygenic photosynthesis in plants takes place in the chloroplasts, a cell organelle of the plastid family (section 1.3).

3.2 A reductant and an oxidant are formed during photosynthesis

In the 1920s Otto Warburg postulated that the energy of light is transferred to CO_2 and that the CO_2, activated in this way, reacts with water to form a carbohydrate, accompanied by the release of oxygen. According to this hypothesis, the oxygen released by photosynthesis was derived from the CO_2. In 1931 this hypothesis was opposed by Cornelis van Niel by postulating that during photosynthesis a **reductant** is formed, which then reacts with CO_2. The so-called van Niel equation describes photosynthesis in the following way:

$$CO_2 + 2H_2A \xrightarrow{\text{Light}} [CH_2O] + H_2O + 2A$$

($[CH_2O]$ = carbohydrate). He proposed that a substance H_2A is split by light energy into a reducing compound (H) and an oxidizing compound (A). For oxygenic photosynthesis of cyanobacteria or plants, it can be rewritten as:

$$CO_2 + 2H_2O \xrightarrow{\text{Light}} [CH_2O] + H_2O + O_2$$

In this equation the oxygen released during photosynthesis is derived from water.

In 1937 Robert Hill in Cambridge proved that a reductant is actually formed in the course of photosynthesis. He was the first to succeed in isolating chloroplasts with some photosynthetic activity, which, however, had no intact envelope membranes and consisted only of thylakoid membranes. When these chloroplasts were illuminated in the presence of Fe^{3+} compounds [initially ferrioxalate, later ferricyanide ($[Fe(CN)_6]^{3-}$)], oxygen evolved accompanied by reduction of the $\mathbf{Fe^{3+}}$-compounds to the $\mathbf{Fe^{2+}}$ form.

$$H_2O + 2Fe^{+++} \xrightarrow{\text{Light}} 2H^+ + 1/2O_2 + 2Fe^{++}$$

Since CO_2 was not involved in this **"Hill reaction,"** this experiment proved that the photochemical splitting of water can be separated from the reduction of the CO_2. The total reaction of photosynthetic CO_2 assimilation can be divided into two partial reactions:

1. The so-called **light reaction**, in which water is split by photon energy to yield reductive power (in the form of NADPH) and chemical energy (in the form of ATP); and

2. The so-called **dark reaction** (Chapter 6), in which CO_2 is assimilated at the expense of this reductive power and of ATP.

In 1952 the Dutchman Louis Duysens made a very important observation that helped explain the photosynthesis mechanism. When illuminating isolated membranes of the purple bacterium *Rhodospirillum rubrum* with short light pulses, he found a decrease in light absorption at 890 nm, which was immediately reversed when the bacteria were darkened again. The same **"bleaching"** effect was found at 870 nm in the purple bacterium *Rhodobacter sphaeroides*. Later, Bessil Kok (United States) and Horst Witt (Germany) also found similar pigment bleaching at 700 nm and 680 nm in chloroplasts. This bleaching was attributed to the **primary reaction of photosynthesis**, and the corresponding pigments of the reaction centers were named P_{870} (*Rb. sphaeroides*) and P_{680} and P_{700} (chloroplasts). When an oxidant (e.g., $[Fe(CN)_6]^{3-}$) was added, this bleaching effect could also be achieved in the dark. These results indicated that these absorption changes of the pigments were due to a **redox reaction**. This was the first indication that chlorophyll can be oxidized. Electron spin resonance measurements revealed that **radicals** are formed during this "bleaching." "Bleaching" could also be observed at the very low temperature of 1°K. This showed that in the electron transfer leading to the formation of radicals, the reaction partners are located so close to each other that thermal oscillation of the reaction partners (normally the precondition for a chemical reaction) is not required for this redox reaction. Spectroscopic measurements indicated that the reaction partner of this primary redox reaction are two closely adjacent chlorophyll molecules arranged as a pair, called a **"special pair**."

3.3 The basic structure of a photosynthetic reaction center has been resolved by X-ray structure analysis

The reaction centers of purple bacteria proved to be especially suitable objects for explaining the structure and function of the photosynthetic machinery. It was a great step forward when in 1970 Roderick Clayton (United States) developed a method for isolating reaction centers from purple bacteria. Analysis of the components of the reaction centers of the different purple bacteria (shown in Table 3.1 for the reaction center of *Rb. sphaeroides* as an example) revealed that the reaction centers had the same

Table 3.1: Composition of the reaction center from *Rhodobacter sphaeroides* (P_{870})

	Molecular mass
1 subunit L	21 kDa
1 subunit M	24 kDa
1 subunit H	28 kDa
4 bacteriochlorophyll-*a*	
2 bacteriopheophytin-*a*	
2 ubiquinone	
1 non-heme-Fe-protein	
1 carotenoid	

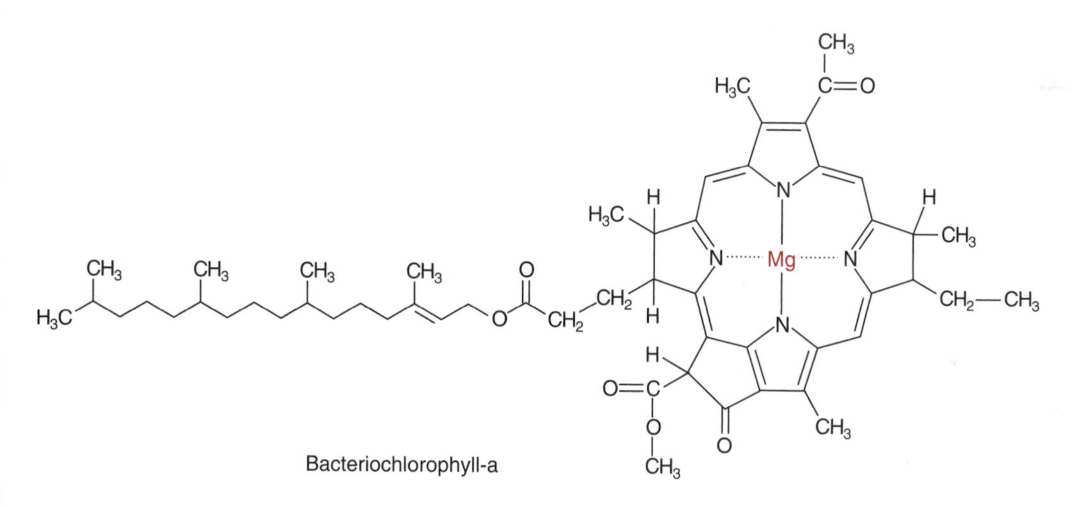

Figure 3.4
Bacteriochlorophyll-*a*.

basic structure in all the purple bacteria investigated. The minimum structure consists of the three subunits L, M, and H (light, medium, and heavy). Subunits L and M are peptides with a similar amino acid sequence. They are homologous. The reaction center of *Rb. sphaeroides* contains four bacteriochlorophyll-*a* (Bchl-*a*, Fig. 3.4) and two bacteriopheophytin-*a* (BPhe-*a*). Pheophytins differ from chlorophylls in that they lack magnesium as the central atom. In addition, the reaction center contains an iron atom that is not a part of a heme. It is therefore called a *non-heme iron*. Furthermore, the reaction center contains two molecules ubiquinone (Fig. 3.5), which are designated as Q_A and Q_B. Q_A is tightly bound to the reaction center, whereas Q_B is only loosely associated with it.

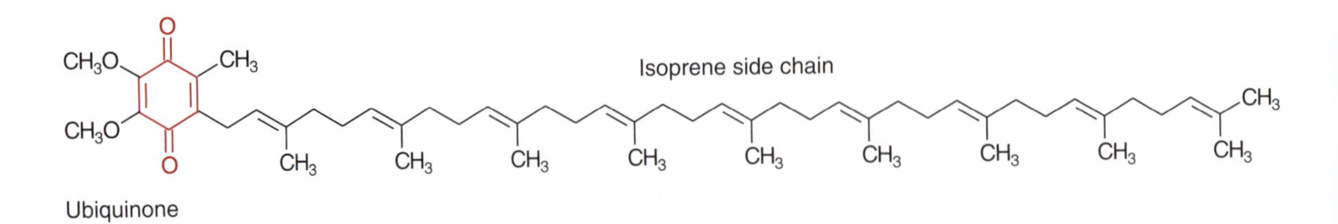

Figure 3.5 Ubiquinone. The long isoprenoid side chain gives the substance a lipophilic character.

X-ray structure analysis of the photosynthetic reaction center

If ordered crystals can be prepared from a protein, it is possible to analyze the spherical structure of the protein molecule by **X-ray structure analysis.** [In a similar way a structural analysis can also be obtained from crystalline molecular layers by using electron cryomicroscopy (section 2.4), but in this method the experimental expenditure is particularly high.] In X-ray structure analysis, a protein crystal is irradiated by an X-ray source. The electrons of the atoms in the molecule cause a scattering of X-rays. Diffraction is observed when the irradiation passes through a regular repeating structure. The corresponding diffraction pattern, consisting of many single reflections, is measured by an X-ray film positioned behind the crystal or by an alternative detector. The principle is demonstrated in Figure 3.6. To obtain as many reflections as possible, the crystal, mounted in a capillary, is rotated. From a few dozen to up to several hundred exposures are required for one set of data, depending on the form of the crystal and the size of the crystal lattice. To evaluate a new protein structure, several sets of data are required in which the protein has been changed by the incorporation or binding of a heavy metal ion. With the help of elaborate computer programs, it is possible to reconstruct the spherical structure of the exposed protein

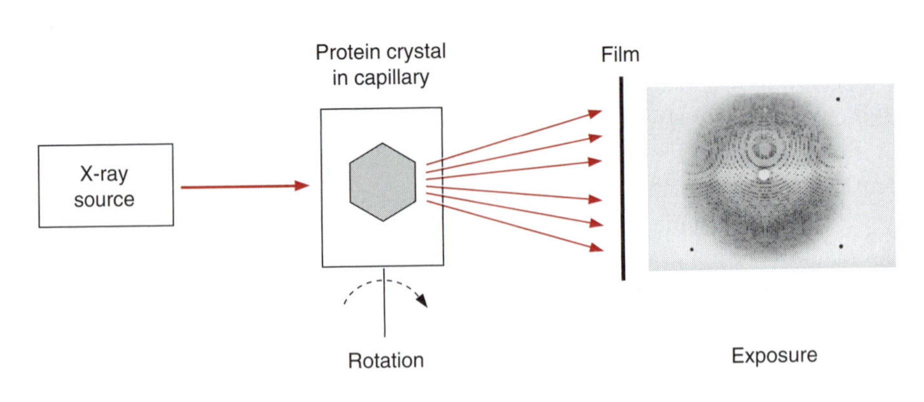

Figure 3.6 Scheme of X-ray structural analysis of a protein crystal. A capillary containing the crystal is made to rotate slowly and the diffraction pattern is monitored on an X-ray film. Nowadays much more sensitive detector systems (image platers) are used instead of films. The diffraction pattern shown was obtained by the structural analysis of the reaction center of *Rb. Sphaeroides.* (By courtesy of H. Michel, Frankfurt.)

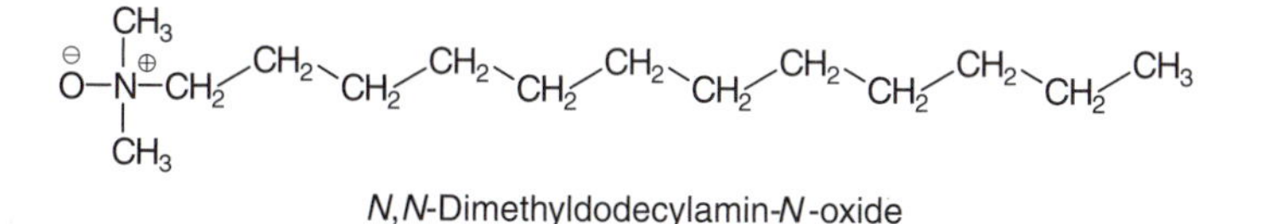

Figure 3.7 The detergent *N,N′*-dimethyldodecylamine-*N*-oxide.

molecules by applying the rules for scattering X-rays by atoms of various electron density.

X-ray structure analysis requires a high technical expenditure and is very time-consuming, but the actual limiting factor in the elucidation of a spherical structure is usually the preparation of **suitable single crystals**. Until 1980 it was thought to be impossible to prepare crystals suitable for X-ray structure analysis from hydrophobic membrane proteins. The application of the detergent **N,N′-dimethyldodecylamine-N-oxide** (Fig. 3.7) was a great step forward in helping to solve this problem. This detergent forms water-soluble protein-detergent micelles with membrane proteins, which can then be made to crystallize when ammonium sulfate or polyethylene glycol is added. The micelles form a regular lattice in these crystals (Fig. 3.8). The protein in the crystal remains in its native state since the hydrophobic regions of the membrane protein, which normally border on the hydrophobic membrane, are covered by the hydrophobic chains of the detergent.

Using this procedure, Hartmut Michel from Munich succeeded in obtaining crystals from the reaction center of the purple bacterium *Rhodopseudomonas viridis* and, together with his colleague Johann Deisenhofer from the department of Robert Huber, performed an X-ray structure analysis of these crystals. The immense amount of time invested in these investigations is illustrated by the fact that the evaluation of the stored data sets alone took two and a half years. With the X-ray structure analysis of a photosynthetic reaction center, for the first time the three-dimensional structure of a membrane protein had been elucidated. For this work, Michel, Deisenhofer, and Huber in 1988 were awarded the Nobel Prize in Chemistry. Using the same method, the reaction center of *Rb. sphaeroides* was analyzed and it turned out that the basic structure of the two reaction centers are astonishingly similar.

The reaction center of *Rhodopseudomonas viridis* has a symmetric structure

Figure 3.9 shows the three-dimensional structure of the reaction center of the purple bacterium *Rhodopseudomonas viridis.* The molecule has a cylindrical shape and is about 8 nm long. The homologous subunits L (red) and

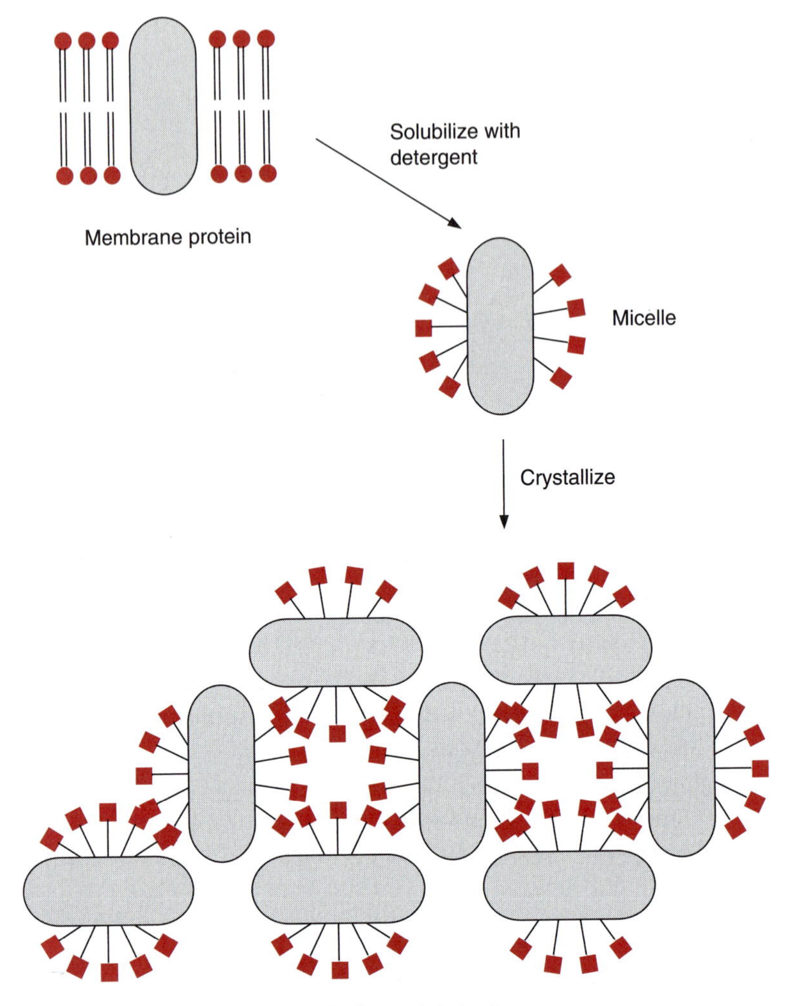

Figure 3.8 A detergent micelle is formed after solubilization of a membrane protein by a detergent. The hydrophobic region of the membrane proteins, the membrane lipids, and the detergent are marked black and the hydrophilic regions are red. Crystal structures can be formed by association of the hydrophilic regions of the detergent micelle.

M (black) are arranged symmetrically and enclose the chlorophyll and pheophytin molecules. The H subunit is attached like a lid to the lower part of the cylinder.

In the same projection as in Figure 3.9, Figure 3.10 shows the location of the chromophores in the protein molecule. All the chromophores are positioned as pairs divided by a symmetry axis. Two Bchl-*a* molecules (D_M, D_L) can be recognized in the upper part of the structure. The two tetrapyrrole rings are so close (0.3 nm) that their orbitals overlap in the excited state. This confirmed the actual existence of the **"special pair"** of chlorophyll molecules, postulated earlier from spectroscopic investigations, as the site of the primary redox process of photosynthesis. The chromophores are arranged

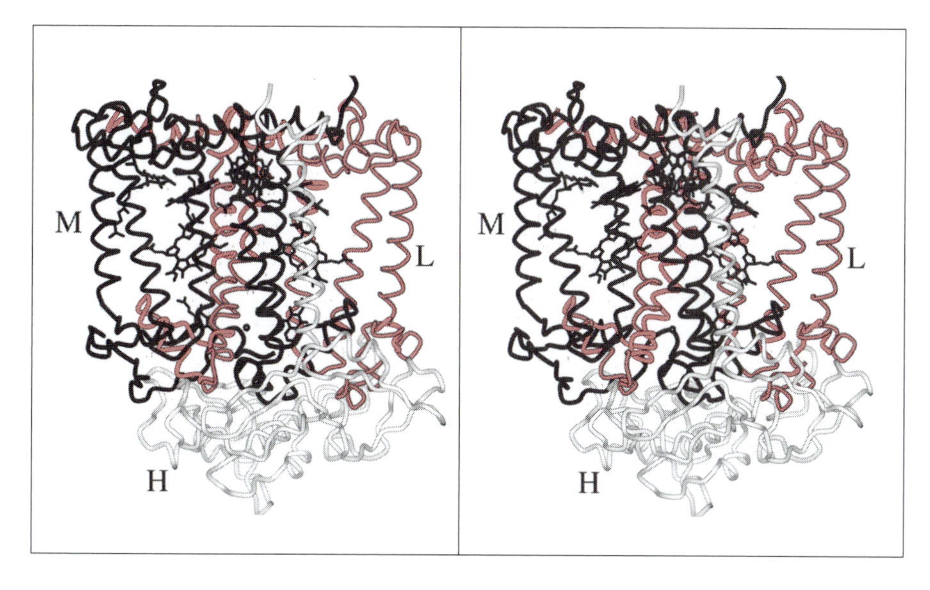

Figure 3.9 Stereo pair of the three-dimensional structure of the reaction center of *Rp. viridis*. The peptide chain of subunit L is marked red and that of subunit M is black. The polypeptide chains are shown as bands and the chromophores (chlorophylls, pheophytins) and quinones are shown as wire models. The upper part of the reaction center borders on the periplasmatic compartment and the lower part on the cytoplasm. (By courtesy of H. Michel, and R. C. R. D. Lancaster, Frankfurt.) *How to look at a stereo picture, see legend Figure 1.29.*

below this chlorophyll pair in two nearly identical branches, each containing a Bchl-*a* molecule (B_A, B_B) as monomer, followed in each branch by a bacteriopheophytin (Φ_A, Φ_B). Whereas the chlorophyll pair (D_M, D_L) is bound by both subunits L and M, the chlorophyll B_A and the pheophytin Φ_A are associated with subunit L, and B_B and Φ_B with subunit M. The quinone ring of Q_A is bound via hydrogen bonds and hydrophobic interaction to subunit M, whereas the loosely associated Q_B is bound to subunit L.

3.4 How does a reaction center function?

The analysis of the structure and extensive kinetic investigations allowed a detailed description of the function of the bacterial reaction center. The kinetic investigations included measurements by absorption and fluorescence spectroscopy after light flashes in the range of less than 10^{-13} s, as well as measurements of nuclear spin and electron spin resonance. Although the reaction center shows a symmetry with two almost identical branches of chromophores, electron transfer proceeds only along the branch on the right in Figure 3.10 (the L side). The chlorophyll monomer (B_B) on the M side is in close contact with a **carotenoid molecule**, which abolishes a harmful **triplet state** of chlorophylls in the reaction center (sections 2.3 and 3.7). The function of the pheophytin (Φ_B) on the M side and of the non-heme iron is not yet fully understood.

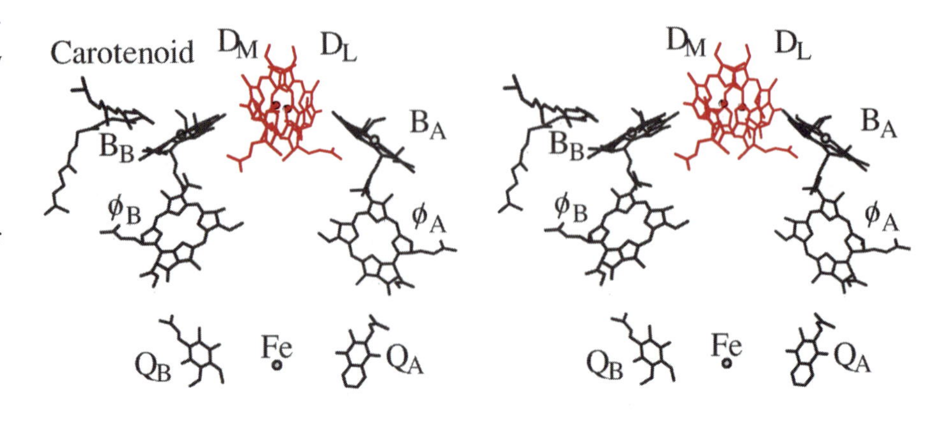

Figure 3.10 Stereo pair of the three-dimensional array of chromophores and quinones in the reaction center of *Rp viridis.* The projection corresponds to the structure shown in Figure 3.9. The Bchl-*a*-pair D_MD_L (see text) is marked red. (By courtesy of H. Michel and R. C. R. D. Lancaster, Frankfurt. The production of the Figures 3.9 and 3.10 was made by P. Kraulis, Uppsala, with the program MOLSCRIPT.)

Figure 3.11 shows the scheme of the reaction center in which the reaction partners are arranged according to their electrochemical potential. The primary reaction with the exciton provided by the antenna (section 2.4) excites the chlorophyll pair. This primary excitation state has only a very short half-life, a charge separation occurs, and, as a result of the large potential difference, within picoseconds an electron is removed to reduce bacteriopheophytin (BPhe).

$$(\text{Bchl})_2 + 1\ \text{Exciton} \longrightarrow (\text{Bchl})_2^*$$

$$(\text{Bchl})_2^* + \text{BPhe} \longrightarrow (\text{Bchl})_2^{+} + \text{BPhe}^{-}$$

The electron is probably transferred first to the Bchl-monomer (B_A) and then to the pheophytin molecule (Φ_A). The second electron transfer proceeds with a half time of 0.9 picoseconds, about four times as fast as the electron transfer to B_A. The **pheophytin radical** has a tendency to return to the ground state by a return of the translocated electron to the Bchl-monomer (B_A). To prevent this, within 200 picoseconds a high potential difference withdraws the electron from the pheophytin radical to a quinone (Q_A) (Fig. 3.11). The **semiquinone radical** thus formed, in response to a further potential difference, transfers its electron to the loosely bound ubiquinone Q_B. In this way first **ubisemiquinone** is formed and then **ubihydroquinone** is formed after a second electron transfer (Fig. 3.12). In contrast to the very labile radical intermediates of the pathway described so far, ubihydroquinone is a **stable reductant**. However, this stability has its price. For the formation of ubihy-

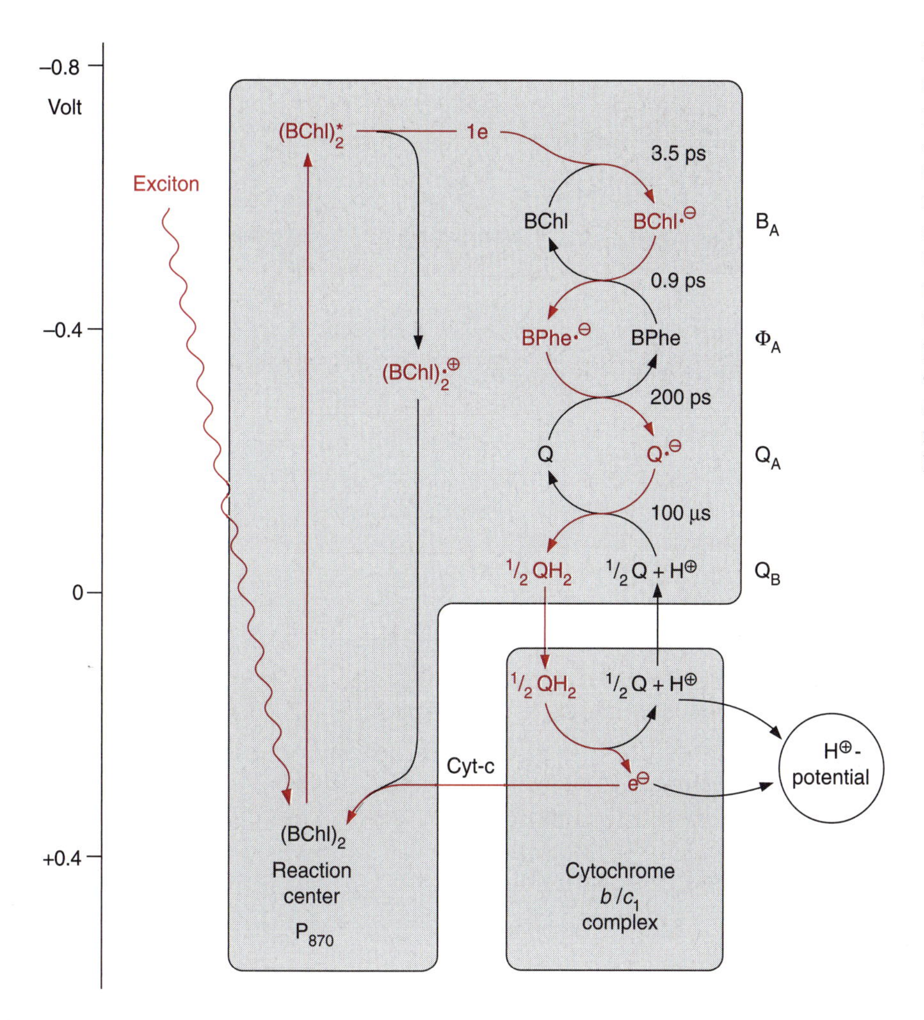

Figure 3.11 Scheme of cyclic electron transport in photosynthesis of *Rb. sphaeroides*. The excited state symbolized by a star results in a charge separation; an electron is transferred via pheophytin, the quinones Q_A, Q_B, and the cyt-*b*/*c* complex to the positively charged chlorophyll radical. Q: quinone, $Q^{\cdot -}$: semiquinone radical, QH_2: hydroquinone.

droquinone as a first stable product from the primary excitation state of the chlorophyll, more than half of the exciton energy is dissipated as **heat**.

Ubiquinone (Fig. 3.5) contains a hydrophobic isoprenoid side chain, which makes it very soluble in the lipid phase of the photosynthetic membrane. The same function of an isoprenoid side chain has already been discussed in the case of chlorophyll (section 2.2). In contrast to chlorophyll, pheophytin, and Q_A, which are all tightly bound to proteins, the ubihydroquinone Q_B is only loosely associated with the reaction center and can be exchanged for another ubiquinone. Ubihydroquinone remains in the membrane phase, is able to diffuse rapidly along the membrane, and functions as a transport metabolite for reducing equivalents in the membrane phase. It feeds the electrons into the **cytochrome-b/c$_1$ complex**, also located in the membrane, and the electrons are transferred back through this complex via

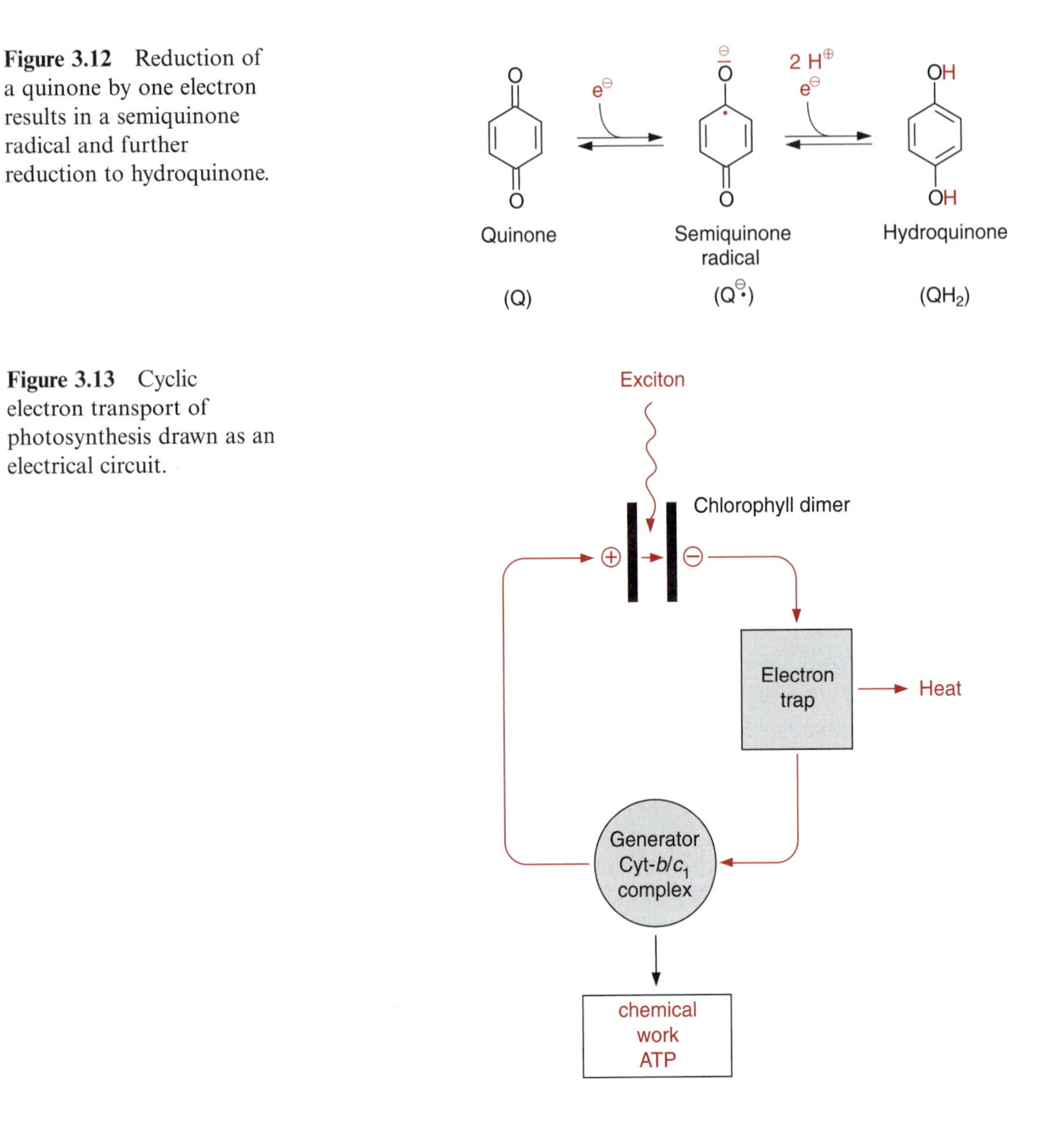

Figure 3.12 Reduction of a quinone by one electron results in a semiquinone radical and further reduction to hydroquinone.

Figure 3.13 Cyclic electron transport of photosynthesis drawn as an electrical circuit.

cytochrome-*c* to the reaction center. Energy is conserved in this electron transport to form a proton potential (section 4.1), which is used for ATP-synthesis. The structure and mechanism of the cytochrome-b/c_1 complex and of ATP-synthase will be described in section 3.7 and Chapter 4, respectively.

In summary, the cyclic electron transport of the purple bacteria may resemble a simple electric circuit as shown in Figure 3.13. The chlorophyll pair and pheophytin, between which an electron is transferred by light energy, may be regarded as the two plates of a capacitor between which a voltage is generated, driving a flux of electrons, a current. A very large part of electron energy is dissipated as heat by a voltage drop via a resistor. This resistor functions as an electron trap, withdrawing the electrons rapidly from the capacitor. A generator utilizes the remaining voltage to produce chemical energy.

3.5 Two photosynthetic reaction centers are arranged in tandem in photosynthesis of algae and plants

A **quantum requirement** (photons absorbed per molecule O_2 produced) of about eight has been determined for photosynthetic water splitting by green algae (section 2.4). Instead of the term *quantum requirement*, one often uses the reciprocal term, **quantum yield** (molecules of O_2 produced per photon absorbed). Investigations into the dependence of quantum yield on the color of irradiated light (action spectrum) revealed that the quantum yield dropped very sharply when the algae were illuminated with red light above a wavelength of 680 nm (Fig. 3.14). At first this effect, named **"red drop,"** remained unexplained since algae contain chlorophyll, which absorbs light at 700 nm. Robert Emerson and coworkers solved this problem in 1957 when they observed in an experiment that the quantum yield in the spectral range above 680 nm increased dramatically when orange light of 650 nm was irradiated together with red light. When the algae were irradiated with the light of the two colors simultaneously, the quantum yield was higher than the sum of the yields obtained when algae were irradiated separately with the light of each wavelength.

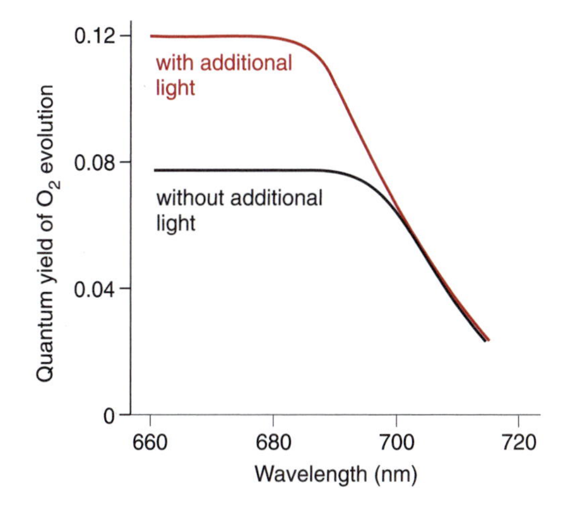

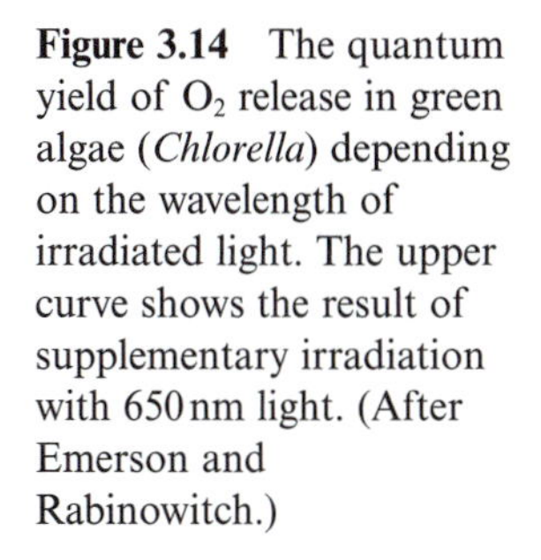
Figure 3.14 The quantum yield of O_2 release in green algae (*Chlorella*) depending on the wavelength of irradiated light. The upper curve shows the result of supplementary irradiation with 650 nm light. (After Emerson and Rabinowitch.)

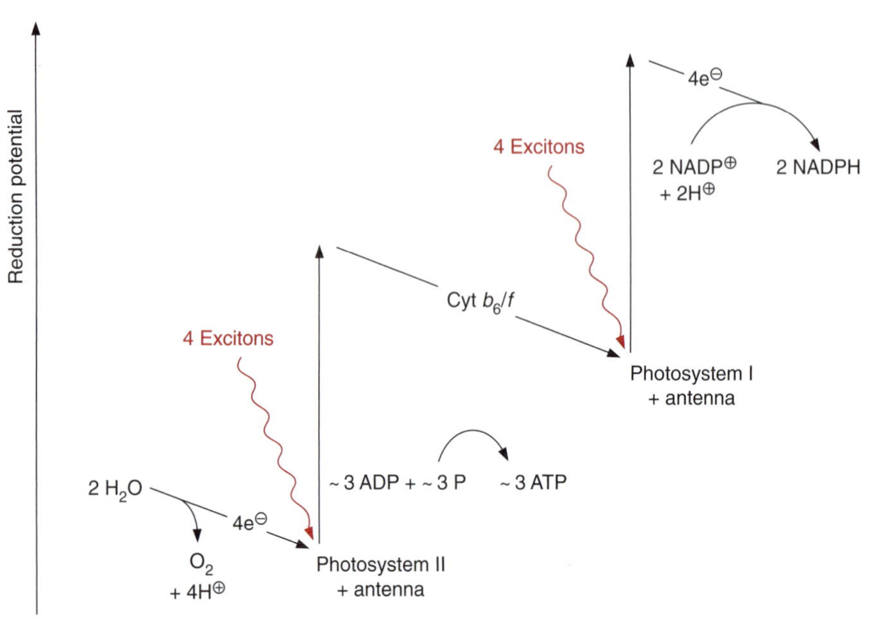

Figure 3.15 The Z scheme of photosynthesis in plants. Electrons are transferred by two photosystems arranged in tandem from water to NADP and during this ATP is formed. The amount of ATP formed is not known but is probably between two and three per four excitons captured at each reaction center (section 4.4).

This **Emerson effect** led to the conclusion that two different reaction centers are involved in photosynthesis of green algae (and also of cyanobacteria and higher plants). In 1960 Robert Hill and Fay Bendall postulated a reaction scheme (Fig. 3.15) in which **two reaction centers are arranged in tandem** and connected by an electron transport chain containing cytochromes-b_6 and -*f* (cytochrome-*f* is a cytochrome of the C type; see section 3.7). Light energy of 700 nm was sufficient for the excitation of the one reaction center, whereas excitation of the other reaction center required light of higher energy with a wavelength up to 680 nm. A reaction diagram according to the redox potentials shows a zigzag, leading to the name **Z scheme**. The numbering of the two photosystems corresponds to the sequence of their discovery. **Photosystem II** (PS II) can use light up to a wavelength of **680 nm**, whereas **photosystem I** (PS I) can utilize light with a wavelength up to **700 nm**. The sequence of the two photosystems makes it possible that at PS II a very **strong oxidant** is generated for oxidation of water and at PS I a very **strong reductant** for reduction of NADP (see also Fig. 3.3).

Figure 3.16 gives an overview of electron transport through the photosynthetic complexes; the carriers of electron transport are drawn according to their electric potential as in Figure 3.11. Figure 3.17 shows how the photosynthetic complexes are arranged in the thylakoid membrane. There is a potential difference of about 1.2 volt between oxidation of water and reduction of the NADP. The two absorbed photons of 680 and 700 nm together correspond to a total potential difference of 3.45 volt (see section 2.2, equa-

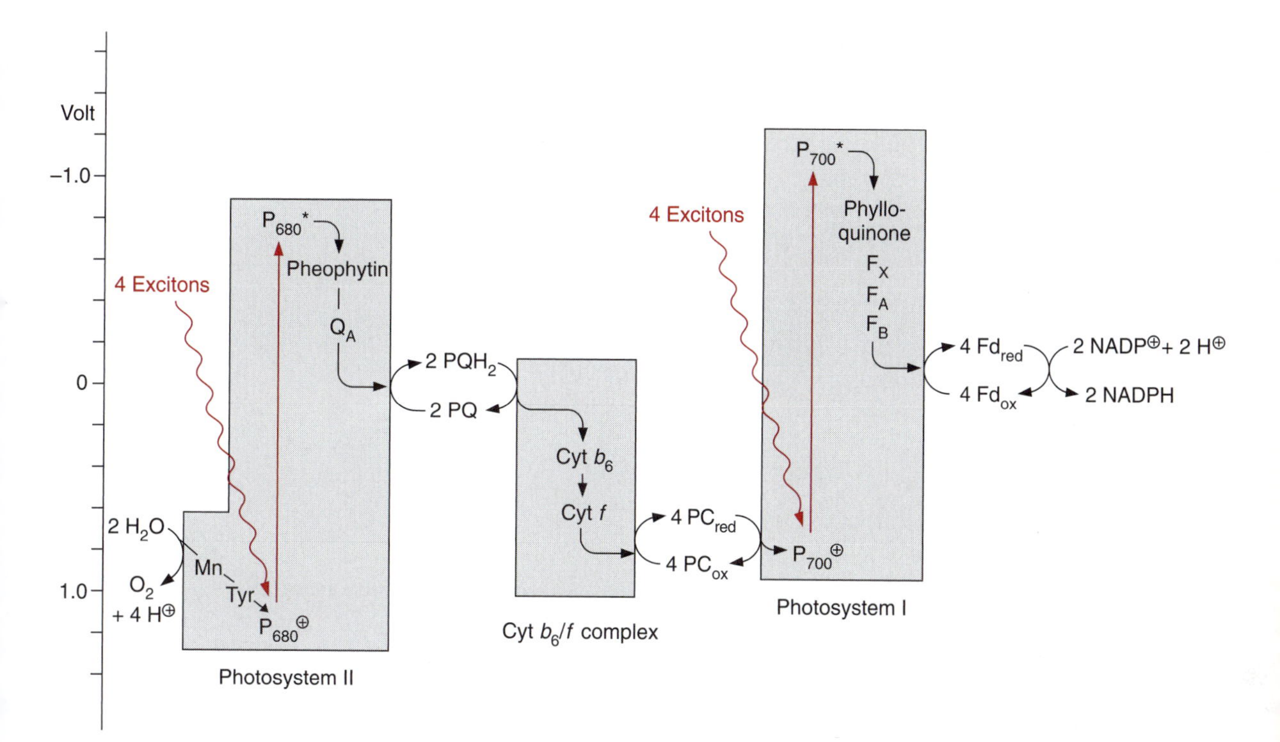

Figure 3.16 Scheme of noncyclic electron transport in plants. The redox components are placed according to their midpoint redox potential and their association with the three complexes involved in the electron transport. A star symbolizes an excited state. The electron transport between the photosystem II complex and the cyt-b_6/f complex occurs by plastohydroquinone (PQH_2), which is oxidized by the cyt-b_6/f complex to plastoquinone (PQ). The electrons are transferred from the cyt-b_6/f complex to photosystem I by plastocyanin (PC). This reaction scheme is also valid for cyanobacteria with the exception that instead of plastocyanin, cytochrome-*c* is involved in the second electron transfer. For details, see Figures 3.18 and 3.31.

tion 2.7). Thus, only about one-third of the energy of the photons absorbed by the two photosystems is used to transfer electrons from water to NADP. In addition to this, about one-eighth of the light energy absorbed by the two photosystems is conserved by pumping protons into the lumen of the thylakoids via PS II and the cytochrome-b_6/f complex (Fig. 3.17). This proton transport leads to the formation of a proton gradient between the lumen and the stroma space. An H^+-ATP synthase, also located in the thylakoid membrane, uses the energy of the proton gradient for synthesis of ATP.

Thus about half the energy of the light absorbed by the two photosystems is not used for chemical work but is dissipated as heat. The significance of this loss of energy in the form of heat in photosynthetic electron transport has been discussed in the previous section.

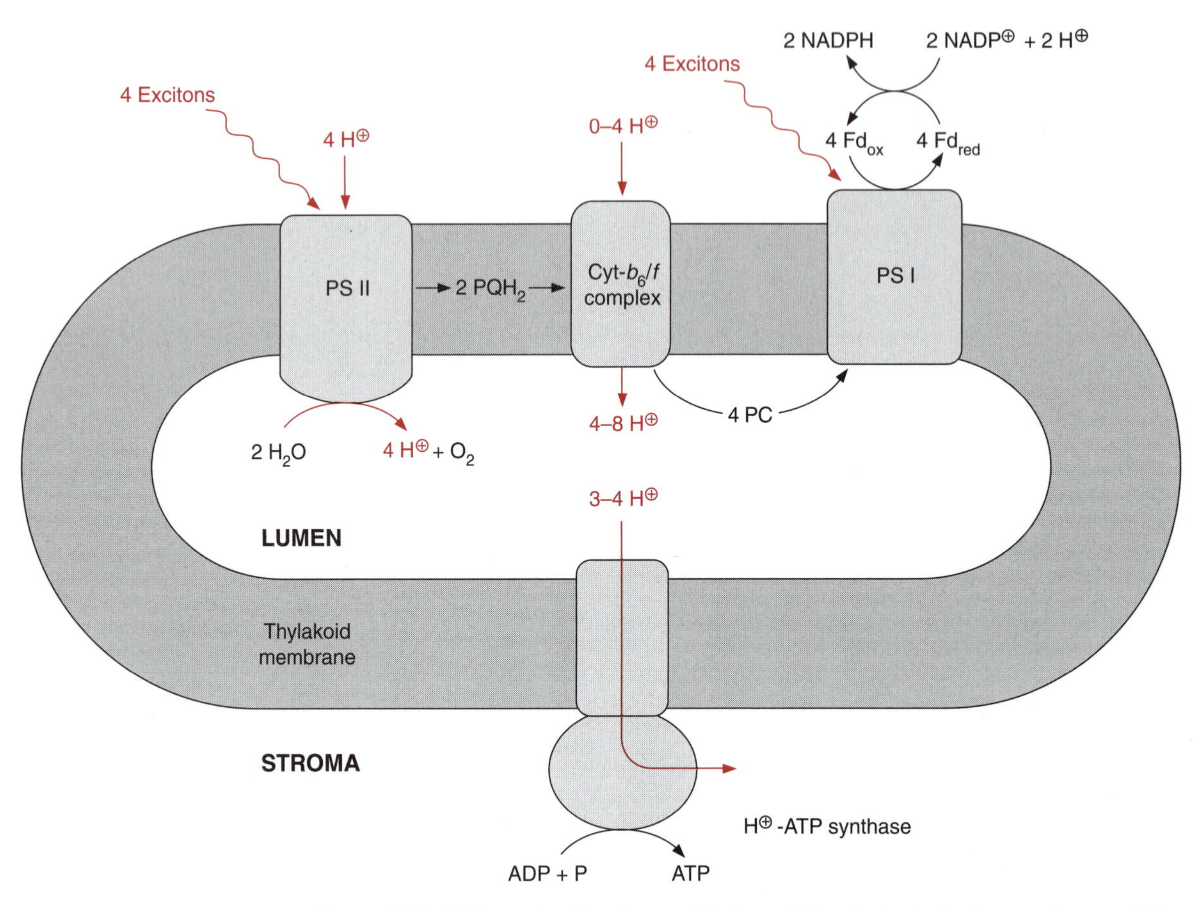

Figure 3.17 Scheme showing the positioning of the photosynthetic complexes and the H^+-ATP synthase in the thylakoid membrane. Transport of electrons between PS II and the cytochrome-b_6/*f* complex is mediated by plastohydroquinone (PQH_2) and that between the cytochrome-b_6/*f* complex and PS I by plastocyanin (PC). Splitting of the water occurs on the luminal side of the membrane, and the formation of NADPH and ATP occurs on the stromal side. The electrochemical gradient of protons pumped into the lumen drives ATP synthesis. The number of protons transported to the lumen during electron transport and the proton requirement of ATP synthesis is not known (section 4.4).

3.6 Water is split by photosystem II

Recently the groups of Horst Witt and Wolfgang Saenger (both in Berlin) resolved the three-dimensional structure of PS II by X-ray structure analysis of crystals from the PS II of the thermophilic cyanobacteria *Synechococcus elongatis*. This work revealed that **PS II**, and as later shown **PS I**, are constructed after the same basic principles as the reaction centers of purple bacteria discussed in section 3.4. This, and the sequence analyses, clearly demonstrate that all **these photosystems have a common origin**. Thus PS II also has a chl-*a* pair in the center, although the distance between the two molecules is so large that probably only one of the two chl-*a* molecules reacts with the exciton. Two arms, each with one chl-*a* and one pheophytin molecule, are connected with this central pair as in the purple bacteria shown in Figure 3.10. Also in the cyanobacteria, only one of these arms appears to be involved in the electron transport.

Upon the excitation of P_{680} by an exciton, one electron is released and transferred via the chl-*a*-monomer to pheophytin and from there to a tightly bound plastoquinone (Q_A), thus forming a semiquinone radical (Fig. 3.18).

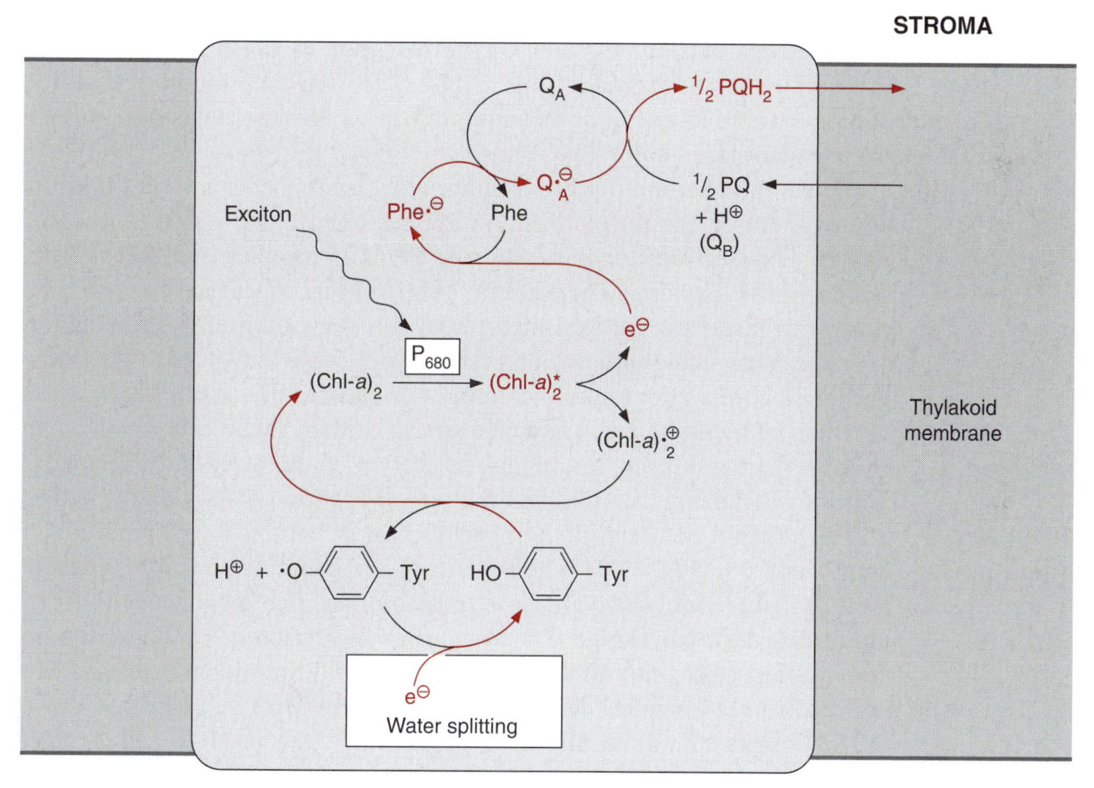

Figure 3.18 Reaction scheme of photosynthetic electron transport in the photosystem II complex. Excitation by a photon results in the release of one electron. The remaining positively charged chlorophyll radical is reduced by a tyrosine residue and the latter by a cluster of probably four manganese atoms involved in the oxidation of water (Fig. 3.20). The negatively charged chlorophyll radical transfers its electron via chl-*a* (not shown) and pheophytin and a quinone Q, of which the entire structure is not yet known, finally to plastoquinone.

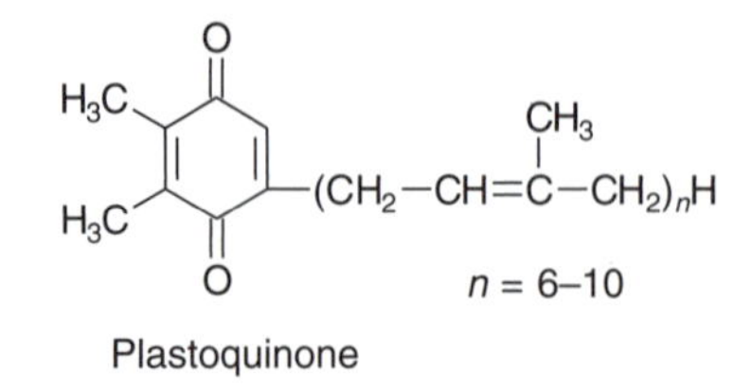

Figure 3.19 Plastoquinone.

The electron is then further transferred to a loosely bound plastoquinone (Q_B). This plastoquinone (PQ) (Fig. 3.19) accepts two electrons and two protons one after the other and is thus reduced to hydroquinone (PQH_2). The hydroquinone is released from the photosynthesis complex and may be regarded as the final product of photosystem II. This sequence, consisting of a transfer of a single electron between (chl-*a*)$_2$ and Q_A and the transfer of two electrons between Q_A and Q_B, corresponds to the reaction sequence shown for *Rb. sphaeroides* (Fig. 3.11). The only difference is that the quinones are ubiquinone or menaquinone in bacteria and plastoquinone in photosystem II.

However, the similarity between the reaction sequence in PS II and the photosystem of the purple bacteria applies only to the electron acceptor region. The electron donor function in PS II is completely different from that in purple bacteria. The electron deficit in (chl-*a*)$_2^{\dot{+}}$ caused by noncyclic electron transport is compensated for by electrons derived from oxidation of water. **Manganese cations** and a **tyrosine** residue are involved in the transport of electrons from water to chlorophyll. The (chl-*a*)$_2^{\dot{+}}$ radical with a redox potential of about +1.1 volt is such a strong oxidant that it can withdraw an electron from a tyrosine residue in the protein of the reaction center and a tyrosine radical remains. This reactive tyrosine residue is often designated as Z. The electron deficit in the tyrosine radical is restored by oxidation of a manganese ion (Fig. 3.20). The PS II complex contains several manganese ions, probably four, which are close to each other. This arrangement of Mn ions is called the **Mn cluster**. The Mn cluster depicts a redox system that can take up four electrons and release them again. During this process the Mn ions probably change between the oxidation state Mn^{3+} and Mn^{4+}.

To liberate one molecule of O_2 from water, the reaction center must withdraw four electrons and thus capture four excitons. The time differences between the capture of the single excitons in the reaction center depends on the intensity of illumination. If oxidation of water were to proceed stepwise, **oxygen radicals** could be formed as intermediary products, especially at low light intensities. Oxygen radicals have a destructive effect on biomolecules such as lipids and proteins (section 3.10). The water splitting machinery of the Mn clusters minimizes the formation of oxygen radical intermediates by supplying the reaction center via tyrosine with four electrons one after the

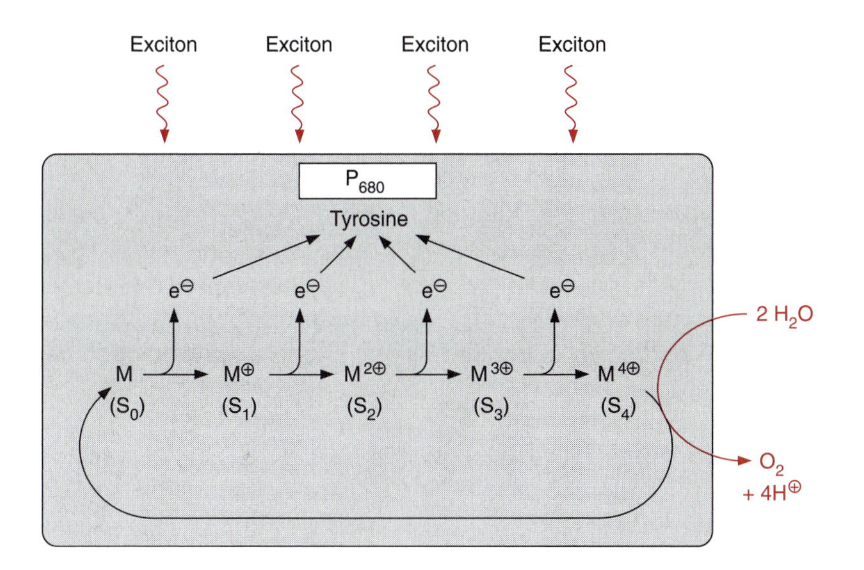

Figure 3.20 A scheme showing the mechanism of water splitting by photosystem II. M means a cluster of probably four manganese atoms. The different manganese atoms are present in different oxidation states. The cluster functions as a redox unit and feeds, one after the other, a total of four electrons into the reaction center of PS II. The deficit of these four electrons is compensated for by splitting of 2 H_2O to O_2 and 4 H^+. M means $(4\,Mn)^{n+}$, M^+ means $(4\,Mn)^{(n+1)+}$, and so on.

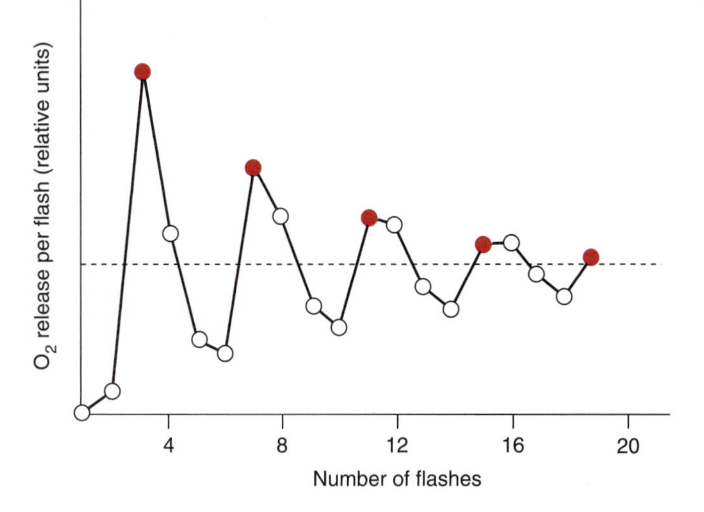

Figure 3.21 Yield of the oxygen released by chloroplasts as a function of the number of light pulses. The chloroplasts, previously kept in the dark, were illuminated by light pulses of 2 µs duration, interrupted by pauses of 0.3 s. (After Forbush, Kok, and McGoild, 1971.)

other (Fig. 3.20). The Mn cluster is transformed during this transfer from the ground oxidation state stepwise to four different oxidation states (these have been designated as S_0 and S_1–S_4).

Experiments by Pierre Joliot and Bessel Kok presented evidence that the water splitting apparatus can be in five different oxidation states (Fig. 3.21). When chloroplasts kept in the dark were illuminated by a series of light pulses, an oscillation of the oxygen release was observed. Whereas after the first two light pulses almost no O_2 was released, the O_2 release was maximal after three pulses and then after a further four pulses, and so on. An increasing number of light pulses, however, dampened the oscillation more and more. This can be explained by some pulses not causing excitation of PS II and thus desynchronizing oscillation. In darkened chloroplasts the water splitting apparatus is apparently in the S_1 state. After the fourth oxidation state (S_4) has been reached, O_2 is released in one reaction and the Mn cluster returns to its ground oxidation state (S_0). In this reaction, protons from water are released to the lumen of the thylakoids. The formal description of this reaction is:

$$2H_2O \longrightarrow 4H^+ + 2O^{2-}$$

$$2O^{2-} + M^{4+} \longrightarrow O_2 + M$$

Figuratively speaking, the four electrons needed in the reaction center are loaned in advance by the Mn cluster and then repaid at one stroke by oxidation of water to one oxygen molecule. In this way the Mn cluster minimizes the formation of oxygen radicals by photosystem II. Despite this safety device, probably enough oxygen radicals are formed in the PS II complex to have some damaging effect on the proteins of the complex, the consequences of which will be discussed in section 3.10.

Photosystem II complex is very similar to the reaction center in purple bacteria

Photosystem II is a complex consisting of at least 17 different subunits (Table 3.2), only two of which are involved in the actual reaction center. The exact function of several subunits is not yet known. For this reason the scheme of the PS II complex shown in Figure 3.22 contains only those subunits for which functions are known. The PS II complex is surrounded by an antenna consisting of light harvesting complexes (Fig. 2.13).

The center of the PS II complex is a heterodimer consisting of the subunits $\mathbf{D_1}$ and $\mathbf{D_2}$ with six chl-*a*, two pheophytin, two plastoquinone, and one to two carotenoid molecules bound to it. The D_1 and the D_2 protein are homologous to each other and also to the L proteins and M proteins from the reaction center of the purple bacteria (section 3.4). As in purple bacteria, only the pheophytin molecule bound to the D_1 protein of PS II is involved in electron transport. On the other hand, Q_A is bound to the D_2 protein, whereas Q_B is bound to the D_1 protein. The Mn cluster is probably enclosed by both the D_1 and D_2 proteins. The tyrosine that is reactive in

Table 3.2: Protein components of photosystem II (list not complete)

Protein	Molecular mass (kDa)	Localization	Encoded in	Function
D_1	32	In membrane	Chloroplast	Binding of P_{680}, Pheo, Q_B, Tyr, Mn-cluster
D_2	34	"	"	Binding of P_{680}, Pheo, Q_A, Mn-cluster
CP_{47}	47	"	"	Core-antenna, binds peripheral antennae LHC
CP_{43}	43	"	"	" "
Cyt-$b_{559\alpha}$	9	"	"	Binds heme, protection of PS II against light damage
Cyt-$b_{559\beta}$	4	"	"	"
Manganese-stabilizing protein (MSP)	33	Peripheral: lumen	Nucleus	Stabilization of Mn-cluster
P	23	"	"	?
Q	16	"	"	?
R	10	Peripheral:stroma	"	?

(After Vermaas, 1993).

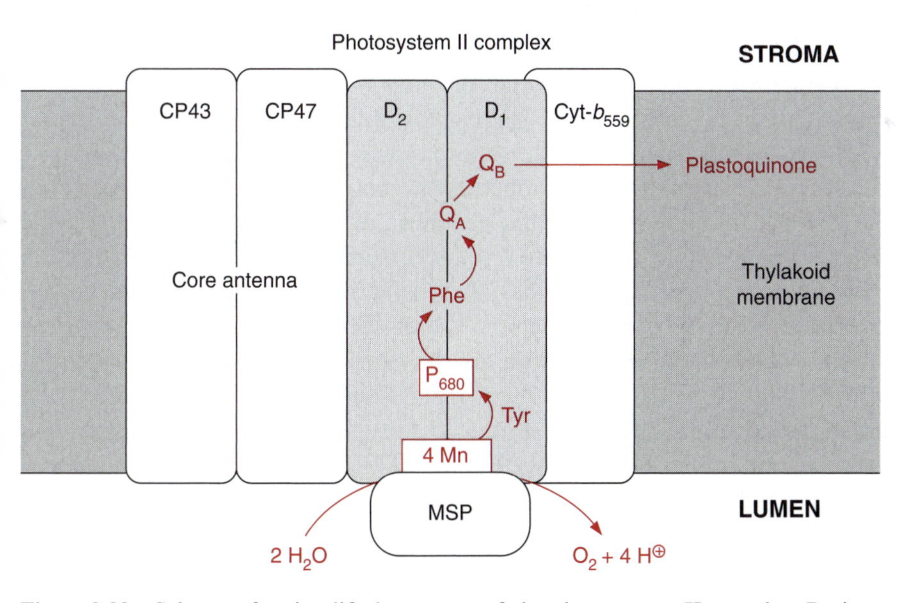

Figure 3.22 Scheme of a simplified structure of the photosystem II complex. Basis for this scheme are the structural analysis carried out by the collaborating groups of Witt and Saenger (Berlin) and also investigations of the binding of quinone to the subunits D_1 and D_2, which are homologous to the subunits L and M in purple bacteria. It appears that the structure of PS II and the structure of the reaction centers in purple bacteria share the same basic features. See also Table 3.2. The two core antennae CP_{43} and CP_{47} flank both sides of the D_1-D_2 complex (not shown in figure).

electron transfer is a constituent of D_1. The subunit MSP (manganese stabilizing protein) stabilizes the Mn cluster. The two subunits CP 43 and CP 47 (CP means chlorophyll protein) each bind about 15 chlorophyll molecules and form the **core complex of the antenna** shown in Figure 2.10. CP43 and CP47 flank both sides of the D_1-D_2 complex. Cyt-b_{559} does not seem to be involved in the electron transport of PS II; possibly its function is to protect the PS II complex from light damage.

The D_1 protein of the PS II complex has a high turnover; it is constantly being resynthesized. It seems that the D_1 protein wears out during its function, perhaps through damage by oxygen radicals, which still occurs despite all the protection mechanisms. It has been estimated that the D_1 protein is replaced after 10^6 to 10^7 catalytic cycles of the reaction center of PS II.

A number of substances that are similar in their structure to plastoquinone can block the plastoquinone binding site at the D_1 protein, causing inhibition of photosynthesis. Such substances are used as weed killers (herbicides). Before the effect of these substances is discussed in detail, some general aspects of the application of herbicides shall be introduced.

Mechanized agriculture usually necessitates the use of herbicides

Herbicides account for about half the money spent worldwide on substances for plant protection. The high cost of labor is one of the main reasons for using herbicides in agriculture. It is cheaper and faster to keep a field free of weeds by using herbicides rather than manual labor. Weed control in agriculture is necessary not only to decrease harvest losses by weed competition, but also because weeds hinder the operation of harvesting machinery; fields free of weeds are a prerequisite for a mechanized agriculture. One uses as herbicides various substances that block vital reactions of plant metabolism and have a low toxicity for animals and therefore for humans. A large number of herbicides (examples will be given at the end of this section) inhibit photosystem II by being antagonists to plastoquinone. For this the herbicide molecule has to be bound to most of the many photosynthetic reaction centers. To be effective, 125 to 4,000 g of these herbicides have to be applied per hectare.

In an attempt to reduce the amount of herbicides needed to be effective, new herbicides have been developed that inhibit certain biosynthetic processes such as the synthesis of fatty acids, certain amino acids, carotenoids, or chlorophyll. There are also herbicides that act as analogues of phytohormones or mitosis inhibitors. Some of these herbicides are effective with amounts as low as 5 g per hectare.

Some herbicides are taken up only by the roots and others by the leaves. For example, to keep the railway tracks free of weeds **nonselective herbicides** are employed, which destroy all vegetation. For such purposes, herbicides that are degraded slowly and are taken up by the roots or emerging shoots are often used. Nonselective herbicides are also used in agriculture (e.g., to combat weeds in citrus plantations), but in this case, herbicides that are taken up only by the leaves of herbaceous plants are applied at ground level. Especially interesting are those **selective herbicides** that combat only weeds and affect cultivars as little as possible (sections 12.2 and 15.3). Selectivity can be caused by various factors (e.g., by differences between the uptake of the herbicide in different plants, between the sensitivity of metabolism in different plants toward the herbicide, or between the ability of the plants to detoxify the herbicide). Important mechanisms utilized by plants to detoxify herbicides and other foreign substances (xenobiotics) are the introduction of hydroxyl groups by P-450 monooxygenases (section 18.2) and the formation of glutathione conjugates (section 12.2). Selective herbicides have the advantage that weeds can be destroyed, for example, when it is opportune at a later growth stage of the cultivars and the dead weeds can form a mulch layer conserving water and preventing erosion.

In some cases, the application of herbicides has led to the development of herbicide-resistant plant mutants (section 10.4). Conventional breeding has used such mutants to generate herbicide-resistant cultivars. In contrast to these results, which have come about by chance, genetic engineering makes it possible to generate made-to-measure herbicide-resistant cultivars. This means that selective herbicides that are degraded rapidly and are effective in small amounts can be used. Examples of this will be discussed in sections 10.1 and 10.4.

A large number of herbicides inhibit photosynthesis: the urea derivative DCMU (Diuron, DuPont), the triazine Atrazine (earlier Ciba Geigy), Bentazon (BASF) (Fig. 3.23), and many similar substances function as herbicides by binding to the plastoquinone binding site on the D_1 protein and thus blocking the photosynthetic electron transport. Nowadays, DCMU is not often used, as the dosage required is high and its degradation is slow. It is, however, often used in the laboratory to inhibit photosynthesis (e.g., of leaves or isolated chloroplasts). Atrazine acts selectively: maize plants are relatively insensitive to this herbicide since they have a particularly efficient mechanism for its detoxification (section 12.2). Because of its relatively slow degradation in the soil, the use of Atrazine has been restricted in some countries. In areas where certain herbicides have been used continuously over the years, some weeds have become resistant to these herbicides. In some cases, the resistance can be traced back to mutations in a single amino acid change in the D_1-proteins. These changes do not markedly affect

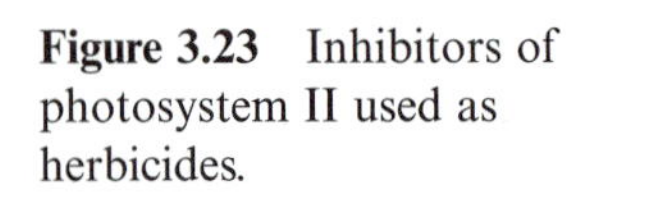

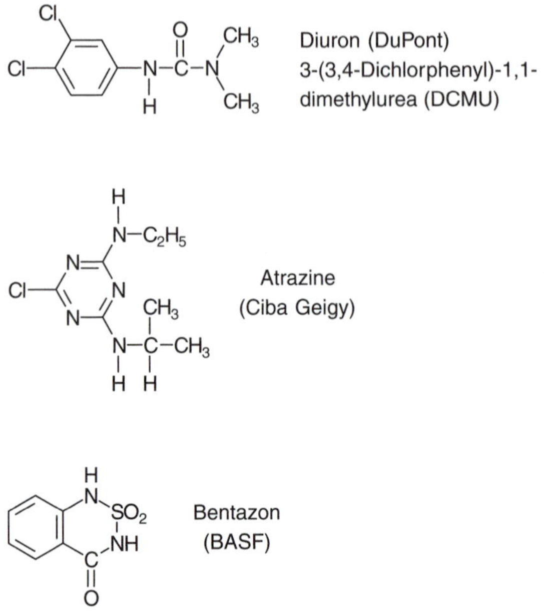

Figure 3.23 Inhibitors of photosystem II used as herbicides.

Diuron (DuPont)
3-(3,4-Dichlorphenyl)-1,1-dimethylurea (DCMU)

Atrazine
(Ciba Geigy)

Bentazon
(BASF)

photosynthesis of these weeds, but they do decrease binding of the herbicides to the D_1-protein.

3.7 The cytochrome-b_6/*f* complex mediates electron transport between photosystem II and photosystem I

Iron atoms in cytochromes and in iron-sulfur centers have a central function as redox carriers

Cytochromes occur in all organisms except a few obligate anaerobes. These are proteins to which one to two **tetrapyrrole** rings are bound. These tetrapyrroles are very similar to the chromophores of chlorophylls. However, chlorophylls contain Mg^{++} as the central atom in the tetrapyrrole, whereas the cytochromes have an iron atom (Fig. 3.24). The tetrapyrrole ring of the cytochromes with iron as the central atom is called the **heme**. The bound iron atom can change between the oxidation states $\mathbf{Fe^{+++}}$ and $\mathbf{Fe^{++}}$ so that cytochromes function as one electron carrier, in contrast to quinones, NAD(P) and FAD, which transfer two electrons together with protons.

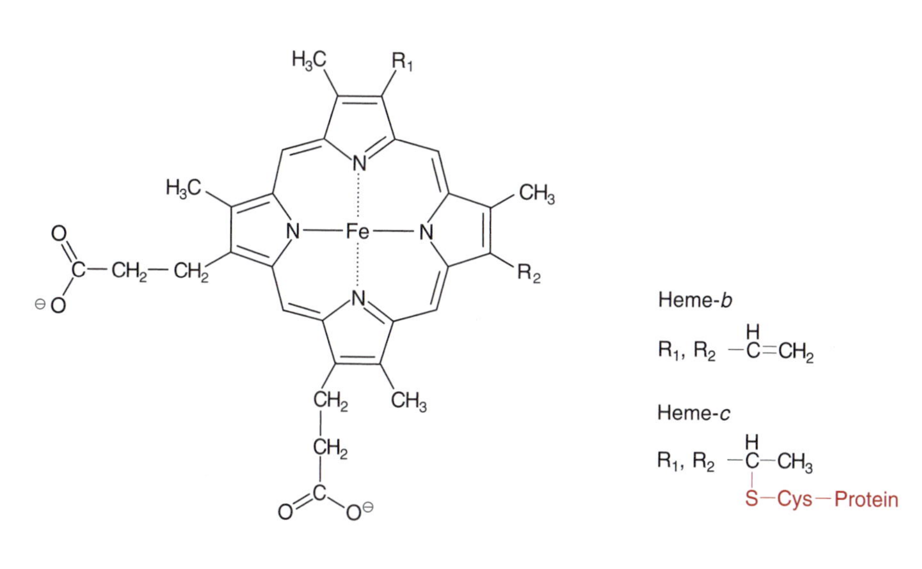

Figure 3.24 Heme-*b* and heme-*c* as prosthetic group of the cytochromes. Heme-*c* is covalently bound to the apoprotein of the cytochrome by the addition of two cysteine residues of the apoprotein to the two vinyl groups of heme-*b*.

Cytochromes are divided into three main groups, the cytochromes-*a*, -*b,* and -*c*. These correspond to heme-*a*, -*b,* and -*c*. Heme-*b* may be regarded as the basic structure (Fig. 3.24). In heme-*c* the SH_2 group of a **cysteine** is added to each of the two vinyl groups of heme-*b*. In this way heme-*c* is covalently bound by a sulfur bridge to the protein of the cytochrome. Such a mode of covalent binding has already been shown for phycocyanin in Figure 2.15, and there is actually a structural relationship between the corresponding apoproteins. In heme-*a* (not shown) an **isoprenoid side chain** consisting of three isoprene units is attached to one of the vinyl groups of heme-*b*. This side chain has the function of a hydrophobic membrane anchor, similar to that found in quinones (Figs. 3.5 and 3.19). Heme-*a* is mentioned here only for the sake of completeness. It plays no role in photosynthesis, but it does have a function in the mitochondrial electron transport chain (section 5.5).

The iron atom in the heme can form up to six coordinative bonds. Four of these bonds are formed with the nitrogen atoms of the tetrapyrrole ring. This ring has a planar structure. The two remaining coordinative bonds to the Fe-atom are formed by two histidine residues, which are positioned vertically to the tetrapyrrole plane (Fig. 3.25). Cyt-*f* (*f* = foliar, in leaves) contains, like cyt-*c*, one heme-*c* and therefore belongs to the *c*-type cytochromes. In cyt-*f* one coordinative bond of the Fe-atom is formed with the terminal amino group of the protein and the other with a histidine residue.

Iron-sulfur centers are of general importance as electron carriers in electron transport chains and thus also in photosynthetic electron transport. Cysteine residues of proteins within iron-sulfur centers (Fig. 3.26) are coordinatively or covalently bound to Fe-atoms. These iron atoms are linked to each other by S-bridges. Upon acidification of the proteins, the sulfur

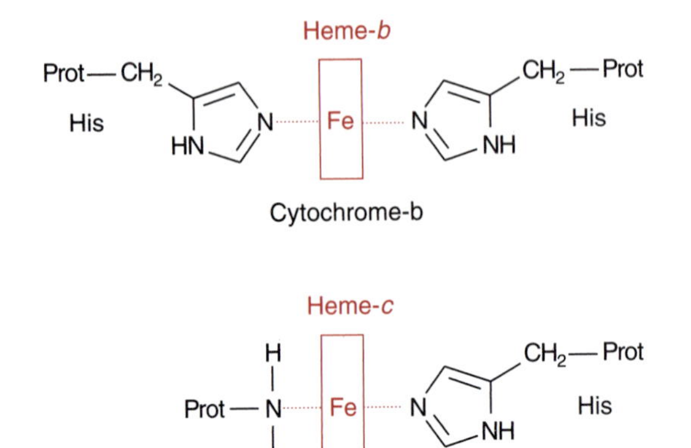

Figure 3.25 Axial ligands of the Fe atoms in the heme groups of cytochrome-*b* and cytochrome-*f*. Of the six possible coordinative bonds of the Fe atom in the heme, four are saturated with the N-atoms present in the planar tetrapyrrole ring. The two remaining coordinative bonds are formed either with two histidine residues of the protein, located vertically to the plane of the tetrapyrrole, or with the terminal amino group and one histidine residue of the protein. Prot = protein.

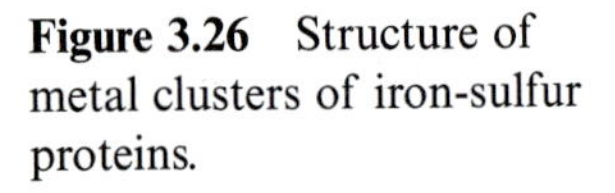

Figure 3.26 Structure of metal clusters of iron-sulfur proteins.

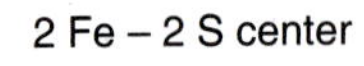

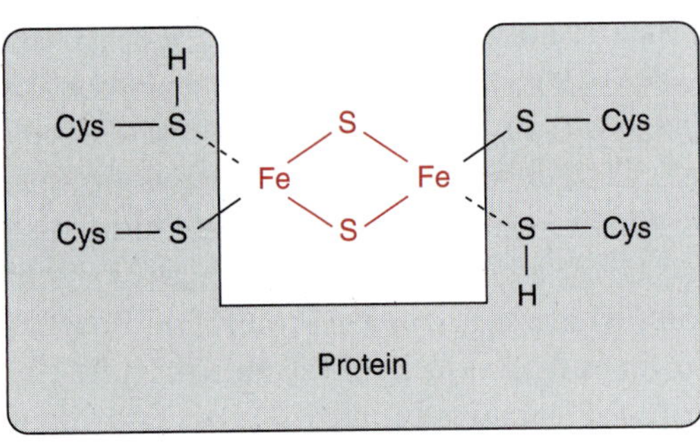

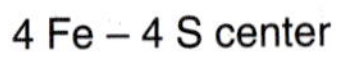

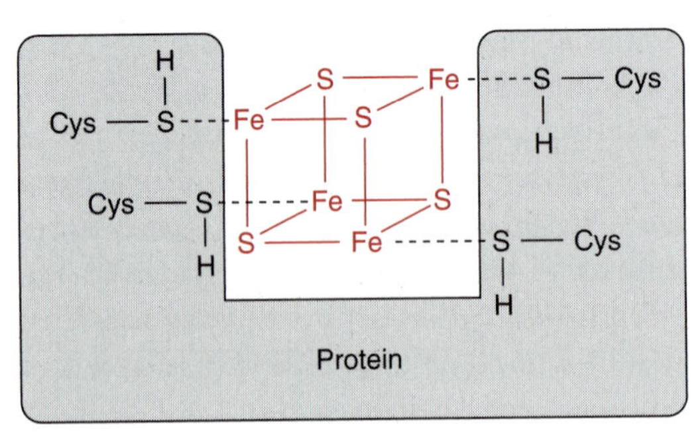

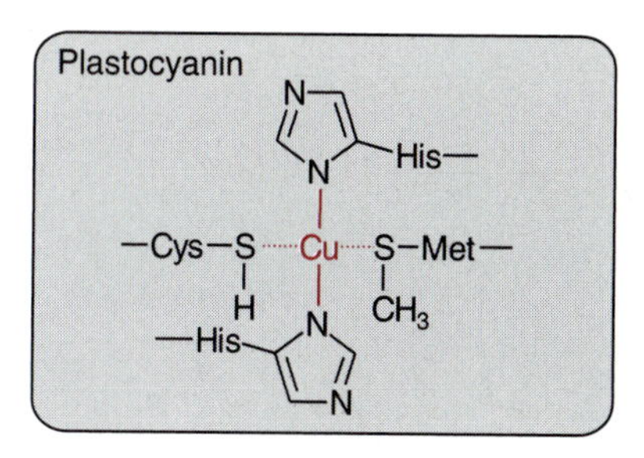

Figure 3.27 Plastocyanin. Two histidine, one methionine, and one cysteine residue of this protein bind one Cu-atom, which changes between the redox states Cu^+ and Cu^{++} by the addition or removal of an electron.

between the Fe-atoms is released as H_2S and for this reason has been called **labile sulfur**. Iron-sulfur centers occur mainly as 2Fe-2S or 4Fe-4S centers. The Fe-atoms in these centers are present in the oxidation states Fe^{++} and Fe^{+++}. Irrespective of the number of Fe-atoms in a center, the oxidized and reduced state of the center differs only by a single charge. For this reason, iron-sulfur centers can take up and transfer only one electron. Various iron-sulfur centers have very different redox potentials, depending on the surrounding protein.

The electron transport by the cytochrome-b_6/f complex is coupled to a proton transport

Plastohydroquinone (PQH_2) formed by PS II diffuses through the lipid phase of the thylakoid membrane and transfers its electrons to the cytochrome-b_6/f complex (Fig. 3.17). This complex then transfers the electrons to **plastocyanin**, which is thus reduced. Therefore the cytochrome-b_6/f complex has also been called **plastohydroquinone-plastocyanin oxidoreductase**. Plastocyanin is a protein with a molecular mass of 10.5 kDa, containing a **copper atom**, which is coordinatively bound to one cysteine, one methionine, and two histidine residues of the protein (Fig. 3.27). This copper atom alternates between the oxidation states Cu^+ and Cu^{++} and thus is able to take up and transfer one electron. Plastocyanin is soluble in water and is located in the thylakoid lumen.

Electron transport through the cyt-b_6/f complex proceeds along a potential difference gradient of about 0.4 V (Fig. 3.16). The energy liberated by the transfer of the electron down this redox gradient is conserved by transporting protons to the thylakoid lumen. The cyt-b_6/f complex is a membrane protein consisting of many subunits. The main components of this complex are four subunits: cyt-b_6, cyt-*f*, an iron-sulfur protein called **Rieske protein** after its discoverer, and a subunit IV. Additionally, there are some smaller peptides and a chlorophyll and a carotenoid of unknown function. The Rieske protein has a 2Fe-2S center with the very positive redox potential of +0.3 V, untypical of such iron-sulfur centers.

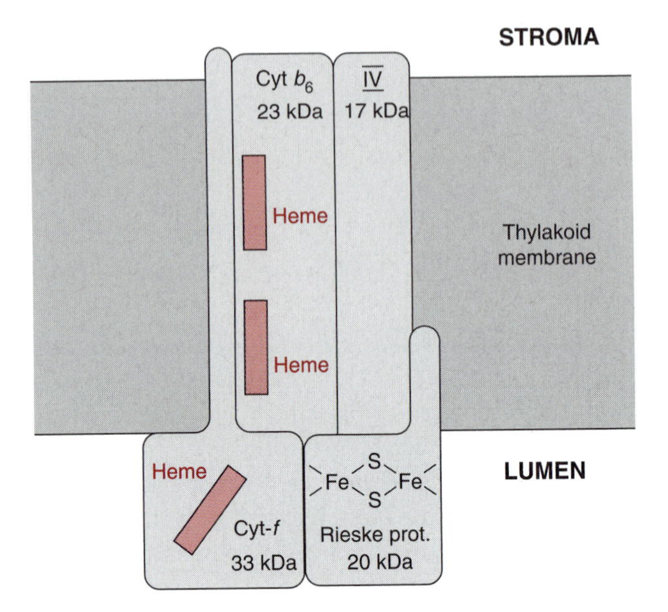

Figure 3.28 Scheme of the structure of the cytochrome-b_6/f complex. The scheme is based on the molecular structures predicted from their amino acid sequences. (After Hauska.)

Table 3.3: Function of cytochrome-*b*/*c* complexes

Purple bacteria	Cyt-b/c_1	Reduction of Cyt-*c*	Proton pump
Green sulfur bacteria	Cyt-b/c_1	"	"
Mitochondria	Cyt-b/c_1	"	"
Cyanobacteria	Cyt-b_6/f	"	"
Chloroplasts	Cyt-b_6/f	Reduction of plastocyanin	"

The cyt-b_6/f complex has an **asymmetric structure** (Fig. 3.28). Cyt-b_6, which contains two molecules of heme-*b*, spans the membrane, as does subunit IV. In the cyt-b_6/f complex, the two heme molecules are placed one above the other and in this way form a redox chain reaching from one side of the membrane to the other. One amino acid chain of cyt-*f* protrudes into the membrane, forming an anchor. The heme-*c*, as the functional group of cyt-*f*, is positioned at the periphery of the membrane on the luminal side. On the same side there is also the Rieske-2Fe-2S-protein, which protrudes only slightly into the membrane.

The cyt-b_6/f complex resembles in its structure the cyt-b/c_1 complex in bacteria and mitochondria (section 5.5). Table 3.3 summarizes the function of these cyt-b_6/f and cyt-b/c_1 complexes. All these complexes possess one iron-sulfur protein. The amino acid sequence of cyt-*b* in the cyt-b/c_1 complex of bacteria and in mitochondria corresponds to the sum of the sequences of cyt-b_6 and the subunit IV in the cyt-b_6/f complex. Apparently a cleavage

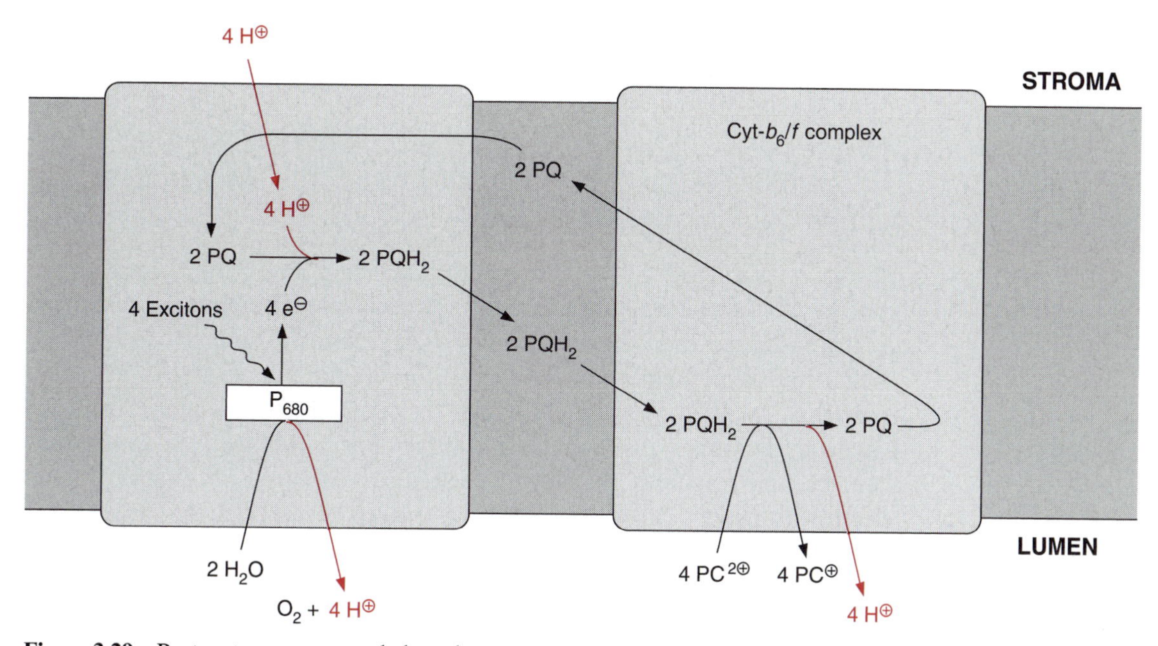

Figure 3.29 Proton transport coupled to electron transport by PS II and the cyt-b_6/f complex in the absence of a Q-cycle. The basis of this scheme is that the oxidation of water in PS II and the oxidation of plastohydroquinone (PQH_2) by cyt-b_6/f occurs at the luminal side of the thylakoid membrane.

of the cyt-*b* gene into the two genes for cyt-b_6 and the subunit IV occurred during evolution. Whereas in plants the cyt-b_6/f complex reduces plastocyanin, the cyt-b/c_1 complex of bacteria and mitochondria reduces cyt-*c*. Cyt-*c* is a very small cytochrome molecule that is water-soluble and, like plastocyanin, transfers redox equivalents from the cyt-b_6/f complex to the next complex along the aqueous phase. In cyanobacteria, which also possess a cyt-b_6/f complex, the electrons are transferred from this complex to photosystem I via cyt-*c* instead of plastocyanin. The great similarity between the cyt-b_6/f complex in plants and the cyt-b/c_1 complexes in bacteria and mitochondria suggests that these complexes have basically similar functions in photosynthesis and in mitochondrial oxidation: they are proton translocators that are driven by a hydroquinone-plastocyanin (or -cyt-*c*) reductase.

The transfer of protons from the stroma to the thylakoid lumen is explained by reduction and oxidation of the quinones taking place on different sides of the membrane. The protons required for reduction of plastoquinone in the PS II complex come from the stromal side, whereas oxidation of plastohydroquinone by plastocyanin, with the resultant release of protons, occurs on the luminal side of the cyt-b_6/f complex (Fig. 3.29). In this way four protons are transferred from the stroma space to the lumen after the capture of four excitons by the PS II complex. In addition to this, four protons are released to the lumen during the splitting of water by PS II.

The number of protons pumped through the cyt-b_6/f complex can be doubled by a Q-cycle

Studies with mitochondria indicated that during electron transport through the cyt-b/c_1 complex, the number of protons transferred per transported electron is larger than shown in Figure 3.29. Peter Mitchell, who established the chemiosmotic hypothesis of energy conservation (section 4.1), also postulated a so-called **Q-cycle**, by which the number of transported protons for each electron transferred through the cyt-b/c_1 complex is doubled. It later became apparent that the Q-cycle also has a role in photosynthetic electron transport.

Figure 3.30 shows the principle of Q-cycle operation in the photosynthesis of chloroplasts. The cyt-b_6/f complex contains two different binding sites for conversion of quinones, one located at the stromal side and the other at the luminal side of the thylakoid membrane. The plastohydroquinone (PQH_2) formed in the PS II complex is oxidized by the Rieske iron-sulfur center at the binding site adjacent to the lumen. Due to its very positive redox potential, the Rieske protein tears off one electron from the plastohydroquinone. Because its redox potential is very negative, the remaining semiquinone is unstable and transfers its electron to the first heme-*b* of the cyt-b_6 (b_p) and from there to the second heme-*b* (b_n), thus raising the redox potential of heme b_n to about −0.1 V. In this way a total of four protons are transported to the thylakoid lumen per two molecules of plastohydroquinone oxidized. Of the two plastoquinone molecules (PQ) formed, only one molecule returns to the PS II complex. The other PQ diffuses through the lipid phase of the membrane to the other binding site on the stromal side and is reduced there by heme-b_n with its high reduction potential via semiquinone to hydroquinone. This is accompanied by the uptake of two protons from the stromal space. The hydroquinone thus regenerated is oxidized, in turn, by the Rieske protein on the luminal side, and so on. In total, the number of transported protons is doubled by the Q-cycle ($1/2 + 1/4 + 1/8 + 1/16 + \ldots + 1/n = 1$). When the Q-cycle is operating fully, the transport of four electrons through the Cyt-b_6/f complex leads in total to the transfer of eight protons from the stroma to the lumen. The function of this Q-cycle in mitochondrial oxidation is now undisputed. In the case of photosynthetic electron transport, however, its function is still a matter of controversy. It is to be expected from the analogy of the cyt-b_6/f complex to the cyt-b/c_1 complex that the Q-cycle also plays an important role in

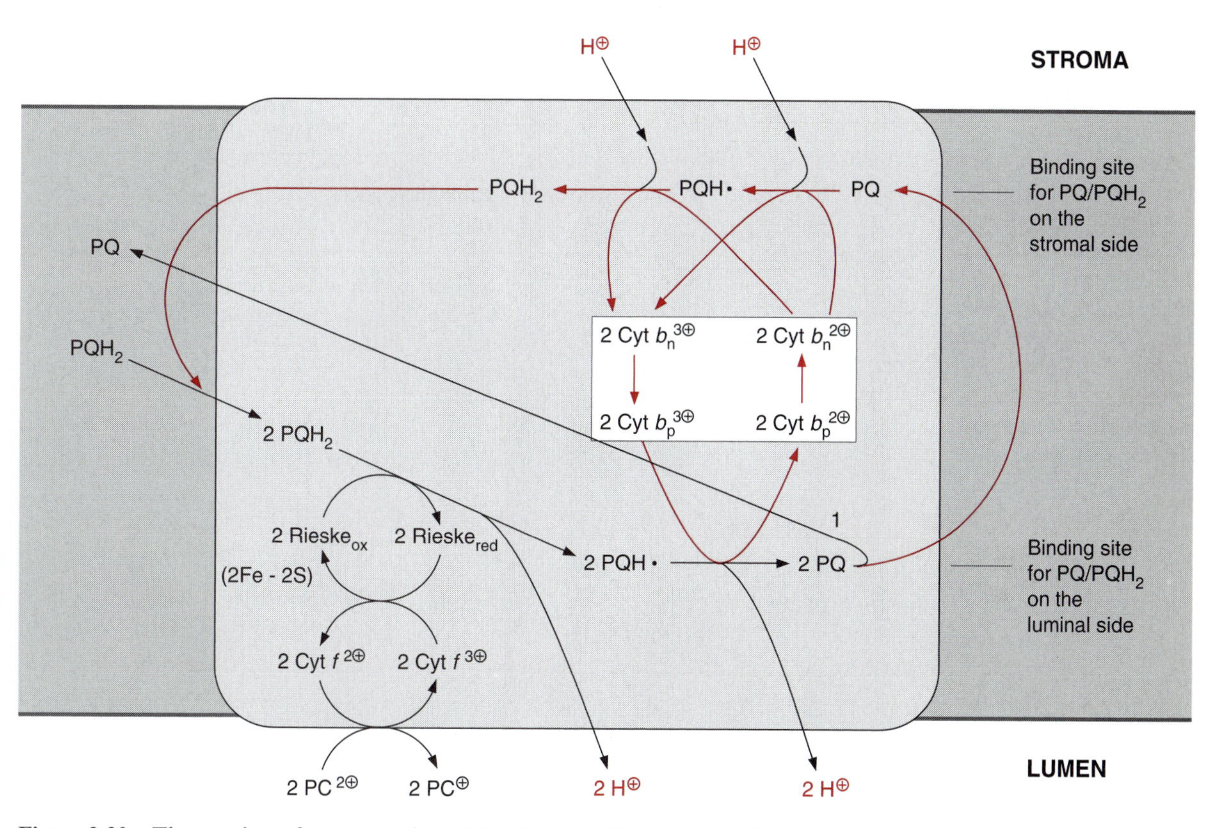

Figure 3.30 The number of protons released by the cyt-b_6/f complex to the lumen is doubled by the Q-cycle. This cycle is based on the finding that the redox reactions of the PQH_2 and PQ occur at two binding sites, one in the lumen and one in the stromal region of the thylakoid membrane. How the cycle functions is explained in the text.

chloroplasts. So far, the functioning of a Q-cycle in plants has been observed mainly under low light conditions. The Q-cycle is perhaps suppressed by a high proton gradient across the thylakoid membrane generated, for instance, by irradiation with high light intensity. In this way the flow of electrons through the Q-cycle could be adjusted to the energy demand of the plant cell.

3.8 Photosystem I reduces NADP

Plastocyanin that has been reduced by the cyt-b_6/f complex diffuses through the lumen of the thylakoids, binds to a positively charged binding site of PS

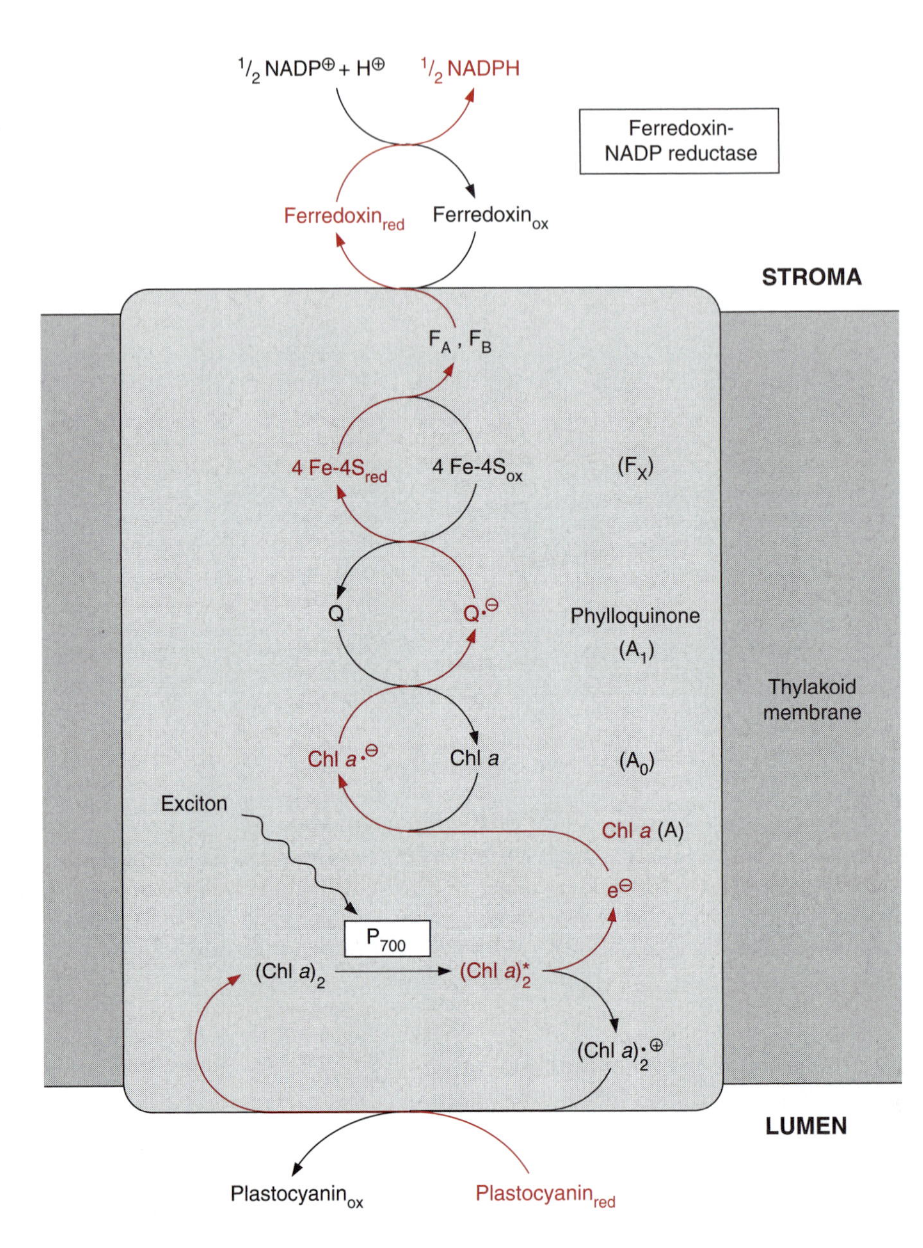

Figure 3.31 Reaction scheme of electron transport in photosystem I. The negatively charged chlorophyll radical formed after excitation of a chlorophyll pair results in reduction of NADP via chl-*a*, phylloquinone, and three iron-sulfur proteins. The electron deficit in the positively charged chlorophyll radical is compensated for by an electron delivered from plastocyanin.

I, transfers its electron, and then diffuses back to the cyt-b_6/f complex in the oxidized form (Fig. 3.31).

The reaction center of PS I with an absorption maximum of 700 nm also contains a chlorophyll pair (chl-a)$_2$ (Fig. 3.31). As in PS II, excitation by a photon results in a charge separation giving (chl-a)$_2^{\dot{+}}$, which is then reduced by plastocyanin. It is assumed that (chl-a)$_2$ transfers its electron to a chl-a monomer (A_0), which then transfers the electron to a strongly bound

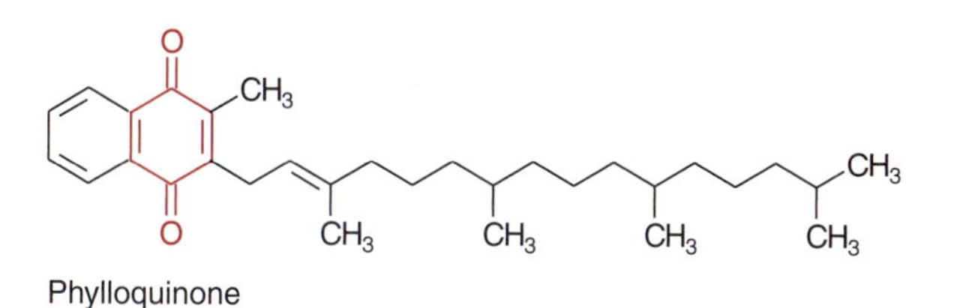

Figure 3.32 Phylloquinone.

Table 3.4: Protein components of photosystem I (list not complete)

Protein	Molecular mass (kDa)	Localization	Encoded in	Function
A	83	In membrane	Chloroplast	Binding of P_{700}, chl-*a*, A_0, A_1, Q F_x, antennae function
B	82	"	"	(as in protein A)
C	9	Peripheral:stroma	"	Binding of F_A, F_B, ferredoxin
D	17	"	Nucleus	"
E	10	"	"	"
F	18	Peripheral:lumen	"	Binding of plastocyanin
H	10	Peripheral:stroma	Nucleus	Binding of phosphorylated LHC II
I	5	In membrane	Chloroplast	?
J	5	"	"	?

(After Andersson and Franzén, 1992, Schubert et al., 1997).

phylloquinone (Q) (Fig. 3.32). Phylloquinone contains the same phytol side chain as chl-*a* and its function corresponds to Q_A in photosystem II. The electron is transferred from the semiquinone form of phylloquinone to an iron-sulfur center named F_X.

F_X is a 4Fe-4S center with a very negative redox potential. It transfers one electron to two further 4Fe-4S centers (F_A, F_B), which in turn reduce **ferredoxin**, a protein with the molecular mass of 11 kDa, containing a 2Fe-2S center. Ferredoxin also takes up and transfers only one electron. The reduction occurs at the stromal side of the thylakoid membrane. For this purpose, the ferredoxin binds at a positively charged binding site on subunit D of PS I (Fig. 3.33). The reduction of $NADP^+$ by ferredoxin, catalyzed by **ferredoxin-NADP reductase**, yields NADPH as an end product of the photosynthetic electron transport.

The PS I complex consists of at least 11 different subunits (Table 3.4). The center of the PS I complex is also a **heterodimer** (as is the center of PS II) consisting of subunits A and B (Fig. 3.33). The molecular masses of **A** and **B** each (82–83 kDa) correspond to about the sum of the molecular masses of D_1 and CP_{43} and D_2 and CP_{47}, respectively, in PS II (Table 3.2).

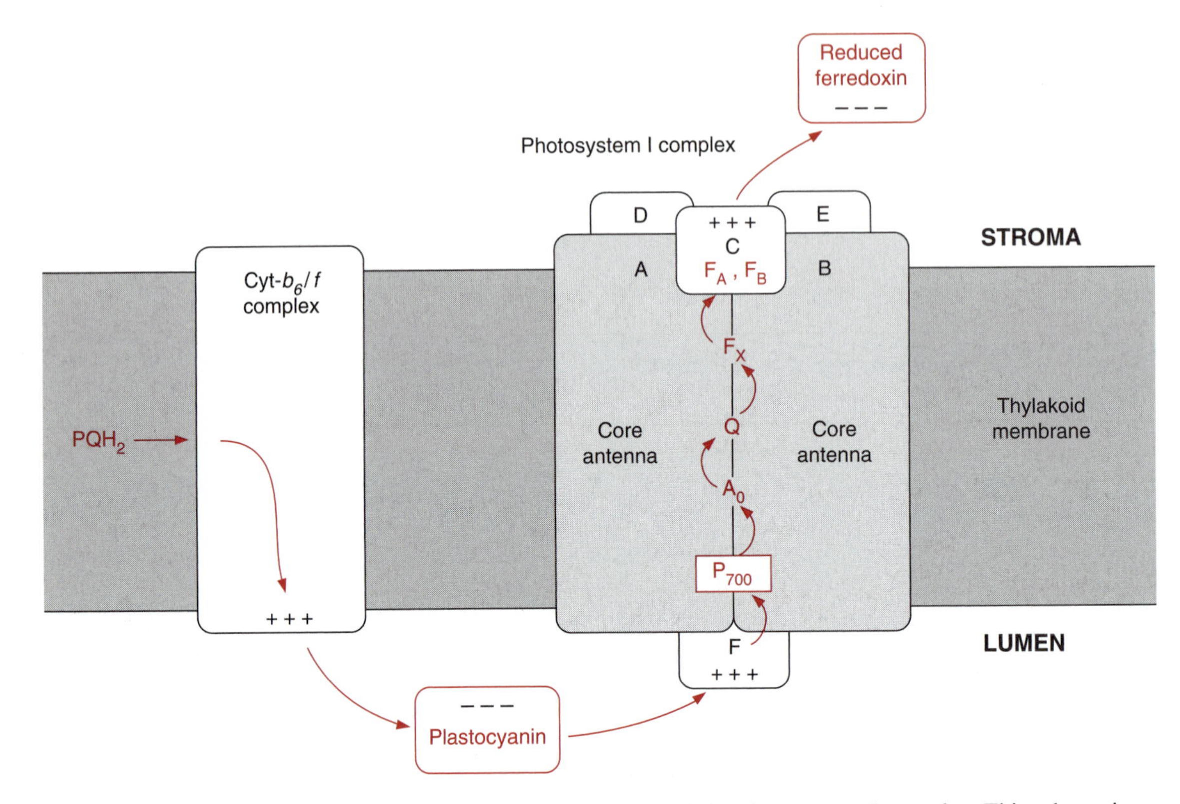

Figure 3.33 Scheme of the structure of the photosystem I complex. This scheme is based on results of X-ray structure analyses. The basic structure of the PSI complex is similar to that of the PSII complex.

In fact, both subunits A and B have a double function. Like D_1 and D_2 in PS II, they bind chromophores (chl-*a*) and redox carriers (phylloquinone, FeX) of the reaction center and, additionally, they contain about 100 chl-*a* molecules as antennae pigments. Thus, the heterodimer of A and B contains the reaction center and also the core antenna. Recently the groups of Saenger and Witt in Berlin resolved the three-dimensional structure of photosystem I of the thermophilic cyanobacterium of *Synechococcus elongatus* by X-ray structural analysis with a resolution of 2.5 Å. The results show that the basic structure of photosystem I, with a central pair of chl-*a* molecules and two branches, each with two chlorophyll molecules, are very similar to photosystem II and to the bacterial photosystem shown in Figure 3.10. But it has not been definitely clarified whether both, or just one of these branches, take part in the electron transport. The FS-centers F_A and F_B are ascribed to subunit C, and subunit F is considered to be the binding site for plastocyanin.

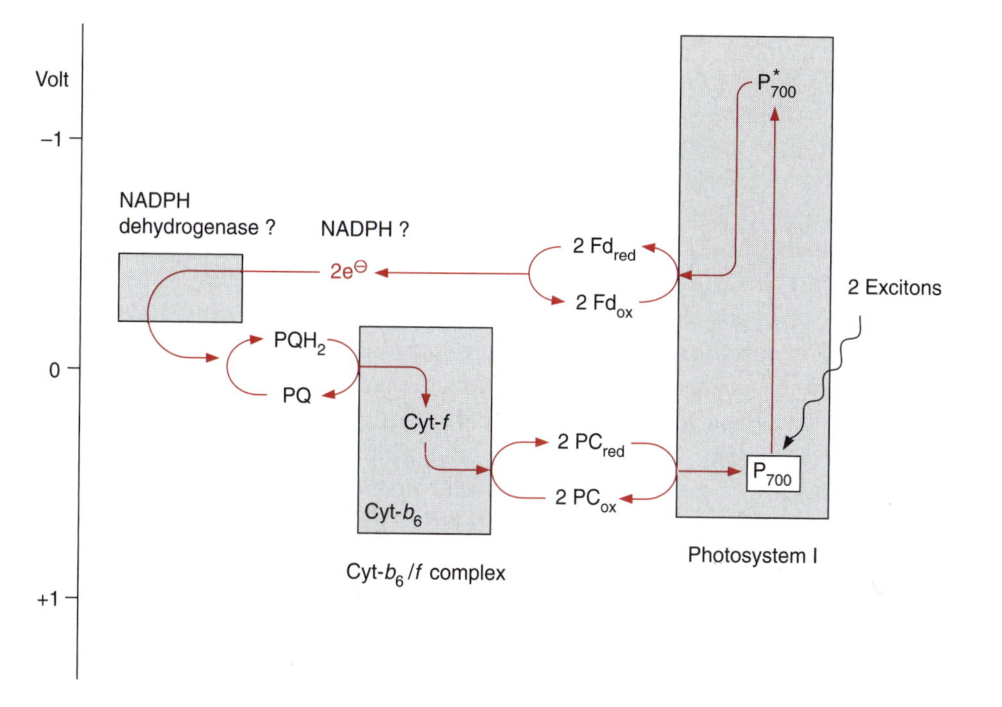

Figure 3.34 Cyclic electron transport between photosystem I and the cyt-b_6/f complex. The path of the electrons from the excited PS I to the cyt-b_6/f complex is still not clear.

In cyclic electron transport by PS I light energy is used for the synthesis of ATP only

Besides the noncyclic electron transport discussed so far, cyclic electron transfer can also take place where the electrons from the excited photosystem I are transferred back to the ground state of PS I, probably via the cyt-b_6/f complex (Fig. 3.34). The energy thus released is used only for the synthesis of ATP, and NADPH is not formed. This is termed **cyclic photophosphorylation.** In intact leaves, and even in isolated intact chloroplasts, it is quite difficult to differentiate experimentally between cyclic and noncyclic photophosphorylation. It has been a matter of debate as to whether and to what extent cyclic photophosphorylation occurs in a leaf under normal physiological conditions. Recent evaluations of the proton stoichiometry of photophosphorylation (see section 4.4) suggest that the yield of ATP in noncyclic electron transport is not sufficient for the requirements of CO_2 assimilation, and therefore cyclic photophosphorylation seems to be required to fill the gap. Moreover, cyclic photophosphorylation must operate at very high rates in the bundle sheath chloroplasts of certain C_4 plants (section 8.4). These cells have a high demand for ATP and they contain high PS I activity but very little PS II. In all likelihood, the cyclic electron flow is governed by the redox state of the acceptor of photosystem I in such a way that an increased reduction of the NADP system, and consequently that of ferredoxin, enhances the diversion of the electrons in the cycle. The function of cyclic electron transport is probably to adjust the rates of ATP and NADPH formation according to demand.

Despite intensive investigations, the pathway of electron flow from PS I to the cyt-b_6/f complex in cyclic electron transport remains unresolved. It has been proposed that cyclic electron transport is structurally separated from the linear electron transport chain in a supercomplex. Most experiments on cyclic electron transport have been carried out with isolated thylakoid membranes that catalyze only cyclic electron transport when redox mediators, such as ferredoxin or flavin adenine mononucleotide (FMN, Fig. 5.16), have been added. Cyclic electron transport is inhibited by the antibiotic **antimycin A**. It is not clear at which site this inhibitor functions. Antimycin A does not inhibit noncyclic electron transport.

Surprisingly, proteins of the NADP-dehydrogenase complex of the mitochondrial respiratory chain (section 5.5) have been identified in the thylakoid membrane of chloroplasts. The function of these proteins in chloroplasts is still not known. The proteins of this complex occur very frequently in chloroplasts from bundle sheath cells of C_4 plants and, as mentioned previously, these cells have little PS II but a particularly high cyclic photophosphorylation activity. These observations raise the possibility that in cyclic electron transport the flow of electrons from NADPH or ferredoxin to plastoquinone proceeds via a complex similar to the mitochondrial NADH dehydrogenase complex. As will be shown in section 5.5, the mitochondrial NADH dehydrogenase complex transfers electrons from NADH to ubiquinone. In cyanobacteria the participation of an NADPH dehydrogenase complex, similar to the mitochondrial NADH dehydrogenase complex, has been proved to participate in cyclic electron transport.

3.9 In the absence of other acceptors electrons can be transferred from photosystem I to oxygen

When ferredoxin is very highly reduced, it is possible that electrons are transferred from PS I to oxygen to form **superoxide radicals** ($\bullet O_2^-$) (Fig. 3.35). This process is called the **Mehler reaction**. The superoxide radical reduces metal ions present in the cell such as Fe^{3+} and Cu^{2+} (M^{n+}):

$$\bullet O_2^- + M^{n+} \rightarrow O_2 + M^{(n-1)+}$$

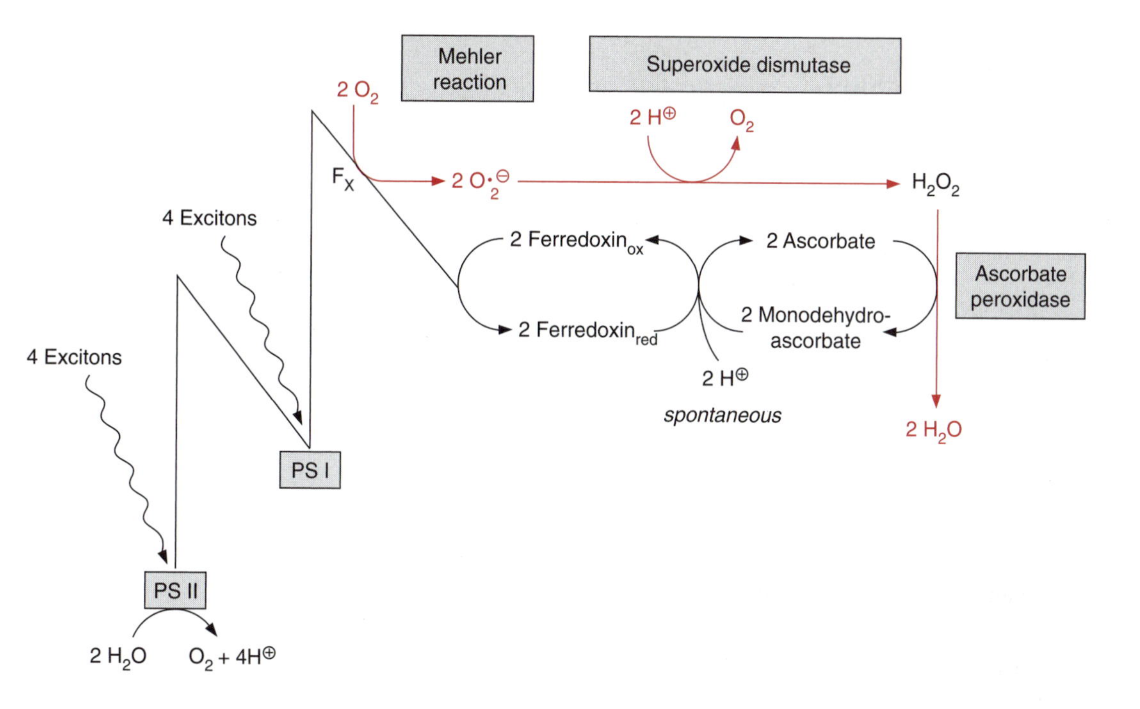

Figure 3.35 A scheme for the Mehler reaction. When ferredoxin becomes highly reduced, electrons are transferred by the Mehler reaction to oxygen, and superoxide is formed. The elimination of this highly aggressive radical involves reactions catalyzed by superoxide dismutase and ascorbate peroxidase.

Superoxide dismutase catalyzes the dismutation of $\bullet O_2^-$ into H_2O_2 and O_2, accompanied by the uptake of two protons:

$$2\bullet O_2^- + 2H^+ \longrightarrow O_2 + H_2O_2$$

$\bullet O_2^-$, H_2O_2 and $\bullet OH$ are summarized as **ROS** (*r*eactive *o*xygen *s*pecies). The metal ions reduced by superoxide react with hydrogen peroxide to form hydroxyl radicals:

$$H_2O_2 + M^{(n-1)+} \longrightarrow OH^- + \bullet OH + M^{n+}$$

The hydroxyl radical ($\bullet OH$) is a very aggressive substance and damages enzymes and lipids by oxidation. The plant cell has no protective enzymes against $\bullet OH$. Therefore it is essential that a reduction of the metal ions be prevented by rapid elimination of $\bullet O_2^-$ by superoxide dismutase. But hydrogen peroxide also has a damaging effect on many enzymes. To prevent such damage, hydrogen peroxide is eliminated by an **ascorbate peroxidase** located in the thylakoid membrane. **Ascorbate**, an important antioxidant in plant cells (Fig. 3.36), is oxidized by this enzyme to the radical **monodehydroascorbate**, which is spontaneously reconverted by photosystem I to ascorbate via reduced ferredoxin. Monodehydroascorbate also can be reduced to

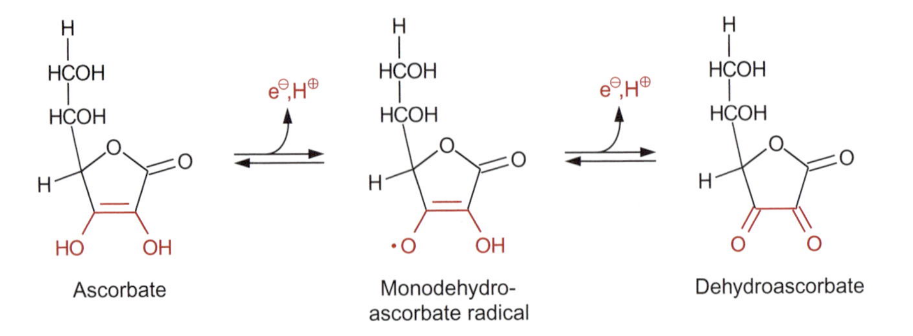

Figure 3.36 The oxidation of ascorbate proceeds via the formation of the monodehydroascorbate radical.

ascorbate by an NAD(P)H-dependent monodehydroascorbate reductase that is present in the chloroplast stroma and the cytosol.

As an alternative to the preceding reaction, two molecules monodehydroascorbate can dismutate to ascorbate and dehydroascorbate. Dehydroascorbate is reconverted to ascorbate by reduction with glutathione in a reaction catalyzed by **dehydroascorbate reductase** present in the stroma (Fig. 3.37). **Glutathione (GSH)** occurs as an antioxidant in all plant cells (section 12.2). It is a tripeptide composed of the amino acids glutamate, cysteine, and glycine (Fig. 3.38). Oxidation of GSH results in the formation of a disulfide (GSSG) between the cysteine residues of two glutathione molecules. Reduction of GSSG is catalyzed by a **glutathione reductase** with NADPH as the reductant (Fig. 3.37).

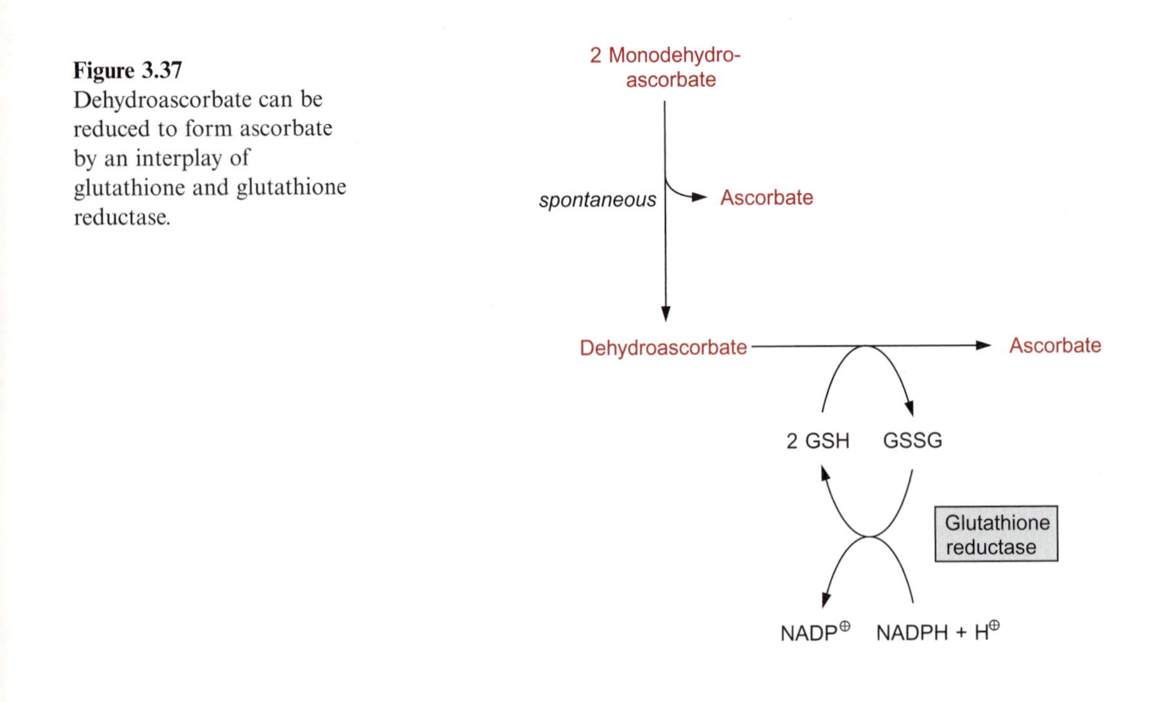

Figure 3.37 Dehydroascorbate can be reduced to form ascorbate by an interplay of glutathione and glutathione reductase.

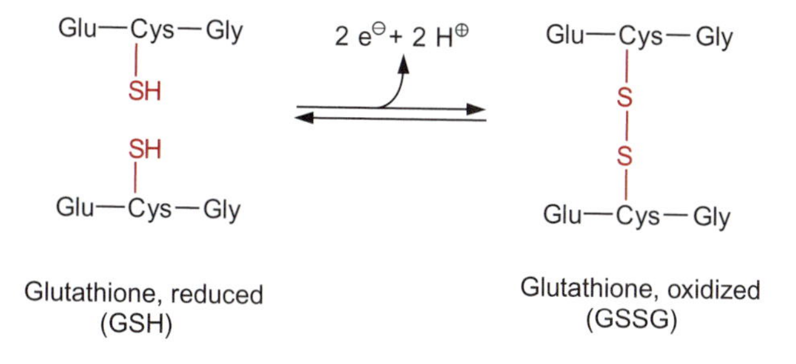

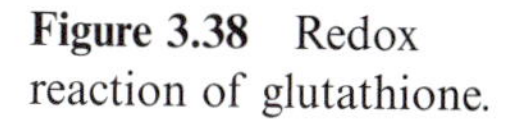
Figure 3.38 Redox reaction of glutathione.

The major function of the Mehler-ascorbate-peroxidase cycle is to dissipate excessive excitation energy of photosystem I as heat. The absorption of a total of eight excitons via PS I results in the formation of two superoxide radicals and two molecules of reduced ferredoxin, the latter serving as a reductant for eliminating H_2O_2 (Fig. 3.35). In a sense, the transfer of electrons to oxygen by the Mehler reaction could be viewed as a reversal of the splitting of water by PS II. As will be discussed in the following section, the Mehler reaction occurs when ferredoxin is very highly reduced. The only gain from this reaction is the generation of a proton gradient from electron transport through photosystem II and the cyt-b_6/f complex. This proton gradient can be used for synthesis of ATP if ADP is present. But since there is usually a shortage in ADP under the conditions of the Mehler reaction, it mostly results in the formation of a high pH gradient. A feature common to the Mehler reaction and cyclic electron transport is that there is no net production of NADPH from reduction of NADP. For this reason, electron transport via the Mehler reaction has been termed **pseudocyclic electron transport**.

Yet another group of antioxidants was recently found in plants, the so-called **peroxiredoxins**. These proteins, containing SH groups as redox carriers, have been known in the animal world for some time. In the model plant *Arabidopsis,* 10 different peroxiredoxin genes have been identified. Peroxiredoxins, being present in chloroplasts as well as in other cell compartments, differ from the aforementioned antioxidants glutathione and ascorbate in that they reduce a remarkably wide spectrum of peroxides, such as $\mathbf{H_2O_2}$, **alkylperoxides,** and **peroxinitrites**. In chloroplasts, oxidized peroxiredoxins are reduced by photosynthetic electron transport of photosystem I with ferredoxin and thioredoxin as intermediates.

Instead of ferredoxin, PS I can also reduce methylviologen. Methylviologen, also called **paraquat,** is used commercially as an herbicide (Fig. 3.39). The herbicidal effect is due to the reduction of oxygen to superoxide radicals by reduced paraquat. Additionally, paraquat competes with dehydroascorbate for the reducing equivalents provided by photosystem I.

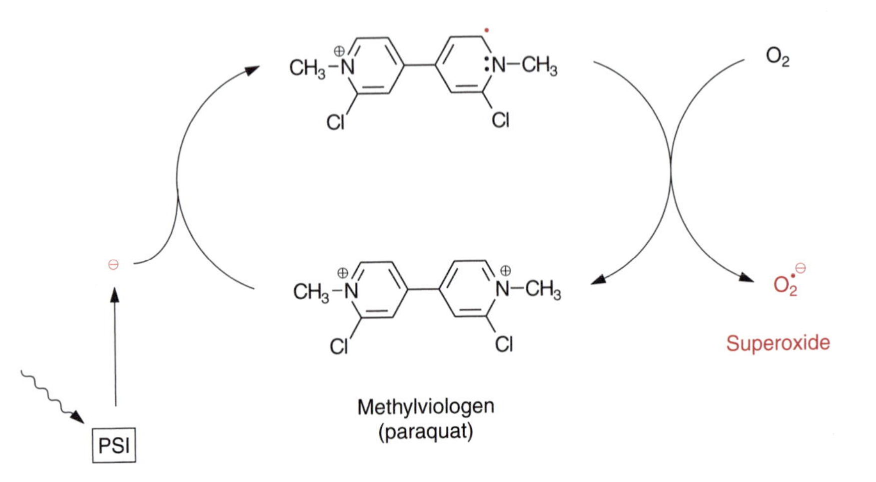

Figure 3.39 Methylviologen, an herbicide also called paraquat, is reduced by the transfer of an electron from the excited PS I to form a radical substance. The latter transfers the electron to oxygen with formation of the aggressive superoxide radical. Paraquat is distributed as an herbicide by ICI under the trade name Gramoxone.

Therefore, in the presence of paraquat, ascorbate is no longer regenerated from dehydroascorbate and the ascorbate peroxidase reaction can no longer proceed. The increased production of superoxide and decreased detoxification of hydrogen peroxide in the presence of paraquat causes severe oxidative damage to mesophyll cells, noticeable by a bleaching of the leaves. In the past, paraquat has been used to destroy marijuana fields.

3.10 Regulatory processes control the distribution of the captured photons between the two photosystems

Linear photosynthetic electron transport through the two photosystems requires the even distribution of the captured excitons between them. As discussed in section 2.4, the excitons are transferred preferentially to that chromophore requiring the least energy for excitation. Photosystem I (P_{700}) requires less energy for excitation than photosystem II (P_{680}). Therefore, in an unrestricted competition between the two photosystems for the excitons, these would be directed mainly to PS I, and therefore distribution of the excitons between the two photosystems must be regulated. The spatial separation of PS I and PS II and their antennae in the thylakoid membrane is an important element in this regulation.

In chloroplasts, the thylakoid membranes are present in two different arrays as **stacked** and **unstacked membranes**. The outer surface of the unstacked membranes has free access to the stromal space; these membranes are called **stromal lamellae** (Fig. 3.40). In the stacked membranes, the neigh-

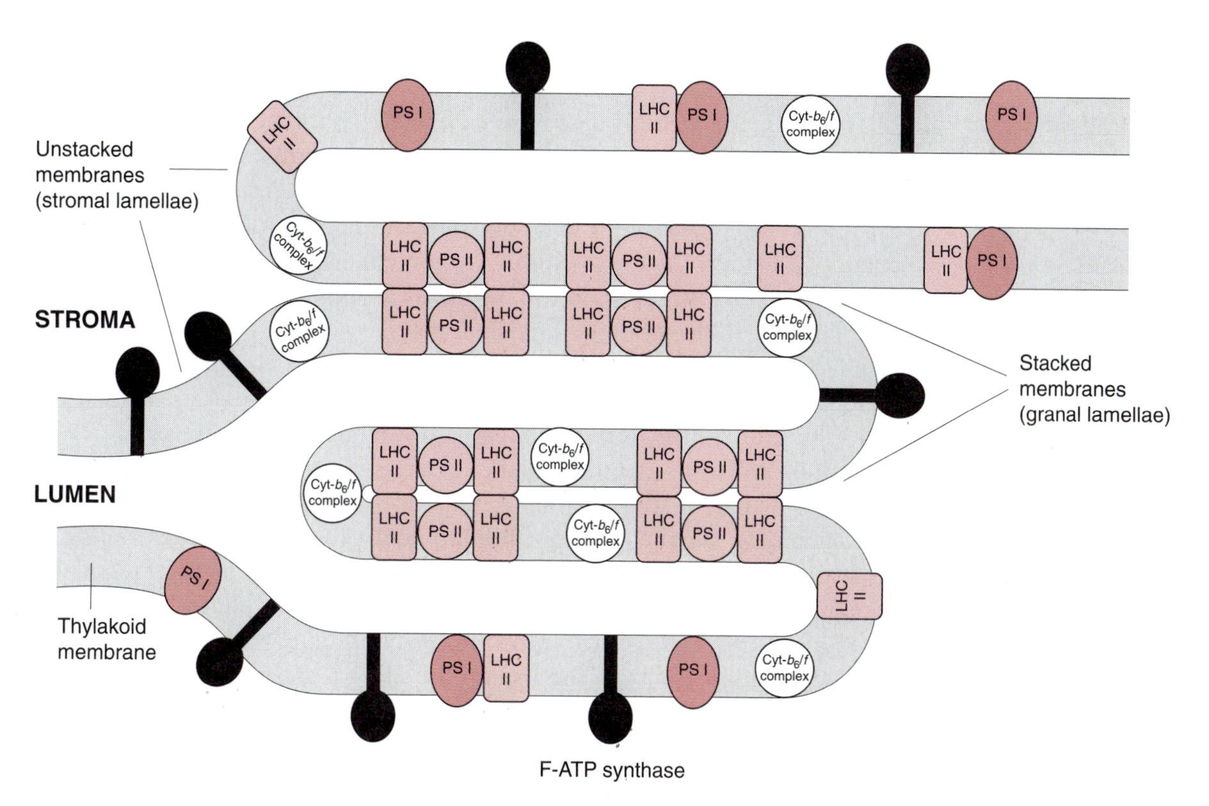

Figure 3.40 Distribution of photosynthetic protein complexes between the stacked and unstacked regions of thylakoid membranes. Stacking is probably caused by light harvesting complexes II (LHC II).

boring thylakoid membranes are in direct contact with each other. These membrane stacks can be seen as grains (grana) in light microscopy and are therefore called **granal lamellae**.

ATP-synthase and the PS I complex (including its light harvesting complexes, which will not be discussed here) are located either in the stromal lamellae or in the outer membrane region of the granal lamellae. Therefore, these proteins have free access to ADP and NADP in the stroma. The PS II complex, on the other hand, is located mainly in the granal lamellae. Peripheral LHC II subunits attached to the PS II complex (section 2.4) contain a protein chain protruding from the membrane, which can probably interact with the LHC II subunit of the adjacent membrane and thus causes tight membrane stacking. In addition to PS II and LHC II, the stacked membranes contain only cyt-b_6/f complexes. Since the proteins of PS I and F-ATP-synthase project into the stroma space, they do not fit into the space between the stacked membranes. Thus the PS II complexes in the stacked membranes are separated spatially from the PS I complexes in the unstacked membranes. It is assumed that this prevents an uncontrolled **spillover of excitons** from PS II to PS I.

However, the spatial separation of the two photosystems and thus the spillover of excitons from PS II to PS I can be regulated. If excitation of PS II is greater than that of PS I, the result is an accumulation of plastohydroquinone, which PS I cannot oxidize rapidly enough via the cyt-b_6/f complex. Under these conditions, a protein kinase is activated, which catalyzes the phosphorylation of hydroxyl groups in threonine residues of peripheral LHC II subunits, causing a change of their conformation. As a result of this, the affinity to PS II is decreased and the LHC II subunits dissociate from the PS II complexes. On the other hand, due to the changed conformation, LHC II subunits can now bind to PS I. Recently the H subunit of PS II was identified as the corresponding binding site. This brings about an increased spillover of excitons from LHC II to PS I. In this way the accumulation of reduced plastoquinone might decrease the excitation of PS II in favor of PS I. A protein phosphatase facilitates the reversal of this regulation. This regulatory process, which has been simplified here, enables the plant to distribute the captured photons optimally between the two photosystems, independent of the spectral quality of the absorbed light.

Excess light energy is eliminated as heat

Plants face the general problem that the energy of irradiated light can be much higher than the demand of photosynthetic metabolism for NADPH and ATP. This is the case with very high light intensities, when the metabolism cannot keep pace. Such a situation arises at low temperatures, when the metabolism is slowed down because of decreased enzyme activities (cold stress) or at high temperatures, when stomata close to prevent loss of water. Excess excitation of the photosystems could result in an excessive reduction of the components of the photosynthetic electron transport.

Very high excitation of photosystem II, recognized by the accumulation of plastohydroquinone, results in damage to the photosynthetic apparatus, termed **photoinhibition**. A major cause for this damage is an overexcitation of the reaction center, by which chlorophyll molecules attain a triplet state, resulting in the formation of aggressive singlet oxygen (section 2.3). The damaging effect of triplet chlorophyll can be demonstrated by placing a small amount of chlorophyll under the human skin, after which illumination causes severe tissue damage. This photodynamic principle is utilized in medicine for selective therapy of skin cancer.

Carotenoids (e.g., β-carotene, Fig. 2.9) are able to convert the triplet state of chlorophyll and the singlet state of oxygen back to the corresponding ground states by forming a triplet carotenoid, which dissipates its energy as heat. In this way carotenoids have an important protective function. This protective function of carotenoids, however, may be unable to cope with excessive excitation of PS II, and the singlet oxygen then has a damaging effect on the PS II complex. The site of this damage could be the D_1 protein of the photosynthetic reaction center in PS II, which even under normal photosynthetic conditions has a high turnover (see section 3.6). When the rate of D1-protein damage exceeds the rate of its resynthesis, the rate of photosynthesis is decreased; in other words, there is **photoinhibition**.

Plants have developed several mechanisms to protect the photosynthetic apparatus from light damage. One mechanism is **chloroplast avoidance movement**, in which chloroplasts move from the cell surface to the side walls of the cells under high light conditions. Another way is to dissipate the energy arising from an excess of excitons as heat. This process is termed **nonphotochemical quenching** of exciton energy. Although our knowledge of this quenching process is still incomplete, it is undisputed that **zeaxanthin** plays an important role in this. Zeaxanthin causes the dissipation of exciton energy to heat by interacting with a chlorophyll-binding protein (CP 22) of photosystem II. Zeaxanthin is formed by the reduction of the diepoxide **violaxanthin**. The reduction proceeds with ascorbate as the reductant and the monoepoxide antheraxanthin is formed as an intermediate. Zeaxanthin can be reconverted to violaxanthin by epoxidation requiring NADPH and O_2 (Fig. 3.41). Formation of zeaxanthin by diepoxidase takes place on the luminal side of the thylakoid membrane at an optimum pH of 5.0, whereas the regeneration of violaxanthin by the epoxidase proceeding at the stromal side of the thylakoid membrane occurs at about pH 7.6. Therefore, the formation of zeaxanthin requires a high pH gradient across the thylakoid membrane. As discussed in connection with the Mehler reaction (section 3.9), a high pH gradient can be an indicator of the high excitation of photosystem II. When there is too much exciton energy, an increased pH gradient initiates zeaxanthin synthesis, dissipating excess exciton energy in the PS II complex as heat. This is how during strong sunlight most plants convert 50% to 70% of all the absorbed photons to heat. The nonphotochemical quenching of exciton energy is the main way for plants to protect themselves from too much light energy. In comparison, the Mehler reaction (section 3.9) and photorespiration (section 7.7) under most conditions play only a minor role in the elimination of excess excitation energy.

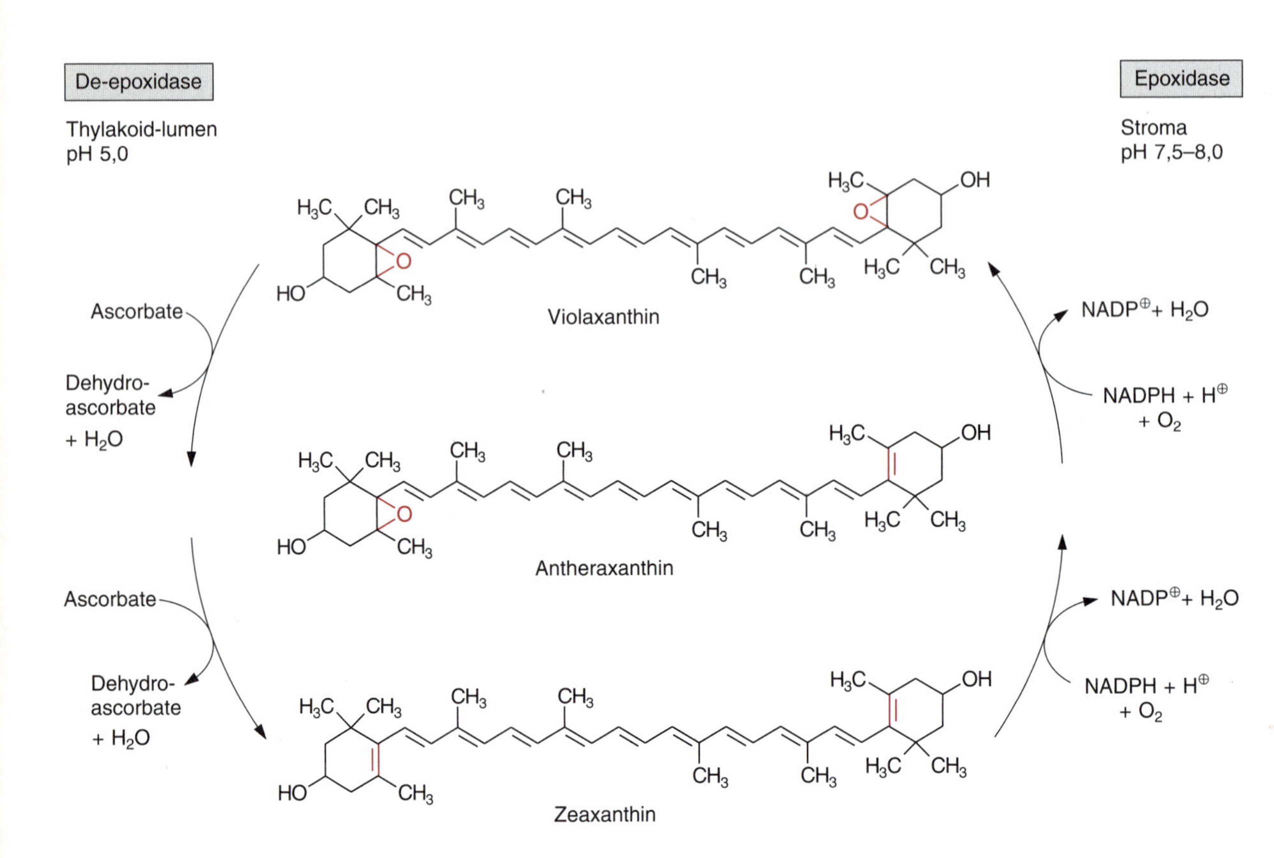

Figure 3.41 The zeaxanthin cycle. (After Demmig-Adams.)

Further reading

Allen, J. F. Cyclic, pseudocyclic and non cyclic photophosphorylation: new links in the chain. Trends Plant Sci 8, 15–19 (2003).

Asada, K. The water-water cycle in chloroplasts: Scavenging of active oxygens and dissipation of excess photons. Annu Rev Plant Physiol Plant Mol Biol 50, 601–639 (1999).

Chitnis, P. R. Photosystem I: Function and physiology. Annu Rev Plant Physiol Plant Mol Biol 52, 593–626 (2001).

Deiner, B. A., Rappaport, F. Structure, dynamics, and energetics of the primary photochemistry of photosystem II of oxigenic photosynthesis. Annu Rev Plant Biol 53, 551–580 (2002).

Deisenhofer, J., Michel, H. Nobel Lecture: The photosynthetic reaction center from the purple bacterium *Rhodopseudomonas viridis*. EMBO J 8, 2149–2169 (1989).

Demmig-Adams, B., Adams III, W. W. The role of the xanthophyll cycle carotenoids in the protection of photosynthesis. Trends Plant Sci 1, 21–26 (1996).

Depège, N., Bellafiore, S., Rochaix, J.-D. Role of chloroplast protein kinase Stt7 in LHCII phosphorylation and state transition in *Chlamydomonas*. Science 299, 1572–1575 (2003).

Dietz, K.-J. Plant peroxiredoxins. Annu Rev Plant Biol 54, 93–107 (2003).

Ermler, U., Fritzsch, G., Buchanan, S.K., Michel, H. Structure of the photosynthetic reaction center from *Rhodobacter sphaeroides* at 2.65 Å resolution: Cofactors and protein-cofactor interactions. Structure 2, 925–936 (1994).

Govindjee. A role for a light-harvesting antenna complex of photosystem II in photoprotection. Plant Cell 14, 1663–1668 (2002).

Govindjee, Gest, H. (eds.). Historical highlights of photosynthesis research I. Photosynth Res 73, 1–308 (2002).

Govindjee, Gest, H. (eds.). Historical highlights of photosynthesis research II. Photosynth Res 79, 1–450 (2003).

Heller, B. A., Holten, D., Kirmaier, C. Control of electron transfer between the L- and M-sides of photosynthetic reaction centers. Science 269, 940–945 (1995).

Horton, P., Ruban, A. V., Walters, R. G. Regulation of light harvesting in green plants. Annu Rev Plant Physiol Plant Mol Biol 47, 655–684 (1996).

Kasahara, M., Kagawa, T., Oikawa, K., Suetsugu, N., Miyao, M., Wada, M. Chloroplast avoidance movement reduces photodamage in plants. Nature 420, 829–832 (2002).

König, J., Baier, M., Horling, F., Kahmannm, U., Harris, G., Schürmann, P. The plant-specific function of 2-Cys peroxiredoxin-mediated detoxification of peroxides in the redox-hierarchy of photosynthetic electron flux. Proc Natl Acad Sci USA 99, 5738–5743 (2002).

Joliot, P., Joliot, A. Cyclic electron transfer in plant leaf. Proc Natl Acad Sci USA 99, 10209–10214 (2002).

Jordan, P., Fromme, P., Witt, H. T., Klukas, O., Saenger, W., Krauß, N. Three-dimensional structure of cyanobacterial photosystem I at 2.5 Å resolution. Nature 411, 909–917 (2001)

Li, X.-P., Björkman, O., Shih, C., Grossman, A. R., Rosenquist, M., Jansson, S., Niyogi, K. K. A pigment-binding protein essential for regulation of photosynthetic light harvesting. Nature 403, 391–395 (2000).

Lunde, C., Jensen, P. E., Haldrup, A., Knoetzel, J., Scheller, H. V. The PSI-H subunit of photosystem I is essential for state transitions in plant photosynthesis. Nature,408, 613–615 (2000).

Melis, A. Photosystem-II damage and repair cycle in chloroplasts: What modulates the rate of photodamage *in vivo*? Trends Plant Sci 4, 1130–135 (1999).

Mittler, R. Oxidative stress, antioxidants and stress tolerance. Trends Plant Sci 7, 405–410 (2002).

Niyogi, K. K. Safety valves for photosynthesis. Curr Opin Plant Biol 3, 455–460 (2000).

Ort, D. R., Yocum, C. F. (eds.). Oxygenic photosynthesis: The light reactions. (Advances in Photosynthesis). Kluwer Academic Publishers, Dordrecht, The Netherlands (1996).

Orr, L., Govindjee. Photosynthesis and the web: 2001. Photosynth Res 68, 1–28 (2001).

Osmond, B., Badger, M., Björkman, O., Leegood, R. Too many photons: Photorespiration, photoinhibition and photooxidation. Trends Plant Sci 2, 119–121 (1997).

Powles, S. B., Holtum, J. A. M. Herbicide resistance in plants. Lewis Publishers, Boca Raton, Ann Arbor, (1994).

Roy, C., Lancaster, D., Michel, H. The coupling of light-induced electron transfer and proton uptake as derived from crystal structures of reaction centers from *Rhodopseudomonas viridis* modified at the binding site of the secondary quinone, Q_B. Structure 5, 1339–1359 (1997).

Schubert, W.-D., Klukas, O., Saenger, W., Witt, H. T., Fromme, P., Krauß, N. A common ancestor for oxygenic and anoxygenic photosynthetic systems: A comparison based on the structural model of photosystem I. J Mol Biol 280, 297–314 (1998).

Vener, A. V., Ohad, I., Andersson, B. Protein phosphorylation and redox sensing in chloroplast thylakoids. Plant Biol 1, 217–223 (1998).

Wada, M., Kagawa, T., Sato, Y. Chloroplast movement. Annu Rev Plant Biol 54, 455–468 (2003).

Werck-Reichhart, D., Hehn, A., Didierjean, L. Cytochromes P450 for engineering herbicide tolerance. Trends Plant Sci 5, 116–123 (2000).

Zouni, A., Witt, H. T., Kern, J., Fromme, P., Krauß, N., Saenger, W., Orth, P. Crystal structure of photosystem II from *Synechococcus elongatus* at 3.8 Å revolution. Nature 409, 739–743 (2001).

4

ATP is generated by photosynthesis

Chapter 3 discussed the transport of protons across a thylakoid membrane by photosynthetic electron transport and how, in this way, a proton gradient is generated. This chapter deals with how this proton gradient is utilized for the synthesis of ATP.

In 1954 Daniel Arnon (Berkeley) discovered that when suspended thylakoid membranes are illuminated, ATP is formed from ADP and inorganic phosphate. This process is called **photophosphorylation**. Further experiments showed that photophosphorylation is coupled to the generation of NADPH. This result was unexpected, as it was then generally believed that the synthesis of ATP in chloroplasts was driven, as in mitochondria, by an electron transport from NADPH to oxygen. It soon became apparent, however, that the mechanism of photophosphorylation coupled to photosynthetic electron transport was very similar to that of ATP synthesis coupled to electron transport of mitochondria, termed **oxidative phosphorylation** (section 5.6).

In 1961 Peter Mitchell (Edinburgh) postulated in his **chemiosmotic hypothesis** that during electron transport-coupled ATP synthesis, a proton gradient is formed, and that it is the **proton motive force** of this gradient that drives the synthesis of ATP. At first this revolutionary hypothesis was strongly opposed by many workers in the field, but in the course of time, experimental results of many researchers supported the chemiosmotic hypothesis, which is now fully accepted. In 1978 Peter Mitchell was awarded the Nobel Prize in Chemistry for this hypothesis.

4.1 A proton gradient serves as an energy-rich intermediate state during ATP synthesis

Let us first ask: How much energy is actually required in order to synthesize ATP?

The free energy for the synthesis of ATP from ADP and phosphate is calculated from the van't Hoff equation:

$$\Delta G = \Delta G^{0'} + RT \ln \frac{[\text{ATP}]}{[\text{ADP}] \cdot [\text{P}]} \tag{4.1}$$

The standard free energy for the synthesis of ATP is:

$$\Delta G^{0'} = +30.5 \text{kJ/mol.}$$

The concentrations of ATP, ADP, and phosphate in the chloroplast stroma are very much dependent on metabolism. Typical concentrations are:

$$\text{ATP} = 2.5 \cdot 10^{-3} \text{ mol/L; ATP} = 0.5 \cdot 10^{-3} \text{ mol/L; P} = 5 \cdot 10^{-3} \text{ mol/L.}$$

When these values are introduced into equation 4.1 (R = 8.32 J/mol·K, T = 298 K), the energy required for synthesis of ATP is evaluated as:

$$\Delta G = +47.8 \text{ kJ/mol.}$$

This value is, of course, variable because it depends on the metabolic conditions. For further considerations an average value of 50 kJ/mol will be employed for ΔG_{ATP}.

The transport of protons across a membrane can have different effects. If the membrane is permeable to counter ions of the proton [e.g., a chloride ion (Fig. 4.1A)], the charge of the proton will be compensated for, since each transported proton will pull a chloride ion across a membrane. This is how a proton concentration gradient can be generated. The free energy for the transport of protons from A to B is:

$$\Delta G = RT \ln \frac{[\text{H}^+]_\text{B}}{[\text{H}^+]_\text{A}} \quad [\text{J mol}^{-1}] \tag{4.2}$$

If the membrane is impermeable for counterions (Fig. 4.1B), a charge compensation for the transported proton is not possible. In this case, the

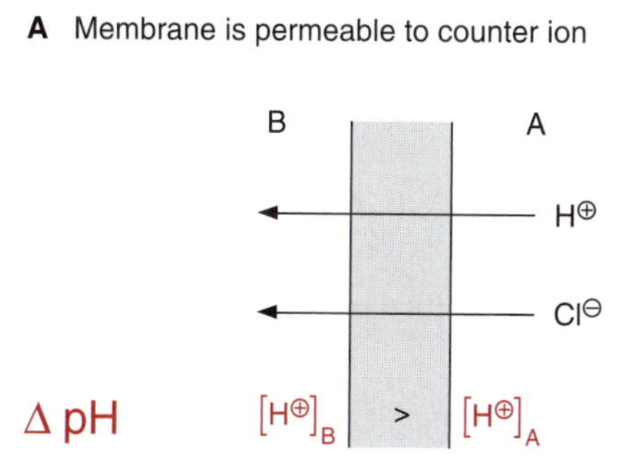

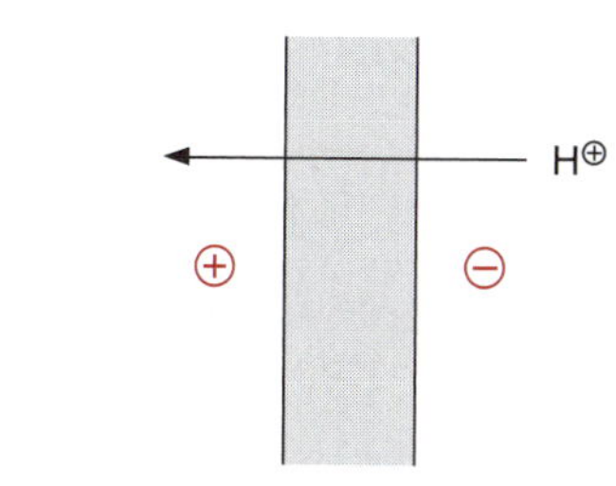

Figure 4.1 A. Transport of protons through a membrane, permeable to a counter ion such as chloride, results in the formation of a proton concentration gradient. B. When the membrane is impermeable to a counter ion, proton transport results in the formation of a membrane potential.

transfer of only a few protons across the membrane results in the formation of a membrane potential $\Delta\Psi$, measured as the voltage difference across the membrane. By convention, $\Delta\Psi$ is positive when a cation is transferred in the direction of the more positive region. Voltage and free energy are connected by the following equation:

$$\Delta G = m\,F \cdot \Delta\Psi \qquad (4.3)$$

where m is the charge of the ion (in the case of a proton 1), and F is the Faraday constant, 96480 V^{-1} J mol^{-1}.

Proton transport across a biological membrane leads to the formation of a proton concentration gradient and a membrane potential. The free energy for the transport of protons from A to B therefore consists of the sum of the free energies for the generation of the H^+ concentration gradient and the membrane potential:

$$\Delta G = RT \ln \frac{[H^+]_B}{[H^+]_A} + F\Delta\Psi \qquad (4.4)$$

In chloroplasts, the energy stored in a proton gradient corresponds to the change of free energy during the flux of protons from the lumen into the stroma.

$$\Delta G = RT \ln \frac{[H^+]_S}{[H^+]_L} + F\Delta\Psi \quad (4.5)$$

where S = stroma, L = lumen, and $\Delta\Psi$ = voltage difference stroma-lumen.

The conversion of the natural logarithm into the decadic logarithm yields:

$$\Delta G = 2.3 \cdot RT \log \frac{[H^+]_S}{[H^+]_L} + F\Delta\Psi \quad (4.6)$$

The logarithmic factor is the negative pH difference between lumen and stroma:

$$\log[H^+]_S - \log[H^+]_L = -\Delta pH$$

A rearrangement yields:

$$\Delta G = -2.3\ RT \cdot \Delta pH + F\Delta\Psi \quad (4.7)$$

At 25°C: $2.3 \cdot RT = 5700\,J\ mol^{-1}$.

Thus: $\Delta G = -5700\Delta pH + F\Delta\Psi \quad [J\ mol^{-1}]$ (4.8)

The expression $\frac{\Delta G}{F}$ is called **proton motive force** (PMF), with unit = volts:

$$\frac{\Delta G}{F} = PMF = -\frac{2.3RT}{F}\Delta pH + \Delta\Psi \quad [V] \quad (4.9)$$

At 25°C: $\frac{2.3RT}{F} = 0.059\,V$

Thus: $PMF = -0.059 \cdot \Delta pH + \Delta\Psi \quad [V]$ (4.10)

Equation 4.10 is of general significance for electron transport-coupled ATP synthesis. In mitochondrial oxidative phosphorylation, the PMF is primarily the result of a membrane potential. In chloroplasts, on the other hand, the membrane potential does not contribute much to the PMF. Here the PMF is almost entirely due to the concentration gradient of protons across the thylakoid membrane. In illuminated chloroplasts, one finds a ΔpH across the thylakoid membrane of about 2.5. Introducing this value into equation 4.8 yields:

$$\Delta G = -14.3\text{kJ/mol}$$

A comparison of this value with ΔG for the formation of ATP (50 kJ/mol) shows that at least four protons are required for the ATP synthesis from ADP and phosphate.

4.2 The electron chemical proton gradient can be dissipated by uncouplers to heat

Photosynthetic electron transport from water to NADP is coupled with photophosphorylation. Electron transport occurs only if ADP and phosphate are present as precursor substances for ATP synthesis. When an **uncoupler** is added, electron transport proceeds at a high rate in the absence of ADP; electron transport is then uncoupled from ATP synthesis. Therefore, in the presence of an uncoupler, ATP synthesis is abolished.

The chemiosmotic hypothesis explains the effect of uncouplers (Fig. 4.2). Uncouplers are amphiphilic substances, soluble in both water and lipids. They are able to permeate the lipid phase of a membrane by **diffusion** and in this way to transfer a proton or an alkali ion across the membrane, thus eliminating a proton concentration gradient or a membrane potential, respectively. In the presence of an uncoupler, due to the absence of the proton gradient, protons are transported by ATP synthase from the stroma to the thylakoid lumen at the expense of ATP, which is hydrolyzed to ADP and phosphate. This is the reason why uncouplers cause an ATP hydrolysis (**ATPase**).

Figure 4.2A shows the effect of the uncoupler carbonylcyanide-*p*-trifluormethoxyphenylhydrazone (**FCCP**), a weak acid. FCCP diffuses in the undissociated (protonated) form from the compartment with a high proton concentration (on the left in Fig. 4.2A), through the membrane into the compartment with a low protein concentration, and dissociates there into a proton and the FCCP anion. The proton remains and the FCCP anion returns by diffusion to the other compartment, where it is protonated again. In this way the presence of FCCP at a concentration of only $7 \cdot 10^{-8}$ mol/L results in complete dissipation of the proton gradient. The substance **SF 6847** (3.5-Di (*tert*-butyl)-4-hydroxybenzyldimalononitril) (Fig. 4.3) has an even higher uncoupler effect. Uncouplers such as FCCP or SF 6847, which transfer protons across a membrane, are called **protonophores**.

In addition to the protonophores, there is a second class of uncouplers, termed **ionophores,** which are able to transfer alkali cations across a

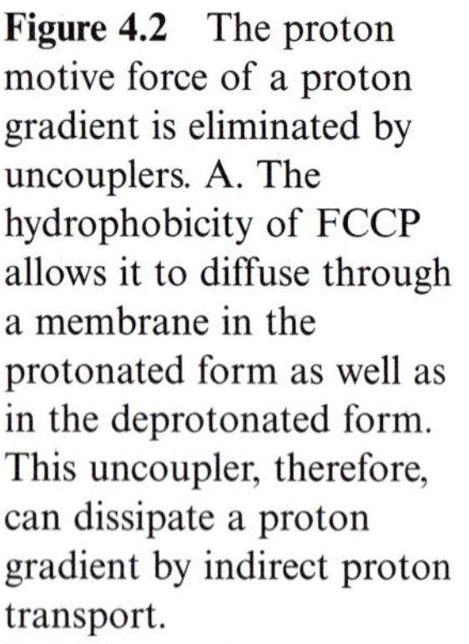

Figure 4.2 The proton motive force of a proton gradient is eliminated by uncouplers. A. The hydrophobicity of FCCP allows it to diffuse through a membrane in the protonated form as well as in the deprotonated form. This uncoupler, therefore, can dissipate a proton gradient by indirect proton transport.
B. Valinomycin, an antibiotic with a cyclic structure, folds to a hydrophobic spherical molecule, which is able to bind K^+ ions in the interior. Loaded with K^+ ions, valinomycin can diffuse through a membrane. In this way valinomycin can eliminate a membrane potential by transferring K^+ ions across a membrane.

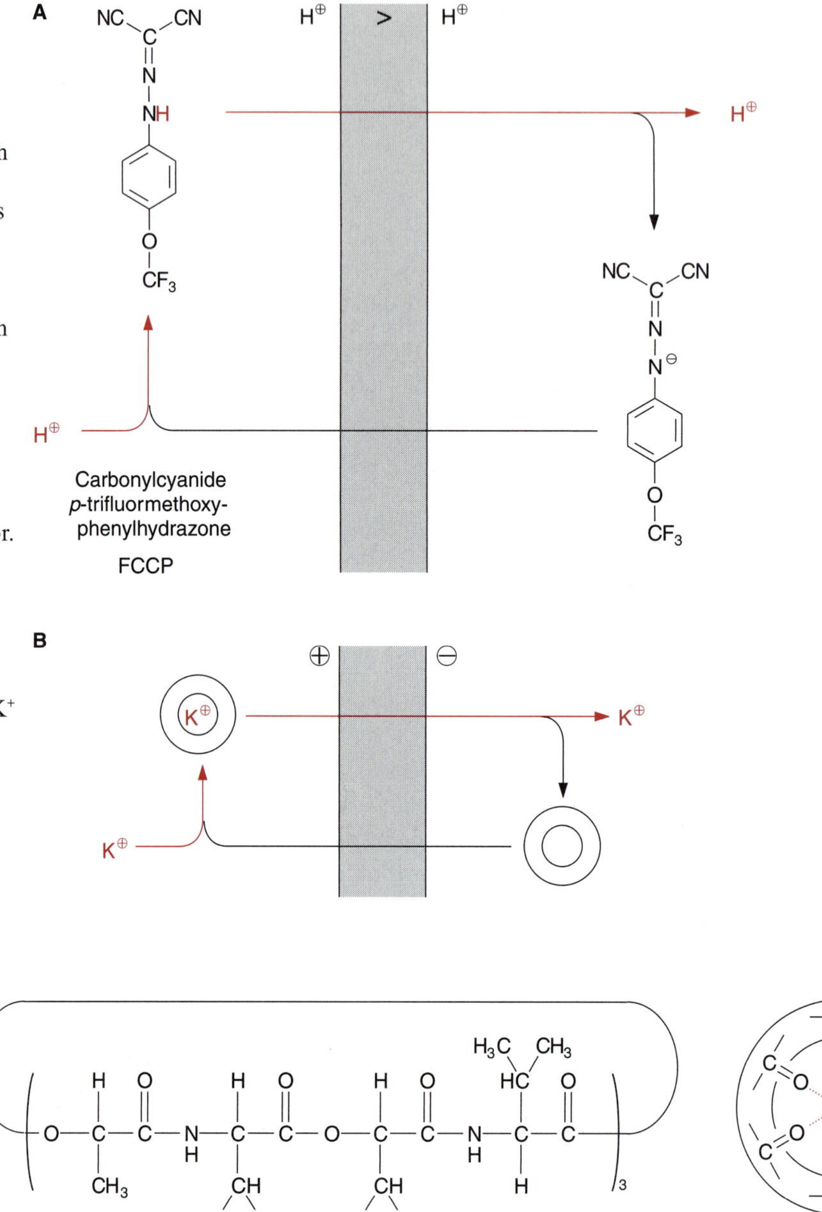

membrane and thus dissipate a membrane potential. **Valinomycin**, an antibiotic from *Streptomyces*, is such an ionophore (Fig. 4.2B). Valinomycin is a cyclic molecule containing the sequence (L-lactate)-(L-valine)-(D-hydroxyisovalerate)-(D-valine) three times. Due to its hydrophobic outer surface, valinomycin is able to diffuse through a membrane. Oxygen atoms directed toward the inside of the valinomycin molecule form the binding site for dehydrated Rb^+-and K^+ ions. Na^+ ions are only very loosely bound. When K^+ ions are present, the addition of valinomycin results in the elimination of the membrane potential. The ionophore **gramicidine**, not discussed here in detail, is also a polypeptide. Gramicidine incorporates into membranes and forms a transmembrane ion channel by which both alkali cations and protons can diffuse through the membrane.

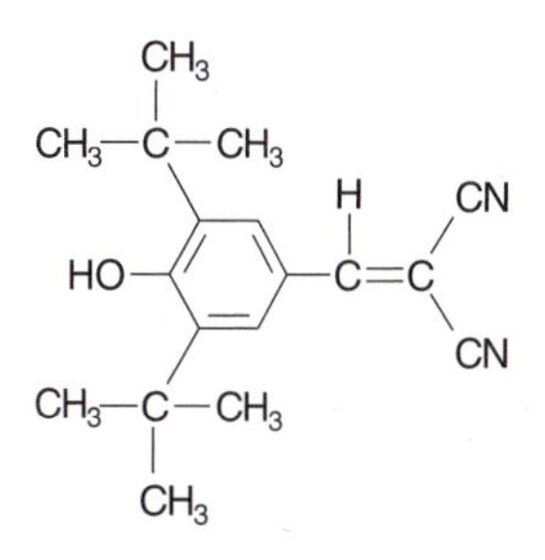

Figure 4.3 Di(*tert*-butyl)-4-hydroxybenzyl malononitrile (SF6847) is an especially effective uncoupler. Only 10^{-9} mol/L of this substance results in the complete dissipation of a proton gradient across a membrane. This uncoupling is based in the permeation of the protonated and deprotonated molecule through the membrane, as shown in Figure 4.2A for FCCP.

The chemiosmotic hypothesis was proved experimentally

In 1966 the American scientist André Jagendorf presented conclusive evidence for the validity of the chemiosmotic hypothesis in chloroplast photophosphorylation (Fig. 4.4). He incubated thylakoid membranes in an acidic medium of pH 4 in order to acidify the thylakoid lumen by unspecific uptake of protons. He added inorganic phosphate and ADP to the thylakoid suspension and then increased the pH of the medium to pH 8 by adding an alkaline buffer. This led to the sudden generation of a proton gradient of $\Delta pH = 4$, and for a short time ATP was found to be synthesized. Since this experiment was carried out in the dark, it presented evidence that synthesis of ATP in chloroplasts can be driven without illumination just by a pH gradient across a thylakoid membrane.

4.3 H^+-ATP synthases from bacteria, chloroplasts, and mitochondria have a common basic structure

How is the energy of the proton gradient utilized to synthesize ATP? A proton coupled ATP synthase (H^+-ATP synthase) is not unique to the chloroplast. It evolved during an early stage of evolution and occurs in its basic structure in bacteria, chloroplasts, and mitochondria. In bacteria this enzyme catalyzes not only ATP synthesis driven by a proton gradient, but also (in a reversal of this reaction) the transport of protons against the concentration gradient at the expense of ATP. This was probably the original function of the enzyme. In some bacteria an ATPase homologous to the H^+-ATP synthase functions as an ATP-dependent Na^+ transporter.

Our present knowledge about the structure and function of the H^+-ATP synthase derives from investigations of mitochondria, chloroplasts, and bacteria. By 1960 progress in electron microscopy led to the detection of small

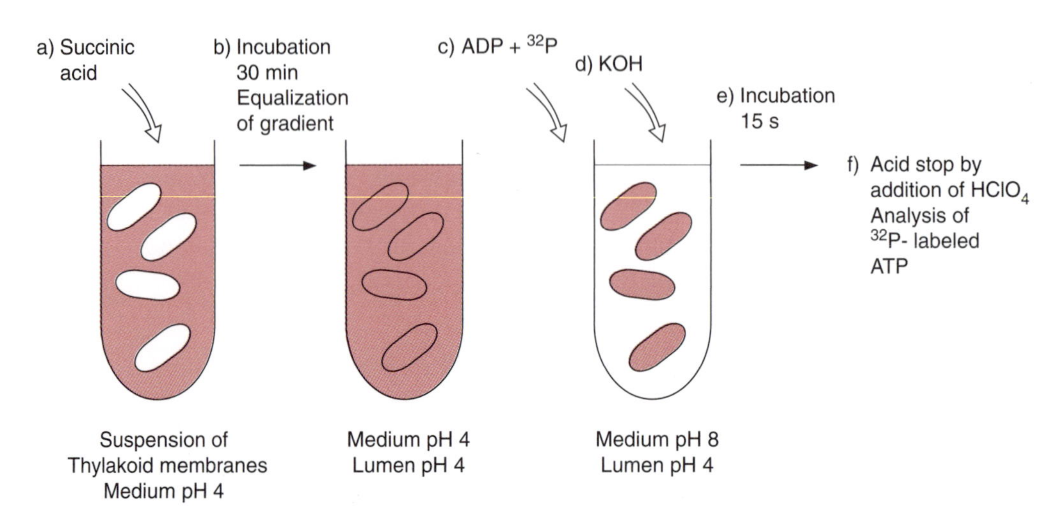

Figure 4.4 Thylakoid membranes can synthesize ATP in the dark by an artificially formed proton gradient. In a suspension of thylakoid membranes, the pH in the medium is lowered to 4.0 by the addition of succinic acid (a). After incubation for about 30 minutes, the pH in the thylakoid lumen is equilibrated with the pH of the medium due to a slow permeation of protons across the membrane (b). The next step is to add ADP and phosphate, the latter being radioactively labeled by the isotope ^{32}P (c). Then the pH in the medium is raised to 8.0 by adding KOH (d). In this way a ΔpH of 4.0 is generated between the thylakoid lumen and the medium, and this gradient drives the synthesis of ATP from ADP and phosphate. After a short time of reaction (e), the mixture is denatured by the addition of perchloric acid, and the amount of radioactively labeled ATP formed in the deproteinized extract is determined. (After Jagendorf, 1966.)

particles, which are attached by stalks to the inner membranes of mitochondria and the thylakoid membranes of chloroplasts. These particles occur only at the matrix or stromal side of the corresponding membranes. By adding urea, Ephraim Racker and coworkers (Cornell University), succeeded in removing these particles from mitochondrial membranes. The particles thus separated catalyzed the hydrolysis of ATP to ADP and phosphate. Racker called them **F_1-ATPase**. Mordechai Avron from Rehovot, Israel, showed that the corresponding particles from chloroplast membranes also have ATPase activity.

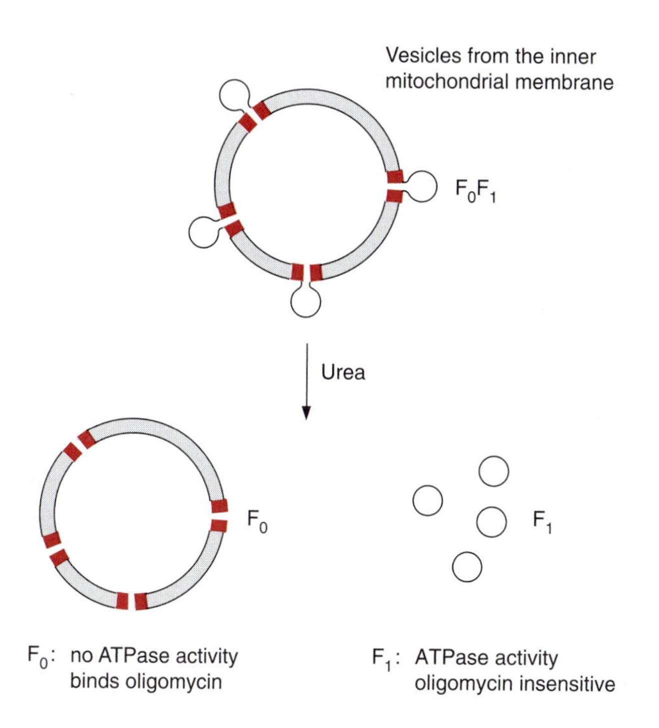

Figure 4.5 Vesicles prepared by ultrasonic treatment of mitochondria contain functionally intact H^+-ATP synthase. The soluble factor F_1 with ATPase function is removed by treatment with urea. The oligomycin binding factor F_o remains in the membrane.

Vesicles containing F_1 particles could be prepared from the inner membrane of mitochondria. These membrane vesicles were able to carry out respiration coupled to ATP synthesis. As in intact mitochondria (section 5.6), the addition of uncouplers resulted in a high ATPase activity. The uncoupler-induced ATPase, as well as ATP synthesis carried out by these vesicles, was found to be inhibited by the antibiotic **oligomycin**. Mitochondrial vesicles where the F_1 particles had been removed showed no ATPase activity but were highly permeable for protons. This proton permeability was eliminated by adding oligomycin. The ATPase activity of the removed F_1 particles, on the other hand, was not affected by oligomycin. These and other experiments showed that the H^+-ATP synthase of the mitochondria consists of two parts:

1. A soluble factor 1 (F_1) that catalyzes the synthesis of ATP; and
2. A membrane-bound factor enabling the flux of protons through the membrane to which oligomycin is bound.

Racker designated this factor $\mathbf{F_o}$ (O, oligomycin) (Fig. 4.5). Basically the same result has also been found for H^+-ATP synthases of chloroplasts and bacteria, with the exception that the H^+-ATP synthase of chloroplasts is not inhibited by oligomycin. Despite this, the membrane part of the chloro-

Table 4.1: Compounds of the F-ATP-synthase from chloroplasts. Nomenclature as in *E. coli* F-ATP synthase

Subunits	Number in F_0F_1-molecule	Molecular mass (kDa)	Encoded in
F_1:α	3	55	Plastid genome
β	3	54	Plastid genome"
γ	1	36	Plastid genome
δ	1	21	Nuclear genome
ε	1	15	Plastid genome
F_o:a	1	27	Plastid genome
b	1	16	Nuclear genome
b′	1	21	Plastid genome
c	12	8	Plastid genome

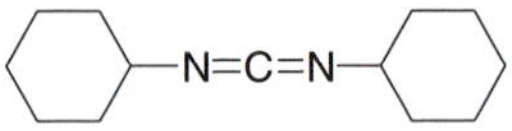

Dicyclohexylcarbodiimide
DCCD

Figure 4.6 Dicyclohexylcarbodiimide (DCCD), an inhibitor of the F_o part of F-ATP synthase.

plastic ATP synthase is also designated as F_o. The H^+-ATP synthases of chloroplasts, mitochondria, and bacteria, as well as the corresponding H^+- and Na^+-ATPases of bacteria, are collectively termed **F-ATP synthases** or **F-ATPases**. The terms F_oF_1-ATP synthase and F_oF_1-ATPase are also used.

F_1, after removal from the membrane, is a soluble oligomeric protein with the composition $\alpha_3\beta_3\gamma\delta\varepsilon$ (Table 4.1). This composition has been found in chloroplasts, bacteria, and mitochondria.

F_o is a strongly hydrophobic protein complex that can be removed from the membrane only by detergents. Dicyclohexylcarbodimide (**DCCD**) (Fig. 4.6) binds to the F_o embedded in the membrane, and thus closes the proton channel. In chloroplasts four different subunits have been detected as the main constituents of F_o and are named a, b, b′, and c. Subunit c, probably occurring in the chloroplastic F_o in 12 copies, contains two transmembrane helices and is the binding site for DCCD. The c subunits appear to form a cylinder, which spans the membrane. In the membrane on the outside of the cylinder, the subunits a, b, and b′ are arranged, whereby b and b′ are in contact with the F_1 part via subunit δ. Subunits γ and ε form the central connection between F_1 and F_o.

Whereas the structure of the F_o part shown in Figure 4.7 is still somewhat hypothetical, the structure of the F_1 part has been thoroughly investi-

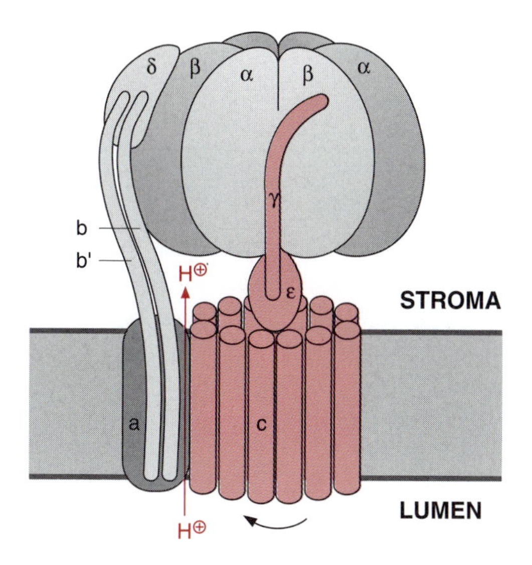

Figure 4.7 Scheme of the structure of an F-ATP synthase. The structure of the F_1 subunit concurs with the results of X-ray analysis discussed in the text. (After Junge.)

gated. The F_1 particles are so small that details of their structure are not visible on a single electron micrograph. However, details of the structure can be resolved if a very large number of F_1 images obtained by electron microscopy are subjected to a **computer-aided image analysis**. Figure 4.8 shows an average image of an F-ATP synthase from chloroplasts. In the side projection, the stalk connecting the F part with the membrane can be recognized. Using more refined picture analysis (not shown here), two stems, one thick and the other thin, were found recently between F_1 and F_2. In the vertical projection, a hexagonal array is to be seen, corresponding to an alternating arrangement of α- and β-subunits. Investigations of the isolated F_1 protein showed that an F_oF_1 protein has three catalytic binding sites for ADP or ATP. One of these binding sites is occupied by very tightly bound ATP, which is released only when energy is supplied from the proton gradient.

X-ray structure analysis of the F_1 part of ATP synthase yields an insight into the machinery of ATP synthesis

In 1994 the group of John Walker in Cambridge (England) succeeded in analyzing the three-dimensional structure of the F_1 part of ATP synthase. Crystals of F_1 from beef heart mitochondria were used for this analysis. Prior to crystallization, the F_1 preparation was loaded with a mixture of ADP and an ATP analogue (5′adenylyl-imidodiphosphate, AMP-PNP). This ATP analogue differs from ATP in that the last two phosphate residues

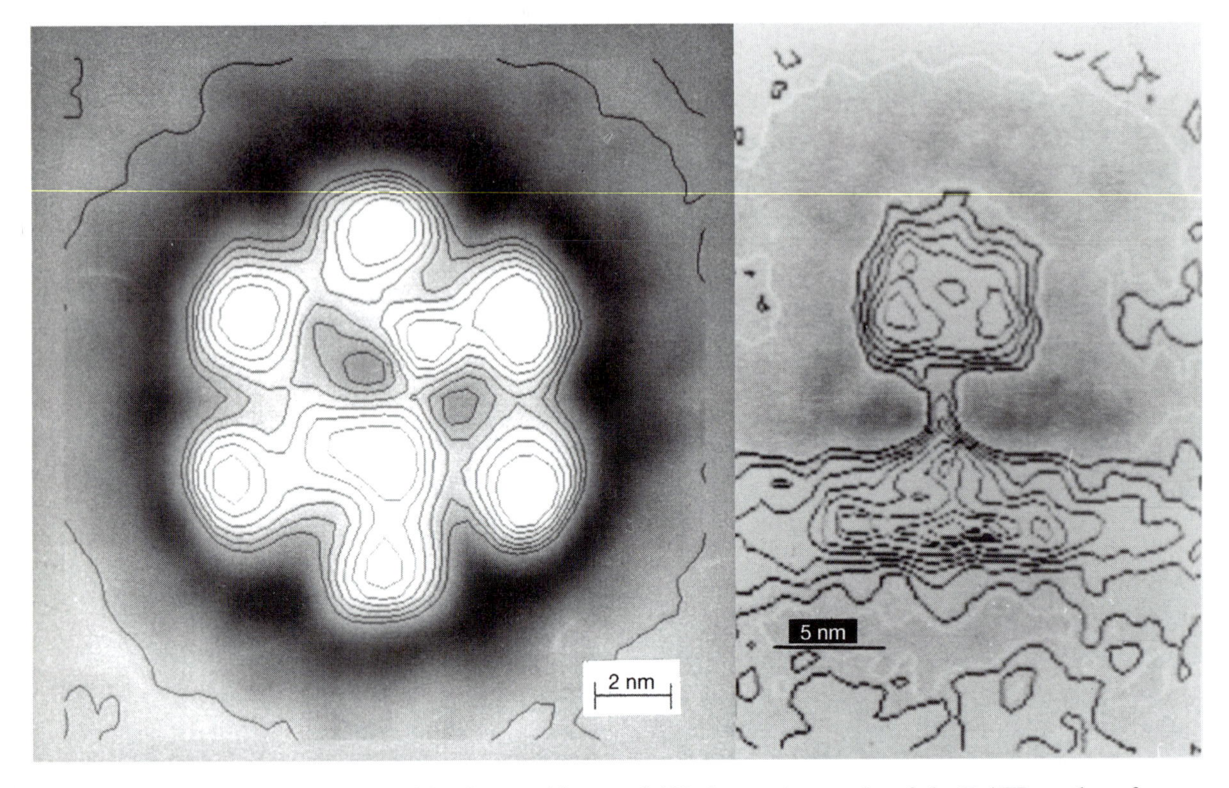

Figure 4.8 Averaged image of 483 electromicrographs of the F-ATP synthase from spinach chloroplasts. A. Vertical projection of the F_1 part. A hexameric structure reflects the alternating ($\alpha\beta$)-subunits. B. Side projection, showing the stalk connecting the F_1 part with the membrane. (By P. Graeber, Stuttgart.)

are connected by an N-atom. It binds to the ATP binding site as ATP, but is not hydrolyzed by ATPase. The structural analysis confirmed the alternating arrangement of the α- and β-subunits (Figs. 4.7 and 4.9). One α- and one β-subunit form a unit with a binding site for one adenine nucleotide. The β-subunit is primarily involved in catalysis of ATP synthesis. In the F_1 crystal investigated, one ($\alpha\beta$)-unit contained one ADP, the second the ATP-analogue, whereas the third ($\alpha\beta$)-subunit was vacant. These differences in nucleotide binding were accompanied by differences in the conformation of the three β-subunits, shown in Figure 4.9 in a schematic drawing. The γ-subunit is arranged **asymmetrically**; protrudes through the center of the F_1 part, and is bent to the side of the ($\alpha\beta$)-unit loaded with ADP (Figs. 4.7 and 4.9). This asymmetry gives an insight into the function of the F_1 part of ATP synthase. Some general considerations about ATP synthase shall be made before dealing with this in more detail.

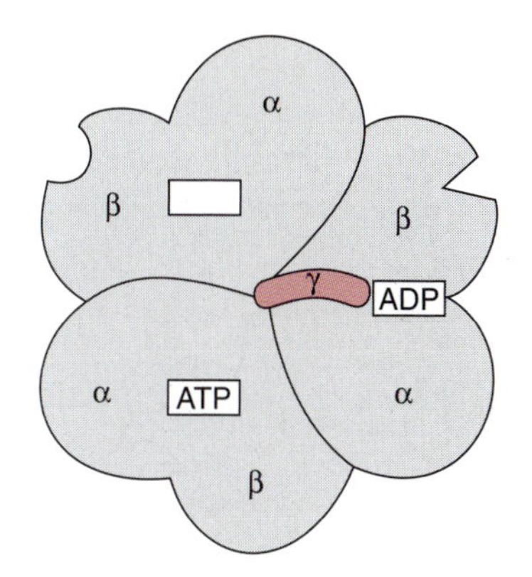

Figure 4.9 Scheme of the vertical projection of the F_1 part of the F-ATP synthase. The enzyme contains three nucleotide binding sites, each consisting of an α-subunit and a β-subunit. Each of the three β-subunits occurs in a different conformation. The γ-subunit in the center, vertical to the viewer, is bent to the α- and β-subunit loaded with ADP. This representation corresponds to the results of X-ray structure analysis by Walker and coworkers mentioned in the text.

4.4 The synthesis of ATP is effected by a conformation change of the protein

For the reaction:

$$ATP + H_2O \rightleftarrows ADP + Phosphate$$

the standard free energy is:

$$\Delta G^{o\prime} = -30.5\,kJ/mol.$$

Because of its high free energy of hydrolysis, ATP is regarded as an energy-rich compound. It may be noted, however, that the standard value $\Delta G^{o\prime}$ has been determined for an aqueous solution of 1 mol of ATP, ADP, and phosphate per liter, respectively, corresponding to a water concentration of 55 mol/L. If the concentration of water were only 10^{-4} mol/L, the ΔG for ATP hydrolysis would have a value of +2.2 kJ/mol. This means that at very low concentrations of water the reaction proceeds toward the synthesis of ATP. This example demonstrates that **in the absence of water the synthesis of ATP does not require the uptake of energy**.

The catalytic site of an enzyme can form a reaction site where water is excluded. Catalytic sites are often located in a hydrophobic area of the enzyme protein in which the substrates are bound in the absence of water. Thus, with ADP and P tightly bound to the enzyme, the synthesis of ATP could proceed spontaneously without requiring energy (Fig. 4.10). This has been proved for H^+-ATP synthase. Since the actual ATP synthesis does proceed without the uptake of energy, the amount of energy required to form ATP from ADP and phosphate in the aqueous phase has to be otherwise consumed (e.g., for the removal of the tightly bound newly synthesized

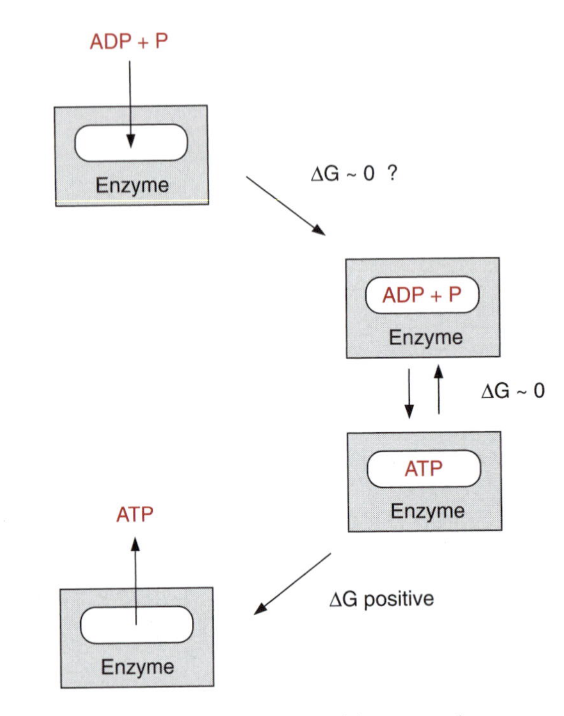

Figure 4.10 In the absence of H_2O, ATP synthesis can occur without the input of energy. In this case, the energy required for ATP synthesis in an aqueous solution has to be spent on binding ADP and P and/or on the release of the newly formed ATP. From available evidence, the latter case is more likely.

ATP from the binding site). This could occur by an energy-dependent conformation change of the protein.

In 1977 Paul Boyer (United States) put forth the hypothesis that the **three identical sites of the F_1 protein alternate in their binding properties** (Fig. 4.11). One of the binding sites is present in the L form, which binds ADP and P loosely but is not catalytically active. A second binding site, T, binds ADP and ATP tightly and is catalytically active. The third binding site, O, is open, binds ADP and ATP only very loosely, and is catalytically inactive. According to this ***"binding change" hypothesis***, the synthesis of ATP proceeds in a cycle. First, ADP and P are bound to the loose binding site, L. A conformation change of the F_1 protein converts site L to a binding site, T, where ATP is synthesized from ADP and phosphate in the absence of water. The ATP formed remains tightly bound. Another conformation change converts the binding site T to an open binding site, O, and the newly formed ATP is released. A crucial point of this hypothesis is that with the conformation change of the F_1 protein, driven by the energy of the proton gradient, the conformation of each of the three catalytic sites is converted simultaneously to the next conformation

$$(L \rightarrow T \qquad T \rightarrow O \qquad O \rightarrow L).$$

The results of X-ray analysis, shown above, support the binding change hypothesis. The evaluated structure clearly shows that the three subunits of

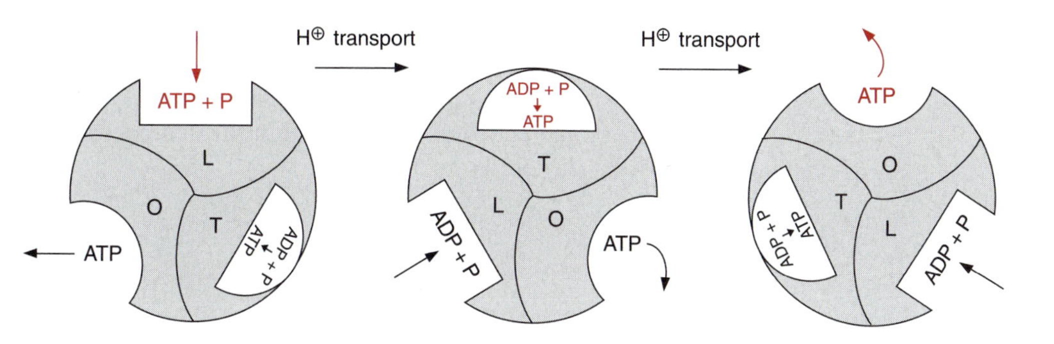

Figure 4.11 ATP synthesis by the binding change mechanism as proposed by Boyer. The central feature of this postulated mechanism is that synthesis of ATP proceeds in the F_1 complex by three nucleotide binding sites, which occur in three different conformations: conformation L binds ADP and P loosely, T binds ADP and P tightly and catalyzes the ATP formation; the ATP thus formed is tightly bound. The open form, O, releases the newly formed ATP. The flux of protons through the F-ATP synthase, as driven by the proton motive force, results in a concerted conformation change of the three binding sites, probably as a rotation.

F_1—one free, one loaded with ADP, and one with the ATP analogue AMP-PNP—have different conformations. Paul Boyer and John Walker were awarded the 1997 Nobel Prize for these results.

Further investigations showed that the central γ subunit rotated. The γ- and ε-subunits of F_1 form together with the 12 c-subunits of F_o a rotor (marked red in Fig. 4.7). In a stator consisting of subunits-$(\alpha\beta)_3$, δ, a, b, b′, this rotor rotates, by which the conformations of each of the catalytic centers shown in Fig. 4.11 is changed. The velocity of rotation of the F-ATP-synthase in chloroplasts has been estimated to be about 160 revolutions per second. It has still not been fully explained how the proton transport drives this rotation. Recent structure analyses suggest that the rotation is the result of an interplay between **subunit a**, which is regarded as part of the **stator**, and **subunit c**, attributed to the **rotor**. According to a preliminary model, subunit a contains a **proton channel**, which is open to the lumen side but closed in the middle. It is assumed that via this channel protons from the thylakoid lumen reach a binding site of subunit c (possibly the carboxyl group of an aspartyl residue). A rotation of the rotor by one step through **thermal movement** shifts the binding site with the proton to another proton channel, which opens to the opposite direction and in this way allows the release of the proton to the stroma space. According to this model, the transfer of a proton from the lumen to the stroma space would cause the **progressive rotation of the rotor** for one c-subunit step. The proton gradient existing between the thylakoid lumen and the stroma space would render the resulting rotation irreversible.

As discussed previously, several bacteria contain an **F-ATP synthase** that is driven by an Na^+ gradient. The model of a proton driven rotor described previously gives an explanation for this by assuming that in the Na^+ F-ATP synthase the subunit c binds Na^+ ions and the two partial ion channels conduct Na^+.

It is still unclear how many c-subunits make up the rotor. To date, investigations of the number of c-subunits per F-ATP synthase molecule yielded values of 12 to 14 (chloroplasts), 10 (yeast mitochondria), and 12 (*E. coli.*). These results, however, remain uncertain. Studies with *E. coli* indicated that the number of c-subunits per rotor might be variable in one and the same organism, depending on its metabolism. In view of this, the general validity of these numbers must be treated with caution.

In photosynthetic electron transport the stoichiometry between the formation of NADPH and ATP is still a matter of debate

According to the model discussed previously, in chloroplasts with 14 c subunits per rotor, a complete rotation would require 14 protons. Since three ATP molecules are formed during one rotation, this would correspond to an H^+/ATP stoichiometry of 4.7. Independent measurements indicated that in chloroplasts at least four protons are necessary for the synthesis of one ATP, which would be similar to the proton stoichiometry of the rotor model. It is still not clear to what extent the Q-cycle plays a part in proton transport. In linear (noncyclic) electron transport, for each NADPH formed without a Q-cycle, four protons (PS II: $2H^+$, Cyt-b_6/f complex: $2H^+$), and with a Q-cycle (Cyt-b_6/f complex: $4H^+$), six protons are transported into the lumen (section 3.7). With a stoichiometry H^+/ATP = 4.7, in noncyclic electron transport with the Q-cycle operating for each NADPH 1.3 ATP would be generated, whereas without a Q-cycle just 0.9. If these stoichiometries hold, noncyclic photophosphorylation would not be sufficient to meet the demands of CO_2 assimilation by the Calvin cycle (ATP/NADPH = 1.5, see Chapter 6) and cyclic photophosphorylation (section 3.8) would be required as well. The question concerning the stoichiometry of photophosphorylation is still not finally answered.

H^+-ATP synthase of chloroplasts is regulated by light

H^+-ATP synthase catalyzes a reaction that is in principle reversible. In chloroplasts, a pH gradient across the thylakoid membrane is generated only during illumination. In darkness, therefore, due to the reversibility of ATP synthesis, one would expect that the ATP synthase then operates in the opposite direction by transporting protons into the thylakoid lumen at the expense of ATP. In order to avoid such a costly reversal, chloroplast ATP synthase is subject to **strict regulation**. This is achieved in two ways. If the pH gradient across the thylakoid membrane decreases below a threshold

3M SelfCheck™ System

Customer name: Tsirigoti, Christina

Title: Mims' pathogenesis of infectious disease / Anthony A. Nash, Robert G. Dalziel, J. Ross Fitzgerald.

ID: 30114015730271

Due: 28-09-15

Title: The molecular life of plants / Russell Jones ... [et al.].

ID: 30114015426649

Due: 28-09-15

Title: Plant biochemistry / by Hans-Walter Heldt in cooperation with Fiona Heldt.

ID: 30114013124972

Due: 28-09-15

Total items: 3

08/06/2015 19:20

Overdue: 0

Thank you for using the

3M SelfCheck™ System.

value, the catalytic sites of the β-subunits are instantaneously switched off, and they are switched on again when the pH gradient is restored upon illumination. The mechanism of this is not yet understood. Furthermore, chloroplast ATP synthase is regulated by **thiol modulation**. By this process, described in detail in section 6.6, a disulfide bond in the γ-subunit of F_1 is reduced in the light by **ferredoxin** to two SH groups, as mediated by reduced thioredoxin. The reduction of the γ-subunit causes the activation of the catalytic centers in the β-subunits. In this way illumination switches F-ATP synthase on. Upon darkening, the two SH groups are deoxidized by oxygen from air to disulfide, and as a result of this, the catalytic centers in the β-subunits are switched off. The simultaneous action of the two regulatory mechanisms allows an efficient control of ATP synthase in chloroplasts.

V-ATPase is related to the F-ATP synthase

A proton transporting V-ATPase is conserved in all eukaryotes. In plants, V-ATPases are located not only in vacuoles (from which the name comes), but also in plasma membranes and membranes of the endoplasmic reticulum and the Golgi apparatus. Genes for at least 12 different subunits have been identified in *Arabidopsis thaliana.* Major functions of this pump are to acidify the vacuole and to generate proton gradients across membranes for driving the transport of ions. V-ATPase also has a role in stomatal closure by the guard cells. The V-ATPase resembles the F-ATP synthase in its basic structure, but it is more complex. It consists of several proteins embedded in the membrane, similar to the F_o part of the F-ATPase, to which a spherical part (similar to F_1) is attached by a stalk and protrudes into the cytosol. The spherical part consists of 3A- and 3B-subunits, which are arranged alternately like the (αβ)-subunits of F-ATP synthase. F-ATP synthase and V-ATPase are derived from a common ancestor. The V-ATPase pumps two protons per molecule of ATP consumed and is able to generate titratable proton concentrations of up to 1.4 mol/L within the vacuoles (section 8.5).

Vacuolar membranes also contain an **H^+-pyrophosphatase**, which, upon the hydrolysis of one molecule pyrophosphate to phosphate, pumps one proton into the vacuole, but it does not reach such high proton gradients as the V-ATPase. H^+-pyrophosphatase probably consists only of a single protein with 16 transmembrane helices. The division of labor between H^+-pyrophosphatase and V-ATPase in transporting protons into the vacuole has not yet been resolved. It remains to be elucidated why there are two enzymes transporting H^+ across the vacuolar membrane. Plasma membranes contain a proton transporting **P-ATPase**, which will be discussed in section 8.2.

Further reading

Abrahams, J. P., Leslie, A. G. W., Lutter, R., Walker, J. E. Structure at 2.8 Å resolution of F_1-ATPase from bovine heart mitochondria. Nature 370, 621–628 (1994).

Boekema, E. J., van Heel, M., Gräber, P. Structure of the ATP synthase from chloroplasts studied by electron microscopy and image processing. Biochim Biophys Acta 933, 365–371 (1988).

Boyer, P. D. The binding change mechanism for ATP synthase—Some probabilities and possibilities. Biochim Biophys Acta 1149, 215–250 (1993).

Dimroth, P. Wirkung des F_0-Motors der ATP-Synthase. Biospektrum 5, 374–380 (1999).

Drozdowicz, Y. M., Rea, P. A. Vacuolar H^+ pyrophosphatases: from the evolutionary backwaters into the mainstream. Trends Plant Sci 6, 206–211 (2001).

Junge, W., Pänke, O, Chrepanov, A, Gumbiowski, K, Engelbrecht, S. Intersubunit rotation and elastic power transmission in FoF1-ATPase. FEBS Lett 504, 152–160 (2001).

Kramer, D. M., Cruz, J. A., Kanazawa, A. Balancing the central roles of the thylakoid proton gradient. Trends Plant Sci 8, 27–32 (2003).

Leigh, R. A., Gordon-Weeks. R., Steele, S. H., Korenkov, V. D. The H^+-pumping inorganic pyrophosphatase of the vacuolar membrane of higher plants. In Blatt, M. R., Leigh, R. A., Sanders, D. (eds.). Membranes for Experimental Biology, Cambridge University Press, Cambridge, pp. 61–74 (1994).

Lüttge, U., Ratajczak, R. The physiology, biochemistry and molecular biology of the plant vacuolar ATPase. In R. A. Leigh, D. Sanders (eds.). The Plant Vacuole, Advances in Botanical Research 25, Academic Press, New York, pp. 253–296 (1997).

Noji, H., Yasuda, R., Yoshida, M., Kinosita Jr., K. Direct observation of the rotation of F_1-ATPase. Nature 386, 299–302 (1997).

Rastogi, V. K., Girvin, M. E. Structural changes linked to proton translocation by subunit *c* of the ATP synthase. Nature 402, 263–268 (1999).

Sambongi, Y., Iko, Y., Tanabe, M., Omote, H., Iwamoto-Kihara, A., Ueda, I., Yanagida, T., Wada, Y., Futai, M. Mechanical rotation of the c subunit oligomer in ATP synthase (F_0F_1): Direct observation. Science 286, 1722–1724 (1999).

Seelert, H., Poetsch, A., Dencher, N. A., Engel, A., Stahlberg, H., Müller, D. J. Proton-powered turbine of a plant motor. Nature 405, 418–419 (2000).

Stock, D., Leslie, A. G. W., Walker, J. E. Molecular architecture of the rotary motor in ATP synthase. Science 286, 1700–1705 (1999).

Sze, H., Schumacher, K., Mueller, M. L., Padmanaban, S., Taiz, L. A simple nomenclature for a complex proton pump: *VHA* genes encode the vacuolar H^+-ATPase. Trends Plant Sci 7, 157–161 (2002).

5

Mitochondria are the power station of the cell

In the process of biological oxidation, substrates such as carbohydrates are oxidized to form water and CO_2. Biological oxidation can be seen as a reversal of the photosynthesis process. It evolved only after the oxygen in the atmosphere had been accumulated by photosynthesis. Both biological oxidation and photosynthesis serve the purpose of generating energy in the form of ATP. Biological oxidation involves a transport of electrons through a mitochondrial electron transport chain, which is in part similar to the photosynthetic electron transport discussed in Chapter 3. The present chapter will show that the machinery of mitochondrial electron transport is also constructed from modules. Of its three complexes, the middle one has the same basic structure as the cytochrome-b_6/f complex of the chloroplasts. Just as in photosynthesis, in mitochondrial oxidation electron transport and ATP synthesis are coupled to each other via the formation of a proton gradient. The synthesis of ATP proceeds by an F-ATP synthase, which was described in Chapter 4.

5.1 Biological oxidation is preceded by a degradation of substrates to form bound hydrogen and CO_2

The total reaction of biological oxidation is equivalent to combustion. In contrast to combustion, however, biological oxidation proceeds in a sequence of partial reactions, which allow the major part of the free energy to be utilized for synthesis of ATP.

The principle of biological oxidation was formulated in 1932 by the Nobel Prize winner Heinrich Wieland:

$$XH_2 + \frac{1}{2}O_2 \longrightarrow X + H_2O$$

First, hydrogen is removed from substrate XH_2 and afterward oxidized to water. Thus, during oxidation, carbohydrates $[CH_2O]_n$ are first degraded by reaction with water to form CO_2 and bound hydrogen [H], and the latter is then oxidized to water:

$$[CH_2O] + H_2O \longrightarrow CO_2 + 4[H]$$

$$4[H] + O_2 \longrightarrow 2H_2O$$

In 1934 Otto Warburg (winner of the 1931 Nobel Prize in Medicine) showed that the transfer of hydrogen from substrates to the site of oxidation occurs as bound hydrogen in the form of **NADH**. From results of studies with homogenates from pigeon muscle in 1937, Hans Krebs formulated the **citrate cycle** (also called the Krebs cycle) as a mechanism for substrate degradation, yielding the reducing equivalents for the reduction of oxygen via biological oxidation. In 1953 he was awarded the Nobel Prize in Medicine for this discovery. The operation of the citrate cycle will be discussed in detail in section 5.3.

5.2 Mitochondria are the sites of cell respiration

Light microscopic studies of many different cells showed that they contain small granules, similar in appearance to bacteria. At the beginning of the last century, the botanist C. Benda named these granules **mitochondria**, which means threadlike bodies. For a long time, however, the function of these mitochondria remained unclear.

As early as 1913, Otto Warburg realized that cell respiration involves the function of granular cell constituents. He succeeded in isolating a protein from yeast that he termed "Atmungsferment" (respiratory ferment), which catalyzes the oxidation by oxygen. He also showed that iron atoms are involved in this catalysis. In 1925 David Keilin from Cambridge (England) discovered the cytochromes and their participation in cell respiration. Using a manual spectroscope, he identified the cytochromes-*a*, -a_3, -*b*, and -*c*

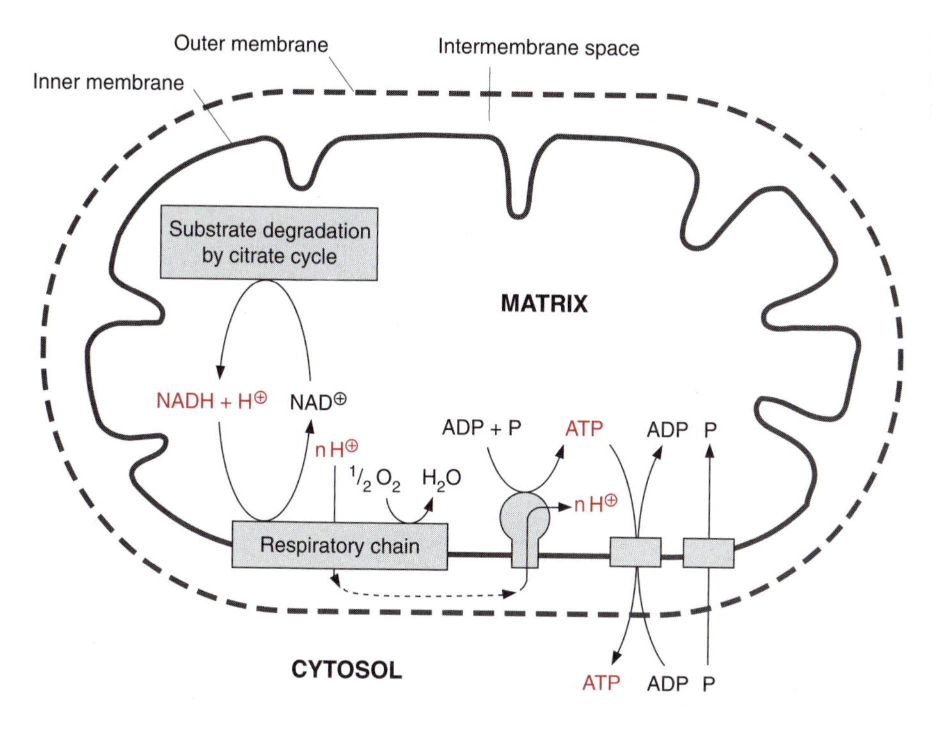

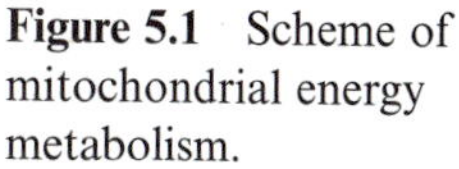
Figure 5.1 Scheme of mitochondrial energy metabolism.

(Fig. 3.24). In 1928 Otto Warburg showed that his **"Atmungsferment"** contained **cytochrome-a_3**. A further milestone in the clarification of cell respiration was reached in 1937, when Hermann Kalckar observed that the formation of ATP in aerobic systems depends on the consumption of oxygen. The interplay between cell respiration and ATP synthesis, termed **oxidative phosphorylation**, was now apparent. In 1948 Eugene Kennedy and Albert Lehninger showed that mitochondria contain the enzymes of the citrate cycle and oxidative phosphorylation. These findings demonstrated the function of the mitochondria as the **power station of the cell**.

Mitochondria form a separated metabolic compartment

Like plastids, mitochondria also form a separated metabolic compartment. The structure of the mitochondria is discussed in section 1.4. Figure 5.1 provides an overview of mitochondrial metabolism. The degradation of substrates to CO_2 and hydrogen (the latter bound to the transport metabolite NADH) takes place in the mitochondrial matrix. NADH thus formed diffuses through the matrix to the mitochondrial inner membrane and is oxidized there by the **respiratory chain**. The respiratory chain consists of a sequence of redox reactions by which electrons are transferred from NADH to oxygen. As in photosynthetic electron transport, in mitochondrial elec-

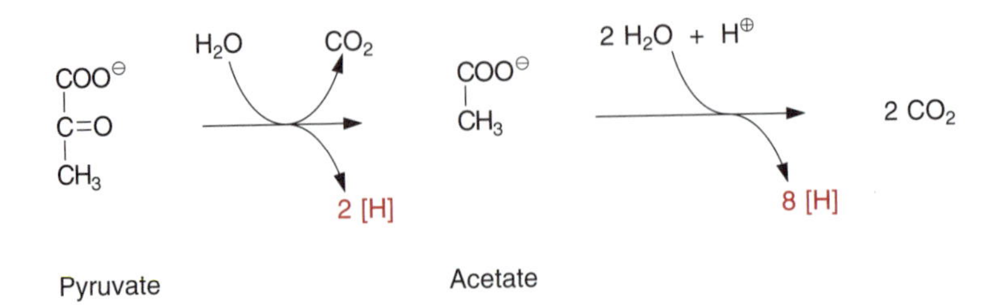

Figure 5.2 Overall reaction of the oxidation of pyruvate by mitochondria. The acetate is formed as acetyl coenzyme A. [H] means bound hydrogen in the form of NADH and $FADH_2$, respectively.

tron transport the released energy is used to generate a proton gradient, which in turn drives the synthesis of ATP. ATP thus formed is exported from the mitochondria and provides the energy required for cell metabolism. This is the universal function of mitochondria in all eukaryotic cells.

5.3 Degradation of substrates for biological oxidation takes place in the matrix compartment

Pyruvate, which is formed by the glycolytic catabolism of carbohydrates in the cytosol, is the starting compound for substrate degradation by the citrate cycle (Fig. 5.2). Pyruvate is first oxidized to acetate (in the form of acetyl coenzyme A), which is then completely degraded to CO_2 by the citrate cycle, yielding 10 reducing equivalents [H] to be oxidized by the respiratory chain to generate ATP. Figure 5.3 shows the reactions of the citrate cycle.

Pyruvate is oxidized by a multienzyme complex

Pyruvate oxidation is catalyzed by the **pyruvate dehydrogenase complex**, a multienzyme complex located in the mitochondrial matrix. It consists of three different catalytic subunits: **pyruvate dehydrogenase, dihydrolipoyl transacetylase,** and **dihydrolipoyl dehydrogenase** (Fig. 5.4). The pyruvate dehydrogenase subunit contains **thiamine pyrophosphate** (**TPP**, Fig. 5.5A) as the prosthetic group. The reactive group of TPP is the thiazole ring. Due to the presence of a positively charged N-atom, the thiazole ring contains an acidic C-atom. After dissociation of a proton, a carbanion is formed, which is able to bind to the carbonyl group of the pyruvate. The positively charged N-atom of the thiazole ring enhances the decarboxylation of the bound pyruvate to form hydroxethyl-TPP (Fig. 5.4). The hydroxethyl group is now transferred to lipoic acid.

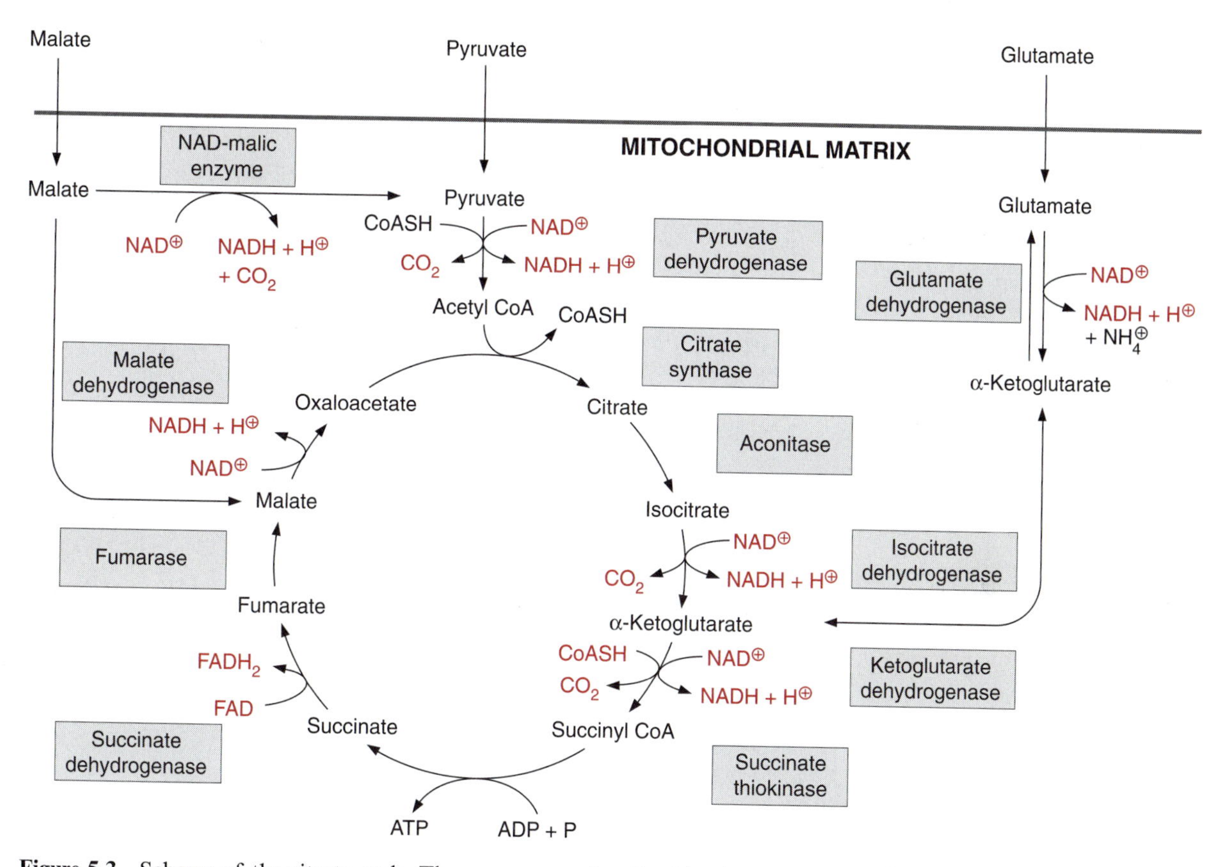

Figure 5.3 Scheme of the citrate cycle. The enzymes are localized in the mitochondrial matrix, with the exception of succinate dehydrogenase, which is located in the inner mitochondrial membrane. As a special feature, plant mitochondria contain NAD malic-enzyme in the mitochondrial matrix. Therefore plant mitochondria are able to oxidize malate via the citrate cycle also in the absence of pyruvate. Glutamate dehydrogenase enables mitochondria to oxidize glutamate.

Lipoic acid is the prosthetic group of the dihydrolipoyl transacetylase subunit. It is covalently bound by its carboxyl group to a lysine residue of the enzyme protein via an amide bond (Fig. 5.5B). The lipoic acid residue is attached to the protein virtually by a long chain and so is able to react with the various reaction sites of the multienzyme complex. Lipoic acid contains two S-atoms linked by a disulfide bond. When the hydroxyethyl residue is transferred to the lipoic acid residue, lipoic acid is reduced to dihydrolipoic acid and the hydroxyethyl residue is oxidized to an acetyl residue. The latter is attached to the dihydrolipoic acid by a thioester bond. This is how energy released during oxidation of the carbonyl group is conserved to generate an energy-rich thioester. The acetyl residue is now transferred by dihydrolipoyl

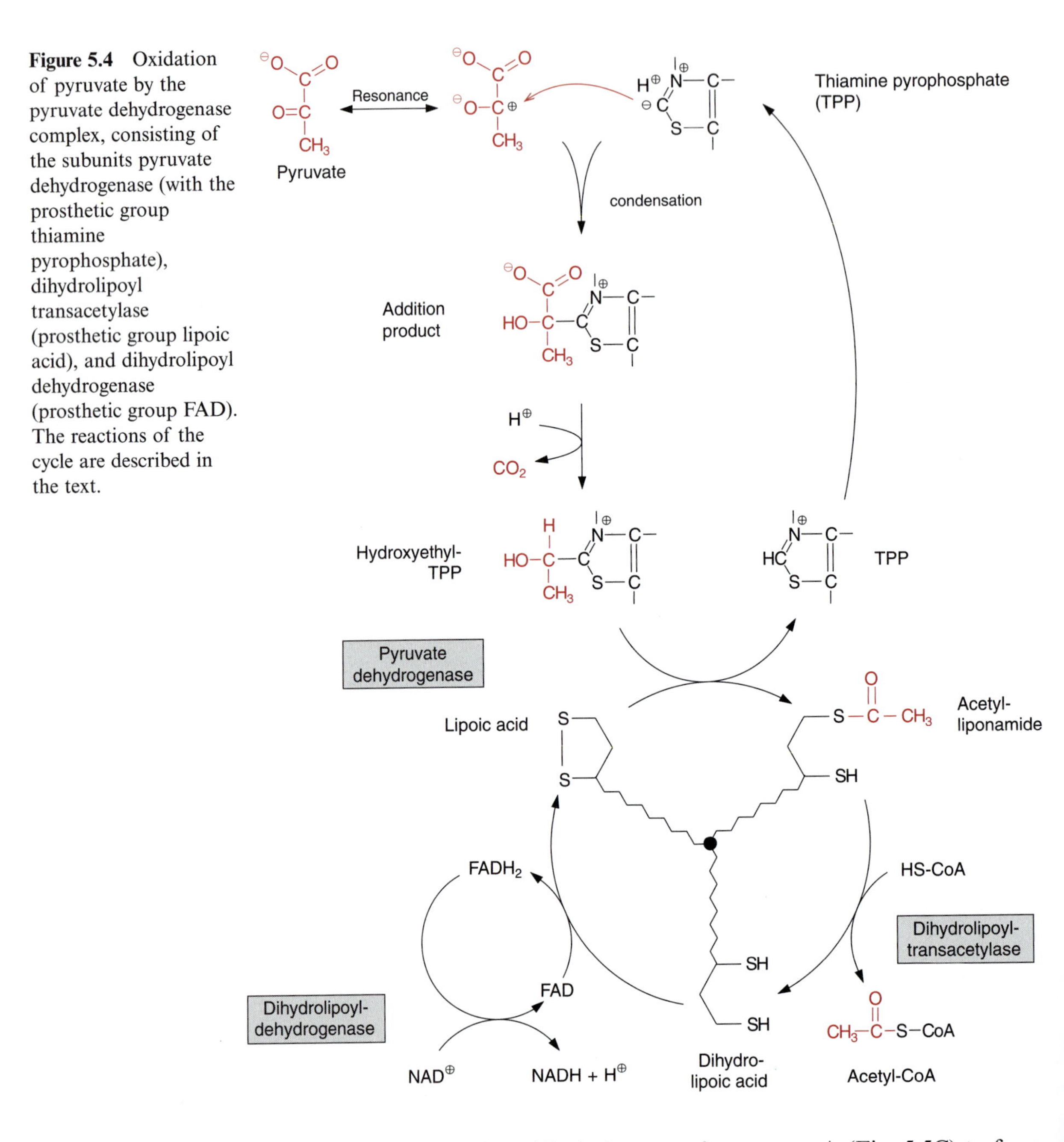

Figure 5.4 Oxidation of pyruvate by the pyruvate dehydrogenase complex, consisting of the subunits pyruvate dehydrogenase (with the prosthetic group thiamine pyrophosphate), dihydrolipoyl transacetylase (prosthetic group lipoic acid), and dihydrolipoyl dehydrogenase (prosthetic group FAD). The reactions of the cycle are described in the text.

transacetylase to the sulfhydryl group of coenzyme A (Fig. 5.5C) to form acetyl coenzyme A. **Acetyl CoA**—also called active acetic acid—was discovered by Feodor Lynen from Munich (1984 Nobel Prize in Medicine). Dihydrolipoic acid is reoxidized to lipoic acid by dihydrolipoyl dehydrogenase and NAD^+ is reduced to NADH via **FAD** (see Fig. 5.16). It should

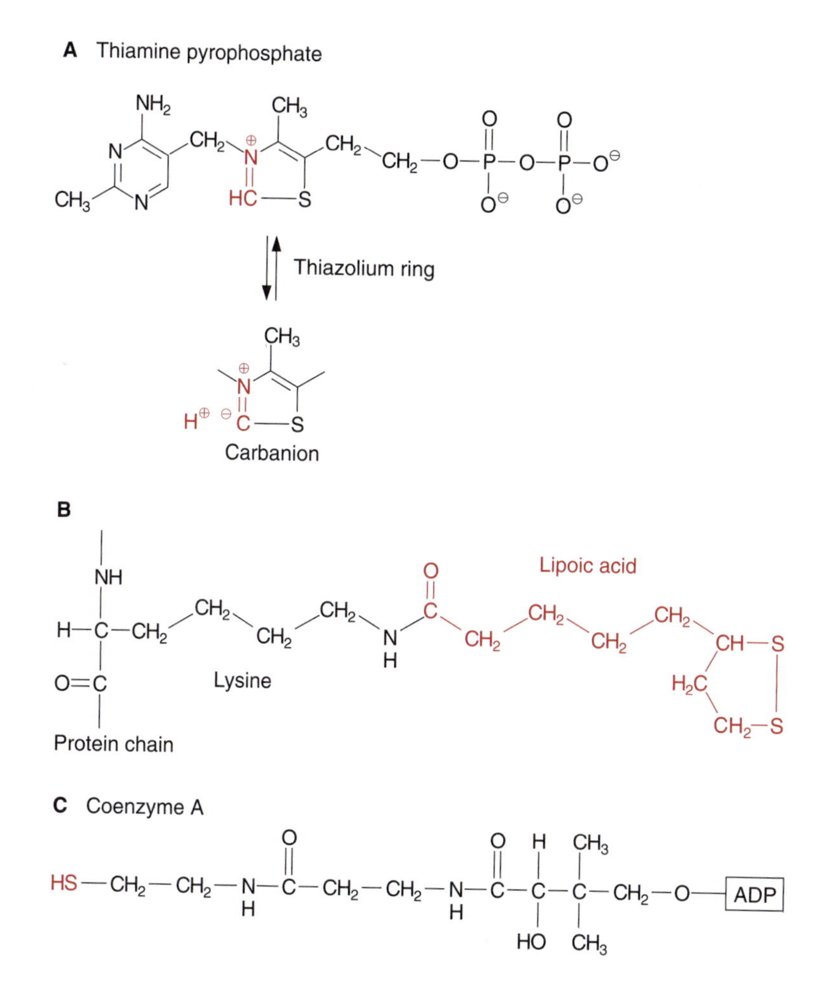

Figure 5.5 Reaction partners of pyruvate oxidation:
A. Thiamine pyrophosphate, B. Lipoic amide, C. Coenzyme A.

be noted that a pyruvate dehydrogenase complex is also found in the chloroplasts, but its function in lipid biosynthesis will be discussed in section 15.3.

Acetate is completely oxidized in the citrate cycle

Acetyl coenzyme A now enters the actual citrate cycle and condenses with oxaloacetate to citrate (Fig. 5.6). This reaction is catalyzed by the enzyme **citrate synthase**. The thioester group promotes the removal of a proton of the acetyl residue, and the carbanion thus formed binds to the carbonyl carbon of oxaloacetate. Subsequent release of CoA-SH makes the reaction irreversible. The enzyme **aconitase** (Fig. 5.7) catalyzes the reversible isomerization of citrate to isocitrate. In this reaction, first water is released, and the *cis*-aconitate thus formed remains bound to the enzyme and is

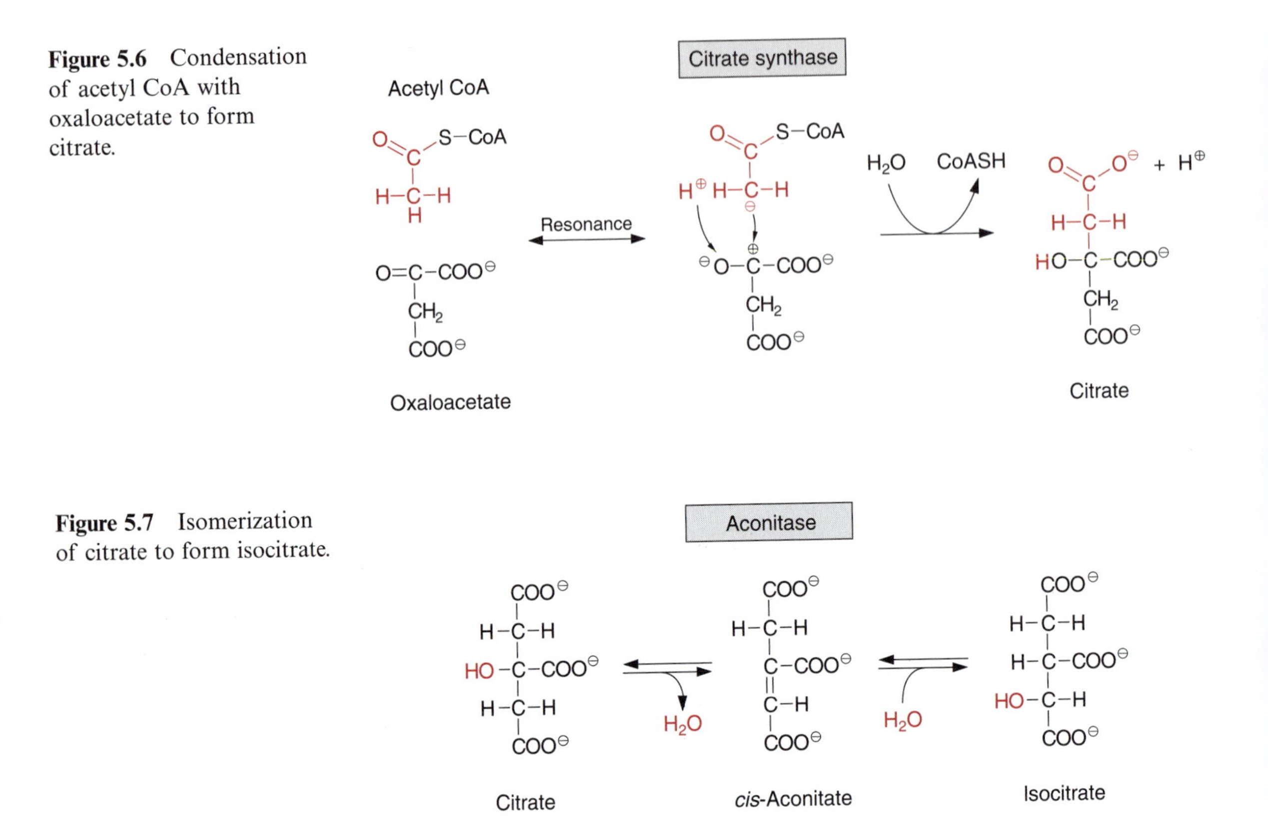

Figure 5.6 Condensation of acetyl CoA with oxaloacetate to form citrate.

Figure 5.7 Isomerization of citrate to form isocitrate.

converted to isocitrate by the addition of water. In addition to the mitochondrial aconitase, there is an isoenzyme of aconitase in the cytosol of plant cells.

Oxidation of isocitrate to α-ketoglutarate by **NAD isocitrate dehydrogenase** (Fig. 5.8) results in the formation of NADH. Oxalosuccinate is formed as an intermediate, which, still tightly bound to the enzyme, is decarboxylated to **α-ketoglutarate** (also termed 2-oxo-glutarate). This decarboxylation makes the oxidation of isocitrate irreversible. Besides NAD-isocitrate dehydrogenase, mitochondria also contain an NADP-dependent enzyme. NADP-isocitrate dehydrogenases also occur in the chloroplast stroma and in the cytosol. The function of the latter enzyme will be discussed in section 10.4.

Oxidation of α-ketoglutarate to succinyl-CoA (Fig. 5.8) is catalyzed by the ***α*-ketoglutarate dehydrogenase multienzyme complex**, also involving thiamine pyrophosphate, lipoic acid, and FAD, analogously to the oxidation of pyruvate to acetyl CoA described previously.

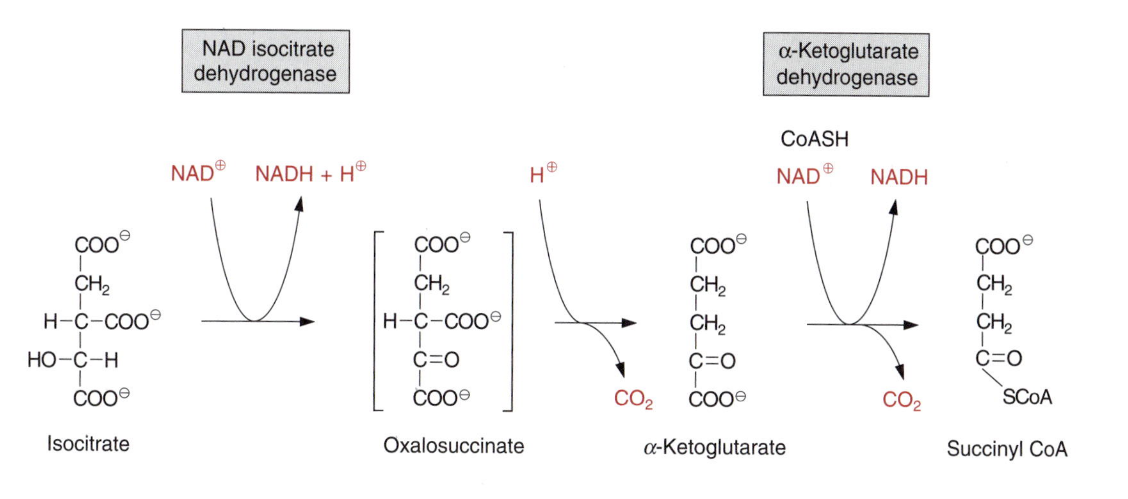

Figure 5.8 Oxidation of isocitrate to form succinyl CoA.

The thioester bond of the succinyl CoA is rich in energy. In the **succinate thiokinase** reaction, the free energy released upon the hydrolysis of this thioester is utilized to form ATP (Fig. 5.9). It may be noted that in animal metabolism the mitochondrial succinate thiokinase reaction yields GTP. The succinate formed is oxidized by **succinate dehydrogenase** to form fumarate. Succinate dehydrogenase is the only enzyme of the citrate cycle that is located not in the matrix, but in the mitochondrial inner membrane, with its succinate binding site accessible from the matrix (section 5.5). Reducing equivalents derived from succinate oxidation are transferred to ubiquinone. Catalyzed by **fumarase**, water reacts by *trans*-addition with the C-C double bond of fumarate to form L-malate. This is a reversible reaction (Fig. 5.9). Oxidation of malate by **malate dehydrogenase,** yielding oxaloacetate and NADH, is the final step in the citrate cycle (Fig. 5.9). The reaction equilibrium of this reversible reaction lies strongly toward the educt malate.

$$\frac{[\text{NADH}]\cdot[\text{oxaloacetate}]}{[\text{NAD}^+]\cdot[\text{malate}]} = 2.8\cdot10^{-5}(\text{pH } 7)$$

Because of this equilibrium, it is essential for the operation of the citrate cycle that the citrate synthase reaction be irreversible. In this way oxaloacetate can be withdrawn from the malate dehydrogenase equilibrium to react further in the cycle. Isoenzymes of malate dehydrogenase also occur outside the mitochondria. Both the cytosol and the peroxisomal matrix contain NAD-malate dehydrogenases, and the chloroplast stroma contains NADP-malate dehydrogenase. These enzymes will be discussed in Chapter 7.

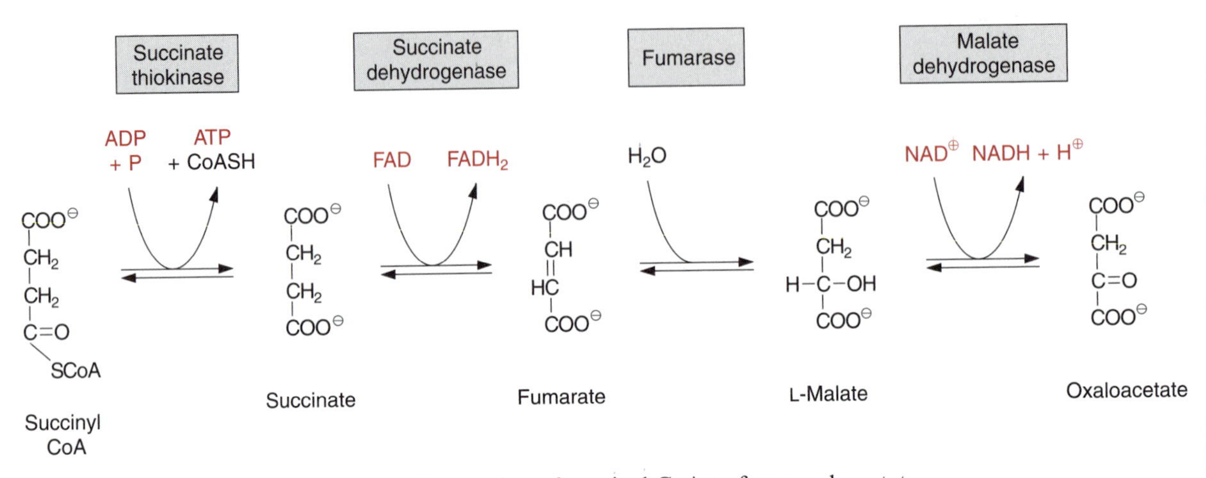

Figure 5.9 Conversion of succinyl CoA to form oxaloacetate.

A loss of intermediates of the citrate cycle is replenished by anaplerotic reactions

The citrate cycle can proceed only when the oxaloacetate required as acceptor for the acetyl residue is fully regenerated. Section 10.4 describes how citrate and α-ketoglutarate are withdrawn from the citrate cycle to synthesize the carbon skeletons of amino acids in the course of nitrate assimilation. It is necessary, therefore, to replenish the loss of citrate cycle intermediates by **anaplerotic reactions**. In contrast to mitochondria from animal tissues, plant mitochondria are able to transport oxaloacetate into the chloroplasts via a specific translocator of the inner membrane (section 5.8). Therefore, the citrate cycle can be replenished by the uptake of oxaloacetate, which has been formed by phosphoenolpyruvate carboxylase in the cytosol (section 8.2). Oxaloacetate can also be delivered by oxidation of malate in the mitochondria. Malate is stored in the vacuole (sections 1.2, 8.2, and 8.5) and is an important substrate for mitochondrial respiration. A special feature of plant mitochondria is that malate is oxidized to pyruvate with the reduction of NAD and the release of CO_2 via **NAD-malic enzyme** in the matrix (Fig. 5.10). Thus an interplay of malate dehydrogenase and NAD malic-enzyme allows citrate to be formed from malate without the operation of the complete citrate cycle (Fig. 5.3). It may be noted that an **NADP-dependent malic** enzyme is present in the chloroplasts, especially in C_4 plants (section 8.4).

Another important substrate of mitochondrial oxidation is **glutamate**, which is one of the main products of nitrate assimilation (section 10.1) and, besides sucrose, is the most highly concentrated organic compound in the

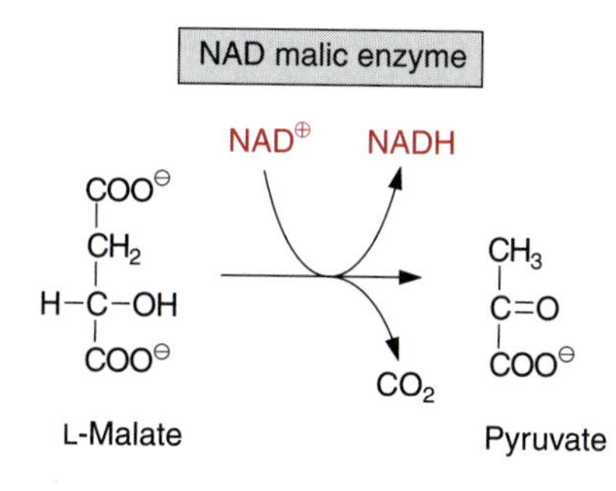

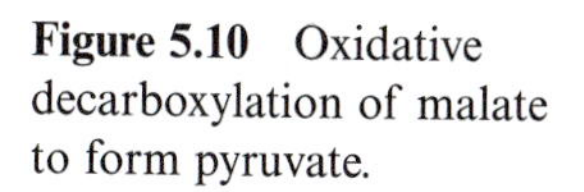
Figure 5.10 Oxidative decarboxylation of malate to form pyruvate.

cytosol of many plant cells. Glutamate oxidation, accompanied by formation of NADH, is catalyzed by **glutamate dehydrogenase** located in the mitochondrial matrix (Fig. 5.11). This enzyme also reacts with NADP. NADP-glutamate dehydrogenase activity is also found in plastids, although its function there is not clear.

Glycine is the main substrate of respiration in the mitochondria from mesophyll cells of illuminated leaves. The oxidation of glycine as a partial reaction of the photorespiratory pathway will be discussed in section 7.1.

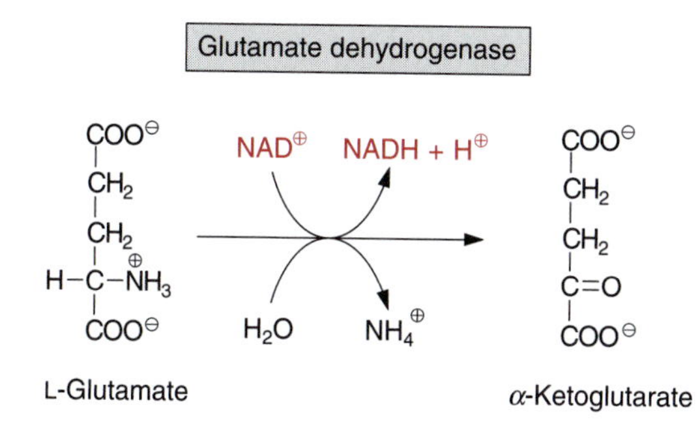

Figure 5.11 Oxidation of glutamate.

5.4 How much energy can be gained by the oxidation of NADH?

Let us first ask ourselves how much energy is released during mitochondrial respiration or, to be more exact, how large is the difference in free energy in the mitochondrial redox processes? To answer this question, one must know the differences of the potentials of the redox pairs involved. This potential difference can be calculated by the Nernst equation:

$$E = E^{0'} + \frac{RT}{nF} \ln \frac{\text{oxidized substance}}{\text{reduced substance}} \tag{5.1}$$

where $E^{0'}$ = standard potential at pH 7, 25°C; R (gas constant) = 8.31 J · K^{-1} · mol^{-1}; T = 298 K; n is the number of electrons transferred; and (Faraday constant) = 96480 JV^{-1} · s · mol^{-1}.

The standard potential for the redox pair $NAD^+/NADH$ is:

$$E^{0'} = -0.320\,\text{V}.$$

Under certain metabolic conditions, a $NAD^+/NADH$ ratio of 3 was found in mitochondria from leaves. The introduction of this value into equation 5.1 yields:

$$E_{\text{NAD+/NADH}} = -0.320 + \frac{RT}{2F} \ln 3 = -0.306\,\text{V} \tag{5.2}$$

The standard potential for the redox pair H_2O/O_2 is :

$$E^{0'} = +0.815\,\text{V}. \quad ([\text{H}_2\text{O}] \text{ in water } 55\,\text{mol/L})$$

The partial pressure of the oxygen in the air is introduced for the evaluation of the actual potential for $[O_2]$.

$$E_{\text{H}_2\text{O/O}_2} = 0.815 + \frac{RT}{2F} \ln \sqrt{p}\,\text{O}_2 \tag{5.3}$$

The partial pressure of the oxygen in the air (pO_2) amounts to 0.2. Introducing this value into equation 5.3 yields:

$$E_{\text{H}_2\text{O/O}_2} = 0.805\,\text{V}$$

The difference of the potentials amounts to:

$$\Delta E = E_{\text{H}_2\text{O/O}_2} - E_{\text{NAD+/NADH}} = +1.11\,\text{V} \tag{5.4}$$

The free energy (ΔG) is related to ΔE as follows:

$$\Delta G = -nF\Delta E \tag{5.5}$$

Two electrons are transferred in the reaction. The introduction of ΔE into equation 5.5 shows that the change of free energy during the oxidation of NADH by the respiratory chain amounts to:

$\Delta G = -214 \text{ kJ mol}^{-1}$.

How much energy is required for the formation of ATP? It has been calculated in section 4.1 that the synthesis of ATP under the metabolic conditions in the chloroplasts requires a change of free energy of $\Delta G \approx +50 \text{ kJ mol}^{-1}$. This value also applies approximately to the ATP provided by the mitochondria for the cytosol.

The free energy released with the oxidation of NADH would therefore be sufficient to generate four molecules of ATP, but in fact the amount of ATP formed by NADH oxidation is much lower (section 5.6).

5.5 The mitochondrial respiratory chain shares common features with the photosynthetic electron transport chain

The photosynthesis of cyanobacteria led to the accumulation of oxygen in the early atmosphere and thus formed the basis for the oxidative metabolism of mitochondria. Many cyanobacteria can satisfy their ATP demand both by photosynthesis and by oxidative metabolism. Cyanobacteria contain a photosynthetic electron transport chain that consists of three modules (complexes), namely, photosystem II, the cyt-b_6/f complex, and photosystem I (Chapter 3, Fig. 5.12). These complexes are located in the

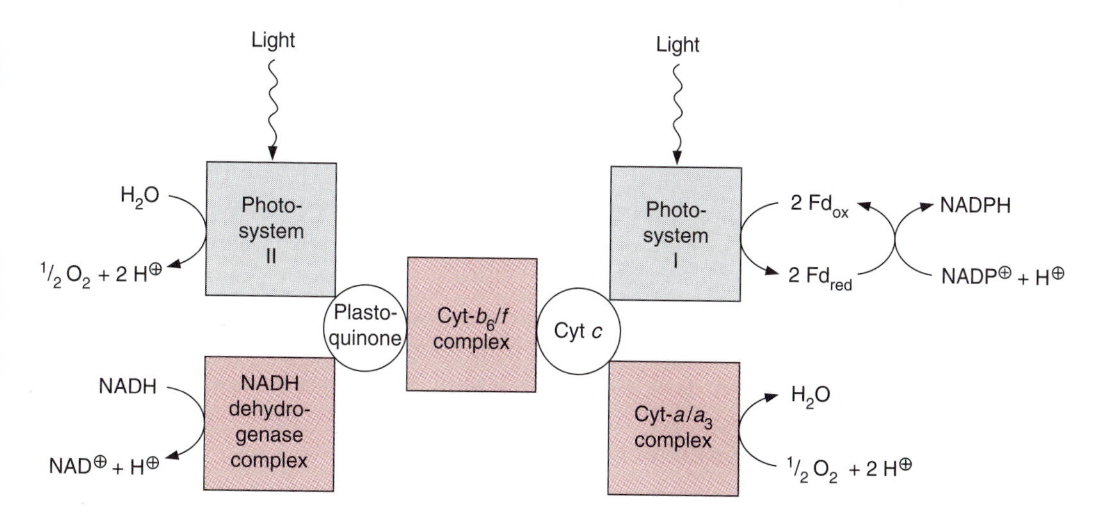

Figure 5.12 Scheme of photosynthetic and oxidative electron transport in cyanobacteria. In both electron transport chains the cytochrome-b_6/f complex functions as the central complex.

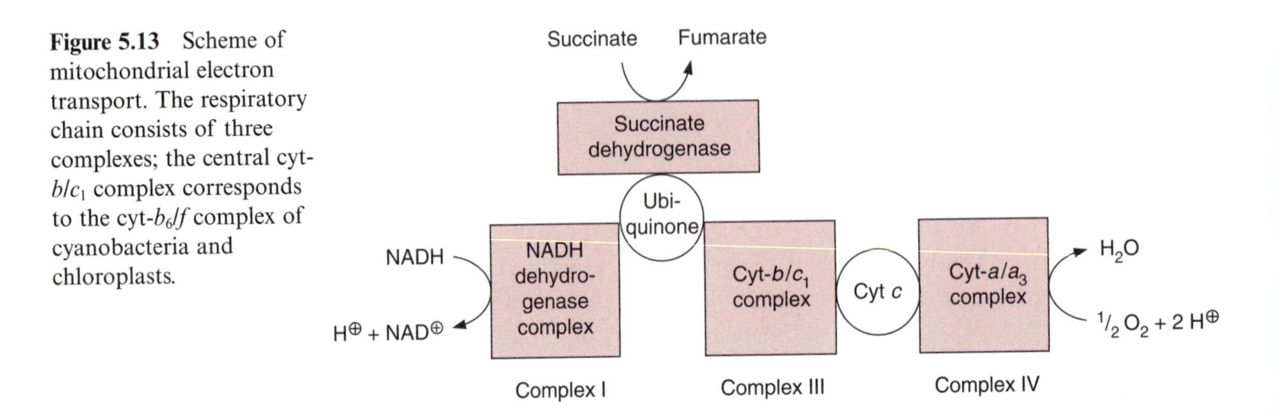

Figure 5.13 Scheme of mitochondrial electron transport. The respiratory chain consists of three complexes; the central cyt-b/c_1 complex corresponds to the cyt-b_6/f complex of cyanobacteria and chloroplasts.

inner membrane of cyanobacteria, where, however, there are also the enzymes of the respiratory electron transport chain. This respiratory chain consists once more of three modules: an **NADH dehydrogenase complex**, catalyzing the oxidation of NADH; the same **cyt-b_6/f complex** that is also part of the photosynthetic electron transport chain; and a **cyt-a/a_3 complex**, by which oxygen is reduced to water. Plastoquinone feeds the electrons into the cyt-b_6/f complex not only in photosynthesis (section 3.7), but also in the respiratory chain of the cyanobacteria. Likewise, cytochrome-*c* mediates the electron transport from the cyt-b_6/f complex to photosystem I as well as to the cyt-a/a_3 complex. The relationship between photosynthetic and oxidative electron transport in cyanobacteria is obvious; both electron transport chains possess the same module as their middle part, the cyt-b_6/f complex. Section 3.7 described how in the cyt-b_6/f complex the energy released by electron transport is used to form a proton gradient. The function of the cyt-b_6/f complex in respiration and photosynthesis shows that the basic principle of energy conservation in photosynthetic and oxidative electron transport is the same.

The mitochondrial respiratory chain is analogous to the respiratory chain of cyanobacteria (Fig. 5.13), but with ubiquinone instead of plastoquinone as redox carrier and slightly different cytochromes. The mitochondria contain a related cyt-b/c_1 complex instead of a cyt-b_6/f complex. Both Cyt-*c* and cyt-*f* contain heme-*c*.

Figure 5.13 shows succinate dehydrogenase as an example of another electron acceptor of the mitochondrial respiratory chain. This enzyme (historically termed complex II) catalyzes the oxidation of succinate to fumarate, a partial reaction of the citrate cycle (Fig. 5.9).

The complexes of the mitochondrial respiratory chain

The subdivision of the respiratory chain into several complexes goes back to the work of Youssef Hatefi, who in 1962, when working with beef heart mitochondria, succeeded in isolating four different complexes, which he termed complexes I–IV. It has been shown recently for plant mitochondria that complex I and III together form a supercomplex. In the complexes I, III, and IV, the electron transport is accompanied by a decrease in the redox potential (Fig. 5.14); the energy thus released is used to form a proton gradient.

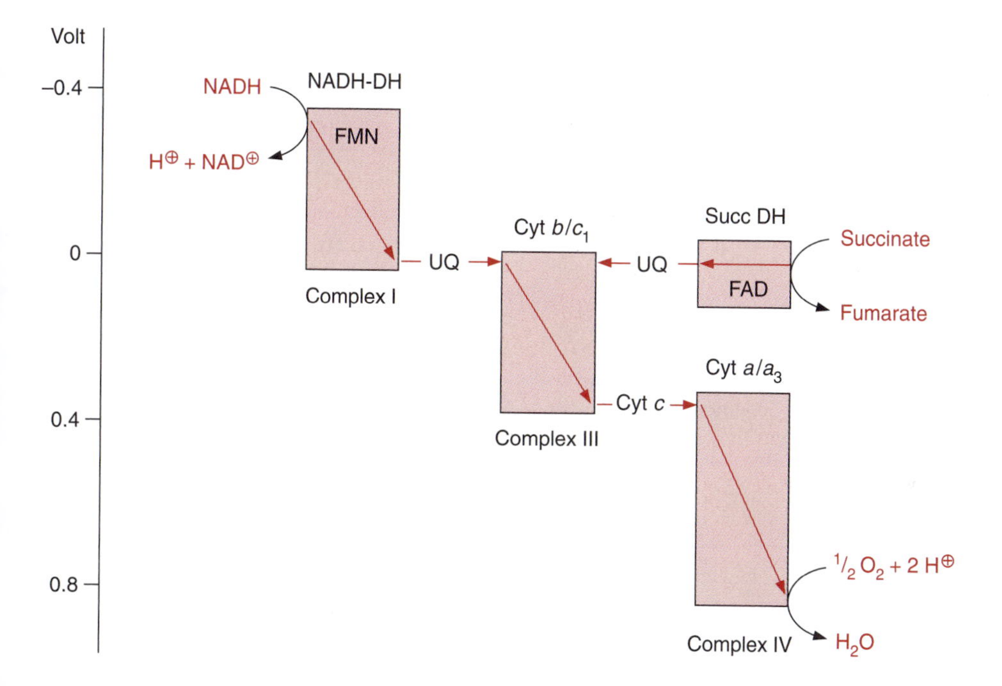

Figure 5.14 Scheme of the complexes of the respiratory chain arranged according to their redox potentials.

The **NADH dehydrogenase complex** (complex I) (Fig. 5.15) feeds the respiratory chain with the electrons from NADH, formed from the degradation of substrates in the matrix. The electrons are transferred to ubiquinone via a flavin adenine mononucleotide and several iron-sulfur centers. Complex I has the most complicated structure of all the mitochondrial electron transport complexes. Made up of more than 40 different subunits (of which, depending on the organism, seven to nine are encoded in the mitochondria), it consists of one part that is embedded in the membrane (**membrane part**) and a **peripheral part** that protrudes into the matrix space. The peripheral part contains the binding site for NADH, a bound flavin mononucleotide (FMN) (Fig. 5.16) and at least three Fe-S-centers

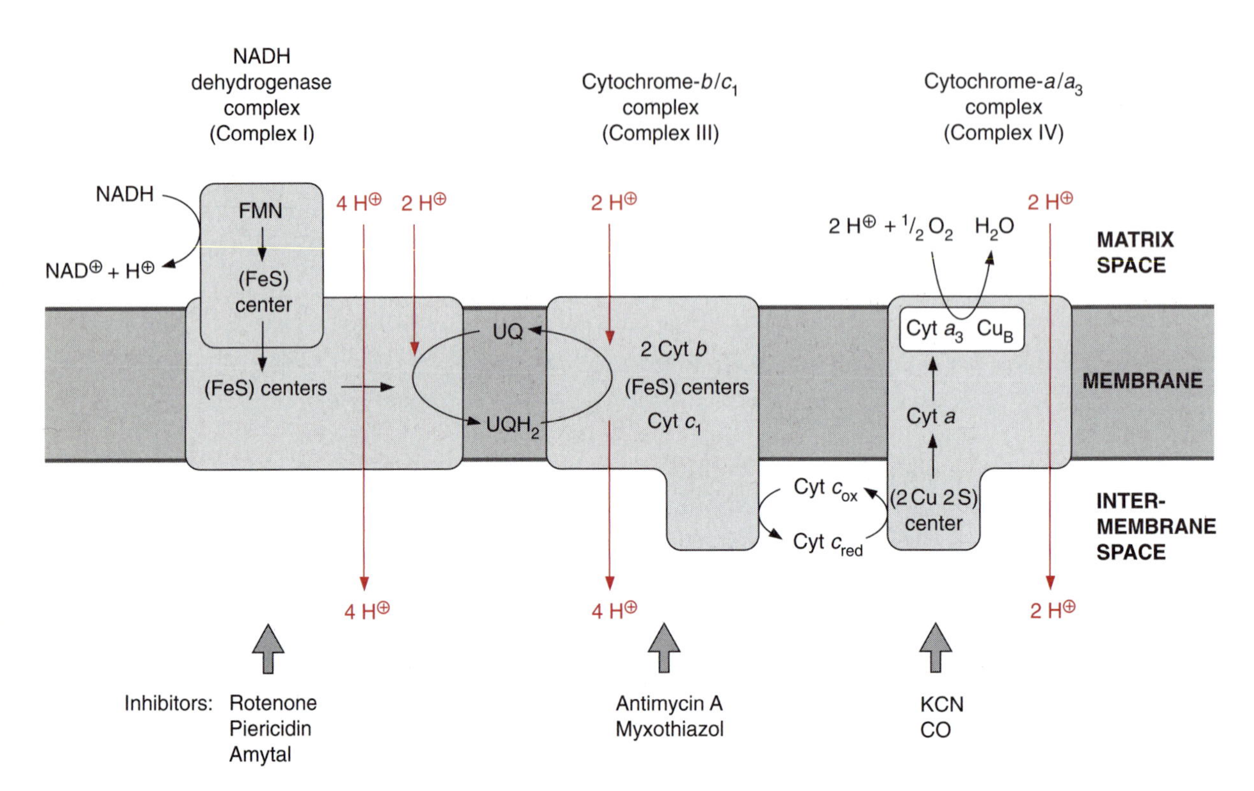

Figure 5.15 Scheme of the location of the respiratory chain complexes I, III, and IV in the mitochondrial inner membrane.

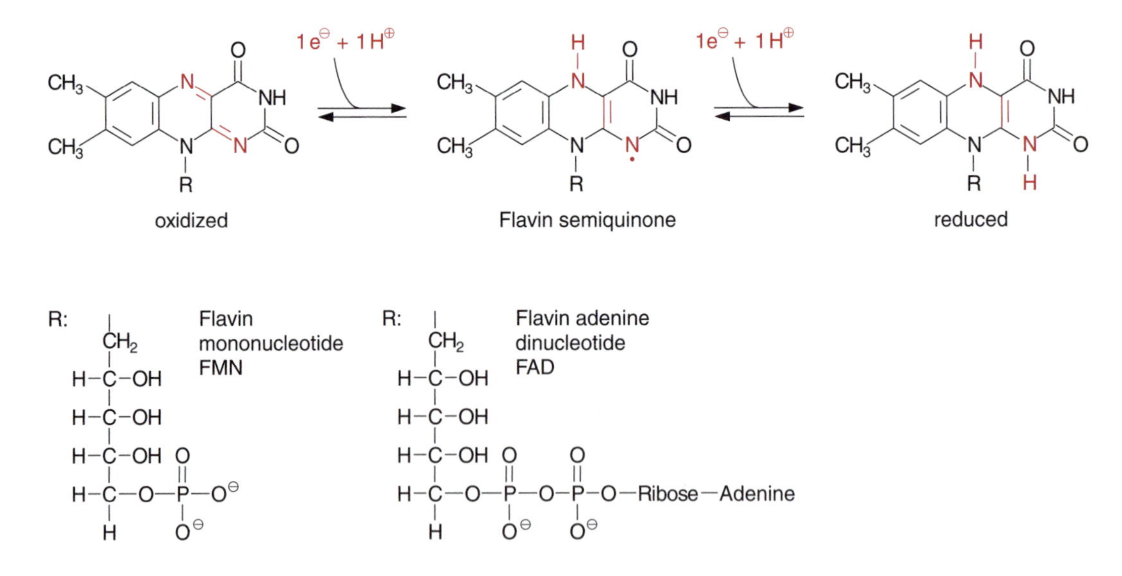

Figure 5.16 FMN, FAD, reduced and oxidized form.

(Fig. 3.26). The membrane part contains a further Fe-S-center, as well as the binding site for ubiquinone. The electron transport is inhibited by a variety of poisons deriving from plants and bacteria, such as **rotenone** (which protects plants from being eaten by animals); the antibiotic **piericidin A**; and **amytal**, a barbiturate. The electron transport catalyzed by complex I is reversible. It is therefore possible for electrons to be transferred from ubiquinone to NAD, driven by the proton motive force of the proton gradient. In this way purple bacteria are provided with NADH by means of a homologous NADH dehydrogenase complex (see Fig. 3.1).

Succinate dehydrogenase (complex II) (Fig. 5.9) in plants consists of seven subunits. It contains a flavin adenine nucleotide (FAD, Fig. 5.16) as the electron acceptor; several Fe-S-centers (Fig. 3.26) as redox carriers; and one cytochrome-*b,* the function of which is not known. Electron transport by succinate dehydrogenase to ubiquinone proceeds with no major decrease in the redox potential, so no energy is gained in the electron transport from succinate to ubiquinone.

Ubiquinone reduced by the NADH dehydrogenase complex or succinate dehydrogenase is oxidized by the **cyt-b/c$_1$ complex** (complex III) (Fig. 5.15). In mitochondria this complex consists of 11 subunits, only one of which (the cyt-*b* subunit) is encoded in the mitochondria. The cyt-b/c_1 complex is very similar in structure and function to the cyt-b_6/f complex of chloroplasts, (section 3.7). Electrons are transferred by the cyt-b/c_1 complex to cyt-*c,* which is bound to the outer surface of the inner membrane. Several antibiotics, such as **antimycin A** and **myxothiazol,** inhibit the electron transport by the cyt-b/c_1 complex.

Due to its positive charge, reduced cyt-*c* diffuses along the negatively charged surface of the inner membrane to the **cyt-a/a$_3$ complex** (Fig. 5.15), also termed **complex IV** or **cytochrome oxidase**. The cyt-a/a_3 complex contains 13 different subunits, three of which are encoded in the mitochondria. Recently the three-dimensional structure of the cyt-a/a_3 complex has been resolved by X-ray structure analysis of these complexes from beef heart mitochondria and from *Paracoccus denitrificans*. The complex has a large hydrophilic region that protrudes into the intermembrane space and contains the binding site for cyt-*c*. In the oxidation of cyt-*c,* the electrons are transferred to a **copper sulfur cluster** containing two Cu atoms called Cu_A. These two Cu atoms are linked by two S-atoms of cysteine side chains (Fig. 5.17). This copper-sulfur cluster probably takes up one electron and transfers it via cyt-*a* to a **binuclear center**, consisting of cyt-a_3 and a Cu atom (Cu_B), bound to histidine. This binuclear center functions as a redox unit in which the Fe-atom of the cyt-a_3, together with Cu_B, take up two electrons.

$$[Fe^{+++} \cdot Cu_B^{++}] + 2e^- \rightarrow [Fe^{++} \cdot Cu_B^{+}]$$

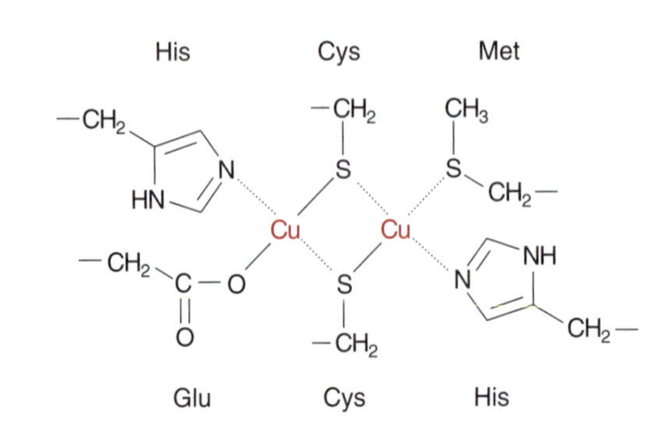

Figure 5.17 In a copper-sulfur cluster of the cytochrome-a/a_3 complex, termed Cu_A, a Cu^{2+}- and a Cu^+-ion are linked by two cysteine residues of the protein and bound further to the protein by two histidines, one glutamate and one methionine residue. Cu_A probably transfers one electron. The structure of this novel redox cluster was revealed by X-ray structural analysis of the cytochrome-a/a_3 complex carried out simultaneously in Frankfurt, Germany, (Iwata et al., 1995) and in Osaka, Japan (Tsukihara et al., 1995).

In contrast to cyt-*a* and the other cytochromes of the respiratory chain, in cyt-a_3 the sixth coordination position of the Fe-atom is not saturated by an amino acid of the protein (Fig. 5.18). This free coordination position as well as Cu_B are the binding site for the oxygen molecule, which is reduced to water by the uptake of four electrons:

$$O_2 + 4e^- \rightarrow 2O^{2-} + 4H^+ \rightarrow 2H_2O$$

Cu_B probably has an important function in electron-driven proton transport, which is discussed in the next section. Instead of O_2, also CO and CN^- can be very tightly bound to the free coordination position of the cyt-a_3, resulting in the inhibition of respiration. Therefore both carbon monoxide and prussic acid (HCN) are very potent poisons.

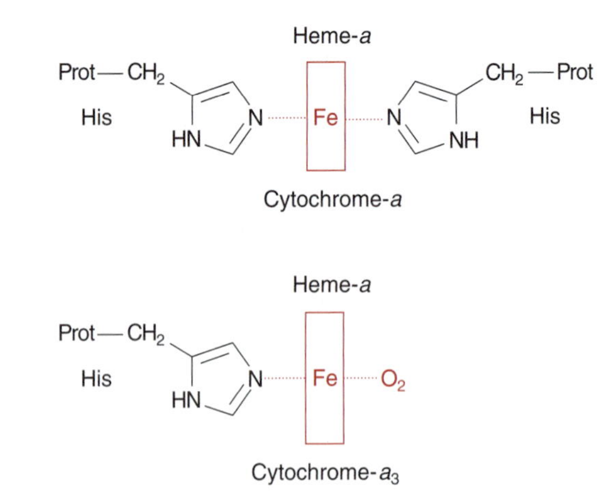

Figure 5.18 Axial ligands of the Fe-atoms in the heme groups of cytochrome-*a* and -a_3. Of the six coordinative bonds formed by the Fe-atom present in the heme, four are saturated by the N atoms present in the planar tetrapyrrole ring. Whereas in cytochrome-*a* the two remaining coordination positions of the central Fe-atom bind to histidine residues of the protein, positioned at either side vertically to the plane of the tetrapyrrole, in cytochrome-a_3 one of these coordination positions is free and functions as binding site for the O_2 molecule.

5.6 Electron transport of the respiratory chain is coupled to the synthesis of ATP via proton transport

The electron transport of the respiratory chain is coupled to the formation of ATP. This is illustrated in the experiment of Figure 5.19, in which the velocity of respiration in a mitochondrial suspension was determined by measuring the decrease of the oxygen concentration in the suspension medium. The addition of a substrate alone (e.g., malate) causes only a minor increase in respiration. The subsequent addition of a limited amount of ADP results in a considerable acceleration of respiration. After some time, however, respiration returns to the lower rate prior to the addition of ADP, as the ADP has been completely converted to ATP. Respiration in the presence of ADP is called **active respiration**, whereas that after ADP is consumed is called **controlled respiration**. As the ADP added to the mitochondria is completely converted to ATP, the amount of ATP formed with the oxidation of a certain substrate can be determined from the ratio of ADP added to oxygen consumed (**ADP/O**). An ADP/O of about 2.5 is determined for substrates oxidized in the mitochondria via the formation of NADH (e.g., malate), and of about 1.6 for succinate, from which the redox equivalents

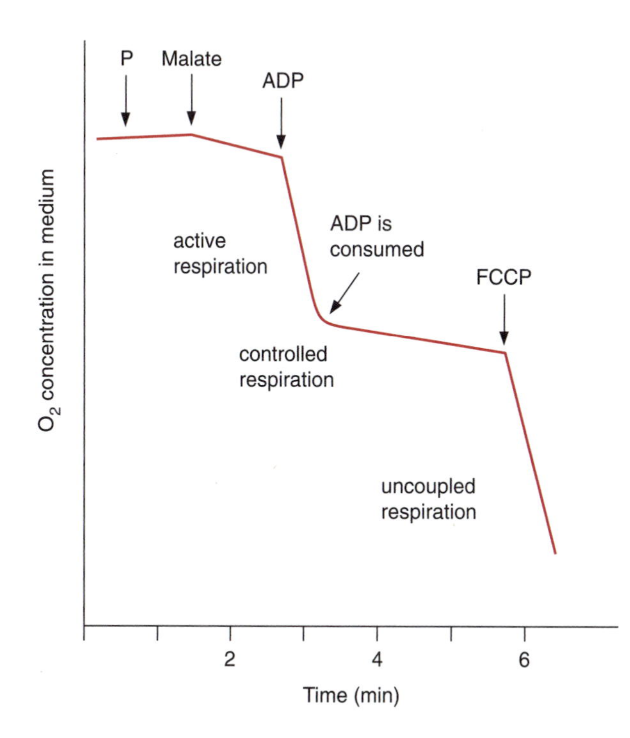

Figure 5.19 Registration of oxygen consumption by isolated mitochondria. Phosphate and malate are added one after the other to mitochondria suspended in a buffered osmotic. Addition of ADP results in a high rate of respiration. The subsequent decrease of oxygen consumption indicates that the conversion of the added ADP into ATP is completed. Upon the addition of an uncoupler, a high respiration rate is attained without ADP: the respiration is now uncoupled.

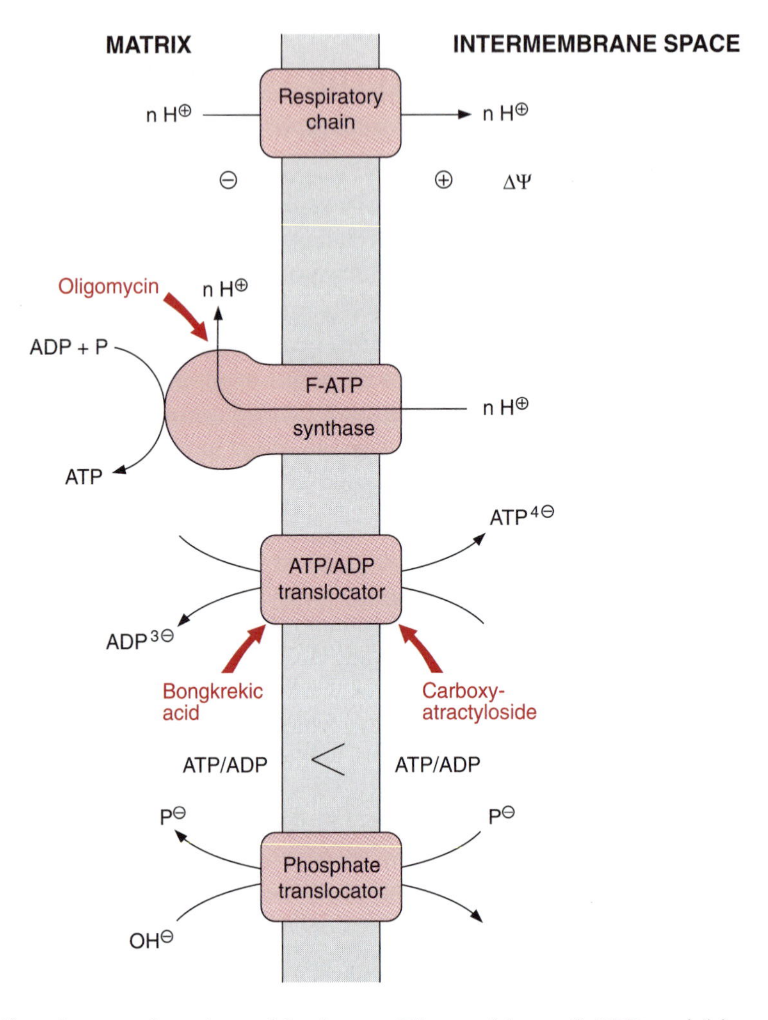

Figure 5.20 ATP synthesis by mitochondria requires an uptake of phosphate by the phosphate translocator in counter-exchange for OH^- ions, and an electrogenic exchange of ADP for ATP, as catalyzed by the ADP/ATP translocator. Due to the membrane potential generated by electron transport of the respiratory chain, ADP is preferentially transported inward and ATP outward, and as a result of this the ATP/ADP ratio in the cytosol is higher than in the mitochondrial matrix. ATP/ADP transport is inhibited by carboxyactractyloside (binding from the outside) and bongkrekic acid (binding from the matrix side). F-ATP synthase of the mitochondria is inhibited by oligomycin.

are directly transferred to ubiquinone. The problem of ATP stoichiometry of respiration will be discussed at the end of this section.

Like photosynthetic electron transport (Chapter 4), the electron transport of the respiratory chain is accompanied by the generation of a proton motive force (Fig. 5.15), which in turn drives the synthesis of ATP (Fig. 5.20). Therefore substances such as FCCP (Fig. 4.2) function as uncouplers of mitochondrial as well as photosynthetic electron transport. Figure 5.19 shows that the addition of the uncoupler FCCP results in high stimulation of respiration. As discussed in section 4.2, the uncoupling function of the FCCP is due to a short circuit of protons across a membrane, resulting in the elimination of the proton gradient. The respiration is then uncoupled from ATP synthesis and the energy set free during electron transport is dissipated as heat.

To match respiration to the energy demand of the cell, it is regulated by an overlapping of two different mechanisms. The classic mechanism of **respiratory control** is based on the fact that when the ATP/ADP ratio increases, the proton motive force also increases, which in turn causes a decrease of electron transport by the respiratory chain. In addition, it was found recently that ATP also impedes the electron transport by binding to a subunit of cytochrome oxidase, which results in a decrease of its activity.

Mitochondrial proton transport results in the formation of a membrane potential

Mitochondria, in contrast to chloroplasts, have no closed thylakoid space for forming a proton gradient. Instead, in mitochondrial electron transport, protons are transported from the matrix to the intermembrane space, which is connected to the cytosol by pores [formed by porines (Fig. 1.30)]. The formation in chloroplasts of a proton gradient of $\Delta pH = 2.5$ in the light results in a decrease of pH in the thylakoid lumen from about pH 7.5 to pH 5.0. If in the case of mitochondrial oxidation such a strong acidification were to occur in the cytosol, it would have a grave effect on the activity of the cytosolic enzymes. In fact, during mitochondrial controlled respiration the ΔpH across the inner membrane is only about 0.2. This is because mitochondrial proton transport leads primarily to the formation of a membrane potential ($\Delta\Psi \approx 200$ mV). Mitochondria are unable to generate a larger proton gradient, as their inner membrane is impermeable for anions, such as chloride. As shown in Figure 4.1, a proton concentration gradient can be formed only when the charge of the transported protons is compensated for by the diffusion of a counter anion.

Our knowledge of the mechanism of coupling between mitochondrial electron transport and transport of protons is still incomplete, despite intensive research for more than 30 years. Four protons are probably taken up during the transport of two electrons from the NADH dehydrogenase complex to ubiquinone on the matrix side and released again into the intermembrane space by the cyt-b/c_1 complex (Fig. 5.15). It is generally accepted that in mitochondria the cyt-b/c_1 complex catalyzes a **Q-cycle** (Fig. 3.30) by which, when two electrons are transported, two additional protons are transported out of the matrix space into the intermembrane space. The cyt-a/a_3 complex transports two protons per two electrons. The three-dimensional structure of the cyt-a/a_3 complex determined recently indicates that the binuclear center from cytochrome-a_3 and Cu_b is involved in this proton transport. If these stoichiometries are correct, altogether 10 protons would be transported during the oxidation of NADH and only six with the oxidation of succinate.

Mitochondrial ATP synthesis serves the energy demand of the cytosol

The energy of the proton gradient is used in the mitochondria for ATP synthesis by an F-ATP-synthase (Fig. 5.20), which has the same basic structure as the F-ATP-synthase of chloroplasts (section 4.3). However, there are differences regarding the inhibition by **oligomycin**, an antibiotic from *Streptomyces*. Whereas the mitochondrial F-ATP-synthase is very strongly inhibited by oligomycin, due to the presence of an oligomycin binding protein, the chloroplast enzyme is insensitive to this inhibitor. Despite these differences, the mechanism of ATP synthesis appears to be identical for both ATP synthases. Also, the proton stoichiometry in mitochondrial ATP synthesis has not been resolved unequivocally. In the event that it is correct that the rotor of the F-ATP-synthase in mitochondria has 10 c-subunits, then, according to the mechanism for ATP synthesis discussed in section 4.4, 3.3 protons would be required for the synthesis of 1 mol of ATP. This rate corresponds more or less with previous independent investigations.

In contrast to chloroplasts, which synthesize ATP essentially for their own consumption, the ATP in mitochondria is synthesized mainly for export to the cytosol. This requires the uptake of ADP and phosphate from the cytosol into the mitochondria and the release of the ATP formed there. The uptake of phosphate proceeds by the **phosphate translocator** in a counter-exchange for OH^- ions, whereas the uptake of ADP and the release of ATP is mediated by the **ATP/ADP translocator** (Fig. 5.20). The mitochondrial ATP/ADP translocator is inhibited by **carboxyatractyloside**, a glucoside from the thistle *Atractylis gumnifera*, and by **bonkrekic acid**, an antibiotic from the bacterium *Cocovenerans*, growing on coconuts. Both substances are deadly poisons.

The ATP/ADP translocator catalyzes a strict **counter-exchange**; for each ATP or ADP transported out of the chloroplasts, an ADP or ATP is transported inward. Since the transported ATP contains one negative charge more than the ADP, the transport is electrogenic. Due to the membrane potential generated by the proton transport of the respiratory chain, there is a preference for ADP to be taken up and ATP to be transported outward. The result of this asymmetric transport of ADP and ATP is that the ADP/ATP ratio outside the mitochondria is much higher than in the matrix. In this way mitochondrial ATP synthesis maintains a high ATP/ADP ratio in the cytosol. With the exchange of ADP for ATP, one negative charge is transferred from the matrix to the outside, which is in turn compensated for by the transport of a proton in the other direction. This is why protons from the proton gradient are required not only for ATP synthesis as such, but also for export of the synthesized ATP from the mitochondria.

Let us return to the stoichiometry between the transported protons and the ATP formation during respiration. It is customary to speak of three coupling sites of the respiratory chain, which correspond to the complexes I, III, and IV. Textbooks often state that when NADH is oxidized by the mitochondrial respiratory chain, one molecule of ATP is formed per coupling site, and as a result of this, the ADP/O quotient for oxidation of NADH amounts to three, and that for succinate to two. However, considerably lower values have been found in experiments with isolated mitochondria. The attempt was made to explain this discrepancy by assuming that owing to a proton leakage of the membrane, the theoretical ADP/O values were not attained in the isolated mitochondria. It appears now that even in theory these whole numbers for ADP/O values are incorrect. Probably 10 protons are transported upon the oxidation of NADH. In the event that 3.3 protons are required for the synthesis of ATP and another one for its export from the mitochondria, the resulting ADP/O would be 2.3. With isolated mitochondria, values of about 2.5 have been obtained experimentally.

At the beginning of this chapter, the change in free energy during the oxidation of NADH was evaluated as $-214\ kJ\ mol^{-1}$ and for the synthesis of ATP as about $+50\ kJ\ mol^{-1}$. An ADP/O of 2.3 for the respiration of NADH-dependent substrates would mean that about 54% of the free energy released during oxidation is used for the synthesis of ATP. However, these values must still be treated with caution.

5.7 Plant mitochondria have special metabolic functions

The function of the mitochondria as the power station of the cell, as discussed so far, is relevant for all mitochondria, from unicellular organisms to animals and plants. In plant cells performing photosynthesis, the role of the mitochondria as a supplier of energy is not restricted to the dark phase; the mitochondria provide the cytosol with ATP also during photosynthesis.

In addition, plant mitochondria fulfill special functions. The mitochondrial matrix contains enzymes for the oxidation of glycine to serine, an important step in the photorespiratory pathway (section 7.1):

$$2\ \text{glycine} + NAD^+ + H_2O \longrightarrow \text{serine} + NADH + CO_2 + NH_4^+$$

The NADH generated from glycine oxidation is the main fuel for mitochondrial ATP generation during photosynthesis. Another important role of plant mitochondria is the conversion of oxaloacetate and pyruvate to form citrate, a precursor for the synthesis of α-ketoglutarate. This pathway is important for providing the carbon skeletons for amino acid synthesis during nitrate assimilation (Fig. 10.11).

Mitochondria can oxidize surplus NADH without forming ATP

In mitochondrial electron transport, the participation of flavins, ubisemiquinones, and other electron carriers leads to the formation of superoxide radicals, H_2O_2, and hydroxyl radicals (summarized as **ROS**, *r*eactive *o*xygen *s*pecies) as by-products. These by-products cause severe cell damage. Since the formation of ROS is especially high, when the components of the respiratory chain are highly reduced, there is a necessity to avoid an overreduction of the respiratory chain. On the other hand, it is essential for a plant that glycine, formed in large quantities by the photorespiratory cycle (section 7.1), is converted by mitochondrial oxidation even when the cell requires no ATP. Plant mitochondria have **overflow mechanisms**, which oxidize surplus NADH without synthesis of ATP in order to prevent an overreduction of the respiratory chain (Fig. 5.21). The inner mitochondrial membrane contains at the matrix side an **alternative NADH dehydrogenase**, which transfers electrons from NADH to ubiquinone, without coupling to proton transport. This pathway is not inhibited by rotenone. However, oxidation of NADH via this rotenone-insensitive pathway proceeds only when the NADH/NAD^+ quotient in the matrix is exceptionally high. In addition, the matrix side of the mitochondrial inner membrane contains an alternative NADPH dehydrogenase, which is not shown in Figure 5.21.

Moreover, mitochondria possess an **alternative oxidase** by which electrons can be transferred directly from ubiquinone to oxygen; this pathway too is not coupled to proton transport. This alternative oxidation is insensitive to **antimycin-A** and **KCN** (inhibitors of complex III and II, respectively), but is inhibited by *s*alicyl*h*ydrox*am*ate (**SHAM**). Recent results show that the alternative oxidase is a membrane protein consisting of two equal subunits (each 36 kDa). From the amino acid sequence it can be predicted that each subunit forms two transmembrane helices. The two subunits together form a **di-iron oxo-center** (like in the fatty acid desaturase, Fig. 15.16), which catalyzes the oxidation of ubiquinone by oxygen.

Electron transport via the alternative oxidase can be understood as a short circuit. It occurs only when the mitochondrial ubiquinone pool is reduced to a very high degree. When metabolites in the mitochondria are in excess, the interplay of the alternative NADH dehydrogenase and the alternative oxidase leads to their elimination by oxidation without accompany-

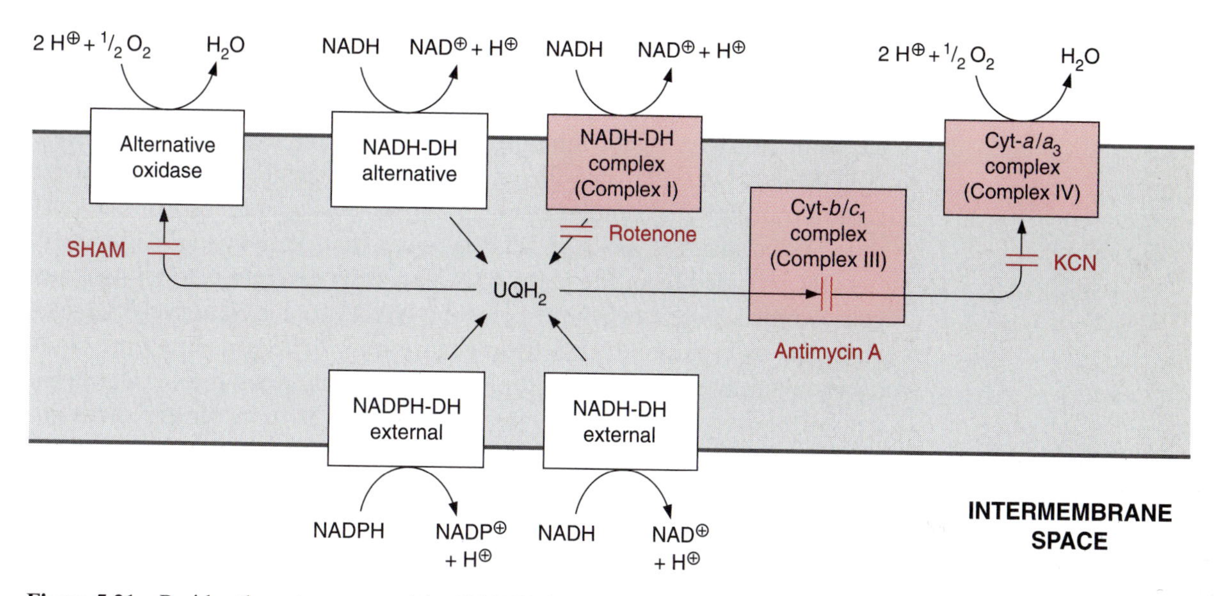

Figure 5.21 Besides the rotenone-sensitive NADH dehydrogenase (NADH DH), there are further dehydrogenases that transfer electrons to ubiquinone without accompanying proton transport. There also exists an alternative NADPH dehydrogenase that is directed to the matrix side, not shown here. An alternative oxidase enables the oxidation of ubihydroquinone (UQH_2). This pathway is insensitive to the inhibitors antimycin A and KCN, but it is inhibited by salicylhydroxamate (SHAM).

ing ATP synthesis, and the oxidation energy is dissipated as heat. The capacity of the alternative oxidase in the mitochondria from different plant tissues is variable and also depends on the developmental state. An especially high alternative oxidase activity has been found in the spadix of the voodoo lily *Sauromatum guttatum*, which uses the alternative oxidase to heat the spadix by which volatile amine compounds are emitted, which produce a nasty smell like carrion or dung. This strong stench attracts insects from far and wide. The formation of the alternative oxidase is synchronized in these spadices with the beginning of flowering.

NADH and NADPH from the cytosol can be oxidized by the respiratory chain of plant mitochondria

In contrast to mitochondria from animal tissues, plant mitochondria can also oxidize cytosolic NADH and in some cases cytosolic NADPH. Oxidation of this external NADH and NADPH proceeds via two specific dehydrogenases of the inner membrane, of which the substrate binding site is directed toward the intermembrane space. As in the case of succinate dehydrogenase, the electrons from external NADH and NADPH dehydrogenase are fed into the respiratory chain at the site of **ubiquinone**, and therefore this electron transport is not inhibited by rotenone. As oxidation of external NADH and NADPH (like the oxidation of succinate) does not involve a proton transport by complex I (Fig. 5.21), in the oxidation of external pyridine nucleotides the yield of ATP is lower than that in the oxidation of NADH provided from the matrix. Oxidation by external NADH dehydrogenase proceeds only when the cytosolic NAD^+ pool is excessively reduced. Also, the external NADH dehydrogenase may be regarded as part of an overflow mechanism, which comes into action only when the NADH in the cytosol is overreduced. As discussed in section 3.10, in certain situations photosynthesis may produce a surplus of reducing power, which is hazardous for a cell. The plant cell has the capacity to eliminate excessive reducing power by making use of the external NADH dehydrogenase, the alternative dehydrogenase for internal NADH from the matrix, and the alternative oxidase mentioned earlier.

5.8 Compartmentation of mitochondrial metabolism requires specific membrane translocators

The mitochondrial inner membrane is impermeable for metabolites. Specific translocators enable a specific transport of metabolites between the mitochondrial matrix and the cytosol in a counter-exchange mode (Fig. 5.22). The role of the ATP/ADP and the phosphate translocators (Fig. 5.20) has been discussed in section 5.6. Malate and succinate are transported into the mitochondria in counter-exchange for phosphate by a **dicarboxylate translocator**. This transport is inhibited by **butylmalonate**. α-Ketoglutarate, citrate, and oxaloacetate are transported in counter-exchange for malate. By these translocators, substrates can be fed into the citrate cycle. Glutamate is transported in counter-exchange for aspartate, and pyruvate in counter-exchange for OH^- ions. Although these translocators all occur in plant mitochondria, most of our present knowledge about them is based on studies with mitochondria from animal tissues. A comparison of the amino acid sequences known for the ATP/ADP, phosphate, citrate, and glutamate/aspartate

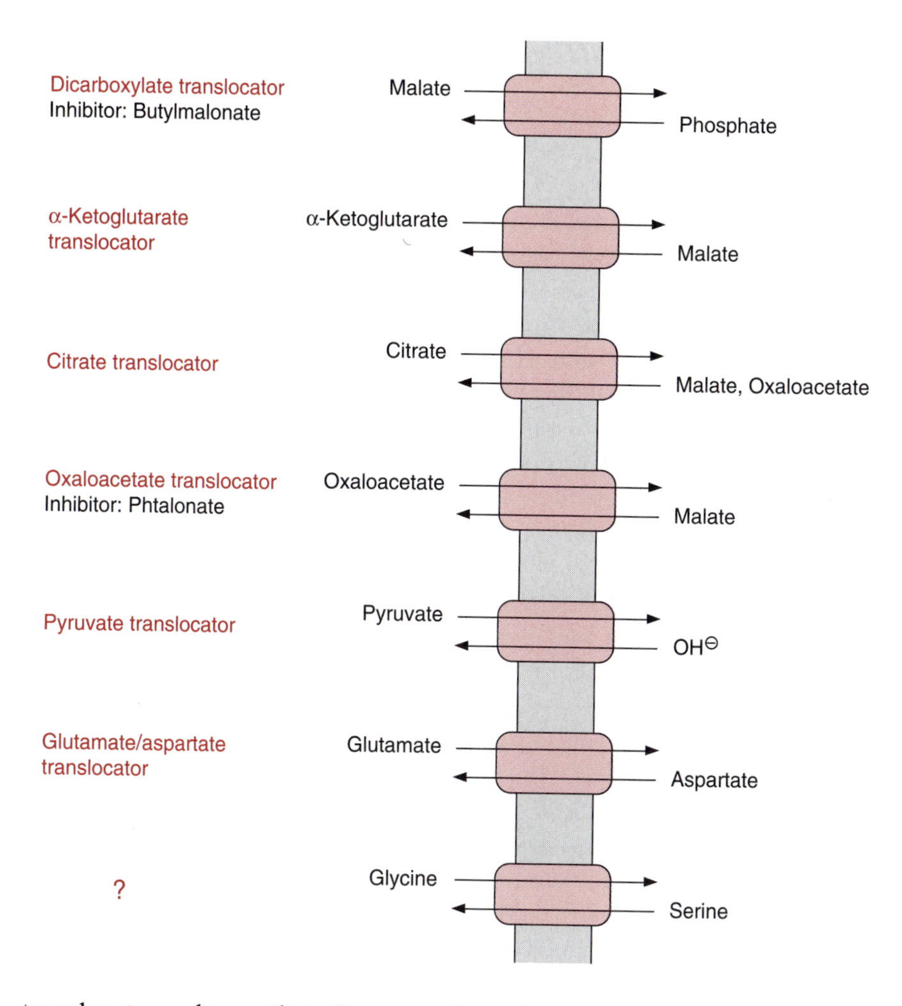

Figure 5.22 Important translocators of the inner mitochondrial membrane. The phosphate- and the ATP/ADP-translocator have been already shown in Figure 5.20.

translocators shows that these are homologous; the proteins of these translocators represent a family deriving from a common ancestor. As mentioned in section 1.9, all these translocators are composed of $2 \cdot 6$ transmembrane helices.

The malate-oxaloacetate translocator is a special feature of plant mitochondria and has an important function in the malate-oxaloacetate cycle described in section 7.3. It also transports citrate and is involved in providing the carbon skeletons for nitrate assimilation (Fig. 10.11). The oxaloacetate translocator and, to a lesser extent, the α-ketoglutarate translocator are inhibited by the dicarboxylate phthalonate. The transport of glycine and serine, involved in the photorespiratory pathway (section 7.1), has not yet been characterized. Although final proof is still lacking, it is to be expected that this transport is mediated by one or two mitochondrial translocators.

Further reading

Atkin, O. K., Millar, A. H., Gardeström, P., Day, D. A. Photosynthesis, carbohydrate metabolism and respiration in leaves of higher plants. In Photosynthesis: Physiology and Metabolism. R. C. Leegood, T. D. Sharkey, S. von Caemmerer (eds.), pp. 153–175. Kluwer Academic Publishers, Dordrecht, The Netherlands (2000).

Calhoun, M. W., Thomas, J., Gennis, R. B. The cytochrome oxidase superfamily of redox-driven proton pumps Trends Biol Sci 19, 325–330 (1994).

Eubel, H., Jaensch, L., Braun, H.-P. New insights into the respiratory chain of plant mitochondria. Supercomplexes and a unique composition of complex II. Plant Physiol 133, 274–286 (2003).

Iwata, S., Ostermeier, C., Ludwig, B., Michel, H. Structure at 2.8 Å resolution of cytochrome C oxidase from *Paracoccus denitrificans*. Nature 376, 660–669 (1995).

Kadenbach, B., Arnold, A. A second mechanism of respiratory control. FEBS Lett 447, 131–134 (1999).

Krämer, R., Palmieri, F. Metabolite carriers in mitochondria in L. Ernster (ed.). Molecular Mechanisms in Bioenergetics, pp. 359–384. Elsevier, Amsterdam (1992).

Krömer, S. Respiration during photosynthesis. Annu Rev Plant Physiol Plant Mol Biol 46, 45–70 (1995).

Laloi, M. Plant mitochondrial carriers: An overview. Cellular and Molecular Life Sciences. BirkhauserVerlag, Basel, 56, pp. 918–944 (1999).

Lambers, H., van der Plaas, L. H. W. Molecular, Biochemical and Physiological Aspects of Plant Respiration. SPB Academic Publishing, Den Haag, The Netherlands (1992).

McIntosh, L. Molecular biology of the alternative oxidase. Plant Physiol 105, 781–786 (1994).

Millenaar, F. F., Lambers, H. The alternative oxidase: *In vivo* regulation and function. Plant Biol 5, 2–15 (2003).

Møller I. M., Gardeström, P., Glimelius, K., Glaser, K. (Hrsg.) Plant Mitochondria from Gene to Function. Backhuys Publishers, Leiden, The Netherlands (1998).

Møller I. M. Plant mitochondria and oxidative stress: electron transport, NADPH turnover, and metabolism of reactive oxygen species. Annu Rev Plant Physiol Plant Mol Biol 52, 561–591 (2001).

Møller I. M. A new dawn for plant mitochondrial NAD(P)H dehydrogenases. Trends Plant Sci 7, 235–237 (2002).

Pebay-Peyroula, E., Dahout-Gonzalez, C., Kahn, R., Trezeguet, V., Lauquin, G. J.-M., Brandolin, G. Structure of mitochondrial ATP/ADP carrier in complex with carboxyatractyloside. Nature 426, 39–44 (2003).

Raghavendra, A. S., Padmasree, K. Beneficial interaction of mitochondrial metabolism with photosynthetic carbon assimilation. Trends Plant Sci 8, 546–553 (2003).

Saraste, Matti. Oxidative phosphorylation at the *fin de siècle.* Science 283, 1488–1493 (1999).

Vanlerberghe, G. C., Ordog, S. H. Alternative oxidase: Integrating carbon metabolism and electron transport in plant respiration. In: C. H. Foyer and G. Noctor (eds.). Advances in Photosynthesis: Photosynthetic Assimilation and Associated Carbon Metabolism, pp. 173–191. Kluwer Academic Publishers, Dordrecht, The Netherlands (2002).

Verkhovsky, M. I., Jassaitis, A., Verkhovskaya, M. L., Morgan, J. E., Wikström. M. Proton translocation by cytochrome *c* oxidase. Nature 400, 480–483 (1999).

Xia, D., Yu, C.-A., Kim, H., Xia, J.-Z., Kachurin, A. M., Zhang, L., Yu, L., Deisenhofer, J. Crystal structure of the cytochrome bc_1 complex from bovine heart mitochondria. Science 277, 60–66 (1997).

Yoshikawa, S., Shinzawa-Ito, K., Nakashima, R., Yaono, R., Yamashita, E., Inoue, N., Yao, M., Jie Fei, M., Peters Libeu, C., Mizushima, T., Yamaguchi, H., Tomizaki, T., Tsukihara, T. Redox-coupled crystal structural changes in bovine heart cytochrome c oxidase. Science 280, 1723–1729 (1998).

Zhang, Z., Huang, L., Shulmeister, V. M., Chi, Y.-I., Kim, K. K., Hung, L.-W., Crofts, A. R., Berry, E. A., Kim, S. H. Electron transfer by domain movement in cytochrome bc_1. Nature 392, 677–684 (1998).

6

The Calvin cycle catalyzes photosynthetic CO_2 assimilation

Chapters 3 and 4 showed how the electron transport chain and the ATP synthase of the thylakoid membrane use the energy from light to provide reducing equivalents in the form of NADPH, and chemical energy in the form of ATP. This chapter will describe how this NADPH and ATP are used for CO_2 assimilation.

6.1 CO_2 assimilation proceeds via the dark reaction of photosynthesis

It is relatively simple to isolate chloroplasts with intact envelope from leaves (see section 1.7). When these chloroplasts are added to an isotonic medium containing an osmoticum, a buffer, bicarbonate, and inorganic phosphate, and the light is switched on, one observes the generation of oxygen. Water is split to oxygen by the light reaction described in Chapter 3, and the resulting reducing equivalents are used for CO_2 assimilation (Fig. 6.1). With intact chloroplasts, there is no oxygen evolution in the absence of CO_2 or phosphate, demonstrating that the light reaction in the intact chloroplasts is coupled to CO_2 assimilation and the product of this assimilation contains phosphate. The main assimilation product of the chloroplasts is **dihydroxy-acetone phosphate**, a **triose phosphate**. Figure 6.2 shows that the synthesis of triose phosphate from CO_2 requires energy in the form of ATP and reducing equivalents in the form of NADPH, which have been provided by the **light reaction of photosynthesis**. The reaction chain for the formation of triose phosphate from CO_2, ATP, and NADPH was formerly called the **dark**

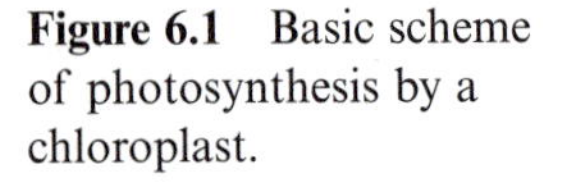
Figure 6.1 Basic scheme of photosynthesis by a chloroplast.

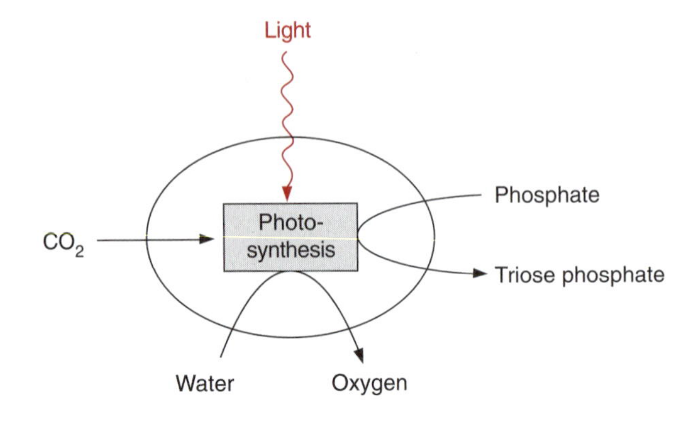

reaction of photosynthesis, as it requires no light *per se* and theoretically it should also be able to proceed in the dark. The fact is, however, that in leaves this reaction does not proceed during darkness, since some of the enzymes of the reaction chain, due to regulatory processes, are active only during illumination (section 6.6).

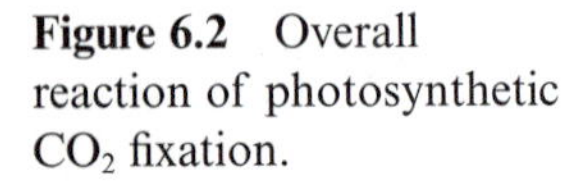
Figure 6.2 Overall reaction of photosynthetic CO_2 fixation.

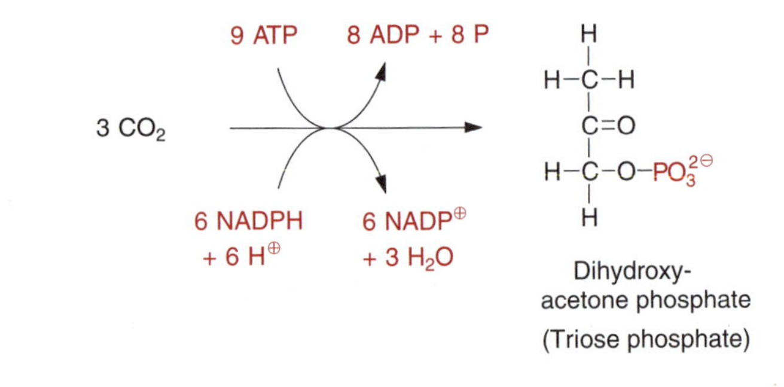

Between 1946 and 1953 Melvin Calvin and his collaborators, Andrew Benson and James Bassham, in Berkeley, California, resolved the mechanism of photosynthetic CO_2 assimilation. In 1961 Calvin was awarded the Nobel Prize in Chemistry for this fundamental discovery. A prerequisite for the elucidation of the CO_2 fixation pathway was the discovery in 1940 of the radioactive carbon isotope ^{14}C, which, as a by-product of nuclear reactors, was available in larger amounts in the United States after 1945. Calvin chose the green alga *Chlorella* for his investigations. He added radioactively labeled CO_2 to illuminated algal suspensions, killed the algae after a short incubation period by adding hot ethanol, and, using paper chromatography, analyzed the radioactively labeled products of CO_2 fixation. By successively shorten-

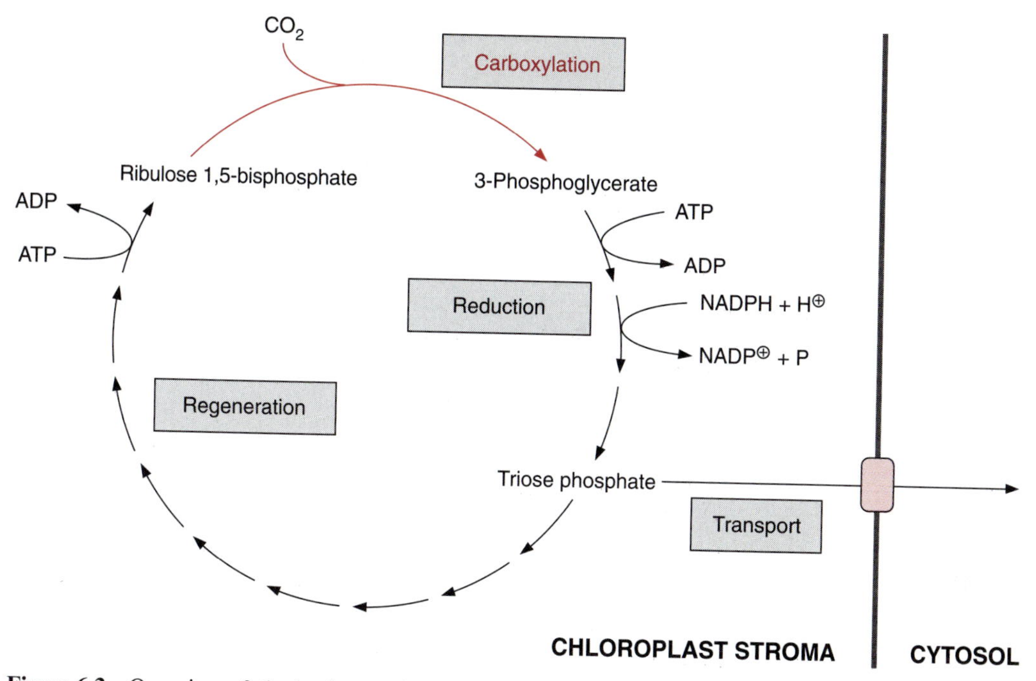

Figure 6.3 Overview of the basic reactions of the Calvin cycle without stoichiometries.

ing the incubation time, he was able to show that 3-phosphoglycerate was formed as the first stable product of CO_2 fixation. More detailed studies revealed that CO_2 fixation proceeds by a cyclic process, which has been named **Calvin cycle** after its discoverer. **Reductive pentose phosphate pathway** is another term that will be used in some sections of this book. This name derives from the fact that a reduction occurs and pentoses are formed in the cycle.

The Calvin cycle can be subdivided into three sections:

1. The **carboxylation** of the C_5 sugar ribulose 1,5-bisphosphate leading to the formation of two molecules 3-phosphoglycerate;
2. The **reduction** of the 3-phosphoglycerate to triose phosphate; and
3. The **regeneration** of the CO_2 acceptor ribulose 1,5-bisphosphate from triose phosphate (Fig. 6.3).

As a product of photosynthesis, triose phosphate is exported from the chloroplasts by specific transport. However, most of the triose phosphate remains in the chloroplasts to regenerate ribulose 1,5-bisphosphate. These reactions will be discussed in detail in the following sections.

6.2 Ribulose bisphosphate carboxylase catalyzes the fixation of CO_2

The key reaction for photosynthetic CO_2 assimilation is the binding of atmospheric CO_2 to the acceptor ribulose 1,5-bisphosphate (RuBP) to form two molecules of 3-phosphoglycerate. The reaction is very exergonic ($\Delta G^{o\prime}$ –35 kJ/mol) and therefore virtually irreversible. It is catalyzed by the enzyme ribulose bisphosphate carboxylase/oxygenase (abbreviated **RubisCO**), so called because the same enzyme also catalyzes a side-reaction in which the ribulose bisphosphate reacts with O_2 (Fig. 6.4).

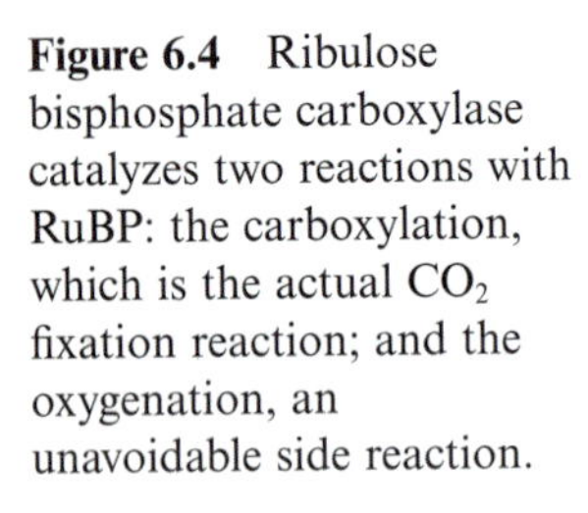

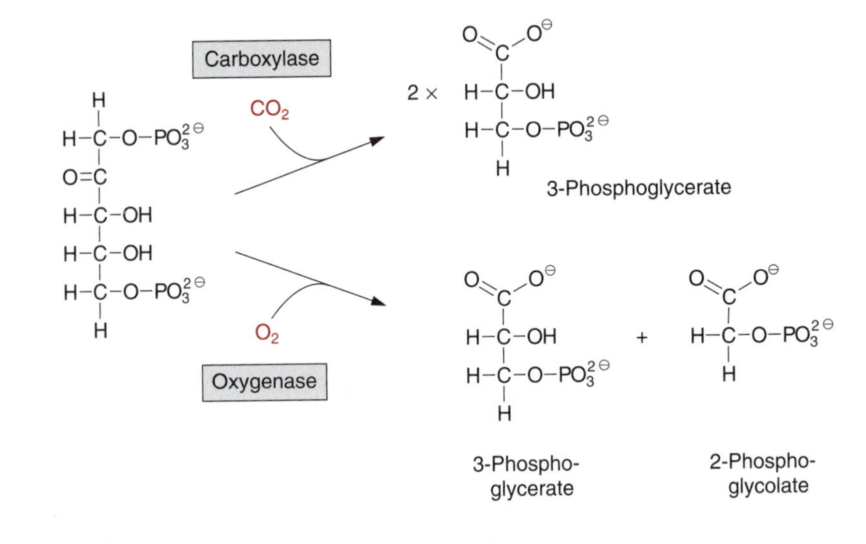

Figure 6.4 Ribulose bisphosphate carboxylase catalyzes two reactions with RuBP: the carboxylation, which is the actual CO_2 fixation reaction; and the oxygenation, an unavoidable side reaction.

Figure 6.5 shows the reaction sequence of the **carboxylase reaction**. Keto-enol isomerization of RuBP yields an enediol, which reacts with CO_2 to form the intermediate 2-carboxy 3-ketoarabinitol 1,5-bisphosphate, which is cleaved to two molecules of 3-phosphoglycerate. In the **oxygenase reaction,** an unavoidable by-reaction, probably O_2 reacts in a similar way as CO_2 with the enediol to form a peroxide as an intermediate. In a subsequent cleavage of the O_2 adduct, one atom of the O_2 molecule is released in the form of water and the other is incorporated into the carbonyl group of 2-phosphoglycolate (Fig. 6.6). The final products of the oxygenase reaction are 2-phosphoglycolate and 3-phosphoglycerate.

Ribulose bisphosphate-carboxylase/oxygenase **is the only enzyme that enables the fixation of atmospheric CO_2 for the formation of biomass.** This enzyme is therefore a prerequisite for the existence of the present life on

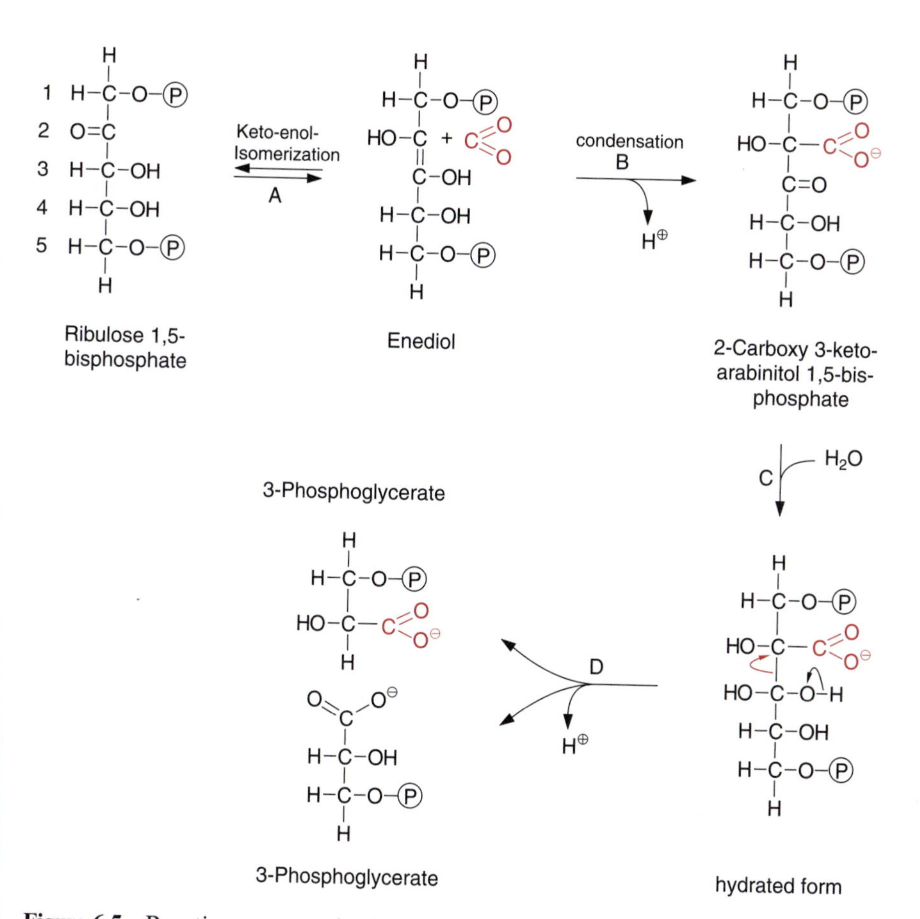

Figure 6.5 Reaction sequence in the carboxylation of RuBP by RubisCO. For the sake of simplicity, —PO_3^{2-} is symbolized as —P. An enediol, formed by keto-enol-isomerization of the carbonyl group of the RuBP (A), allows the nucleophilic reaction of CO_2 with the C-2 atom of RuBP by which 2-carboxy 3-ketoarabinitol 1,5-bisphosphate (B) is formed. After hydration (C), the bond between C-2 and C-3 is cleaved and two molecules of 3-phosphoglycerate are formed (D).

earth. In plants and cyanobacteria it consists of eight identical **large subunits** (depending on the species of a molecular mass of 51–58 kDa) and eight identical **small subunits** (molecular mass 12–18 kDa). **With its 16 subunits, RubisCO is one of the largest enzymes in nature**. In plants the genetic information for the large subunit is encoded in the plastid genome and for the small subunit in the nucleus. Each large subunit contains one catalytic center. The function of the small subunits is not yet fully understood. It has been proposed that the eight small subunits stabilize the complex of the eight large subunits. The small subunit apparently is not essential for the process of CO_2 fixation *per se*. RubisCO occurs in some phototrophic purple

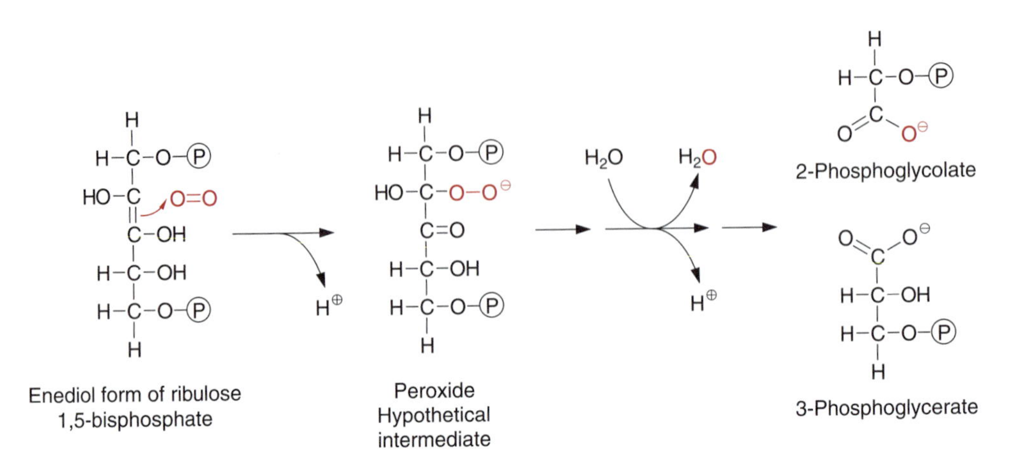

Figure 6.6 Part of the reaction sequence in the oxygenation of RuBP as catalyzed by RubisCO.

bacteria as a dimer only of large subunits, and the catalytic properties of the corresponding bacterial enzymes are not basically different from those in plants. The one difference in these bacterial enzymes, which consist of only two large subunits, is that the ratio between oxygenase and carboxylase activity is higher than in the plant enzyme, which is made up of eight large and eight small subunits.

The oxygenation of ribulose bisphosphate: a costly side-reaction

Although the CO_2 concentration required for half saturation of the enzyme (K_M [CO_2]) is much lower than that of O_2 (K_M [O_2]) (Table 6.1), the velocity of the oxygenase reaction is very high since the concentration of O_2 in air amounts to 21% and that of CO_2 to only 0.035%. Moreover, the CO_2 concentration in the gaseous space of the leaves can be considerably lower than the CO_2 concentration in the outside air. For these reasons, the ratio of oxygenation to carboxylation during photosynthesis of a leaf at 25°C is mostly in the range of **1 : 4** to **1 : 2**. *This means that every third to fifth ribulose 1,5-bisphosphate molecule is consumed in the side-reaction.* When the temperature rises, the CO_2/O_2 specificity (Table 6.1) decreases, and as a consequence, the ratio of oxygenation to carboxylation increases. On the other hand, a rise in the CO_2 concentration in the atmosphere lowers oxygenation, which in many cases leads to higher plant growth.

It will be shown in Chapter 7 that recycling of the by-product 2-phosphoglycolate, produced in very large amounts, is a very costly process

Table 6.1: Kinetic properties of ribulose bisphosphate carboxylase/oxygenase (RubisCO) at 25°C

Substrate concentrations at half saturation of the enzyme

$K_M[CO_2]$: 9 μmol/L*
$K_M[O_2]$: 535 μmol/L*
$K_M(RuBP)$: 28 μmol/L

Maximal turnover (related to one subunit)

$Kcat\ [CO_2]\ 3.3\,s^{-1}$
$Kcat\ [O_2]\ 2.4\,s^{-1}$

$$CO_2/O_2 \text{ specificity} = \left(\frac{Kcat[CO_2]}{K_M[CO_2]} \Big/ \frac{Kcat[O_2]}{K_M[O_2]} \right) = 82:$$

* For comparison:
In equilibrium with air (0.035% = 350 ppm CO_2, 21% O_2) the concentrations in water at 25°C amount to

CO_2 : 11 μmol/L
O_2 : 253 μmol/L

(Data from Woodrow and Berry, 1988).

for plants. This recycling process requires a metabolic chain with more than 10 enzymatic reactions distributed over three different organelles (chloroplasts, peroxisomes, and mitochondria), as well as very high energy consumption. Section 7.5 describes in detail that about a **third of the photons absorbed** during photosynthesis of a leaf are consumed to reverse the consequences of oxygenation.

Apparently evolution has not been successful in eliminating this costly side-reaction of ribulose bisphosphate carboxylase. Between cyanobacteria and higher plants the ratio in the activities of carboxylase and oxygenase of RubisCO is increased by a factor of less than two. It seems as if we are confronted here with a case in which the evolutionary refinement of a key process of life has reached a limit set by the chemistry of the reaction. The reason is probably that early evolution of the RubisCO occurred at a time when there was no oxygen in the atmosphere. A comparison of the RubisCO proteins from different organisms leads to the conclusion that RubisCO was already present about three and a half billion years ago, when the first chemolithotrophic bacteria evolved . When more than one and a half billion years later, due to photosynthesis, oxygen appeared in the atmosphere in higher concentrations, the complexity of the RubisCO protein probably made it too difficult to change the catalytic center to eliminate oxygenase activity. Experimental results support this notion. A large number of experiments, in which genetic engineering was employed to obtain site-specific mutations of the amino acid sequence in the region of the active center of

RubisCO, were unable to improve the ratio between the activities of carboxylation to oxygenation. The only chance of lowering oxygenation by molecular engineering may lie in simultaneously exchanging several amino acids in the catalytic binding site of RubisCO, which would be an extremely unlikely event in the process of evolution. Section 7.7 will show how plants make a virtue of necessity, and use the energy-consuming oxygenation to eliminate surplus NADPH and ATP produced by the light reaction.

Ribulose bisphosphate carboxylase/oxygenase: special features

The catalysis of the carboxylation of RuBP by RubisCO is very slow (Table 6.1): the turnover number for each subunit amounts to 3.3 s^{-1}. This means that at substrate saturation only about three molecules of CO_2 and RuBP are converted per second at a catalytic site of RubisCO. In comparison, the turnover numbers of dehydrogenases and carbonic anhydrase are on the order of 10^3 s^{-1} and 10^5 s^{-1}. Because of the extremely low turnover number of RubisCO, very large amounts of enzyme protein are required to catalyze the fluxes required for photosynthesis. RubisCO can amount to **50% of the total soluble proteins** in leaves. The wide distribution of plants makes RubisCO by far the **most abundant protein on earth**. The concentration of the catalytic large subunits in the chloroplast stroma is as high as 4–10 × 10^{-3} mol/L. A comparison of this value with the aqueous concentration of CO_2 in equilibrium with air (at 25°C about $11 \cdot 10^{-6}$ mol/L) shows the abnormal situation in which the concentration of an enzyme is up to 1,000 times higher than the concentration of the substrate CO_2 and is at a similar concentration as the substrate RuBP.

Activation of ribulose bisphosphate carboxylase/oxygenase

All the large subunits of RubisCO contain a lysine in position 201 of their sequence of about 470 amino acids. RubisCO is active only when the ε-amino group of this lysine reacts with CO_2 to form a **carbamate** (carbonic acid amide), to which an Mg^{++} ion is bound (Fig. 6.7). The activation is due to a change in the conformation of the protein of the large subunit. The active conformation is stabilized by the complex formation with Mg^{++}. This carbamylation is a prerequisite for the activity of all known RubisCO proteins. It should be noted that the CO_2 bound as carbamate is different from the CO_2 that is a substrate of the carboxylation reaction of RubisCO.

The activation of RubisCO requires ATP and is catalyzed by the enzyme **RubisCO activase**. The noncarbamylated, inactive form of RubisCO binds

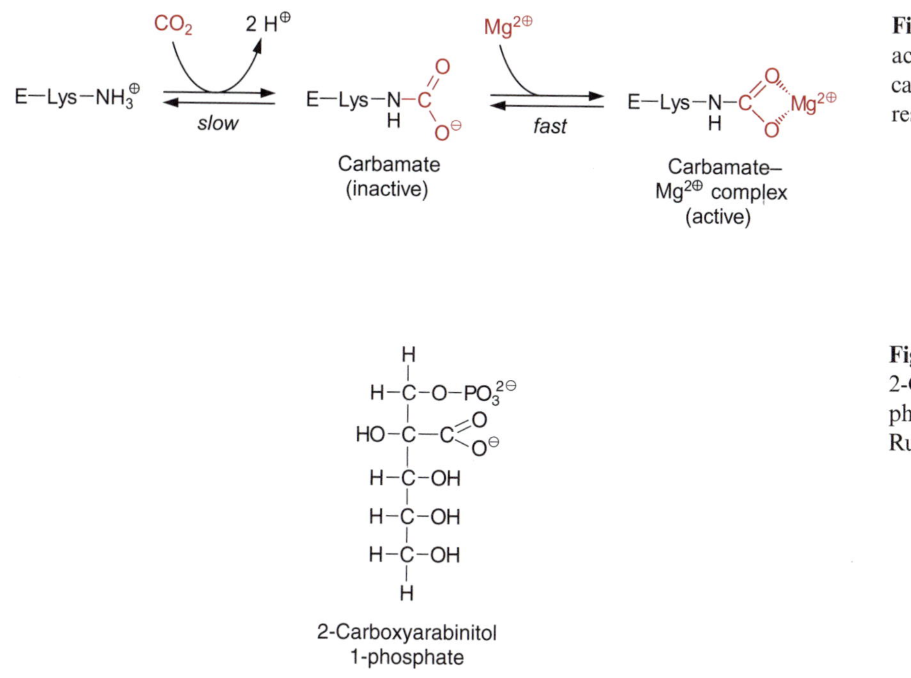

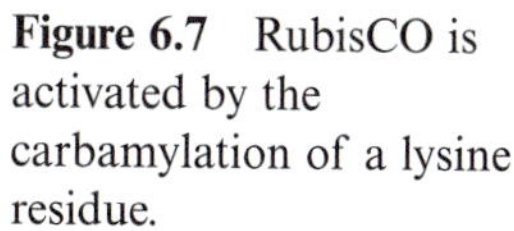

Figure 6.7 RubisCO is activated by the carbamylation of a lysine residue.

Figure 6.8 2-Carboxyarabinitol 1-phosphate, an inhibitor of RubisCO.

RuBP very tightly, resulting in a blockage of the enzyme. Upon the consumption of ATP, the activase releases the tightly bound RuBP and thus enables the carbamylation of the free enzyme. The regulation of RubisCO activase is discussed in section 6.6.

RubisCO is inhibited by several hexose phosphates and by 3-phosphoglycerate, which all bind to the active site instead of RuBP. A very strong inhibitor is **2-carboxyarabinitol 1-phosphate** (CA1P) (Fig. 6.8). This substance has a structure very similar to that of 2-carboxy 3-ketoarabinitol 1,5-bisphosphate (Fig. 6.5), which is an intermediate of the carboxylation reaction. CA1P has a 1,000-fold higher affinity than RuBP for the RuBP binding site of RubisCO. In a number of species, CA1P accumulates in the leaves during the night, blocking a large number of the binding sites of RubisCO and thus inactivating the enzyme. During the day, CA1P is released by RubisCO activase and is degraded by a specific phosphatase, which hydrolyzes the phosphate residue from CA1P and thus eliminates the effect of the RubisCO inhibitor. CA1P is synthesized from fructose1.6-bisphosphate with the hexosephosphates hamamelose bisphosphate and hamamelose monophosphate as intermediary substances. Since CA1P is not formed in all plants, its role in the regulation of RubisCO is still a matter of debate.

6.3 The reduction of 3-phosphoglycerate yields triose phosphate

For the synthesis of dihydroxyacetone phosphate, the carboxylation product 3-phosphoglycerate is phosphorylated to 1,3-bisphosphoglycerate by the enzyme **phosphoglycerate kinase**. In this reaction, with the consumption of ATP, a mixed anhydride is formed between the new phosphate residue and the carboxyl group (Fig. 6.9). As the free energy for the hydrolysis of this anhydride is similarly high to that of the phosphate anhydride in ATP, the phosphoglycerate kinase reaction is reversible. An isoenzyme of the chloroplast phosphoglycerate kinase is also involved in the glycolytic pathway proceeding in the cytosol, where it catalyzes the formation of ATP from ADP and 1,3-bisphosphoglycerate.

The reduction of 1,3-bisphosphoglycerate to D-glyceraldehyde 3-phosphate is catalyzed by the enzyme **glyceraldehyde phosphate dehydrogenase** (Fig. 6.9). A thioester is formed as an intermediate of this reaction by the exchange of the phosphate residue at the carboxyl group for an SH-group of a cysteine residue in the active center of the enzyme (Fig. 6.10). The free energy for the hydrolysis of the so formed thioester is similarly high to that of the anhydride ("energy-rich bond"). When a thioester is reduced, a thio-semiacetal is formed which has a low free energy of hydrolysis.

Through the catalysis of phosphoglycerate kinase and glyceraldehyde phosphate dehydrogenase, the large difference in redox potentials between

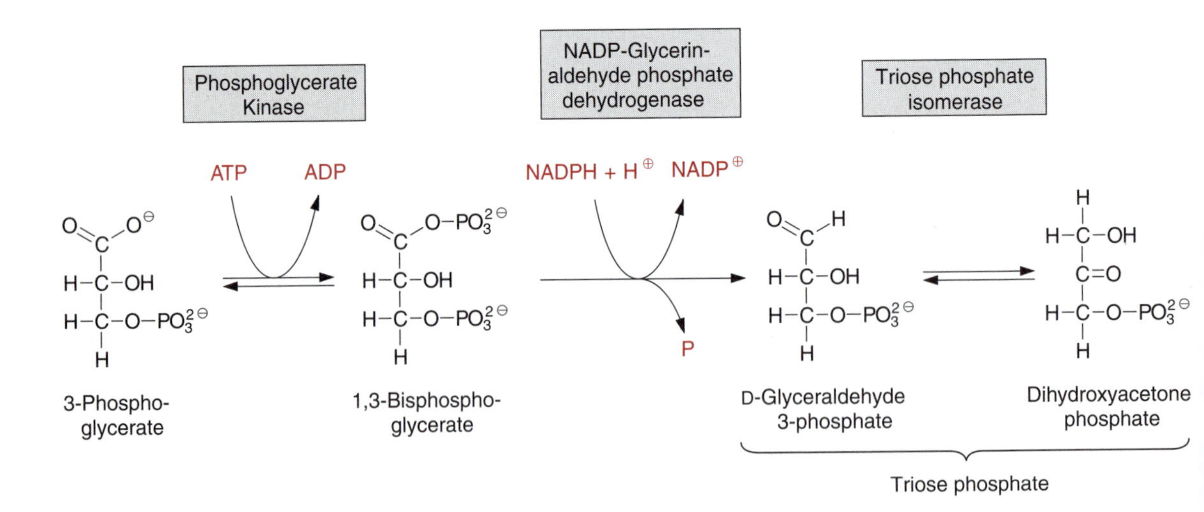

Figure 6.9 Conversion of 3-phosphoglycerate to triose phosphate.

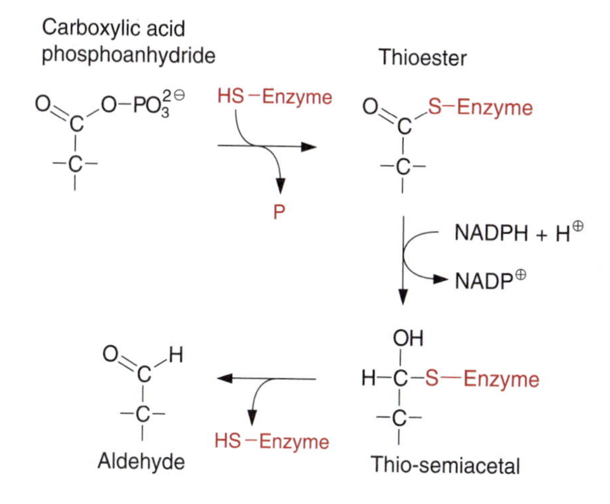

Figure 6.10 Reaction sequence catalyzed by glyceraldehyde phosphate dehydrogenase. H-S-enzyme symbolizes the sulfhydryl group of a cysteine residue in the active center of the enzyme.

the aldehyde and the carboxylate in the reduction of 3-phosphoglycerate to glyceraldehyde phosphate is overcome by the consumption of ATP. It is therefore a reversible reaction. A glyceraldehyde phosphate dehydrogenase in the cytosol catalyzes the conversion of glyceraldehyde phosphate to 1,3-bisphosphoglycerate. In contrast to the cytosolic enzyme, which mainly catalyzes the oxidation of glyceraldehyde phosphate with NAD^+ as hydrogen acceptor, the chloroplast enzyme uses NADPH as a hydrogen donor.

This is an example of the different roles that the NADH/NAD^+ and NADPH/$NADP^+$ systems play in the metabolism of eukaryotic cells. Whereas the NADH system is specialized in collecting reducing equivalents to be oxidized for the synthesis of ATP, the NADPH system mainly gathers reducing equivalents to be donated to synthetic processes. Figuratively speaking, the NADH system has been compared with a **hydrogen low pressure line** through which reducing equivalents are pumped off for oxidation to generate energy, and the NADPH system as a **hydrogen high pressure line** through which reducing equivalents are pressed into synthesis processes. Usually the reduced/oxidized ratio is about 100 times higher for the NADPH system than for the NADH system. The relatively high degree of reduction of the NADPH system in chloroplasts (about 50–60% reduced) allows the very efficient reduction of 1,3-bisphosphoglycerate to glyceraldehyde-3-phosphate.

Triose phosphate isomerase catalyzes the isomerization of glyceraldehyde phosphate to dihydroxyacetone phosphate. This conversion of an aldose to a ketose proceeds via an 1,2-enediol as intermediate and is basically similar to the reaction catalyzed by ribose phosphate isomerase. The equilibrium of the reaction lies toward the ketone. Triose phosphate as a collective term, therefore, consists of about 96% of dihydroxyacetone phosphate and only 4% glyceraldehyde phosphate.

6.4 Ribulose bisphosphate is regenerated from triose phosphate

From the fixation of three molecules of CO_2 in the Calvin cycle, six molecules of phosphoglycerate are formed and are converted to six molecules of triose phosphate (Fig. 6.11). Of these, only one molecule of triose phosphate is the actual gain, which is provided to the cell for various biosynthetic processes. The remaining five triose phosphates are needed to regenerate three molecules of ribulose bisphosphate so that the Calvin cycle can continue. Figure 6.12 shows the metabolic pathway for the conversion of the five triose phosphates (white boxes) to three pentose phosphates (red boxes).

The two trioses dihydroxyacetone phosphate and glyceraldehyde phosphate are condensed in a reversible reaction to fructose 1,6-bisphosphate, as catalyzed by the enzyme **aldolase** (Fig. 6.13). Figure 6.14 shows the reaction mechanism. As an intermediate of this reaction, a protonated Schiff base is formed between a lysine residue in the active center of the enzyme and the keto group of the dihydroxyacetone phosphate. This Schiff base enhances the release of a proton from the C-3 position and enables a nucleophilic reaction with the C-atom of the aldehyde group of glyceraldehyde phosphate. Fructose 1,6-bisphosphate is hydrolyzed in an irreversible reaction to fructose 6-phosphate, as catalyzed by **fructose 1,6 bisphosphatase** (Fig. 6.15).

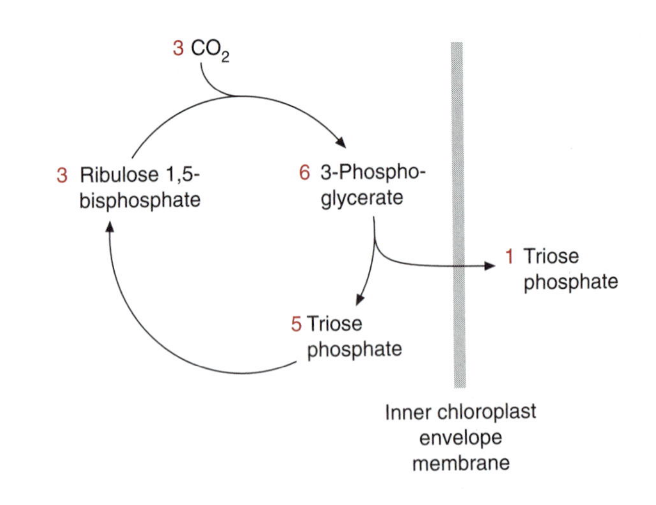

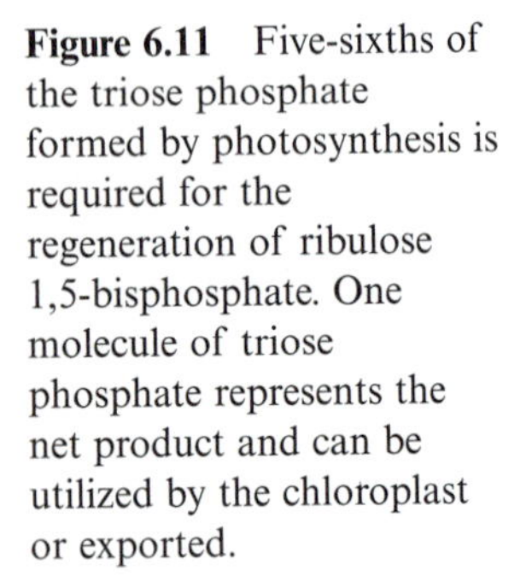

Figure 6.11 Five-sixths of the triose phosphate formed by photosynthesis is required for the regeneration of ribulose 1,5-bisphosphate. One molecule of triose phosphate represents the net product and can be utilized by the chloroplast or exported.

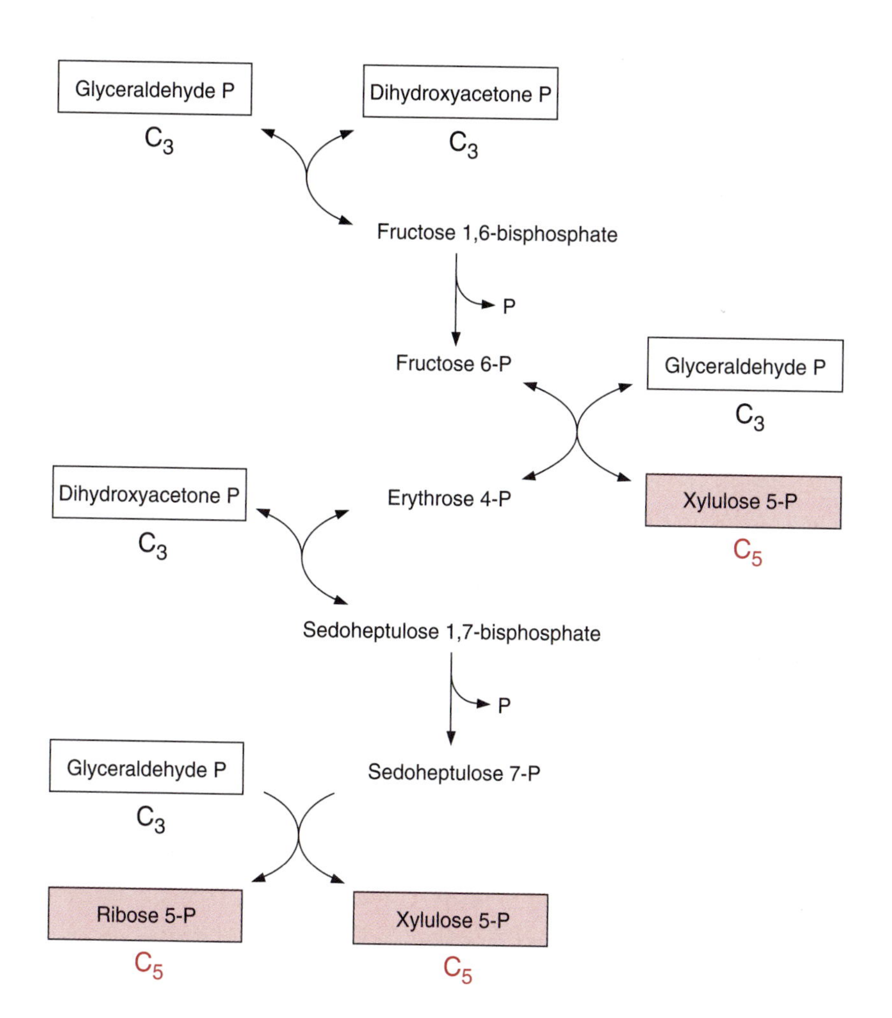

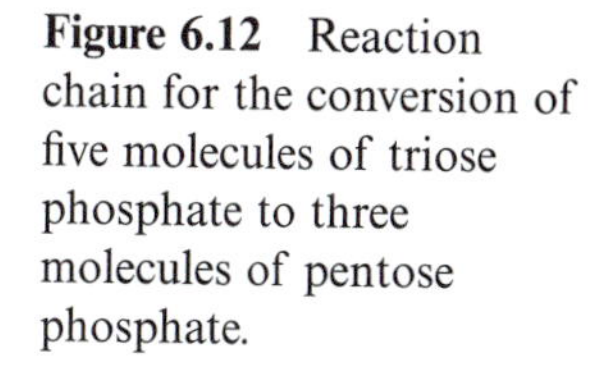

Figure 6.12 Reaction chain for the conversion of five molecules of triose phosphate to three molecules of pentose phosphate.

The enzyme **transketolase** transfers a carbohydrate residue with two carbon atoms from fructose 6-phosphate to glyceraldehyde 3-phosphate yielding, xylulose 5-phosphate, and erythrose 4-phosphate in a reversible reaction (Fig. 6.16). In this reaction, thiamine pyrophosphate (Fig. 5.5), already discussed as a reaction partner of pyruvate oxidation (section 5.3), is involved as the prosthetic group (Fig. 6.17).

Once more **aldolase** (Fig. 6.13) catalyzes a condensation, this time of erythrose 4-phosphate with dihydroxyacetone phosphate to form sedoheptulose 1,7-bisphosphate. Subsequently, the enzyme **sedoheptulose 1,7-bisphosphatase** catalyzes the irreversible hydrolysis of sedoheptulose 1,7-bisphosphate. This reaction is similar to the hydrolysis of fructose 1,6-bisphosphate, although the two reactions are catalyzed by different enzymes. Again, a carbohydrate residue with two C-atoms is transferred

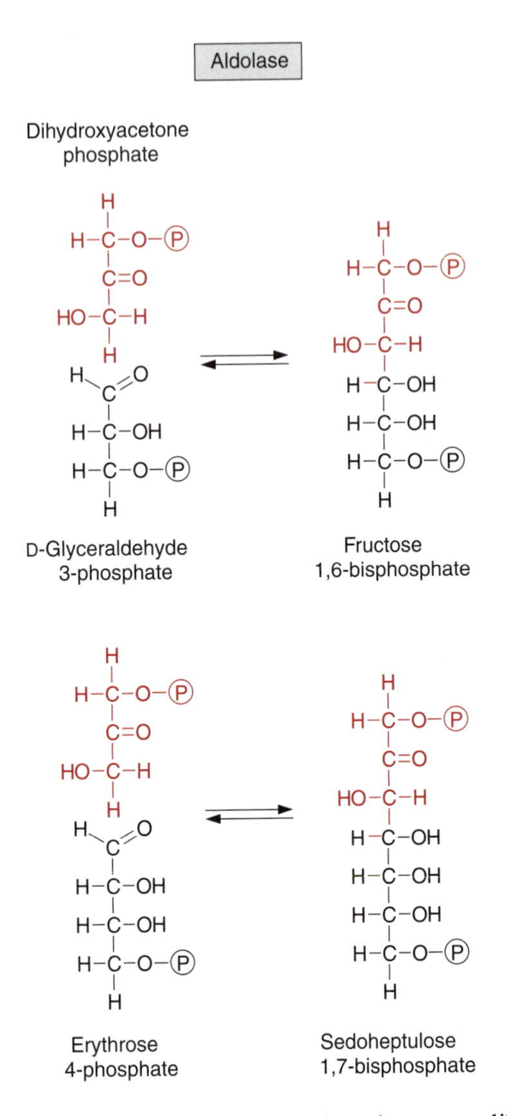

Figure 6.13 Aldolase catalyzes the condensation of dihydroxyacetone phosphate with the aldoses glyceraldehyde 3-phosphate or erythrose 4-phosphate.

by **transketolase** from sedoheptulose 7-phosphate to dihydroxyacetone phosphate and this forms ribose 5-phosphate and xylulose 5-phosphate (Fig. 6.16).

The three pentose phosphates formed are now converted to ribulose 5-phosphate (Fig. 6.18). The conversion of xylulose 5-phosphate is catalyzed by **ribulose phosphate epimerase**; this reaction proceeds via a keto-enol isomerization with a 2,3-enediol as the intermediary product. The conversion of the aldose ribose 5-phosphate to the ketose ribulose 5-phosphate is catalyzed by **ribose phosphate isomerase**, again via an enediol as intermediate, although in the 1,2-position. The three molecules of ribulose 5-phosphate formed in this way are converted to the CO_2 acceptor ribulose

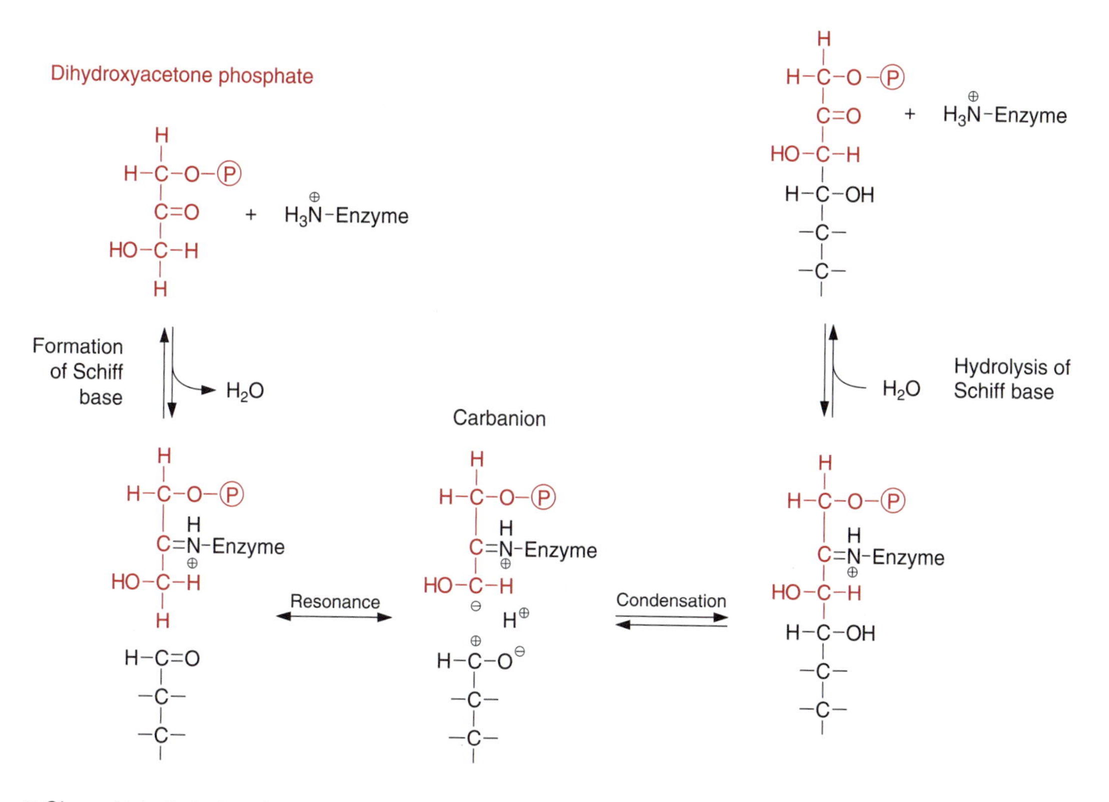

Figure 6.14 Pathway of the aldolase reaction. Dihydroxyacetone phosphate forms a Schiff base with the terminal amino group of a lysine residue of the enzyme protein. The positive charge at the nitrogen atom favors the release of a proton at C-3, and thus a carbanion is formed. In one mesomeric form of the glyceraldehyde phosphate, the C-atom of the aldehyde group is positively charged. This enables condensation between this C-atom and the negatively charged C-3 of the dihydroxyacetone phosphate. After condensation, the Schiff base is cleaved again and fructose 1,6-bisphosphate is released. Sedoheptulose 1,7-bisphosphate is formed by the same enzyme from reaction with erythrose 4-phosphate. The aldolase reaction is reversible.

1,5-bisphosphate upon consumption of ATP by **ribulose-phosphate kinase** (Fig. 6.19). This kinase reaction is irreversible, since a phosphate is converted from the "energy-rich" anhydride in the ATP to a phosphate ester with a low free energy of hydrolysis.

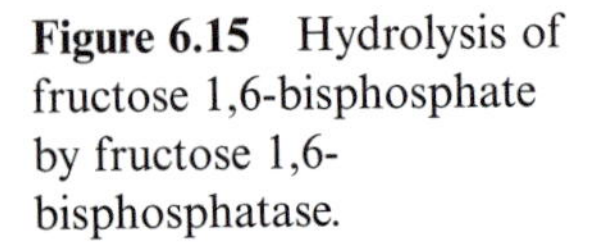
Figure 6.15 Hydrolysis of fructose 1,6-bisphosphate by fructose 1,6-bisphosphatase.

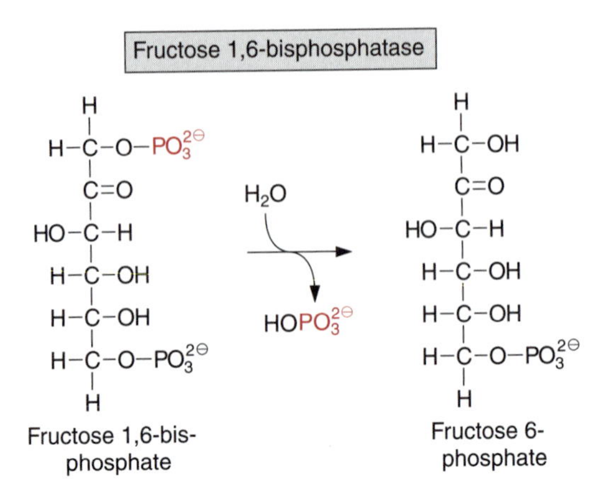

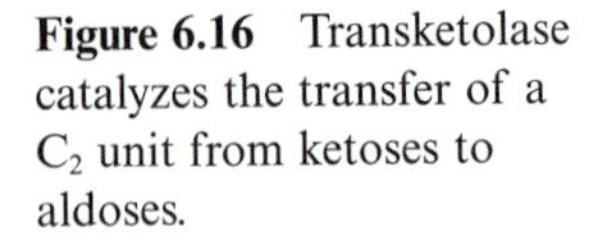
Figure 6.16 Transketolase catalyzes the transfer of a C_2 unit from ketoses to aldoses.

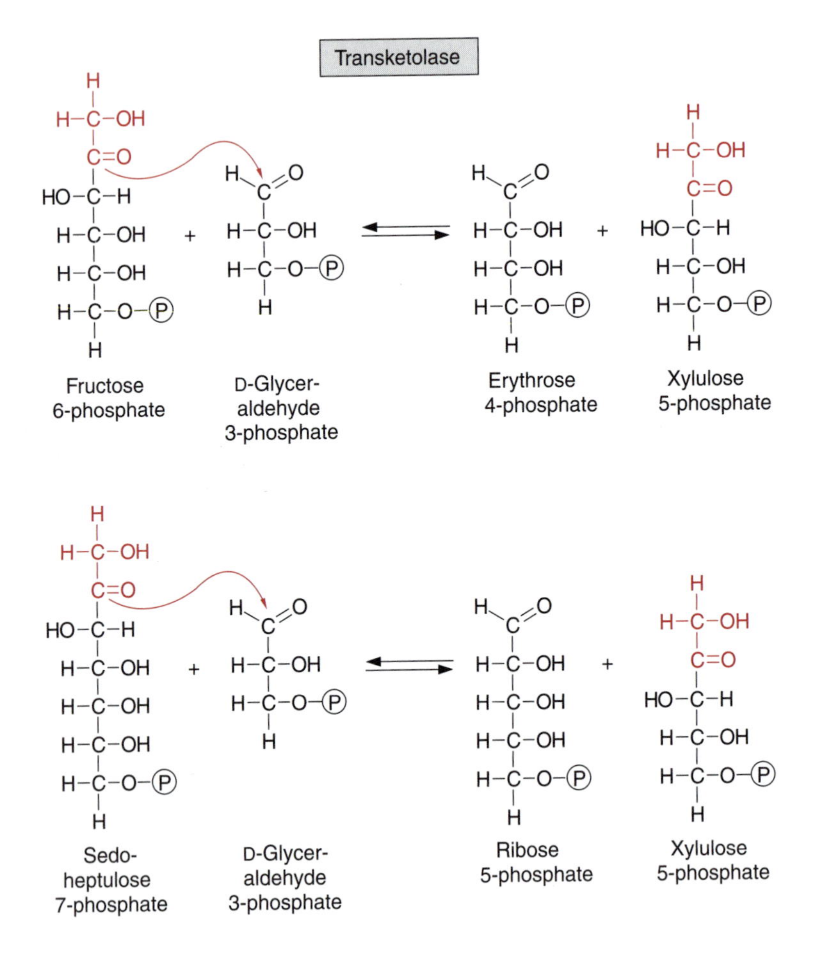

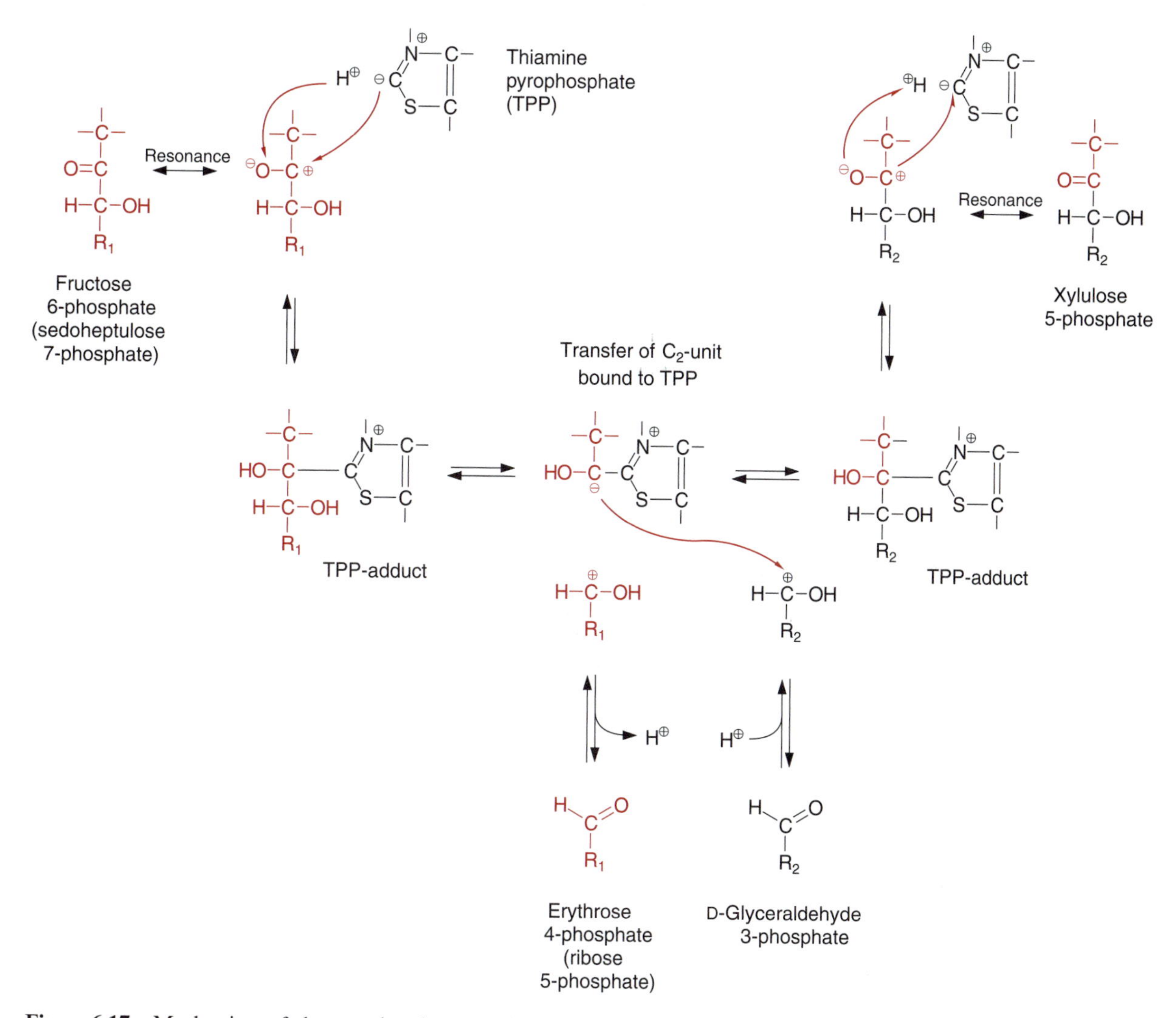

Figure 6.17 Mechanism of the transketolase reaction. The enzyme contains as a prosthetic group thiamine pyrophosphate with a thiazole ring as the reactive component. The positive charge of the N-atom in this ring enhances the release of a proton at the neighboring C-atom, resulting in a negatively charged C-atom (carbanion), to which the partially positively charged C-atom of the keto group of the substrate is bound. The positively charged N-atom of the thiazole favors the cleavage of the carbon chain, and the carbon atom in position 2 becomes a carbanion. The reaction mechanism is basically the same as that of the aldolase reaction in Figure 6.14. The C_2 carbohydrate moiety bound to the thiazole is transferred to the C-1 position of the glyceraldehyde 3-phosphate.

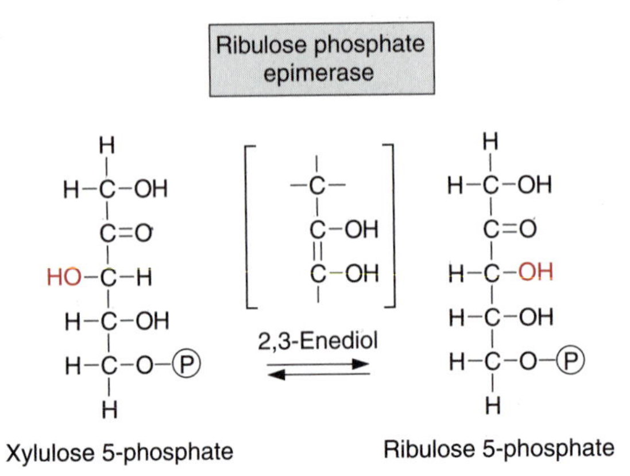

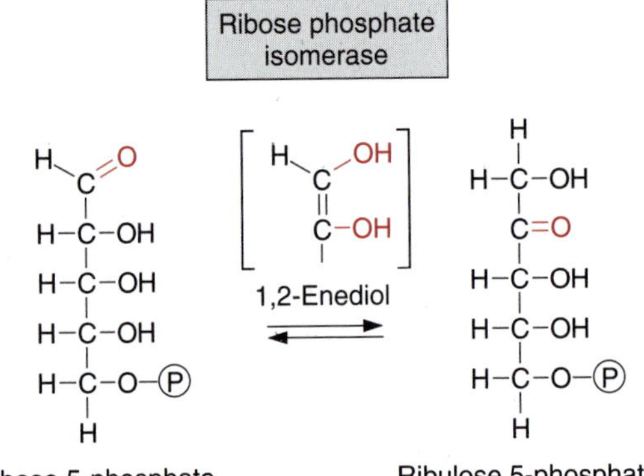

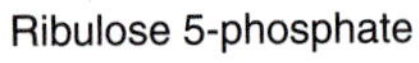

Figure 6.18 The conversion of xylulose 5-phosphate and ribose 5-phosphate to ribulose 5-phosphate. In both cases a *cis*-enediol is formed as the intermediate.

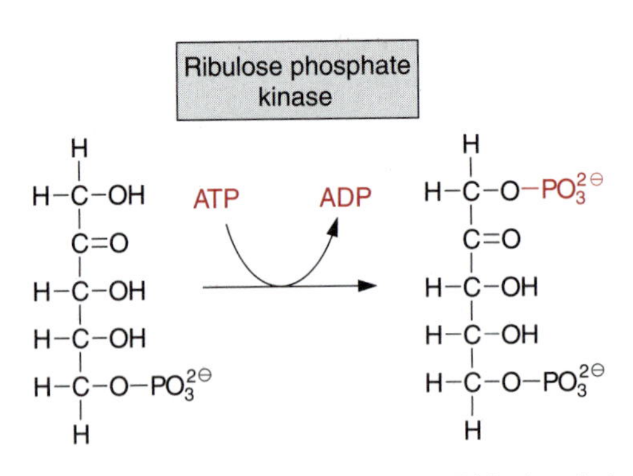

Figure 6.19 Ribulose phosphate kinase catalyzes the irreversible formation of ribulose 1,5-bisphosphate.

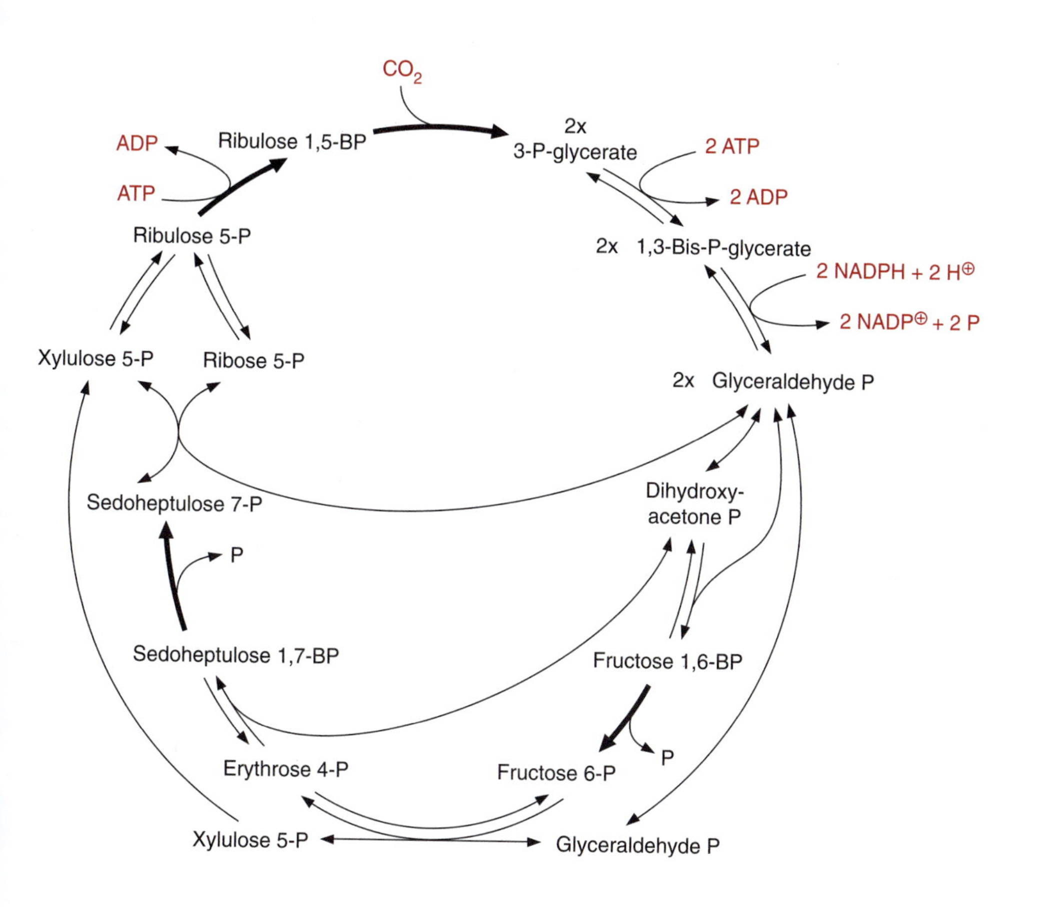

Figure 6.20 The Calvin cycle (reductive pentose phosphate pathway). P, phosphate; BP, bisphosphate.

The scheme in Figure 6.20 gives a summary of the various reactions of the Calvin cycle. There are four irreversible steps in the cycle, marked by bold arrows: carboxylation, hydrolysis of fructose bisphosphate, hydrolysis of sedoheptulose bisphosphate, and phosphorylation of ribulose 5-phosphate. Fixation of one molecule of CO_2 requires in total two molecules of NADPH and three molecules of ATP.

6.5 Besides the reductive pentose phosphate pathway there is also an oxidative pentose phosphate pathway

Besides the reductive pentose phosphate pathway discussed in the preceding section, the chloroplasts also contain the enzymes of an oxidative pentose phosphate pathway. This pathway, which occurs in both the plant and

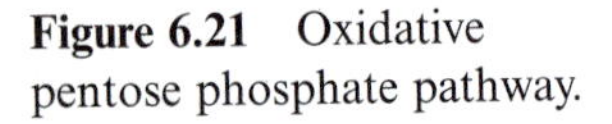
Figure 6.21 Oxidative pentose phosphate pathway.

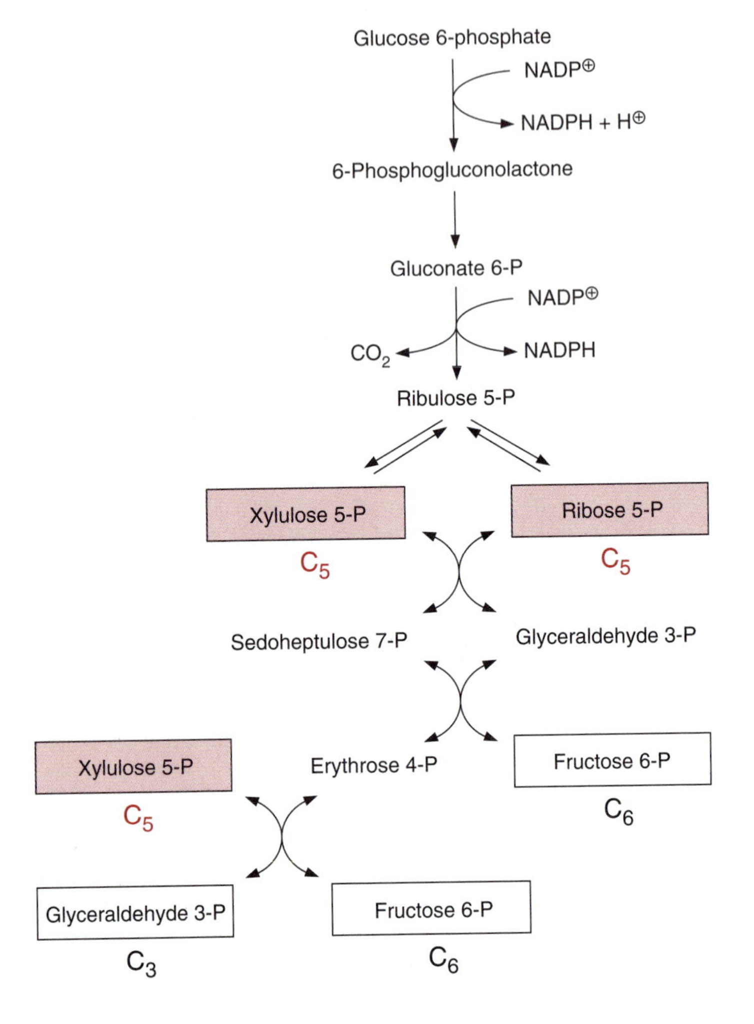

animal kingdoms, oxidizes a **hexose phosphate** to a **pentose phosphate** with the release of one molecule of CO_2. This pathway provides **NADPH** as "high pressure hydrogen" for biosynthetic processes. Figure 6.21 shows the pathway. Glucose 6-phosphate is first oxidized by **glucose 6-phosphate dehydrogenase** to 6-phosphogluconolactone (Fig. 6.22). This reaction is highly exergonic and therefore is not reversible. 6-Phosphogluconolactone, an internal ester, is hydrolyzed by **lactonase**. The gluconate 6-phosphate thus formed is oxidized by the enzyme **gluconate 6-phosphate dehydrogenase** to ribulose 5-phosphate. In this reaction, CO_2 is released and NADPH is formed.

In the oxidative pathway, xylulose 5-phosphate is formed from ribulose 5-phosphate by **ribulose phosphate epimerase** and from ribose 5-phosphate

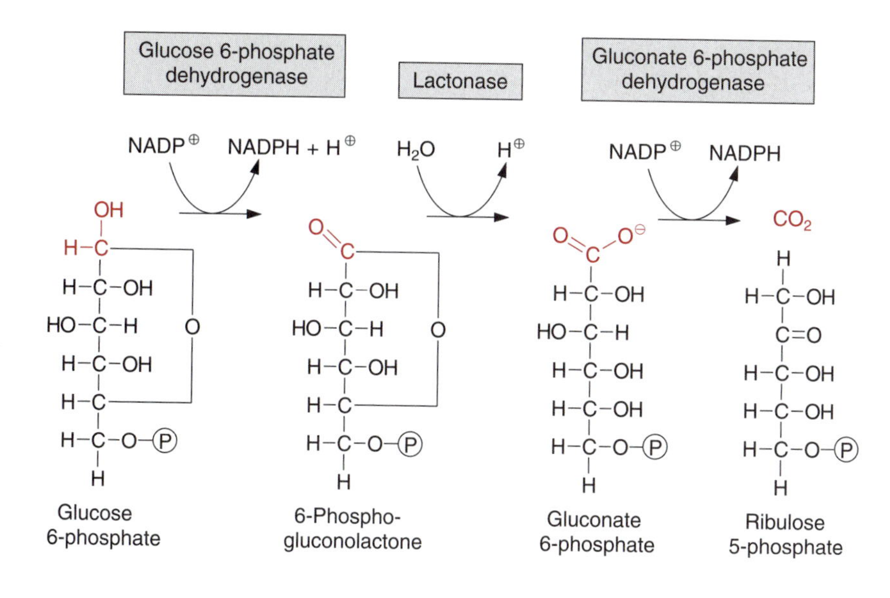

Figure 6.22 The two oxidation reactions of the pentose phosphate pathway.

by **ribose-phosphate isomerase**. These two products are then converted by **transketolase** to sedoheptulose 7-phosphate and glyceraldehyde 3-phosphate. This reaction sequence is a reversal of the reductive pentose phosphate pathway. The further reaction sequence is a special feature of the oxidative pathway: **Transaldolase** transfers a nonphosphorylated C_3 residue from sedoheptulose 7-phosphate to glyceraldehyde 3-phosphate, forming fructose 6-phosphate and erythrose 4-phosphate (Fig. 6.23). The reaction mechanism is basically the same as in the aldolase reaction (Fig. 6.13), with the only difference being that after the cleavage of the C-C bond, the remaining C_3 residue continues to be bound to the enzyme via a Schiff base, until it is transferred. Erythrose 4-phosphate reacts with another xylulose 5-phosphate via **transketolase** to form glyceraldehyde 3-phosphate and fructose 6-phosphate. In this way two hexose phosphates and one triose phosphate are formed from three pentose phosphates:

$$3\,C_5\text{-P} \rightleftarrows 2\,C_6\text{-P} + 1\,C_3\text{-P}$$

This reaction chain is reversible. It allows the cell to provide ribose 5-phosphate for nucleotide biosynthesis even when no NADPH is required.

In the oxidative pathway, two molecules of NADPH are gained from the oxidation of glucose 6-phosphate with the release of one molecule of CO_2, whereas in the reductive pathway the fixation of one molecule of CO_2 requires not only two molecules of NADPH but also three molecules of ATP (Fig. 6.24). This expenditure of energy makes it possible for the reductive

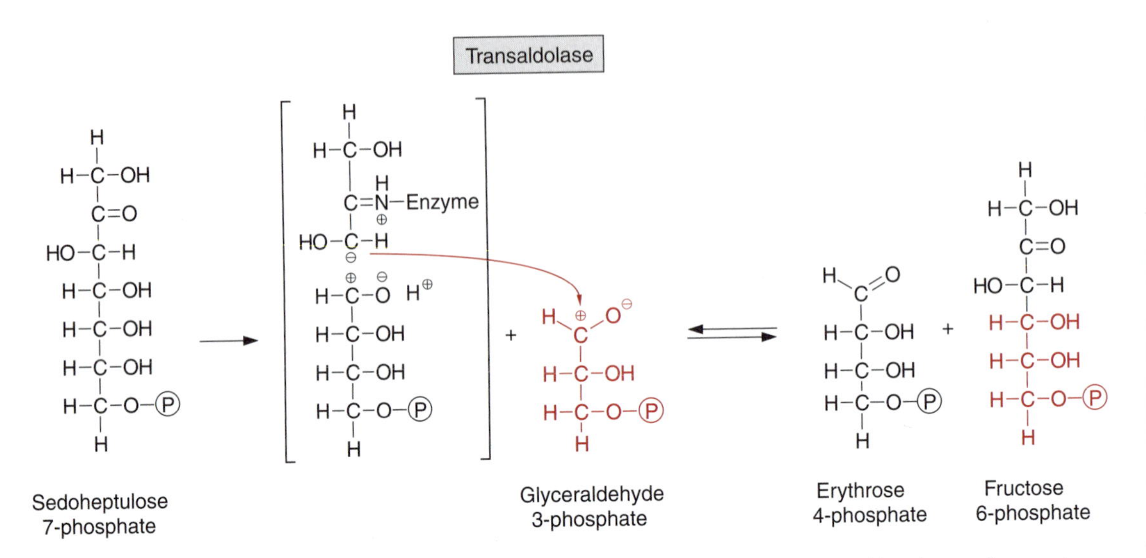

Figure 6.23 Transaldolase catalyzes the transfer of the C_3 residue from a ketone to an aldehyde. The reaction is reversible. The reaction mechanism is the same as with aldolase, with the difference that after the cleavage of the C-C bond the C_3-residue remains bound to the enzyme and is released only after the transfer to glyceraldehyde phosphate.

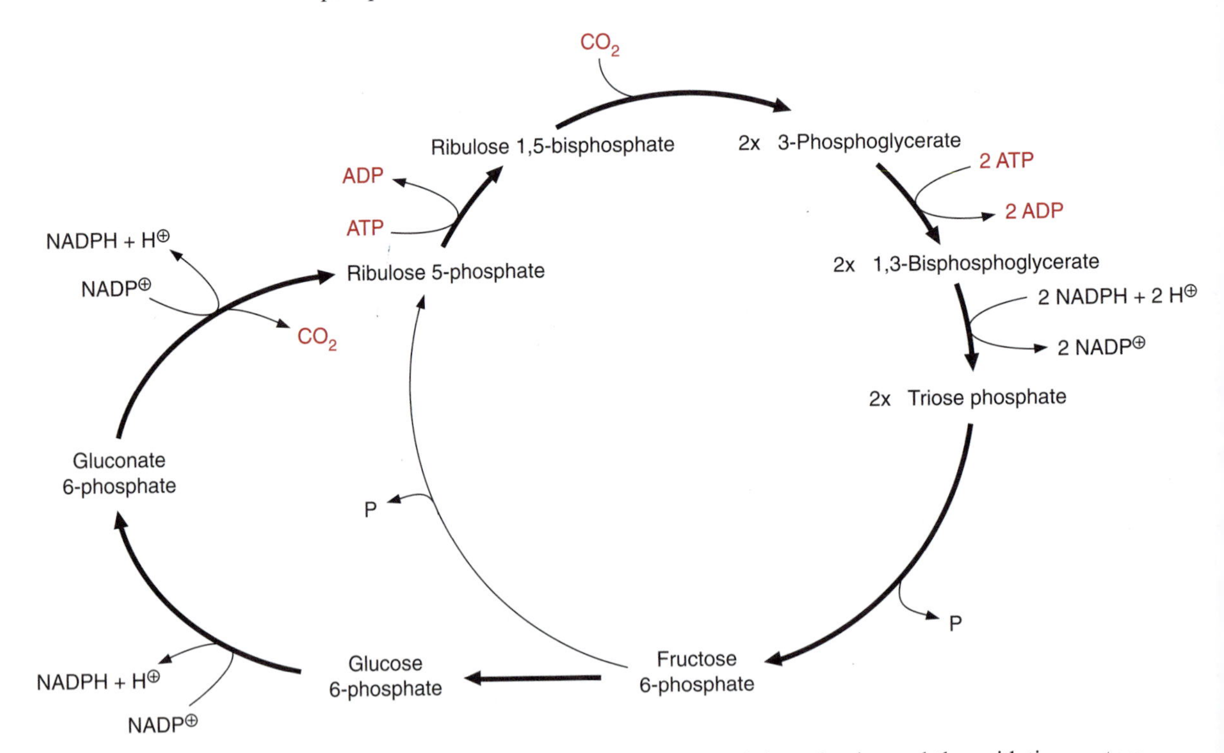

Figure 6.24 A simultaneous operation of the reductive and the oxidative pentose phosphate pathway would result in a futile, ATP wasting cycle.

pentose phosphate pathway to proceed in the opposite direction to the oxidative pathway with a very high flux rate.

6.6 Reductive and oxidative pentose phosphate pathways are regulated

The enzymes of the reductive as well as the oxidative pentose phosphate pathways are located in the chloroplast stroma (Fig. 6.24). A simultaneous operation of both metabolic pathways, in which one molecule of CO_2 is reduced to a carbohydrate residue at the expense of three ATP and two NADPH, which would then be reoxidized by the oxidative pathway to CO_2, yielding two molecules of NADPH, would represent a futile cycle in which three molecules of ATP are wasted in each turn. This is prevented by metabolic regulation, which ensures that key enzymes of the reductive pentose phosphate pathway are active only during illumination and switched off in darkness, whereas the key enzyme of the oxidative pentose phosphate pathway is active only in the dark.

Reduced thioredoxins transmit the signal for "illumination" to enzyme proteins

An important signal for the state "illumination" is provided by photosynthetic electron transport as reducing equivalents in the form of reduced thioredoxin (Fig. 6.25). These reducing equivalents are transferred from ferredoxin to thioredoxin by the enzyme **ferredoxin-thioredoxin reductase**, an iron sulfur protein of the 4Fe4S type.

Thioredoxins form a family of small proteins, consisting of about 100 amino acids, which contain as a reactive group the sequence **Cys-Gly-Pro-Cys**, located at the periphery of the protein. Due to the neighboring cysteine groups, the thioredoxin can be present in two redox states: the reduced thioredoxin with **two SH-groups** and the oxidized thioredoxin in which the two cysteines are linked by a **disulfide (S-S) bond**.

Thioredoxins are found in all living organisms from the archaebacteria to plants and animals. They function as **protein disulfide oxido-reductases**, in reducing disulfide bonds in target proteins to the SH form and reoxidizing them again to the S-S form. Despite their small size, they have a relatively high substrate specificity. Thioredoxins participate as redox carriers in the reduction of high as well as low molecular substances [e.g., the reduction of

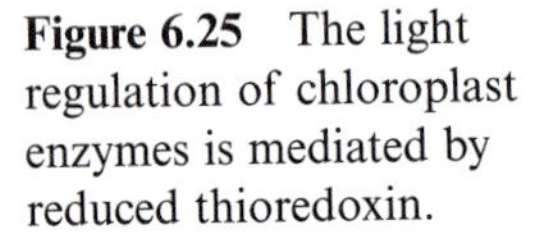

Figure 6.25 The light regulation of chloroplast enzymes is mediated by reduced thioredoxin.

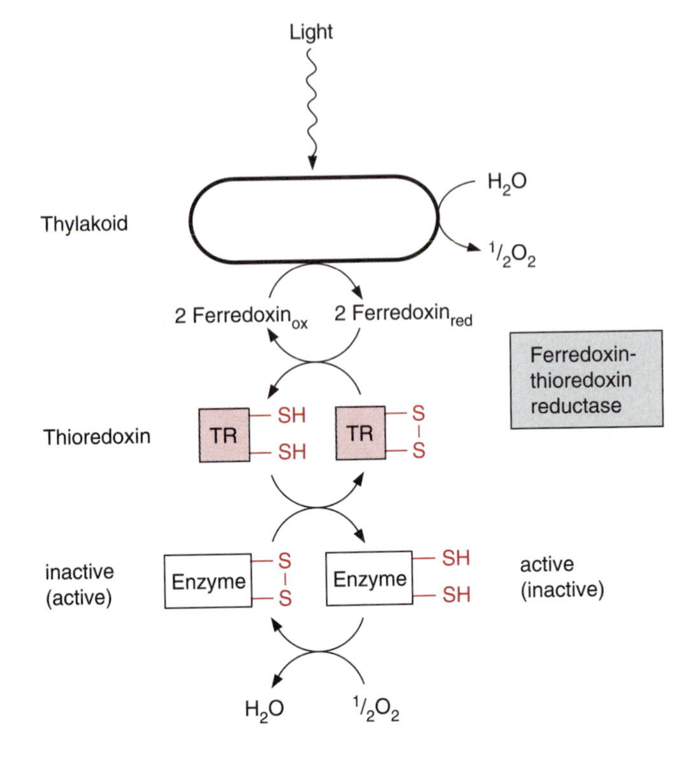

ribonucleotides to deoxyribonucleotides; the reduction of sulfate, a process occurring in plants and microorganisms (section 12.1); and the reductive activation of seed proteins during germination]. An abundance of other processes are known in which thioredoxins play an essential role, for instance, the assembly of bacteriophages and hormone action or the blood-clotting process in animals. The involvement of thioredoxins in the light regulation of chloroplast enzymes may be looked upon as a special function in addition to their general metabolic functions.

The chloroplast enzymes **ribulose phosphate kinase**, **sedoheptulose 1,7-bisphosphatase** and the chloroplast isoform of **fructose 1,6-bisphosphatase** are converted from an inactive state to an active state by reduced thioredoxin and are thus switched on by light. This also applies to other chloroplast enzymes such as **NADP-malate-dehydrogenase** (section 7.3) and **F-ATPase** (section 4.4). Reduced thioredoxin also converts **RubisCO-activase** (section 6.2) and **NADP glyceraldehyde phosphate dehydrogenase** contained in the chloroplasts from a less active state to a more active state. On the other hand, reduced thioredoxin inactivates **glucose 6-phosphate dehydrogenase**, the first enzyme of the oxidative pentose phosphate pathway.

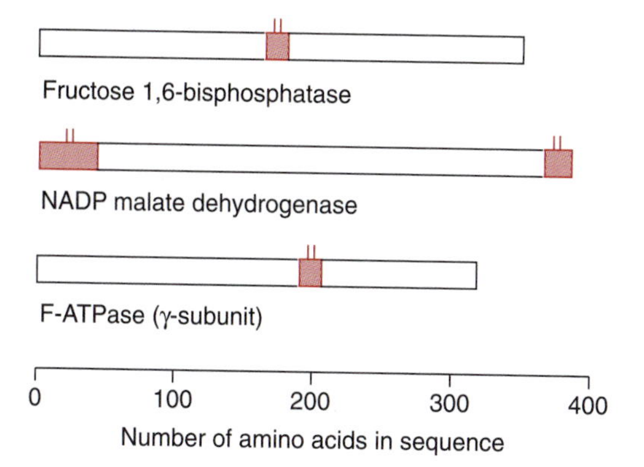

Figure 6.26 In contrast to the non-plastid isoenzymes, several thioredoxin-modulated chloroplast enzymes contain additional sections in the sequence (marked red) in which two cysteine residues are located (After Scheibe 1990).

The thioredoxin modulated activation of chloroplast enzymes releases an inbuilt blockage

Important knowledge of the mechanism of thioredoxin action on the chloroplast enzymes has been obtained from comparison with the corresponding isoenzymes from other cellular compartments. Isoenzymes of chloroplast fructose 1,6-bisphosphatase and malate dehydrogenase exist in the cytosol and are not regulated by thioredoxin. This also applies to F-ATPase in the mitochondria. A comparison of the amino acid sequences shows that at least in some cases the chloroplast isoenzymes possess **additional sections** at the end or in an inner region of their sequence, containing **two cysteine residues** (Fig. 6.26). The SH-groups of these cysteine residues can be converted by oxidation to a disulfide and represent the substrate for the protein disulfide oxidoreductase activity of thioredoxin.

The isoenzymes that are not regulated by thioredoxin, and therefore do not contain these additional sequences, often have a higher activity than the isoenzymes regulated by thioredoxin. An exchange of the cysteine residues involved in the regulation for other amino acids by genetic engineering (Chapter 22) resulted in enzymes that were fully active in the absence of reduced thioredoxin. Under oxidizing conditions, the enzymes regulated by thioredoxin are forced by the formation of a disulfide bridge into a conformation in which the catalytic center is inactivated. The reduction of this disulfide bridge by thioredoxin releases this blockage and the enzyme protein is converted to a relaxed conformation in which the catalytic center is active.

The light activation discussed so far is not an all or nothing effect. It is due to a continuous change between the thioredoxin-mediated reduction of the enzyme protein and its simultaneous oxidation by oxygen. The degree of activation of the enzyme depends on the rate of reduction. This is due not only to the degree of the reduction of thioredoxin (and thus to the degree of reduction of ferredoxin), but also to the presence of metabolites. Thus the reductive activation of fructose- and sedoheptulose bisphosphatase is enhanced by the corresponding bisphosphates. These effectors cause a decrease in the redox potential of the SH groups in the corresponding enzymes, which enhances the reduction of the disulfide group by thioredoxin. In this way the activity of these enzymes increases when the concentration of their substrates rises. On the other hand, the reductive activation of NADP malate dehydrogenase is decreased by the presence of $NADP^+$. This has the effect that the enzyme is active only at a high NADPH/NADP ratio. In contrast, the oxidative activation of glucose 6-phosphate dehydrogenase is enhanced by NADP and this increases the activity of the oxidative pentose phosphate pathway when there is a demand for NADPH.

An abundance of further regulatory processes ensures that the various steps of the reductive pentose phosphate pathway are matched

An additional light regulation of the Calvin cycle is based on the effect of light-dependent changes of the proton and Mg^{++} concentrations in the stroma on the activity of chloroplast enzymes. When isolated chloroplasts are illuminated, the acidification of the thylakoid space (Chapter 3) is accompanied by an alkalization and an increase in the Mg^{++} concentration in the stroma. During a darklight transition, the pH in the stroma may change from about pH 7.2 to 8.0. CO_2 fixation by isolated chloroplasts shows an optimum at about pH 8.0 with a sharp decline toward the acidic range. An almost identical pH dependence is found with the light-activated enzymes **fructose 1,6-bisphosphatase** and **sedoheptulose 1,7-bisphosphatase**. Moreover, the catalytic activity of both these enzymes is increased by the light-dependent increase in the Mg^{++} concentration in the stroma. The light activation of these enzymes by the thioredoxin system, together with the increase in enzyme activity by light-induced changes of the pH and Mg^{++} concentration in the stroma, of which each alone results in an extensive inactivation of the corresponding enzymes during darkness, is a very efficient system for switching these enzymes on and off, according to demand.

The activities of several stromal enzymes are also regulated by metabolite levels. The chloroplast **fructose 1,6-bisphosphatase** and **sedoheptulose 1,7-bisphosphatase** are inhibited by their corresponding products, fructose 6-phosphate and sedoheptulose 7-phosphate, respectively. This can decrease the activity of these enzymes when their products accumulate. **Ribulose phosphate kinase** is inhibited by 3-phosphoglycerate and also by ADP. Inhibition

Dai, S., Schwendtmayer, C., Schürmann, P., Ramaswamy, S., Eklund H. Redox signaling in chloroplasts: cleavage of disulfides by an iron-sulfur cluster. Science 287, 655–658 (2000).

Edwards, G., Walker, D. A. *C_3, C_4*; Mechanisms and Cellular and Environmental Regulation of Photosynthesis. Blackwell Scientific Publications, Oxfor, UK (1983).

Furbank, R. T., Taylor. W. C. Regulation of photosynthesis in C_3 and C_4 plants. Plant Cell 7, 797–807 (1995).

Flügge, U.-I. Phosphate translocators in plastids. Annu Rev Plant Physiol Plant Mol Biol 50, 27–46 (1999).

Gutteridge, S., Gatenby, A. RubisCO synthesis, assembly, mechanism and regulation. Plant Cell 7, 809–819 (1995).

Kelly, G., J. Photosynthesis: carbon metabolism from DNA to deoxyribose. Prog Bot 62, 238–265 (2001).

Roy, H., Andrews, T. J. RubisCO: Assembly and mechanism. In Photosynthesis: Physiology and Metabolism. R. C. Leegood, T. D. Sharkey, S. von Caemmerer (eds.), pp. 53–83. Kluwer Academic Publishers, Dordrecht, The Netherlands (2000).

Ruelland, E., Miginiac-Maslow, M. Regulation of chloroplast enzyme activities by thioredoxins: Activation or relief from inhibition? Trends Plant Sci 4, 136–141 (1999).

Schnarrenberger, C., Flechner, A., Martin, W. Enzymatic evidence for a complete oxidative pentose pathway in chloroplasts and an incomplete pathway in the cytosol of spinach leaves. Plant Physiol 108, 609–614 (1995).

Schürmann P., Jacquot, J.-P. Plant thioredoxin systems revisited. Annu Rev Plant Physiol Plant Mol Biol 51, 371–400 (2000).

Spreitzer,R. J., Salvucci, M. E. RubisCO: structure, regulatory interactions and possibilities for a better enzyme. Annu Rev Plant Biol 53, 449–475 (2002).

Stitt, M. Metabolic regulation of photosynthesis. In: Advances in `Photosynthesis, Vol. 5. Environmental Stress and Photosynthesis. N. Baker (eds.), pp. 151–190. Academic Press, New York (1997).

Von Caemmerer, S., Quick, W. P. RubisCO: Physiology in vivo. In Photosynthesis: Physiology and Metabolism. R. C. Leegood, T. D. Sharkey, S. von Caemmerer (eds.), pp. 85–113. Kluwer Academic Publishers, Dordrecht, The Netherlands (2000).

Zhang, N., Kallis, R. P. Ewy, R. G., Portis, A. R. Light modulation of Rubisco in *Arabidopsis* requires a capacity for redox regulation of the larger Rubisco activase isoform. Proc Natl Acad Sci USA 5, 3330–3334 (2002).

7

In the photorespiratory pathway phosphoglycolate formed by the oxygenase activity of RubisCo is recycled

Section 6.2 described how large amounts of 2-phosphoglycolate are formed as a by-product during CO_2 fixation by RubisCO, due to the oxygenase activity of this enzyme. In the photorespiratory pathway, discovered in 1972 by the American scientist Edward Tolbert, the by-product 2-phosphoglycolate is recycled to ribulose 1,5-bisphosphate. The term *photorespiration* indicates that it involves oxygen consumption occurring in the light, which is accompanied by the release of CO_2. Whereas in mitochondrial respiration (cell respiration; Chapter 5), the oxidation of substrates to CO_2 serves the purpose of producing ATP, in the case of photorespiration ATP is consumed.

7.1 Ribulose 1,5-bisphosphate is recovered by recycling 2-phosphoglycolate

Figure 7.1 gives an overview of the reactions of the photorespiratory pathway and their localization. Recycling of 2-phosphoglycolate begins with the hydrolytic release of phosphate by **phosphoglycolate phosphatase** present in the chloroplast stroma (Fig. 7.2). The resultant glycolate leaves the chloroplasts by a specific translocator located in the inner envelope membrane and enters the peroxisomes via nonspecific pores in the peroxisomal boundary membrane, probably formed by a **porin** (section 1.11).

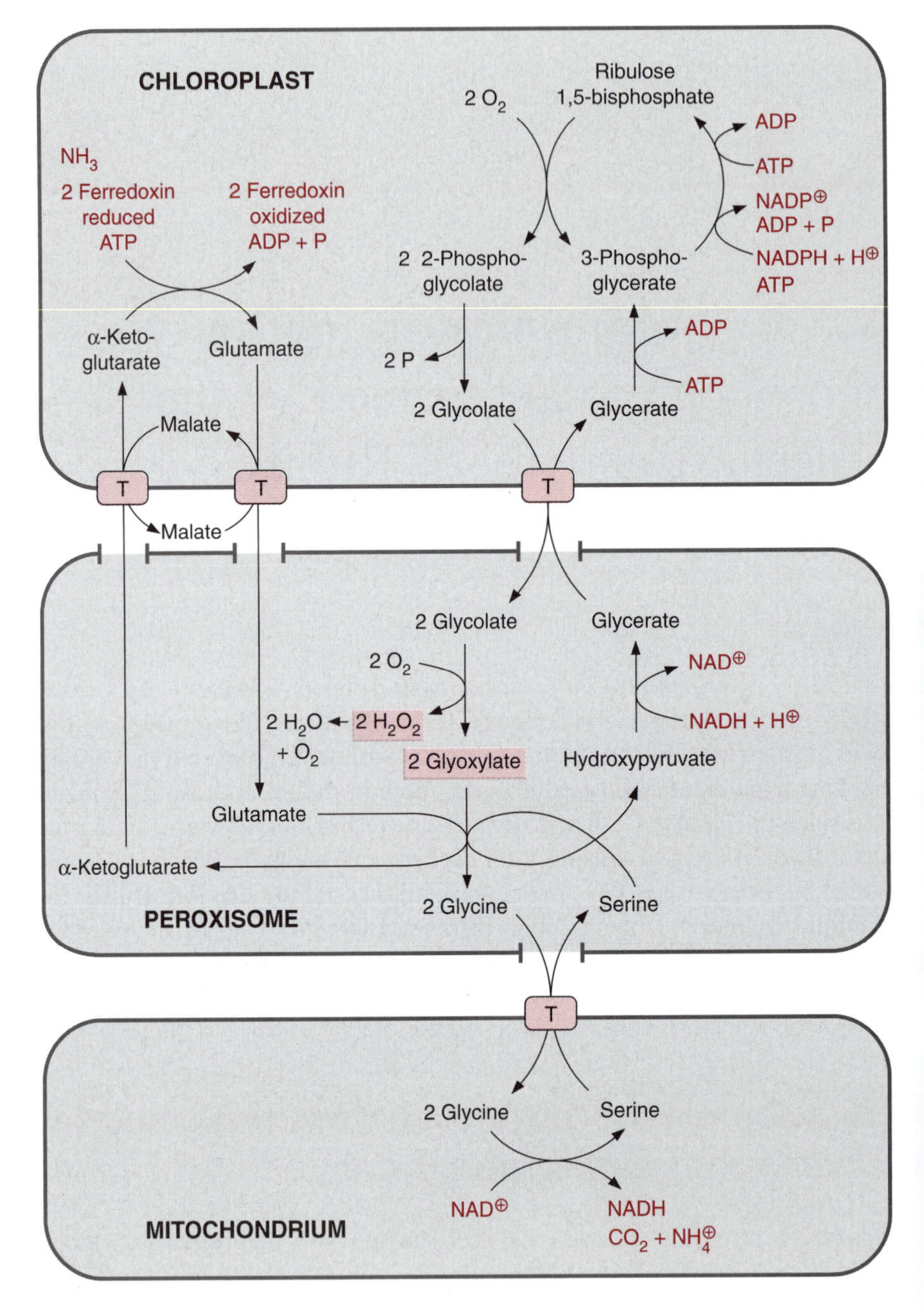

Figure 7.1
Compartmentation of the photorespiratory pathway. This scheme does not show the outer membranes of the chloroplasts and mitochondria, which are nonspecifically permeable for metabolites, due to the presence of porins. T = translocator. Translocators for glycine and serine have not been identified yet.

In the peroxisomes the alcoholic group of glycolate is oxidized to a carbonyl group in an irreversible reaction catalyzed by **glycolate oxidase**, resulting in the formation of glyoxylate. The reducing equivalents are transferred to molecular oxygen forming H_2O_2 (Fig. 7.2). Like other H_2O_2 forming

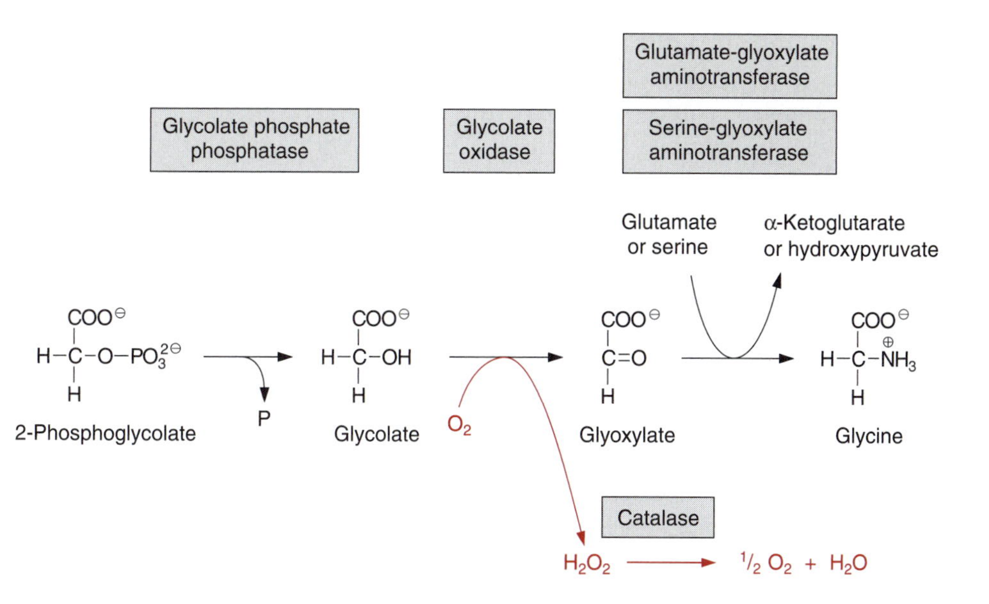

Figure 7.2 Reaction sequence for the conversion of 2-phosphoglycolate to form glycine.

oxidases, glycolate oxidase contains a flavin mononucleotide cofactor (FMN, Fig. 5.16) as redox mediator between glycolate and oxygen. The H_2O_2 formed is converted to water and oxygen by the enzyme **catalase** present in the peroxisomes. Thus, in total, 0.5 mol of O_2 is consumed for the oxidation of one mole of glycolate to glyoxylate.

The glyoxylate formed is converted to the amino acid glycine by two different reactions proceeding in the peroxisome simultaneously at a 1 : 1 ratio. The enzyme **glutamate-glyoxylate aminotransferase** catalyzes the transfer of an amino group from the donor glutamate to glyoxylate. This enzyme also reacts with alanine as the amino donor. In the other reaction, the enzyme **serine glyoxylate aminotransferase** catalyzes the transamination of glyoxylate by serine. These two enzymes, like other aminotransferases, (e.g. glutamate-oxaloacetate aminotransferase, see section 10.4) contain bound pyridoxal phosphate with an aldehyde function as reactive group (Fig. 7.3). Figure 7.4 shows the reaction sequence.

The glycine thus formed, leaves the peroxisomes via pores and is transported into the mitochondria. Although this transport has not yet been characterized in detail, one would expect it to proceed via a specific translocator. In the mitochondria two molecules of glycine are oxidized yielding one molecule of serine with release of CO_2 and NH_4^+ and a transfer of reducing equivalents to NAD^+ (Fig. 7.5). The oxidation of glycine is catalyzed by the **glycine decarboxylase-serine hydroxymethyltransferase complex**. This is a multi-enzyme complex, consisting of four different subunits (Fig. 7.7), which shows a great similarity to the pyruvate dehydroge-

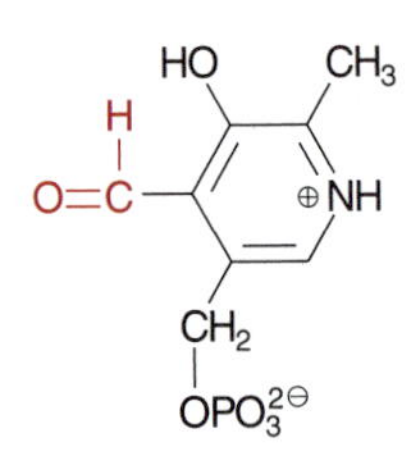

Figure 7.3 Pyridoxal phosphate.

nase complex described in section 5.3. The so-called **H-protein** with the prosthetic group lipoic acid amide (Fig. 5.5) represents the center of the glycine decarboxylase complex. Around this center are positioned the pyridoxal phosphate-containing **P-protein**, the **T-protein** with a tetrahydrofolate (Fig. 7.6) as a prosthetic group, and the **L-protein**, also named dihydrolipoate-dehydrogenase, which is identical to the dihydrolipoate-dehydrogenase of the pyruvate and α-ketoglutarate dehydrogenase complex (Figs. 5.4 and 5.8). Since the disulfide group of the lipoic acid amide in the H-protein is located at the end of a flexible polypeptide chain (see also Fig. 5.4), it is able to react with the three other subunits. Figure 7.7 shows the reaction sequence. The enzyme **serine hydroxymethyltransferase**, which is in close proximity to the glycine decarboxylase complex, catalyzes the transfer of the formyl residue to another molecule of glycine to form serine.

The NADH produced in the mitochondrial matrix from glycine oxidation can be oxidized by the mitochondrial respiratory chain in order to generate ATP. Alternatively, these reducing equivalents can be exported from the mitochondria to other cell compartments, as will be discussed in section 7.3. The capacity for glycine oxidation in the mitochondria of green plant cells is very high. The glycine decarboxylase complex of the mitochondria

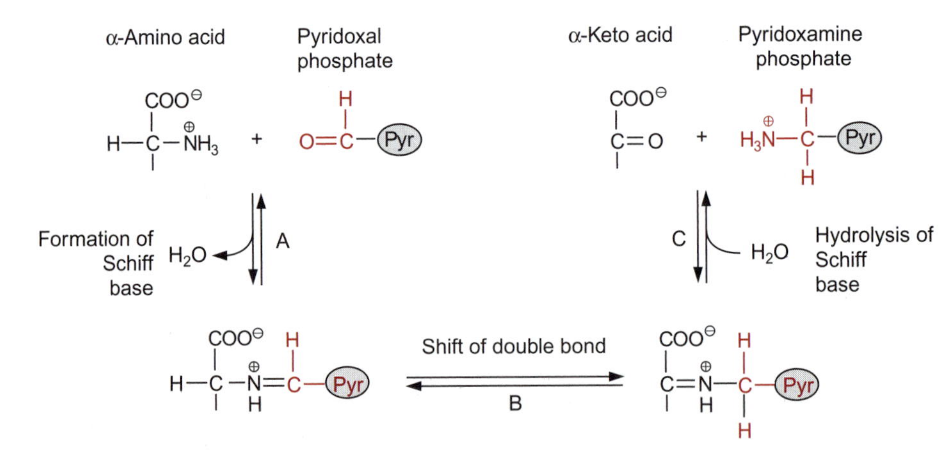

Figure 7.4 Reaction sequence of the aminotransferase reaction. The aldehyde group of pyridoxal phosphate forms a Schiff base with the α-amino group of the amino acid (in this case glutamate or serine) (A), which is subsequently converted to an isomeric form (B) by a base-catalyzed movement of a proton. Hydrolysis of the isomeric Schiff base results in the formation of an α-ketoacid (α-ketoglutarate or hydroxypyruvate), and pyridoxamine remains (C). The amino group of this pyridoxamine then forms a Schiff base with another α-ketoacid (in this case glyoxylate), and glycine is formed by a reversal of the steps C, B, and A. Pyridoxal is thus regenerated and is available for the next reaction cycle.

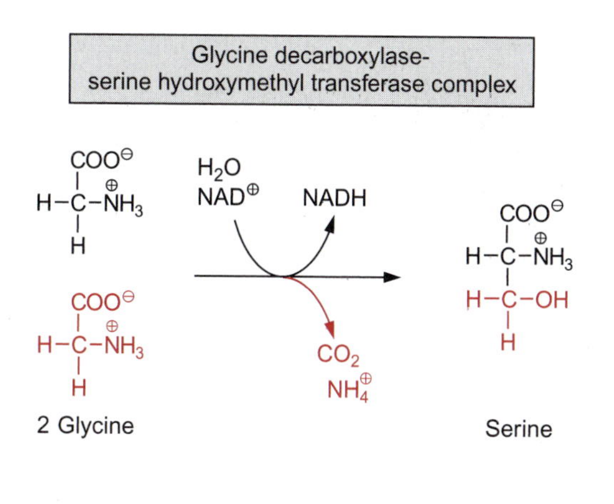

Figure 7.5 Overall reaction scheme for the conversion of two molecules of glycine to form one molecule of serine as catalyzed by the glycine-decarboxylase complex.

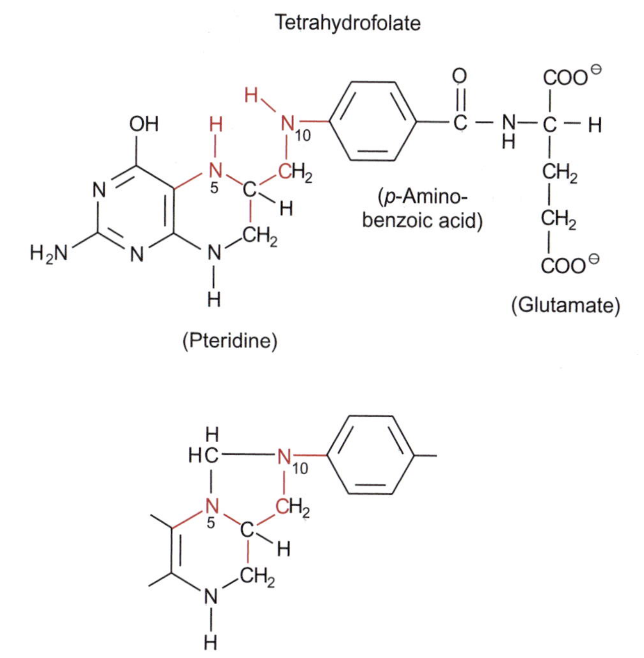

Figure 7.6 Tetrahydrofolate.

can amount to 30% to 50% of the total content of soluble proteins. In mitochondria of non-green plant cells, however, the proteins of glycine oxidation are present only in very low amounts or are absent.

Serine probably leaves the mitochondria via a specific translocator, possibly the same translocator by which glycine is taken up. After entering the peroxisomes through pores, serine is converted to hydroxypyruvate by the enzyme **serine-glyoxylate aminotransferase** mentioned previously (Fig. 7.8).

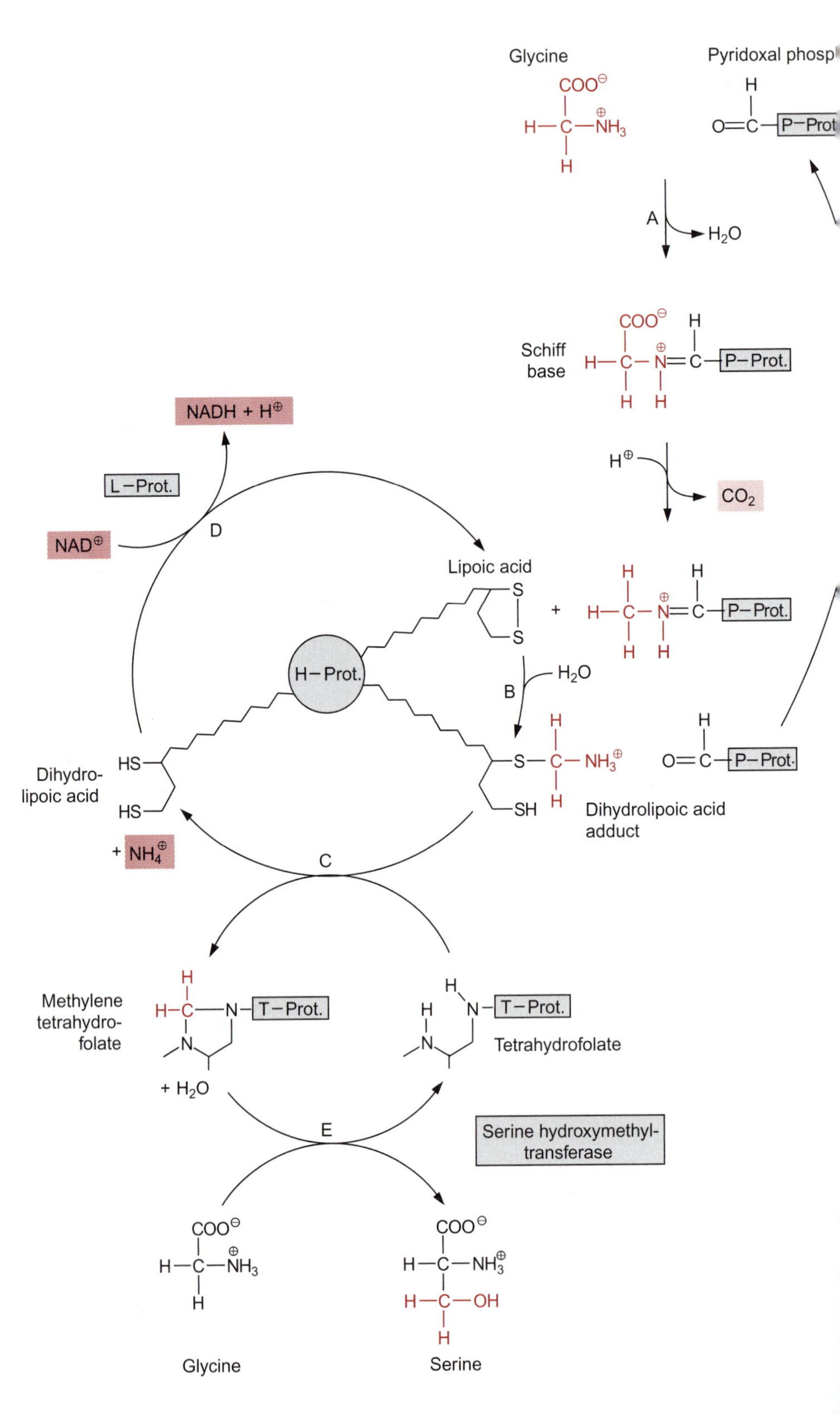

Figure 7.7 Reaction sequence for the interconversion of two molecules of glycine to form one molecule of serine. The amino group of glycine reacts first with the aldehyde group of pyridoxal in the P-protein to form a Schiff base (A). The glycine residue is then decarboxylated and transferred from the P-protein to the lipoic acid residue of the H-protein (B). This is the actual oxidation step: the C_1 residue is oxidized to a formyl group and the lipoic acid residue is reduced to dihydrolipoic acid. The dihydrolipoic acid adduct now reacts with the T-protein, the formyl residue is transferred to tetrahydrofolate, and the dihydrolipoic acid residue remains (C). The dihydrolipoic acid is reoxidized via the L-protein (dihydrolipoate dehydrogenase) to lipoic acid and the reducing equivalents are transferred to NAD (D). A new reaction cycle can then begin. The formyl residue bound to tetrahydrofolate is transferred to a second molecule of glycine by serine-hydroxymethyl transferase and serine is formed (E).

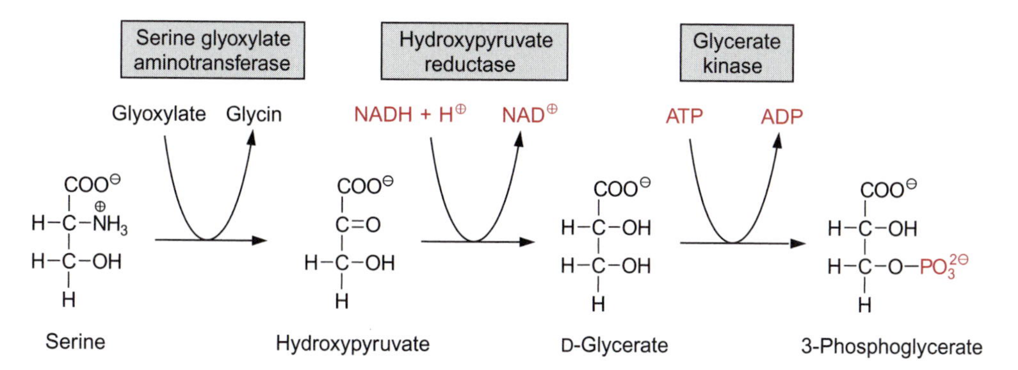

Figure 7.8 Reaction sequence for the conversion of serine to form 3-phosphoglycerate.

At the expense of NADH, hydroxypyruvate is reduced by **hydroxypyruvate reductase** to glycerate, which is released from the peroxisomes and imported into the chloroplasts.

The uptake into the chloroplasts proceeds by the same translocator as that catalyzing the release of glycolate from the chloroplasts (**glycolate-glycerate translocator**). This translocator facilitates a glycolate-glycerate counter-exchange as well as a co-transport of just glycolate with a proton. In this way, the translocator enables the export of two molecules of glycolate from the chloroplasts in exchange for the import of one molecule of glycerate. Glycerate is converted to 3-phosphoglycerate, consuming ATP by **glycerate kinase** present in the chloroplast stroma. Finally, 3-phosphoglycerate is reconverted to ribulose 1,5-bisphosphate via the reductive pentose phosphate pathway (section 6.3, 6.4). These reactions complete the recycling of 2-phosphoglycolate.

7.2 The NH_4^+ released in the photorespiratory pathway is refixed in the chloroplasts

Nitrogen is an important plant nutrient. Nitrogen supply is often a limiting factor in plant growth. It is therefore necessary for the economy of plant metabolism that the ammonium, which is released at very high rates in the photorespiratory pathway during glycine oxidation, is completely refixed. This refixation occurs in the chloroplasts. It is catalyzed by the same enzymes

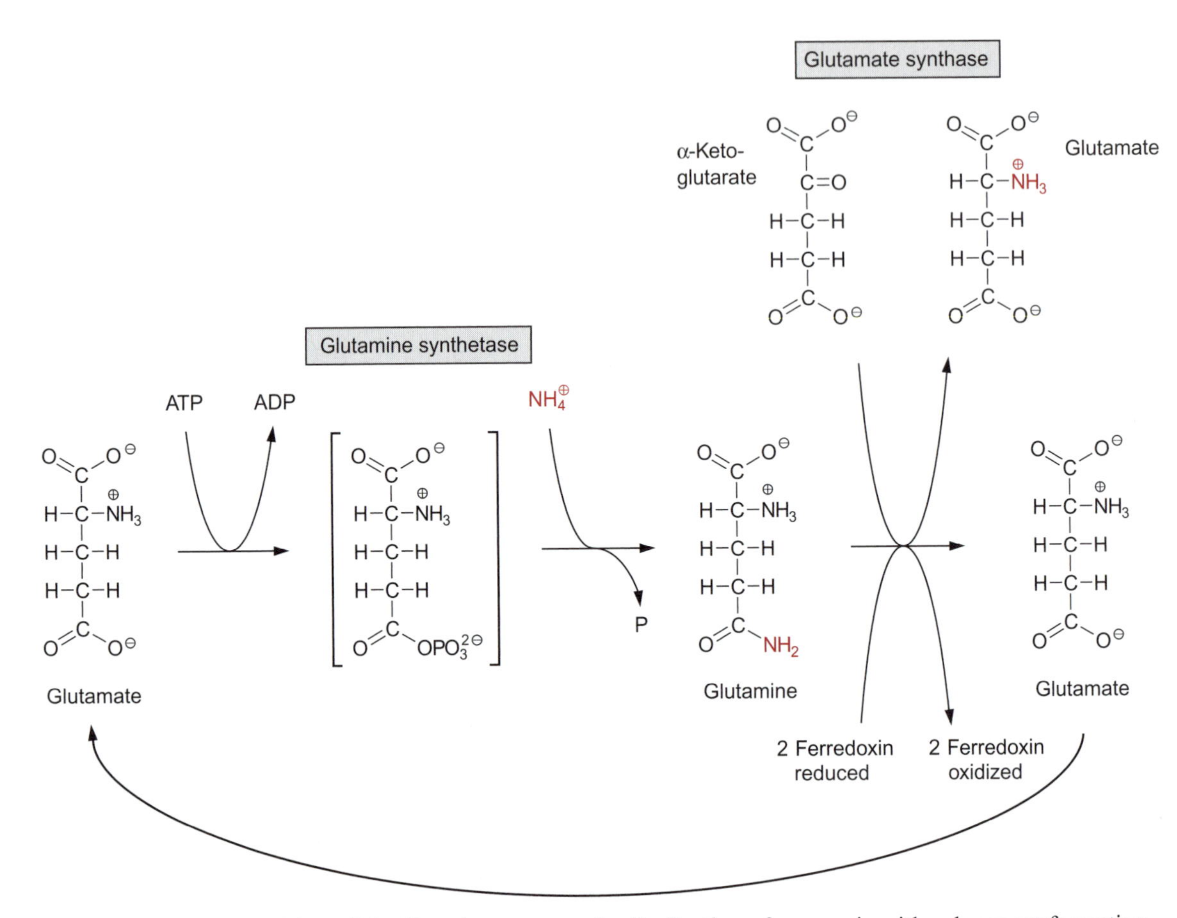

Figure 7.9 Reaction sequence for the fixation of ammonia with subsequent formation of glutamate from α-ketoglutarate.

that participate in nitrate assimilation (Chapter 10). However, the rate of NH_4^+ refixation in the photorespiratory pathway is 5 to 10 times higher than the rate of NH_4^+ fixation in nitrate assimilation.

In a plant cell, chloroplasts and mitochondria are mostly in close proximity to each other. The NH_4^+ formed during oxidation of glycine passes through the inner membrane of the mitochondria and the chloroplasts. Whether this passage occurs by simple diffusion or is facilitated by specific translocators or ion channels is still a matter of debate. The enzyme **glutamine synthetase**, present in the chloroplast stroma, catalyzes the transfer of an ammonium ion to the δ-carboxyl group of glutamate (Fig. 7.9) to form glutamine. This reaction is driven by the conversion of one molecule of ATP to ADP and P. In an intermediary step, the δ-carboxyl group is activated by reaction with ATP to form a carboxy-phosphate anhydride. Glutamine synthetase has a high affinity for NH_4^+ and catalyzes an essentially irreversible reaction. This enzyme has a key role in the fixation of NH_4^+ not only in plants, but also in bacteria and animals.

The nitrogen fixed as amide in glutamine is transferred by reductive amination to α-ketoglutarate (Fig. 7.9). In this reaction, catalyzed by **glutamate synthase**, two molecules of glutamate are formed. The reducing equivalents are provided by reduced ferredoxin, which is a product of photosynthetic electron transport (see section 3.8). In green plant cells, glutamate synthase is located exclusively in the chloroplasts.

Of the two glutamate molecules thus formed in the chloroplasts, one is exported by the **glutamate-malate translocator** in exchange for malate and, after entering the peroxisomes, is available as a reaction partner for the transamination of glyoxylate. The α-ketoglutarate thus formed is re-imported from the peroxisomes into the chloroplasts by a **malate-α-ketoglutarate translocator**, again in counter-exchange for malate.

7.3 For the reduction of hydroxypyruvate, peroxisomes have to be provided with external reducing equivalents

NADH is required as reductant for the conversion of hydroxypyruvate to glycerate in the peroxisomes. Since leaf peroxisomes have no metabolic pathway capable of delivering NADH at the very high rates required, peroxisomes are dependent on the supply of reducing equivalents from outside.

Reducing equivalents are taken up into the peroxisomes via a malate-oxaloacetate shuttle

The cytosolic NADH system of a leaf cell is oxidized to such an extent ($NADH/NAD^+ = 10^{-3}$), that the concentration of NADH in the cytosol is only about 10^{-6} mol/L. This very low concentration would not allow a diffusion gradient to build up, which is large enough to drive the necessary high diffusive fluxes of reducing equivalents in the form of NADH into the peroxisomes. Instead, the reducing equivalents are imported indirectly into the peroxisomes via the uptake of malate and the subsequent release of oxaloacetate (this is termed a **malate-oxaloacetate shuttle**) (Fig. 7.10).

Malate dehydrogenase (Fig. 5.9), which catalyzes the oxidation of malate to oxaloacetate in a reversible reaction, has a key function in this shuttle.

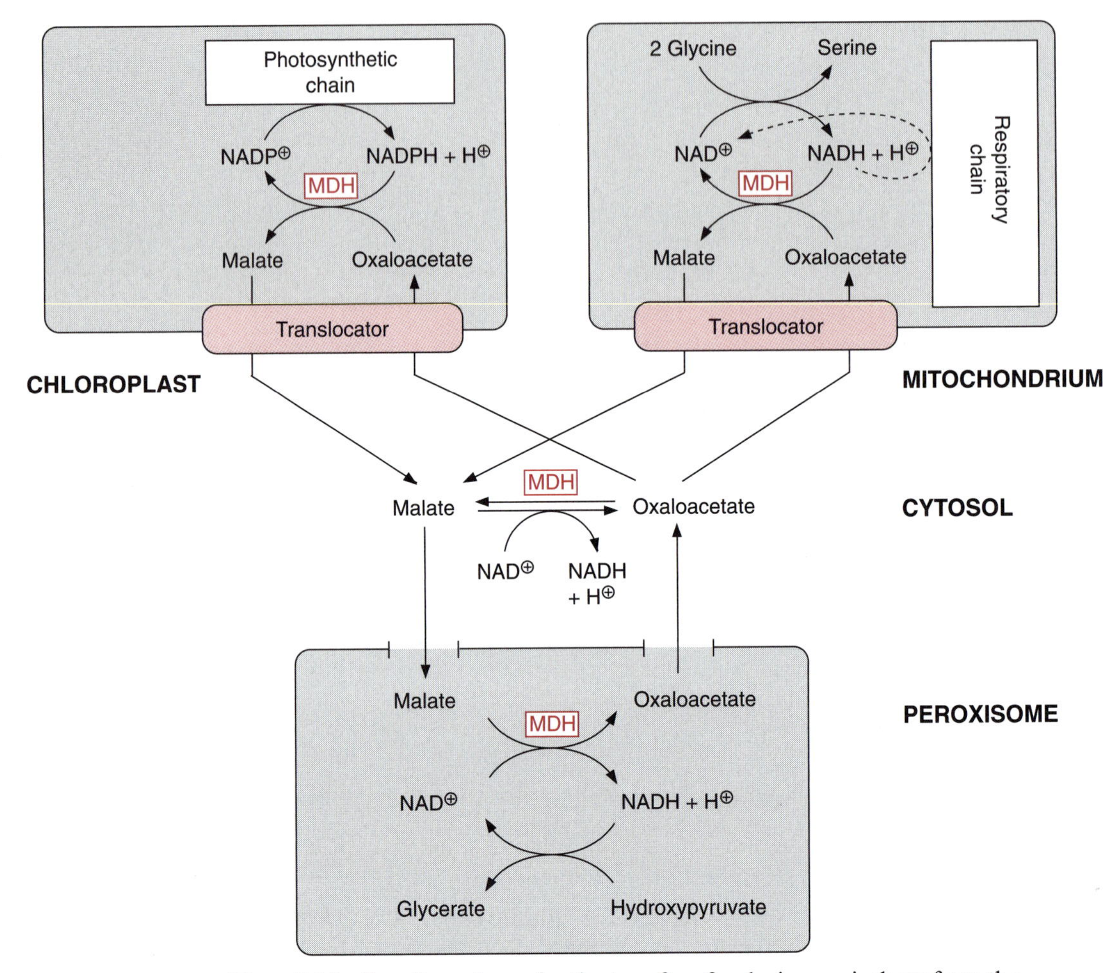

Figure 7.10 Reaction scheme for the transfer of reducing equivalents from the chloroplasts and the mitochondria to the peroxisomes. MDH: malate-dehydrogenase.

High malate dehydrogenase activity is found in the cytosol as well as in chloroplasts, mitochondria, and peroxisomes. The malate dehydrogenases in the various cell compartments are considered to be isoenzymes. They show some differences in their structure and are encoded by different, but homologous genes. Apparently, these are all related proteins, which have derived in the course of evolution from a common precursor. Whereas the NADH system is the redox partner for malate dehydrogenases in the cytosol, mitochondria and peroxisomes, the chloroplast isoenzyme reacts with the NADPH system.

Mitochondria export reducing equivalents via a malate oxaloacetate shuttle

In contrast to mitochondria from animal tissues, where the inner membrane is impermeable for oxaloacetate, plant mitochondria contain in their inner membrane a specific **malate-oxaloacetate translocator**, which transports malate and oxaloacetate in a counter-exchange mode. Since the activity of malate dehydrogenase in the mitochondrial matrix is very high, NADH formed in mitochondria during glycine oxidation can be captured to reduce oxaloacetate to form malate, for export by the **malate-oxaloacetate shuttle**. This shuttle has a high capacity. As can be seen from Figure 7.1, the amount of NADH generated in the mitochondria from glycine oxidation is equal to that of NADH required for the reduction of hydroxypyruvate in the peroxisomes. If all the oxaloacetate formed in the peroxisomes should reach the mitochondria, the NADH generated from glycine oxidation would be totally consumed in the formation of malate and would be no longer available to support ATP synthesis by the respiratory chain. However, mitochondrial ATP synthesis is required during photosynthesis to supply energy to the cytosol of mesophyll cells. In fact, mitochondria deliver only about half the reducing equivalents required for peroxisomal hydroxypyruvate reduction, with the remaining portion being provided by the chloroplasts (Fig. 7.10). Thus only about half of the NADH formed during glycine oxidation is captured by the malate-oxaloacetate shuttle for export, and the remaining NADH is oxidized by the respiratory chain for synthesis of ATP.

A "malate valve" controls the export of reducing equivalents from the chloroplasts

Chloroplasts are also able to export reducing equivalents by a malate-oxaloacetate shuttle. Malate and oxaloacetate are transported across the chloroplast inner envelope membrane via a specific translocator, in a counter-exchange mode. Despite the high activity of the chloroplast malate-oxaloacetate shuttle, a high gradient exists between the chloroplast and cytosolic redox systems: The ratio $NADPH/NADP^+$ in chloroplasts is more than 100 times higher than the corresponding $NADH/NAD^+$ ratio in the cytosol. Whereas malate dehydrogenases usually catalyze a reversible equilibrium reaction, the reduction of oxaloacetate by chloroplast malate dehydrogenase is virtually irreversible and does not reach equilibrium. This is due to a regulation of chloroplast malate dehydrogenase.

Section 6.6 described how chloroplast malate dehydrogenase is activated by **thioredoxin** and is therefore active only in the light. In addition to this,

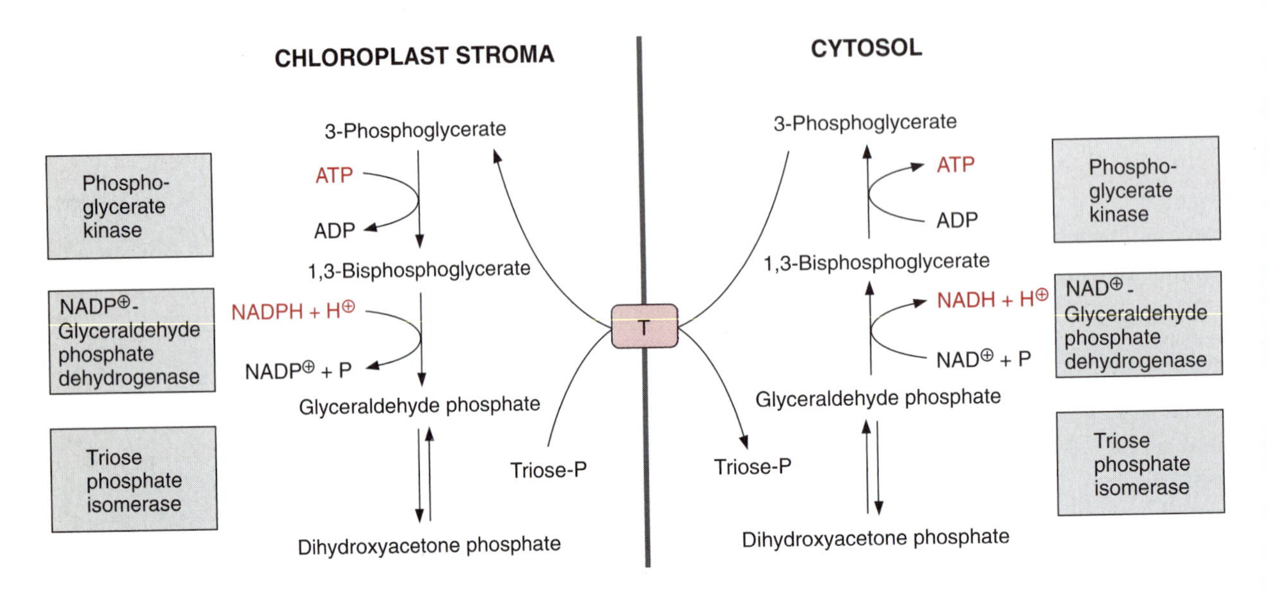

Figure 7.11 Triose phosphate-3-phosphoglycerate shuttle operating between the chloroplast stroma and the cytosol. In the chloroplast stroma triose phosphate is formed from 3-phosphoglycerate at the expense of NADPH and ATP. Triose phosphate is transported by the triose phosphate-phosphate translocator across the inner envelope membrane in exchange for 3-phosphoglycerate. In the cytosol, triose phosphate is reconverted to 3-phosphoglycerate with the generation of NADPH and ATP.

increasing concentrations of $NADP^+$ inhibit the reductive activation of the enzyme by thioredoxin. $NADP^+$ increases the redox potential of the regulatory SH-groups of malate dehydrogenase, and, as a result of this, the reductive activation of the enzyme by thioredoxin is lowered. Thus a decrease in the $NADP^+$ concentration, which corresponds to an increase in the reduction of the $NADPH/NADP^+$ system, switches on chloroplast malate dehydrogenase. This allows the enzyme to function like a **valve**, through which **excessive reducing** equivalents can be released by the chloroplasts to prevent harmful overreduction of the redox carriers of the photosynthetic electron transport chain. At the same time, this valve makes it possible for the chloroplasts to provide reducing equivalents for the reduction of hydroxypyruvate in the peroxisomes and also for other processes [e.g., nitrate reduction in the cytosol (section 10.1)].

An alternative way for exporting reducing equivalents from chloroplasts to the cytosol is the **triose phosphate-3-phosphoglycerate shuttle** (Fig. 7.11). By this shuttle, together with NADH also ATP is delivered to the cytosolic compartment.

7.4 The peroxisomal matrix is a special compartment for the disposal of toxic products

The question arises, why besides the chloroplasts, are two other organelles involved in the recycling process of 2-phosphoglycolate? That mitochondria are the site of the conversion of glycine to serine appears to have the advantage that the respiratory chain can utilize the resultant NADH for the synthesis of ATP. During the conversion of glycolate to glycine, two toxic intermediates are formed: **glyoxylate** and **H_2O_2**. In isolated chloroplasts, photosynthesis is completely inhibited by the addition of low concentrations of H_2O_2 or glyoxylate. The inhibitory effect of H_2O_2 is due to the oxidation of SH-groups in thioredoxin-activated enzymes of the reductive pentose phosphate pathway (section 6.6), resulting in their inactivation. Glyoxylate, a very reactive carbonyl compound, also has a strong inhibitory effect on thioredoxin activated enzymes by also reacting with their SH-groups. Glyoxylate also inhibits RubisCO. Compartmentalization of the conversion of glycolate to glycine in the peroxisomes serves the purpose of eliminating the toxic intermediate products glyoxylate and H_2O_2 at the site of their formation, so that they do not invade other cell compartments.

How is such a compartmentalization possible? Compartmentalization of metabolic processes in other cell compartments (e.g., the chloroplast stroma or the mitochondrial matrix), is due to separating membranes, which contain specific translocators for the passage of certain metabolites, but are impermeable to metabolic intermediates present in these different compartments. This principle, however, does not apply to the compartmentalization of glycolate oxidation, since membranes are normally quite permeable to H_2O_2 as well as to glyoxylate. For this reason, a boundary membrane would be unable to prevent these substances from escaping from the peroxisomes.

The very efficient compartmentalization of the conversion in the peroxisomes of glycolate to glycine, and of serine to glycerate is due to specific properties of the **peroxisomal matrix**. When the boundary membrane of chloroplasts or mitochondria is disrupted (e.g., by suspending the organelles for a short time in pure water to cause an osmotic shock), the proteins of the stroma or the matrix, respectively, which are all soluble, are released from these disrupted organelles. However, in peroxisomes, after disruption of the boundary membrane, the peroxisomal matrix proteins remain aggregated in the form of particles of a size similar to peroxisomes, and the compartmentalization of the peroxisomal reactions is maintained. Glyoxylate, H_2O_2, and hydroxypyruvate, intermediates of peroxisomal metabolism, are not released from these particles in the course of glycolate oxidation. Apparently, the enzymes of the photorespiratory pathway are arranged in the peroxisomal matrix in the form of a **multienzyme complex**, by which the product of one enzymatic reaction binds immediately to the enzyme of the following reaction and is therefore not released.

This process, termed **metabolite channeling**, probably occurs not only in the peroxisomal matrix but also may occur in a similar orderly manner in other metabolic pathways [e.g., the Calvin cycle in the chloroplast stroma, (Chapter 6)]. It seems to be a special feature of the peroxisomes, however, that such an enzyme complex remains intact after disruption of the boundary membrane. This may be a protective function to avoid the escape of glycolate oxidase after eventual damage to the peroxisomal membrane. An escape of glycolate oxidase would result in glycolate being oxidized outside the peroxisomes and poisoning the cell because of the accumulation of the products glyoxylate and H_2O_2 in the cytosol.

For any glyoxylate and hydroxypyruvate leaking out of the peroxisomes despite metabolite channeling, rescue enzymes that are present in the cytosol use NADPH to convert glyoxylate to glycolate (**NADPH-glyoxylate-reductase**) and hydroxypyruvate to glycerate (**NADPH-hydroxypyruvate reductase**). Moreover, glyoxylate can be also eliminated by an NADPH-glyoxylate-reductase present in the chloroplasts.

7.5 How high are the costs of the ribulose bisphosphate oxygenase reaction for the plant?

On the basis of the metabolic schemes in Figures 6.20 and 7.1, the expenditure in ATP and NADPH (respectively the equivalent of two reduced ferredoxins) for oxygenation and carboxylation of RuBP by RubisCO is listed in Table 7.1. The data illustrate that the consumption of ATP and NADPH, required to compensate the consequences of oxygenation, is much higher than the ATP and NADPH expenditure for carboxylation. Whereas in CO_2 fixation the conversion of CO_2 to triose phosphate requires three molecules of ATP and two molecules of NADPH, the oxygenation of RuBP costs in total five molecules of ATP and three molecules of NADPH per molecule O_2. Table 7.2 shows the additional expenditure in ATP and NADPH at various ratios of carboxylation to oxygenation. In the leaf, where the carboxylation : oxygenation ratio is usually between two and four, the additional expenditure of NADPH and ATP to compensate for the oxygenation is more than 50% of the corresponding expenditure for CO_2 fix-

Table 7.1: Expenditure of ATP and NADPH for carboxylation of ribulose 1,5-bisphosphate (CO_2 assimilation) in comparison to the corresponding expenditure for oxygenation

	Expenditure (mol) ATP	NADPH or 2 reduced Ferrredoxin
Carboxylation:		
Fixation of 1 mol CO_2		
1 CO_2 → 0.33 triose phosphate	**3**	**2**
Oxygenation:		
2 ribulose 1,5-bisphosphate + 2 O_2 → 2 3-phosphoglycerate + 2 2-phosphoglycolate		
2 2-phosphoglycolate → 3-phosphoglycerate + 1 CO_2	2	1
1 CO_2 → 0.33 triose phosphate	3	2
3 3-phosphoglycerate → 3 triose phosphate	3	3
3.33 triose phosphate → 2 ribulose 1,5-bisphosphate	2	
	Σ 10	Σ 6
Oxygenation by 1 mol O_2:	**5**	**3**

Table 7.2: Additional consumption for RuBP oxygenation as related to the consumption for CO_2 fixation

Ratio Carboxylation/oxygenation	Additional consumption ATP	NADPH
2	83%	75%
4	42%	38%

ation. **Thus the oxygenase by-reaction of RubisCO costs the plant more than one-third of the captured photons.**

7.6 There is no net CO_2 fixation at the compensation point

At a carboxylation : oxygenation ratio of 1/2 there is no net CO_2 fixation, as the amount of CO_2 fixed by carboxylation is equal to the amount of CO_2 released by the photorespiratory pathway as a result of oxygenation. One can simulate this situation experimentally by illuminating a plant in a closed chamber. Due to photosynthesis, the CO_2 concentration decreases until it reaches a concentration at which the fixation of CO_2 and the release of CO_2 are counterbalanced. This state is termed the **compensation point**. Although the release of CO_2 is caused not only by the photorespiratory pathway but also by other reactions (e.g., the citrate cycle in mitochondria), the latter sources of CO_2 release are negligible compared with the photorespiratory pathway. For the plants discussed so far, designated as **C_3 plants** (this term is derived from the fact that the first carboxylation product is the C_3 compound 3-phosphoglycerate), the CO_2 concentration in air at the compensation point, depending on the species and temperature, is in the range of 35 to 70 ppm., which corresponds to 10% to 20% of the CO_2 concentration in the atmosphere. In the aqueous phase, where RubisCO is present, this corresponds at 25°C to a CO_2 concentration of $1–2 \times 10^{-6}$ mol/L. For **C_4 plants**, discussed in section 8.4, the CO_2 concentration at the compensation point is only about 5 ppm. How these plants manage to have such a low compensation point in comparison with C_3 plants will be discussed in detail in section 8.4.

With a plant kept in a closed system, it is possible to decrease the CO_2 concentration to a value below the compensation point by trapping CO_2 with KOH. In this case, upon illumination, oxygenation by RubisCO and the accompanying photorespiratory pathway results in a net release of CO_2 at the expense of the plant matter, which is degraded to produce carbohydrates to allow the regeneration of ribulose 1,5-bisphosphate. In such a situation, illumination of a plant causes its consumption.

7.7 The photorespiratory pathway, although energy-consuming, may also have a useful function for the plant

Due to the high ATP and NADPH consumption during photorespiration, photosynthetic metabolism proceeds at full speed at the compensation point, although there is no net CO_2 fixation. Such a situation arises when in leaves exposed to full light, the stomata are closed because of water shortage (section 8.1) and therefore CO_2 cannot be taken up. An overreduction and an overenergization of the photosynthetic electron transport carriers can cause severe damage to them (section 3.10). The plant utilizes the energy-consuming photorespiratory pathway to eliminate ATP and NADPH, which have been produced by light reactions, but which cannot be used for CO_2

assimilation. Photorespiration, the unavoidable by-reaction of photosynthesis, is thus utilized by the plant for its protection. It is feasible, therefore, that lowering the oxygenase reaction of RubisCO by molecular engineering (Chapter 22), as attempted by many researchers, although still without success, not only may lead to the plant using energy more efficiently, but also at the same time may increase its vulnerability to excessive illumination or shortage of water (see Chapter 8).

Further reading

Baker, A., Graham, I. A. eds. Plant Peroxisomes. Kluwer Academic Publishers, Dordrecht, The Netherlands (2002).

Douce, R., Heldt, H. W. Photorespiration. In Photosynthesis: Physiology and Metabolism. R. C. Leegood, T. D. Sharkey, S., von Caemmerer (eds.), pp. 115–136. Kluwer Academic Publishers, Dordrecht, The Netherlands (2000).

Douce, R., Bourrguignon, J., Neuburger, M., Rébeillé, F. The glycine decarboxylase system: A fascinating complex. Trends Plant Sci 6, 167–176 (2001).

Givan, C. V., Kleczkowski, L. A. The enzymic reduction of glyoxylate and hydroxypyruvate in leaves of higher plants. Plant Physiol 100, 552–556 (1992).

Husic, D. W., Husic H. D., Tolbert, N. E. The oxidative photosynthetic carbon cycle or C_2 cycle. CRC Critical Reviews in Plant Sciences, Vol. 1, pp. 45–100, CRC Press, Boca Raton (1987).

Husic, D. W., Husic H. D. The oxidative photosynthetic carbon (C_2) cycle: An update of unanswered questions. Rev Plant Biochem and Biotechnol. 1, 33–56 (2002).

Kozaki, A., Takeba, G. Photorespiration protects C_3 plants from photooxidation. Nature 384, 557–560 (1996).

Reumann, S., Maier, E., Benz, R., Heldt, H. W. The membrane of leaf peroxisomes contains a porin-like channel. J Biol Chem 270, 17559–17565 (1995).

Reumann, S. The structural properties of plant peroxisomes and their metabolic significance. Biol Chem 381, 639–648 (2000).

Tolbert, N. E. Microbodies—peroxisomes and glyoxysomes. In The Biochemistry of Plants, P. K. Stumpf and E. E. Conn (eds.), Vol. 1, pp. 359–388, Academic Press, New York (1980).

8

Photosynthesis implies the consumption of water

This chapter deals with the fact that photosynthesis is unavoidably linked with a substantial loss of water and therefore is often limited by the lack of it. Biochemical mechanisms that enable certain plants living in hot and dry habitats to reduce their water requirement will be described.

8.1 The uptake of CO_2 into the leaf is accompanied by an escape of water vapor

Since CO_2 assimilation is linked with a high water demand, plants require an ample water supply for their growth. A C_3 plant growing in temperate climates **requires 700 to 1,300 mol of H_2O for the fixation of 1 mol of CO_2**. In this assessment the water consumption for photosynthetic water oxidation is negligible in quantitative terms. Water demand is dictated by the fact that water escaping from the leaves in the form of water vapor has to be replenished by water taken up through the roots. Thus during photosynthesis there is a steady flow of water, termed the **transpiration stream**, from the roots via the xylem vessels into the leaves.

The loss of water during photosynthesis is unavoidable, as the uptake of CO_2 into the leaves requires openings in the leaf surface, termed **stomata**. The stomata open up to allow the diffusion of CO_2 from the atmosphere into the air space within the leaf, but at the same time water vapor escapes through the open stomata (Fig. 8.1). As the water vapor concentration in

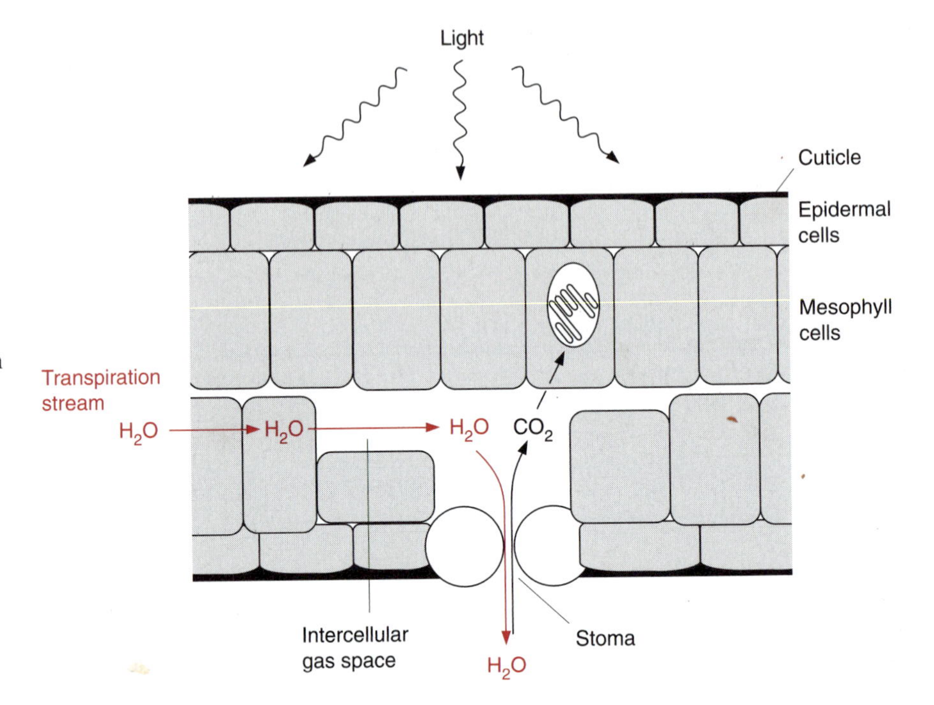

Figure 8.1 Diagram of a cross section of a leaf. The stomata are often located on the lower surface of the leaf. CO_2 diffuses through the stomata into the intercellular air space and thus reaches the mesophyll cells carrying out photosynthesis. Water escapes from the cells into the atmosphere by diffusion in the form of water vapor. This scheme is simplified. In reality, a leaf is formed mostly from several cell layers, and the intercellular gas space is much smaller than in the diagram.

the air space of a leaf in equilibrium with the cell water (31,000 ppm, 25°C) is higher by two orders of magnitude than the CO_2 concentration in air (350 ppm), the escape of a very high amount of water vapor during the influx of CO_2 is inevitable. To minimize this water loss from the leaves, the opening of the stomata is regulated. Thus, when there is a rise in the atmospheric CO_2 concentration, plants require less water. Opening and closing of the stomata is caused by biochemical processes and will be discussed in the next section.

Even when the water supply is adequate, plants open their stomata just enough to provide CO_2 for photosynthesis. During water shortage, plants prevent dehydration by closing their stomata partially or completely, which results in a slowing down or even cessation of CO_2 assimilation. Therefore water shortage is often a decisive factor in limiting plant growth, especially in the warmer and drier regions of our planet, where a large number of plants have evolved a strategy for decreasing water loss during photosynthesis. In CO_2 fixation the first product is 3-phosphoglycerate, a compound with three carbon atoms; hence the name **C_3-plants** (see section 6.2). Some plants save water by first producing the C_4-compound oxaloacetate from fixing CO_2 and are therefore named **C_4-plants**.

8.2 Stomata regulate the gas exchange of a leaf

Stomata are formed by two **guard cells**, which are often surrounded by subsidiary cells. Figures 8.2 and 8.3 show a closed and an open stomatal pore. The pore is opened by the increase in osmotic pressure in the guard cells, resulting in the uptake of water. The corresponding increase in the cell volume **inflates the guard** cells and the **pore opens**.

In order to study the mechanism of the opening process, the guard cells must be isolated. Biochemical and physiological studies are difficult, as the guard cells are very small and can be isolated with only low yields. Although guard cells can be regarded as one of the most thoroughly investigated plant cells, our knowledge of the mechanism of stomatal closure is still incomplete.

Malate plays an important role in guard cell metabolism

The increase in osmotic pressure in guard cells during stomatal opening is mainly due to an accumulation of **potassium salts**. The corresponding anions are usually **malate**, but, depending on the species, sometimes also **chloride**. Figure 8.4 shows a scheme of the metabolic reactions occurring during the opening process with malate as the main anion. An **H^+-P-ATPase** pumps protons across the plasma membrane into the extracellular compartment. The H^+-P-ATPase, which is entirely different from the F-ATPase and V-ATPase (sections 4.3, 4.4), is of the same type as the Na^+/K^+-ATPase in animal cells. An aspartyl residue of the P-ATPase protein is phosphorylated during the transport process (hence the name **P**-ATPase). The potential difference generated by the H^+-P-ATPase drives the influx of K^+ ions into the guard cells via a **K^+ channel**. This channel is open only at a negative voltage (section 1.10) and allows only an inwardly directed flux. For this reason, it is called a **K^+ inward channel**. Most of the K^+ ions taken up into the cell are transported into the vacuole, but the mechanism of this transport is still not known. Probably a vacuolar H^+-ATPase (**V-ATPase**; see section 4.4) is involved, pumping protons into the vacuoles, which could then be exchanged for K^+ ions via a vacuolar potassium channel.

Accumulation of cations in the vacuole leads to the formation of a potential difference across the vacuolar membrane, driving the influx of malate via a channel specific for organic anions. Malate is provided by glycolytic degradation of the starch stored in the chloroplasts. As described in section 9.1, this degradation yields triose phosphate, which is released from the chloroplasts to the cytosol in exchange for inorganic phosphate via the **triose**

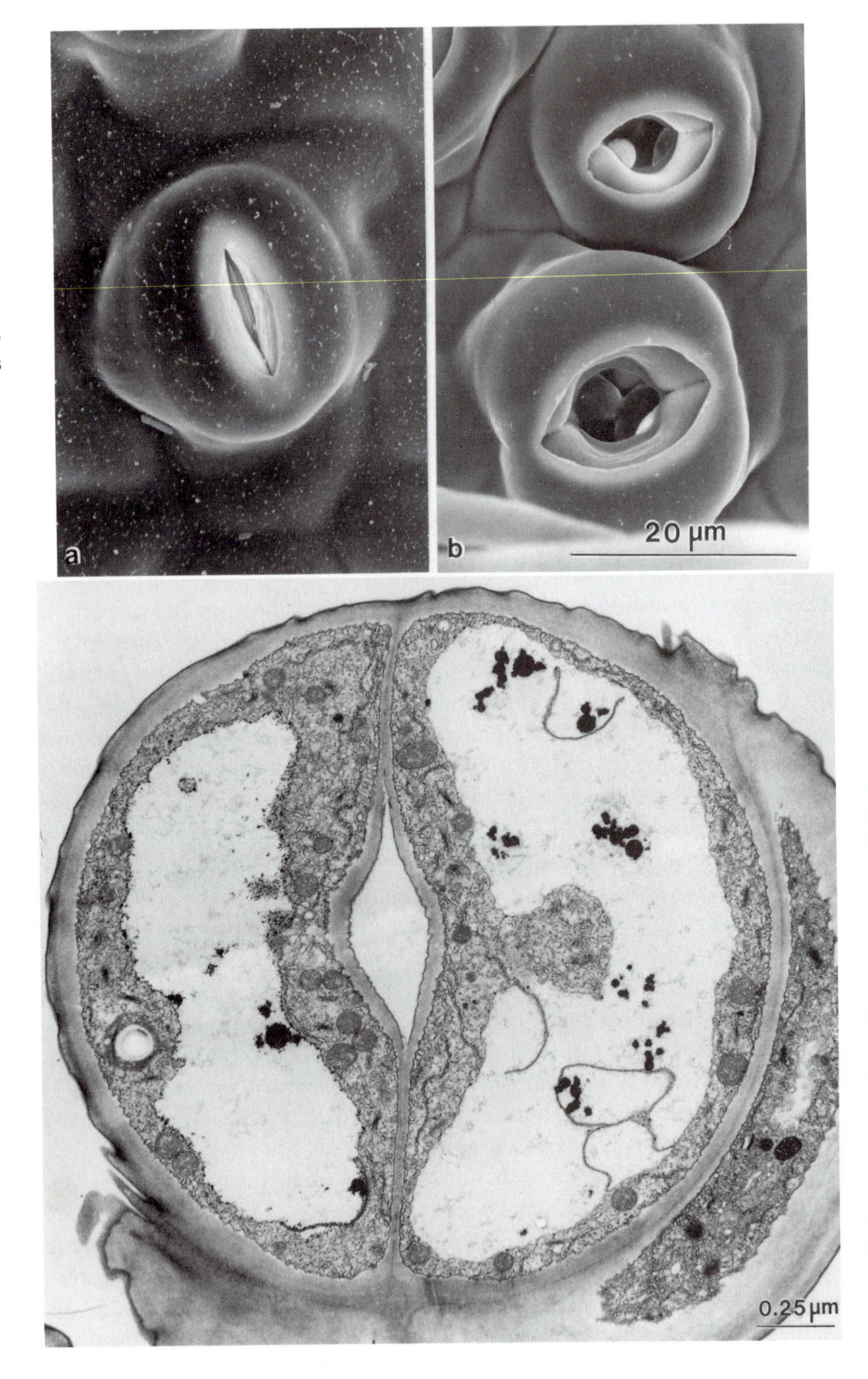

Figure 8.2 Scanning electron micrograph of stomata from the lower epidermis of hazel nut leaves in (a) closed state, and (b) open. By R. S. Harrison-Murray and C. M. Clay, Wellesbourne). Traverse section of a pair of guard cells from a tobacco leaf. The large central vacuole and the gap between the two guard cells can be seen. (By D. G. Robinson, Heidelberg.)

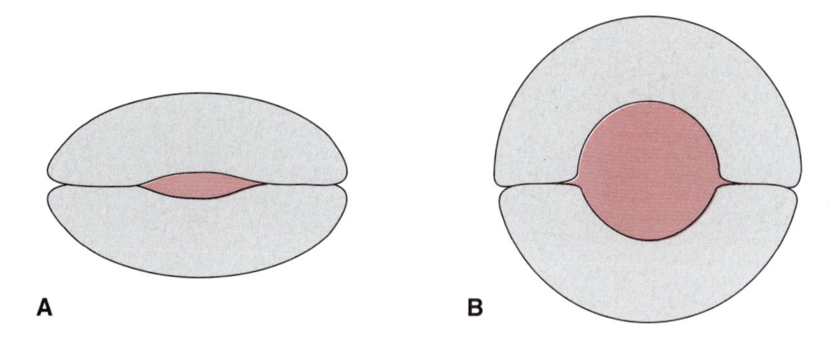

Figure 8.3 Diagram of a stoma formed from the two guard cells, (A) closed and (B) open state.

phosphate-phosphate translocator (section 1.9) and is subsequently converted to phosphoenolpyruvate (see Fig. 10.11). Phosphoenolpyruvate reacts with HCO_3^- to form oxaloacetate in a reaction catalyzed by the enzyme **phosphoenolpyruvate carboxylase** (Fig. 8.5), in which the high energy enol ester bond is cleaved, making the reaction irreversible. The oxaloacetate formed is transported via a specific translocator to the chloroplasts and reduced to malate via **NADP-malate dehydrogenase** (Fig. 8.4). The malate is then released to the cytosol, probably by the same translocator as that which transports oxaloacetate.

During stomatal closure most of the malate is released from the guard cells. As the guard cells contain only very low activities of RubisCO, they are unable to fix CO_2 in significant amounts. The starch is regenerated from glucose, which is taken up into the guard cells. In contrast to chloroplasts from mesophyll cells, the guard cell chloroplasts have a **glucose 6-phosphate-phosphate translocator**, which transports not only glucose 6-phosphate and phosphate, but also triose phosphate and 3-phosphoglycerate. This translocator is also to be found in plastids from non-green tissues, such as roots (section 13.3).

Complex regulation governs stomatal opening

A number of parameters are known to influence the stomatal opening in a very complex way. The opening is regulated by light via a blue light receptor (section 19.9). An important factor is the CO_2 concentration in the intercellular air space, although the nature of the CO_2 sensor is not known. At micromolar concentrations, **abscisic acid (ABA)** (section 19.6) causes the closure of the stomata. If, due to lack of water, the water potential sinks below a critical mark, a higher ABA synthesis comes into force. The effect of ABA on the stomatal opening is dependent on the intercellular CO_2 concentration and on the presence of the signal substance nitric oxide (NO) (see also section 19.9). The binding of ABA to a membrane receptor triggers one

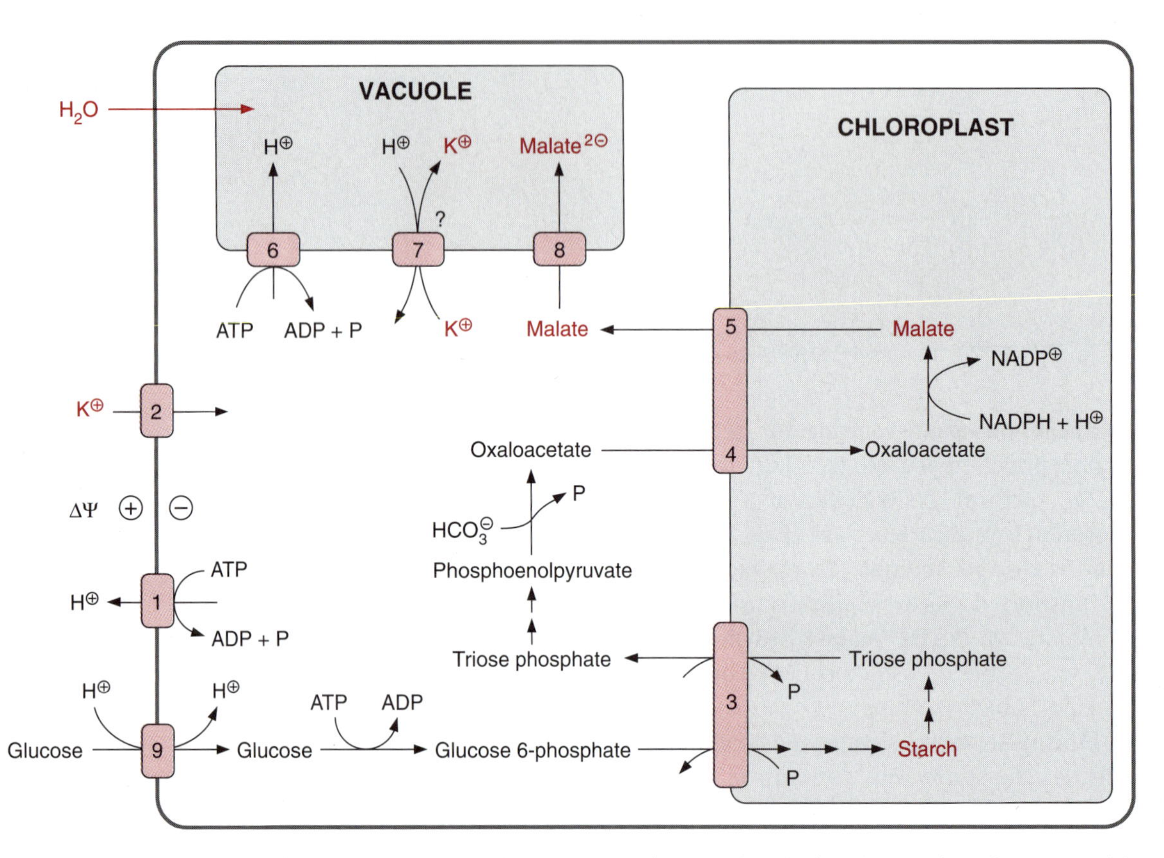

Figure 8.4 Diagram of the processes in operation during the opening of stomata with malate as the main anion. The proton transport by H^+-P-ATPase (1) of the plasma membrane of the guard cell results in an increase in the proton potential and in a hyperpolarization. This opens the voltage-dependent K^+ inward channel (2) and the proton potential drives the influx of potassium ions through this channel. Starch degradation occurs simultaneously in the chloroplasts yielding triose phosphate, which is then released from the chloroplasts via the triose phosphate-phosphate translocator (3) and converted in the cytosol to oxaloacetate. Oxaloacetate is transported into the chloroplasts (4) and is converted to malate by reduction. This malate is transported from the chloroplast to the cytosol, possibly via the same translocator responsible for the influx of oxaloacetate (5). The mechanism of the accumulation of potassium malate in the guard cell vacuoles is not yet known in detail. Protons are transported into the vacuole (6), probably by a H^+V-ATPase, and these protons are exchanged for potassium ions (7). The electric potential difference formed by the H^+V-ATPase drives the influx of malate ions via a malate channel (8). The accumulation of potassium malate increases the osmotic potential in the vacuole and results in an influx of water. For resynthesis of starch, glucose is taken up into the guard cells via an H^+-symport (9), where it is converted in the cytosol to glucose 6-phosphate, which is then transported into the chloroplast via a glucose-phosphate-phosphate translocator (3).

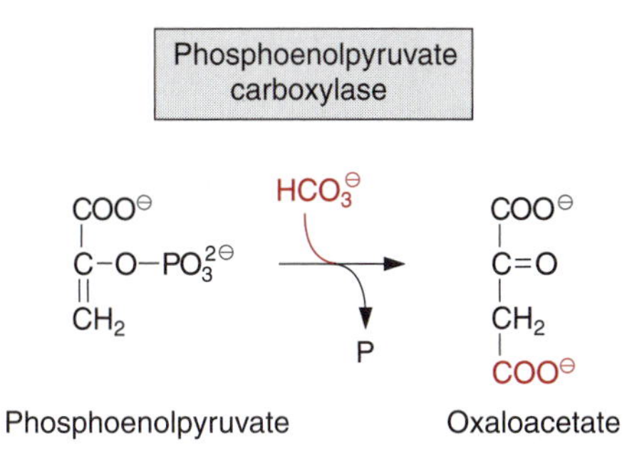

Figure 8.5 Phosphoenolpyruvate carboxylase.

or several signal cascades, at the end of which the opening of ion channels is controlled. There is strong evidence that protein kinases, cyclic ADP ribose, and inositol trisphosphate participate in these signal cascades, which open Ca^{++} channels of the plasma membrane and of internal Ca^{++} stores, such as the endoplasmic reticulum. The resulting Ca^{++} ions in the cytosol function as secondary messengers (section 19.6). By these cascades, ABA activates anion channels in the guard cells, resulting in an efflux of anions. This causes depolarization of the plasma membrane and thus leads to the opening of **K^+ outward channels** (section 1.10). NO regulates these Ca^{++}-sensitive ion channels by promoting Ca^{++} release from intracellular stores to raise the cytosolic free Ca^{++} concentration. The resulting release of K^+-, malate^{2-}-, and Cl^--ions from the guard cells by the joint effect of ABA and NO lowers the osmotic pressure, which leads to a decrease in the guard cell volume and hence to a closure of the stomata. The introduction of the patch clamp technique (section 1.10) has brought important insights into the role of specific ion channels in the stomatal opening process. NO is probably formed by nitric oxide-synthase (section 19.9) or via reduction of nitrite (NO_2^-), as catalyzed by nitrate reductase (section 10.1). In the guard cells, nitrate reductase is induced by ABA. Apparently the interaction of ABA and NO in controlling stomatal opening is very complex.

8.3 The diffusive flux of CO_2 into a plant cell

The movement of CO_2 from the atmosphere to the catalytic center of RubisCO—through the stomata, the intercellular air space, across the plasma membrane, the chloroplast envelope, and the chloroplast stroma—proceeds by diffusion.

According to a simple derivation of the Fick law, the diffusive flux, I, over a certain distance is:

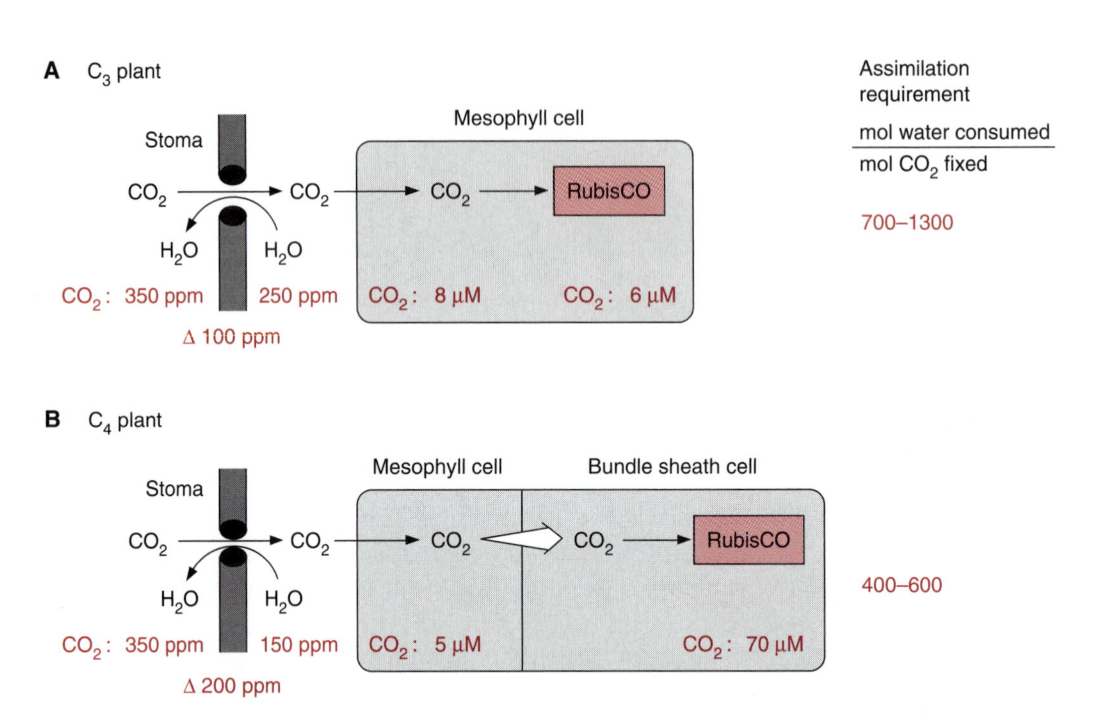

Figure 8.6 Reaction diagram for the uptake of CO_2 in C_3 (A) and C_4 (B) plants. This scheme shows typical stomatal resistances for C_3 and C_4 plants. The values for the CO_2 concentration in the vicinity of RubisCO are taken from von Caemmerer and Evans (C_3 plants) and Hatch (C_4 plants).

$$I = \frac{\Delta C}{R}$$

where I is defined as the amount of a substance diffusing per unit of time and surface area; ΔC, the diffusion gradient, is the difference of concentrations between start and endpoint; and R is the diffusion resistance. R of CO_2 is 10^4 times larger in water than in air.

In Figure 8.6A a model illustrates the diffusive flux of CO_2 into a leaf of a C_3 plant with a limited water supply. The control of the aperture of the stomata leads to a stomatal diffusion resistance, by which a diffusion gradient of 100 ppm is maintained. The resultant CO_2 concentration of 250 ppm in the intercellular air space is in equilibrium with the CO_2 concentration in an aqueous solution of 8×10^{-6} mol/L (8 µM). In water saturated with air containing 350 ppm CO_2, the equilibrium concentration of the dissolved CO_2 is 11.5 µM at 25°C.

Since the chloroplasts are positioned at the inner surface of the mesophyll cells (see Fig. 1.1), within the mesophyll cell the major distance for the

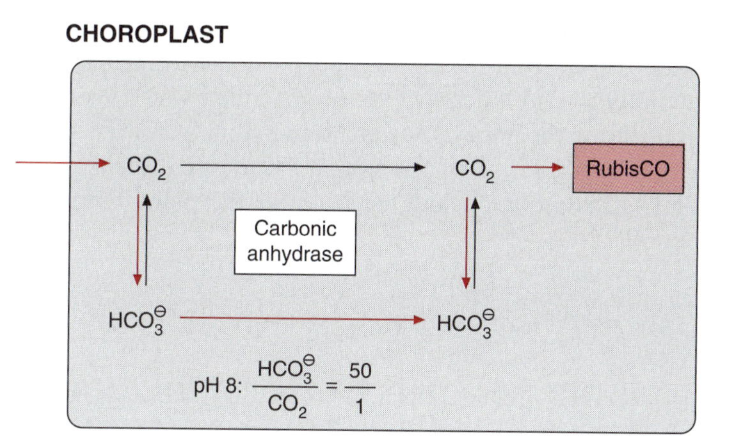

Figure 8.7 Carbonic anhydrase catalyzes the rapid equilibration of CO_2 with HCO_3^- and thus increases the diffusion gradient and hence the diffusive flux of the inorganic carbon across the chloroplast stroma. The example is based on the assumption that the pH value is 8.0. Dissociation constant $[HCO_3^-]\cdot[H^+] / [CO_2] = 5 \times 10^{-7}$.

diffusion of CO_2 to the reaction site of RubisCO is the passage through the chloroplast stroma. To facilitate this diffusive flux, the stroma contains high activities of **carbonic anhydrase**. This enzyme allows the CO_2 entering the chloroplast stroma, after crossing the envelope, to equilibrate with **HCO_3^-** (Fig. 8.7). At pH 8.0, 8 μM CO_2 is in equilibrium with 400 μM HCO_3^- (25°C). Thus, in the presence of carbonic anhydrase the gradient for the diffusive movement of HCO_3^- is 50 times higher than that of CO_2. As the diffusion resistance for HCO_3^- is only about 20% higher than that of CO_2, the diffusive flux of HCO_3^- in the presence of carbonic anhydrase is about 40 times higher than that of CO_2. Due to the presence of carbonic anhydrase in the stroma, the diffusive flux of CO_2 from the intercellular air space to the site of RubisCO in the stroma results in a decrease in CO_2 concentration of only about 2 μM. At the site of RubisCO, a CO_2 concentration of about 6 μM has been measured. In equilibrium with air, the O_2 concentration at the carboxylation site is 250 μM. This results in a carboxylation : oxygenation ratio of about 2.5.

Let us turn our attention again to Figure 8.6. Since CO_2 and O_2 are competitors for the active site of RubisCO, and the CO_2 concentration in the atmosphere is very low as compared with the O_2 concentration, the concentration decrease of CO_2 during the diffusive flux from the atmosphere to the active site of carboxylation is still a limiting factor for efficient CO_2 fixation by RubisCO. This also may be a reason why this enzyme has such a high concentration (see section 6.2). Naturally, the stomatal resistance could be decreased by increasing the aperture of the stomata (e.g., by a factor of two). In this case, with still the same diffusive flux, the CO_2 concentration in the intercellular air space would be increased from 250 to 300 ppm, and the ratio of carboxylation to oxygenation by RubisCO would be increased accordingly. The price, however, for such a reduction of the stomatal diffusion resistance would be a doubling of the water loss. Since the diffusive efflux of water vapor from the leaves is proportional to the diffusion gradient, air humidity is also a decisive factor governing water loss. These considerations illustrate the important function of stomata for the gas exchange of the leaves. The regulation of the stomatal aperture determines how high the rate of CO_2 assimilation may be, without the plant losing too much essential water.

8.4 C_4 plants perform CO_2 assimilation with less water consumption than C_3 plants

In equilibrium with fluid water, the density of water vapor increases exponentially with the temperature. A temperature increase from 20°C to 30°C leads to an almost doubling of water vapor density. Therefore, at high temperatures, the problem of water loss during CO_2 assimilation becomes very serious for plants. C_4 plants found a way to decrease this water loss considerably. At around 25°C these plants use only 400 to 600 mol of H_2O for the fixation of one mol of CO_2, which is just about half the water consumption of C_3 plants, and this difference is even greater at higher temperatures. C_4 plants grow mostly in warm areas that are often also dry; they include important crop plants such as maize, sugarcane, and millet. The principle by which these C_4 plants save water can be demonstrated by comparing the models of C_3 and C_4 plants in Figure 8.6. By doubling the stomatal resistance prevailing in C_3 plants, the C_4 plant can decrease the diffusive efflux of water vapor by 50%.

In order to maintain the same diffusive flux of CO_2 as in C_3 plants at this increased stomatal resistance, in accordance with Fick's law, the diffusion gradient has to be increased by a factor of two. This means that at 350 ppm CO_2 in the air, the CO_2 concentration in the intercellular air space would be only 150 ppm, which is in equilibrium with 5 μM CO_2 in water. At such low CO_2 concentrations C_3 plants would be approaching the compensation point (section 7.6), and therefore the rate of net CO_2 fixation by RubisCO would be very low.

Under these conditions, the crucial factor for the maintenance of CO_2 assimilation in C_4 plants is the presence of a **pumping mechanism** that elevates the concentration of CO_2 at the carboxylation site from 5 μM to about 70 μM. This pumping requires two compartments and the input of energy. However, the energy costs may be recovered, since with this high CO_2 concentration at the carboxylation site, the oxygenase reaction is eliminated to a great extent and the loss of energy connected with the photorespiratory pathway is largely decreased (section 7.5). For this reason, C_4 metabolism does not necessarily imply a higher energy demand; in fact, at higher temperatures C_4 photosynthesis is more efficient than C_3 photosynthesis. This is due to the fact that the oxygenase activity of RubisCO increases more rapidly than the carboxylase activity at increasing temperature. *This is why in warm climates C4 plants with, not only their reduced water demand, but also their suppression of photorespiration, have an advantage over C_3 plants.*

The discovery of C_4 metabolism was stimulated by an unexplained experimental result: After Melvin Calvin and Andrew Benson had established that 3-phosphoglycerate is the primary product of CO_2 assimilation by plants, Hugo Kortschak and colleagues studied the incorporation of radioactively labeled CO_2 during photosynthesis of sugarcane leaves at a sugarcane research institute in Hawaii. The result was surprising. The primary fixation product was not, as expected, 3-phosphoglycerate, but the **C_4** compounds **malate** and **aspartate**. This result questioned whether the then fully accepted Calvin cycle was universally valid for CO_2 assimilation. Perhaps Kortschak was reluctant to raise these doubts, and his results remained unpublished for almost 10 years. It is interesting to note that during this time and without knowing these results, Yuri Karpilov in the former Soviet Union observed similar radioactive CO_2 fixation into C_4 compounds during photosynthesis in maize.

Following the publication of these puzzling results, Hal Hatch and Roger Slack in Australia set out to solve the riddle by systematic studies. They found that the incorporation of CO_2 in malate was a reaction preceding the CO_2 fixation by the Calvin cycle and that this first carboxylation reaction was part of a CO_2 concentration mechanism, the function of which was elucidated by the two researchers by 1970. Earlier, this process was known as the Hatch-Slack pathway. However, they used the term *C_4 dicarboxylic acid pathway of photosynthesis* and this was later abbreviated to the **C_4 pathway** or **C_4 photosynthesis**.

The CO_2 pump in C_4 plants

The requirement of two different compartments for pumping CO_2 from a low to a high concentration is reflected in the leaf anatomy of C_4 plants. The leaves of C_4 plants show a so-called **Kranz-anatomy** (Fig. 8.8). The vascular bundles, containing the sieve tubes and the xylem vessels, are surrounded by a sheath of cells (**bundle sheath cells**), and these are encircled by **mesophyll cells**. The latter are in contact with the intercellular air space of the leaves. In 1884 the German botanist Gustav Haberland described in his textbook *Physiologische Pflanzenanatomie (Physiological Plant Anatomy)* that the

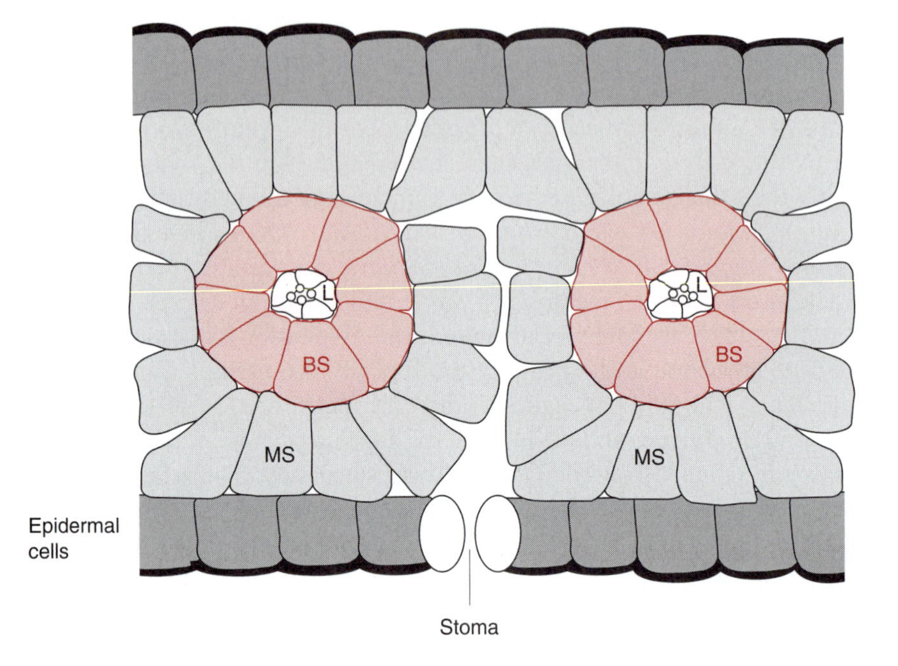

Figure 8.8 Characteristic leaf anatomy of a C_4 plant. Schematic diagram. V = Vascular bundle; BS = bundle sheath cells; MS = mesophyll cells.

assimilatory cells in several plants, including sugarcane and millet, are arranged in what he termed a Kranz (wreath)-type mode. With remarkable foresight, he suggested that this special anatomy may indicate a division of labor between the chloroplasts of the mesophyll and bundle sheath cells.

Mesophyll and bundle sheath cells are separated by a cell wall, which in some instances contains a **suberin layer**, which is probably gas-impermeable. Suberin is a polymer of phenolic compounds that are impregnated with wax (section 18.3). The border between the mesophyll and bundle sheath cells is penetrated by a large number of **plasmodesmata** (section 1.1). These plasmodesmata enable the diffusive flux of metabolites between the mesophyll and bundle sheath cells.

The CO_2 pumping of C_4 metabolism does not rely on the specific function of a membrane transporter but is due to a prefixation of CO_2, after conversion to HCO_3^-, by reaction with phosphoenolpyruvate to form oxaloacetate in the mesophyll cells. After the conversion of this oxaloacetate to malate, the malate diffuses through the plasmodesmata into the bundle sheath cells, where CO_2 is released as a substrate for RubisCO. Figure 8.9 shows a simplified scheme of this process. The formation of the CO_2 gradient between the two compartments by this pumping process is due to the fact that the prefixation of CO_2 and its subsequent release are catalyzed by two different reactions, each of which is virtually irreversible. As a crucial

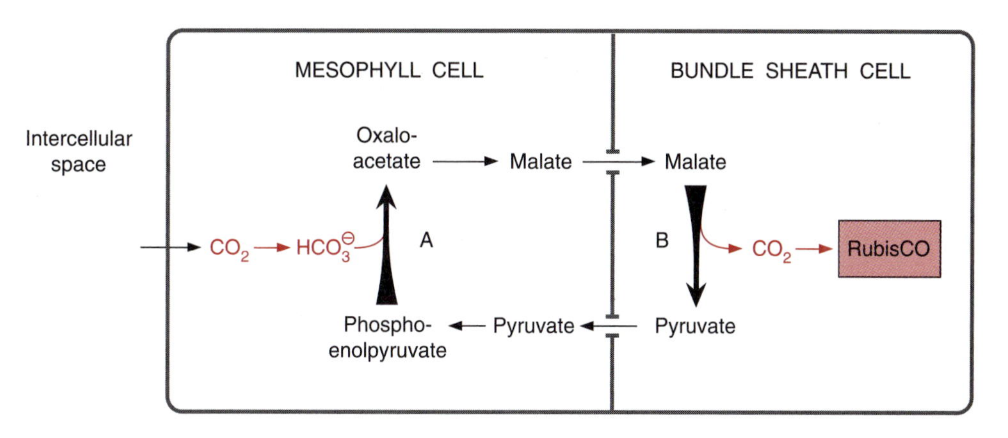

Figure 8.9 Principle of C_4 metabolism.

feature of C_4 metabolism, RubisCO is located exclusively in the bundle sheath chloroplasts.

The reaction of HCO_3^- with phosphoenolpyruvate is catalyzed by the enzyme **phosphoenolpyruvate carboxylase**. This enzyme has already been discussed when dealing with the metabolism of guard cells (Figs. 8.4 and 8.5). The reaction is strongly exergonic and therefore irreversible. As the enzyme has a very high affinity for HCO_3^-, micromolar concentrations of HCO_3^- are fixed very efficiently. The formation of HCO_3^- from CO_2 is facilitated by carbonic anhydrase present in the cytosol of the mesophyll cells.

The release of CO_2 in the bundle sheath cells occurs in three different ways (Fig. 8.10). In most C_4 species, decarboxylation of malate with an accompanying oxidation to pyruvate is catalyzed by **malic enzyme**. In one group of these species termed **NADP-malic enzyme type** plants, release of CO_2 occurs in the bundle sheath chloroplasts and oxidation of malate to pyruvate is coupled with the reduction of $NADP^+$. In other plants, termed **NAD-malic enzyme type**, decarboxylation takes place in the mitochondria of the bundle sheath cells and is accompanied by the reduction of NAD^+. In the **phosphoenolpyruvate carboxykinase type** plants, oxaloacetate is decarboxylated in the cytosol of the bundle sheath cells. ATP is required for this reaction and phosphoenolpyruvate is formed as a product.

The following discusses the metabolism and its compartmentation in the three different types of C_4 plants in more detail.

C_4 metabolism of the NADP-malic enzyme type plants

This group includes the important crop plants maize and sugarcane. Figure 8.11 shows the reaction chain and its compartmentalization. The oxaloacetate arising from the carboxylation of phosphoenolpyruvate is transported via a specific translocator into the chloroplasts where it is reduced by

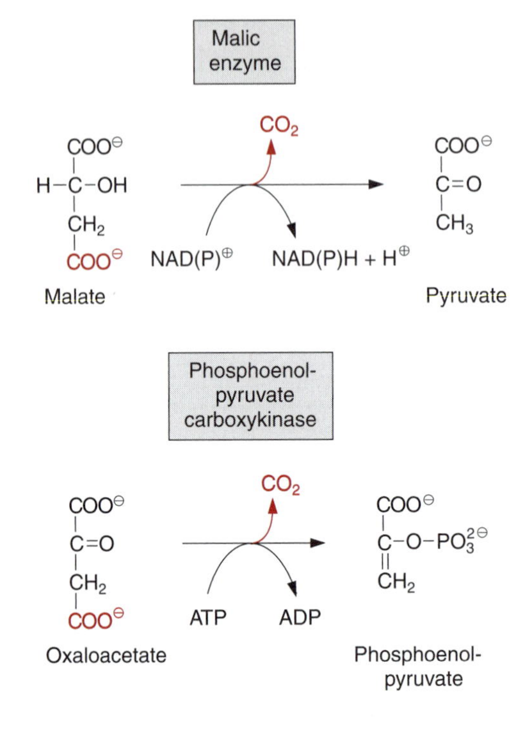

Figure 8.10 Reactions by which CO_2 prefixed in C_4 metabolism in mesophyll cells can be released in bundle sheath cells.

NADP-malate dehydrogenase to malate, which is subsequently transported into the cytosol. (The reduction of oxaloacetate in the chloroplasts has been discussed in section 7.3 in connection with photorespiratory metabolism). Malate diffuses via plasmodesmata from the mesophyll to the bundle sheath cells. The diffusive flux of malate between the two cells requires a diffusion gradient of about 2×10^{-3} mol/L. The malic enzyme present in the bundle sheath cells causes the conversion of malate to pyruvate and CO_2, and the CO_2 is fixed by RubisCO.

The remaining pyruvate is exported by a specific translocator from the bundle sheath chloroplasts, diffuses through the plasmodesmata into the mesophyll cells, where it is transported by another specific translocator into the chloroplasts. The enzyme **pyruvate-phosphate dikinase** in the mesophyll chloroplasts converts pyruvate to phosphoenolpyruvate by a rather unusual reaction (Fig. 8.12). The name *dikinase* means an enzyme that catalyzes a twofold phosphorylation. In a reversible reaction, one phosphate residue is transferred from ATP to pyruvate and a second one to phosphate, converting it to pyrophosphate. A **pyrophosphatase** present in the chloroplast stroma immediately hydrolyzes the newly formed pyrophosphate and thus makes this reaction irreversible. In this way pyruvate is transformed upon the consumption of two energy-rich phosphates of ATP (which is converted to

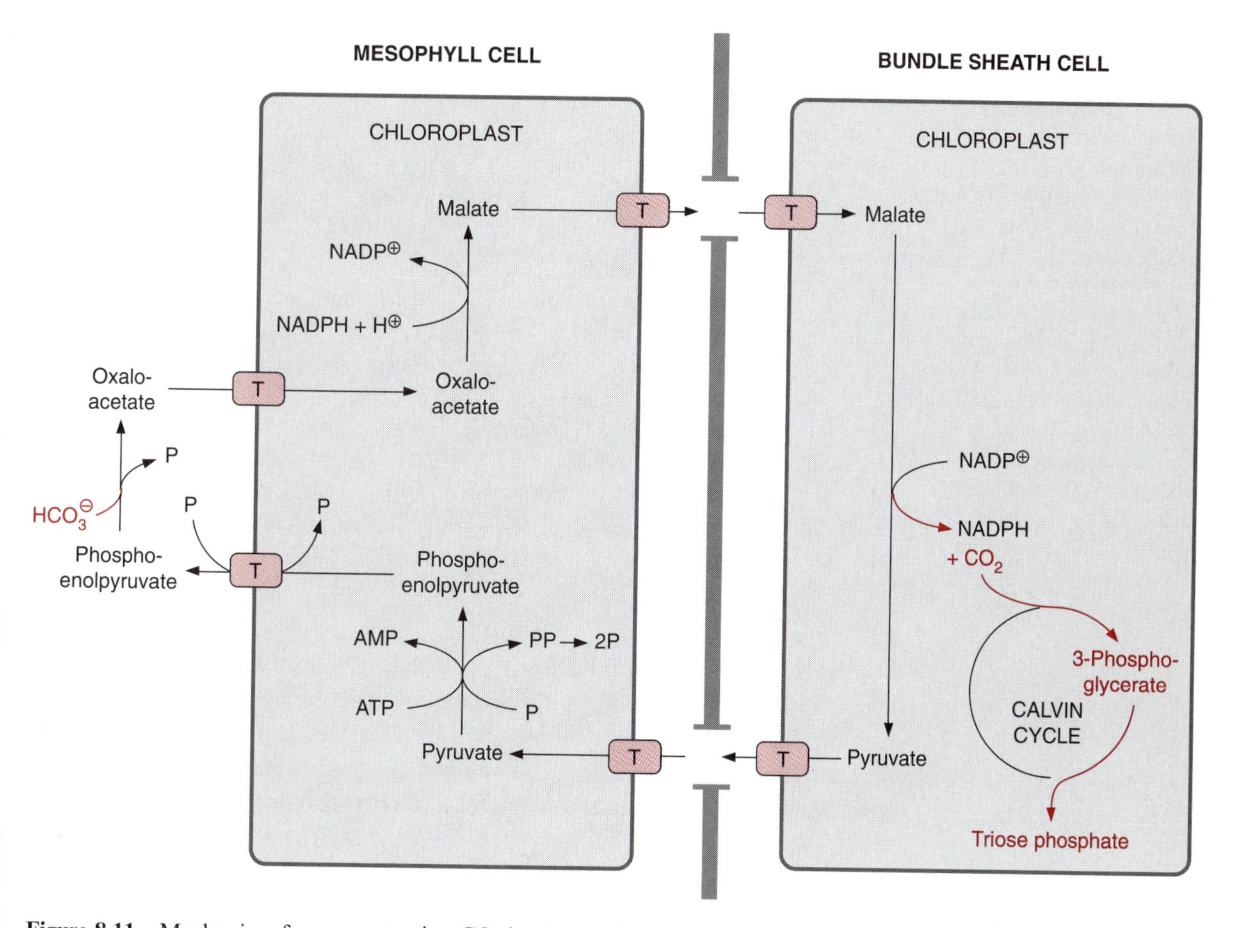

Figure 8.11 Mechanism for concentrating CO_2 in plants of the C_4-NADP-malic enzyme type (e.g., maize). In the cytosol of the mesophyll cells, HCO_3^- is fixed by reaction with phosphoenolpyruvate, and the oxaloacetate formed is reduced in the chloroplast to form malate. After leaving the chloroplasts, malate diffuses into the bundle sheath cells, where it is oxidatively decarboxylated, leading to the formation of pyruvate, CO_2, and NADPH. The pyruvate formed is phosphorylated to phosphoenolpyruvate in the chloroplasts of the mesophyll cells. The transport across the chloroplast membranes proceeds by specific translocators. The diffusive flux between the mesophyll and the bundle sheath cells proceeds through plasmodesmata. The transport of oxaloacetate into the mesophyll chloroplasts and the subsequent release of malate from the chloroplasts is probably facilitated by the same translocator. T = translocator.

AMP) irreversibly into phosphoenolpyruvate. The latter is exported in exchange for inorganic phosphate from the chloroplasts via a **phospho-enolpyruvate-phosphate translocator**.

With the high CO_2 gradient between bundle sheath and mesophyll cells, the question arises why does most of the CO_2 not leak out before it is fixed

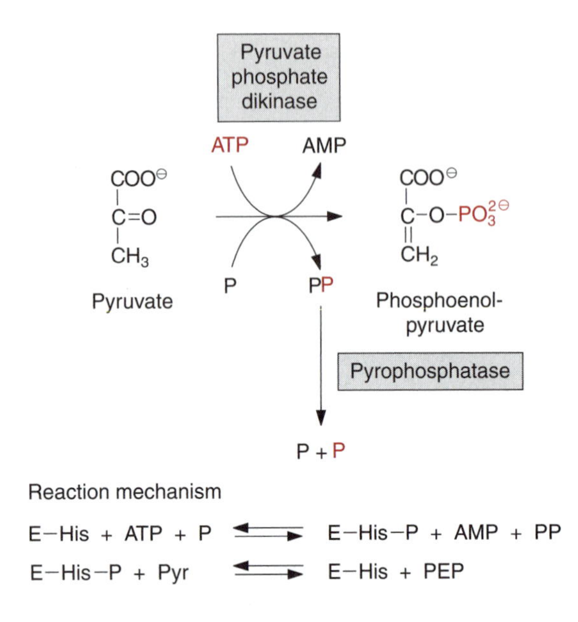

Figure 8.12 Pyruvate-phosphate dikinase. In this reaction, one phosphate residue is transferred from ATP to inorganic phosphate, resulting in the formation of pyrophosphate, and a second phosphate residue is transferred to a histidine residue at the catalytic site of the enzyme. In this way a phosphor amide (R-H-N-PO_3^{2-}) is formed as an intermediate, and this phosphate residue is then transferred to pyruvate, resulting in the formation of phosphoenolpyruvate.

by RubisCO? As the bundle sheath chloroplasts, in contrast to those from mesophyll cells (see Fig. 8.7), do not contain carbonic anhydrase, the diffusion of CO_2 through the stroma of bundle sheath cells proceeds more slowly than in the mesophyll cells. The suberin layer between the cells of some plants probably prevents the leakage of CO_2 through the cell wall, and in this case there would be only a diffusive loss through plasmodesmata. The portion of CO_2 concentrated in the bundle sheath cells that is lost by diffusion back to the mesophyll cells is estimated at 10% to 30% in different species.

In maize leaves, the chloroplasts from mesophyll cells differ in their structure from those of bundle sheath cells. Mesophyll chloroplasts have many grana, whereas bundle sheath chloroplasts contain mainly stroma lamellae, with only very few granal stacks and little photosystem II activity (section 3.10). The major function of the bundle sheath chloroplasts is to provide ATP by **cyclic photophosphorylation** via photosystem I. NADPH required for the reductive pentose phosphate pathway (Calvin cycle) is provided mainly by the linear electron transport in the mesophyll cells. This NADPH is delivered in part via the oxidative decarboxylation of malate (by NADP-malic enzyme), but this reducing power is actually provided by the mesophyll cells for the reduction of oxaloacetate. The other part of NADPH required is transferred along with ATP from the mesophyll chloroplasts to the bundle sheath chloroplasts by a **triose phosphate-3-phosphoglycerate shuttle** via the triose phosphate-phosphate translocators of the inner envelope membranes of the corresponding chloroplasts (Fig. 8.13).

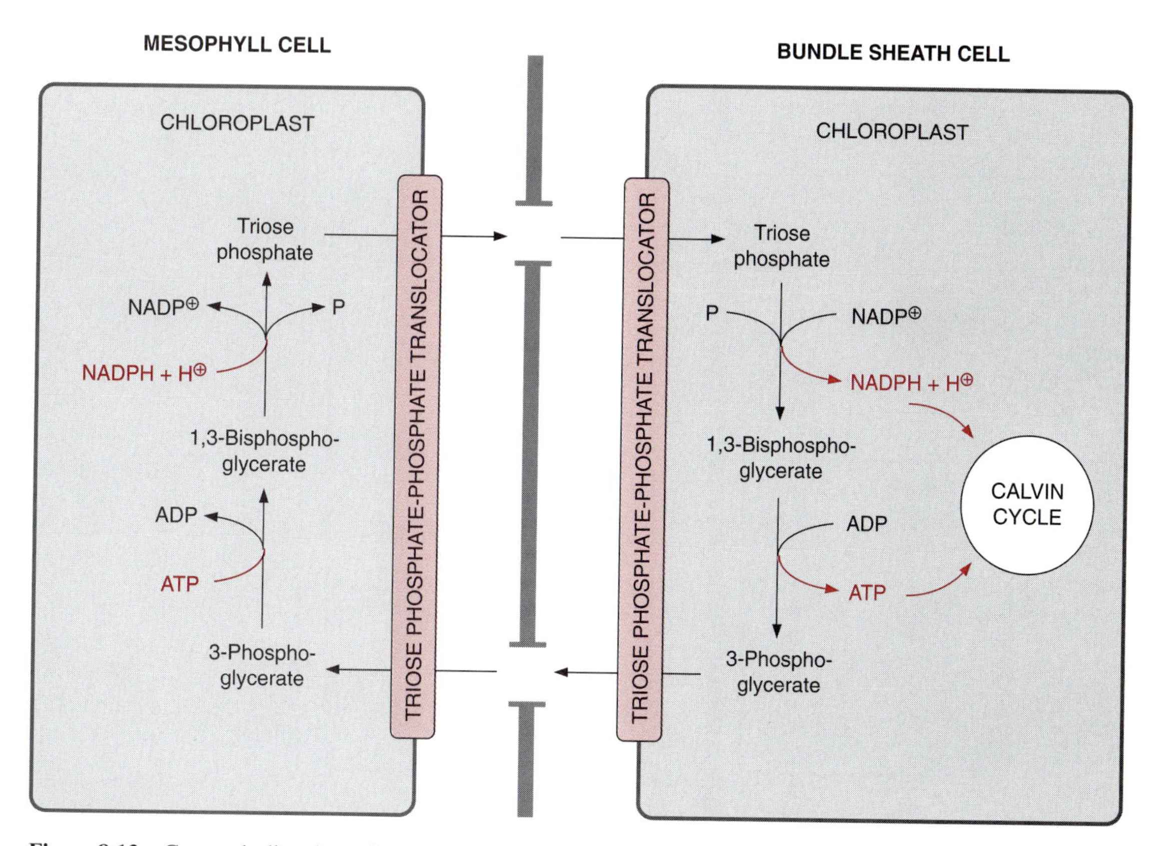

Figure 8.13 C_4 metabolism in maize. Indirect transfer of NADPH and ATP from the mesophyll chloroplast to the bundle sheath chloroplast via a triose phosphate-3-phosphoglycerate shuttle. In the chloroplasts of mesophyll cells, 3-phosphoglycerate is reduced to triose phosphate at the expense of ATP and NADPH. In the bundle sheath chloroplasts, triose phosphate is reconverted to phosphoglycerate, leading to the formation of NADPH and ATP. Transport across the chloroplast membranes proceeds by counter-exchange via triose phosphate-phosphate translocators.

C_4 metabolism of the NAD-malic enzyme type

The metabolism of the NAD-malic enzyme type, shown in the metabolic scheme of Figure 8.14, is found in a large number of species including millet. Here the oxaloacetate formed by phosphoenolpyruvate carboxylase is converted to aspartate by transamination via **glutamate-aspartate aminotransferase**. Since the oxaloacetate concentration in the cell is below 0.1×10^{-3} mol/L, oxaloacetate cannot form a high enough diffusion gradient for the necessary diffusive flux into the bundle sheath cells. Because of the high concentration of glutamate in a cell, the transamination of oxaloacetate yields aspartate concentrations in a range between 5 and 10×10^{-3} mol/L.

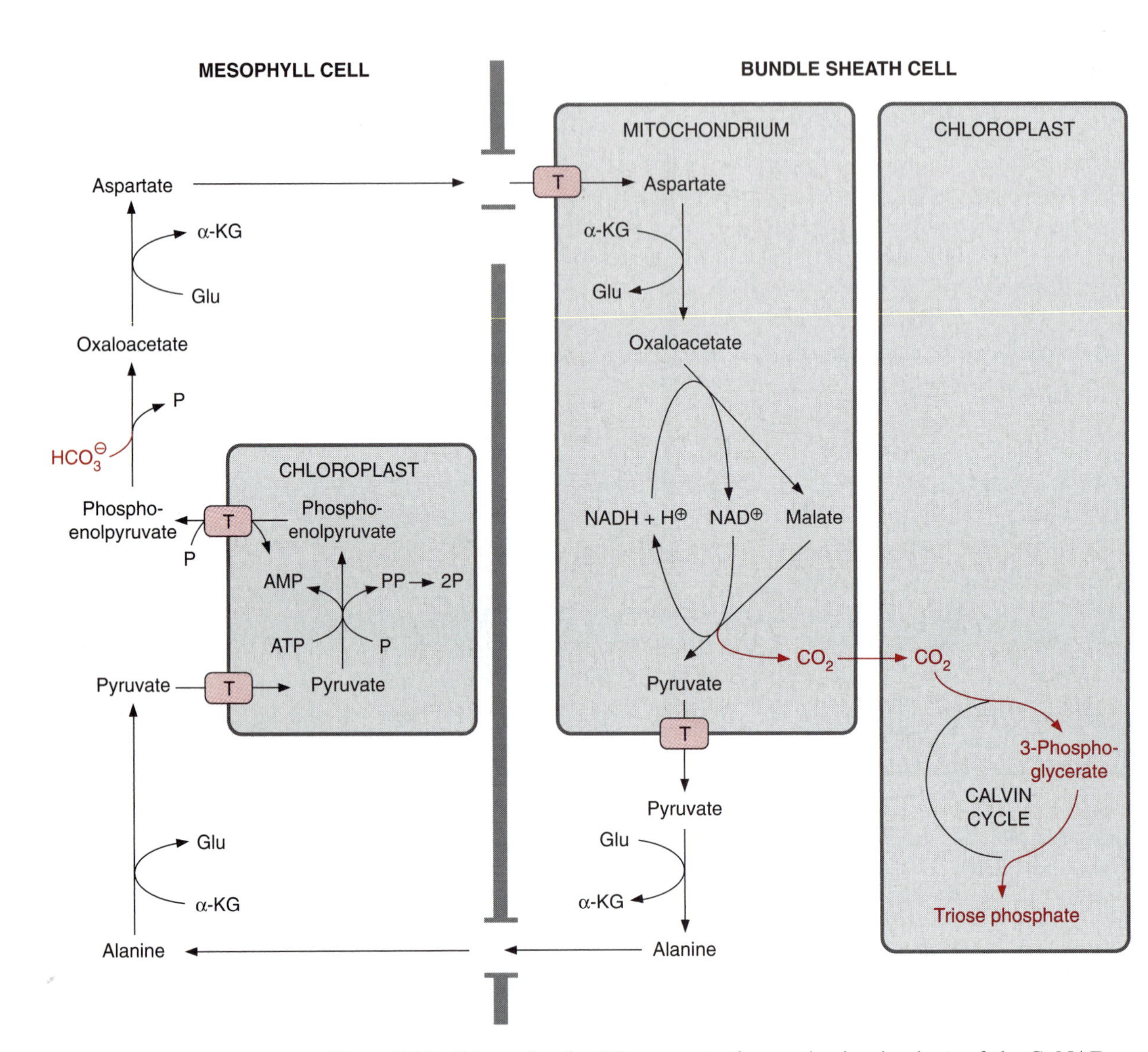

Figure 8.14 Scheme for the CO_2 concentrating mechanism in plants of the C_4 NAD-malic enzyme type. In contrast to C_4 metabolism described in Figure 8.11, oxaloacetate is transaminated in the cytosol to aspartate. After diffusion through the plasmodesmata, aspartate is transported into the mitochondria of the bundle sheath cells by a specific translocator and is reconverted there to oxaloacetate and the latter reduced to malate. In the mitochondria, malate is oxidized by NAD-malic enzyme, giving CO_2 and pyruvate. In the cytosol, pyruvate is transaminated to alanine, which then diffuses into the mesophyll cells. The CO_2 released from the mitochondria diffuses into the bundle sheath chloroplasts, which are in close contact with the mitochondria, and this CO_2 serves as substrate for RubisCO. Abbreviations: Glu = glutamate; α-Kg = α-ketoglutarate; P = phosphate; PP = pyrophosphate; T = translocator.

This is why aspartate is very suitable for supporting a diffusive flux between the mesophyll and bundle sheath cells.

After diffusing into the bundle sheath cells, aspartate is transported by a translocator into the mitochondria. An isoenzyme of glutamate-aspartate aminotransferase present in the mitochondria catalyzes the conversion of aspartate to oxaloacetate, which is then transformed by **NAD-malate dehydrogenase** to malate. This malate is decarboxylated by **NAD-malic enzyme** to pyruvate and the NAD formed during the malate dehydrogenase reaction is reduced to NADH again. CO_2 thus released in the mitochondria then diffuses into the chloroplasts, where it is available for assimilation via RubisCO. The pyruvate leaves the mitochondria via the pyruvate translocator and is converted in the cytosol by an alanine glutamate amino transferase to alanine. Since in the equilibrium of this reaction the alanine concentration is much higher than that of pyruvate, a high diffusive flux of alanine into the mesophyll cells is possible. In the mesophyll cells, alanine is transformed to pyruvate by an isoenzyme of the aminotransferase mentioned previously. Pyruvate is transported into the chloroplasts, where it is converted to phosphoenolpyruvate by pyruvate phosphate dikinase in the same way as in the chloroplasts of the NADP-malic enzyme type.

All NADH formed by malic enzyme in the mitochondria is sequestered for the reduction of oxaloacetate, and thus there are no reducing equivalents left to be oxidized by the respiratory chain (Fig. 8.14). To enable mitochondrial oxidative phosphorylation to form ATP, some of the oxaloacetate formed in the mesophyll cells by phosphoenolpyruvate carboxylase is reduced in the mesophyll chloroplasts to malate, as in the NADP-malic enzyme type metabolism. This malate diffuses into the bundle sheath cells, is taken up by the mitochondria, and is oxidized there by malic enzyme to yield NADH. ATP is generated from oxidation of this NADH by the respiratory chain. This pathway also operates in the phosphoenolpyruvate carboxykinase type metabolism, described below.

C_4 metabolism of the phosphoenolpyruvate carboxykinase type

This type of metabolism is found in several of the fast-growing tropical grasses used as forage crops. Figure 8.15 shows a scheme of the metabolism. As in the NAD-malic enzyme type, oxaloacetate is converted in the mesophyll cells to aspartate and the latter diffuses into the bundle sheath cells, where the oxaloacetate is regenerated via an aminotransferase in the cytosol. In the cytosol the oxaloacetate is converted to phosphoenolpyruvate at the expense of ATP via phosphoenolpyruvate carboxykinase. The CO_2 released

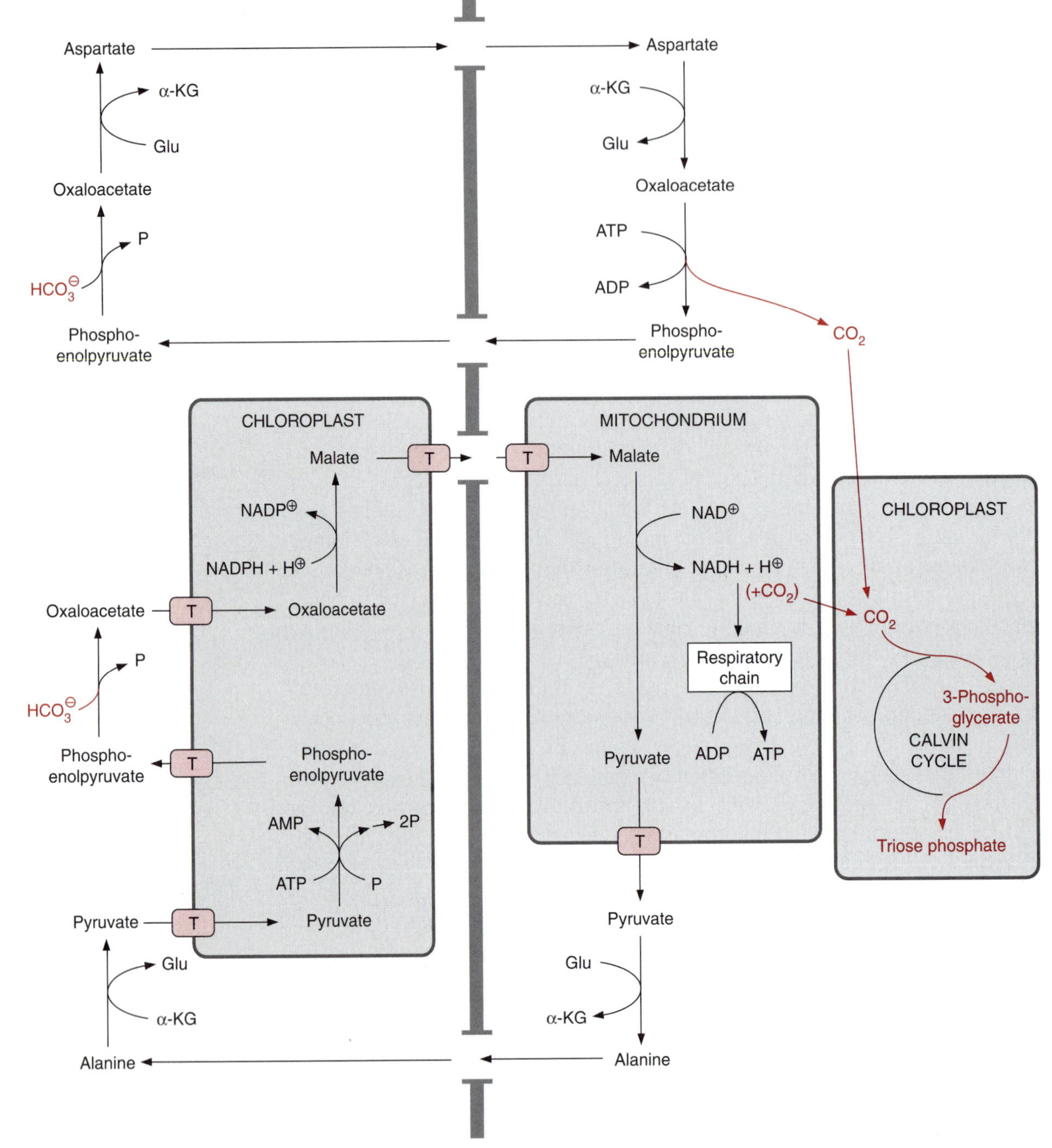

Figure 8.15 Scheme of the CO_2 concentrating mechanism in plants of the C_4-phosphoenolpyruvate carboxykinase type. In contrast to C_4 metabolism described in Figure 8.14, oxaloacetate is formed from aspartate in the cytosol of the bundle sheath cells, and is then decarboxylated to phosphoenolpyruvate and CO_2 via the enzyme phosphoenolpyruvate carboxykinase. Phosphoenolpyruvate diffuses back into the mesophyll cells. Simultaneously, as in Figure 8.11, some malate formed in the mesophyll cells diffuses into the bundle sheath cells and is oxidized there by an NAD-malic enzyme in the mitochondria. The NADH thus formed serves as a substrate for the formation of ATP by mitochondrial oxidative phosphorylation. This ATP is transported to the cytosol to be used for phosphoenolpyruvate carboxykinase reaction. The CO_2 released in the mitochondria, together with the CO_2 released by phosphoenolpyruvate carboxykinase in the cytosol, serves as a substrate for the RubisCO in the bundle sheath chloroplasts. T = translocator.

in this reaction diffuses into the chloroplasts and the remaining phosphoenolpyruvate diffuses back into the mesophyll cells. In this C_4 type, the ATP demand for pumping CO_2 into the bundle sheath compartment is due especially to ATP consumption by the **phosphoenolpyruvate carboxykinase** reaction (Fig. 8.10). The mitochondria provide the ATP required for phosphoenolpyruvate carboxykinase reaction by oxidizing malate via NAD-malic enzyme. This malate originates from mesophyll cells as in the NADP-malic enzyme type (Fig. 8.15, lower part). Thus, in the C_4-phosphoenolpyruvate carboxykinase type plants, a minor portion of the CO_2 is released in the mitochondria and the bulk is released in the cytosol.

Kranz-anatomy with its mesophyll and bundle sheath cells is not an obligatory requirement for C_4 metabolism

In individual cases, the spatial separation of the prefixation of CO_2 by PEP carboxylase and the final fixation by RubisCO can also be achieved in other ways. Recently it was demonstrated in a species of *Chenopodiacae* that its C_4 metabolism takes place in uniform, extended cells where in the cytoplasm at the peripheral end there is a cell layer with PEP carboxylase and at the proximal end chloroplasts are located containing RubisCO. Although this is a special case, it illustrates the variability of the C_4 system.

Enzymes of C_4 metabolism are regulated by light

Phosphoenolpyruvate carboxylase (PEP carboxylase), the key enzyme of C_4 metabolism, is highly regulated. In a darkened leaf, this enzyme is present in a state of low activity. In this state, the affinity of the enzyme to its substrate phosphoenolpyruvate is very low and it is inhibited by low concentrations of malate. Therefore, during the dark phase the enzyme in the leaf is practically inactive. Upon illumination of the leaf, a serine protein kinase (see also Figs. 9.18 and 10.9) is activated, which phosphorylates the hydroxyl group of a serine residue in PEP carboxylase. The enzyme can be inactivated again by hydrolysis of the phosphate group by a protein serine phosphatase. The activated phosphorylated enzyme is also inhibited by malate, but in this case very much higher concentrations of malate are required for its inhibition than for the nonphosphorylated less active enzyme. The rate of the irreversible carboxylation of phosphoenolpyruvate can be adjusted in such a way through a feedback inhibition by malate that a certain malate level is maintained in the mesophyll cell. It is not yet known which signal turns on the activity of the protein kinase when the leaves are illuminated.

NADP-malate dehydrogenase is activated by light via reduction by thioredoxin as described in section 6.6.

Pyruvate-phosphate dikinase (Fig. 8.12) is also subject to dark/light regulation. It is inactivated in the dark by phosphorylation of a threonine residue. This phosphorylation is rather unusual, as it requires ADP rather than ATP as phosphate donor. The enzyme is activated in the light by the phosphorolytic cleavage of the threonine phosphate group. Thus, the regulation of pyruvate phosphate dikinase proceeds in a completely different way from the regulation of PEP carboxylase.

Products of C_4 metabolism can be identified by mass spectrometry

Measuring the distribution of the ^{12}C and the ^{13}C isotopes in a photosynthetic product (e.g., sucrose) can reveal whether it has been formed by C_3 or C_4 metabolism. ^{12}C and ^{13}C occur as natural carbon isotopes in the CO_2 of the atmosphere in the ratio of 98.89% and 1.11%, respectively. RubisCO reacts with $^{12}CO_2$ more rapidly than with $^{13}CO_2$. This is due to a kinetic isotope effect. For this reason, the ratio $^{13}C/^{12}C$ is lower in the products of C_3 photosynthesis than in the atmosphere. The ratio ^{13}C to ^{12}C can be determined by mass spectrometry and is expressed as a $\delta^{13}C$ value.

$$\delta^{13}C[‰] = \left(\frac{^{13}C/^{12}C \text{ sample}}{^{13}C/^{12}C \text{ in standard}} - 1 \right) \times 1000$$

As a standard, one uses the distribution of the two isotopes in a defined limestone. In products of C_3 photosynthesis $\delta^{13}C$ values of 28‰ are found. In the PEP carboxylase reaction, the preference for ^{12}C over ^{13}C is less pronounced. Since in C_4 plants practically all the CO_2, that had been prefixed by PEP carboxylase, reacts further in the compartment of the bundle sheath cells with RubisCO, the photosynthesis of C_4 plants has a $\delta^{13}C$ value in the range of 14‰ only. Therefore it is possible to determine by mass spectrometric analysis of the ^{13}C to ^{12}C ratio, whether, for instance, sucrose has been formed by sugar beet (C_3 metabolism) or by sugarcane (C_4 metabolism).

C_4 plants include important crop plants but also many of the worst weeds

In C_4 metabolism ATP is consumed to concentrate the CO_2 in the bundle sheath cells. This avoids the loss of energy incurred by photorespiration in C_3 plants. The ratio of oxygenation versus carboxylation by RubisCO increases with the temperature (section 6.2). At a low temperature, with resultant low photorespiratory activity, the C_3 plants are at an advantage. Under these circumstances, C_4 plants offer no benefit and very few C_4 plants occur as wild plants in a temperate climate. However, at temperatures of about 25°C or above, the C_4 plants are at an advantage as, under these conditions, the energy consumption for C_4 photosynthesis (measured as a quantum requirement of CO_2 fixation) is lower than in C_3 plants. As indicated previously, this is due largely to increased photorespiration resulting from an increase in the oxygenase reaction of RubisCO. A further advantage of C_4 plants is that, because of the high CO_2 concentration in the bundle sheath chloroplasts, they need less RubisCO. Since RubisCO is the main protein of leaves, C_4 plants require less nitrogen for growth than C_3 plants. Last, but not least, C_4 plants require less water. In warmer climates these advantages make C_4 plants very suitable as crop plants. Of the 12 most rapidly growing crop or pasture plants, 11 are C_4 plants. It has been estimated that about 20% of global photosynthesis by terrestrial plants comes from C_4 plants. One disadvantage, however, is that many C_4 crop plants, such as maize, millet, and sugarcane, are very sensitive to chilling, and this restricts them to the warmer areas. Especially persistent weeds are members of the C_4 plants, including 8 of the 10 worldwide worst specimens [e.g., Bermuda grass (*Cynodon dactylon*), and barnyard grass (*Echinochloa crusgalli*)].

8.5 Crassulacean acid metabolism makes it possible for plants to survive even during a very severe water shortage

Many plants growing in very dry and often hot habitats have developed a strategy not only for surviving periods of severe water shortage, but also for carrying out photosynthesis under such conditions. The succulent ornamental plant *Kalanchoe* and all the cacti are examples of such plants, as are plants that grow as epiphytes in tropical rain forests, including half of the orchids. These plants solve the problem of water loss during photosynthesis by opening their stomata only during the night, when it is cool and air humidity is comparatively high. Through the open stomata during the night, CO_2 is taken up, and it is fixed in an acid, which is stored until the following day, when the acid is degraded to release the CO_2. This CO_2 feeds it into the Calvin cycle, which can now proceed while the stomata are closed. Figure 8.16 shows the basic scheme of this process. Note the strong similarity of

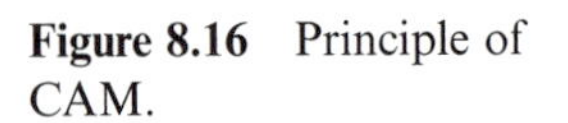

Figure 8.16 Principle of CAM.

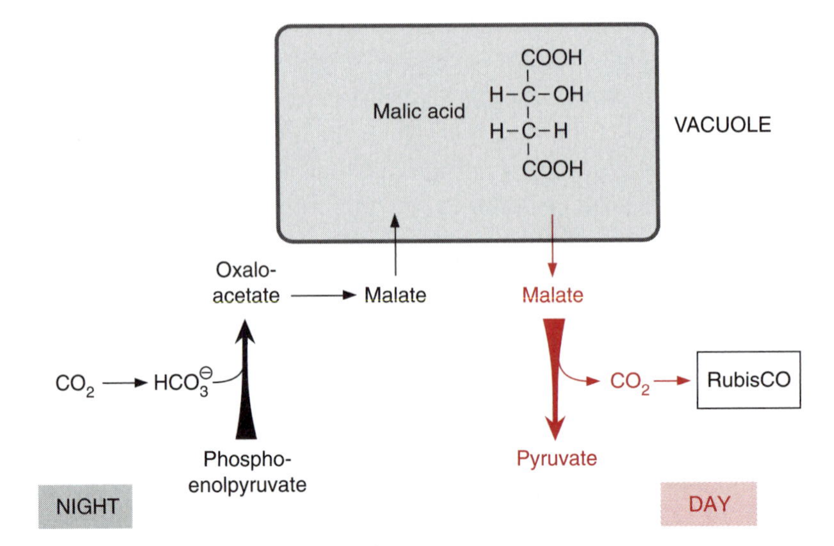

this scheme with the basic scheme of C_4 metabolism in Figure 8.9. The scheme in Figure 8.16 differs from that of C_4 metabolism only in that carboxylation and decarboxylation are separated in time instead of being separated spatially. As this metabolism has first been elucidated in *Crassulaceae* and involves the storage of an acid, it has been named **crassulacean acid metabolism** (abbreviated **CAM**). Important CAM crop plants are pineapples and the agave sisal, yielding natural fibers.

The first observations concerning CAM metabolism were made at the beginning of the nineteenth century. In 1804 the French scientist de Saussure observed that upon illumination and in the absence of CO_2, branches of the cactus *Opuntia* produced oxygen. He concluded that these plants consumed their own matter to produce CO_2, which was then used for CO_2 assimilation. An English gentleman, Benjamin Heyne, noticed in his garden in India that the leaves of the then very popular ornamental plant *Bryophyllum calycinum* had an herby taste in the afternoon, whereas in the morning the taste was as acid as sorrel. He found this observation so remarkable that, after his return to England in 1813, he communicated it in a letter to the Linnean Society.

CO_2 fixed during the night is stored in the form of malic acid

Nocturnal fixation of CO_2 is brought about by reaction with phosphoenolpyruvate, as catalyzed by **phosphoenolpyruvate carboxylase**, in the same way as in C_4 metabolism and in the metabolism of guard cells. In many CAM plants the required phosphoenolpyruvate is generated from a degradation of starch, but in other plants soluble carbohydrates, such as sucrose (section

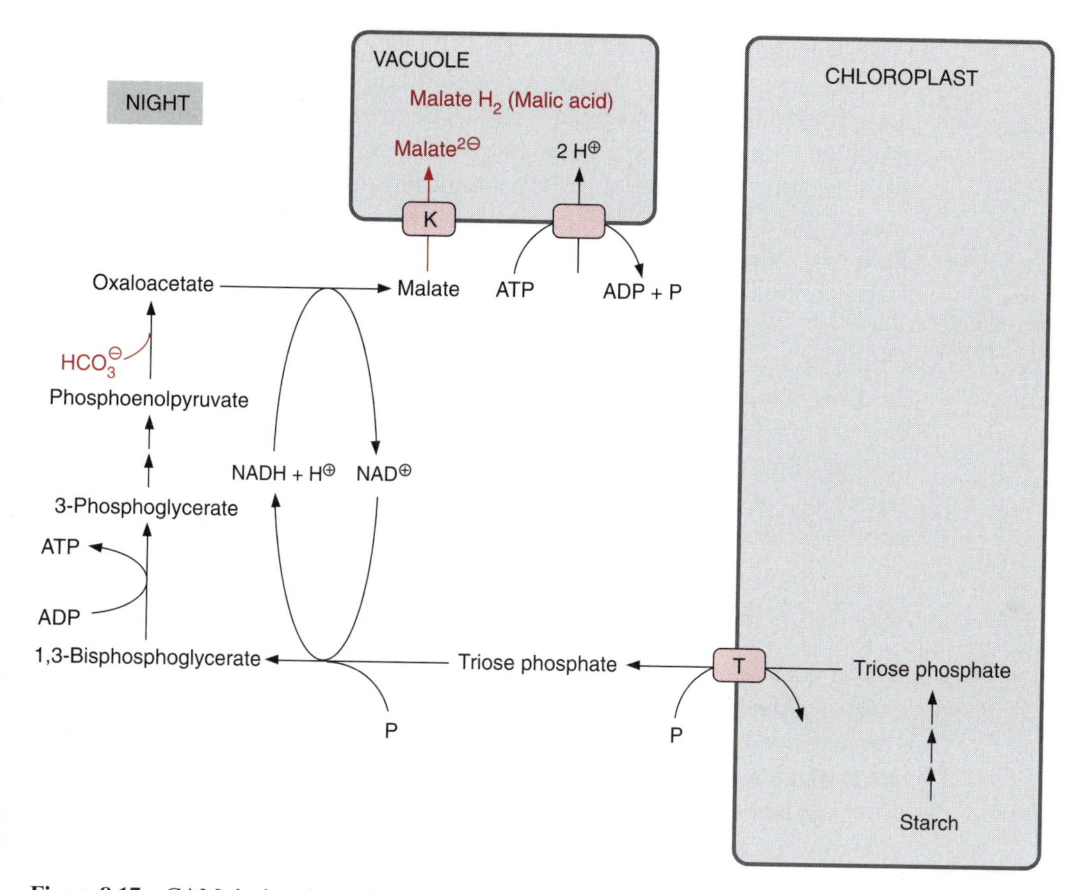

Figure 8.17 CAM during the night. Degradation of starch in the chloroplasts provides triose phosphate, which is converted along with the generation of NADH and ATP to phosphoenolpyruvate, the acceptor for HCO_3^-. The oxaloacetate is reduced in the cytosol to malate. An H^+-V-ATPase in the vacuolar membrane drives the accumulation of malate anions in the vacuole, where they are stored as malic acid. T = translocator; C = channel.

9.2) and fructanes (section 9.5), also may serve as carbon stores. Figure 8.17 shows a metabolic scheme of CAM metabolism with starch as carbon reserve. The starch located in the chloroplasts is degraded to triose phosphate, (section 9.1), which is then exported via the triose phosphate-phosphate translocator and is converted to phosphoenolpyruvate in the cytosol.

The oxaloacetate formed from the prefixation of CO_2 is reduced in the cytosol to malate via NAD-malate dehydrogenase. The NADH required for this is provided by the oxidation of triose phosphate in the cytosol. Malate is pumped at the expense of energy into the vacuoles. As described for the guard cells (section 8.2), the energy-dependent step in this pumping process is the transport of protons by the H^+-V-ATPase, (section 4.4) located in the vacuolar membrane. In contrast to the guard cells, these transported protons are not exchanged for potassium ions. The malate that, by uptake through a malate channel driven by the proton potential, has accumulated in the vacuoles, is stored there as malic acid and makes the vacuolar content very acidic (about pH 3) during the night. The two carboxyl groups of malic acid have pK-values of 3.4 and of 5.1, respectively. Thus at pH 3 malic acid is largely undissociated and the osmotic pressure deriving from the accumulation of malic acid is only about one-third of the osmotic pressure produced by the accumulation of potassium malate ($2K^+ + Mal^{2-}$) in the guard cells. In other words, at a certain osmotic pressure, three times as much malate can be stored as malic acid than as potassium malate. In order to gain a high storage capacity, most CAM plants have unusually large vacuoles and are succulent. The ATP required for CAM metabolism is generated by mitochondrial oxidative phosphorylation from the oxidation of malate.

Photosynthesis proceeds with closed stomata

The malate stored in the vacuoles during the night is released during the day by a regulated efflux through the malate channel. In CAM, as in C_4 metabolism, different plants release CO_2 in various ways: via **NADP-malic enzyme**, **NAD-malic enzyme** or also via **phosphoenolpyruvate carboxykinase**.

CAM of the **NADP-malic enzyme type** is described in Figure 8.18. A specific translocator takes up malate into the chloroplasts, where it is decarboxylated to give pyruvate, NADPH, and CO_2. The latter reacts as substrate with RubisCO and the pyruvate is converted via pyruvate phosphate dikinase to phosphoenolpyruvate (see also Figs. 8.11, 8.12, 8.14, and 8.15). Since plastids are normally unable to convert phosphoenolpyruvate to 3-phosphoglycerate (still to be investigated for CAM chloroplasts), the phosphoenolpyruvate is exported in exchange for 3-phosphoglycerate (probably catalyzed by two different translocators) as shown in Figure 8.18. As in C_4 plants, CAM chloroplasts contain, in addition to a triose phosphate-phosphate translocator (transporting in a counter-exchange triose phosphate, phosphate, and 3-phosphoglycerate), a phosphoenolpyruvate-phosphate translocator (catalyzing a counter-exchange for phosphate). The 3-phosphoglycerate taken up into the chloroplasts is fed into the Calvin cycle. The triose phosphate thus formed is used primarily for resynthesis of the starch consumed during the previous night. Only a small surplus of triose phosphate remains and this is the actual gain of CAM photosynthesis.

Since in CAM photosynthesis proceeds with closed stomata, the water requirement for CO_2 assimilation (compare Fig. 8.6) amounts to only 5% to10% of the water needed for the photosynthesis of C_3 plants. Since the

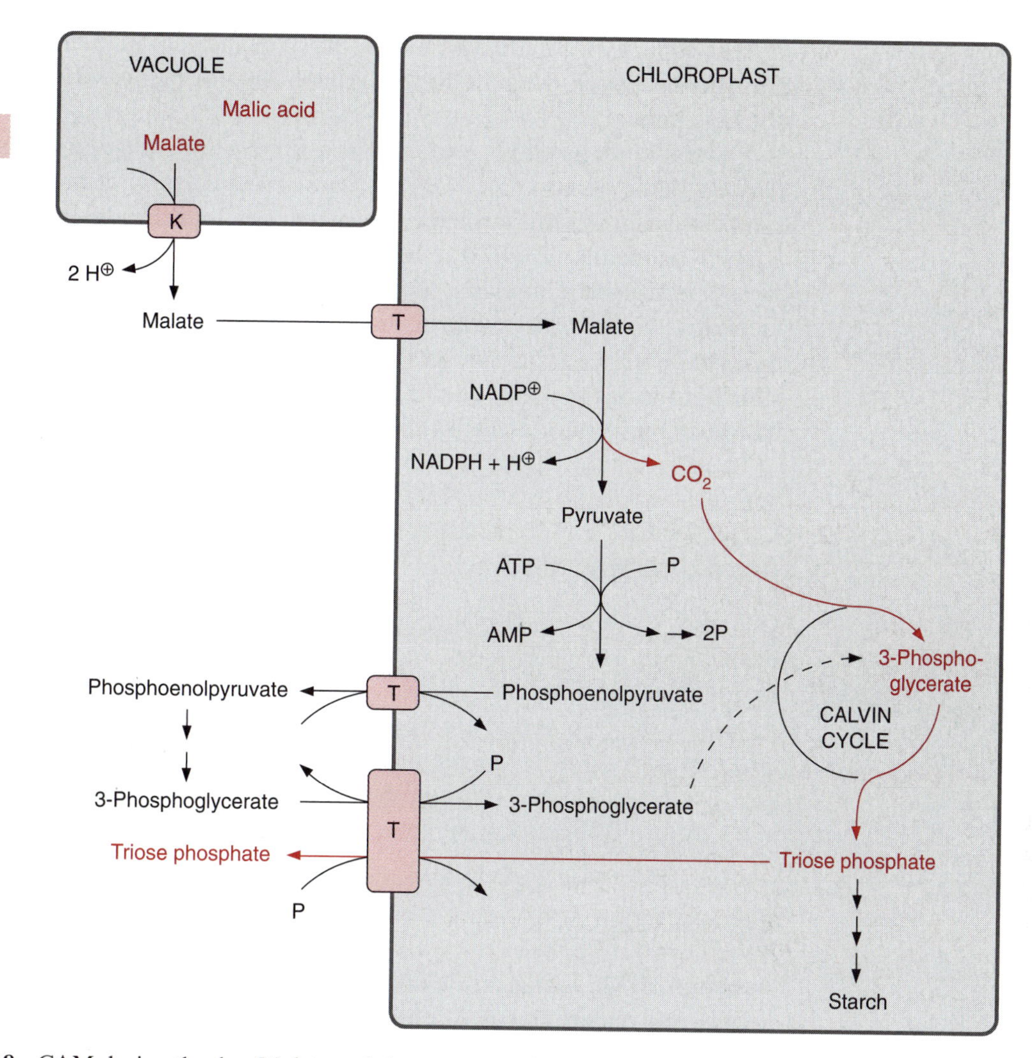

Figure 8.18 CAM during the day. Malate and the accompanying protons are released from the vacuole by a mechanism that is not yet known in detail. In the example given, malate is oxidized in the chloroplasts to pyruvate, yielding CO_2 for the CO_2 fixation by RubisCO. Pyruvate is converted via pyruvate phosphate dikinase to phosphoenolpyruvate, which is probably converted in the cytosol to 3-phosphoglycerate. After transport into the chloroplasts, 3-phosphoglycerate is converted to triose phosphate, which is used mainly for the regeneration of starch. The transport of phosphoenolpyruvate, 3-phosphoglycerate, triose phosphate and phosphate proceeds via the triose phosphate phosphate translocator. T = translocator; C = channel.

storage capacity for malate is limited, the daily increase in biomass in CAM plants is usually very low. Thus the growth rate for plants relying solely on CAM is limited.

Quite frequently plants use CAM as a strategy for surviving extended dry periods. Some plants (e.g., *Mesembryanthemum*) perform normal C_3 photosynthesis when water is available, but switch to CAM during drought or salt stress by inducing the corresponding enzymes. It is possible to determine by mass spectrometric analysis of the $^{13}C/^{12}C$ ratio (section 8.4) whether a facultative CAM plant performs C_3 metabolism or CAM. During extreme drought, cacti can survive for a long time without even opening their stomata during the night. Under these conditions, they can conserve carbon by refixing respiratory CO_2 by CAM photosynthesis.

C_4 as well as CAM metabolism has been developed several times during evolution

One finds C_4 and CAM plants in many unrelated families of monocot and dicot plants. This shows that C_4 metabolism and CAM have both evolved many times independently from C_3 precursors. Thus over 30 evolutionary origins of the C_4 pathway are recognized. As the structural elements and the enzymes of C_4 and CAM plants are also present in C_3 plants (e.g., in the guard cells of stomata), the conversion of C_3 plants to C_4 and CAM plants seems to involve relatively simple evolutionary processes.

Further reading

Assmann, S. M., Wang, X.-Q. From milliseconds to millions of years: Guard cells and environmental responses. Curr Opin Plant Biol 4, 421–428 (2001).

Badger, M. R., Price, D. G. The role of carbonic anhydrase in photosynthesis. Annu Rev Plant Physiol Plant Mol Biol 45, 369–392 (1995).

Barkla, B. J., Pantoja, O. Physiology of ion transport across the tonoplast of higher plants. Annu Rev Plant Physiol Plant Mol Biol 47, 159–184 (1996).

Blatt, M. R. Ca^{2+} signaling and control of guard-cell volume in stomatal movements. Curr Opin Plant Biol 3, 196–204 (2000).

Borland, A. M., Maxwell, K., Griffith, H. Ecophysiology of plants with crassulacean acid metabolism. In Photosynthesis: Physiology and Metabolism. R. C. Leegood, T. D. Sharkey, S. von Caemmerer (eds.), pp. 583–605. Academic Publishers, Dordrecht, The Netherlands (2000).

Chollet, R., Vidal J., O'Leary, M. H. Phosphoenolpyruvate carboxylase: a ubiquitous, highly regulated enzyme in plants. Annu Rev Plant Physiol Plant Mol Biol 47, 273–298 (1996).

Desikan R., Griffiths, R., Hancock, J., Neill, S. A new role for an old enzyme: Nitrate reductase-mediated nitric oxide generation is required for abscisic acid-induced stomatal closure in *Arabidopsis thaliana.* Proc Natl Acad Sci USA 99, 16314–16318 (2002).

Edwards, G., Walker, D. A. C_3, C_4: Mechanisms and Cellular and Environmental Regulation of Photosynthesis. Blackwell Scientific Publications, Oxford, London (1983).

Edwards, G. E., Furbank, R. T., Hatch, M. D., Osmond, C. B. What does it take to be C_4? Lessons from the evolution of C_4 photosynthesis. Plant Physiol 125, 46–49 (2001).

Evans, J. R., Loreto, F. Acquisition and diffusion of CO_2 in higher plants. In Photosynthesis: Physiology and Metabolism. R. C. Leegood, T. D. Sharkey, S. von Caemmerer (eds.), pp. 321–351. Academic Publishers, Dordrecht, The Netherlands (2000).

Fischer, K., Kammerer, B., Gutensohn, M., Arbinger, B., Weber, A., Häusler, R. E., Flügge, U.-I. A new class of plastidic phosphate translocators: A putative link between primary and secondary metabolism by the phosphoenolpyruvate/phosphate antiporter. Plant Cell 9, 453–462 (1997).

Freitag, H., Stichler, W. *Bienertia cycloptera* Bunge ex Boiss., Chenopodiaceae, another C_4 plant without Kranz tissues. Plant Biol 4, 121–132 (2002).

Furbank, R. T., Hatch, M. D., Jenkins, C. L. D. C_4 Photosynthesis: Mechanism and regulation. In Photosynthesis: Physiology and Metabolism. R. C. Leegood, T. D. Sharkey, S. von Caemmerer (eds.), pp. 435–457. Academic Publishers, Dordrecht, The Netherlands (2000).

Garcia-Mata, C., Gay, R., Sokolovski, S., Hills, A., Lamattina, L., Blatt, M. R. Nitric oxide regulates K^+ and Cl^- channels in guard cells through a subset of abscisic acid-evoked signaling pathways. Proc Natl Acad Sci USA 100, 1116–11121 (2003).

Hatch, M. D. C_4 Photosynthesis: An unlikely process full of surprises. Plant Cell Physiol 33, 333–342 (1992).

Kramer, P. J., Boyer, J. S. Water Relations of Plants and Soils. Academic Press, New York (1995).

Leegood, R. C. Transport during C_4 photosynthesis. In Photosynthesis: Physiology and Metabolism. R. C. Leegood, T. D. Sharkey, S. von Caemmerer (eds.), pp. 459–469. Academic Publishers, Dordrecht, The Netherlands (2000).

Mellilo, J. M., McGuire, A. D., Kicklighter, D. W., Moore III, B., Vorosmarty, C. J., Schloss, A. L. Global climate change and terrestrial net primary production. Nature 363, 234–240 (1993).

Mott, K. A., Buckley, T. N. Patchy stomatal conductance: emergent collective behavior of stomata. Trends Plant Sci 5, 258–262 (2000).

Nimmo, H. G. The regulation of phosphoenolpyruvate carboxylase in CAM plants. Trends Plant Sci 5, 75–80 (2000).

Sage, R. F. Environmental and evolutionary preconditions for the origin and diversification of the C_4 photosynthetic syndrome. Plant Biol 3, 202–213 (2001).

Schroeder, J. I., Allen G. J., Hugouvieux, V., Kwak, J. M., Waner, D. Guard cell signal transduction. Annu Rev Plant Physiol Plant Mol Biol 52, 627–658 (2001).

Thiel, G., Wolf, A. H. Operation of K^+-channels in stomatal movement. Trends Plant Sci 2, 339–345 (1997).

Von Caemmerer, S., Evans, J. R. Determination of the average partial pressure of CO_2 in the leaves of several C_3 plants. Australian J Plant Physiol 18, 287–305 (1991).

Vidal, J., Bakrim, N., Hodges, M. The regulation of plant phosphoenolpyruvate carboxylase by reversible phosphorylation. In C. H. Foyer and G. Noctor (eds.). Advances in Photosynthesis: Photosynthetic Assimilation and Associated Carbon Met abolism, pp. 135–150. Kluwer Academic Publishers, Dordrecht, The Netherlands (2002).

Voznesenskaya, E. V., Franceschi, V. R., Kiirats, O., Freitag, H., Edwards, G. E. Kranz anatomy is not essential for terrestrial C_4 plant photosynthesis. Nature 414, 543–546 (2001).

9

Polysaccharides are storage and transport forms of carbohydrates produced by photosynthesis

In higher plants, photosynthesis in the leaves provides substrates, such as carbohydrates, for the various heterotrophic plant tissues (e.g., the roots). Substrates delivered from the leaves are oxidized in the root cells by the large number of mitochondria present there. The ATP thus generated is required for driving the ion pumps of the roots by which mineral nutrients are taken up from the surrounding soil. Therefore respiratory metabolism of the roots, supported by photosynthesis of the leaves, is essential for plants. The plant dies when the roots are not sufficiently aerated and there is not enough oxygen available for their respiration.

Various plant parts are supplied with carbohydrates via the sieve tubes (Chapter 13). A major transport form is the disaccharide sucrose, but in some plants also tri- and tetrasaccharides or sugar alcohols. Since the synthesis of carbohydrates by photosynthesis occurs only during the day, these carbohydrates have to be stored in the leaves to ensure their continued supply to the rest of the plant during the night or during unfavorable weather conditions. Moreover, plants need to build up carbohydrate stores to tide them over the winter or dry periods, and also as a reserve in seeds for the growth of the following generation. For this purpose, carbohydrates are stored primarily in the form of high molecular weight polysaccharides, in particular as starch or fructans, but also as low molecular weight oligosaccharides.

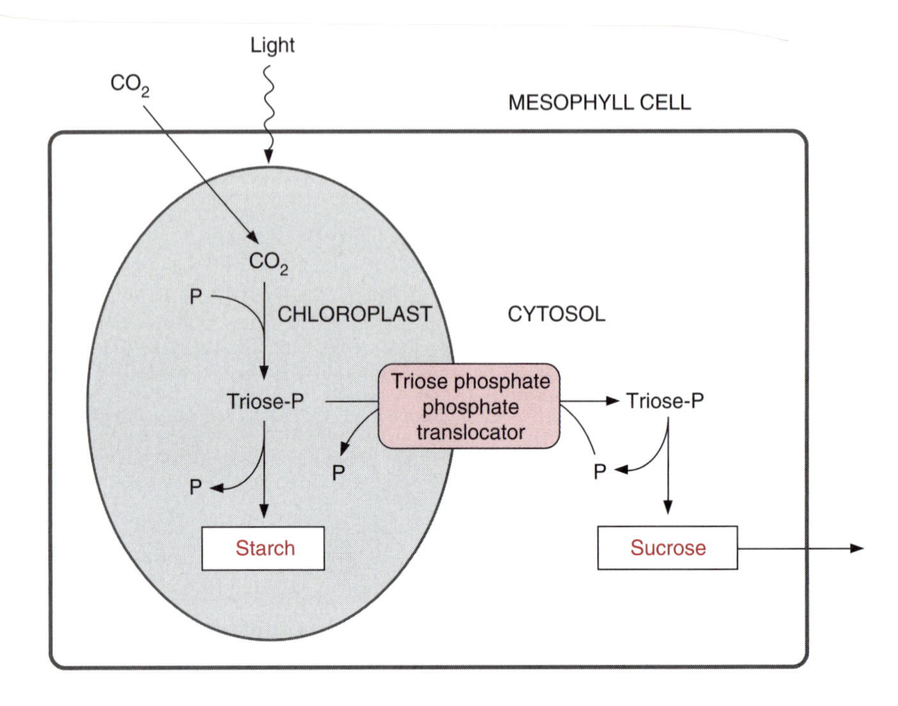

Figure 9.1 Triose phosphate, the product of photosynthetic CO_2 fixation, is either converted in the chloroplasts to starch or, after transport out of the chloroplasts, transformed to sucrose and in the latter form exported from the mesophyll cells.

Starch and sucrose are the main products of CO_2 assimilation in many plants

In most crop plants (e.g., cereals, potato, sugar beet, and rapeseed), carbohydrates are stored in the leaves as **starch** and exported as **sucrose** to other parts of the plants such as the roots or growing seeds. CO_2 assimilation in the chloroplasts yields triose phosphate, which is transported by the **triose phosphate-phosphate translocator** (section 1.9) in counter-exchange for phosphate into the cytosol, where it is converted to sucrose, accompanied by the release of inorganic phosphate (Fig. 9.1). The return of this phosphate is essential, since if there were a phosphate deficiency in the chloroplasts, photosynthesis would come to a stop. Part of the triose phosphate generated by photosynthesis is converted in the chloroplasts to starch, serving primarily as a reserve for the following night period.

9.1 Large quantities of carbohydrate can be stored as starch in the cell

Glucose is a relatively unstable compound since its aldehyde group can be spontaneously oxidized to a carboxyl group. Therefore glucose is not a

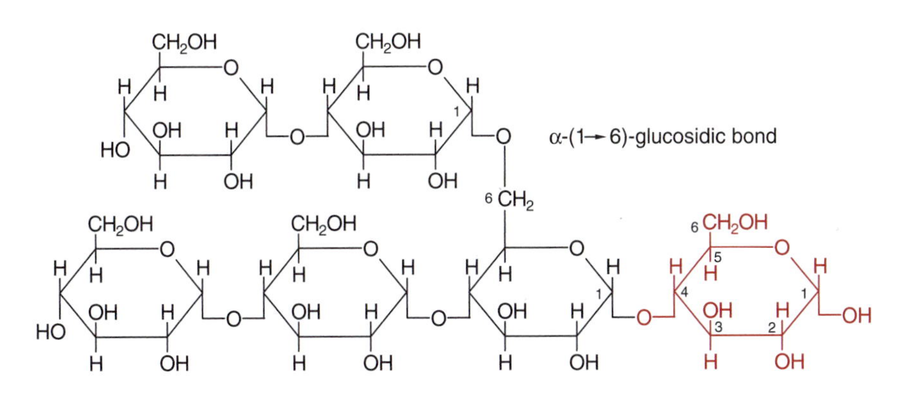

Figure 9.2 The glucose molecules in starch are connected by ($\alpha 1 \rightarrow 4$)- and ($\alpha 1 \rightarrow 6$)-glycosidic linkages to form a polyglucan. Only the glucose residue colored red contains a reducing group.

suitable carbohydrate storage compound. Moreover, for osmotic reasons, the cell has a limited storage capacity for monosaccharides. By polymerization of glucose to the osmotically inert starch, large quantities of glucose molecules can be deposited in a cell without effecting an increase in the osmotic pressure of the cell sap. This may be illustrated by an example: At the end of the day the starch content in potato leaves may amount to 10^{-4} mol glucose units per mg of chlorophyll. If this amount was dissolved as free glucose in the aqueous phase of the mesophyll cell, it would yield a glucose concentration of 0.25 mol/L. Such an accumulation of glucose would result in an increase by more than 50% in the osmotic pressure of the cell sap.

The glucose molecules in starch are primarily connected by ($\alpha 1 \rightarrow 4$)-glycosidic linkages (Fig. 9.2). These linkages protect the aldehyde groups of the glucose molecules against oxidation; only the first glucose molecule, colored red in Figure 9.2, is unprotected. In this way long glucose chains are formed that can be branched by ($\alpha 1 \rightarrow 6$)-glycosidic linkages. Branched starch molecules contain many terminal glucose residues at which the starch molecule can be enlarged.

In plants the formation of starch is restricted to plastids (section 1.3), namely, **chloroplasts** in leaves and green fruits and **leucoplast***s* in heterotrophic tissues. Starch is deposited in the plastids in the form of **starch granules** (Fig. 9.3). The starch granules in a leaf are very large at the end of the day and are usually degraded extensively during the following night. This starch is called **transitory starch**. In contrast, the starch in storage organs (e.g., seeds or tubers) is deposited for longer time periods, and therefore is called **reserve starch**. In cereals the reserve starch often represents 65% to 75% and in potato tubers even 80% of the dry weight.

Starch granules consist primarily of **amylopectin**, an **amylose**, and in some cases **phytoglycogen** (Table 9.1). Starch granules contain the enzymes

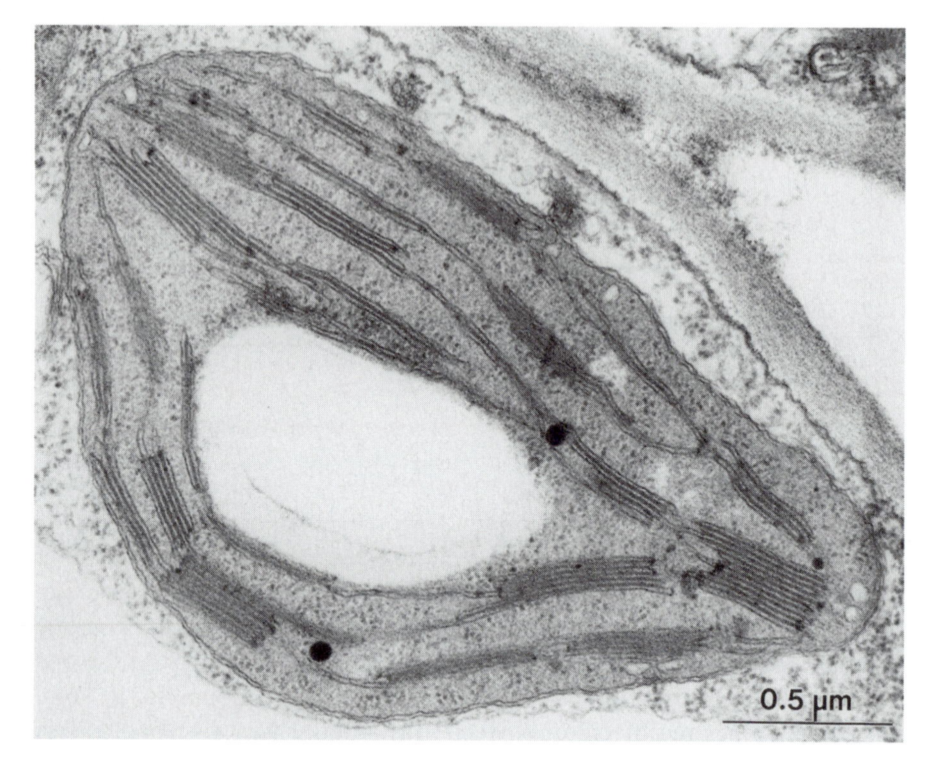

Figure 9.3 Transitory starch in a chloroplast of a mesophyll cell in a tobacco leaf at the end of the day. The starch granule in chloroplasts appears as a large white spot. (By D. G. Robinson, Heidelberg.)

for starch synthesis and degradation. These enzymes are present in several isoforms, some of which are bound to the starch granules whereas others are soluble. Amylose consists mainly of unbranched chains of about 1,000 glucose molecules. Amylopectin, with 10^4 to 10^5 glucose molecules, is much larger than amylose and has a branching point at every 20 to 25 glucose residues (Fig. 9.4). Results of X-ray structure analysis (section 3.3) show that a starch granule is constructed of concentric layers, consisting alternately of amylopectin and amylose. The amylopectin molecules are arranged in a radial fashion (Fig. 9.5). The reducing glucose (colored red in Figs. 9.2 and

Table 9.1: Constituents of plant starch

	Number of glucose residues	Number of glucose residues per branching	Absorption maximum of the glucan iodine complex
Amylose	10^3		660 nm
Amylopectin	10^4–10^5	20–25	530–550 nm
Phytoglycogen	10^5	10–15	430–450 nm

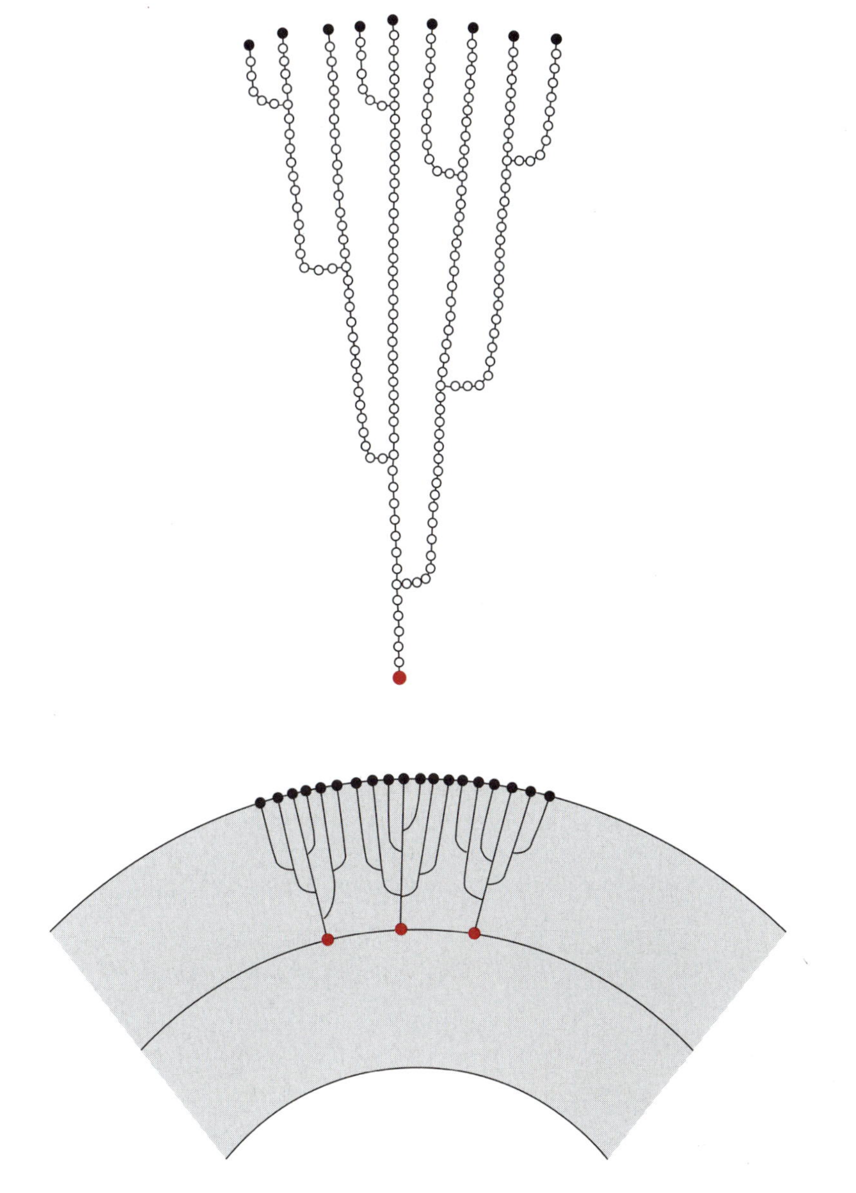

Figure 9.4 The polyglucan chains in amylopectin contain a branch point at every 20 to 25 glucose residues. Neighboring chains are arranged in an ordered structure. The glucose residue, colored red at the beginning of the chain, contains a reducing group. The groups colored black at the end of the branches are the acceptors for the addition of further glucose residues by starch synthase.

Figure 9.5 In a starch granule the amylopectin and amylose molecules are arranged alternately in layers. Compare with Figure 9.4.

9.4) is directed toward the inside, and glucose residues at the ends of the branches (colored black) are directed toward the outside. Neighboring branches of amylopectin form double helices that are packed in a crystalline array. Amylose, on the other hand, is probably present in an amorphous form. Some starch granules contain phytoglycogen, a particularly highly branched starch.

A starch granule usually contains 20% to 30% amylose, 70% to 80% amylopectin, and in some cases up to 20% phytoglycogen. Wrinkled peas, which Gregor Mendel used in his classic breeding experiments, have an amylase content of up to 80%. In the so-called waxy maize mutants, the starch granules consist almost entirely of amylopectin. On the other hand, the starch of the maize variety amylomaize consists of 50% amylose. Transgenic potato plants that contain only amylopectin in their tubers have been generated. A uniform starch content in potato tubers is of importance for the use of starch as a multipurpose raw material in the chemical industry.

Amylose, amylopectin, and phytoglycogen form blue- to violet-colored complexes with iodine molecules (Table 9.1). This makes it very easy to detect starch in a leaf by a simple iodine test.

Starch is synthesized via ADP-glucose

Fructose 6-phosphate, an intermediate of the Calvin cycle, is the precursor for starch synthesis in chloroplasts (Fig. 9.6). Fructose 6-phosphate is converted by hexose phosphate isomerase to glucose 6-phosphate, and a *cis*-enediol is formed as an intermediate of this reaction. Phosphoglucomutase transfers the phosphate residue from the 6-position of glucose to the 1-position. A crucial step for starch synthesis is the activation of glucose 1-phosphate by reaction with ATP to **ADP-glucose**, accompanied by the release of pyrophosphate. This reaction, catalyzed by the enzyme **ADP-glucose pyrophosphorylase** (Fig. 9.7), is reversible. The high activity of

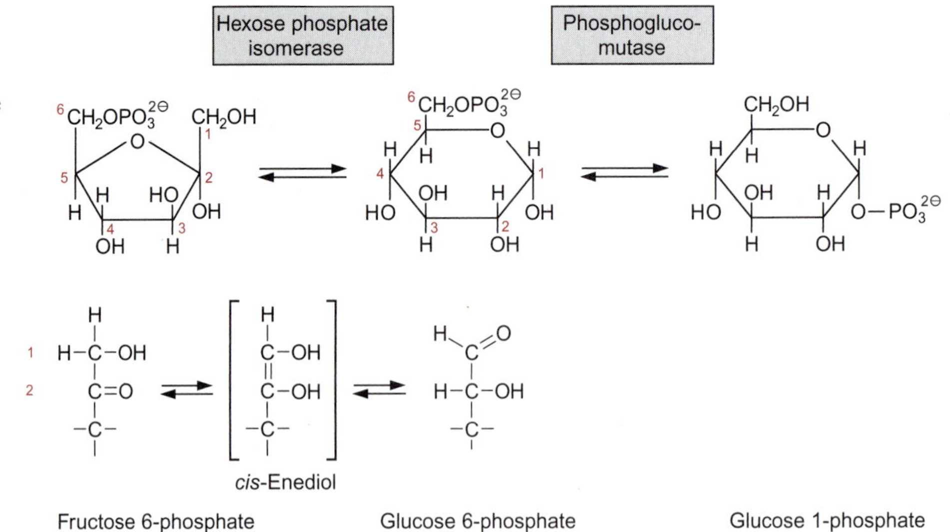

Figure 9.6 Conversion of fructose 6-phosphate to glucose 1-phosphate. In the hexose phosphate isomerase reaction, a *cis*-enediol is formed as an intermediary product.

pyrophosphatase in the chloroplast stroma, however, ensures that the pyrophosphate formed is immediately hydrolyzed to phosphate and thus withdrawn from equilibrium. Therefore the formation of ADP-glucose is an **irreversible process** and is very suitable for regulating starch synthesis. The American biochemist Jack Preiss, who has studied the properties of ADP-glucose pyrophosphorylase in detail, found that this enzyme is allosterically activated by 3-phosphoglycerate and inhibited by phosphate. The significance of this regulation will be discussed at the end of this section. The glucose residue is transferred by **starch synthases** from ADP-glucose to the OH-group in the 4-position of the terminal glucose molecule in the polysaccharide chain of starch (Fig. 9.7). The deposition of glucose residues in a starch grain proceeds by an interplay of several isoenzymes of starch synthase.

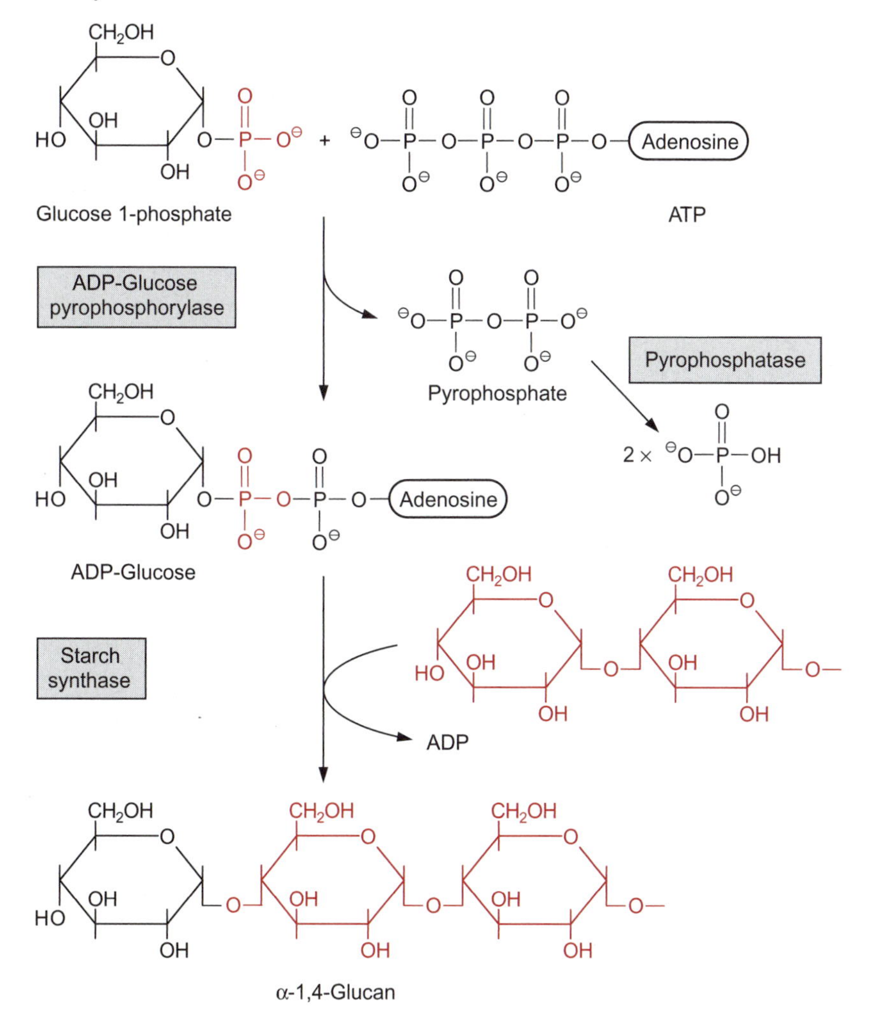

Figure 9.7 Biosynthesis of starch. Glucose 1-phosphate reacts with ATP to give ADP-glucose. The pyrophosphate formed is hydrolyzed by pyrophosphatase and in this way the formation of ADP-glucose becomes irreversible. The glucose activated by ADP is transferred by starch synthase to a terminal glucose residue in the glucan chain.

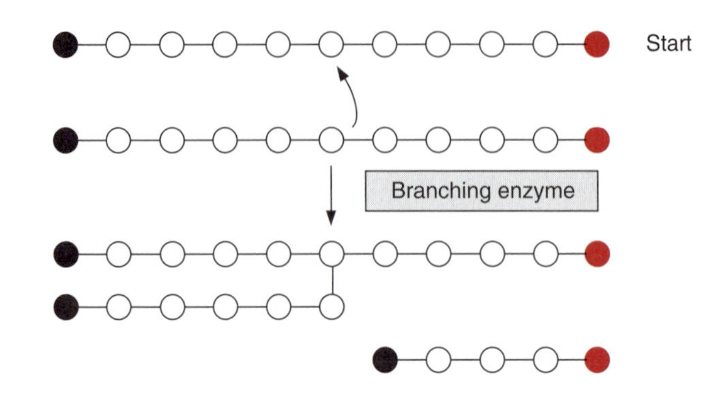

Figure 9.8 In a polyglucan chain an ($\alpha 1 \rightarrow 4$)-linkage is cleaved by the branching enzyme and the disconnected chain fragment is linked to a neighboring chain by a ($\alpha 1 \rightarrow 6$)-glycosidic linkage.

Branches are formed by a **branching enzyme**. At certain chain lengths, the polysaccharide chain is cleaved at the ($\alpha 1 \rightarrow 4$) glycosidic bond (Fig. 9.8) and the chain fragment thus separated is connected via a newly formed ($\alpha 1 \rightarrow 6$) bond to a neighboring chain. These chains are elongated further by starch synthase until a new branch develops. In the course of starch synthesis, branches are also cleaved again by a **debranching enzyme**, which will be discussed at a later point. It is assumed that the activities of the branching and the debranching enzymes determine the degree of branching in starch. The wrinkled peas, with the high amylose content already mentioned, are the result of a decrease in the activity of the branching enzyme in these plants, leading on the whole to a lowered starch content.

Degradation of starch proceeds in two different ways

Degradation of starch proceeds in two basically different reactions (Fig. 9.9). **Amylases** catalyze a hydrolytic cleavage of ($\alpha 1 \rightarrow 4$) glycosidic bonds. Different amylases attack the starch molecule at different sites (Fig. 9.10). **Exoamylases** hydrolyze starch at the end of the molecules. **β-Amylase** is an important amylase that splits off two glucose residues in the form of the disaccharide maltose from the end of the starch molecule (Fig. 9.11). The enzyme is named after its product β-maltose, in which the OH-group in the 1-position is present in the β-configuration. Amylases that hydrolyze starch in the interior of the glucan chain (**endoamylases**) produce cleavage products in which the OH-group in the 1-position is in α-configuration and are therefore named **α-amylases**. ($\alpha 1 \rightarrow 6$) Glycosidic bonds at the branch points are hydrolyzed by **debranching enzymes**.

Phosphorylases (Fig. 9.9) cleave ($\alpha 1 \rightarrow 4$) bonds phosphorolytically, resulting in the formation of glucose 1-phosphate. Since the energy of the glycosidic bond is used here to form a phosphate ester, only one molecule of ATP is consumed for the storage of a glucose residue as starch,

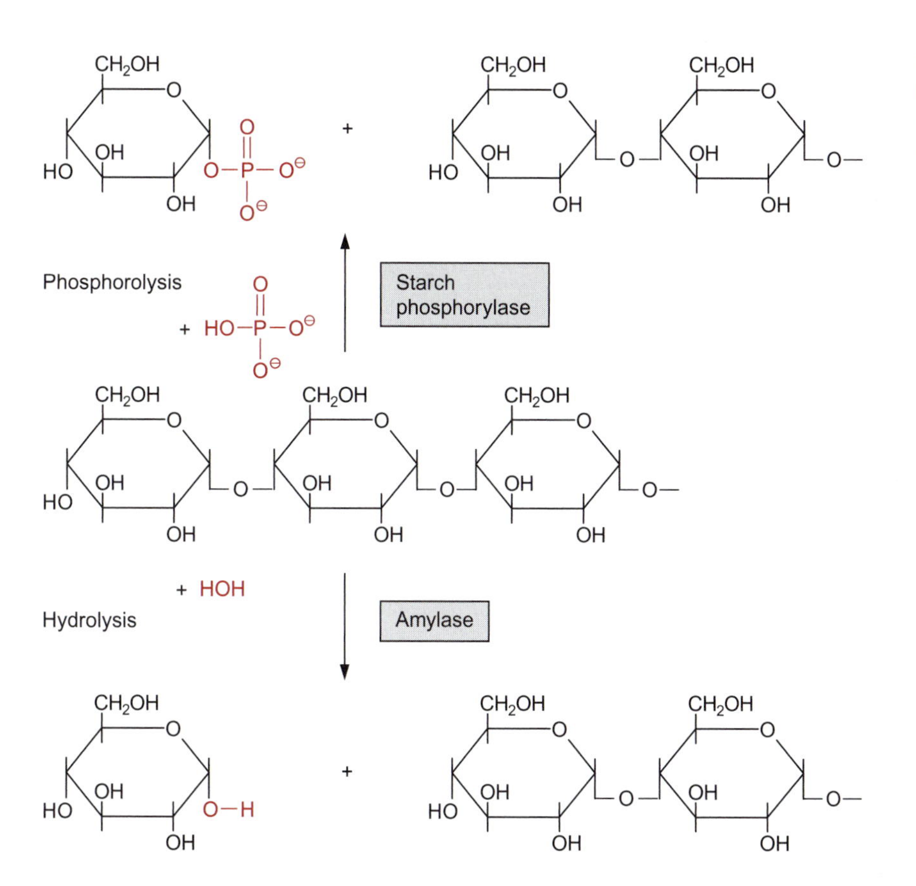

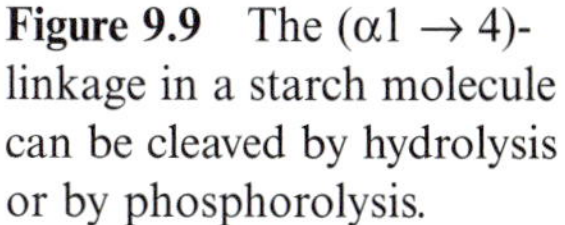
Figure 9.9 The (α1 → 4)-linkage in a starch molecule can be cleaved by hydrolysis or by phosphorolysis.

whereas two molecules of ATP are needed when the starch is mobilized by amylases.

Mobilization of chloroplast transitory starch involves both phosphorolytic and hydrolytic degradation. The phosphorylases alone are not capable of degrading large branched starch molecules. Degradation of transitory starch begins with α-amylases cleaving large starch molecules into smaller fragments, which are then degraded further not only by amylases but also by phosphorylases. In plant tissue an **α-glucan-water dikinase** recently was discovered, through which glucose residues in starch molecules are phosphorylated by ATP. The phosphorylation follows the type of a dikinase reaction (see also Fig. 8.12) in which three substrates, an α-polyglucan, ATP, and H_2O are converted into the three products α-polyglucan-P, AMP, and phosphate.

$$\alpha\text{-polyglucan} + \text{ATP} + \text{H}_2\text{O} \xrightarrow{\alpha\text{-glucan-water-kinase}} \alpha\text{-polyglucan-P} + \text{AMP} + \text{P}$$

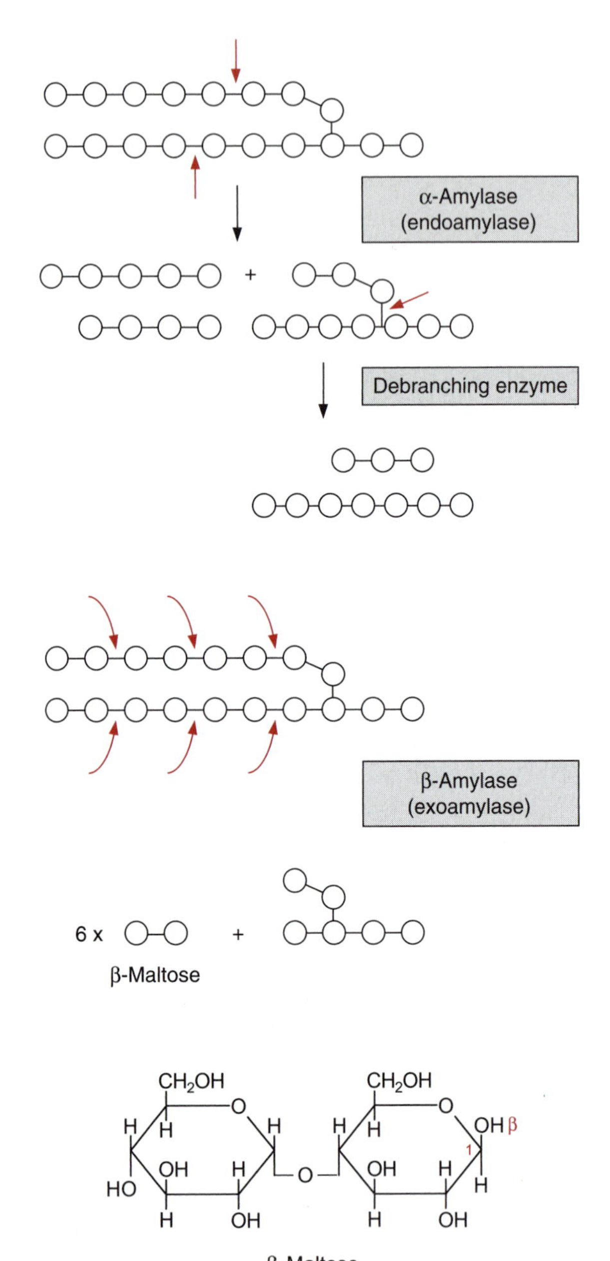

Figure 9.10 (α1 → 4)-Glycosidic linkages in the interior of the starch molecule are hydrolyzed by α-amylases. The debranching enzyme hydrolyses (α1 → 6)-linkages. β-Amylases release the disaccharide β-maltose by hydrolysis of (α1 → 4)-linkages successively from the end of the starch molecules.

Figure 9.11 In the disaccharide β-maltose, the OH-group in position 1 is in the β-configuration.

This enzyme appears to play an important role in starch degradation. *Arabidopsis* mutants devoid of this enzyme showed impaired starch degradation. The exact relationship between phosphorylation and starch degradation remains to be elucidated.

Phosphorylases also occur in the cytosol. It is feasible that small oligosaccharides, products of hydrolytic starch degradation, are released from the chloroplast to the cytosol to be degraded completely by the phosphorylases present there. Our knowledge about this is still incomplete.

Surplus photosynthesis products can be stored temporarily in chloroplasts by starch synthesis

Figure 9.12 outlines the synthesis and degradation of transitory starch in chloroplasts. The regulation of ADP-glucose pyrophosphorylase by **3-phosphoglycerate** (3-PGA) and **phosphate** (P) mentioned earlier, enables the flux of carbohydrates into starch to be regulated. The activity of the enzyme is governed by the 3-PGA/P concentration ratio. 3-PGA is a major metabolite in the chloroplast stroma. Due to the equilibrium of the reactions catalyzed by phosphoglycerate kinase and glyceraldehyde phosphate dehydrogenase (section 6.3), the stromal PGA concentration is much higher than that of triose phosphate. Since in the chloroplast stroma the total amount of phosphate and phosphorylated intermediates of the Calvin cycle is kept virtually constant by the counter-exchange of the triose phosphate-phosphate translocator (section 1.9), an increase in the concentration of 3-PGA results in a decrease in the concentration of phosphate. The 3-PGA/P ratio is therefore a very sensitive indicator of the metabolite status in the chloroplast stroma. When a decrease in sucrose synthesis brings about a decrease in phosphate liberation in the cytosol, the chloroplasts suffer from phosphate deficiency, which limits their photosynthesis (Fig. 9.1). In such a situation, the PGA/P quotient increases, leading in turn to an increase in starch synthesis, by which phosphate is released, thus allowing photosynthesis to continue. Starch functions here as a buffer. Assimilates that are not utilized for synthesis of sucrose or other substances are deposited temporarily in the chloroplasts as transitory starch. Moreover, starch synthesis is programmed in such a way (by a mechanism largely unknown) that starch is deposited each day for use during the following night.

So far very little is known about the regulation of transitory starch degradation. It probably is stimulated by an increase in the stromal phosphate concentration, but the mechanism for this is still unclear. An increase in the stromal phosphate concentration indicates a shortage of substrates. Glucose arising from the **hydrolytic degradation** of starch is exported from the chloroplasts via a **glucose translocator** (Fig. 9.12). This glucose is phosphorylated by ATP to glucose-6-phosphate via a hexokinase bound to the chloroplast outer envelope membrane, before being released to the cytosol.

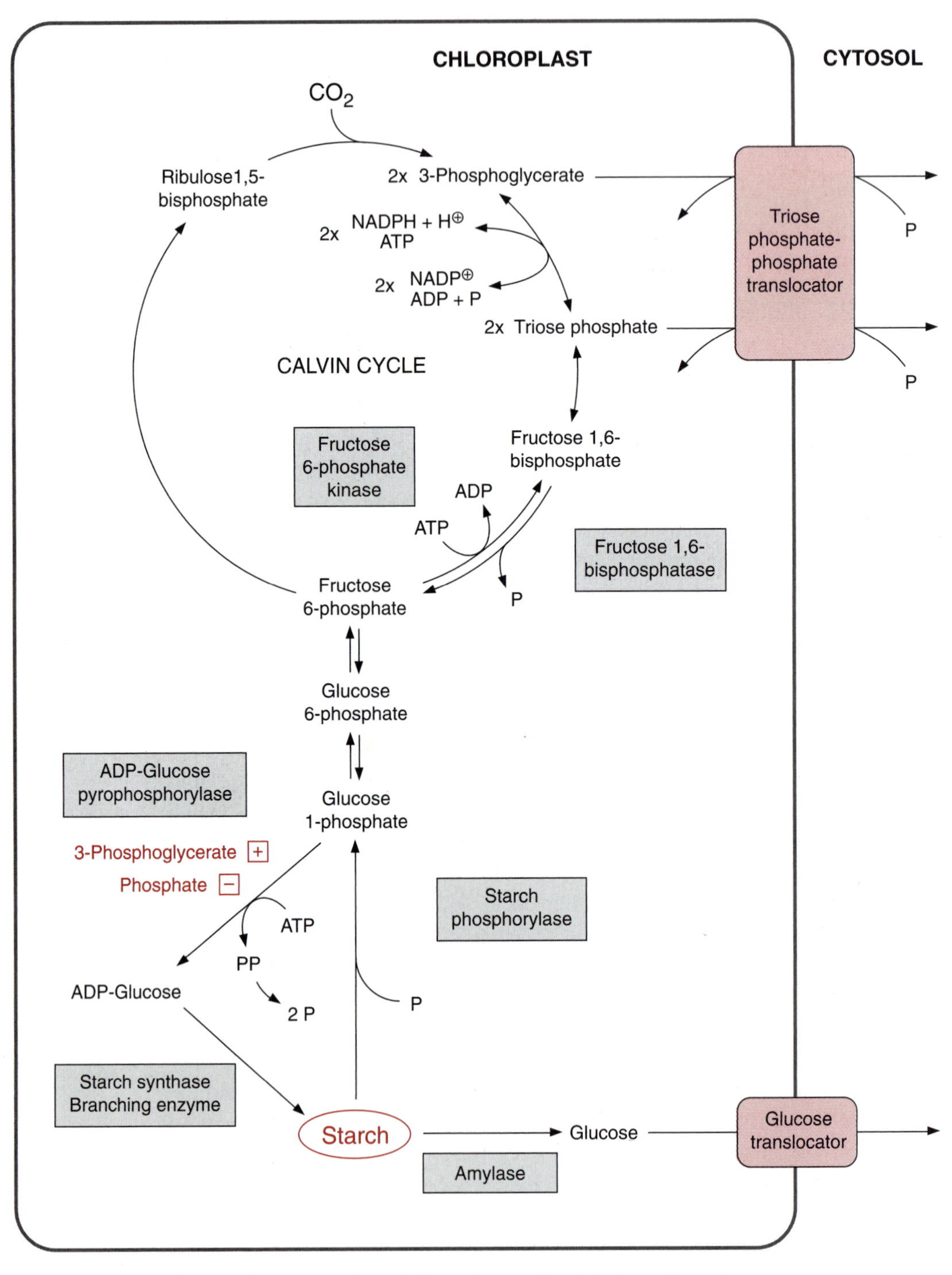

Figure 9.12 Synthesis and degradation of starch in a chloroplast.

Glucose 1-phosphate, derived from **phosphorolytic starch degradation**, is converted in a reversal of the starch synthesis pathway to fructose 6-phosphate, and the latter to fructose 1,6-bisphosphate by fructose 6-phosphate kinase. Triose phosphate formed from fructose 1,6-bisphosphate by aldolase is released from the chloroplasts via the **triose phosphate-phosphate translocator**. Part of the triose phosphate is oxidized within the chloroplasts to 3-phosphoglycerate and is subsequently exported also via the triose phosphate-phosphate translocator. This translocator as well as the glucose translocator are thus involved in the mobilization of the chloroplast transitory starch.

9.2 Sucrose synthesis takes place in the cytosol

The synthesis of sucrose, a disaccharide of glucose and fructose (Fig. 9.13), takes place in the cytosol of the mesophyll cells. As in starch synthesis, the glucose residue is activated as nucleoside diphosphate-glucose, although in this case via UDP-glucose pyrophosphorylase:

$$\text{glucose 1-phosphate} + \text{UTP} \xrightleftharpoons{\text{UDP–glucose pyrophosphorylase}} \text{UDP-glucose} + \text{PP}$$

In contrast to the chloroplast stroma, the cytosol of mesophyll cells does not contain pyrophosphatase to withdraw pyrophosphate from the equilibrium, and therefore the UDP-glucose pyrophosphorylase reaction is reversible. **Sucrose phosphate synthase** (abbreviated SPS, Fig 9.13) catalyzes the transfer of the glucose residue from UDP-glucose to fructose 6-phosphate forming sucrose 6-phosphate. **Sucrose phosphate phosphatase**, forming an enzyme complex together with SPS, hydrolyzes sucrose 6-phosphate, thus withdrawing it from the sucrose phosphate synthase reaction equilibrium. Therefore, the overall reaction of sucrose synthesis is an irreversible process.

In addition to sucrose phosphate synthase, plants also contain a **sucrose synthase**:

$$\text{sucrose} + \text{UDP} \xrightleftharpoons{\text{Sucrose synthase}} \text{UDP-glucose} + \text{fructose}$$

This reaction is reversible. It is not involved in sucrose synthesis but in the utilization of sucrose by catalyzing the formation of UDP-glucose and fructose from UDP and sucrose. This enzyme occurs primarily in nonphoto-

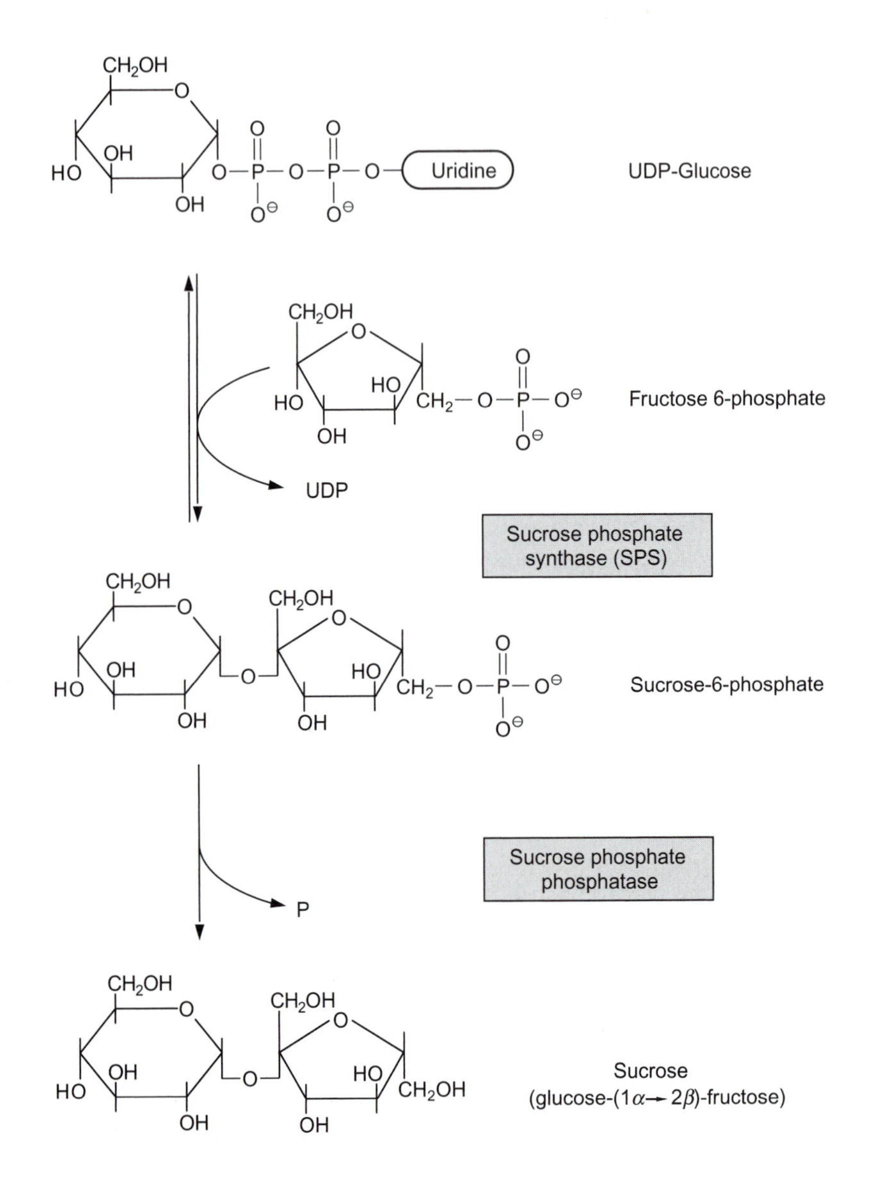

Figure 9.13 Synthesis of sucrose. The glucose activated by UDP is transferred to fructose 6-phosphate. The total reaction becomes irreversible by hydrolysis of the formed sucrose 6-phosphate.

synthetic tissues. It is involved in sucrose breakdown for the synthesis of starch in amyloplasts of storage tissue such as potato tubers (section 13.3). It also plays a role in the synthesis of cellulose and callose, where the sucrose synthase, otherwise soluble, is membrane-bound (see section 9.6).

9.3 The utilization of the photosynthesis product triose phosphate is strictly regulated

As shown in Figure 6.11, five-sixths of the triose phosphate generated in the Calvin cycle is required for the regeneration of the CO_2 acceptor ribulose bisphosphate. Therefore a maximum of one-sixth of the triose phosphate formed is available for export from the chloroplasts. In fact, due to photorespiration (Chapter 7), the portion of available triose phosphate is only about one-eighth of the triose phosphate formed in the chloroplasts. If more triose phosphate were withdrawn from the Calvin cycle, the CO_2 acceptor ribulose bisphosphate could no longer be regenerated and the Calvin cycle would collapse. Therefore it is crucial for the functioning of the Calvin cycle that the withdrawal of triose phosphate does not exceed this limit. On the other hand, photosynthesis of chloroplasts can proceed only if its product triose phosphate is utilized (e.g., for synthesis of sucrose), and in this phosphate is released. A phosphate deficiency would result in a decrease or even a total cessation of photosynthesis. Thus it is important for a plant that when the rate of photosynthesis is increased (e.g., by strong sunlight), the increase in the formation of assimilation products is matched by a corresponding increase in its utilization.

Therefore the utilization of the triose phosphate generated by photosynthesis should be regulated in such a way that as much as possible is utilized without exceeding the set limit and so ensuring the regeneration of the CO_2 acceptor ribulose bisphosphate.

Fructose 1,6-bisphosphatase functions as an entrance valve for the pathway of sucrose synthesis

In mesophyll cells, sucrose synthesis is normally the main consumer of triose phosphate generated by CO_2 fixation. The withdrawal of triose phosphate from chloroplasts for the synthesis of sucrose is not regulated at the translocation step. Due to the hydrolysis of fructose 1,6-bisphosphate and sucrose 6-phosphate, the total reaction of sucrose synthesis (Fig. 9.14) is an irreversible process, which, owing to the high enzymatic activities, has a high synthetic capacity. Sucrose synthesis must be strictly regulated to ensure that not more than the permitted amount of triose phosphate (see preceding) is withdrawn from the Calvin cycle

The first irreversible step of sucrose synthesis is catalyzed by the **cytosolic fructose 1,6-bisphosphatase**. This reaction is an important control point

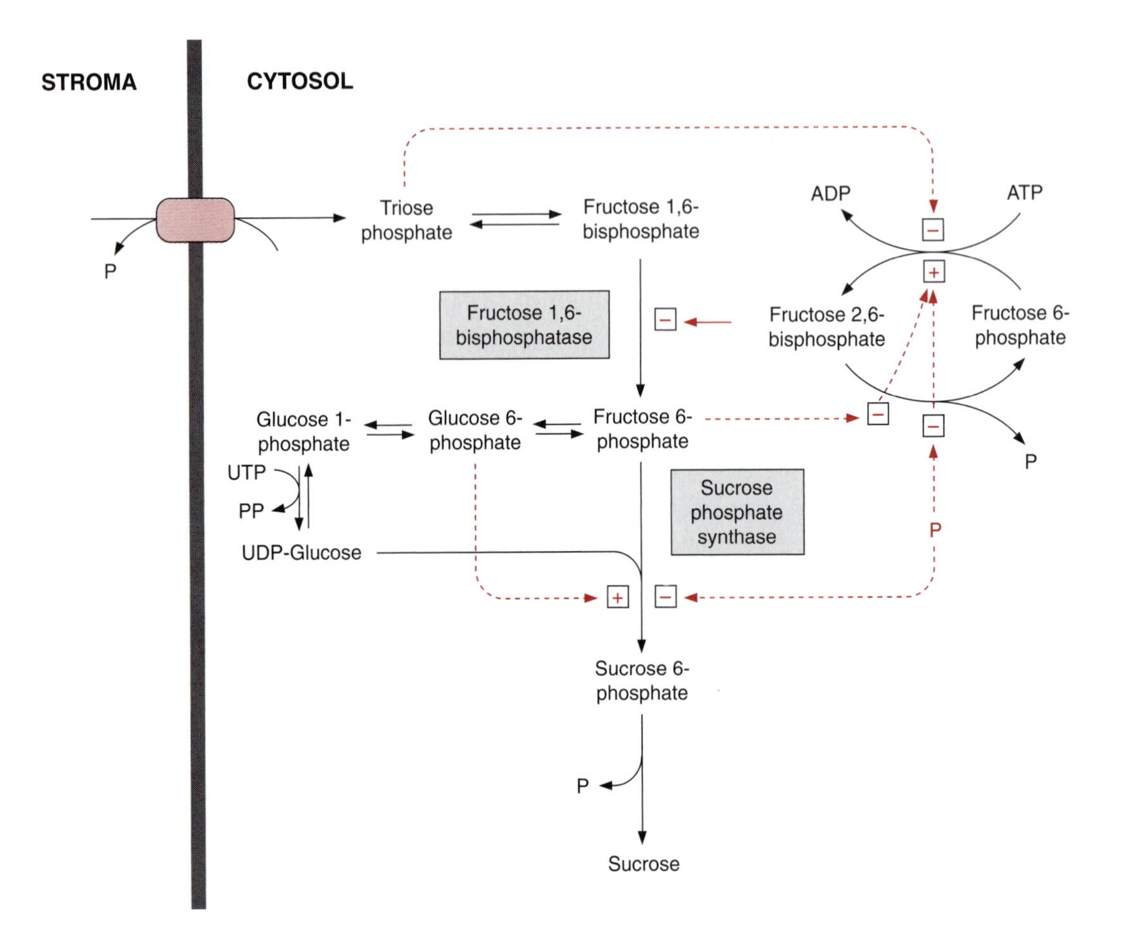

Figure 9.14 Conversion of triose phosphate into sucrose. The dashed red lines represent the regulation of single reactions by metabolites, (–) means inhibition, (+) activation. The effect of the regulatory substance fructose 2,6-bisphosphate is explained in detail in Figure 9.15.

and is the entrance valve where triose phosphate is withdrawn for the synthesis of sucrose. Figure 9.15 shows how this valve functions. An important role is played here by **fructose 2,6-bisphosphate** (Fru2,6BP), a regulatory substance that differs from the metabolic fructose 1,6-bisphosphate only in the positioning of one phosphate group (Fig. 9.16).

Fru2,6BP was discovered to be a potent activator of ATP-dependent fructose 6-phosphate kinase and an inhibitor of fructose 1,6-bisphosphatase in liver. Later it became apparent that Fru2,6BP has a general function in controlling glycolysis and gluconeogenesis in animals, plants, and fungi. It is a powerful regulator of cytosolic fructose 1,6-bisphosphatase in mesophyll cells and just micromolar concentrations of Fru2,6BP, as may occur in the

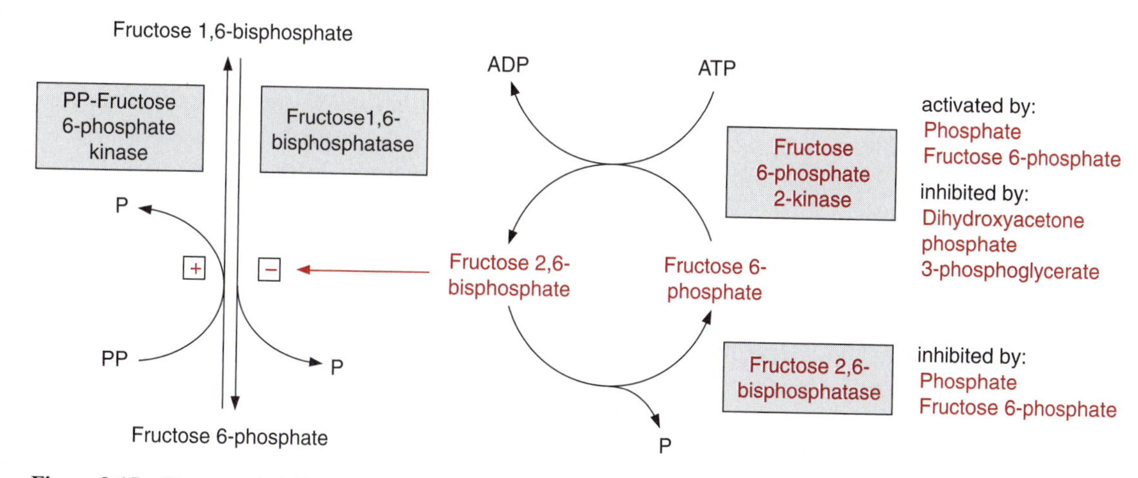

Figure 9.15 Fructose 1,6-bisphosphatase represents the entrance valve for the conversion of the CO_2 assimilate into sucrose. The enzyme is inhibited by the regulatory substance fructose-2,6-bisphosphate (Fru2,6BP). The pyrophosphate-dependent fructose 6-phosphate kinase, which synthesizes fructose 1,6-bisphosphate from fructose 6-phosphate, with the consumption of pyrophosphate, is active only in the presence of the regulatory substance Fru2,6BP. The concentration of this Fru2,6BP is adjusted by continuous synthesis and degradation. The enzymes catalyzing Fru2,6BP synthesis and degradation are regulated by metabolites. In this way the presence of triose phosphate and 3-phosphoglycerate decreases the concentration of Fru2,6BP and thus increases the activity of fructose 1,6-bisphosphatase.

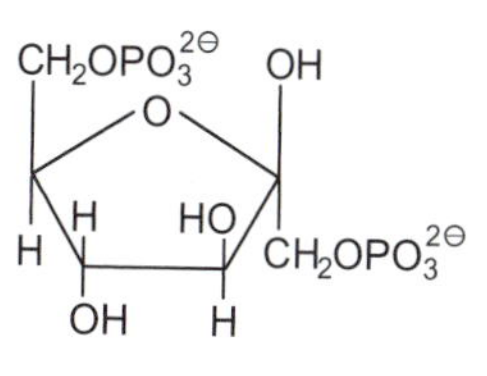

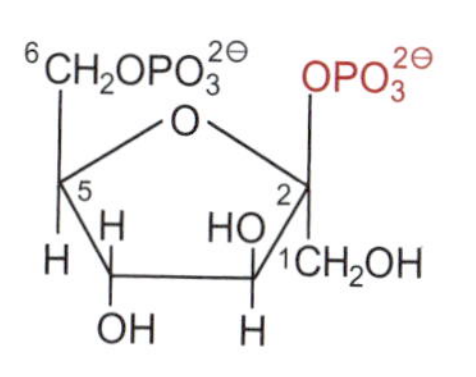

Figure 9.16 The regulatory substance fructose 2,6-bisphosphate differs from the metabolite fructose 1,6-bisphosphate only in the position of one phosphate group.

cytosol of mesophyll cells, result in a large decrease in the affinity of the enzyme toward its substrate fructose 1,6-bisphosphate. On the other hand, Fru2,6BP activates a **pyrophosphate-dependent fructose 6-phosphate kinase** found in the cytosol of plant cells. Without Fru2,6BP this enzyme is inactive. The pyrophosphate-dependent fructose 6-phosphate kinase can utilize pyrophosphate, which is formed in the UDP glucose pyrophosphorylase reaction.

Fru2,6BP is formed from fructose 6-phosphate by a specific kinase (**fructose 6-phosphate 2-kinase**) and is degraded hydrolytically by a specific phosphatase (**fructose 2,6-bisphosphatase**) to fructose 6-phosphate again. The level of the regulatory substance Fru2,6BP is adjusted by regulation of the relative rates of synthesis and degradation. Triose phosphate and 3-phosphoglycerate inhibit the synthesis of Fru2,6BP, whereas fructose 6-phosphate and phosphate stimulate synthesis and decrease hydrolysis. In this way an increase in the triose phosphate concentration results in a decrease in the level of Fru2,6BP and thus in an increased affinity of the cytosolic fructose 1,6-bisphosphatase toward its substrate fructose 1,6-bisphosphate. Moreover, due to the equilibrium catalyzed by cytosolic aldolase, an increase in the triose phosphate concentration results in an increase in the concentration of fructose 1,6-bisphosphate. The simultaneous increase in substrate concentration and substrate affinity has the effect that the rate of sucrose synthesis increases in a sigmoidal fashion with rising triose phosphate concentrations (Fig. 9.17). In this way the rate of sucrose synthesis can be adjusted effectively to the supply of triose phosphate.

The principle of regulation can be compared with an overflow valve. A certain threshold concentration of triose phosphate has to be exceeded for an appreciable metabolite flux via fructose 1,6-bisphosphatase to occur. This ensures that the triose phosphate level in chloroplasts does not decrease below the minimum level required for the Calvin cycle to function. When this threshold is overstepped, a further increase in triose phosphate results in a large increase in enzyme activity, whereby the surplus triose phosphate can be channeled very efficiently into sucrose synthesis.

Cytosolic fructose 1,6-bisphosphatase adjusts its activity, as shown above, not only to the substrate supply, but also to the demand for its product. With an increase in fructose 6-phosphate, the level of the regulatory substance Fru2,6BP is increased by stimulation of fructose 6-phosphate 2-kinase and simultaneous inhibition of fructose 2,6-bisphosphatase,

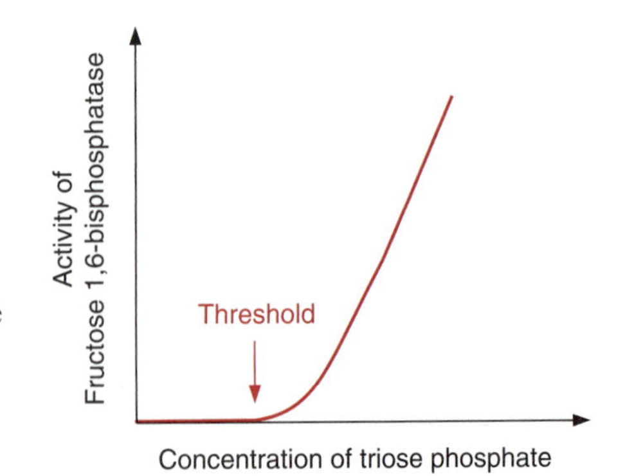

Figure 9.17 Cytosolic fructose 1,6-bisphosphatase acts as an entrance valve to adjust the synthesis of sucrose to the supply of triose phosphate. Increasing triose phosphate leads, via aldolase, to an increase in the substrate fructose 1,6-bisphosphate (Fig. 9.14), and in parallel (Fig. 9.15) to a decrease in the concentration of the regulatory substance fructose-2,6-bisphosphate. As a consequence of these two synergistic effects, the activity of fructose 1,6-bisphosphatase increases with the concentration of triose phosphate in a sigmoidal manner. The enzyme becomes active only after the triose phosphate concentration reaches a threshold concentration and then responds in its activity to increasing triose phosphate concentrations.

whereby the activity of cytosolic fructose 1,6-bisphosphatase is reduced (Fig. 9.15).

Sucrose phosphate synthase is regulated not only by metabolites but also by covalent modification

Also, **sucrose phosphate synthase** (Fig. 9.14) is subject to strict metabolic control. This enzyme is activated by glucose 6-phosphate and is inhibited by phosphate. Due to hexose phosphate isomerase, the activator glucose 6-phosphate is in equilibrium with fructose 6-phosphate. In this equilibrium, the concentration of glucose 6-phosphate greatly exceeds the concentration of the substrate fructose 6-phosphate. In equilibrium, therefore, the change in the concentration of the substrate results in a much larger change in the concentration of the activator. In this way the activity of the enzyme is adjusted effectively to the supply of the substrate.

Moreover, the activity of sucrose phosphate synthase is altered by a covalent modification of the enzyme. The enzyme contains at **position 158** a **serine residue**, of which the OH-group is phosphorylated by a special protein kinase, termed **sucrose phosphate synthase kinase** (SPS kinase) and is dephosphorylated by the corresponding **SPS phosphatase** (Fig. 9.18). The SPS phosphatase is inhibited by **okadaic acid**, an inhibitor of protein phos-

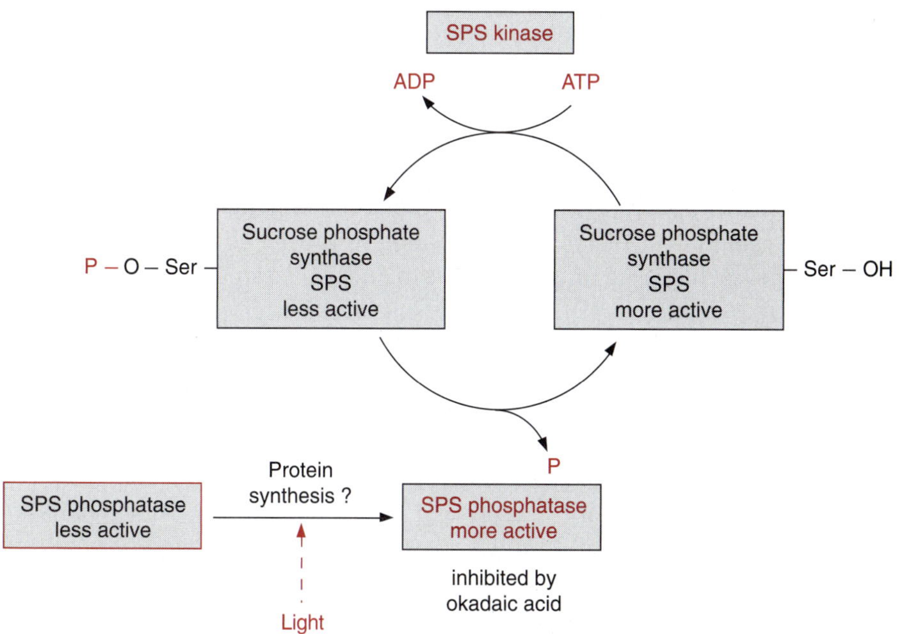

Figure 9.18 Sucrose-phosphate synthase (abbreviated SPS) is converted to a less active form by phosphorylation of a serine residue via a special protein kinase (SPS kinase). The hydrolysis of the phosphate residue by SPS phosphatase results in an increase in the activity. The activity of SPS phosphatase is increased by illumination, probably via a *de novo* synthesis of the enzyme protein. (After Huber, 1996.)

phatases of the so-called 2A-type (not discussed in more detail here). The activity of SPS kinase is probably regulated by metabolites, such as glucose-6-phosphate.

The phosphorylated form of sucrose phosphate synthase is less active than the dephosphorylated form. The activity of the enzyme is adjusted by the relative rates of its phosphorylation and dephosphorylation. Illumination of a leaf increases the activity of SPS phosphatase and thus the sucrose phosphate synthase is converted into the more active form. The mechanism for this is still not fully known. It is debated that the decrease of SPS phosphatase activity observed during darkness is due to a lowered rate of the synthesis of the SPS phosphatase. In **position 424** sucrose phosphate synthase contains a **second serine residue**, which is phosphorylated by another protein kinase (activated by osmotic stress), resulting in an activation of SPS. This demonstrates the complexity of SPS regulation. The phosphorylation of one serine residue by the corresponding protein kinase causes an inhibition, while the phosphorylation of another serine residue by a different protein kinase leads to activation. Moreover, SPS has a **third phosphorylation site**, to which a **14.3.3 protein** is bound (similarly as in the case of nitrate reductase, section 10.3). The physiological role of this binding remains to be resolved.

Partitioning of assimilates between sucrose and starch is due to the interplay of several regulatory mechanisms

The preceding section discussed various regulatory processes involved in the regulation of sucrose synthesis. Metabolites acting as enzyme inhibitors or activators can adjust the rate of sucrose synthesis immediately to the prevailing metabolic conditions in the cell. Such an immediate response is called **fine control**. The covalent modification of enzymes, influenced by diurnal factors and probably also by phytohormones (Chapter 19), results in a **general regulation of metabolism** according to the metabolic demand of the plant. This includes partitioning of assimilates between sucrose, starch, and amino acids (Chapter 10). Thus, slowing down sucrose synthesis, which results in an increase in triose phosphate and also of 3-phosphoglycerate, can lead to an increase in the rate of **starch synthesis** (Fig. 9.12). As already discussed, during the day a large part of the photoassimilates is deposited temporarily in the chloroplasts of leaves as transitory starch, to be converted during the following night to sucrose and delivered to other parts of the plant. However, in some plants, such as barley, during the day large quantities of the photoassimilates are stored as sucrose in the leaves. Therefore during darkness the rate of sucrose synthesis varies in leaves of different plants.

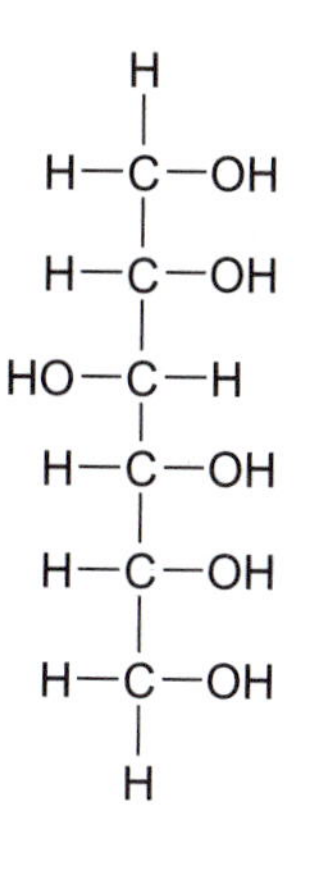

Figure 9.19 In some plants assimilated CO_2 is exported from the leaves in the form of the sugar alcohol sorbitol.

9.4 In some plants assimilates from the leaves are exported as sugar alcohols or oligosaccharides of the raffinose family

Sucrose is not the main transport form in all plants for the translocation of assimilates from the leaves to other parts of the plant. In some plants photoassimilates are translocated as **sugar alcohols**. In the *Rosaceae* (these include orchard trees in temperate regions), the assimilates are translocated from the leaves in the form of **sorbitol** (Fig. 9.19). Other plants, such as squash, several deciduous trees, (e.g., lime, hazelnut, elm), and olive trees, translocate in their sieve tubes oligosaccharides of the **raffinose family**, in which sucrose is linked by a glycosidic bond to one or more galactose molecules (Fig. 9.20). Oligosaccharides of the raffinose family include **raffinose** with one, **stachyose** with two, and **verbascose** with three galactose residues. These also function as storage compounds and, for example, in pea and bean seeds, make up 5 to 15% of the dry matter. Humans do not have the enzymes that catalyze the hydrolysis of α-galactosides and are therefore unable to digest oligosaccharides of the raffinose family. When these sugars are ingested, they are decomposed in the last section of the intestines by anaerobic bacteria to form digestive gases.

The galactose required for raffinose synthesis is formed by epimerization of UDP-glucose (Fig. 9.21). **UDP-glucose epimerase** catalyzes the oxidation

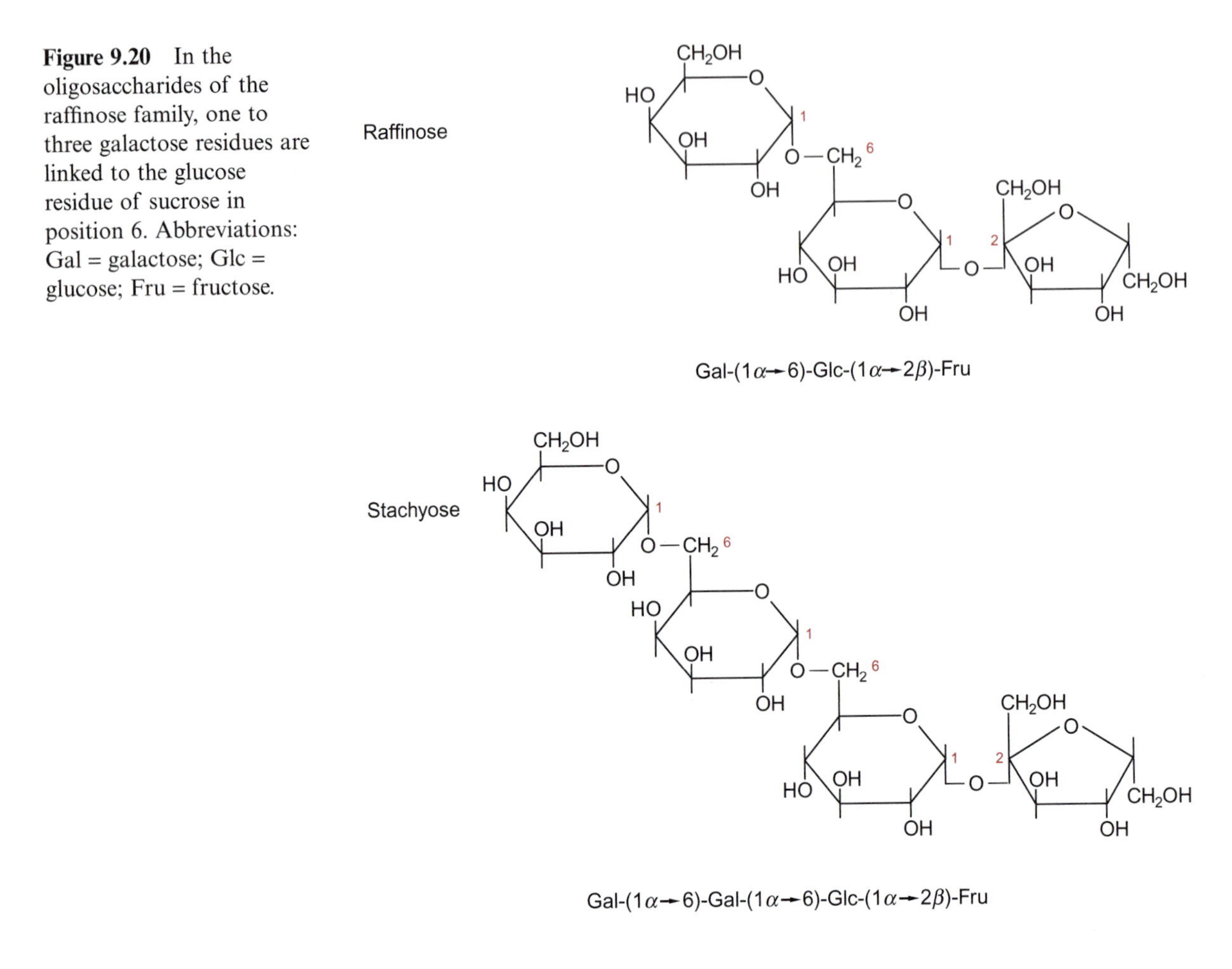

Figure 9.20 In the oligosaccharides of the raffinose family, one to three galactose residues are linked to the glucose residue of sucrose in position 6. Abbreviations: Gal = galactose; Glc = glucose; Fru = fructose.

of the OH-group in position 4 of the glucose molecule by NAD, which is tightly bound to the enzyme. A subsequent reduction results in the formation of glucose as well as galactose residues in an epimerase equilibrium. The galactose residue is transferred by a transferase to the cyclic alcohol myo-inositol producing galactinol. **Myo-inositol-galactosyl-transferases** catalyze the transfer of the galactose residue from galactinol to sucrose, to form raffinose, and correspondingly also stachyose and verbascose.

$$\text{sucrose} + \text{galactinol} \rightarrow \text{raffinose} + \text{myo-inositol}$$
$$\text{raffinose} + \text{galactinol} \rightarrow \text{stachyose} + \text{myo-inositol}$$
$$\text{stachyose} + \text{galactinol} \rightarrow \text{verbascose} + \text{myo-inositol}$$

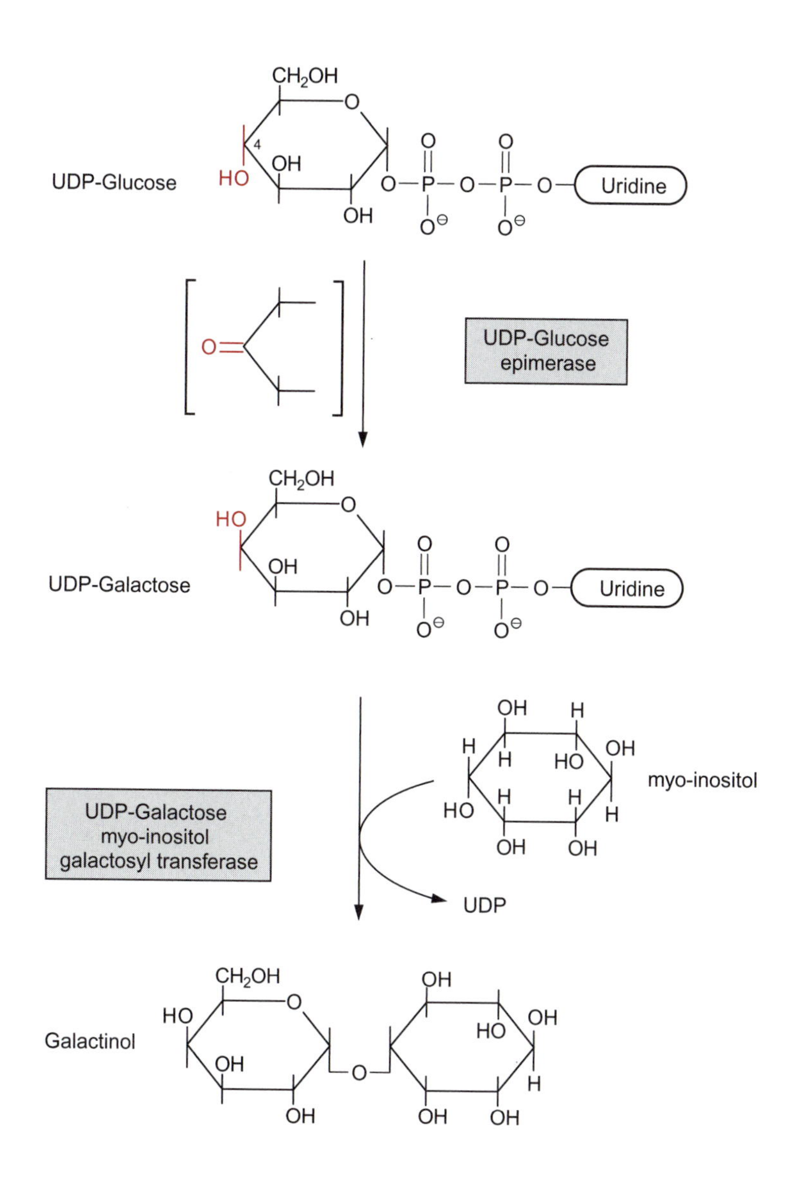

Figure 9.21 Synthesis of galactinol as an intermediate in raffinose synthesis from UDP-glucose and myo-inositol. The epimerization of UDP-glucose to UDP-galactose proceeds via the formation of a keto group as intermediate in position 4.

9.5 Fructans are deposited as storage substances in the vacuole

In addition to starch, many plants use **fructans** as carbohydrate storage compounds. Whereas starch is an insoluble polyglucose formed in the plastids, fructans are soluble polyfructoses that are synthesized and stored in the vacuole. They were first found in the tubers of ornamental flowers such as dahlias. Fructans are stored, often in the leaves and stems, of many grasses from temperate climates, such as wheat and barley. Fructans are also the major carbohydrate found in onions and, like the raffinose sugars, cannot be digested by humans. Because of their sweet taste, fructans are used as natural calorie-free sweeteners. Fructans are also used in the food industry as a replacement for fat.

The precursor for the polysaccharide chain of fructans is a sucrose molecule to which additional fructose molecules are attached by glycosidic linkages. The basic structure of a fructan in which sucrose is linked with one additional fructose molecule to a trisaccharide is called **kestose**. Figure 9.22 shows three major types of fructans.

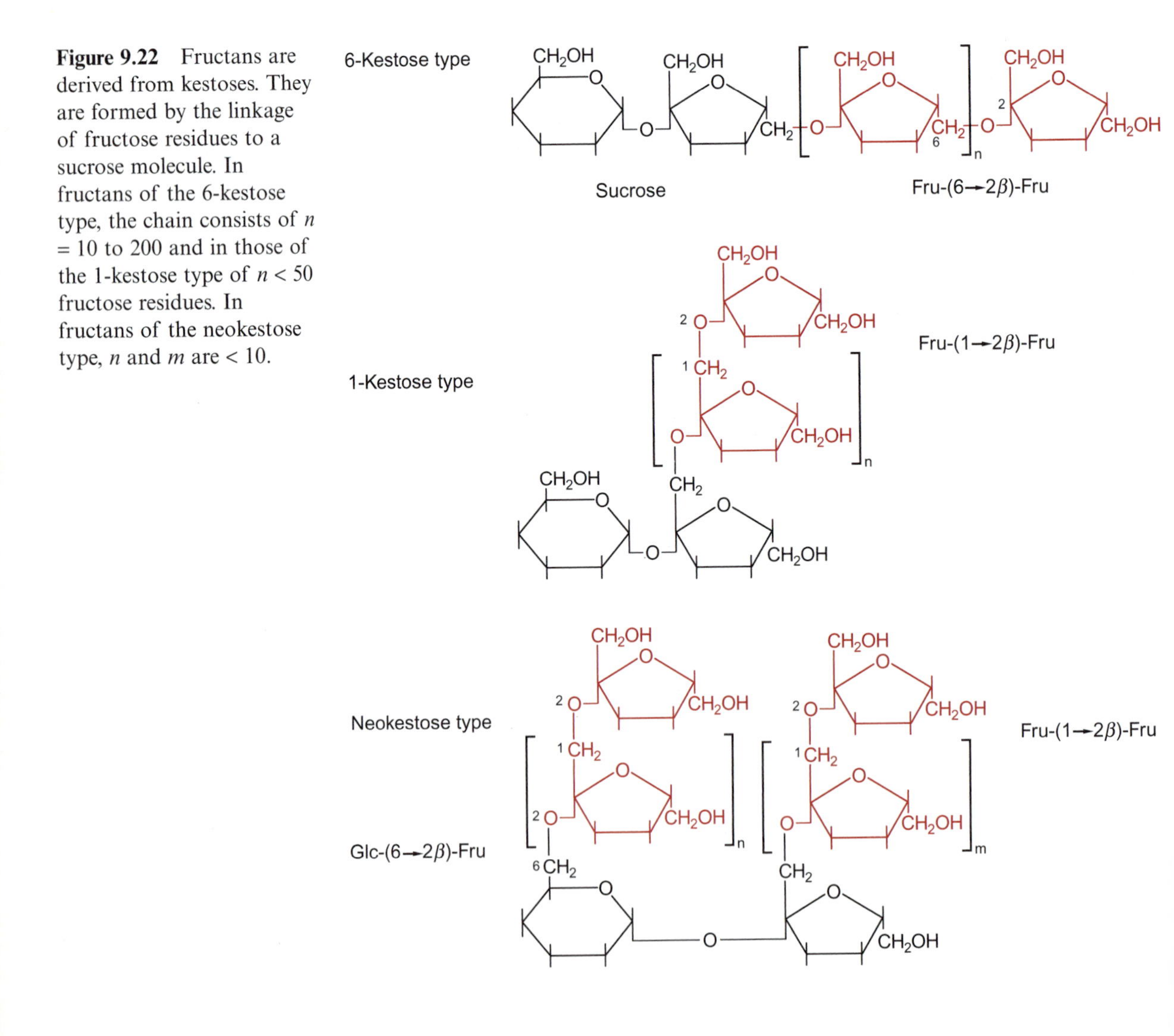

Figure 9.22 Fructans are derived from kestoses. They are formed by the linkage of fructose residues to a sucrose molecule. In fructans of the 6-kestose type, the chain consists of $n = 10$ to 200 and in those of the 1-kestose type of $n < 50$ fructose residues. In fructans of the neokestose type, n and m are < 10.

In fructans of the **6-kestose type**, the fructose residue of sucrose in position 6 is glycosidically linked with another fructose in the 2β-position. Chains of different lengths (10–200 fructose residues) are formed by (6 → 2β)-linkages with further fructose residues. These fructans are also called **levan type fructans** and are often found in grasses.

The fructose residues in fructans of the **1-kestose type** are linked to the sucrose molecule and to each other by (1 → 2β) glycosidic linkages. These fructans, also called **inulin type fructans**, consist of up to 50 fructose molecules. Inulin is found in dahlia tubers.

In fructans of the **neokestose type**, two polyfructose chains are connected to sucrose, one as in 1-kestose via (1 → 2β) glycosidic linkage with the fructose moiety, and the other in (6 → 2β) glycosidic linkage with the glucose residue of the sucrose molecule. The fructans of the neokestose type are the smallest fructans with only 5 to 10 fructose residues. Branched fructans in which the fructose molecules are connected by both (1 → 2β)- and (6 → 2β)-glycosidic linkages, are to be found in wheat and barley and are called **graminanes**.

Although fructans appear to have an important function in the metabolism of many plants, our knowledge of their function and metabolism is still fragmentary. Fructan synthesis occurs in the vacuoles and sucrose is the precursor for its synthesis. The fructose moiety of a sucrose molecule is transferred by a **sucrose-sucrose-fructosyl transferase** to a second sucrose molecule, resulting in the formation of a 1-kestose with a glucose molecule remaining (Fig. 9.23A). Additional fructose residues are transferred not from another sucrose molecule but from another kestose molecule for the elongation of the kestose chain (Fig. 9.23B). The enzyme **fructan-fructan 1-fructosyl transferase** preferentially transfers the fructose residue from a trisaccharide to a kestose with a longer chain. Correspondingly, the formation of 6-kestoses is catalyzed by a **fructan-fructan 6-fructosyl transferase**. For the formation of neokestoses, a fructose residue is transferred via a **6-glucose-fructosyl transferase** from a 1-kestose to the glucose residue of sucrose (Fig. 9.23C). The trisaccharide thus formed is a precursor for further chain elongation as shown in Figure 9.23B.

The degradation of fructans proceeds by the successive hydrolysis of fructose residues from the end of the fructan chain by exo-hydrolytic enzymes.

In many grasses, fructans are accumulated for a certain time period in the leaves and in the stems. They can comprise up to 30% of the dry matter. Often carbohydrates are accumulated as fructans before the onset of flowering as a reserve for rapid seed growth after pollination of the flowers. Plants in marginal habitats, where periods of positive CO_2 balance are succeeded by periods in which adequate photosynthesis is not possible, use

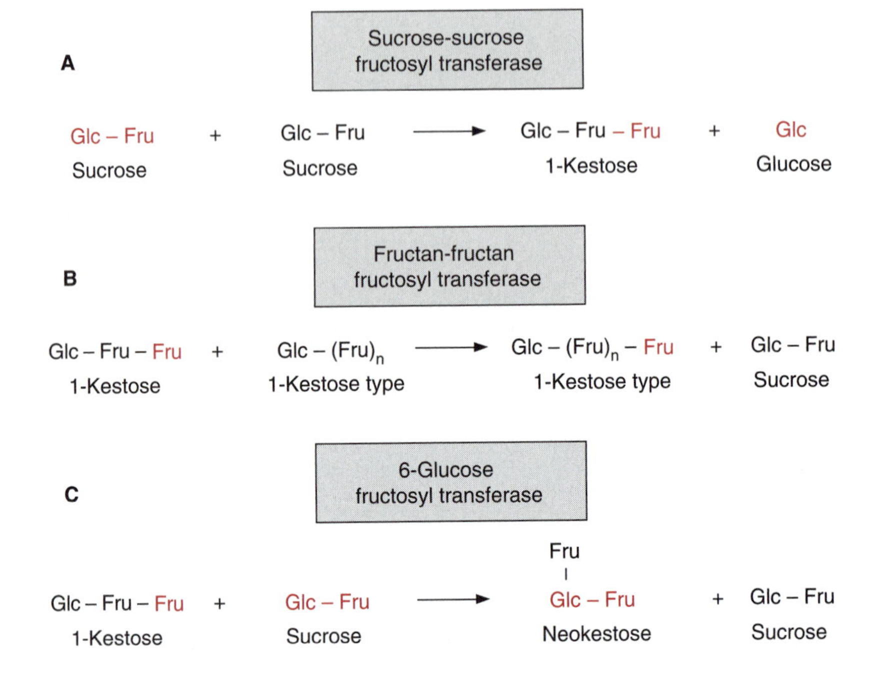

Figure 9.23 Sucrose is precursor for synthesis of kestoses. Three important reactions of the kestose biosynthesis pathway proceeding in the vacuole are shown.

fructans as a reserve to survive unfavorable conditions. Thus in many plants fructans are formed when these are subjected to water or cold stress.

Plants that accumulate fructans usually also store sucrose and starch in their leaves. Fructan is then an additional store. Figure 9.24 shows a simplified scheme of the formation of fructans as alternative storage compounds in a leaf. For the synthesis of fructans, sucrose is first synthesized in the cytosol (see Fig. 9.14 for details). The UDP-glucose required is formed from the glucose molecule that is released from the vacuole in the course of fructan synthesis and is phosphorylated by hexokinase. Thus the conversion of fructose 6-phosphate, generated by photosynthesis, into fructan needs altogether two ATP equivalents per molecule.

The large size of the leaf vacuoles, often comprising about 80% of the total leaf volume, offers the plant a very advantageous storage capacity for carbohydrates in the form of fructan. Thus, in a leaf, on top of the diurnal carbohydrate stores in the form of transitory starch and sucrose, an additional carbohydrate reserve can be maintained for such purposes as rapid seed production or the endurance of unfavorable growth conditions.

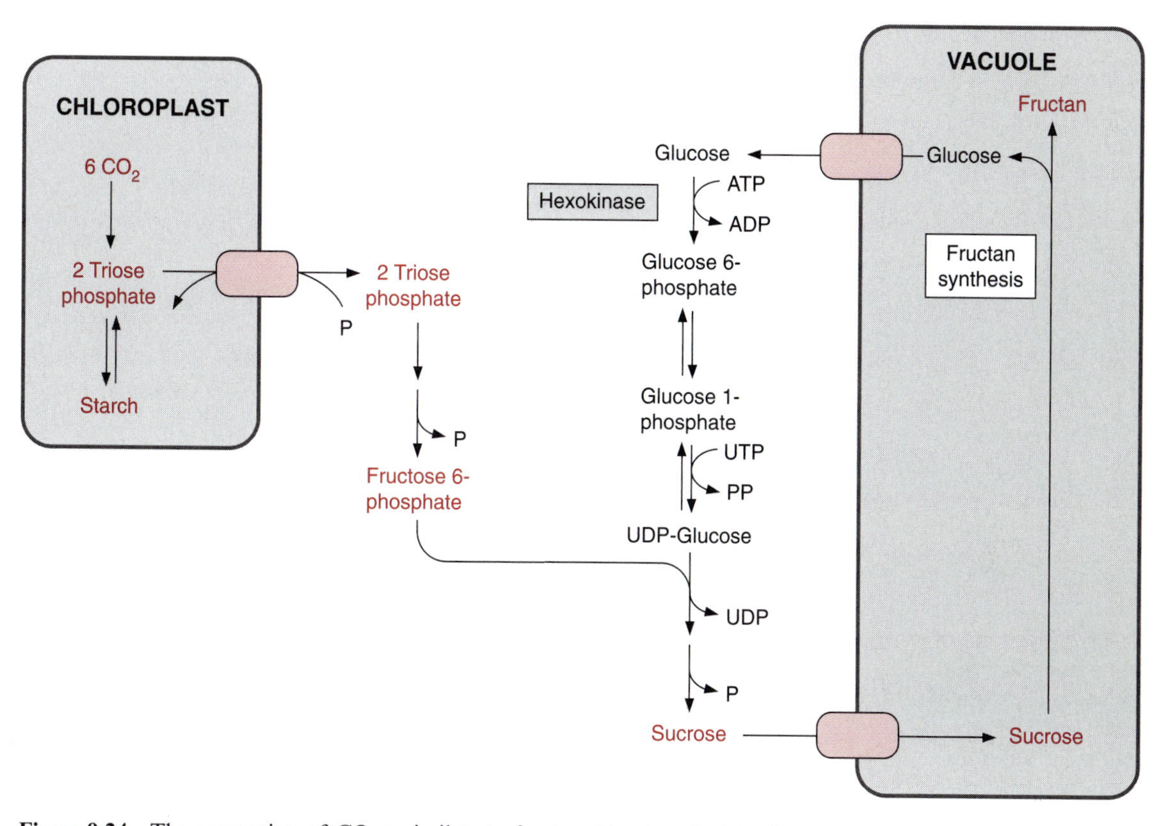

Figure 9.24 The conversion of CO_2 assimilate to fructan. Fructose 6-phosphate, which is provided as a product of photosynthesis in the cytosol, is first converted to sucrose. The glucose required for this is formed as a by-product in the synthesis of fructan in the vacuole (see Fig. 9.23). Phosphorylation is catalyzed by a hexokinase present in the cytosol. The entry of sucrose into the vacuole and the release of glucose are facilitated by different translocators.

9.6 Cellulose is synthesized by enzymes located in the plasma membrane

Cellulose, an important cell constituent (section 1.1), is a glucan in which the glucose residues are linked by ($\beta1 \rightarrow 4$)-glycosidic bonds forming a very long chain (Fig. 9.25). The synthesis of cellulose is catalyzed by **cellulose synthase** located in the plasma membrane. The required glucose molecules are delivered as **UDP-glucose** from the cytosol of the cell, and the newly synthesized cellulose chain is excreted into the extracellular compartment (Fig. 9.26). It has been shown in cotton-producing cells—a useful system for studying cellulose synthesis—that UDP-glucose is supplied by a membrane-

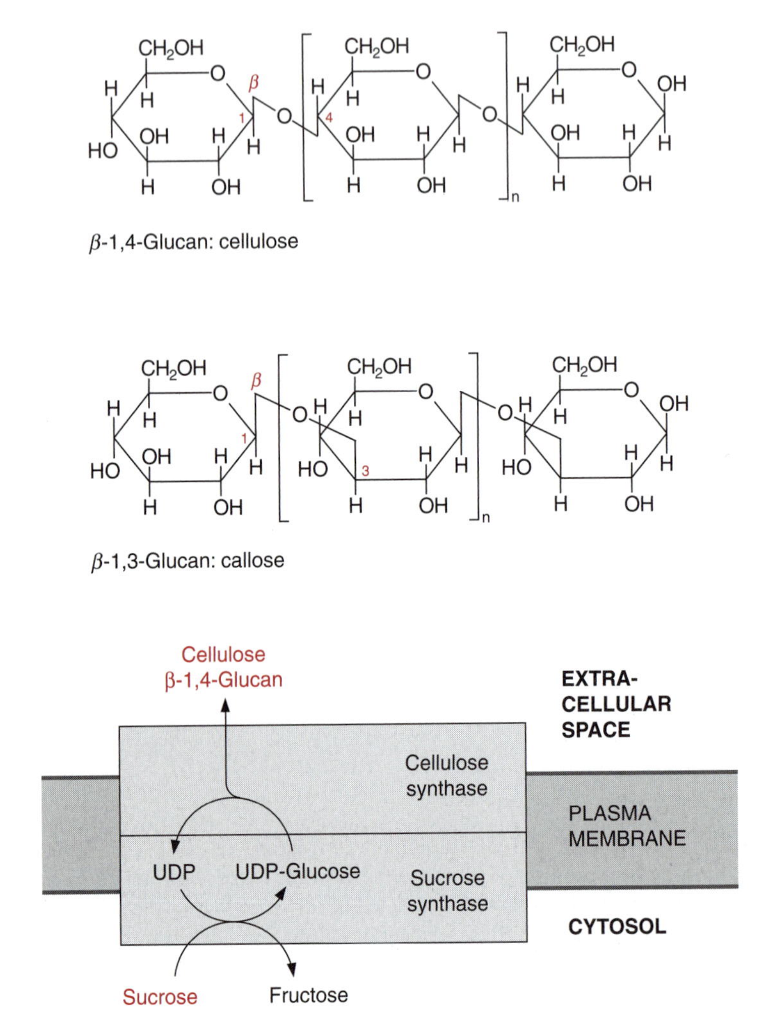

Figure 9.25 Cellulose and callose.

Figure 9.26 Synthesis of β-1,4-glucan chains by a membrane-bound cellulose synthase. The UDP-glucose required is formed from sucrose by a membrane-bound sucrose synthase.

bound **sucrose synthase** (see section 9.2) from sucrose provided by the cytosol. The UDP-glucose formed is transferred directly to the cellulose synthase. According to recent results, the synthesis of cellulose starts with the transfer of a glucose residue from UDP glucose to **sitosterol** (Fig. 15.3), a plasma membrane lipid. The glucose residue is bound to the hydroxyl group of the membrane lipid via a glycosidic linkage and acts as a primer for the cellulose synthesis, thus anchoring the growing cellulose chain to the membrane. Cellulose never occurs in single chains but always in a crystalline array of many chains called a **microfibril** (section 1.1). It is assumed that, due to the many neighboring cellulose synthases in the membrane, all the β-1,4-glucan chains of a microfibril are synthesized simultaneously and spontaneously form a microfibril.

Toroser, D., Huber, S. C. Protein phosphorylation as a mechanism for osmotic-stress activation of sucrose-phosphate synthase in spinach leaves. Plant Physiol 114, 947–955 (1997).
Wiese, A., Gröner, F., Sonnewald, U., Deppner, H., Lerchl, J., Hebbeker, U., Flügge, U.-I., Weber, A. Spinach hexokinase I is located in the outer envelope membrane of plastids. FEBS Lett 461, 13–18 (1999).
Williamson, R. E., Burn, J. E., Hocart, C. H. Towards the mechanism of cellulose synthesis. Trends Plant Sci 7, 461–467 (2002).
Winter, H., Huber, S. C. Regulation of sucrose metabolism in higher plants: Localization and regulation of activity of key enzymes. Crit Rev Plant Sci 19, 31–67 (2000).
Zimmermann, M. H., Ziegler, H. List of sugars and sugar alcohols in sieve-tube exudates. In (Zimmermann, M.H.; Milburn, J.A. (eds.)) Encyclopedia of Plant Physiology, Springer Verlag, Heidelberg, Vol. 1, 480–503 (1975).

10

Nitrate assimilation is essential for the synthesis of organic matter

Living matter contains a large amount of nitrogen incorporated in proteins, nucleic acids, and many other biomolecules. This organic nitrogen is present in oxidation state –III (as in NH_3). During autotrophic growth the nitrogen demand for the formation of cellular matter is met by inorganic nitrogen in two alternative ways:

1. Fixation of molecular nitrogen from air; or
2. Assimilation of the nitrate or ammonia contained in water or soil.

Only some bacteria, including cyanobacteria, are able to fix N_2 from air. Some plants enter a symbiosis with N_2-fixing bacteria, which supply them with organic bound nitrogen. Chapter 11 deals with this in detail. However, about 99% of the organic nitrogen in the biosphere is derived from the assimilation of nitrate. NH_4^+ is formed as an end product of the degradation of organic matter, primarily by the metabolism of animals and bacteria, and is oxidized to nitrate again by nitrifying bacteria in the soil. Thus a continuous cycle exists between the nitrate in the soil and the organic nitrogen in the plants growing on it. NH_4^+ accumulates only in poorly aerated soils with insufficient drainage, where, due to lack of oxygen, nitrifying bacteria cannot grow. Mass animal production can lead to a high ammonia input into the soil, from manure as well as from the air. If it is available, many plants can also utilize NH_4^+ instead of nitrate as a nitrogen source.

10.1 The reduction of nitrate to NH_3 proceeds in two partial reactions

Nitrate is assimilated in the leaves and also in the roots. In most fully grown herbaceous plants, nitrate assimilation occurs primarily in the leaves, although nitrate assimilation in the roots often plays a major role at an early growth state of these plants. In contrast, many woody plants (e.g., trees, shrubs), as well as legumes such as soybean, assimilate nitrate mainly in the roots.

The transport of nitrate into the root cells proceeds as symport with two protons (Fig. 10.1). A proton gradient across the plasma membrane, generated by a H^+-P-ATPase (section 8.2) drives the uptake of nitrate against a concentration gradient. The ATP required for the formation of the proton gradient is mostly provided by **mitochondrial respiration**. When inhibitors or uncouplers of respiration abolish mitochondrial ATP synthesis in the roots, nitrate uptake normally comes to a stop. Root cells contain several **nitrate transporters** in their plasma membrane; among these are a transporter with a relatively low affinity (half saturation >500 μM nitrate) and a transporter with a very high affinity (half saturation 20–100 μM nitrate), where the latter is induced only when required by metabolism. In this way the capacity of nitrate uptake into the roots is adjusted to the environmental conditions. The efficiency of the nitrate uptake systems makes it possible that plants can grow when the external nitrate concentration is as low as 10 μM.

The nitrate taken up into the root cells can be stored there temporarily in the vacuole. As discussed in section 10.2, nitrate is reduced to NH_4^+ in the epidermal and cortical cells of the root. This NH_4^+ is mainly used for the synthesis of **glutamine** and **asparagine** (collectively named amide in Fig. 10.1). These two amino acids can be transported to the leaves via the **xylem vessels**. However, when the capacity for nitrate assimilation in the roots is exhausted, nitrate is released from the roots into the xylem vessels and is carried by the transpiration stream to the leaves. There it is taken up into the mesophyll cells, probably also by a proton symport. Large quantities of nitrate can be stored in a leaf by uptake into the vacuole. Sometimes this vacuolar store is emptied by nitrate assimilation during the day and replenished during the night. Thus, in spinach leaves, for instance, the highest nitrate content is found in the early morning.

The nitrate in the mesophyll cells is reduced to nitrite by **nitrate reductase** present in the **cytosol** and then to NH_4^+ by **nitrite reductase** in the **chloroplasts** (Fig. 10.1).

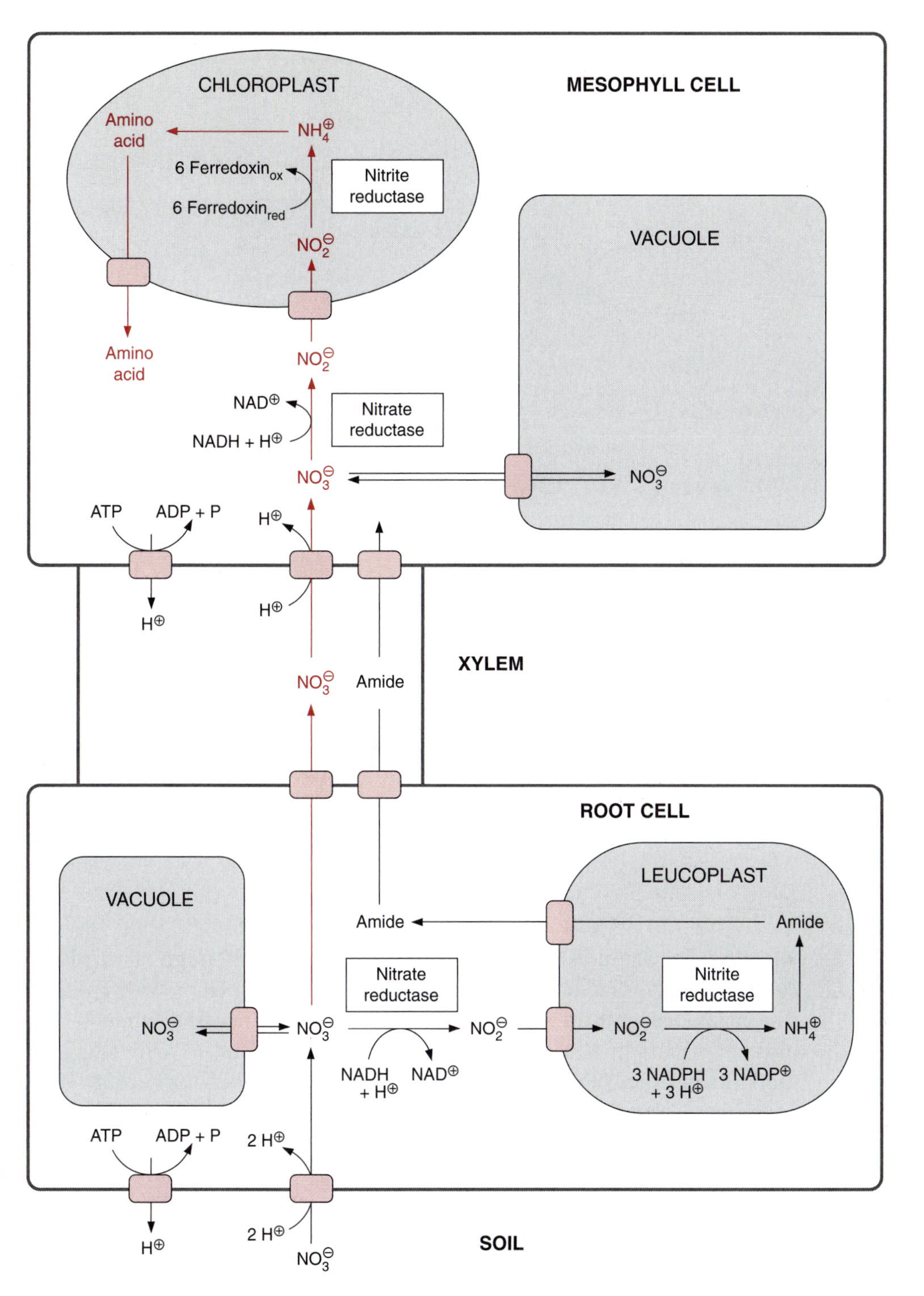

Figure 10.1 Nitrate assimilation in the roots and leaves of a plant. Nitrate is taken up from the soil by the root. It can be stored in the vacuoles of the root cells or assimilated in the cells of the root epidermis and the cortex. Surplus nitrate is carried via the xylem vessels to the mesophyll cells, where nitrate can be temporarily stored in the vacuole. Nitrate is reduced to nitrite in the cytosol and then nitrite is reduced further in the chloroplasts to NH_4^+, from which amino acids are formed. H^+ transport out of the cells of the root and the mesophyll proceeds via a H^+-P-ATPase.

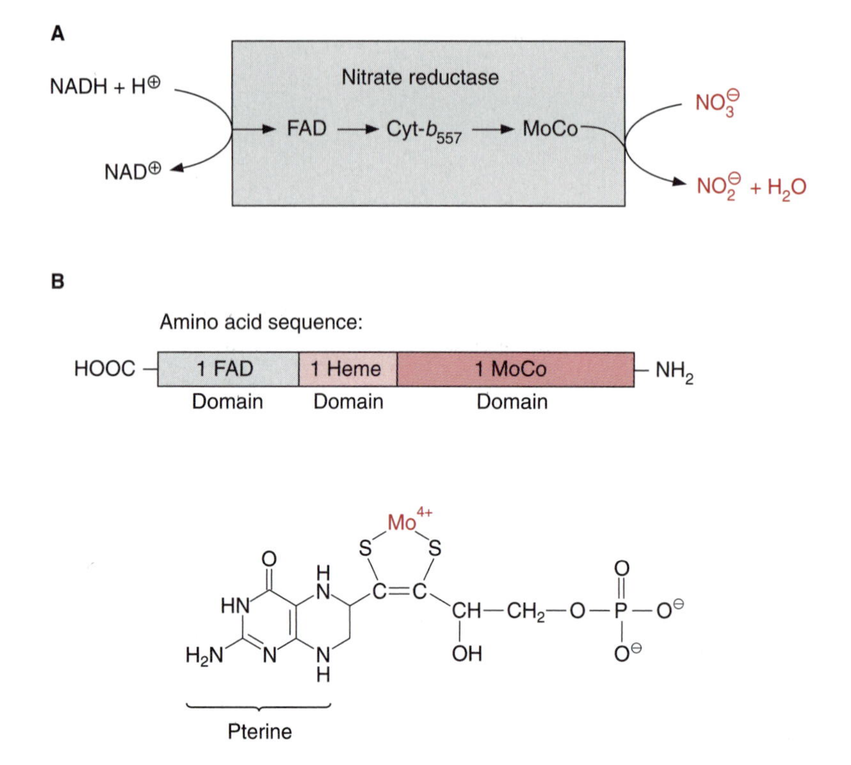

Figure 10.2 A.) Nitrate reductase transfers electrons from NADH to nitrate. B.) The enzyme contains three domains where FAD, heme, and the molybdenum cofactor (MoCo) are bound.

Figure 10.3 The molybdenum cofactor (MoCo).

Nitrate is reduced to nitrite in the cytosol

Nitrate reduction uses mostly NADH as reductant, although some plants contain a nitrate reductase reacting with NADPH as well as with NADH. The **nitrate reductase** of higher plants consists of two identical subunits. The molecular mass of each subunit varies from 99 to 104 kDa, depending on the species. Each subunit contains an electron transport chain (Fig. 10.2) consisting of one **flavin adenine dinucleotide** molecule (FAD), one heme of the **cytochrome-*b*** type (cyt-b_{557}), and one cofactor containing molybdenum (Fig. 10.3). This cofactor is a **pterin** with a side chain to which the molybdenum is attached by two sulfur bonds and is called the **molybdenum cofactor**, abbreviated **MoCo**. The bound Mo atom probably changes between oxidation states +IV and +VI. The three redox carriers of nitrate reductase are each covalently bound to the subunit of the enzyme. The protein chain of the subunit can be cleaved by limited proteolysis into three domains, each of which contains only one of the redox carriers. These separated domains, as well as the holoenzyme, are able to catalyze, via their redox carriers, electron transport to artificial electron acceptors [e.g., from NADPH to Fe^{+++} ions via the FAD domain or from reduced methylviologen (Fig. 3.39) to

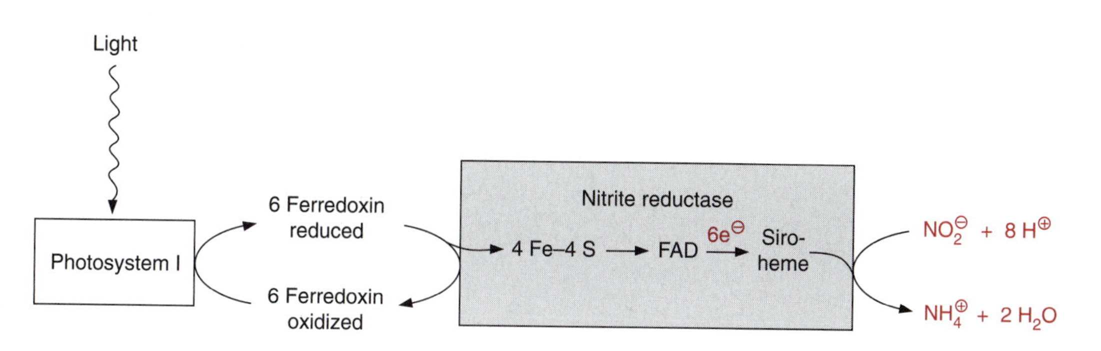

Figure 10.4 Nitrite reductase in chloroplasts transfers electrons from ferredoxin to nitrite. Reduction of ferredoxin by photosystem I is shown in Figure 3.16.

nitrate via the Mo domain]. Nitrite reductase is also able to reduce nitrite to nitric oxide (NO), which is an important signal substance in signal cascades (section 8.2, 19.1). Moreover, nitrate reductase reduces chlorate (ClO_3^-) to chlorite (ClO_2^-). The latter is a very strong oxidant and therefore is highly toxic to plant cells. In the past chlorate was used as an inexpensive nonselective herbicide for keeping railway tracks free of vegetation.

The reduction of nitrite to ammonia proceeds in the plastids

The reduction of nitrite to ammonia requires the uptake of six electrons. This reaction is catalyzed by only one enzyme, the **nitrite reductase** (Fig. 10.4), which is located exclusively in plastids. This enzyme utilizes reduced ferredoxin as electron donor, which is supplied by photosystem I as a product of photosynthetic electron transport (Fig. 3.31). To a much lesser extent, the ferredoxin required for nitrite reduction in a leaf can also be provided during darkness via reduction by NADPH, which is generated by the oxidative pentose phosphate pathway present in chloroplasts and leucoplasts (Figs. 6.21 and 10.8).

Nitrite reductase contains a covalently bound **4Fe-4S cluster** (see Figure 3.26), one molecule of **FAD**, and one siroheme. **Siroheme** (Fig. 10.5) is a cyclic tetrapyrrole with one Fe-atom in the center. Its structure is different from that of heme as it contains additional acetyl and propionyl residues deriving from pyrrole synthesis (see section 10.5).

The 4Fe-4S cluster, FAD, and siroheme form an electron transport chain by which electrons are transferred from ferredoxin to nitrite. Nitrite reductase has a very high affinity for nitrite. The capacity for nitrite reduction in the chloroplasts is much greater than that for nitrate reduction in the cytosol. Therefore all nitrite formed by nitrate reductase can be completely converted to ammonia. This is important since nitrite is toxic to the cell. It forms diazo

Figure 10.5 Siroheme.

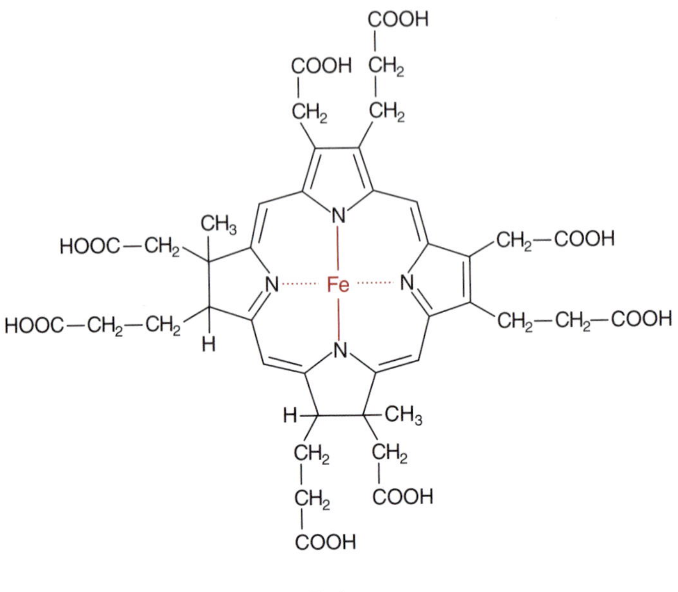

compounds with amino groups of nucleobases (R–NH_2), which are converted into alcohols with the release of nitrogen.

$$R-NH_2+NO_2^- \rightarrow [R-N=N-OH+OH^-] \rightarrow R-OH+N_2+OH^-$$

Thus, for instance, cytosine can be converted to uracil. This reaction can lead to mutations in nucleic acids. The very efficient reduction of nitrite by chloroplast nitrite reductase prevents nitrite from accumulating in the cell.

The fixation of NH_4^+ proceeds in the same way as in photorespiration

Glutamine synthetase in the chloroplasts transfers the newly formed NH_4^+ at the expense of ATP to glutamate, forming glutamine (Fig. 10.6). The activity of glutamine synthetase and its affinity for NH_4^+ ($Km \approx 5 \cdot 10^{-6}$ mol/L) are so high that the NH_4^+ produced by nitrite reductase is taken up completely. The same reaction also fixes the NH_4^+ released during photorespiration (see Fig. 7.9). Because of the high rate of photorespiration, the amount of NH_4^+ produced by the oxidation of glycine is about 5 to 10 times higher than the amount of NH_4^+ generated by nitrate assimilation. Thus only a minor proportion of glutamine synthesis in the leaves is actually involved in nitrate assimilation. Leaves also contain an isoenzyme of glutamine synthetase in their cytosol.

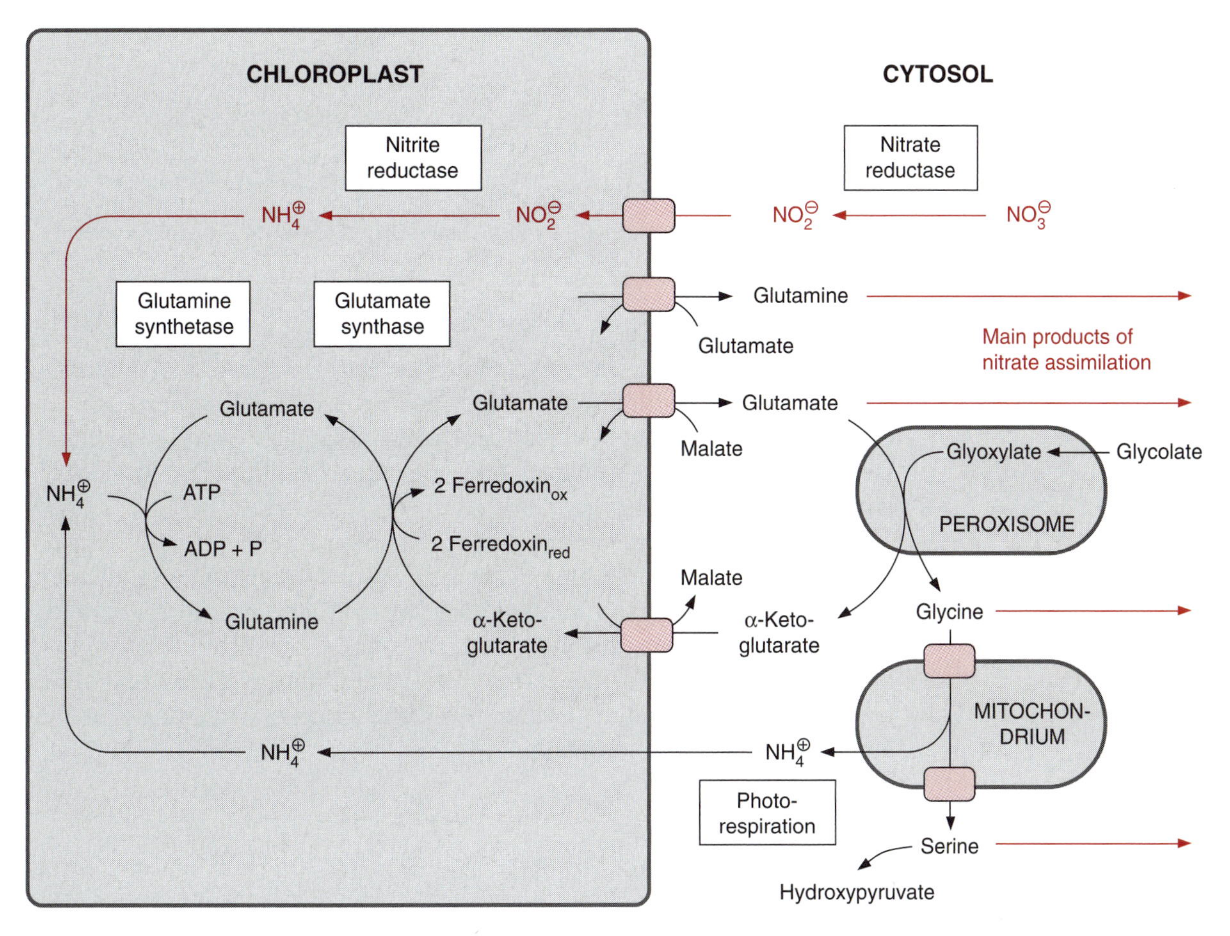

Figure 10.6 Compartmentation of partial reactions of nitrate assimilation and the photorespiratory pathway in mesophyll cells. NH_4^+ formed in the photorespiratory pathway is colored black and NH_4^+ formed by nitrate assimilation is colored red. The main products of nitrate assimilation are marked with a red arrow.

Glufosinate (Fig. 10.7), a substrate analogue of glutamate, inhibits glutamine synthesis. Plants in which the addition of glufosinate has inhibited the synthesis of glutamine accumulate toxic levels of ammonia and die off. NH_4^+-glufosinate is distributed as an herbicide (section 3.6) under the trade name Liberty (Aventis). It has the advantage that it is degraded rapidly in the soil, leaving behind no toxic degradation products. Recently glufosinate-resistant crop plants have been generated by genetic engineering, enabling the use of glufosinate as a selective herbicide for eliminating weeds in growing cultures (section 22.6).

The glutamine formed in the chloroplasts is converted via **glutamate synthase** (also called glutamine-oxoglutarate amino transferase, abbreviated

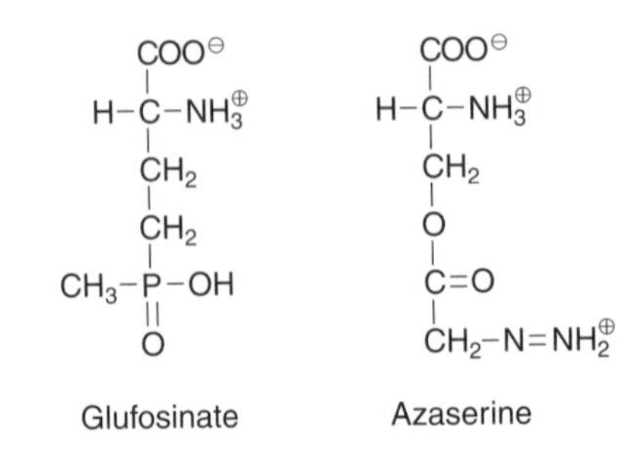

Figure 10.7 Glufosinate (also called phosphinotricin) is a substrate analogue of glutamate and a strong inhibitor of glutamine synthetase. Ammonium glufosinate is an herbicide (Liberty, Aventis). Azaserine is a substrate analogue of glutamate and an inhibitor of glutamate synthase.

GOGAT), by reaction with α-ketoglutarate to two molecules of glutamate (see also Fig. 7.9) with ferredoxin as reductant. Some chloroplasts and leucoplasts also contain an NADPH-dependent glutamate synthase. Glutamate synthases are inhibited by the substrate analogue **azaserine** (Fig. 10.7), which is toxic to plants.

α-Ketoglutarate, which is required for the glutamate synthase reaction, is transported into the chloroplasts by a specific translocator in counter-exchange for malate, and the glutamate formed is transported out of the chloroplasts into the cytosol by another translocator, also in exchange for malate (Fig. 10.6). A further translocator in the chloroplast envelope transports glutamine in counter-exchange for glutamate, enabling the export of glutamine from the chloroplasts.

10.2 Nitrate assimilation also takes place in the roots

As mentioned, nitrate assimilation occurs in part, and in some species even mainly, in the roots. NH_4^+ taken up from the soil is normally fixed in the roots. The reduction of nitrate and nitrite as well as the fixation of NH_4^+ proceeds in the root cells in an analogous way to the mesophyll cells. However, in the root cells the necessary reducing equivalents are supplied exclusively by oxidation of carbohydrates. The reduction of nitrite and the subsequent fixation of NH_4^+ (Fig. 10.8) occur in the leucoplasts, a differentiated form of plastids (section 1.3).

The oxidative pentose phosphate pathway provides reducing equivalents for nitrite reduction in leucoplasts

The reducing equivalents required for the reduction of nitrite and the formation of glutamate are provided in leucoplasts by oxidation of glucose 6-

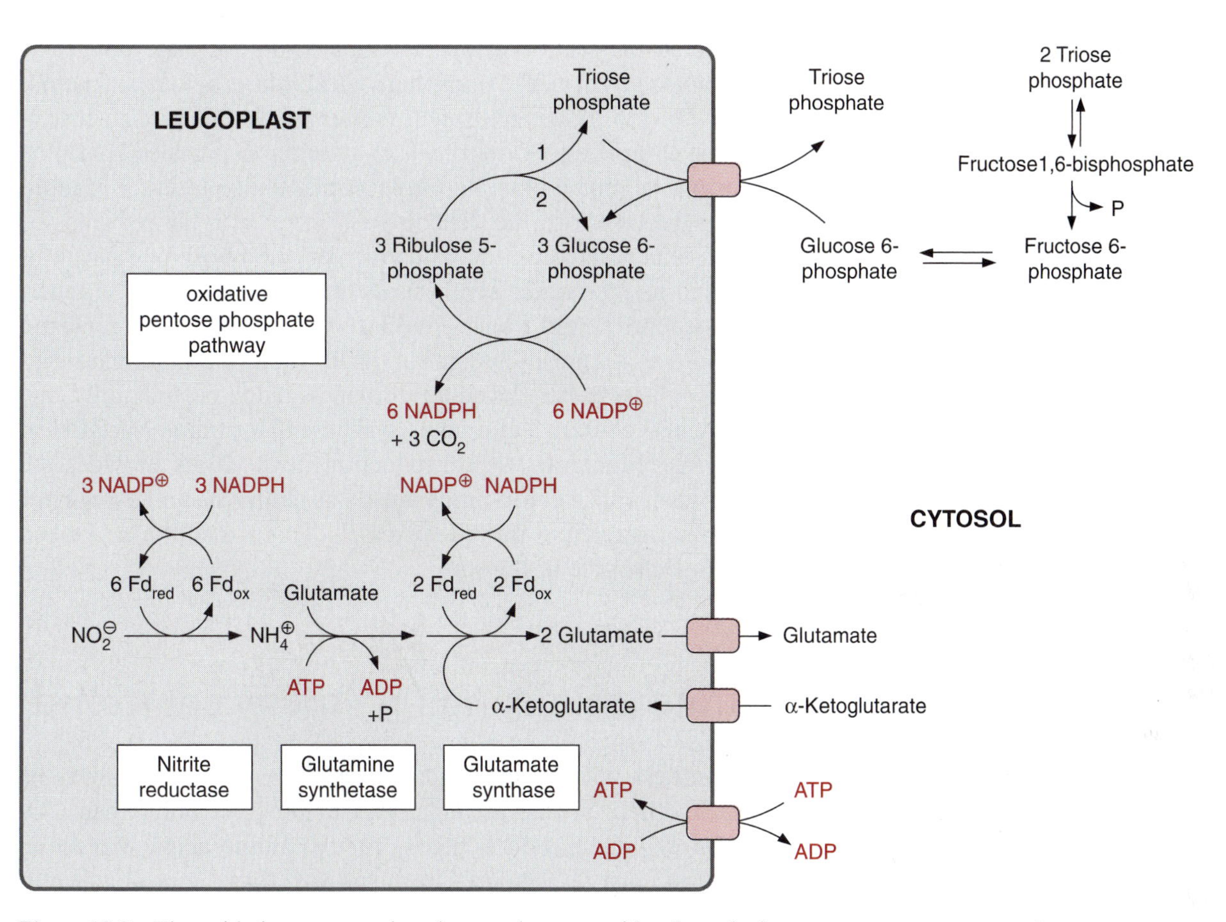

Figure 10.8 The oxidative pentose phosphate pathway provides the reducing equivalents for nitrite reduction in plastids (leucoplasts) from non-green tissues In some plastids, glucose 1-phosphate is transported in counter-exchange for triose phosphate or phosphate. Fd, ferredoxin.

phosphate via the oxidative pentose phosphate pathway discussed in section 6.5 (Fig. 10.8). The uptake of glucose 6-phosphate proceeds in counter-exchange for triose phosphate. The **glucose 6-phosphate-phosphate translocator** of leucoplasts differs from the triose phosphate-phosphate translocator of chloroplasts in transporting glucose 6-phosphate in addition to phosphate, triose phosphate, and 3-phosphoglycerate. In the oxidative pentose phosphate pathway, three molecules of glucose 6-phosphate are converted to three molecules of ribulose 5-phosphate with the release of three molecules of CO_2, yielding six molecules of NADPH. The subsequent reactions yield one molecule of triose phosphate and two molecules of fructose 6-phosphate; the latter are reconverted to glucose 6-phosphate via hexose phosphate isomerase. In the cytosol, glucose 6-phosphate is regenerated from two molecules of triose phosphate via aldolase, cytosolic fructose 1,6-bisphosphatase, and hexose phosphate isomerase. In this way glucose 6-phosphate can be completely oxidized to CO_2 in order to produce NADPH.

As in chloroplasts, nitrite reduction in leucoplasts also requires reduced ferredoxin as reductant. In the leucoplasts, ferredoxin is reduced by NADPH, which is generated by the oxidative pentose phosphate pathway. The ATP required for glutamine synthesis in the leucoplasts can be generated by the mitochondria and transported into the leucoplasts by a plastid ATP translocator in counter-exchange for ADP. Also, the glutamate synthase of the leucoplasts uses reduced ferredoxin as redox partner, although some leucoplasts also contain a glutamate synthase that utilizes NADPH or NADH directly as reductant. Nitrate reduction in the roots provides the shoot with organic nitrogen compounds mostly as **glutamine** and **asparagine** via the transpiration stream in the xylem vessels. This is also the case when NH_4^+ is the nitrogen source in the soil.

10.3 Nitrate assimilation is strictly controlled

During photosynthesis, CO_2 assimilation and nitrate assimilation have to be matched to each other. Nitrate assimilation can progress only when CO_2 assimilation provides the carbon skeletons for the amino acids. Moreover, nitrate assimilation must be regulated in such a way that the production of amino acids does not exceed demand. Finally, it is important that nitrate reduction does not proceed faster than nitrite reduction, since otherwise toxic levels of nitrite (section 10.1) would accumulate in the cells. Under certain conditions such a dangerous accumulation of nitrite can indeed occur in roots when excessive moisture makes the soil anaerobic. Flooded roots are able to discharge nitrite into water, avoiding the buildup of toxic levels of nitrite, but this escape route is not open to leaves, making the strict control of nitrate reduction there especially important.

The NADH required for nitrate reduction in the cytosol can also be provided during darkness (e.g., by glycolytic degradation of glucose). However, reduction of nitrite and fixation of NH_4^+ in the chloroplasts depends largely on photosynthesis providing reducing equivalents and ATP. The oxidative pentose phosphate pathway can provide only very limited amounts of reducing equivalents in the dark. Therefore, during darkness nitrate reduction in the leaves has to be slowed down or even switched off to prevent an accumulation of nitrite. This illustrates how essential it is for a plant to regulate the activity of nitrate reductase, the entrance step of nitrate assimilation.

The synthesis of the nitrate reductase protein is regulated at the level of gene expression

Nitrate reductase is an exceptionally short-lived protein. Its half-life is only a few hours. The rate of *de novo* synthesis of this enzyme is therefore very high. Thus, by regulating its synthesis, the activity of nitrate reductase can be altered within hours.

Various factors control the synthesis of the enzyme at the level of gene expression. Nitrate and light stimulate the enzyme synthesis. The question as to whether a phytochrome or a blue light receptor (Chapter 19) is involved in this effect of light has not been fully resolved. The synthesis of the nitrate reductase protein is stimulated by glucose and other carbohydrates and is inhibited by NH_4^+, glutamine, and other amino acids (Fig. 10.9). Sensors seem to be present in the cell that adjust the capacity of nitrate reductase both to the demand for amino acids and to the supply of carbon skeletons from CO_2 assimilation for its synthesis via regulation of gene expression.

Nitrate reductase is also regulated by reversible covalent modification

The regulation of *de novo* synthesis of nitrate reductase (NR) allows regulation of the enzyme activity within a time span of hours. This would not be sufficient to prevent an accumulation of nitrite in the plants during darkening or sudden shading of the plant. Rapid inactivation of nitrate reductase in the time span of minutes occurs via **phosphorylation of the nitrate reductase protein** (Fig. 10.9). Upon darkening, a **serine** residue, which is located in the nitrate reductase protein between the heme and the MoCo domain, is phosphorylated by a specific protein kinase termed **nitrate reductase kinase**. This protein kinase is inhibited by the photosynthesis product triose phosphate and other phosphate esters and is stimulated by Ca^{++}-ions, a messenger substance of many signal chains (section 19.1). The phosphorylated nitrate reductase binds an **inhibitor protein**. This interrupts the electron transport between cytochrome b_{557} and the MoCo domain (Fig. 10.2) and thus inhibits nitrate reductase. The **nitrate reductase phosphatase** hydrolyzes the enzyme's serine phosphate and this causes the inhibitor protein to be released from the enzyme and thus to lose its effect; and the enzyme is active again. **Okadaic acid** inhibits nitrate reductase phosphatase and in this way also inhibits the reactivation of nitrate reductase. It is still not clear which effectors of the nitrate reductase kinase and phosphatase signal the state "illumination." The inhibition of nitrate reductase kinase by triose phosphate and other phosphate esters ensures that nitrate reductase

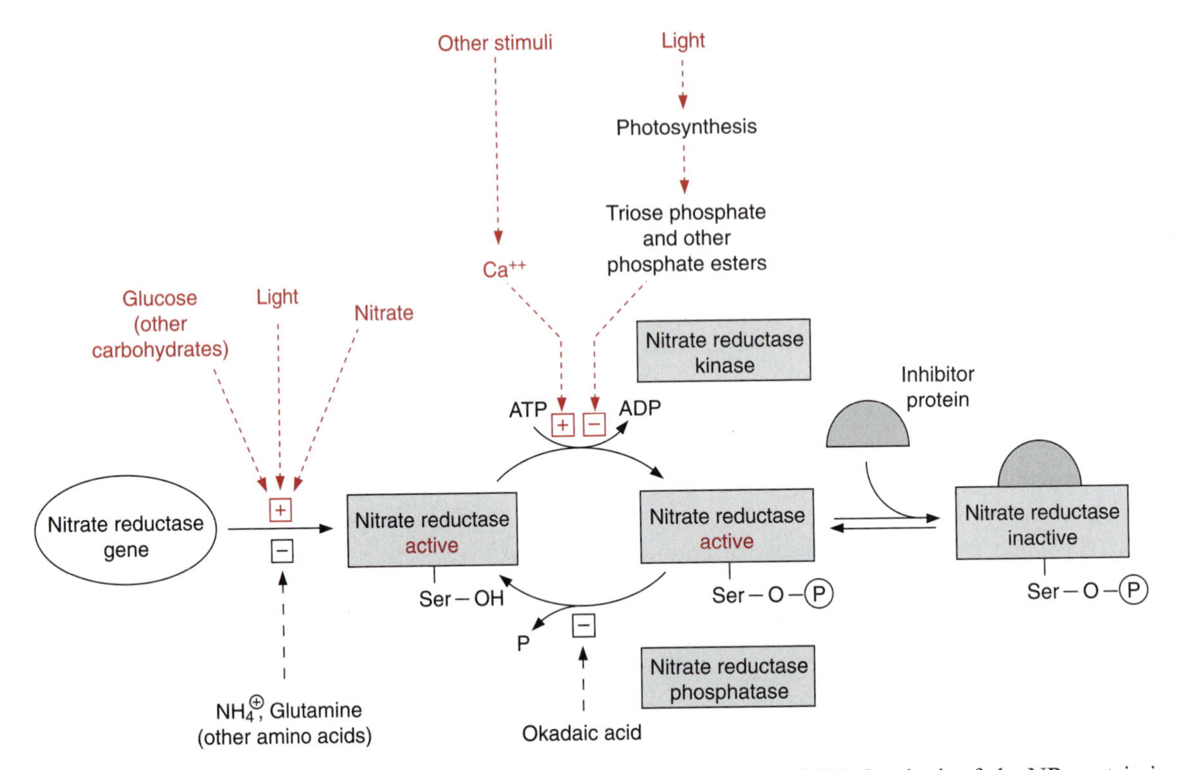

Figure 10.9 Regulation of nitrate reductase (NR). Synthesis of the NR protein is stimulated by carbohydrates (perhaps glucose or its metabolic products) and light [+], and inhibited by glutamine or other amino acids [–]. The newly formed NR protein is degraded within a time span of a few hours. Nitrate reductase is inhibited by phosphorylation of a serine residue and the subsequent interaction with an inhibitor protein. After hydrolytic liberation of the phosphate residue by a protein phosphatase, the inhibitor is dissociated and nitrate reductase regains its full activity. The activity of the nitrate reductase kinase is inhibited by triose phosphate and other phosphate esters—products of photosynthesis in the light—and in this way nitrate reductase is active in the light. Through the effect of Ca^{++} on nitrate reductase kinase other still not identified factors may modulate the activity of nitrate reductase. Okadaic acid, an inhibitor of protein phosphatases, counteracts the activation of nitrate reductase. (After Huber *et al.*, 1996.)

is active only when CO_2 fixation is operating for delivery of the carbon skeletons for amino acid synthesis, which is discussed in the next section.

14-3-3 Proteins are important metabolic regulators

It was discovered that the nitrate reductase inhibitor protein belongs to a family of regulatory proteins called **14-3-3 proteins**, which are widely spread throughout the animal and plant worlds. 14-3-3 proteins bind to a specific binding site of the **target protein with six amino acids**, containing a serine phosphate in position 4. This was proved for nitrate reductase in an experiment, in which the serine in the 14-3-3 protein binding site of nitrate reductase was exchanged for arginine via mutagenesis; the altered nitrate reductase was no longer inactivated by phosphorylation.14-3-3 proteins bind to a variety of proteins and change their activity. They form a large family of **multifunctional regulatory proteins**, many isoforms of which occur in a single plant. Thus 14-3-3 proteins regulate in plants the activity of the **H^+-P-ATPase** (section 8.2) of the plasma membrane. 14-3-3 Proteins regulate the function of transcription factors (section 20.2) and protein transport into chloroplasts (section 21.3). There are indications that 14-3-3 proteins are involved in the regulation of signal transduction (section 19.1), as they bind to various protein kinases, and play a role in defense processes against biotic and abiotic stress. The elucidation of these various functions of 14.3.3 proteins is at present a very hot topic in research.

This important function of the 14-3-3 proteins in metabolic regulation is exploited by the pathogenic fungus *Fusicoccum* to attack plants. This fungus forms the substance **fusicoccin**, which binds specifically to the 14-3-3 protein binding sites of various proteins and thus cancels the regulatory function of 14-3-3 proteins. This can disrupt the metabolism to such an extent that the plant finally dies. This attack proceeds in a subtle way. When *Fusicoccum amygdalis* infects peach or almond trees, at first only a few leaves are affected. In these leaves the fungus excretes fusicoccin into the apoplasts, from which it is spread via the transpiration stream through the rest of the plant. Finally, fusicoccin arrives in the guard cells, where it activates a signal chain by which the stomata are kept open all the time. This leads to a very high loss of water; consequently, the leaves wilt, the tree dies, and it becomes a source of nutrition for the fungus.

The regulation of nitrate reductase and sucrose phosphate synthase have great similarities

The mechanism shown here for the regulation of nitrate reduction by phosphorylation of serine residues of the enzyme protein by special protein kinases and protein phosphatases is remarkably similar to the regulation of sucrose phosphate synthase discussed in Chapter 9 (Fig. 9.18). Upon darkening, both enzymes are inactivated by phosphorylation, which in the case of nitrate reductase also requires a binding of an inhibitor protein. Both enzymes are reactivated by protein phosphatases, which are inhibited by okadaic acid. Also, sucrose phosphate synthase has a binding site for 14-3-3 proteins, but its significance for regulation is not yet clear. Although many details are still not known, it is obvious that the basic mechanisms for the rapid light regulation of sucrose phosphate synthase and nitrate reductase are similar.

10.4 The end product of nitrate assimilation is a whole spectrum of amino acids

As described in Chapter 13, the carbohydrates formed as the product of CO_2 assimilation are transported from the leaves via the sieve tubes to various parts of the plants only in defined transport forms, such as sucrose, sugar alcohols (e.g., sorbitol), or raffinoses, depending on the species. There are no such special transport forms for the products of nitrate assimilation. All amino acids present in the mesophyll cells are exported via the sieve tubes. Therefore the sum of amino acids can be regarded as the final product of nitrate assimilation. Synthesis of these amino acids takes place mainly in the chloroplasts. The pattern of the amino acids synthesized varies largely, depending on the species and the metabolic conditions. In most cases **glutamate** and **glutamine** represent the major portion of the synthesized amino acids. Glutamate is exported from the chloroplasts in exchange for malate and glutamine in exchange for glutamate (Fig. 10.6). Also, **serine** and **glycine**, which are formed as intermediate products in the photorespiratory cycle, represent a considerable portion of the total amino acids present in the mesophyll cells. Large amounts of **alanine** are often formed in C_4 plants.

CO_2 assimilation provides the carbon skeletons to synthesize the end products of nitrate assimilation

CO_2 assimilation provides the carbon skeletons required for the synthesis of the various amino acids. Figure 10.10 gives an overview of the origin of the carbon skeletons of individual amino acids.

3-Phosphoglycerate is the most important carbon precursor for the synthesis of amino acids. It is generated in the Calvin cycle and is exported from the chloroplasts to the cytosol by the triose phosphate-phosphate translocator in exchange for phosphate (Fig. 10.11). 3-Phosphoglycerate is converted in the cytosol by **phosphoglycerate mutase** and **enolase** to phosphoenolpyruvate (PEP). From PEP two pathways branch off, the reaction via pyruvate kinase leading to **pyruvate**, and via PEP-carboxylase to **oxaloacetate**. Moreover, PEP together with erythrose 4-phosphate is the precursor

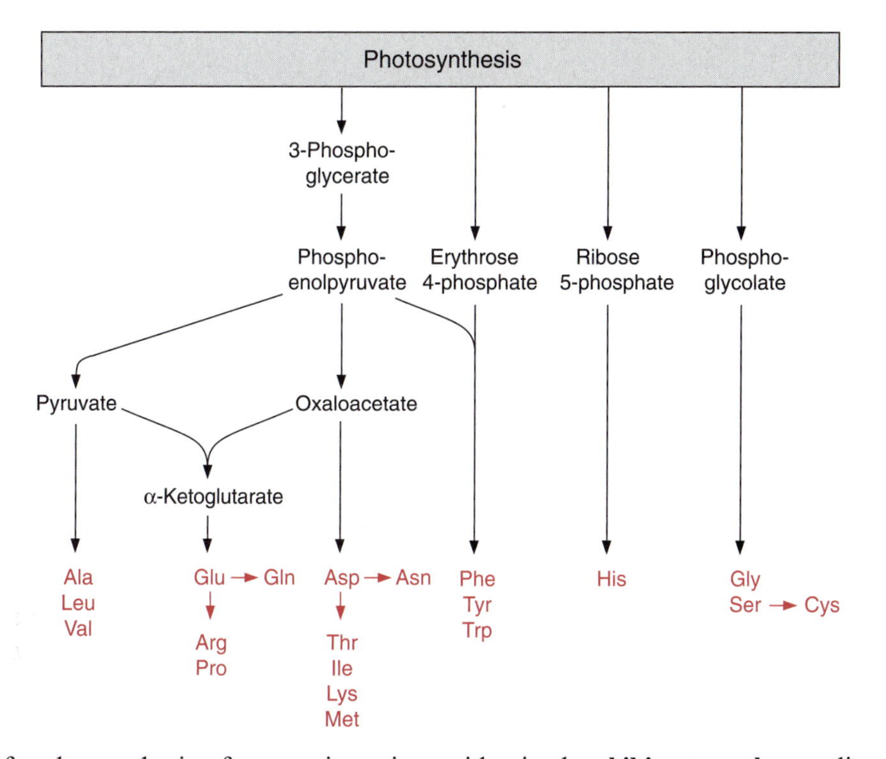

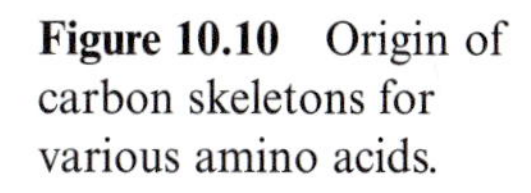
Figure 10.10 Origin of carbon skeletons for various amino acids.

for the synthesis of aromatic amino acids via the **shikimate pathway**, discussed later in this chapter. Since the shikimate pathway is located in the chloroplasts, the required PEP is transported via a specific **PEP-phosphate translocator** into the chloroplasts.

The PEP-carboxylase reaction has already been discussed in conjunction with the metabolism of stomatal cells (section 8.2) and C_4 and CAM metabolism (sections 8.4 and 8.5). Oxaloacetate formed by PEP-carboxylase has two functions in nitrate assimilation:

1. It is converted by transamination to aspartate, which is the precursor for the synthesis of five other amino acids (asparagine, threonine, isoleucine, lysine, and methionine).
2. Together with pyruvate it is the precursor for the formation of α-ketoglutarate, which is converted by transamination to glutamate, being the precursor of three other amino acids (glutamine, arginine, and proline).

Glycolate formed by photorespiration is the precursor for the formation of glycine and serine (see Fig. 7.1), and from the latter cysteine is formed (Chapter 12). In non-green cells, serine and glycine can also be formed from

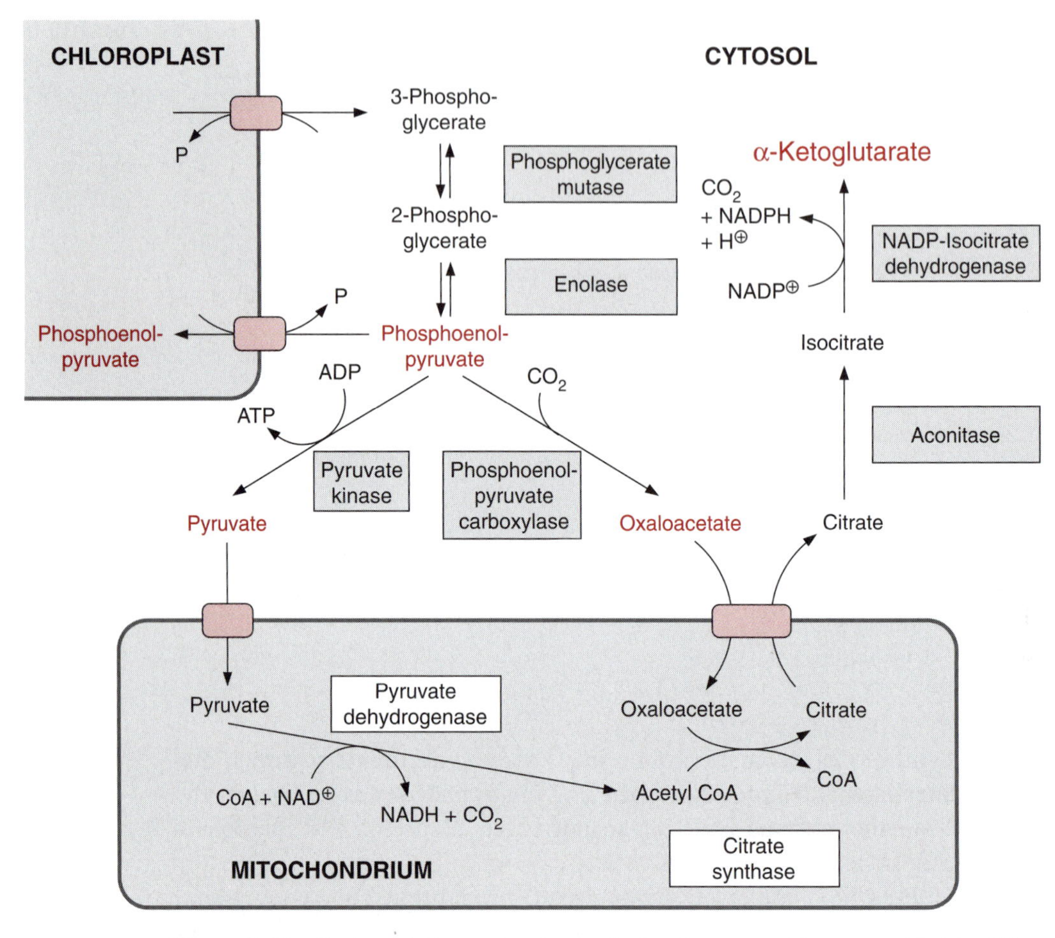

Figure 10.11 Carbon skeletons for the synthesis of amino acids are provided by CO_2 assimilation. Important precursors for amino acid synthesis are colored red.

3-phosphoglycerate. Details of this are not discussed here. Ribose 5-phosphate is the precursor for the synthesis of histidine. This pathway has not yet been fully resolved in plants.

The synthesis of glutamate requires the participation of mitochondrial metabolism

Figure 10.6 shows that glutamate is formed from α-ketoglutarate, which can be provided by a partial sequence of the mitochondrial citrate cycle (Fig. 10.11). Pyruvate and oxaloacetate are transported from the cytosol to the mitochondria by specific translocators. Pyruvate is oxidized by **pyruvate**

dehydrogenase (see Fig. 5.4), and the acetyl-CoA thus generated condenses with oxaloacetate to citrate (see Fig. 5.6). This citrate can be converted in the mitochondria via **aconitase** (Fig. 5.7), oxidized further by **NAD-isocitrate dehydrogenase** (Fig. 5.8), and the resultant α-ketoglutarate can be transported into the cytosol by a specific translocator. But often a major part of the citrate produced in the mitochondria is exported to the cytosol and converted there to α-ketoglutarate by cytosolic isoenzymes of aconitase and NADP isocitrate dehydrogenase. In this case, only a short partial sequence of the citrate cycle is involved in the synthesis of α-ketoglutarate from pyruvate and oxaloacetate. Citrate is released from the mitochondria by a specific translocator in exchange for oxaloacetate.

Biosynthesis of proline and arginine

Glutamate is the precursor for the synthesis of **proline** (Fig. 10.12). Its δ-carboxylic group is first converted by a **glutamate kinase** to an energy-rich phosphoric acid anhydride and is then reduced by NADPH to an aldehyde. The accompanying hydrolysis of the energy-rich phosphate drives the reaction, resembling the reduction of 3-phosphoglycerate to glyceraldehyde 3-phosphate in the Calvin cycle. A ring is formed by the condensation of the carbonyl group with the α-amino group. Reduction by NADPH results in the formation of proline.

Besides its role as a protein constituent, proline has a special function as a **protective substance against dehydration damage** in leaves. When exposed to aridity or to a high salt content in the soil (both leading to water stress), many plants accumulate very high amounts of proline in their leaves, in some cases several times the sum of all the other amino acids. It is assumed that the accumulation of proline during water stress is caused by the induction of the synthesis of the enzyme protein of **pyrrolin-5-carboxylate reductase**.

Proline protects a plant against dehydration, because, in contrast to inorganic salts, it has no inhibitory effect on enzymes even at very high concentrations. Therefore proline is classified as a **compatible substance**. Other compatible substances, formed in certain plants in response to water stress, are sugar alcohols such as **mannitol** (Fig. 10.13), and **betains**, consisting of amino acids, such as proline, glycine, and alanine, of which the amino groups are methylated. The latter are termed proline-, glycine-, and alaninebetains. The accumulation of such compatible substances, especially in the cytosol, chloroplasts, and mitochondria, minimizes in these compartments the damaging effect of water shortage or a high salt content of the soil. But these substances also participate as antioxidants in the elimination of reactive oxygen species (**ROS**) (section 3.9). Water shortage and high salt content of the soil causes an inhibition of CO_2 assimilation, resulting in an

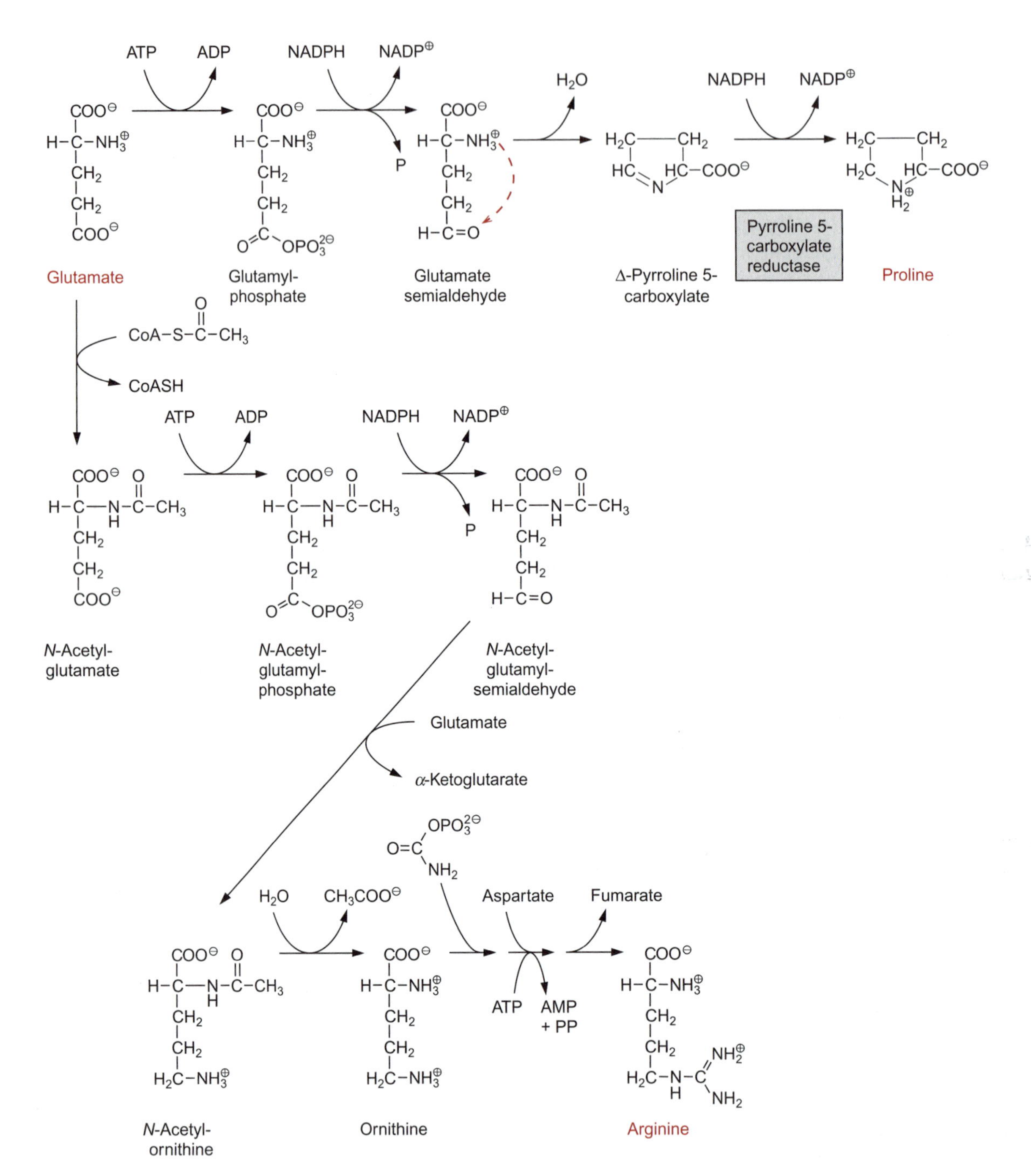

Figure 10.12 Pathway for the formation of amino acids from glutamate.

overreduction of photosynthetic electron transport carriers, which in turn leads to an increased formation of ROS.

In the first step of the synthesis of **arginine**, the α-amino group of glutamate is acetylated by reaction with acetyl-CoA and is thus protected. Subsequently, the δ-carboxylic group is phosphorylated and reduced to a semi-aldehyde in basically the same reaction as in proline synthesis. Here the α-amino group is protected and the formation of a ring is not possible. By transamination with glutamate, the aldehyde group is converted to an amino group, and after cleavage of the acetyl residue, ornithine is formed. The conversion of ornithine to arginine (shown only summarily in Fig. 10.12) proceeds in the same way as in the urea cycle of animals, by condensation with carbamoyl phosphate to citrulline. An amino group is transferred from aspartate to citrulline, resulting in the formation of arginine and fumarate.

Aspartate is the precursor of five amino acids

Aspartate is formed from oxaloacetate by transamination with glutamate by **glutamate-oxaloacetate amino transferase** (Fig. 10.14). The synthesis of **asparagine** from aspartate requires a transitory phosphorylation of the terminal carboxylic group by ATP, as in the synthesis of glutamine. In contrast to glutamine synthesis, however, it is not NH_4^+ but the amide group of glutamine that usually serves as the amino donor in asparagine synthesis. Therefore, the energy expenditure for the amidation of aspartate is twice as high as for the amidation of glutamate. Asparagine is formed to a large extent in the roots (section 10.2), especially when NH_4^+ is the nitrogen source in the soil. Synthesis of asparagine in the leaves often plays only a minor role.

For the synthesis of **lysine**, **isoleucine**, **threonine**, and **methionine**, the first two steps are basically the same as for proline synthesis: After phosphorylation by a kinase, the γ-carboxylic group is reduced to a semi-aldehyde. For the synthesis of lysine (not shown in detail in Fig. 10.14), the semi-aldehyde condenses with pyruvate and, in a sequence of six reactions involving reduction by NADPH and transamination by glutamate, *meso*-2,6-diaminopimelate is formed and from this lysine arises by decarboxylation.

For the synthesis of **threonine**, the semi-aldehyde is further reduced to homoserine. After phosphorylation of the hydroxyl group by homoserine kinase, threonine is formed by isomerization of the hydroxyl group, accompanied by the removal of phosphate. The synthesis of isoleucine from threonine will be discussed in the following paragraph, and the synthesis of methionine in conjunction with sulfur metabolism is discussed in Chapter 12.

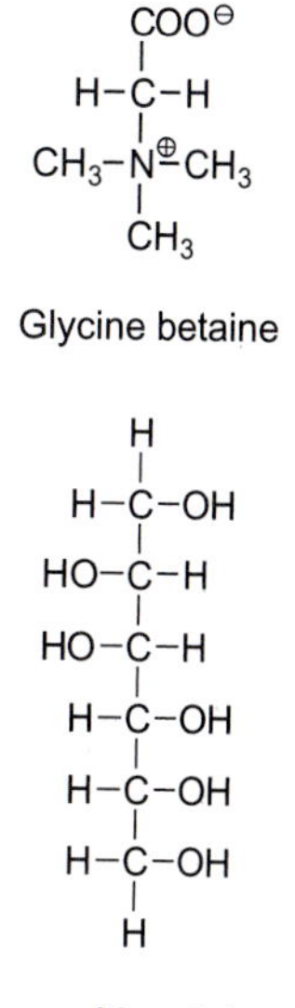

Figure 10.13 Two compatible substances that, like proline, are accumulated in plants as protective agents against desiccation and high salt content in the soil.

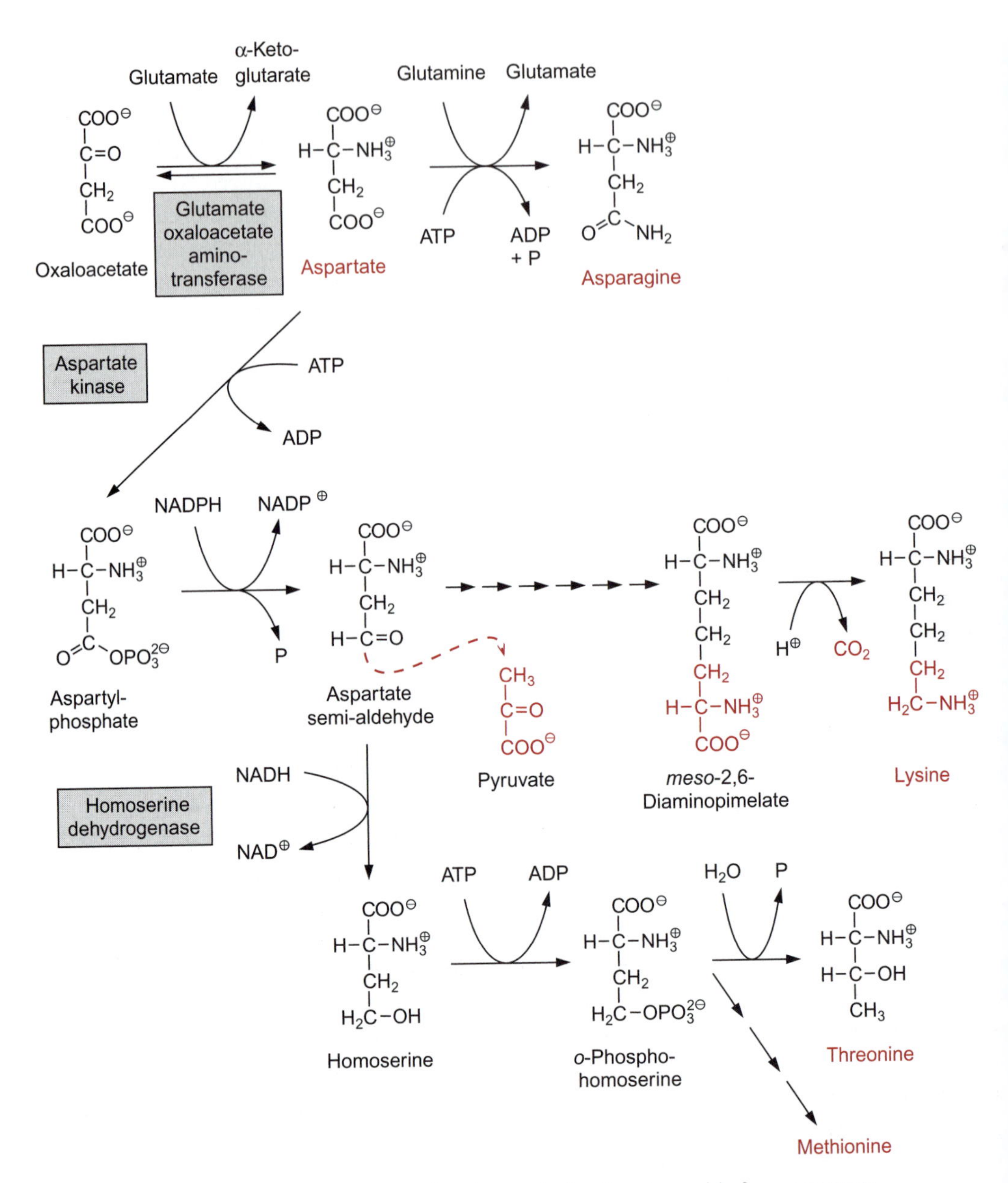

Figure 10.14 The pathway for the synthesis of amino acids from aspartate.

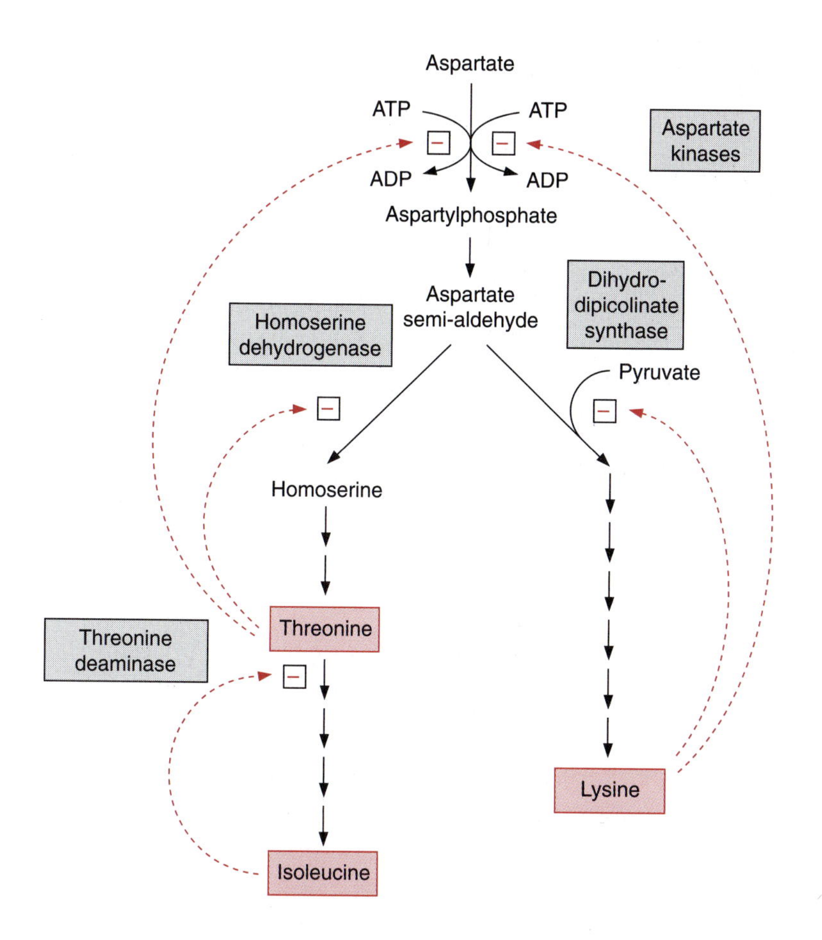

Figure 10.15 Feedback inhibition by end products regulates the entrance enzyme for the synthesis of amino acids from aspartate according to demand. [–] indicates inhibition. Aspartate kinase exists in two isoforms.

Synthesis of amino acids from aspartate is subject to strong feedback control by its end products (Fig. 10.15). Aspartate kinase, the entrance valve for the synthetic pathways, is present in two isoforms. One is inhibited by threonine and the other by lysine. In addition, the reactions of aspartate semi-aldehyde at the branch points of both synthetic pathways are inhibited by the corresponding end products.

Acetolactate synthase participates in the synthesis of hydrophobic amino acids

Pyruvate can be converted by transamination to **alanine** (Fig. 10.16A). This reaction plays a special role in C_4 metabolism (see Figs. 8.14 and 8.15).

Synthesis of **valine** and **leucine** begins with the formation of acetolactate from two molecules of pyruvate. **Acetolactate synthase**, catalyzing this reaction, contains **thiamine pyrophosphate** (TPP) as its prosthetic group. The

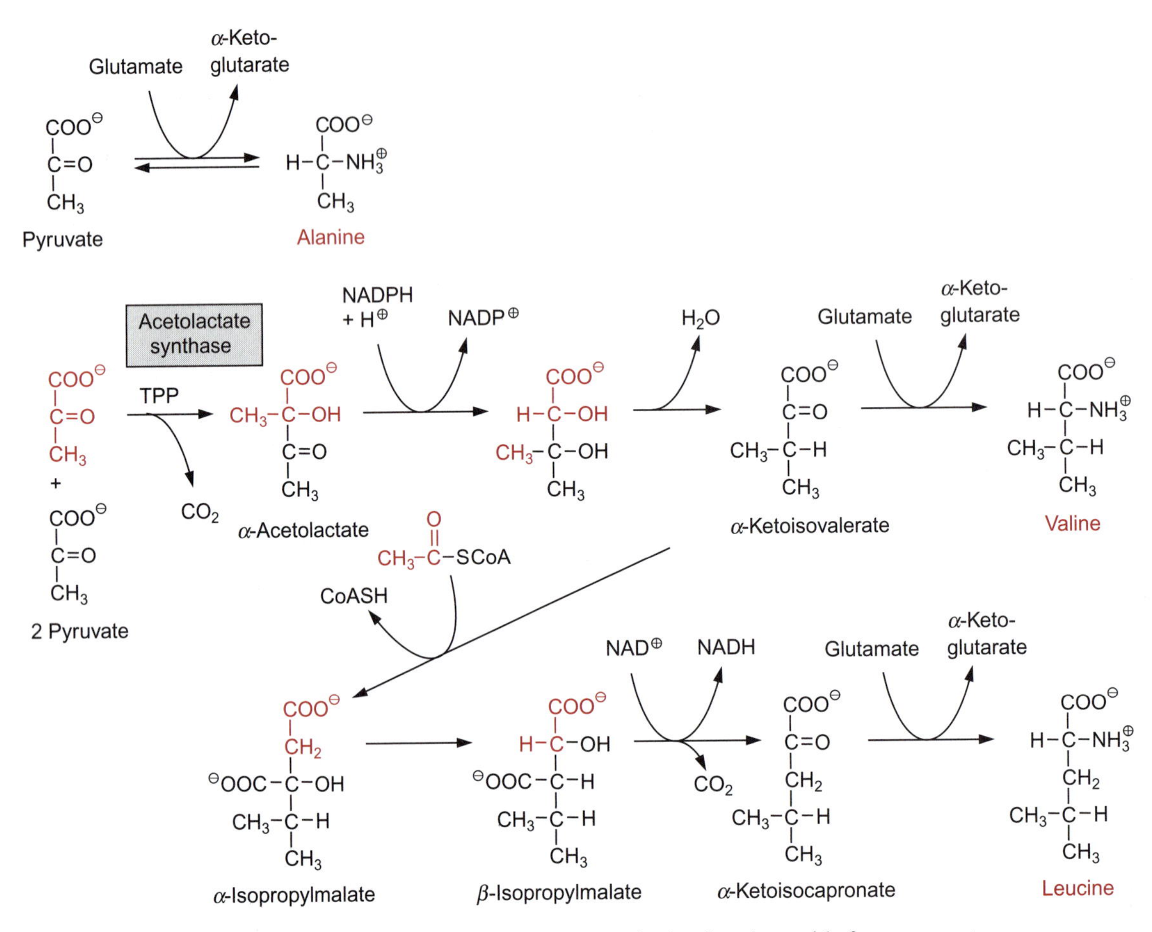

Figure 10.16A Pathway for the synthesis of amino acids from pyruvate.

reaction of TPP with pyruvate yields hydroxyethyl-TPP and CO_2, in the same way as in the pyruvate dehydrogenase reaction (see Figure 5.4). The hydroxyethyl residue is transferred to a second molecule of pyruvate and thus **acetolactate** is formed. Its reduction and rearrangement and the release of water yields α-ketoisovalerate and a subsequent transamination by glutamate produces valine.

The formation of **leucine** from α-ketoisovalerate proceeds in basically the same reaction sequences as for the formation of glutamate from oxaloacetate shown in Figure 5.3. First, acetyl-CoA condenses with α-ketoisovalerate (analogous to the formation of citrate), the product α-isopropylmalate isomerizes (analogous to isocitrate formation), and the β-isopropylmalate thus formed is oxidized by NAD^+ with the release of CO_2 to α-ketoisocapronate (analogous to the synthesis of α-ketoglutarate by isocitrate

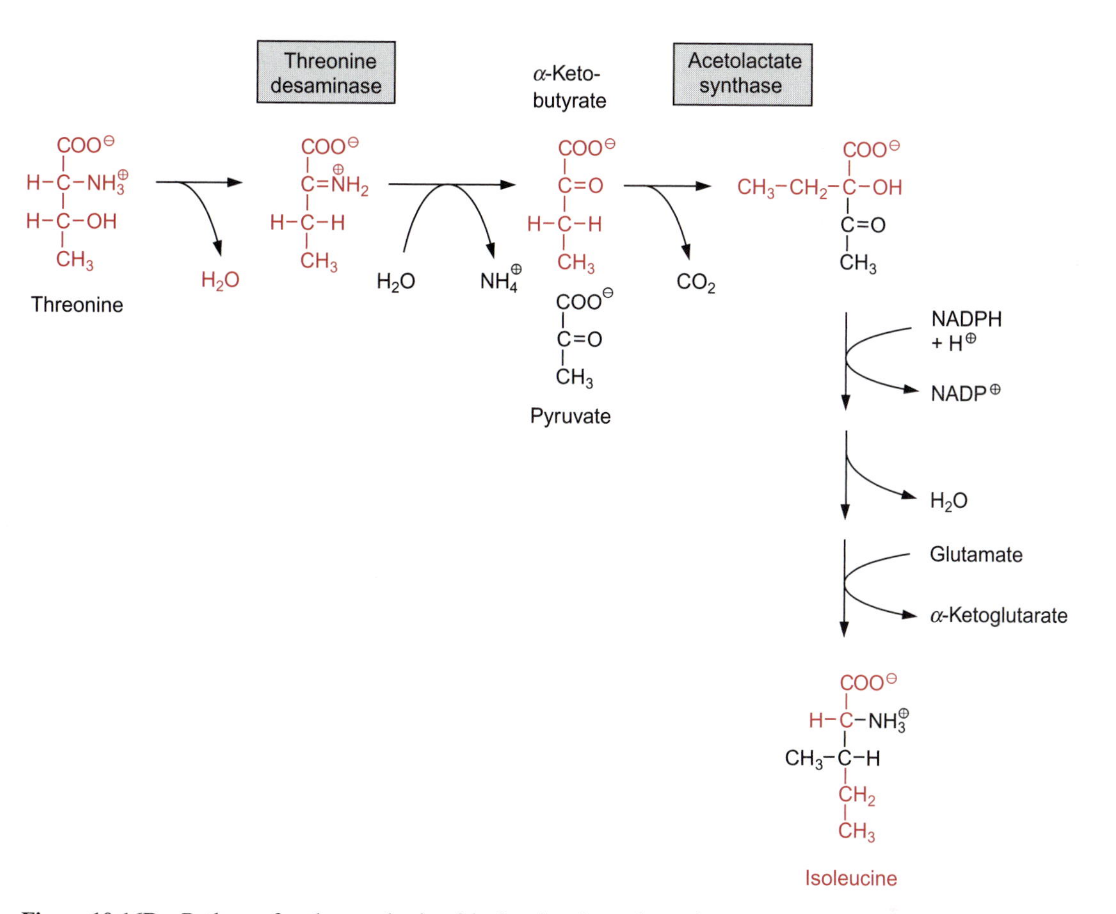

Figure 10.16B Pathway for the synthesis of isoleucine from threonine and pyruvate.

dehydrogenase). Finally, in analogy to the synthesis of glutamate, α-ketoisocapronate is transformed by transamination to leucine.

For the synthesis of **isoleucine** from threonine, the latter is first converted by a deaminase to α-ketobutyrate (Fig. 10.16B). Acetolactate synthase condenses α-ketobutyrate with pyruvate in a reaction analogous to the synthesis of acetolactate from two molecules of pyruvate (Fig. 10.16A). The further reactions in the synthesis of isoleucine correspond to the reaction sequence in the synthesis of valine.

The synthesis of leucine, valine, and isoleucine is also subject to feedback control by the end products. Isopropylmalate synthase is inhibited by leucine (Fig. 10.17) and threonine deaminase is inhibited by isoleucine (Fig. 10.15). The first enzyme, **acetolactate synthase** (ALS), is inhibited by valine and leucine. Sulfonyl ureas, (e.g., chlorsulfurone) and imidazolinones, (e.g., imazethapyr) (Fig. 10.18) are very strong inhibitors of ALS, where they bind

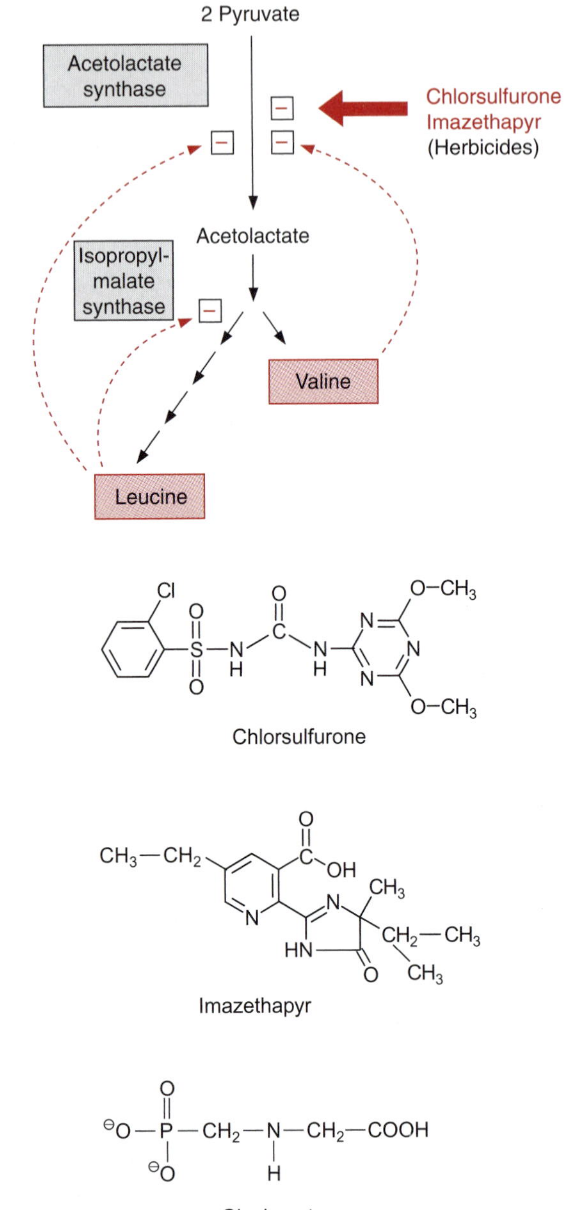

Figure 10.17 Synthesis of valine and leucine is adjusted to demand by the inhibitory effect of both amino acids on acetolactate synthase and the inhibition of isopropyl malate synthase by leucine. The herbicides chlorsulfurone and imazethapyr inhibit acetolactate synthase. [–] indicates inhibition.

Figure 10.18 Herbicides: chlorsulfurone, a sulfonyl urea, (trade name Glean, DuPont) and imazethapyr, an imidazolinone, (trade name Pursuit, ACC) inhibit acetolactate synthase (Fig. 10.16A). Glyphosate (trade name Roundup, Monsanto) inhibits EPSP synthase (Fig. 10.19).

to the pyruvate binding site. A concentration as low as 10^{-9} mol/L of chlorsulfurone is sufficient to inhibit ALS by 50%. Since the pathway for the formation of valine, leucine, and isoleucine is present only in plants and microorganisms, the inhibitors aforementioned are suitable for destroying plants specifically and are therefore used as **herbicides** (section 3.6). Chlorsulfurone (trade name Glean, DuPont) is used as a selective herbicide in the cultivation of cereals, and imazethapyr (Pursuit, American Cyanamide Co.) is used for protecting soybeans. From the application of these herbicides, mutants of maize, soybean, rapeseed, and wheat have naturally evolved, which are resistant to sulfonyl ureas or imidazolinones, or even to both herbicides. In each case, a mutation was found in the gene for acetolactate synthase, making the enzyme insensitive to the herbicides without affecting its enzyme activity. By crossing these mutants with other lines, herbicide-resistant varieties have been bred and are, in part, already commercially cultivated.

Aromatic amino acids are synthesized via the shikimate pathway

Precursors for the formation of aromatic amino acids are erythrose 4-phosphate and phosphoenolpyruvate. These two compounds condense to form cyclic dehydrochinate accompanied by the liberation of both phosphate groups (Fig. 10.19). Following the removal of water and the reduction of the carbonyl group, **shikimate** is formed. After protection of the 3′-hydroxyl group by phosphorylation, the 5′-hydroxyl group of the shikimate reacts with phosphoenolpyruvate to give the enolether 5′-enolpyruvyl shikimate-3-phosphate (**EPSP**) and chorismate is formed from this by the removal of phosphate. Chorismate represents a branch point for two biosynthetic pathways.

1. Tryptophan is formed via four reactions,which are not discussed in detail here.
2. Prephenate is formed by a rearrangement, in which the side chain is transferred to the 1′-position of the ring, and arogenate is formed after transamination of the keto group. Removal of water results in the formation of the third double bond and **phenylalanine** is formed by decarboxylation. Oxidation of arogenate by NAD, accompanied by a decarboxylation, results in the formation of **tyrosine**. According to recent results, the enzymes of the shikimate pathway are located exclusively in the **plastids**. The synthesis of aromatic amino acids is also controlled by the end products at several steps in the pathway (Fig. 10.20).

Glyphosate acts as an herbicide

Glyphosate (Fig. 10.18), a structural analogue of phosphoenolpyruvate, is a very strong inhibitor of **EPSP synthase**. Glyphosate inhibits specifically the

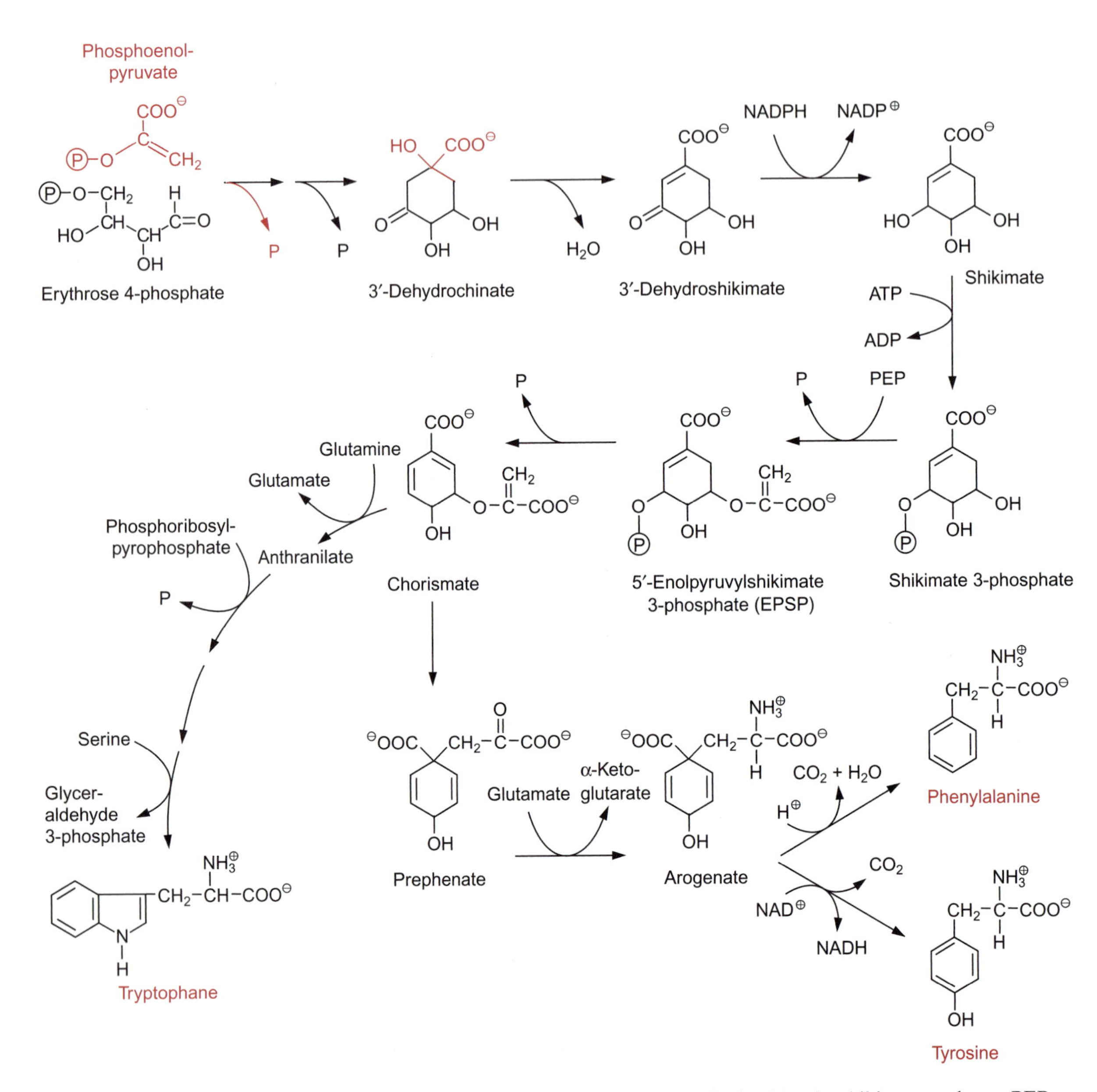

Figure 10.19 Aromatic amino acids are synthesized by the shikimate pathway. PEP = phosphoenolpyruvate.

synthesis of aromatic amino acids but has only a low effect on other phosphoenolpyruvate metabolizing enzymes (e.g., pyruvate kinase or PEP-carboxykinase). Interruption of the shikimate pathway by glyphosate has a lethal effect on plants. Since the shikimate pathway is not present in animals,

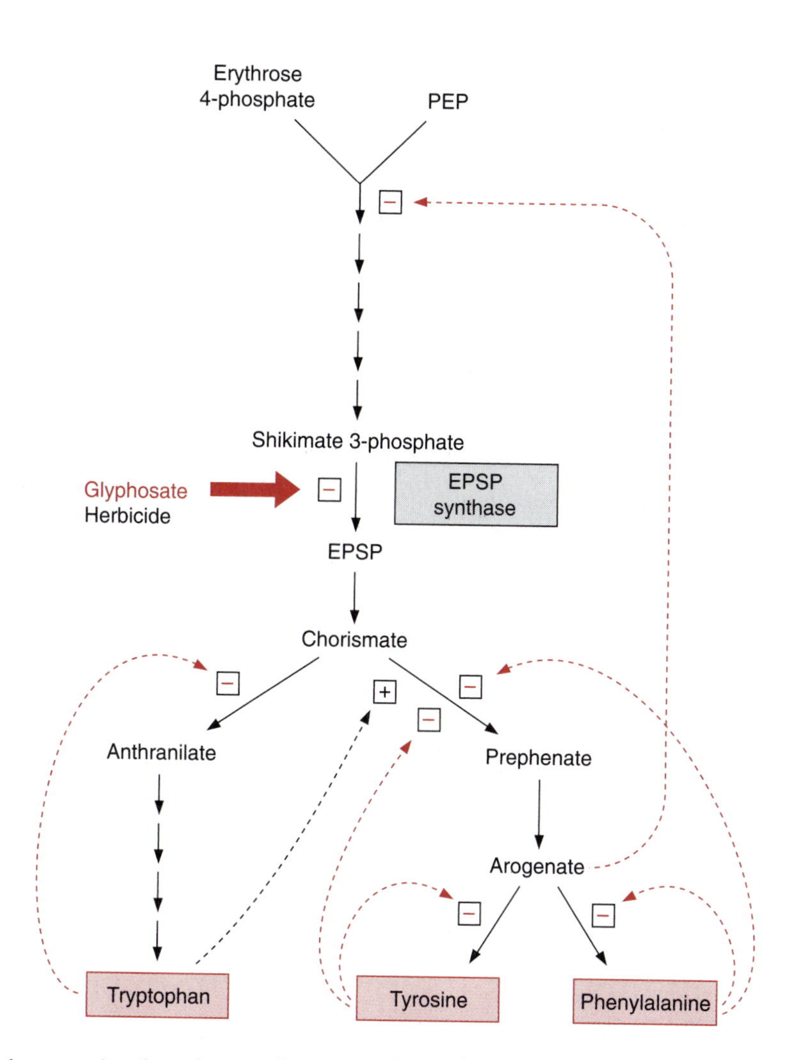

Figure 10.20 Several steps in the synthesis of aromatic amino acids are regulated by product feedback inhibition, thus adjusting the rate of synthesis to demand. Tryptophan stimulates the synthesis of tyrosine and phenylalanine [+]. The herbicide glyphosate (Fig. 10.19) inhibits EPSP synthase [–].

glyphosate (under the trade name Roundup, Monsanto) is used as an herbicide (section 3.6). Due to its simple structure, glyphosate is relatively rapidly degraded by bacteria present in the soil. Glyphosate is the herbicide with the highest sales worldwide. Genetic engineering has been successful in creating glyophosate-resistant crop plants (section 22.6). This makes it possible to combat weeds very efficiently in the presence of such transgenic crop plants.

A large proportion of the total plant matter can be formed by the shikimate pathway

The function of the shikimate pathway is not restricted to the generation of amino acids for protein biosynthesis. It also provides precursors for a large

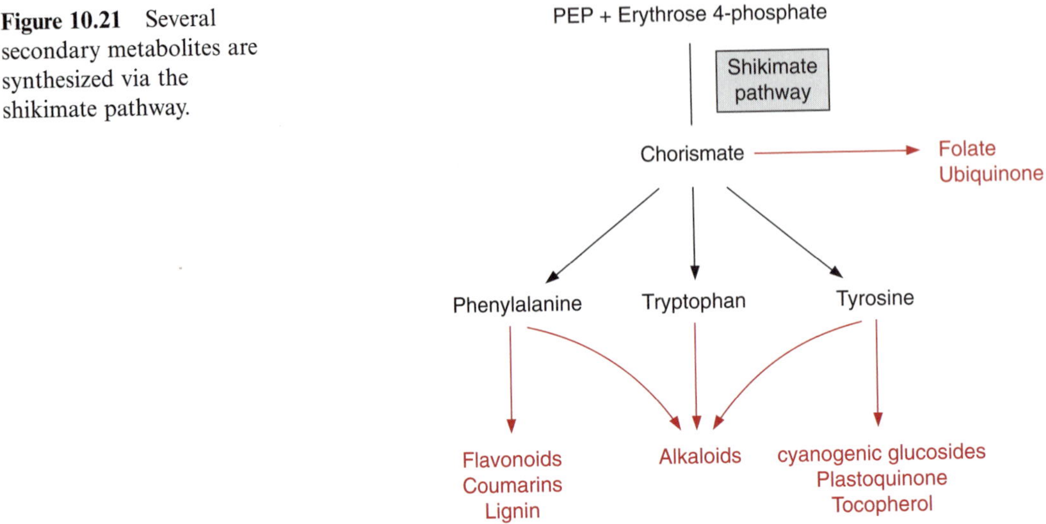

Figure 10.21 Several secondary metabolites are synthesized via the shikimate pathway.

variety of other substances (Fig. 10.21) formed by plants in large quantities, particularly phenylpropanoids such as **flavonoids** and **lignin** (Chapter 18). As the sum of these products can amount to a high proportion of the total cellular matter, in some plants up to 50% of the dry matter, the shikimate pathway can be regarded as one of the main biosynthetic pathways of plants.

10.5 Glutamate is the precursor for synthesis of chlorophylls and cytochromes

Chlorophyll amounts to 1% to 2% of the dry matter of leaves. Its synthesis proceeds in the plastids. As shown in Figure 2.4, chlorophyll consists of a **tetrapyrrole** ring with **magnesium** as the central atom and with a **phytol side chain** as a membrane anchor. **Heme**, likewise a tetrapyrrole, but with **iron** as the central atom, is a constituent of cytochromes and catalase.

Porphobilinogen, a precursor for the synthesis of tetrapyrroles, is formed by the condensation of two molecules of **δ-amino levulinate**. δ-Amino levulinate is synthesized in animals, yeast, and some bacteria from succinyl-CoA and glycine, accompanied by the liberation of CoASH and CO_2. In contrast, the synthesis of δ-amino levulinate in plastids, cyanobacteria, and many eubacteria proceeds by reduction of **glutamate**. As discussed in section 6.3, the difference in redox potentials between a carboxylate and an alde-

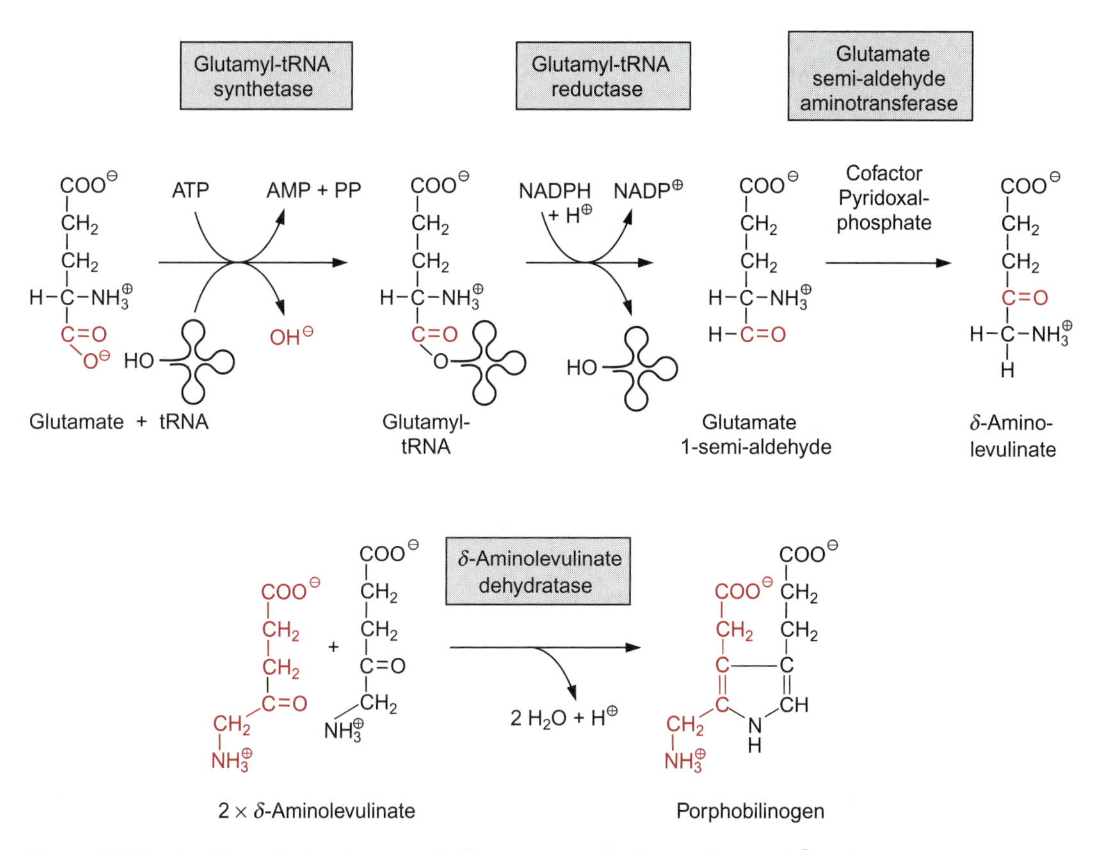

Figure 10.22 In chloroplasts, glutamate is the precursor for the synthesis of δ-amino levulinate, which is condensed to porphobilinogen.

hyde is so high that a reduction of a carboxyl group by NADPH is possible only when this carboxyl group has been previously activated [e.g., as a thioester (Fig. 6.10) or as a mixed phosphoric acid anhydride (Fig. 10.12)]. In the plastid δ-amino levulinate synthesis, glutamate is activated in a very unusual way by a covalent linkage to a **transfer RNA** (tRNA) (Fig. 10.22). This tRNA for glutamate is encoded in the plastids and is involved there in the synthesis of δ-amino levulinate as well as in protein biosynthesis. As in protein biosynthesis (see Fig. 21.1), the linkage of the carboxyl group of glutamate to tRNA is accompanied by consumption of ATP. During reduction of glutamate tRNA by **glutamate tRNA reductase**, tRNA is liberated and in this way the reaction becomes irreversible. The **glutamate 1-semi-aldehyde** thus formed is converted to δ-aminolevulinate by an aminotransferase containing **pyridoxal phosphate** as a prosthetic group. This reaction proceeds according to the same mechanism as the aminotransferase reaction shown in Figure 7.4, with the only difference being that here the amino group (as amino donor) and the keto group (as amino acceptor) are present in the same molecule.

Two molecules of δ-aminolevulinate condense to form porphobilinogen (Fig. 10.22). The open-chain tetrapyrrole hydroxymethylbilan is formed from four molecules of porphobilinogen via **porphobilinogen deaminase** (Fig. 10.23). The enzyme contains a dipyrrole as cofactor, which it produces itself. Uroporphyrinogen III is formed after the exchange of the two side chains on ring d and by closure of the ring. Subsequently, protoporphyrin IX is formed by reaction with decarboxylases and oxidases (not shown in detail). Mg^{++} is incorporated into the tetrapyrrole ring by **magnesium chelatase** and the resultant Mg-protoporphyrin IX is converted by three more enzymes to protochlorophyllide. The tetrapyrrole ring of protochlorophyllide contains the same number of double bonds as protoporphyrin IX. The reduction of one double bond in ring d by NADPH yields chlorophyllide. **Protochlorophyllide reductase**, which catalyzes this reaction, is active only when protochlorophyllide is activated by absorption of light. The transfer of a phytyl chain, activated by pyrophosphate via a prenyl transferase called **chlorophyll synthetase** (see section 17.7), completes the synthesis of the chlorophyll.

Because of the light dependence of protochlorophyllide reductase, the shoot starts greening only when it reaches the light. Also, synthesis of the chlorophyll binding proteins of the light harvesting complexes is light-dependent. The only exceptions are some gymnosperms (e.g., pine), in which protochlorophyllide reduction as well as the synthesis of chlorophyll binding proteins also progresses during darkness. Porphyrins are photooxidized in light, which may lead to photochemical cell damage. It is therefore important that intermediates of chlorophyll biosynthesis do not accumulate. To prevent this, the synthesis of δ-amino levulinate is light-dependent, but the mechanism of this regulation is not yet understood. Moreover, δ-amino levulinate synthesis is subject to feedback inhibition by Mg-protoporphyrin IX and by heme. The end products protochlorophyllide and chlorophyllide inhibit magnesium chelatase (Fig. 10.24). Moreover, intermediates of chlorophyll synthesis control the synthesis of light harvesting proteins (section 2.4) via the regulation of gene expression.

Protophorphyrin is also a precursor for heme synthesis

Incorporation of an iron atom into protoporphyrin IX by a **ferro-chelatase** results in the formation of heme. By assembling the heme with apoproteins, chloroplasts are able to synthesize their own cytochromes. Also, mitochondria possess the enzymes for the biosynthesis of their cytochromes from protoporphyrin IX, but the corresponding enzyme proteins are different from those in the chloroplasts. The precursor protoporphyrin IX is delivered from

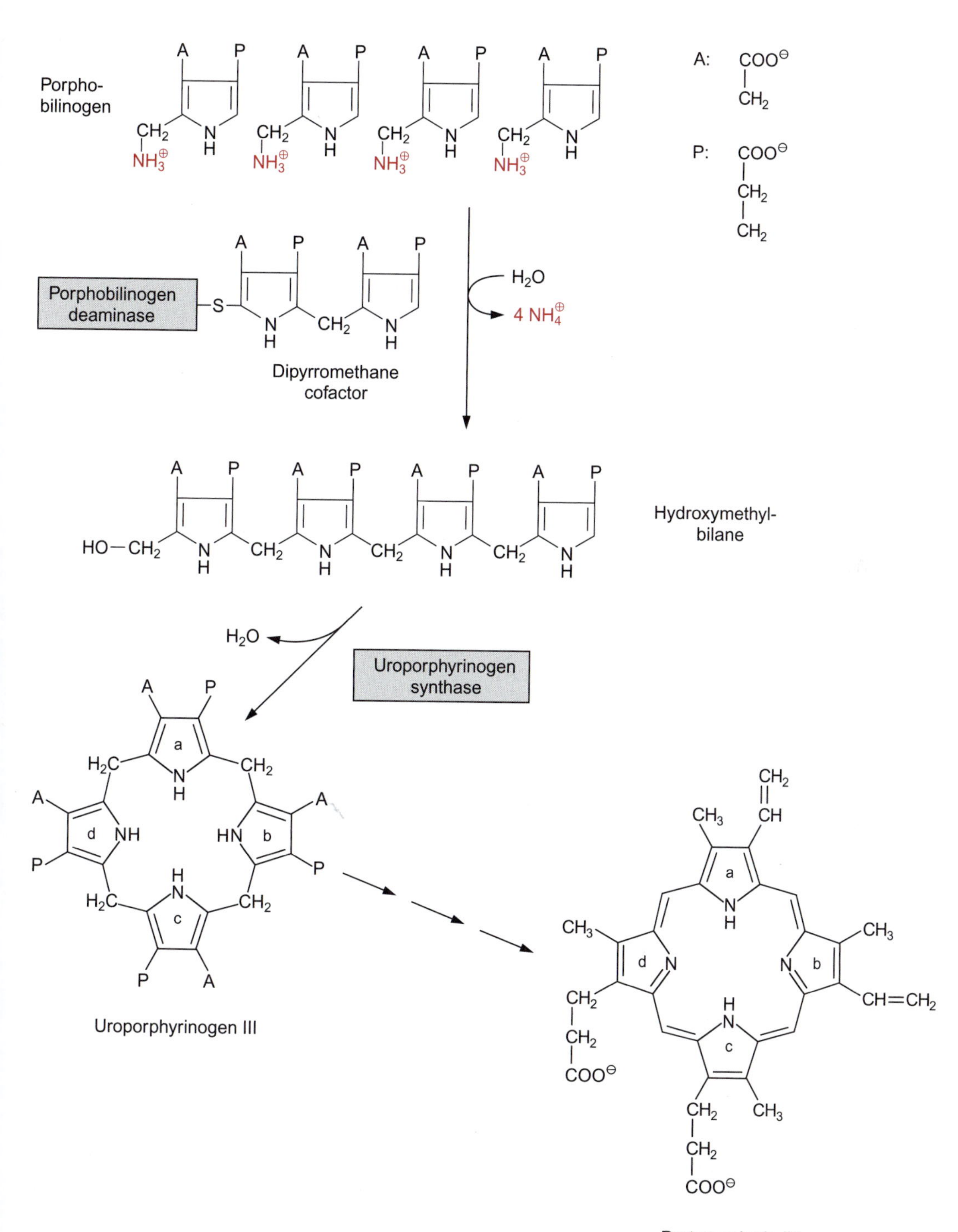

Figure 10.23 Protoporphyrin synthesis.

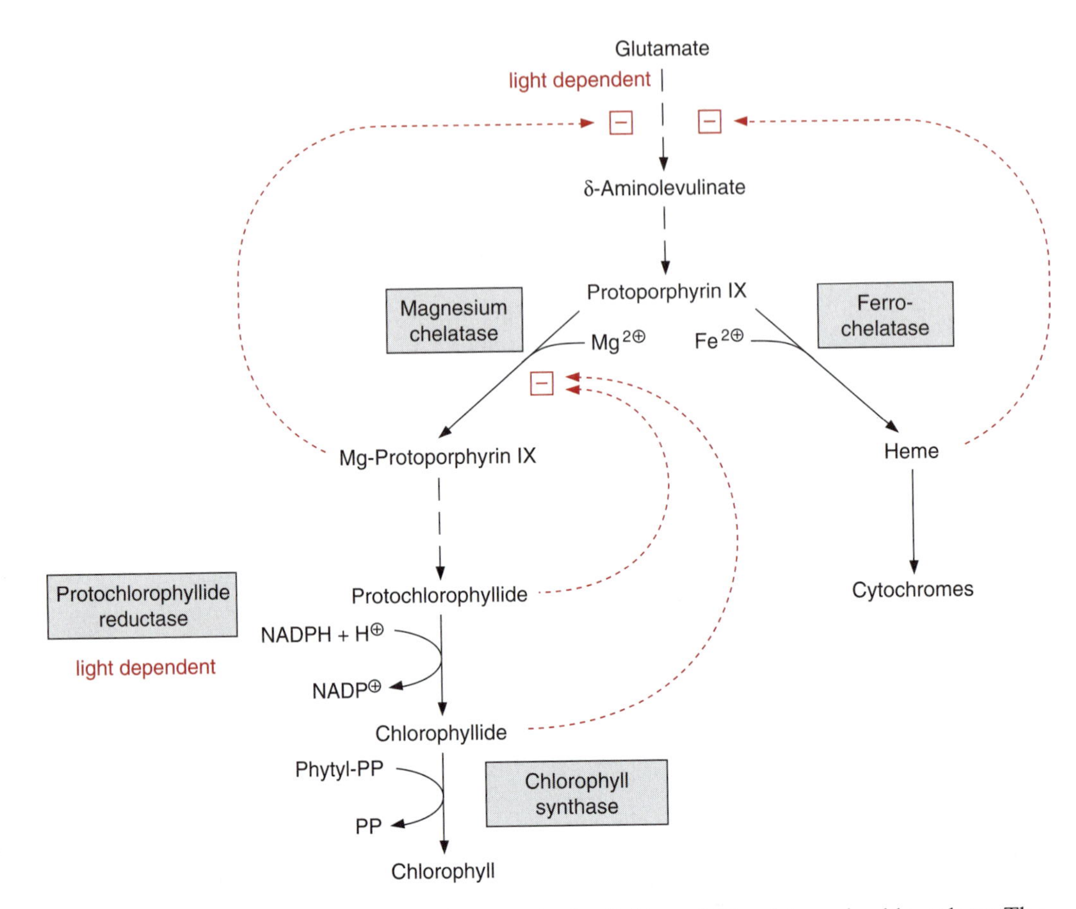

Figure 10.24 Overview of the synthesis of chlorophyll and heme in chloroplasts. The dashed red lines symbolize inhibition of enzymes by products of the biosynthesis chain. Enzymes of the biosynthesis chain from protoporphyrin IX to cytochromes are also present in the mitochondria; these proteins, however, are different from those in the chloroplasts.

the plastids to the mitochondria, probably involving porphyrin transporters in the two organelles. Thus chloroplasts have a central function in the synthesis of tetrapyrrole compounds in a plant.

Further reading

Brusslau, J. A., Peterson, M. P. Tetrapyrrol regulation of nuclear gene expression. Photosynth Res 71, 185–194 (2002).

Campbell, W. H. Nitrate reductase, structure, function and regulation: bridging the gap between biochemistry and physiology. Annu Rev Plant Physiol Plant Mol Biol 50, 277–303 (1999).

Cornah, J. E., Terry, M. J., Smith, A. G. Green or red: What stops the traffic in the tetrapyrrole pathway? Trends Plant Sci 8, 224–230 (2003).

De Boer, B. Fusicoccin—a key to multiple 14-3-3 locks? Trends Plant Sci 2, 60–66 (1997).

Forde, B. G. Local and long-range signaling pathways regulating plant responses to nitrate. Annu Rev Plant Biol 53, 203–224 (2002).

Galili, G. Regulation of lysine and threonine synthesis. Plant Cell 7, 899–906 (1995).

Grimm, B. Novel insights in the control of tetrapyrrole metabolism of higher plants. Plant Biol 1, 245–250 (1998).

Hasegawa, P. M., Bressan, R. A., Zhu, J.-K., Bohnert, H. J. Plant cellular and molecular responses to high salinity. Annu Rev Plant Physiol Plant Mol Biol 51, 463–499 (2000).

Hermann, K. M., Weaver, L. M. The Shikimate pathway. Annu Rev Plant Physiol Plant Mol Biol 50, 473–503 (1999).

Hirel, B., Lea, P. J. The biochemistry, molecular biology and genetic manipulation of primary ammonia assimilation. In: C. H. Foyer and G. Noctor (eds.), Advances in Photosynthesis: Photosynthetic assimilation and associated carbon metabolism, pp, 239–263. Kluwer Academic Publishers, Dordrecht, Netherlands (2002).

Huber, S. C., Bachmann, M., Huber, J. L. Post-translational regulation of nitrate reductase activity: A role for Ca^{++} and 14-3-3 proteins. Trends Plant Sci 1, 432–438 (1996).

Kaiser, W. M., Stoimenova, M., Man, H.-M. What limits nitrogen reduction in leaves. In: C. H. Foyer and G. Noctor (eds.), Advances in Photosynthesis: Photosynthetic assimilation and associated carbon metabolism, pp. 63–70. Kluwer Academic Publishers, Dordrecht, Netherlands (2002).

Kumar, A. M., Schaub, U., Söll, D., Ujwal, M.L: Glutamyl-transfer RNA: At the crossroad between chlorophyll and protein biosynthesis. Trends Plant Sci 1, 371–376 (1996).

Lancien, M., Gadal, P., Hodges, M. Enzyme redundancy and the importance of 2-oxoglutarate in higher plant ammonium assimilation. Plant Physiol 123, 817–824 (2000).

MacKintosh, C. Regulation of cytosolic enzymes in primary metabolism by reversible protein phosphorylation. Plant Biol 1, 224–229 (1998).

McNiel, S. D., Nuccio, M. L., Hanson, A. D. Betaines and related osmoprotectants. Targets for metabolic engineering of stress resistance. Plant Physiol 120, 945–949 (1999).

Roberts, M. R. Regulatory 14-3-3 protein-protein interactions in plant cells. Curr Opin Plant Biol 3, 400–405 (2000).

Roberts, M. R., 14-3-3 proteins find new partners in plant cell signalling. Trends Plant Sci 8, 218–223 (2003).

Schultz, G. Assimilation of non-carbohydrate compounds. In: Photosynthesis: A Comprehensive Treatise, Raghavendra, A. S. Hrsg, Cambridge University Press, Cambridge, England, pp. 183–196 (1998).

Singh, B. K., Shaner, D. L. Biosynthesis of branched chain amino acids: From test tube to field. Plant Cell 7, 935–944 (1995).

Stitt, M. Nitrate regulation of metabolism and growth. Curr Opin Plant Biol 2, 178–186 (1999).

Streatfield, S. J., Weber, A., Kinsman, E. A., Häusler, R. E., Li, J., Post-Breitmiller, D., Kaiser, W. M., Pyke, K. A., Flügge, U.-I., Chory, J. The phosphoenolpyruvate/phosphate translocator is required for phenolic metabolism, palisade cell development, and plastid-dependent nuclear gene expression. Plant Cell 11, 1609–1621 (1999).

Thomashow, M. F. Plant cold acclimation: Freezing tolerance genes and regulatory mechanisms. Annu Rev Plant Physiol Plant Mol Biol 50, 571–599 (1999).

von Wettstein, D., Gough, S., Kannagara, C. G. Chlorophyll biosynthesis. Plant Cell, 7, 1039–1057 (1995).

Williams, L. E., Miller, A. J. Transporters responsible for the uptake and partitioning of nitrogenous solutes. Annu Rev Plant Physiol Plant Mol Biol 52, 659–688 (2001).

11

Nitrogen fixation enables the nitrogen in the air to be used for plant growth

In a closed ecological system, the nitrate required for plant growth is derived from the degradation of the biomass. In contrast to other plant nutrients (e.g., phosphate or sulfate), nitrate cannot be delivered by the weathering of rocks. Smaller amounts of nitrate are generated by lightning and carried into the soil by rain water (in temperate areas about 5 kg N/ha per year). Due to the effects of civilization (e.g., car traffic, mass animal production, etc.), the amount of nitrate, other nitrous oxides and ammonia carried into the soil by rain can be in the range of 15 to 70 kg N/ha per year. Fertilizers are essential for agricultural production to compensate for the nitrogen that is lost by the withdrawal of harvest products. For the cultivation of maize, for instance, per year about 200 kg N/ha have to be added as fertilizers in the form of nitrate or ammonia. Ammonia, the primary product for the synthesis of nitrate fertilizer, is produced from nitrogen and hydrogen by the **Haber-Bosch process**:

$$3H_2 + N_2 \rightarrow 2NH_3 \ (\Delta H - 92{,}6 kJ/mol)$$

Because of the high bond energy of the N≡N triple bond, this synthesis requires a high activation energy and is therefore, despite a catalyzator, carried out at a pressure of several hundred atmospheres and temperatures of 400°C to 500°C. Therefore it involves very high energy costs. The synthesis of nitrogen fertilizer amounts to about one-third of the total energy expenditure for the cultivation of maize. If it were not for the production of nitrogen fertilizer by Haber-Bosch synthesis, large parts of the world's population could no longer be fed. Using "organic cycle" agriculture, one hectare of land can feed about 10 people, whereas with the use of nitrogen fertilizer the amount is increased fourfold.

The majority of cyanobacteria and some bacteria are able to synthesize ammonia from nitrogen in air. A number of plants live in symbiosis with N_2-fixing bacteria, which supply the plant with organic nitrogen. In return, the plants provide these bacteria with metabolites for their nutrition. The symbiosis of legumes with **nodule-inducing bacteria (rhizobia)** is widespread and important for agriculture. **Legumes**, which include soybean, lentil, pea, clover, and lupines, form a large family (***Leguminosae***) with about 20,000 species. A very large part of the legumes have been shown to form a symbiosis with rhizobia. In temperate climates, the cultivation of legumes can lead to an N_2 fixation of 100 to 400 kg N_2/ha per year. Therefore legumes are important as green manure; in crop rotation they are an inexpensive alternative to artificial fertilizers. The symbiosis of the water fern *Azolla* with the cyanobacterium ***Nostoc*** supplies rice fields with nitrogen. N_2-fixing actinomycetes of the genus ***Frankia*** form a symbiosis with woody plants such as the alder or the Australian casuarina. The latter is a pioneer plant on nitrogen-deficient soils.

11.1 Legumes form a symbiosis with nodule-inducing bacteria

Initially it was thought that the nodules of legumes (Fig. 11.1) were caused by a plant disease, until their function in N_2 fixation was recognized by H. Hellriegel and H. Wilfarth in 1888. They found that beans containing these nodules were able to grow without nitrogen fertilizer.

The nodule-inducing bacteria include, among other genera, the genera *Rhizobium*, *Bradyrhizobium*, and *Azorhizobium* and are collectively called **rhizobia**. The rhizobia are strictly aerobic gram-negative rods, which live in the soil and grow heterotrophically in the presence of organic compounds. Some species (*Bradyrhizobium*) are also able to grow autotrophically in the presence of H_2, although at a low growth rate.

The uptake of rhizobia into the host plant is a **controlled infection**. The molecular basis of specificity and recognition is still only partially known. The rhizobia form species-specific nodulation factors (**Nod factors**). These are lipochito-oligosaccharides that acquire a high structural specificity (e.g., by acylation, acetylation, and sulfatation). They are like a security key with many notches and open the house of the specific host with which the rhizobia associate. The Nod factors bind to specific **receptor kinases** of the host,

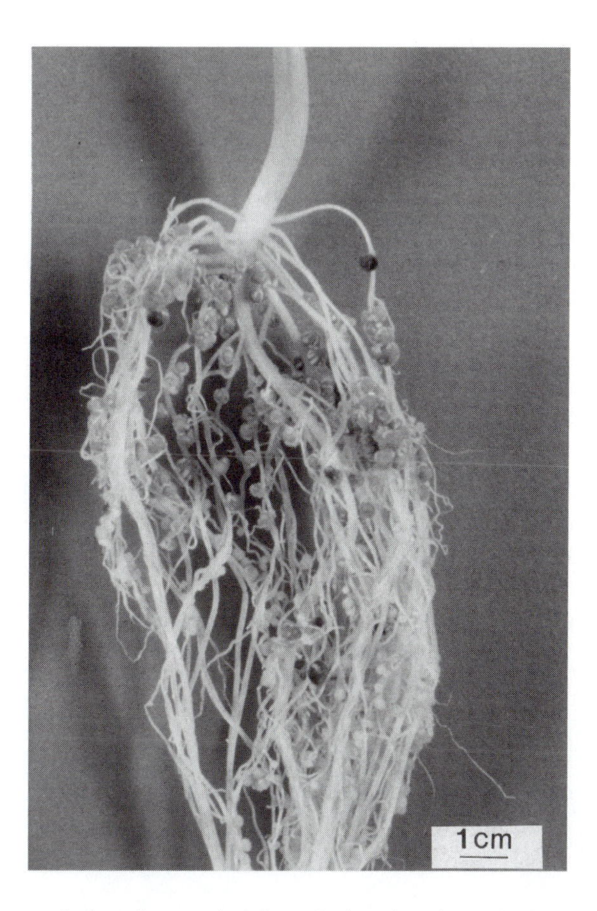

Firgure 11.1 Root system of *Phaseolus vulgaris* (bean) with a dense formation of nodules after infection with *Rhizobium etli.* (By P. Vinuesa-Fleischmann and D. Werner, Marburg.)

which are part of signal transduction chains (section 19.1). In this way the "key" induces the root hair of the host to curl and the root cortex cells to divide, forming the **nodule primordium**. After the root hair has been invaded by the rhizobia, an **infection thread** forms (Fig. 11.2), which extends into the cortex of the roots, forms branches there, and infects the cells of the nodule primordium. A **nodule** thus develops from the infection thread. The morphogenesis of the nodule is of similarly high complexity to that of any other plant organ such as the root or shoot. The nodules are connected with the root via vascular tissues, which supply them with substrates formed by photosynthesis. The bacteria, which have been incorporated into the plant cell, are enclosed by **a peribacteroid membrane** (also called a symbiosome membrane), which is formed by the plant. The incorporated bacteria are thus separated from the cytoplasm of the host cell in a so-called **symbiosome** (Fig. 11.3). In the symbiosome, the rhizobia differentiate to **bacteroids.** The volume of these bacteroids can be 10 times the volume of individual bacteria. Several of these bacteroids are surrounded by a peribacteroid membrane.

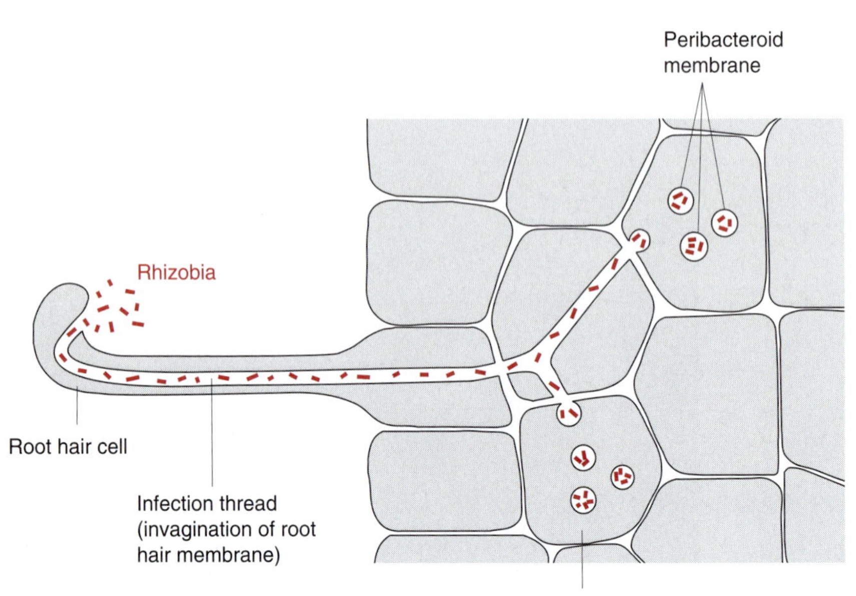

Figure 11.2 Controlled infection of a host cell by rhizobia is induced by an interaction with the root hairs. The rhizobia induce the formation of an infection thread, which is formed by invagination of the root hair cell wall and protrudes into the cells of the root cortex. In this way the rhizobia invaginate the host cell where they are separated by a peribacteroid membrane from the cytosol of the host cells. The rhizobia grow and differentiate into large bacteroids.

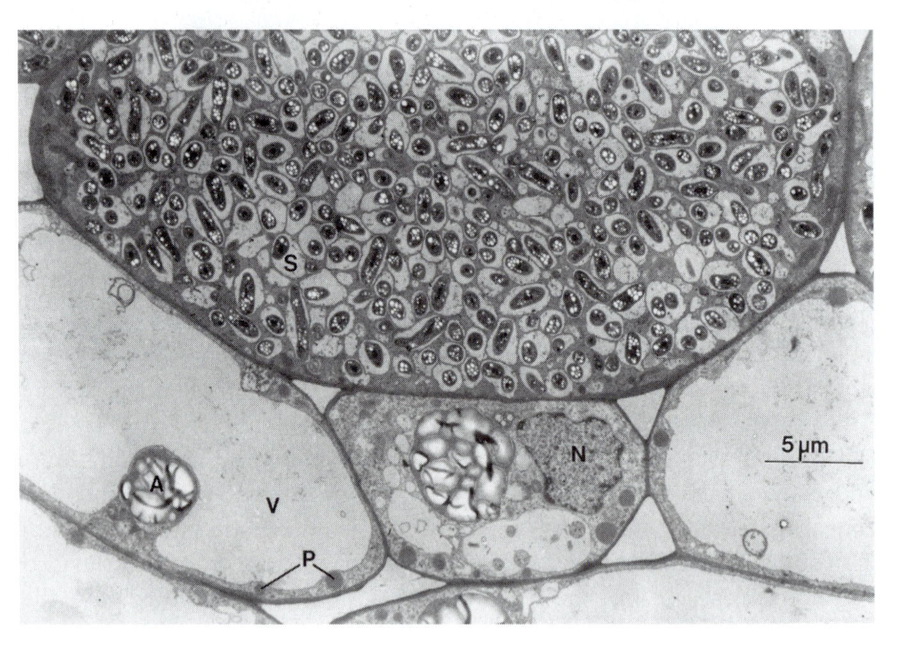

Figure 11.3 Electron microscopic cross section through a nodule of *Glycine max cv. Caloria* (soybean) infected with *Bradyrhizobium japonicum*. The upper large infected cell shows intact symbiosomes (S) with one or two bacteroids per symbiosome. In the lower section, three noninfected cells with nucleus (N), central vacuole (V), amyloplasts (A), and peroxisomes (P) are to be seen. (By E. Mörschel and D. Werner, Marburg.)

Rhizobia possess a respiratory chain with a basic structure corresponding to that of the mitochondrial respiratory chain (see Fig. 5.15). In a *Bradyrhizobium* species, an additional electron transport path is formed during differentiation of the rhizobia to bacteroids. This path branches at the cyt-bc_1 complex of the respiratory chain and conducts electrons to another terminal oxidase, enabling an increased respiratory rate. It is encoded for by symbiosis-specific genes.

The formation of nodules is due to a regulated interplay of the expression of specific bacteria and plant genes

Rhizobia capable of entering a symbiosis contain a large number of genes, which are switched off in the free-living bacteria and are activated only after an interaction with the host, to contribute to the formation of an N_2-fixing nodule. The bacterial genes for proteins required for N_2 fixation are named ***nif*** and ***fix*** **genes**, and those that induce the formation of the nodules are called ***nod*** **genes.**

The host plant signals its readiness to form nodules by excreting several **flavonoids** (section 18.5) as signal compounds. These flavonoids bind to a bacterial protein, which is encoded by a constitutive (which means expressed at all times) ***nod*** **gene**. The protein, to which the flavonoid is bound, activates the transcription of the other *nod* genes. The proteins encoded by these *nod* genes are involved in the synthesis of the Nod factors mentioned previously. Four so-called "general" *nod* genes are present in all rhizobia. In addition, more than 20 other *nod* genes are known, which are responsible for the host's specificity.

Those proteins required especially for the formation of nodules, and which are synthesized by the **host plant** in the course of nodule formation, are called **nodulins**. These nodulins include leghemoglobin (section 11.2), the enzymes of carbohydrate degradation [including sucrose synthase (section 9.2)], enzymes of the citrate cycle and the synthesis of glutamine and asparagine, and, if applicable, also of ureide synthesis. They also include an aquaporin of the peribacteroid membrane. The plant genes encoding these proteins are called **nodulin genes**. One differentiates between **"early"** and **"late" nodulins.** Early nodulins are involved in the process of infection and formation of nodules, where the expression of the corresponding genes is induced in part by signal substances released from the rhizobia. "Late" nodulins are synthesized only after the formation of the nodules. In many cases, nodulins are isoforms of proteins found in other plant tissues.

Metabolic products are exchanged between bacteroids and host cells

The main substrate provided by the host cells to the bacteroids is **malate** (Fig. 11.4), formed from sucrose, which is delivered by the sieve tubes. The

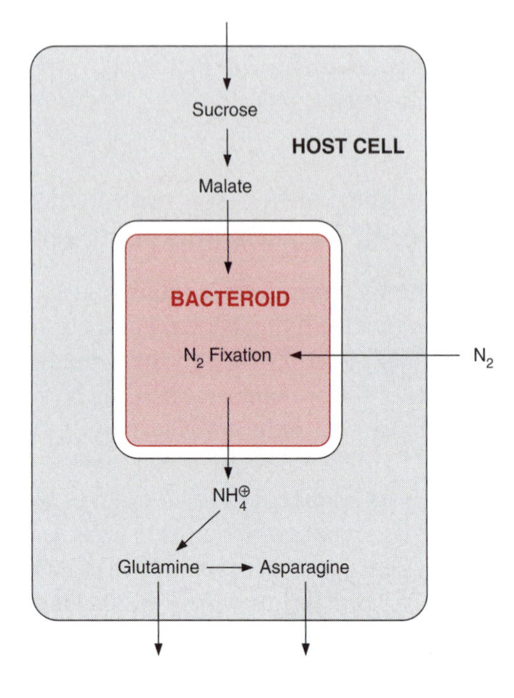

Figure 11.4 Metabolism of infected cells in a root nodule. Glutamine and asparagine are formed as the main products of N_2 fixation (see also Fig. 11.5).

sucrose is metabolized by sucrose synthase (Fig. 13.5), degraded by glycolysis to phosphoenolpyruvate, which is carboxylated to oxaloacetate (see Fig. 10.11), and the latter is reduced to malate. Nodule cells contain high activities of phosphoenolpyruvate carboxylase. NH_4^+ is delivered as a product of N_2 fixation to the host cell, where it is subsequently converted mainly into **glutamine** (Fig. 7.9) and **asparagine** (Fig. 10.14) and then transported via the xylem vessels to the other parts of the plant. It was recently shown that alanine also can be exported from bacteroids.

The nodules of some plants (e.g., those of soybean) export the fixed nitrogen as ureides (urea degradation products), especially **allantoin** and **allantoic acid** (Fig. 11.5). These compounds have a particularly high nitrogen to carbon ratio. The formation of ureides in the host cells requires a complicated synthetic pathway. First, inosine monophosphate is synthesized via the pathway of purine synthesis, which is present in all cells for the synthesis of AMP and GMP, and then it is degraded via xanthine and ureic acid to the ureides mentioned previously.

Malate taken up into the bacteroids is oxidized by the citrate cycle (Fig. 5.3). The reducing equivalents thus generated are the fuel for the fixation of N_2.

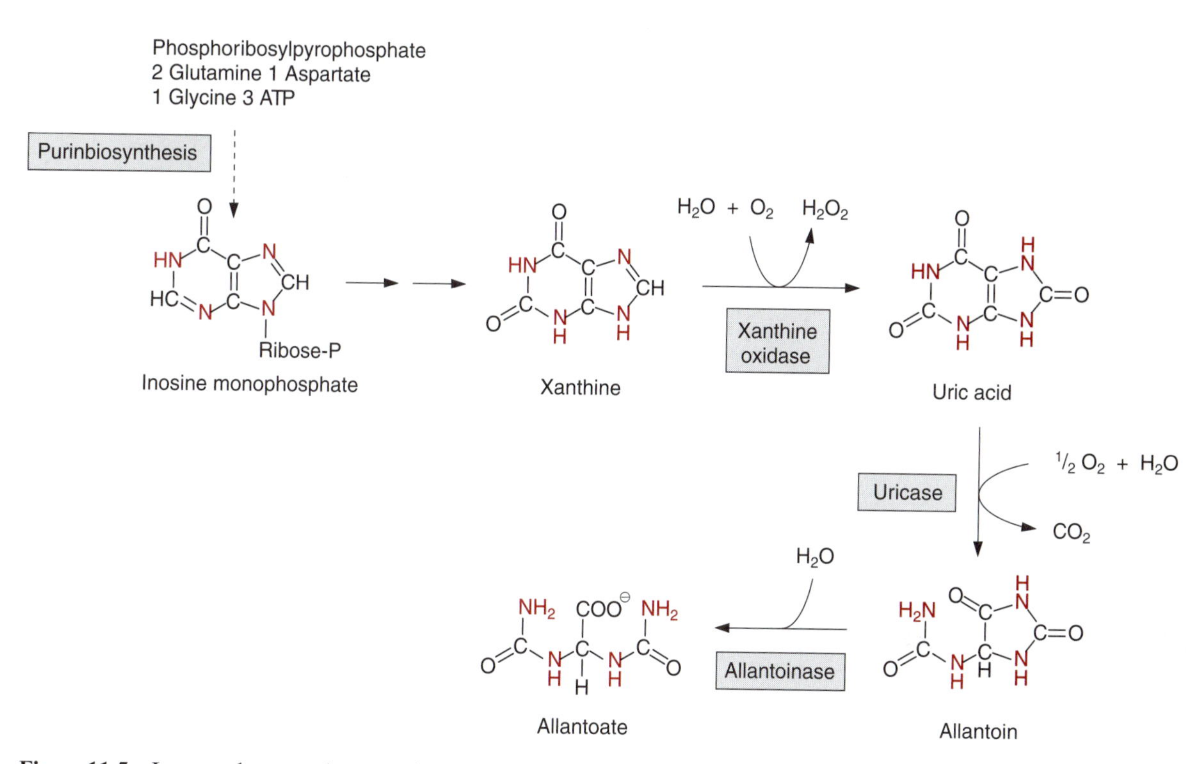

Figure 11.5 In some legumes (e.g., soy bean and cow pea), allantoin and allantoic acid are formed as products of N_2 fixation and are delivered via the roots to the xylem. Their formation proceeds first via inosine monophosphate by the purine synthesis pathway. Inosine monophosphate is oxidized to xanthine and then further to ureic acid. Allantoin and allantoic acid are formed by hydrolysis and opening of the ring.

Nitrogenase reductase delivers electrons for the nitrogenase reaction

Nitrogen fixation is catalyzed by the **nitrogenase complex**, a highly complex system with nitrogenase reductase and nitrogenase as the main components (Fig. 11.6). This complex is highly conserved and is present in the cytoplasm of the bacteroids. From NADH formed in the citrate cycle, electrons are transferred via soluble ferredoxin to **nitrogenase reductase**. The latter is a one-electron carrier, consisting of two identical subunits, which together form a **4Fe-4S cluster** (see Fig. 3.26) and contain two binding sites for ATP. After reduction of nitrogenase reductase, two molecules of ATP bind to it, resulting in a conformational change of the protein, by which the redox potential of the 4Fe-4S cluster is raised from –0.25 to –0.40 V. Following the transfer of an electron to nitrogenase, the two ATP molecules bound to

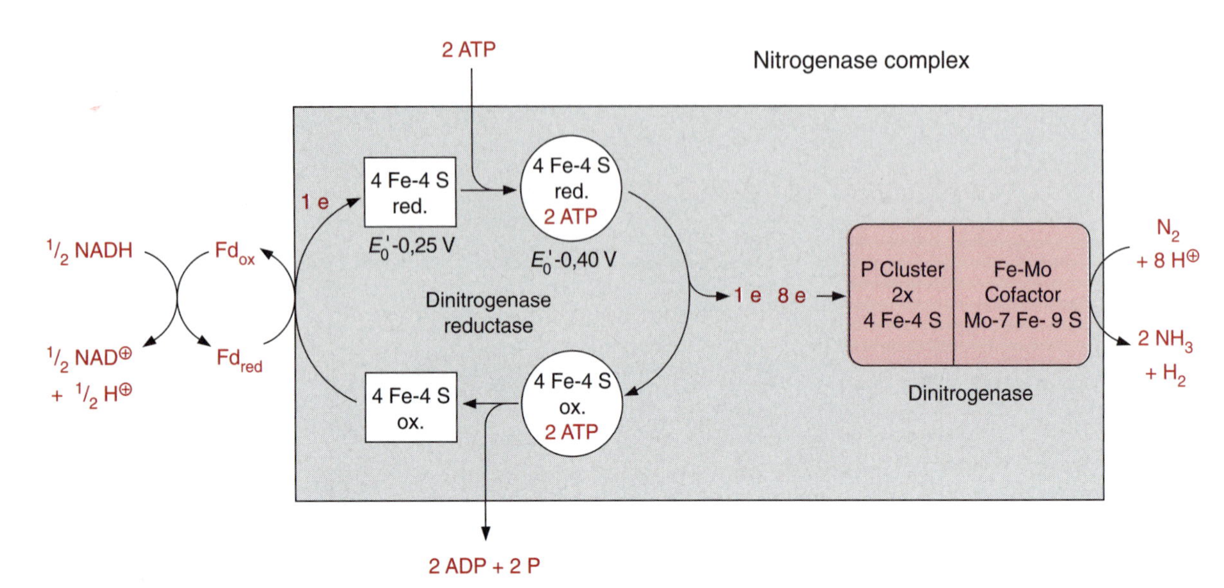

Figure 11.6 The nitrogenase complex consists of the nitrogenase reductase and the nitrogenase. Their structure and function are described in the text. The reduction of one molecule of N_2 is accompanied by the reduction of at least two protons to form molecular hydrogen.

the protein are hydrolyzed to ADP and phosphate, and then released from the protein. As a result, the conformation with the lower redox potential is restored and the enzyme is again ready to take up one electron from ferredoxin. Thus. with the consumption of two molecules of ATP, one electron is transferred from NADH to nitrogenase by nitrogenase reductase.

N_2 as well as H^+ are reduced by nitrogenase

Nitrogenase is an $\alpha_2\beta_2$ tetramer. The α and β subunits have a similar size and are similarly folded. The tetramer contains two catalytic centers, probably reacting independently of each other, and each contains a so-called P cluster, consisting of two **4Fe-4S clusters** and an iron molybdenum cofactor **(FeMoCo)**. FeMoCo is a large redox center made up of Fe_4S_3 and Fe_3MoS_3, which are linked to each other via three inorganic sulfide bridges (Fig. 11.7). A further constituent of the cofactor is **homocitrate**, which is linked via oxygen atoms of the hydroxyl and carboxyl group to molybdenum. Another ligand of molybdenum is the imidazole ring of a histidine residue of the protein. The function of the Mo atom is still unclear. Alternative nitrogenases are known, in which molybdenum is replaced by vanadium or iron, but these nitrogenases are much more unstable than the nitrogenase containing FeMoCo. The Mo atom possibly causes a more favorable geometry

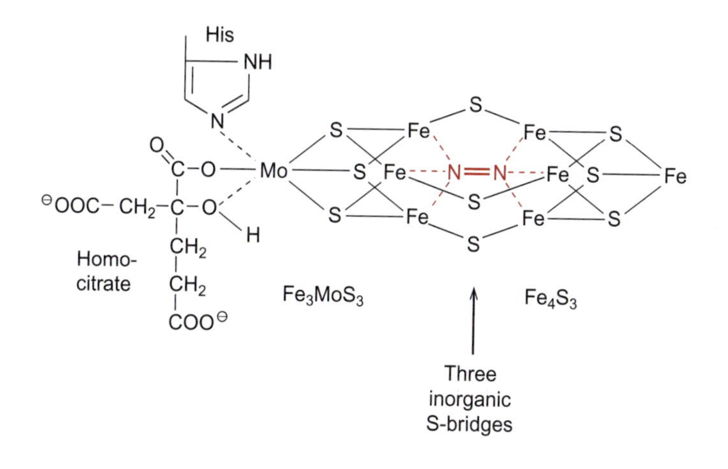

Figure 11.7 The iron-molybdenum cofactor consists of the fragments Fe_4S_3 and $MoFe_3S_3$, which are linked to each other by three inorganic sulfide bridges. In addition, the molybdenum is ligated with homocitrate and the histidine side group of the protein. The cofactor binds one N_2 molecule and reduces it to two molecules of NH_3 by successive uptake of electrons. The position where N_2 is bound in the cofactor has not yet been experimentally proven. (After Karlin 1993.)

and electron structure of the center. It is not yet known how nitrogen reacts with the iron-molybdenum cofactor. One possibility would be that the N_2 molecule is bound in the cavity of the FeMoCo center (Fig. 11.7) and that the electrons required for N_2 fixation are transferred by the P cluster to the FeMoCo center.

Nitrogenase is able to reduce other substrates beside N_2 (e.g., protons, which are reduced to molecular hydrogen):

$$2H^+ + 2e^- \xrightarrow{\textit{Nitrogenase}} H_2$$

During N_2 fixation at least one molecule of hydrogen is formed per N_2 reduced:

$$8H^+ + 8e^- + N_2 \xrightarrow{\textit{Nitrogenase}} 2NH_3 + H_2$$

Thus the balance of N_2 fixation is at least:

$$N_2 + 4NADH + 4H^+ + 16ATP \longrightarrow 2NH_3 + H_2 + 4NAD^+ + 16ADP + 16P$$

In the presence of sufficient concentrations of acetylene, only this is reduced and ethylene is formed:

$$HC \equiv CH + 2e^- + 2H^+ \xrightarrow{\textit{Nitrogenase}} H_2C = CH_2$$

This reaction is used to measure the activity of nitrogenase. Why H_2 evolves during N_2 fixation is not known. It may be part of the catalytic mechanism or a side reaction or a reaction to protect the active center against the inhibitory effect of oxygen. The formation of molecular hydrogen during N_2 fixation can be observed in a clover field.

Many bacteroids, however, possess hydrogenases by which H_2 is reoxidized by electron transport:

$$2H_2 + O_2 \xrightarrow{Hydrogenase} 2H_2O$$

It is questionable, however, whether this reaction is coupled in the bacteroids to the generation of ATP.

11.2 N_2 fixation can proceed only at very low oxygen concentrations

Nitrogenase is extremely sensitive to oxygen. Therefore N_2 fixation can proceed only at very low oxygen concentrations. The nodules form an anaerobic compartment. Since N_2 fixation depends on the uptake of nitrogen from the air, the question arises how is the enzyme protected against the oxygen present in air? The answer is that oxygen, which has diffused together with nitrogen into the nodules, is consumed by the **respiratory chain** contained in the bacteroid membrane. Due to a very high affinity of the bacteroid cytochrome-a/a_3 complex, respiration is still possible with an oxygen concentration of only 10^{-9} mol/L. As described previously, at least a total of 16 molecules of ATP are required for the fixation of one molecule of N_2. Upon oxidation of one molecule of NADH, about 2.5 molecules of ATP are generated by the mitochondrial respiratory chain (section 5.6). In the bacterial respiratory chain, which normally has a lower degree of coupling than that of mitochondria, only about two molecules of ATP may be formed per molecule of NADH oxidized. Thus about four molecules of O_2 have to be consumed for the formation of 16 molecules of ATP (Fig. 11.8). If the bacteroids possess a hydrogenase, due to the oxidation of H_2 formed during N_2 fixation, oxygen consumption is further increased by half an O_2 molecule. Thus during N_2 fixation, for each N_2 molecule at least four O_2 molecules are consumed by bacterial respiration ($O_2/N_2 \geq 4$). In contrast, the O_2/N_2 ratio in air is about 0.25. This comparison shows that air required for N_2 fixation contains in relation to nitrogen far too little oxygen.

The outer layer of the nodules is a considerable **diffusion barrier** for the entry of air. The diffusive resistance is so high that bacteroid respiration is limited by the uptake of oxygen. This leads to the astonishing situation that N_2 fixation is limited by the influx of O_2 for formation of the required ATP. Experiments by the Australian Fraser Bergersen have presented evidence for

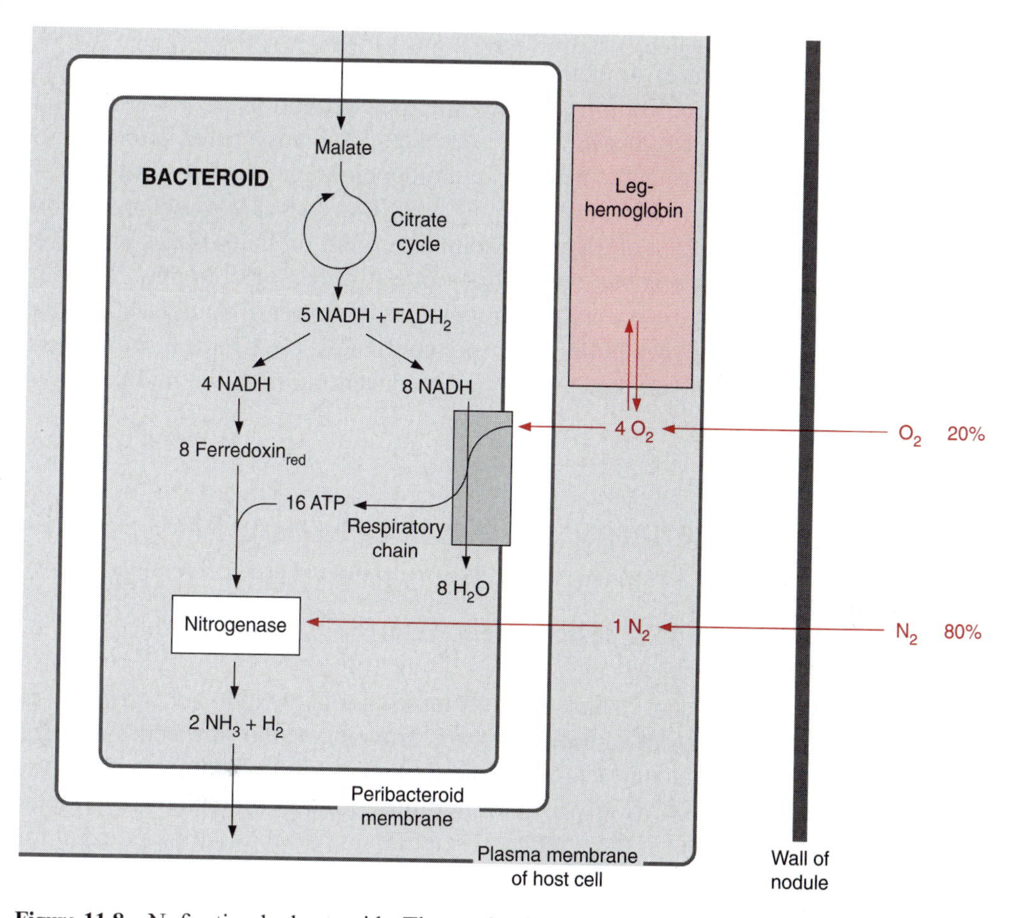

Figure 11.8 N_2 fixation by bacteroids. The total oxidation of malate by the citrate cycle yields five NADH and one $FADH_2$ (see Figure 5.3). The formation of two NH_3 from N_2 and the accompanying reduction of $2H^+$ to H_2 requires at least 16 molecules of ATP. Generation of this ATP by the respiratory chain localized in the bacteroid membrane requires the oxidation of at least eight molecules of NADH. Thus, for each molecule of N_2 fixed, at least four molecules of O_2 are consumed by the oxidation of NADH in the respiratory chain of the bacteroid membrane.

this. He observed with soybean nodules that a doubling of the O_2 content in air (with a corresponding decrease of the N_2 content) resulted in a doubling of the rate of N_2 fixation. But, because of the O_2 sensitivity of the nitrogenase, a further increase in O_2 resulted in a steep decline in N_2 fixation.

Since the bacterial respiratory chain is located in the membrane and nitrogenase in the interior of the bacteroids, O_2 is kept at a safe distance from nitrogenase. The high diffusive resistance for O_2, which, as shown in the experiment, can limit N_2 fixation, ensures that even at low temperatures, at which N_2 fixation and the bacterial respiration are slowed down, oxygen is kept away from the nitrogenase complex.

The cells infected by rhizobia form **leghemoglobin**, which is very similar to the myoglobin of animals, but has a 10-fold higher affinity for oxygen. The oxygen concentration required for half saturation of leghemoglobin amounts to only $10–20 \times 10^{-9}$ mol/L. Leghemoglobin is located in the cytosol of the host cell—outside the peribacteroid membrane—and present there in unusually high concentrations (3×10^{-3} mol/L in soybeans). Leghemoglobin can amount to 25% of the total soluble protein of the nodules and gives them a pink color. It has been proposed that leghemoglobin plays a role in the transport of oxygen within the nodules. However, it is more likely that it serves as an **oxygen buffer** to ensure continuous electron transport in the bacteroids at the very low prevailing O_2 concentration in the nodules.

11.3 The energy costs for utilizing N_2 as a nitrogen source are much higher than for the utilization of NO_3^-

As shown in Figure 11.8, at least six molecules of NADH are consumed in the formation of one molecule of NH_4^+ from molecular nitrogen. Assimilation of nitrate, in contrast, requires only four NAD(P)H equivalents for the formation of NH_4^+ (Fig. 10.1). In addition, it costs the plant much metabolic energy to form the nodules. Therefore it is much more economical for plants, which have the potential to fix N_2 with the help of their symbionts, to satisfy their nitrogen demand by nitrate assimilation. This is why the formation of nodules is regulated. Nodules are formed only when the soil is nitrate-deficient. The advantage of this symbiosis is that legumes and actinorhizal plants can grow in soils with very low nitrogen content, where other plants have no chance.

11.4 Plants improve their nutrition by symbiosis with fungi

Frequently plant growth is limited by the supply of nutrients other than nitrate (e.g., phosphate). Because of its low solubility, the extraction of phosphate by the roots from the soil requires very efficient uptake systems. For this reason, plant roots possess very high affinity transporters, with a half saturation of 1 to 5 μM phosphate, where the phosphate transport is driven by proton symport, similar to the transport of nitrate (Section 10.1). In order to increase the uptake of phosphate, but also of other mineral nutrients (e.g., nitrate and potassium), most plants enter a symbiosis with fungi. Fungi are able to form a mycelium with hyphae, which have a much lower diameter than root hairs and which are therefore well suited to penetrate soil particles and to mobilize their nutrients. The symbiotic fungi (microsymbionts) deliver these nutrients to the plant root (macrosymbiont) and are in turn supplied by the plant with substrates for maintaining their metabolism. The supply of the symbiotic fungi by the roots demands a high amount of assimilates. For this reason, many plants make the establishment of the mycorrhiza dependent on the phosphate availability in the soil. In the case of a high phosphate concentration in the soil, when the plant can do without, it treats the fungus as a pathogen and activates its defense system against fungal infections (see Chapter 16).

The arbuscular mycorrhiza is widespread

The **arbuscular mycorrhiza** has been detected in more than 80% of all plant species. In this symbiosis the fungus penetrates the cortex of plant roots and forms there a network of hyphae, which protrude into cortical cells and form there treelike invaginations, which are termed **arbuscules** (Fig. 11.9) or form **hyphal coils**. The boundary membranes of fungus and host remain intact.

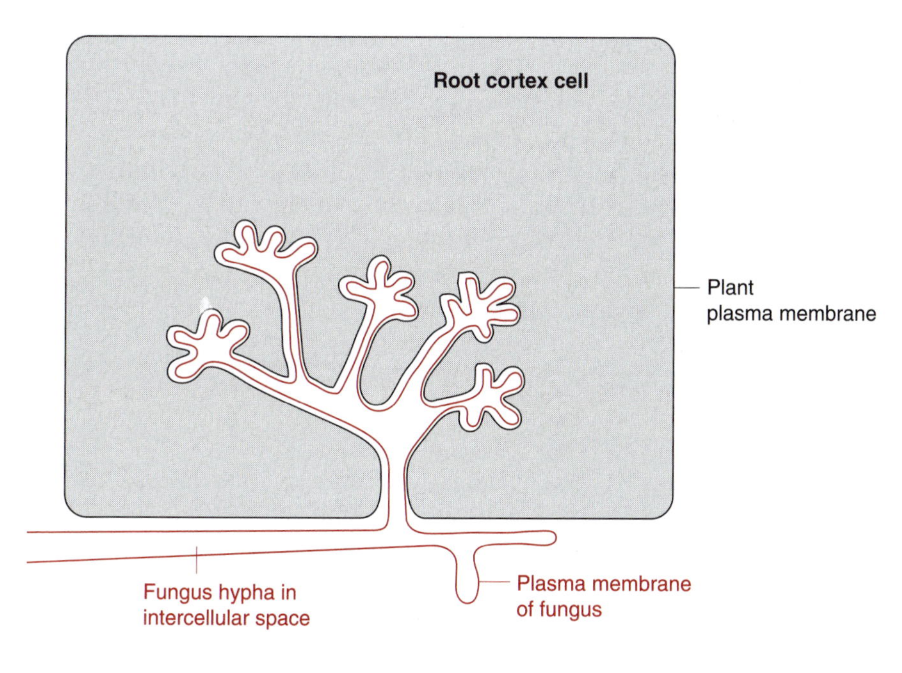

Figure 11.9 Schematic representation of an arbuscule. The hypha of a symbiotic fungus traverses the rhizodermis cells and spreads in the intercellular space of the root cortex. From there treelike invaginations into the inner layer of the cortex are formed. The cell walls of the plant and of the fungus (not shown in the figure) and the plasma membranes remain intact. The large boundary surface between the host and the microsymbiont enables an effective exchange of substances.

The arbuscules form a large surface, enabling an efficient exchange of substances between the fungus and the host. The fungus delivers phosphate, nitrate, K^+-ions, and water, and the host delivers carbohydrates. The arbuscules have a lifetime of only maximally two weeks, but the subsequent degeneration does not damage the corresponding host cell. Therefore, the maintenance of symbiosis requires a constant formation of new arbuscules. The arbuscular mycorrhiza evolved at a very early stage of plant evolution about 450 million years ago. Whereas the number of plant species capable of forming an arbuscular mycorrhiza is very large (about 80% of terrestrial plants), there are only six genera of fungi functioning as microsymbionts. Therefore the symbiosis is rather unspecific.

Ectomycorrhiza supplies trees with nutrients

Many trees in temperate and cool climates form a symbiosis with fungi termed **Ectomycorrhiza**. In this the hyphae of the fungi do not penetrate the cortex cells, but colonize only the surface and the intercellular space of the cortex with a network of hyphae, termed **Hartig net**, which is connected with a very extensive mycel in the soil. Microsymbionts are *Asco-* and *Basidiomycetae* from more than 60 genera, including several mushrooms. The plant root tips colonized by the fungi are thickened and do not form any root hairs. The uptake of nutrients and water is delegated to the microsymbiont, which in turn is served by the plant with substrates to maintain its metabolism, The exchange of substances occurs, as in arbuscular mycorrhiza, via closely neighbored fungal and plant plasma membranes. The ectomycorrhiza also enables a transfer of assimilates between adjacent plants. Ectomycorrhiza is of great importance for the growth of trees, such as beech, oak, and pine, as it increases the uptake of phosphate by a factor of three to five. It has been observed that the formation of ectomycorrhiza is negatively affected when the nitrate content of the soil is high. This may explain the damaging effect of nitrogen input to forests by air pollution.

Other forms of mycorrhiza (e.g., the endomycorrhiza with orchids and *Ericaceae*) will not be discussed here.

11.5 Root nodule symbioses may have evolved from a preexisting pathway for the formation of arbuscular mycorrhiza

There are parallels between the establishing of arbuscular mycorrhiza and of root nodule symbiosis. In both cases, receptor-like kinases (**RLK**, section

19.1) appear to be involved, linked to **signal cascades**, which induce the synthesis of the proteins required for the controlled infection. These signal cascades probably involve G-proteins, MAP-kinases, and Ca^{++} ions as messenger (section 19.1). For several legume species, mutants are known that have lost the ability to establish both root nodule symbiosis and arbuscular mycorrhiza. One of the genes that cause such a defect in different legume species has been identified to encode an RLK, indicating that this RLK has an essential function in the formation of both arbuscular mycorrhiza and root nodule symbiosis. Fungi and bacteria, despite their different natures, apparently induce similar genetic programs upon infection.

Molecular phylogenetic studies have shown that all plants with the ability to enter root nodule symbiosis, rhizobial or actinorhizal, belong to a single clade (named **Eurosid I**), that is, they go back to a common ancestor (although not all descendants of this ancestor are symbiotic). Obviously, this ancestor has acquired a property on the basis of which a bacterial symbiosis could develop. The overlaps in signal transduction pathways lead to the assumption that this property is based on the ability to enter fungal symbiosis. Based on this property, root nodule symbiosis evolved about 50 million years ago, not as a single evolutionary event, but reoccurred about eight times. In order to transfer by genetic engineering the ability to enter a root nodule symbiosis to agriculturally important monocots, such as rice, maize, and wheat, it will be necessary to find out which property of the Eurosid I clade plants allowed the evolution of such symbiosis

Further reading

Christiansen, J., Dean, D. R. Mechanistic feature of the Mo-containing nitrogenase. Annu Rev Plant Physiol Mol Biol 52, 269–295 (2002).

Cohn, J., Day, R. B., Stacey, G. Legume nodule organogenesis. Trends Plant Sci 3, 105–110 (1998).

Harrison, M. J. Molecular and cellular aspects of the arbuscular mycorrhizal symbiosis. Annu Rev Plant Physiol Mol Biol 50, 361–389 (1999).

Hirsch, A. M., Lum, M. R., Downie, J. A. What makes the rhizobia-legume symbiosis so special? Plant Physiol 127, 1484–1492 (2001).

Karlin, K. D. Metalloenzymes, structural motifs and inorganic models. Science, 701–708 (1993).

Kistner, C., Parniske, M. Evolution of signal transduction in intracellular symbiosis. Trends Plant Sci 7, 511–517 (2002).

Lang, S. R. Genes and signals in the *Rhizobium*-legume symbiosis. Plant Physiol 125, 69–72 (2001).

Limpens, E., Franken, C., Smit, P., Willemse, J., Bisseling, T., Geurts, R. LysM domain receptor kinases regulating rhizobial Nod factor-induced infection. Science 302, 630–633 (2003)

Mithöfer, A. Suppression of plant defence in Rhizobia-legume symbiosis. Trends Plant Sci 7, 440–•• (2002).

Mylona, P., Pawlowski, K., Bisseling, T. Symbiotic nitrogen fixation. Plant Cell 7, 869–885 (1995).

Ndakidemi, P. A, Dakora, F. D. Legume seed flavonoids and nitrogenous metabolites as signals and protectants in early seedling development. Funct Plant Biol 30, 729–745 (2003).

Parniske, M., Downie, J. A. Locks, keys and symbioses. Nature 425, 569–570 (2003).

Radutolu, S., Madsen, L. H., Madsen, E. B., Felle, H. H., Gronlund, M., Sato, S., Nakamura, Y., Tabata, S., Sandal, N., Stougaard, J. Plant recognition of symbiotic bacteria requires two LysM receptor-like kinases. Nature 425, 585–591 (2003).

Sanders, I. R. Preference, specificity and cheating in the arbuscular mycorrhizal symbiosis. Trends Plant Sci 8, 143–145 (2003).

Smil, V. Enriching the earth: Fritz Haber, Carl Bosch, and the transformation of world food. Massachussets Institute of Technology Press, Boston (2001).

Smith, P. M. C., Atkins, C. A. Purine biosynthesis. Big in cell division, even bigger in nitrogen assimilation. Plant Physiol 128, 793–802 (2002).

Triplett, E. W. (ed.) Prokaryotic nitrogen fixation. Horizon Scientific Press Wymondham, Norfolk, England (2000).

Zhu, Y.-G., Miller R. M. Carbon cycling by arbuscular mycorrhizal fungi in soil-plant systems. Trends Plant Sci 8, 407–409 (2003).

12

Sulfate assimilation enables the synthesis of sulfur-containing substances

Sulfate is an essential constituent of living matter. In the oxidation state -II, it is contained in the two amino acids cysteine and methionine, in the detoxifying agent glutathione, in various iron sulfur redox clusters, in peroxiredoxins, and in thioredoxins. Plants, bacteria, and fungi are able to synthesize these substances by assimilating sulfate taken up from the environment. Animal metabolism is dependent on nutrients to supply amino acids containing sulfur. Therefore sulfate assimilation of plants is a prerequisite for animal life, just like the carbon and nitrate assimilation discussed previously.

Whereas the plant uses nitrate only in its reduced form for syntheses, sulfur, also in the form of sulfate, is an essential plant constituent. Sulfate is contained in sulfolipids, which comprise about 5% of the lipids of the thylakoid membrane (Chapter 15). In sulfolipids sulfur is attached as sulfonic acid via a C-S bond to a carbohydrate residue of the lipid. The biosynthesis of this sulfonic acid group is, to a great extent, not known.

12.1 Sulfate assimilation proceeds primarily by photosynthesis

Sulfate assimilation in plants occurs primarily in the **chloroplasts** and is then a part of photosynthesis, but it also takes place in the **plastids** of the roots. However, the rate of sulfate assimilation is relatively low, amounting to only about 5% of the rate of nitrate assimilation and only 0.1% to 0.2%

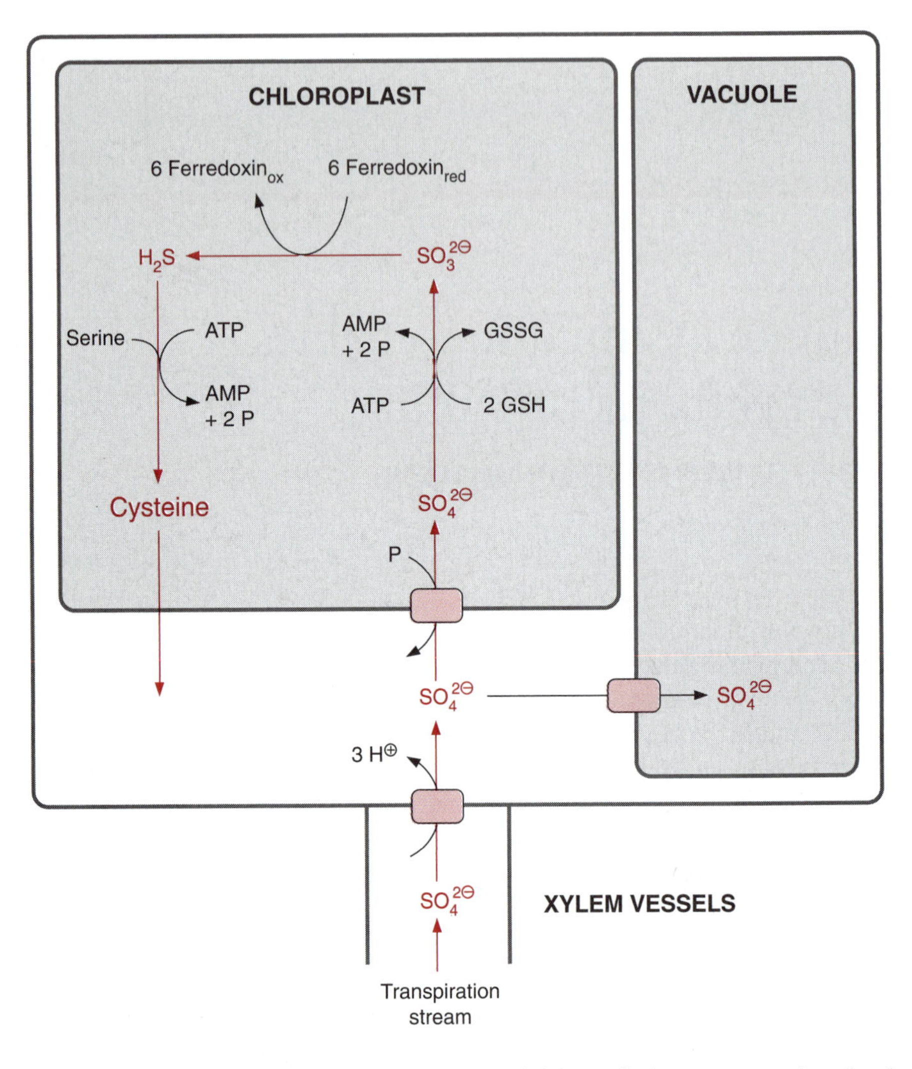

Figure 12.1 Sulfate metabolism in a leaf. Sulfate is carried by the transpiration stream into the leaves and is transported to the mesophyll cells. After being transported to the chloroplast via the phosphate translocator, sulfate is reduced there to H_2S and subsequently converted to cysteine. Sulfate can also be deposited in the vacuole. Simplified scheme. In reality, serine is activated as acetylserine.

of the rate of CO_2 assimilation. The activities of the enzymes involved in sulfate assimilation are minute, making it very difficult to elucidate the reactions involved. Therefore our knowledge about sulfate assimilation is still fragmentary.

Sulfate assimilation has some parallels to nitrogen assimilation

Plants take up **sulfate** via a specific translocator of the roots, in a manner similar to that described for nitrate in Chapter 10. The transpiration stream in the xylem vessels carries the sulfate to the leaves, where it is taken up by a specific translocator, probably a symport with three protons, into the mesophyll cells (Fig. 12.1). Surplus sulfate is transported to the vacuole and is deposited there.

The basic scheme for sulfate assimilation in the mesophyll cells corresponds to that of nitrate assimilation. Sulfate is reduced to **sulfite** by the uptake of two electrons and then by the uptake of another six electrons, to **hydrogen sulfide**:

$$SO_4^{2-} + 2e^- + 2H^+ \rightarrow SO_3^{2-} + H_2O$$

$$SO_3^{2-} + 6e^- + 8H^+ \rightarrow H_2S + 3H_2O$$

Whereas the NH_3 formed during nitrite reduction is fixed by the formation of the amino acid glutamine (Fig. 10.6), the hydrogen sulfide formed during sulfite reduction is fixed to form the amino acid cysteine. A distinguishing difference between nitrate assimilation and sulfate assimilation is that the latter requires a much higher input of energy. This is shown in an overview in Figure 12.1. The reduction of sulfate to sulfite, which in contrast to nitrate reduction occurs in the chloroplasts, requires in total the cleavage of two energy-rich phosphate anhydride bonds, and the fixation of the hydrogen sulfide into cysteine requires another two. Thus the ATP consumption of sulfate assimilation is four times higher than that of nitrate assimilation. Let us now look at the individual reactions.

Sulfate is activated prior to reduction

Sulfate is probably taken up into the chloroplasts in counter-exchange for phosphate. Sulfate cannot be directly reduced in the chloroplasts, because the redox potential of the substrate pair SO_3^{2-}/SO_4^{2-} ($\Delta E^{0\prime} = -517$ mV) is too high. No reductant is available in the chloroplasts that could reduce SO_4^{2-} to SO_3^{2-} in one reaction step. To make the reduction of sulfate possible, the redox potential difference to sulfite is lowered by activation of the sulfate prior to reduction.

As shown in Figure 12.2, activation of sulfate proceeds via the formation of an anhydride bond with the phosphate residue of AMP. Sulfate is exchanged by the enzyme **ATP-sulfurylase** for a pyrophosphate residue of ATP, and AMP-sulfate (**APS**) is thus formed. Since the free energy of the hydrolysis of the sulfate-phosphate anhydride bond ($\Delta G^{o\prime} = -71$ kJ/mol) is very much higher than that of the phosphate-phosphate anhydride bond in ATP ($\Delta G^{o\prime} = -31$ kJ/mol), the equilibrium of the reaction lies far toward ATP. This reaction can proceed only because pyrophosphate is withdrawn from the equilibrium by a high **pyrophosphatase activity** in the chloroplasts.

Sulfate present in the form of APS is reduced by **glutathione** (Figs. 12.5, 3.38) to sulfite. The **APS reductase** involved in this reaction catalyzes not only the reduction, but also the subsequent liberation of sulfite from AMP. The redox potential difference from sulfate to sulfite is lowered, since the

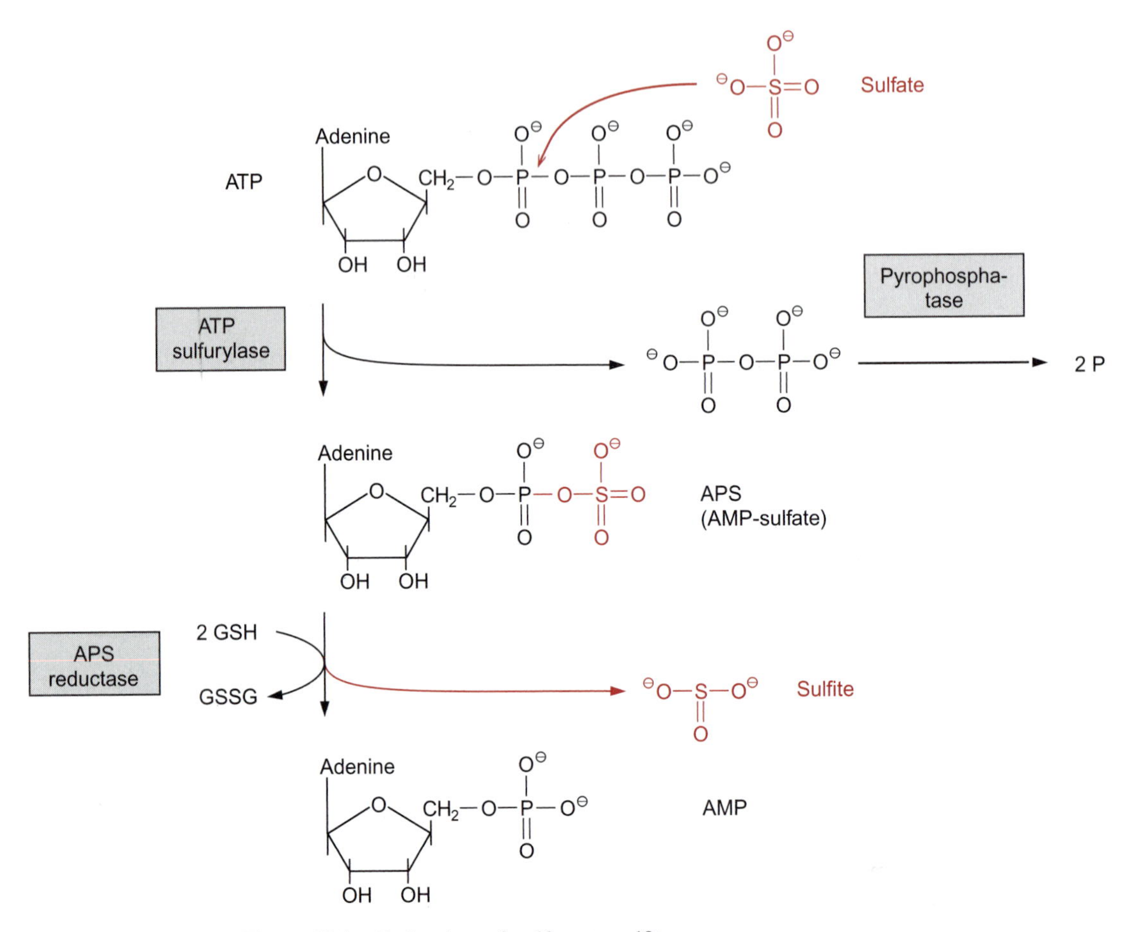

Figure 12.2 Reduction of sulfate to sulfite.

reduction of sulfate is driven by hydrolysis of the very energy-rich sulfite anhydride bond. The mechanism of the APS reductase reaction remains to be elucidated.

Sulfite reductase is similar to nitrite reductase

As in nitrite reduction, six molecules of reduced ferredoxin are required as reductant for the reduction of sulfite in the chloroplasts (Fig. 12.3). The **sulfite reductase** is homologous to the nitrite reductase, it also contains a **siroheme** (Fig. 10.5) and a **4Fe-4S cluster**. The enzyme is half saturated at a sulfite concentration in the range of 10^{-6} mol/L and thus is suitable to reduce efficiently the newly formed sulfite to hydrogen sulfide. The ferredoxin required by sulfite reductase, as in the case of nitrite reductase (Fig. 10.1),

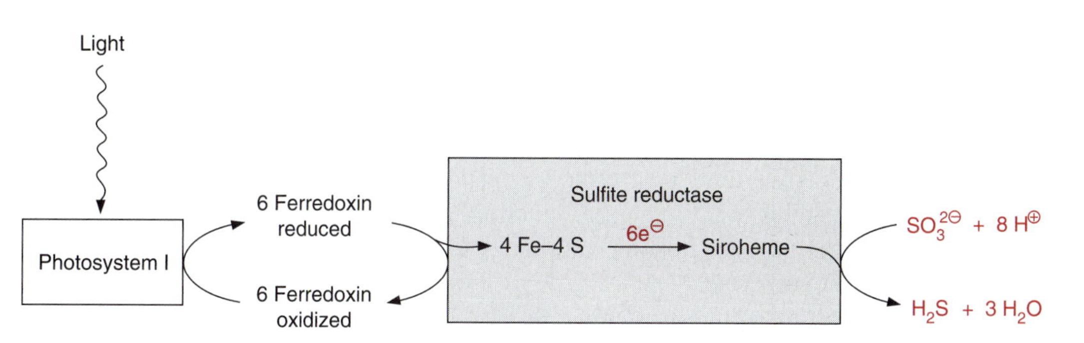

Figure 12.3 Reduction of sulfite to hydrogen sulfide by sulfite reductase in the chloroplasts. The reducing equivalents are delivered via ferredoxin from photosystem I.

can be reduced by NADPH. This makes it possible for sulfite reduction to occur in heterotrophic tissues also.

H_2S is fixed in the form of cysteine

The fixation of the newly formed H_2S requires the activation of **serine**, and for this its hydroxyl group is acetylated by **acetyl-CoA** via a **serine transacetylase** (Fig. 12.4). The latter is formed from acetate and CoA with the consumption of ATP (which is converted to AMP) by the enzyme **acetyl-CoA synthetase**. As pyrophosphate formed in this reaction is hydrolyzed by the pyrophosphatase present in the chloroplasts; the activation of the serine costs the chloroplasts in total two energy-rich phosphates.

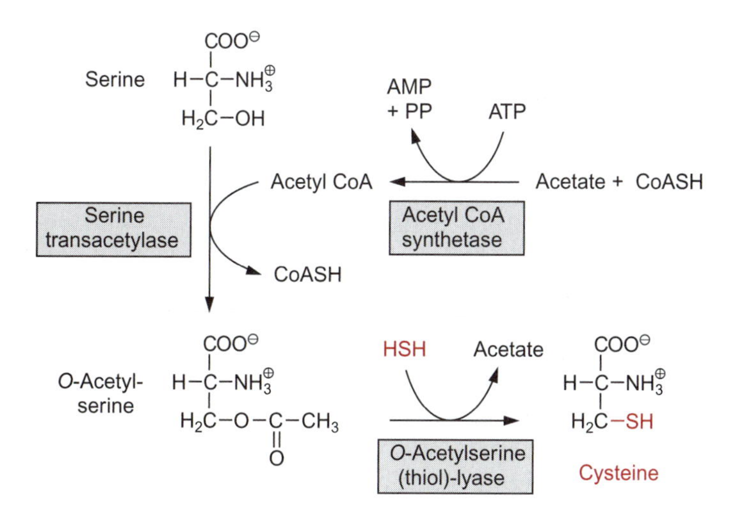

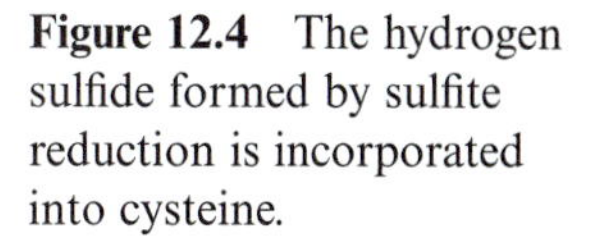

Figure 12.4 The hydrogen sulfide formed by sulfite reduction is incorporated into cysteine.

Fixation of H_2S is catalyzed by the **enzyme *O*-acetyl serine (thiol) lyase**. The enzyme contains pyridoxal phosphate as a prosthetic group and has a high affinity for H_2S and acetyl serine. The incorporation of the SH group can be described as a cleavage of the ester linkage by H-S-H. In this way **cysteine** is formed as the end product of sulfate assimilation.

Cysteine has an essential function in the structure and activity of the catalytic site of enzymes and cannot be replaced there by any other amino acids. Moreover, cysteine residues form iron-sulfur clusters (Fig. 3.26,) and are constituents of thioredoxin (Fig. 6.25).

12.2 Glutathione serves the cell as an antioxidant and is an agent for the detoxification of pollutants

A relatively large proportion of the cysteine produced by the plant is used for synthesis of the tripeptide **glutathione**, (Fig. 12.5). The synthesis of glutathione proceeds via two enzymatic steps: first, an amide linkage between the γ-carboxyl group of glutamate with the amino group of the cysteine is formed by **γ-glutamyl-cysteine-synthetase** accompanied by the hydrolysis of ATP; and second, a peptide bond between the carboxyl group of the cysteine and the amino group of the glycine is produced by **glutathione synthetase**, again with the consumption of ATP. Glutathione, abbreviated **GSH**, is present at relatively high concentrations in all plant cells, where it has various functions. The function of GSH as a reducing agent was discussed in the previous section. As an antioxidant, it protects cell constituents against oxidation. Together with ascorbate, it eliminates the oxygen radicals formed as by-products of photosynthesis (section 3.9). In addition, glu-

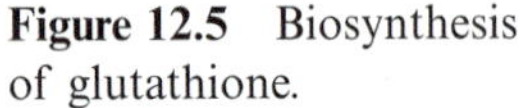
Figure 12.5 Biosynthesis of glutathione.

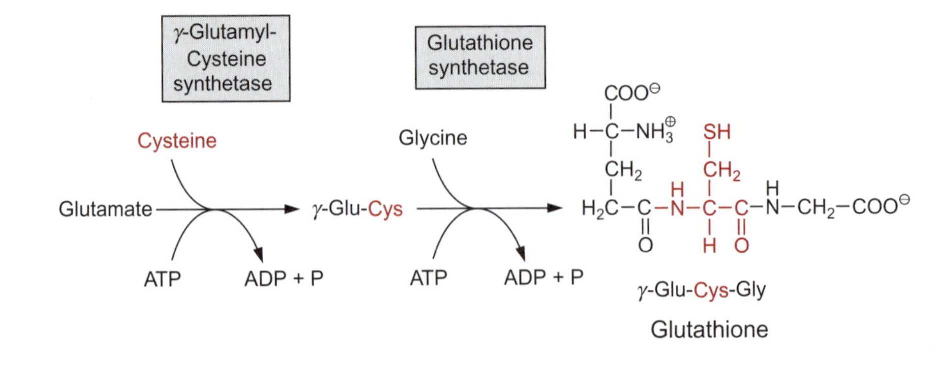

tathione has a protective function for the plant in forming conjugates with xenobiotics and also as a precursor for the synthesis of **phytochelatins**, which are involved in the detoxification of heavy metals. Moreover, glutathione acts as a reserve for organic sulfur. If required, cysteine is released from glutathione by enzymatic degradation.

Xenobiotics are detoxified by conjugation

Toxic substances formed by the plant or which it has taken up (xenobiotics) are detoxified by reaction with glutathione. Catalyzed by **glutathione-S transferases**, the reactive SH group of glutathione can form a thioether by reacting with carbon double bonds, carbonyl groups, and other reactive groups. **Glutathione conjugates** (Fig. 12.6) formed in this way are transported into the vacuole by a specific **glutathione translocator** against a concentration gradient. In contrast to the transport processes so far, where metabolite transport against a gradient proceeds by secondary active transport, the uptake of glutathione conjugates into the vacuole proceeds by an ATP-driven primary active transport (Fig. 1.20). This translocator belongs to the superfamily of the **ABC-transporter** (ATP binding cassette), which is ubiquitous in plants and animals and also occurs in bacteria. In the vacuolar membrane, various ABC transporters with different specificities are present. The conjugates taken up are often modified (e.g., by degradation to a cysteine conjugate) and are finally deposited in this form. In this way plants can also

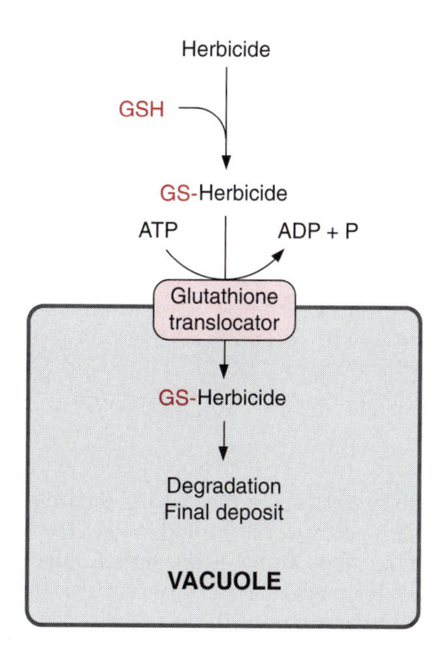

Figure 12.6 Detoxification of a herbicide. Glutathione (GSH) forms a conjugate with the herbicide, which is pumped into the vacuole by a specific glutathione translocator to be finally deposited there after degradation.

detoxify herbicides. Herbicide resistance (e.g., resistance of maize to atrazine) can be due to the activity of a specific glutathione-*S* transferase. In an attempt to develop herbicides that selectively attack weeds and not crop plants, the plant protection industry has produced a variety of different substances that increase the tolerance of crop plants to certain herbicides. These protective substances are called **safeners**. Such safeners, like other xenobiotics, stimulate the increased expression of glutathione-*S* transferase and of the vacuolar glutathione translocator, and in this way the herbicides taken up into the plants are detoxified more rapidly. Formation of glutathione conjugates and their transport into the vacuole is also involved in the deposition of flower pigments (section 18.6).

Phytochelatins protect the plant against heavy metals

Glutathione is also a precursor for the formation of **phytochelatins** (Fig. 12.7). **Phytochelatin synthase**, a transpeptidase, transfers the amino group of glutamate to the carboxyl group of the cysteine of a second glutathione

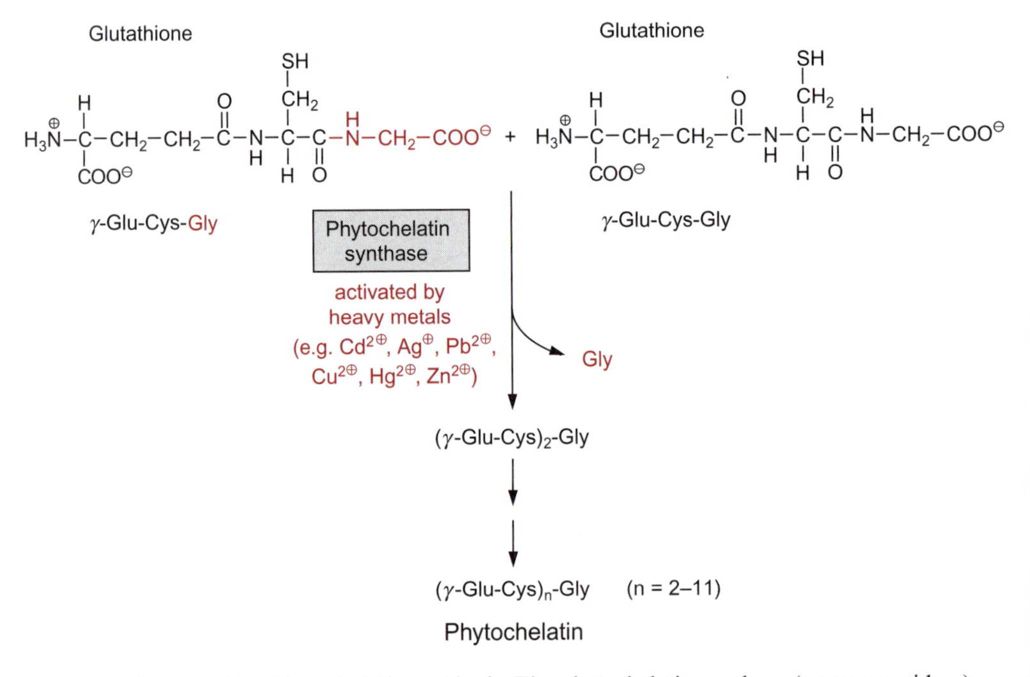

Figure 12.7 Phytochelatin synthesis. The phytochelatin synthase (a transpeptidase) cleaves the peptide bond between the cysteine and the glycine of a glutathione molecule and transfers the α-amino group of the glutamate residue of a second glutathione molecule to the liberated carboxyl group of the cysteine. Long chain phytochelatins are formed by repetition of this reaction.

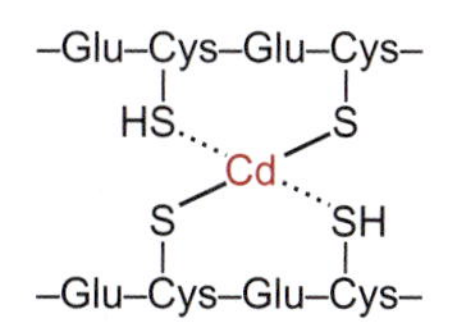

Figure 12.8 Detoxification of heavy metals by phytochelatins: Heavy metals are complexed by the thiol groups of the cysteine and thus rendered harmless.

molecule, accompanied by liberation of one glycine molecule. The repetition of this process results in the formation of chains of up to 11 Glu-Cys residues. Phytochelatins have been found in all plants investigated so far, although sometimes in a modified form as *iso*-phytochelatins, in which glycine is replaced by serine, glutamate, or β-alanine.

Phytochelatins protect plants against toxicity from heavy metals and are storage compounds for Cu^{++} and Zn^{++} . Through the thiol groups of the cysteine residues, they form tight complexes with metal ions such as Cd^{++}, Ag^{+}, Pb^{++}, Cu^{++}, Hg^{++}, and Zn^{++} as well as the nonmetal As^{3+} (Fig. 12.8). The phytochelatin synthase present in the cytosol is activated by the ions of at least one of the heavy metals listed previously. Thus, upon the exposure of plants to heavy metals, within a very short time the phytochelatins required for detoxification are synthesized *de novo* from glutathione. Exposure to heavy metals can therefore lead to a dramatic fall in the glutathione reserves in the cell. The phytochelatins loaded with heavy metals are pumped, in a similar manner to the glutathione conjugates, at the expense of ATP into the vacuoles. Because of the acidic environment in the vacuole, the heavy metals are liberated from the phytochelatins and finally deposited there as sulfides. Phytochelatins are essential to protect plants against heavy metal poisoning. Mutants of *Arabidopsis* have been found with a defect in the phytochelatin synthase, which showed an extreme sensitivity to Cd^{++}.

The capacity of plants to sequester heavy metal ions by binding them to phytochelatins has been utilized in recent times to detoxify soils polluted with heavy metals. On such soils plants are grown, which by breeding or genetic engineering have a particularly high capacity for heavy metal uptake by the roots and of phytochelatin biosynthesis, and in this way are able to extract heavy metals from the polluted ground. This procedure, termed **phytoremediation**, may have a great future, since it is much less costly than other methods to remediate heavy metal polluted soils.

12.3 Methionine is synthesized from cysteine

Cysteine is a precursor for **methionine**, another sulfur-containing amino acid. *O*-Phosphohomoserine, which has already been mentioned as an inter-

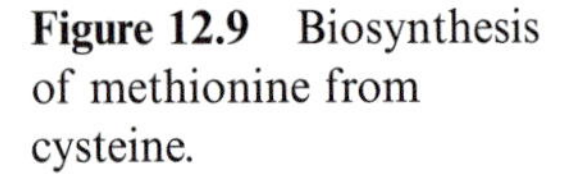

Figure 12.9 Biosynthesis of methionine from cysteine.

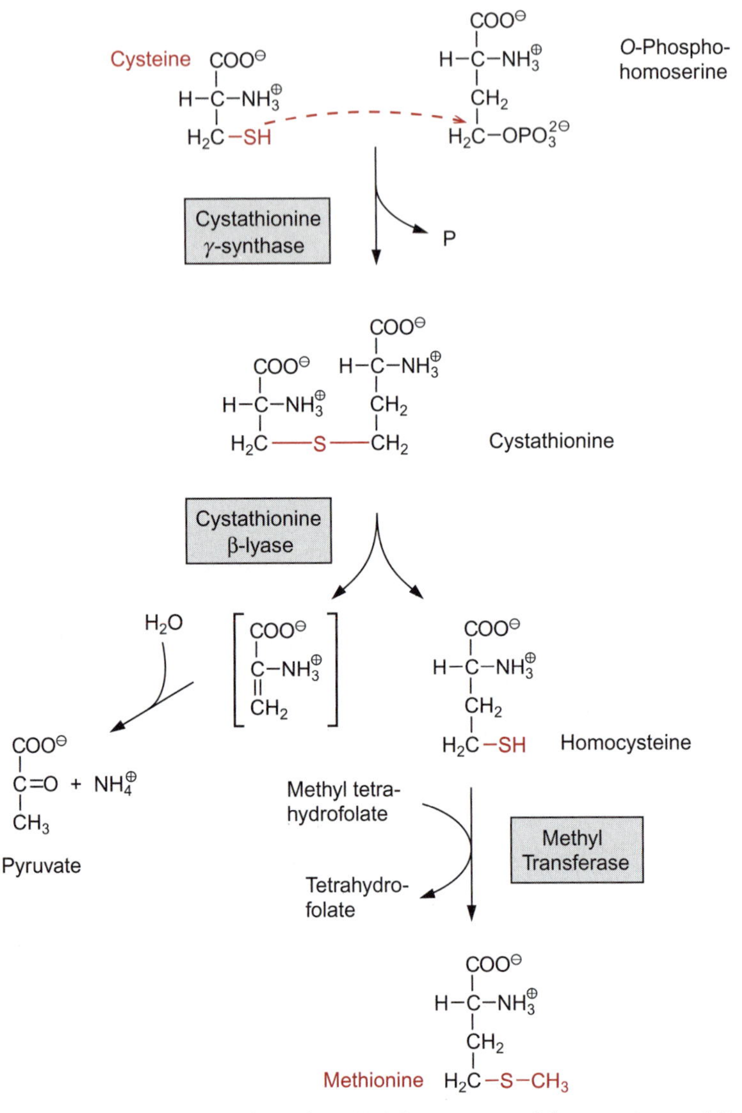

mediate of threonine synthesis (Fig. 10.14), reacts with cysteine, while a phosphate group is liberated to form cystathionine (Fig. 12.9). The thioether is cleaved by **cystathionine-β-lyase** to form homocysteine and an unstable enamine, which spontaneously degrades into pyruvate and NH_4^+. The sulfhydryl group of homocysteine is methylated by methyltetrahydrofolate (**methyl-THF**) (see Fig. 7.6), and thus the end product methionine is formed.

S-Adenosylmethionine is a universal methylation reagent

The origin of the methyl group provided by tetrahydrofolic acid (THF) is not clear. It is possible that it is derived from formate molecules, reacting in

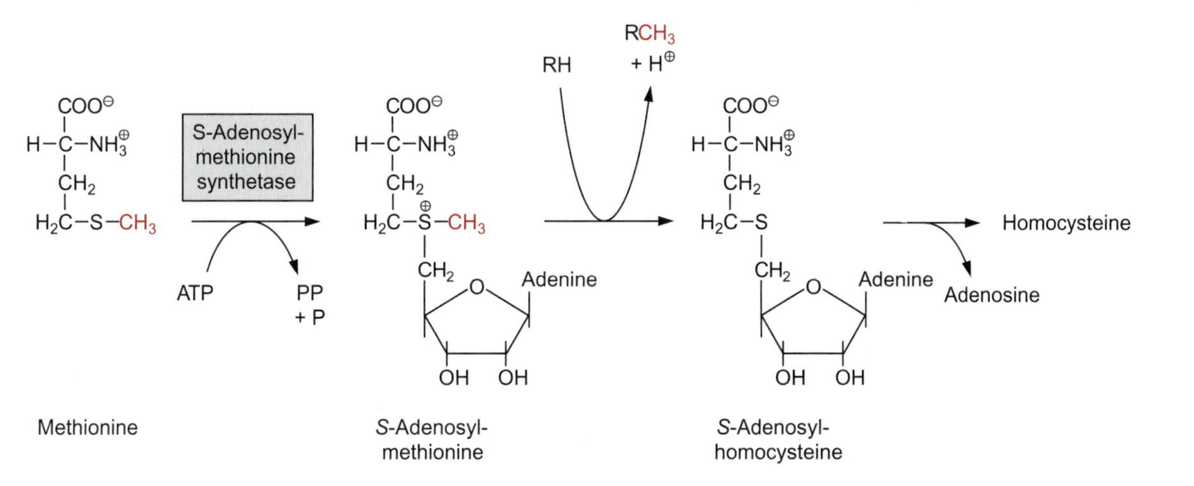

Figure 12.10 *S*-Adenosylmethionine formed from methionine and ATP is a methylating agent.

an ATP-dependent reaction with THF to form formyl-THF, which is reduced by two molecules of NADPH to methyl-THF. Methyl-THF has only a low methyl transfer potential. ***S*-Adenosylmethionine**, however, has a more general role as a methyl donor. It is involved in the methylation of nucleic acids, proteins, carbohydrates, membrane lipids, and many other substances and can therefore be regarded as a universal methylating agent of the cell.

S-Adenosylmethionine is formed by the transfer of an adenosyl residue from ATP to the sulfur atom of methionine, with the release of phosphate and pyrophosphate (Fig. 12.10). The methyl group to which the positively charged *S*-atom is linked is activated and can thus be transferred by corresponding methyl transferases to other acceptors. The remaining *S*-adenosylhomocysteine is hydrolyzed to adenosine and homocysteine and from the latter methionine is recovered by reduction with **methyltetrahydrofolate** (Fig. 12.9). *S*-adenosylmethionine is a precursor for the synthesis of the phytohormone ethylene (section 19.7).

12.4 Excessive concentrations of sulfur dioxide in air are toxic for plants

Sulfur dioxide in the air, which is formed in particularly high amounts during the smelting of ores containing sulfur, but also during the combustion of fossil fuel, can cover the total nutritional sulfur requirement of a plant. In higher concentrations, however, it leads to dramatic damage in plants. Gaseous SO_2 is taken up via the stomata into the leaves, where it is converted to sulfite:

$$SO_2 + OH^- \rightarrow SO_3^{2-} + H^+$$

Plants possess protective mechanisms for removing the sulfite, which has been formed in the leaves. In one of these, sulfite is converted by the sulfite reductase, discussed in section 12.1, to hydrogen sulfide and then further into cysteine. Cysteine formed in increasing amounts can be converted to glutathione. Thus one often finds an accumulation of glutathione in the leaves of SO_2-polluted plants. Excessive hydrogen sulfide can leak out of the leaves through the stomata, although only in small amounts. Alternatively, sulfite can be oxidized, possibly by peroxidases in the leaf, to sulfate. Since this sulfate cannot be removed by transport from the leaves, it is finally deposited in the vacuoles of the leaf cells as K^+ or Mg^{++}-sulfate. When the deposit site is full, the leaves are abscised. This explains in part the toxic effect of SO_2 on pine trees: The early loss of the pine needles of SO_2-polluted trees is to a large extent due to the fact that the capacity of the vacuoles for the final deposition of sulfate is exhausted. In cation-deficient soils, the high cation demand for the final deposition of sulfate can lead to a serious K^+ or Mg^{++} deficiency in leaves or pine needles. The bleaching of pine needles, often observed during SO_2 pollution, is partly attributed to a decreased availability of Mg^{++} ions.

Further reading

Clemens, S., Palmgren, M. G., Kraemer, U. A long way ahead: Understanding and engineering plant metal accumulation. Trends Plant Sci 7, 309–315 (2002).

Cobbett, C., Goldsbrough, P. Phytochelatins and metallothioneins: Roles in heavy metal detoxification and homeostasis. Annu Rev Plant Biol 53, 159–182 (2002).

Cole, J. O. D., Blake-Kalff, M. M. A., Davies, T. G. E. Detoxification of xenobiotics by plants: Chemical modification and vacuolar compartmentation. Trends Plant Sci 2, 144–151 (1997).

Dixon, D. P., Cummins, I., Cole, D. J., Edwards, R. Glutathione-mediated detoxification systems in plants. Plant Biol 1, 258–266 (1998).

Foyer, C. H., Theodoulou, F. L., Delrot, S. The functions of inter- and intracellular glutathione transport systems in plants. Trends Plant Sci 6, 486–492 (2001).

Heber, U., Kaiser, W., Luwe, M., Kindermann, G., Veljovic-Iovanovic, S., Yin, Z-H., Pfanz, H., Slovik, S. Air pollution, photosynthesis and forest decline. Ecol Stud 100, 279–296 (1994).

Hesse, H., Hoefgen, R. Molecular aspects of methionine biosynthesis. Trends Plant Sci 8, 259–262 (2003).

Higgins, C. F., Linton, K. J. The xyz of ABC transporters. Science 293, 1782–1784 (2001).

Howden, R., Goldsbrough, C. R., Anderson, C. R., Cobbett, C. S. Cadmium-sensitive, *cad1* mutants of *Arabidopsis thaliana* are phytochelatin deficient. Plant Physiol 107, 1059–1066 (1995).

Hung, L.-W., Wang, I. X., Nikaido, K., Liu, P.-Q., Ames, G. F.-L., Kim, S.-H. Crystal structure of the ATP-binding subunit of an ABC transporter. Nature 396, 703–707 (1998).

Kreuz, K., Tommasini, R., Martinoia, E. Old enzymes for a new job: Herbicide detoxification in plants. Plant Physiol 111, 349–353 (1996).

Leustek, T., Martin, M. N., Bick, J.-A., Davies, J. P. Pathways and regulation of sulfur metabolism revealed through molecular and genetic studies. Annu Rev Plant Physiol Plant Mol Biol 51, 141–165 (2000).

Ma, L. Q., Komart, K. M., Tu, C., Zhang, W., Cai, Y., Kennelley, E. D. A fern that hyperacculmulates arsenic. Nature 409, 579 (2001).

Saito, K. Regulation of sulfate transport and synthesis of sulfur-containing amino acids. Curr Opin Plant Biol 3, 188–195 (2000).

Zenk, M. H. Heavy metal detoxification in higher plants. Gene 179, 21–30 (1996).

13

Phloem transport distributes photoassimilates to the various sites of consumption and storage

This chapter deals with the export of photoassimilates from the leaves to the other parts of the plant. Besides having the xylem as a long-distance translocation system for transport from the root to the leaves, plants have a second long-distance transport system, the phloem, which exports the photoassimilates formed in the leaves to wherever they are required. The **xylem** and **phloem** together with the parenchyma cells form **vascular bundles** (Fig. 13.1). The xylem (*xylon*, Greek for wood) consists of lignified tubes, which translocate water and dissolved mineral nutrients from the root via the transpiration stream (section 8.1) to the leaves. Several translocation vessels arranged mostly on the outside of the vascular bundles make up the phloem (*phloios*, Greek for bark), which transports photoassimilates from the site of formation (**source**) (e.g., the mesophyll cell of a leaf) to the sites of consumption or storage (**sink**) (e.g., roots, tubers, fruits, or areas of growth). The phloem system thus connects the sink and source tissues.

The phloem contains elongated cells, joined by **sieve plates**, the latter consisting of diagonal cell walls perforated by pores. The single cells are called **sieve elements** and their longitudinal arrangement is called the **sieve tube** (Fig. 13.2). The pores of the sieve plate are widened plasmodesmata lined with **callose** (section 9.6). The sieve elements can be regarded as living cells that have lost their nucleus, Golgi apparatus, and vacuoles, and contain only a few mitochondria, plastids, and some endoplasmic reticulum. The absence of many cell structures normally present in a cell specializes the sieve tubes for the long-distance transport of carbon- and nitrogen-containing metabolites and of various inorganic and organic compounds. In most plants sucrose is the main transport form for carbon, but some plants also trans-

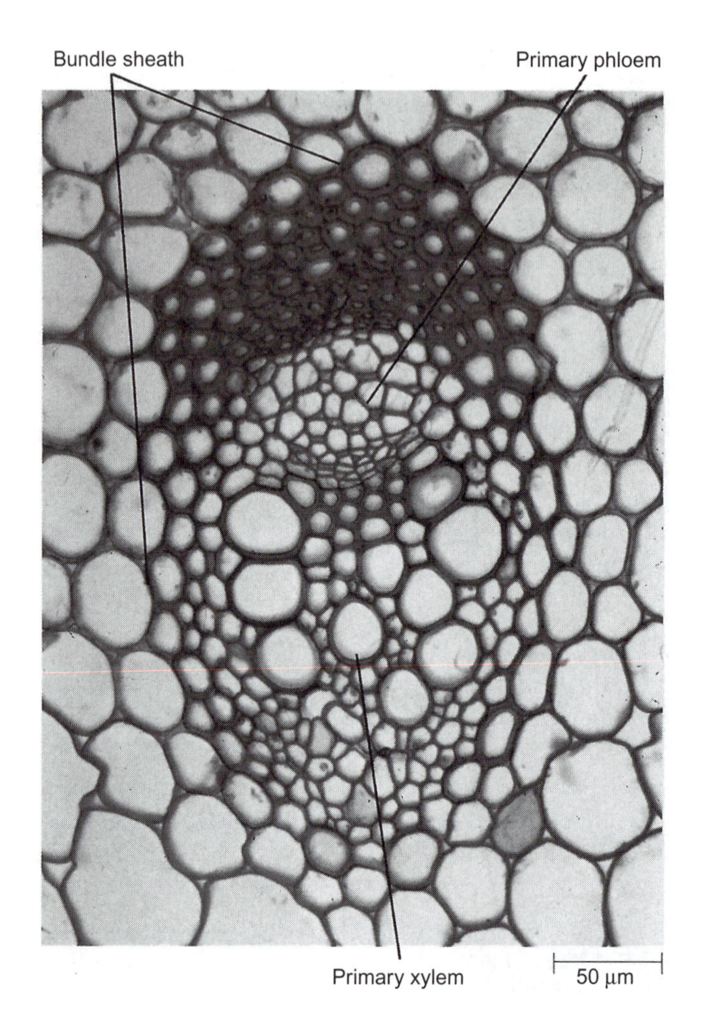

Figure 13.1 Transverse section through a vascular bundle of *Ranunculus* (buttercup), an herbaceous dicot plant. The phloem and xylem are surrounded by bundle sheath cells. (From Raven, Evert, and Curtis, Biologie der Pflanzen, De Gruiter Verlag, Berlin, by permission.)

port oligosaccharides from the raffinose family, or sugar alcohols (section 9.4). Nitrogen is transported in the sieve tubes almost exclusively in the organic form as amino acids. Organic acids, nucleotides, proteins, and phytohormones are present in the phloem sap in lower concentrations. In addition to these organic substances, the sieve tubes transport inorganic ions, mainly K^+ ions.

Companion cells, adjacent to the sieve elements of angiosperms, contain all the constituents of a normal living plant cell, including the nucleus and many mitochondria. Sieve elements and companion cells have developed from a common precursor cell and are connected to each other by numerous **plasmodesmata** (section 1.1). They are an important element of phloem loading. Depending on the kind of phloem loading, the companion cells are named **transfer cells** or **intermediary cells**.

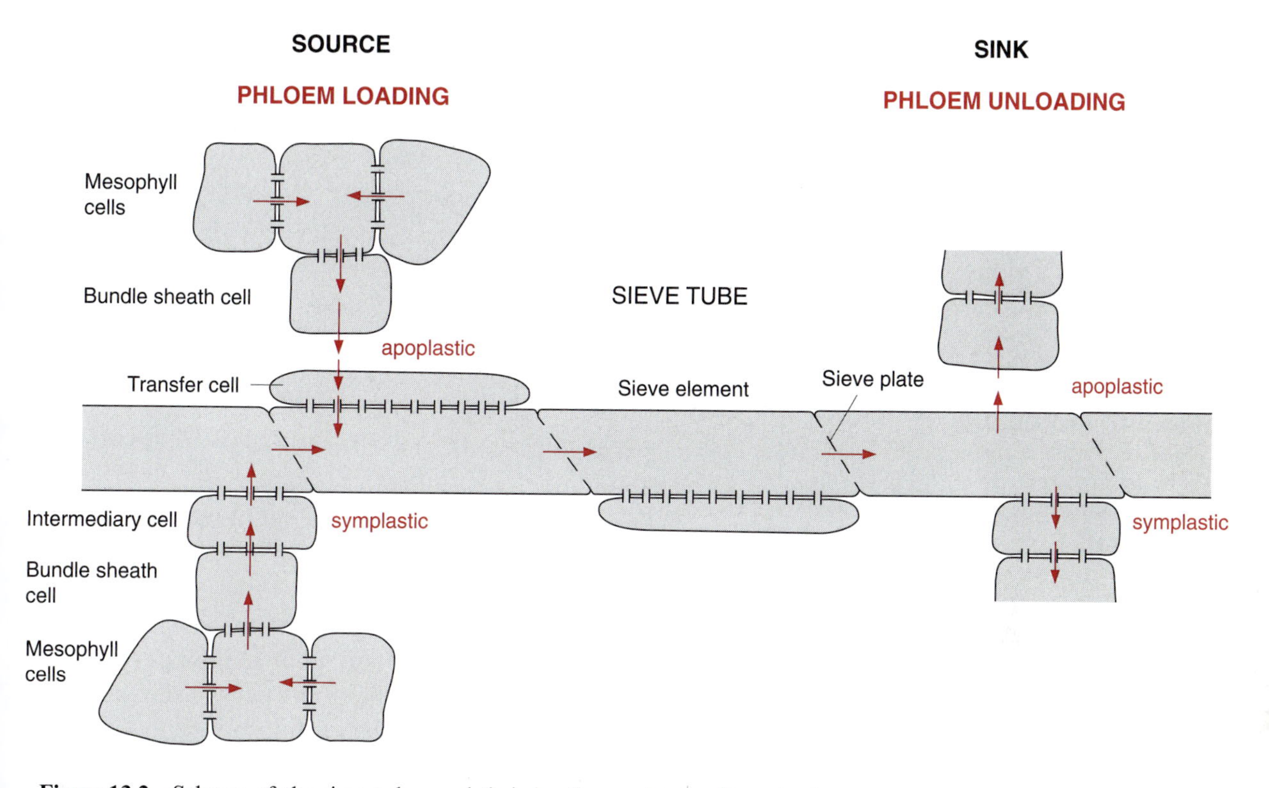

Figure 13.2 Scheme of the sieve tubes and their loading and unloading via the apoplastic and symplastic pathways. The plasmodesmata indicated by the double line allow unhindered diffusion of sugar and amino acids. The structures are not shown to scale. The companion cells participating in apoplastic loading are also called transfer cells. Intermediary cells are specialized companion cells involved in symplastic phloem loading.

13.1 There are two modes of phloem loading

Photoassimilates generated in the mesophyll cells, such as sucrose and various oligosaccharides as well as amino acids, diffuse via plasmodesmata to the **bundle sheath cells.** The further transport of photoassimilates from the bundle sheath cells to the sieve tubes can occur in two different ways:

1. Especially in those plants in which oligosaccharides from the raffinose family (section 9.4) are translocated in the sieve tubes (e.g., squash plants), the bundle sheath cells are connected to specialized companion cells, named **intermediary cells**, and further to the sieve tubes, by a large number of plasmodesmata. Therefore, in these plants the transfer of the photoassimilates to the sieve tubes via plasmodesmata is termed **symplastic phloem loading**.

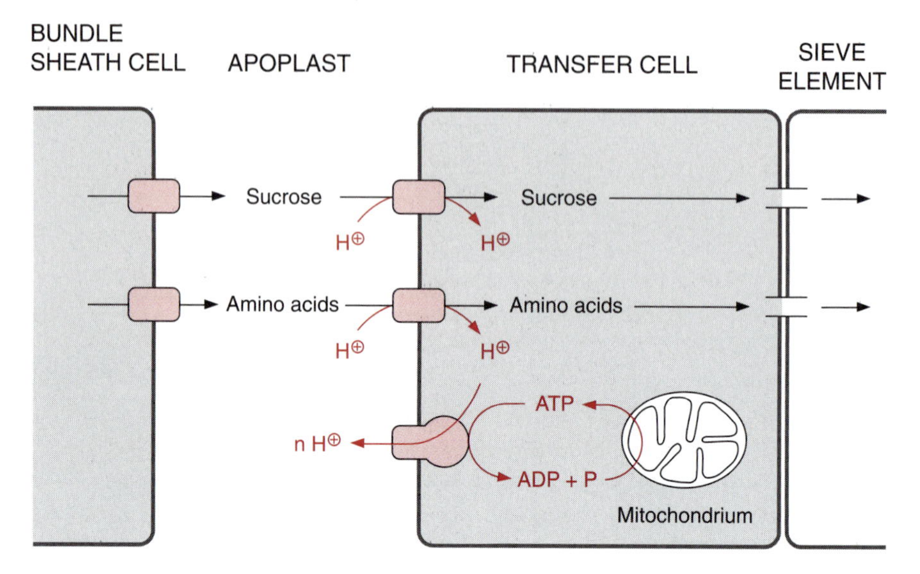

Figure 13.3 Apoplastic phloem loading. Transfer of the photoassimilates from the bundle sheath cells to the sieve tubes. Many observations indicate that active loading takes place in the plasma membrane of the transfer cells and that the subsequent transfer to the sieve elements occurs by diffusion via plasmodesmata. However, recent results indicate that part of the assimilates also can be taken up directly from the apoplasts into the sieve elements.

2. In contrast, in **apoplastic** phloem loading, found for instance in the leaves of cereals, sugar beet, rapeseed, and potato, photoassimilates are first transported from the source cells via the bundle sheath cells to the extracellular compartment, the **apoplast,** and then by active transport into the sieve tube compartment (Fig. 13.3). The translocators mediating the export from the bundle sheath to the apoplast have not yet been characterized. Since the concentration of sucrose and of amino acids in the source cells is very much higher than in the apoplast, this export requires no energy input. The companion cells participating in the apoplastic phloem loading are termed **transfer cells**. The transport of sucrose and amino acids from the apoplasts to the phloem proceeds via a proton symport (Fig. 13.3). This is driven by a proton gradient between the apoplast and the interior of the companion cells and the sieve tubes. The proton gradient is generated by a **H^+-P ATPase** (section 8.2) present in the plasma membrane. The required ATP is produced by mitochondrial oxidation. The **H^+-sucrose translocator** involved in phloem loading has been identified and characterized in several plants: in the vascular bundles of *Plantago major* using specific antibodies, an H^+-sucrose translocator has been localized in the plasma membrane of transfer cells. (Fig. 13.3). On the other hand, an H^+-sucrose translocator has been identified in the plasma membrane of the sieve elements of potato leaves. The substrates for mitochondrial respiration are provided by degradation of sucrose via sucrose synthase (see also Fig. 13.5) to hexose phosphates, which are further degraded by glycolytic metabolism. Another substrate for respiration is glutamate (section 5.3). In many plants this amino acid is present in relatively high concentrations in the phloem sap.

Whereas in the plants with apoplastic phloem loading investigated so far, sucrose is the exclusive transport form for carbohydrates (hexoses are not transported), no special transport form exists for amino nitrogen. In principle, all protein-building amino acids are transported. The proportion of the single amino acids related to the sum of amino acids is very similar in the phloem sap to that in the source cells. The amino acids most frequently found in the phloem sap are glutamate, glutamine, and aspartate, but alanine also is found in some plants. Several amino acid translocators with a broad specificity for various amino acids were identified, which are presumed to participate in the phloem loading.

13.2 Phloem transport proceeds by mass flow

The proton-substrate-co-transport results in very high concentrations of sucrose and amino acids in the sieve tubes. Depending on the plant and on growth conditions, the concentration of sucrose in the phloem sap amounts to 0.6 to 1.5 mol/L, and the sum of the amino acids ranges from 0.05 to 0.5 mol/L. **Aphids** turned out to be useful helpers for obtaining phloem sap samples for such analyses. An aphid, after some attempts, can insert its stylet exactly into a sieve tube. As the phloem sap is under pressure, it flows through the tube of the stylet and is consumed by the aphid (Fig. 13.4). The

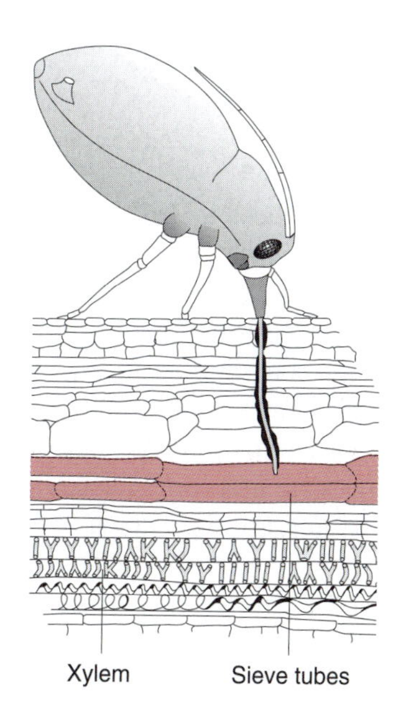

Figure 13.4 Aphids know where to insert their stylet into the sieve tubes and feed themselves in this way from the exuding phloem sap. (Figure by A. F. G. Dixon, Encyclopaedia of Plant Physiology, Vol. 1, Springer-Verlag, by permission.)

aphid takes up more sucrose than it can metabolize and excretes the surplus as honeydew, which is the sticky sugary layer covering aphid-infested house plants. When the stylet of a feeding aphid is severed by a laser beam, the phloem sap exudes from the sieve tube through the stump of the stylet. Although the amount of phloem sap obtained in this way is very low (0.05 to 0.1×10^{-6} L/h), modern techniques make a quantitative assay of the phloem sap in these samples possible.

In plants performing photosynthesis in the presence of radioactively labeled CO_2, phloem transport velocities of 30 to 150 cm/h have been measured. This rapid transport proceeds by **mass flow**, driven, on the one hand, by very efficient pumping of sucrose and amino acids into the sieve tubes and, on the other hand, by their withdrawal at the sites of consumption. This mass flow is driven by many transversal osmotic gradients. The surge of this mass flow carries along substances present at low concentrations, such as **phytohormones**. The direction of mass flow is governed entirely by the consumption of the phloem contents. Depending on what is required, phloem transport can proceed in an upward direction (e.g., from the mature leaf to the growing shoot or flower) or downward into the roots or storage tubers. Since the phloem sap is under high pressure and the phloem is highly branched, wounding the vascular tissue might result in the phloem sap "bleeding". Protective mechanisms prevent this. Due to the presence of substrates in the phloem sap and the enzymes of sucrose synthase and **callose synthase**, which are probably membrane-bound, the sieve pores of damaged sieve tubes are sealed by the formation of **callose** (section 9.6), and damaged sieve tubes are thus put out of action.

13.3 Sink tissues are supplied by phloem unloading

The delivered photosynthate is utilized in the sink tissues to sustain the metabolism, but may also be deposited there as reserves, mainly in the form of starch. There are again two possibilities for phloem unloading (Fig. 13.2). In **symplastic unloading**, the sucrose and amino acids reach the cells of the sink organs directly from the sieve elements via plasmodesmata. In **apoplastic unloading**, the substances are first transported from the sieve tubes to the extracellular compartment and are then taken up into the cells of the sink organs. Electron microscopic investigations of the plasmodesmatal frequency indicate that in vegetative tissues, such as roots or growing shoots, phloem unloading proceeds primarily symplastically, whereas in storage tissues unloading is often, but not always, apoplastic.

Starch is deposited in plastids

In storage tissues, the delivered carbohydrates are mostly converted to starch and stored as such. In apoplastic phloem unloading, this may proceed by two alternative pathways. In the pathway colored red in Figure 13.5, the sucrose is taken up from the apoplast into the storage cells and converted there via **sucrose synthase** and **UDP-glucose-pyrophosphorylase** to fructose and glucose 1-phosphate. In this reaction, pyrophosphate is consumed and UTP is generated. It is still unresolved how the necessary pyrophosphate is formed. **Phosphoglucomutase** converts glucose 1-phosphate to glucose 6-phosphate. Alternatively, the enzyme **invertase** first hydrolyzes sucrose in the apoplast to glucose and fructose, and these two hexoses are then transported into the cell. This pathway is colored black in Figure 13.5. A **fructokinase** and a **hexokinase** (the latter phosphorylating mannose as well as glucose) catalyze the formation of the corresponding hexose phosphates. Glucose 6-phosphate is transported via the **glucose 6-phosphate-phosphate translocator** (see section 8.2) in counter-exchange for phosphate to the amyloplast, where starch is formed via the synthesis of ADP-glucose (section 9.1). Some leucoplasts transport glucose 1-phosphate in counter-exchange for phosphate. In potato tubers, the storage of starch probably proceeds mainly via sucrose synthase. In the taproots of sugar beet, the carbohydrates are stored as sucrose in the vacuoles. In some fruits (e.g., grapes), carbohydrates are stored in the vacuole as glucose.

The glycolysis pathway plays a central role in the utilization of carbohydrates

The carbohydrates delivered by phloem transport to the sink cells are fuel for the energy metabolism and also a carbon source for the synthesis of the cell matter. The **glycolysis pathway,** which is present at least in part in almost all living organisms, has a fundamental role in the utilization of carbohydrates. The enzymes of this pathway not only occur in sink tissues but also are present in all plant cells. Each cell has two sets of glycolytic enzymes, one in the **cytosol** and one in the **plastids**. Some of the plastidic enzymes participate in the Calvin cycle, as discussed in Chapter 6. In the plastids of some plants, the glycolysis pathway is incomplete because one or two enzymes are lacking. The corresponding glycolytic enzymes in the cytosol and in the plastids are isoenzymes encoded by different genes.

Figure 13.6 depicts the glycolysis pathway in the cytosol. Glucose 6-phosphate, deriving from either the degradation of sucrose (Fig. 13.5) or the degradation of starch (Fig. 9.12), is converted in a reversible reaction by **hexose phosphate isomerase** to fructose 6-phosphate. This reaction proceeds

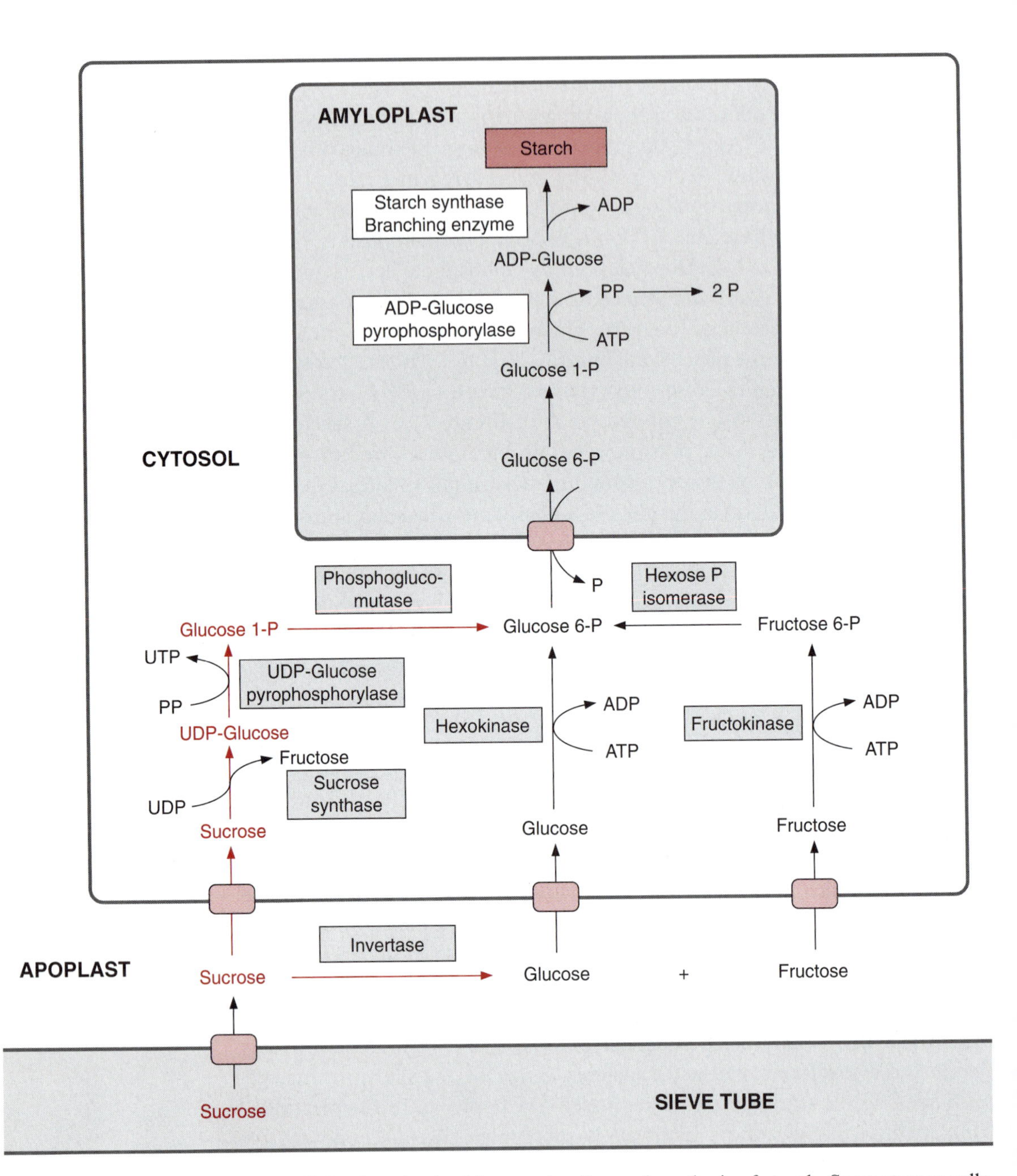

Figure 13.5 Apoplastic phloem unloading and synthesis of starch. Some storage cells take up sucrose, whereas others take up glucose and fructose, which have been formed by the hydrolysis of sucrose catalyzed by invertase. It is not yet known whether glucose and fructose are transported by the same or by different translocators. For details see section 9.1. Some amyloplasts transport glucose-1-phosphate in exchange for phosphate.

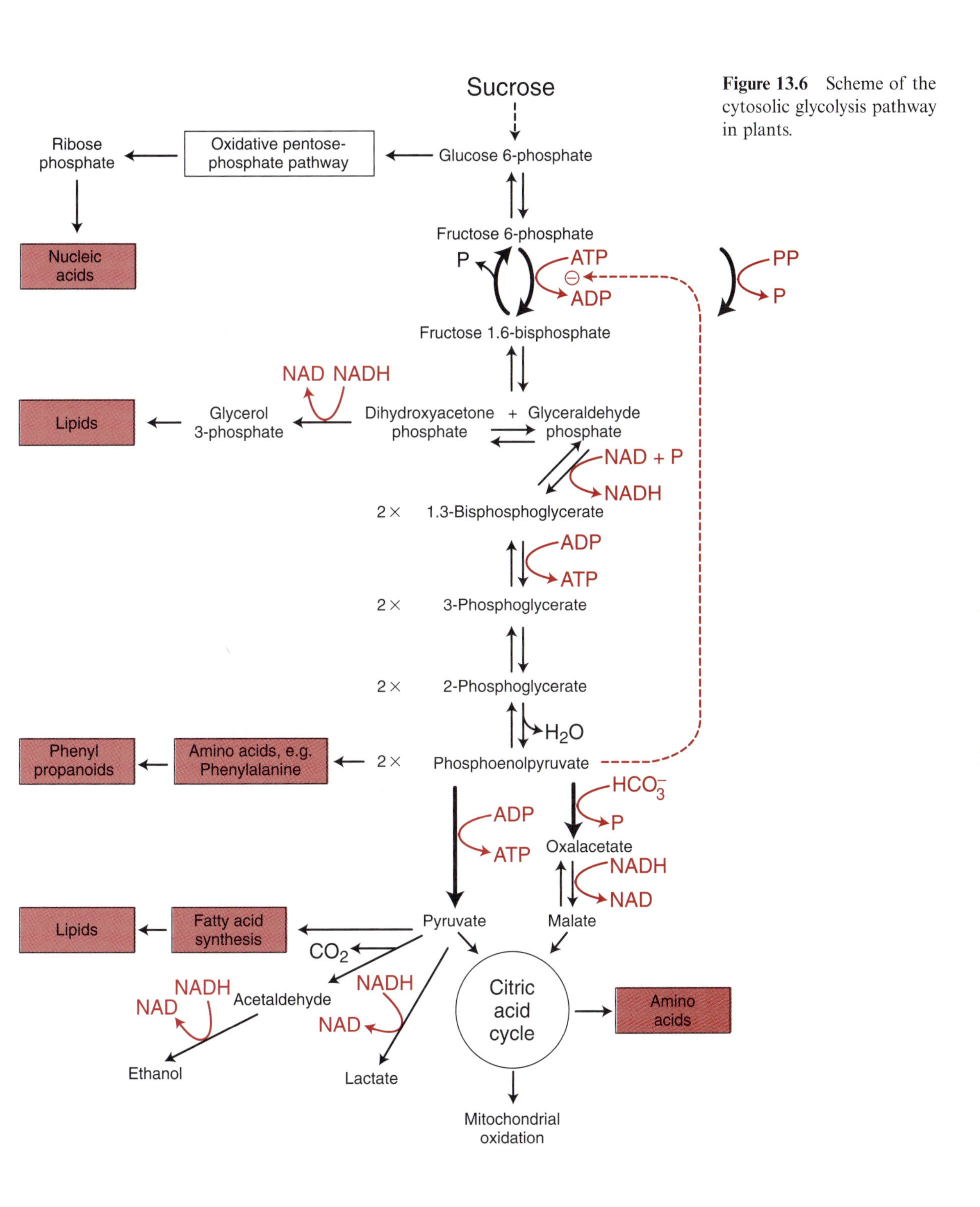

Figure 13.6 Scheme of the cytosolic glycolysis pathway in plants.

in analogy to the isomerization of ribose 5-phosphate (Fig. 6.18). Fructose 6-phosphate is phosphorylated by ATP to fructose 1.6-bisphosphate, as catalyzed by **ATP-phosphofructokinase.** Alternatively, it is also phosphorylated by inorganic pyrophosphate via **pyrophosphate-phosphofructokinase**. The latter enzyme does not occur in plastids and its physiological function is not yet clear. Since in both reactions the free energy for the hydrolysis of the anhydride phosphate donor is much higher than that of the formed phosphate ester, the formation of fructose 1.6-bisphosphate is an irreversible process. For this reason, the conversion of fructose 1.6-bisphosphate to fructose 6-phosphate proceeds via another reaction, namely, the hydrolysis of phosphate, as catalyzed by **fructose 1-6-bisphosphatase** (Figs. 6.15, 9.13). Fructose 1.6-bisphosphate is split in a reversible reaction into glyceraldehyde phosphate and dihydroxyacetone phosphate as catalyzed by **aldolase** (Fig. 6.14). Dihydroxyacetone phosphate is converted to glyceraldehyde phosphate by **triose isomerase**, again in analogy to the isomerization of ribose 5-phosphate (Fig. 6.18).

In the reaction sequence of the glycolysis pathway discussed so far, the hexose phosphate was prepared for the generation of reducing equivalents and of ATP. Glyceraldehyde phosphate is oxidized in a reversible reaction by **glyceraldehyde phosphate dehydrogenase** to 1.3-bisphosphoglycerate yielding the reduction of NAD. This reaction has already been discussed, although in the opposite direction, as part of the Calvin cycle in Figures 6.9 and 6.10. The change in free energy during the oxidation of the aldehyde to a carboxylate is conserved to form a phosphate anhydride, and by the reversible conversion of 1.3 bisphosphoglycerate to 3 phosphoglycerate, as catalyzed by **phosphoglycerate kinase,** this is utilized for the synthesis of ATP (Fig. 6.9). In order to prepare the remaining phosphate group for the synthesis of ATP, 3-phosphoglycerate is first converted by **phosphoglycerate mutase** to 2-phosphoglycerate (in analogy to the phosphogluco mutase reaction, Fig. 9.6) and then H_2O is split off in a reversible reaction catalyzed by **enolase,** yielding phosphoenol pyruvate. In this way a phosphate ester is converted to an enol ester, of which the free energy of hydrolysis is considerably higher than that of the anhydride bond of ATP. Therefore the subsequent conversion of phosphoenolpyruvate to pyruvate coupled to the phosphorylation ADP by **pyruvate kinase** is an irreversible reaction. Alternatively, phosphoenolpyruvate can be converted in the cytosol via **PEP carboxylase** (Fig. 8.5) to oxaloacetate, and the latter can be reduced by **malate dehydrogenase** (Fig. 5.9) to malate. Malate can be converted to pyruvate by malic enzyme, as described in Figure 8.10. Both pyruvate and malate can be fed into the citrate cycle of the mitochondria for the generation of ATP via the respiratory chain.

Under usual aerobic conditions, the glycolysis pathway makes only a minor contribution to the energy demand of a cell. The conversion of glucose 6-phosphate to pyruvate produces just **three molecules** of **ATP**. In contrast, mitochondrial oxidation of pyruvate and of the NADH formed by glyceraldehyde phosphate dehydrogenase yields about **25 molecules** of **ATP** per glucose 6-phosphate (see section 5.6). But in the absence of oxygen, which may occur when roots are flooded or during imbibition of water by germinating seeds, the ATP production by the glycolysis pathway is crucial for maintaining a minimal metabolism. In such a situation, the NADH generated in the glycolysis pathway can be reoxidized by the reduction of pyruvate to lactate, as catalyzed by **lactate dehydrogenase**. In roots and developing seeds, this enzyme is induced by oxygen deficit. The lactate formed is excreted as **lactic acid**. Alternatively, the NADH produced by glycolysis can be consumed in converting pyruvate to **ethanol**, in the same way as in ethanol fermentation by yeast. In this reaction, pyruvate is first decarboxylated by **pyruvate decarboxylase** to acetaldehyde, involving thiamine pyrophosphate as cofactor, similar to Figure 5.4. Subsequently, **alcohol dehydrogenase** catalyzes the reduction of acetaldehyde to ethanol, which is excreted. In plant cells, the activity of alcohol dehydrogenase is largely increased as a response to oxygen deficit. In most plants, ethanol is the main product of anaerobic metabolism, with smaller amounts of lactic acid formed.

Apart from the generation of ATP, the glycolysis pathway provides **precursors for a multitude of cell components**. Here are a few examples: The oxidative pentose phosphate pathway (Fig. 6.21) yields ribose 5-phosphate as a precursor for the synthesis of nucleotides and nucleic acids. The reduction of dihydroxyacetone phosphate by **glycerol phosphate dehydrogenase** yields glycerol 3-phosphate, a precursor for the synthesis of lipids. Phosphoenolpyruvate is the precursor of a number of amino acids (e.g., phenylalanine), which is the precursor for the synthesis of phenylpropanoids, such as lignin (Fig. 18.9) and tannin (Fig. 18.16). Pyruvate is the precursor for the synthesis of fatty acids (Fig.15.7) and hence the synthesis of lipids.

The glycolysis pathway is governed by a very complex regulation, particularly as most of the enzymes are located in the cytosol as well as in the plastids. In both compartments, the phosphorylation of fructose 6-phosphate by **ATP-phosphofructokinase** is inhibited by **phosphoenolpyruvate**, allowing a feedback control of the glycolysis pathway. The phosphorylation by the **pyrophosphate-dependent phosphofructokinase** is activated by **fructose 2.6-bisphosphate**, whereas the hydrolysis of fructose 1.6-bisphosphate by **fructosebisphosphatase** is inhibited by this substance (see Fig. 9.15).

Further reading

Eschrich, W. Funktionelle Pflanzenanatomie. Springer-Verlag, Berlin-Heidelberg (1995).

Fischer, W. N., André, B., Reutsch, D., Krolkiewicz, S., Tegeder, M., Breitkreuz, K., Frommer, W. B. Amino acid transport in plants. Trends Plant Sci 3, 188–195 (1998).

Givan, C. V. Evolving concepts in plant glycolysis: Two centuries of progress. Biol Rev (Cambridge) 74, 277–309 (1999).

Herschbach C., Rennenberg, H. Significance of phloem-translocated organic sulfur compounds for the regulation of sulfur nutrition. Prog Bot 62, 177–193 (2001).

Knop, C., Voitsekhovskaja, O., Lohaus, G. Sucrose transporters in two members of the *Scrophulariaceae* with different types of transport sugar. Planta 213, 80–91 (2001).

Kuhn, C., Barber L., Bürkle, L., Frommer, W.-B. Update on sucrose transport in higher plants. J Exp Bot 50, 935–953 (1999).

Kuehn, C. A comparison of the sucrose transporter systems of different plant species. Plant Biol 5 215–232 (2003).

Lohaus, G., Winter, H., Riens, B., Heldt, H. W. Further studies of the phloem loading process in leaves of barley and spinach. The comparison of metabolite concentrations in the apoplastic compartment with those in the cytosolic compartment and in the sieve tubes. Botanica Acta 108, 270–275 (1995).

Lohaus, G., Fischer, K. Intracellular and intercellular transport of nitrogen and carbon. In: C. H. Foyer and G. Noctor (eds.) Advances in Photosynthesis: Photosynthetic assimilation and associated carbon metabolism, pp. 239–263, Kluwer Academic Publishers, Dordrecht, The Netherlands, (2002).

Oparka, K. J., Santa Cruz, S. The great escape: Phloem transport and unloading of macromolecules. Annu Rev Plant Physiol Plant Mol Biol 51, 323–347 (2000).

Plaxton, W. C. The organization and regulation of plant glycolysis. Annu Rev Plant Physiol Plant Mol Biol 47, 185–214 (1996).

Ruiz-Medrano, R., Xoconostle-Cázares, B., Lucas, W. J. The phloem as a conduit for inter-organ communication. Curr Opin Plant Biol 4, 202–209 (2001).

Sjolund, R. D. The phloem sieve element: A river runs through it. Plant Cell 9, 1137–1146 (1997).

Stadler, R., Brandner, J., Schulz, A., Gahrtz, M., Sauer, N. Phloem loading by the PmSUC2 sucrose carrier from *Plantago major* occurs in companion cells. Plant Cell 7, 1545–1554 (1995).

Stitt, M. Manipulation of carbohydrate partitioning. Curr Biol 5, 137–143 (1994).

Thompson, G. A., Schulz, A. Macromolecular trafficking in the phloem. Trends Plant Sci 4, 354–360 (1999).

Turgeon, R. Phloem loading and plasmodesmata. Trends Plant Sci 1, 418–423 (1996).

van Bel, A. J. E., Ehlers, K., Knoblauch, M. Sieve elements caught in the act. Trends Plant Sci 7, 126–132 (2002).

Williams, L. E., Lemoine, R., Sauer N. Sugar transporters in higher plants—a diversity of roles and complex regulation. Trends Plant Sci 5, 283–290 (2000).

Zimmermann, M. H., Milburn, J. A. (eds.). Phloem Transport. Encyclopaedia of Plant Physiology, Vol. 1, Springer-Verlag, Berlin–Heidelberg, (1975).

14

Products of nitrate assimilation are deposited in plants as storage proteins

Whereas the products of CO_2 assimilation are deposited in plants in the form of oligo- and polysaccharides, as discussed in Chapter 9, the amino acids formed as products of nitrate assimilation are stored as proteins. These are mostly special **storage proteins**, which have no enzymatic activity and are often deposited in the cell within **protein bodies**. Protein bodies are enclosed by a single membrane and are derived from the endomembrane system of the endoplasmic reticulum and the Golgi apparatus or the vacuoles. In potato tubers, storage proteins are also stored in the vacuole.

Storage proteins can be deposited in various plant organs, such as leaves, stems, and roots. They are stored in seeds and tubers and also in the cambium of tree trunks during winter to enable the rapid formation of leaves during seed germination and sprouting. Storage proteins are located in the endosperm in cereal seeds and in the cotyledons of most of the legume seeds. Whereas in cereals the protein content amounts to 10% to 15% of the dry weight, in some legumes (e.g., soybean) it is as high as 40% to 50%. About 85% of these proteins are storage proteins.

Globally, about 70% of the human demand for protein is met by the consumption of seeds, either directly, or indirectly by feeding them to animals for meat production. Therefore plant storage proteins are the basis for human nutrition. However, in many plant storage proteins, the content of certain amino acids essential for the nutrition of humans and animals is too low. In cereals, for example, the storage proteins are deficient in **threonine**, **tryptophan**, and particularly **lysine**, whereas in legumes there is a deficiency of **methionine**. Since these amino acids cannot be synthesized by human metabolism, humans depend on being supplied with them in their food. In

Table 14.1: Some examples of plant storage proteins

Plant	Globulin	Prolamin (incl. Glutelin)	2 S-Protein
Rapeseed			Napin*
Pea, bean	Legumin, Vicilin		
Wheat, rye		Gliadin, Glutenin	
Maize		Zein	
Potato	Patatin		

* Structurally related to prolamins which, according to the solubility properties, are counted as globulins.

humans with an entirely vegetarian diet, such amino acid deficiencies can lead to irreparable physical and mental damage, especially in children. It can also be a serious problem in pig and poultry feed. A target of research in plant genetic engineering is to improve the amino acid composition of the storage proteins of harvest products.

Scientists have long been interested in plant proteins. By 1745 the Italian J. Beccari already had isolated proteins from wheat. In 1924, at the Connecticut Agricultural Experimental Station, T. B. Osborne classified plant proteins according to their solubility properties. He fractionated plant proteins into **albumins** (soluble in pure water), **globulins** (soluble in diluted salt solutions), **glutelins** (soluble in diluted solutions of alkali and acids), and **prolamins** (soluble in aqueous ethanol). When the structures of these proteins were determined later, it turned out that glutelins and prolamins were structurally closely related. Therefore, in more recent literature, glutelins are regarded as members of the group of prolamins. Table 14.1 shows some examples of various plant storage proteins.

14.1 Globulins are the most abundant storage proteins

Storage globulins occur in varying amounts in practically all plants. The most important globulins belong to the **legumin** and **vicilin** groups. Both of these globulins are encoded by a multigene family. These multigene families descend from a common ancestor. Legumin is the main storage protein of leguminous seeds. In broad bean, for instance, 75% of the total storage protein consists of legumin. Legumin is a hexamer with a molecular mass of 300 to 400 kDa. The monomers contain two different peptide chains (α, β), which are linked by a disulfide bridge. The large α-chain usually has a molecular mass of about 35 to 40 kDa, and the small β-chain has a molecular mass of about 20 kDa. Hexamers can be composed of different (α, β) monomers. Some contain methionine, whereas others do not. In the hexamer, the protein molecules are arranged in a very regular package and can be deposited in this form in the protein bodies. Protein molecules, in which some of the protein chains are not properly folded, do not fit into this package and are degraded by peptidases. Although it is relatively easy nowadays to exchange amino acids in a protein by genetic engineering, this has turned out to be difficult in storage proteins, as both the folding and the three-dimensional structure of the molecule may be altered by such exchanges. Recent progress in obtaining crystals enabled the analysis of the three-dimensional protein structure of the precursor trimers as well as of the mature storage proteins. These studies revealed that the stability of the storage proteins toward the proteases in the storage vacuoles is due to the fact that possible cleavage sites are hidden within the protein structure and in this way are protected against proteolysis.

Vicilin shows similarities in its amino acid sequence to legumin, but occurs mostly as a trimer, of which the monomers consist of only one peptide chain. Due to the lack of cysteine, the vicilin monomers are unable to form S-S bridges. In contrast to legumins, vicilins are often glycosylated; they contain carbohydrate residues, such as mannose, glucose, and *N*-acetylglucosamine.

14.2 Prolamins are formed as storage proteins in grasses

Prolamins are contained only in grasses, such as cereals. They are present as a polymorphic mixture of many different subunits of 30 to 90 kDa each. Some of these subunits contain cysteine residues and are linked by S-S bridges. Also in **glutenin**, which occurs in the grains of wheat and rye, monomers are linked by S-S bridges. The glutenin molecules differ in size. The suitability of flour for bread-making depends on the content of high molecular glutenins, and therefore flour from barley, oat, or maize lacking glutenin, is not suitable for baking bread. Since the glutenin content is a critical factor in determining the quality of bread grain, investigations are in progress to improve the glutenin content of bread grain by genetic engineering.

14.3 2S-Proteins are present in seeds of dicot plants

2S-Proteins are also widely distributed storage proteins. They represent a heterogeneous group of proteins, of which the sole definition is their sedimentation coefficient of about 2 svedberg (S). Investigations of their structure have revealed that most 2S-proteins have an interrelated structure and are possibly derived, along with the prolamins, from a common ancestor protein. **Napin**, the predominant storage protein in rapeseed, is an example of a 2S-protein. This protein is of substantial economic importance since, after the oil has been extracted, the remainder of the rapeseed is used as fodder. Napin and other related 2S-proteins consist of two relatively small polypeptide chains of 9 kDa and 12 kDa, which are linked by S-S bridges. So far, little is known about the packing of the prolamins and 2S-proteins in the protein bodies.

14.4 Special proteins protect seeds from being eaten by animals

The protein bodies of some seeds contain other proteins, which, although also acting as storage proteins, protect the seeds from being eaten. To give some examples: The storage protein **vicilin** has a defense function by binding to the chitin matrix of fungi and insects. In some insects, it interferes with the development of the larvae. The seeds of some legumes contain **lectins**, which bind to sugar residues, irrespective of whether these are free sugars or constituents of glycolipids or glycoproteins. When these seeds are consumed by animals, the lectins bind to glycoproteins in the intestine and thus interfere with the absorption of food. The seeds of some legumes and other plants also contain **proteinase inhibitors**, which block the digestion of proteins by inhibiting proteinases in the animal digestive tract. Because of their content of lectins and proteinase inhibitors, many beans and other plant products are suitable for human consumption only after denaturing by cooking. This is one reason why humans have learned to cook. Castor beans contain the extremely toxic protein **ricin**. A few milligrams of it can kill a human. Beans also contain **amylase inhibitors**, which specifically inhibit the hydrolysis of starch by amylases in the digestive tract of certain insects. Using genetic engineering, α-amylase inhibitors from beans have been successfully expressed in the seeds of pea. Whereas the larvae of the pea beetle

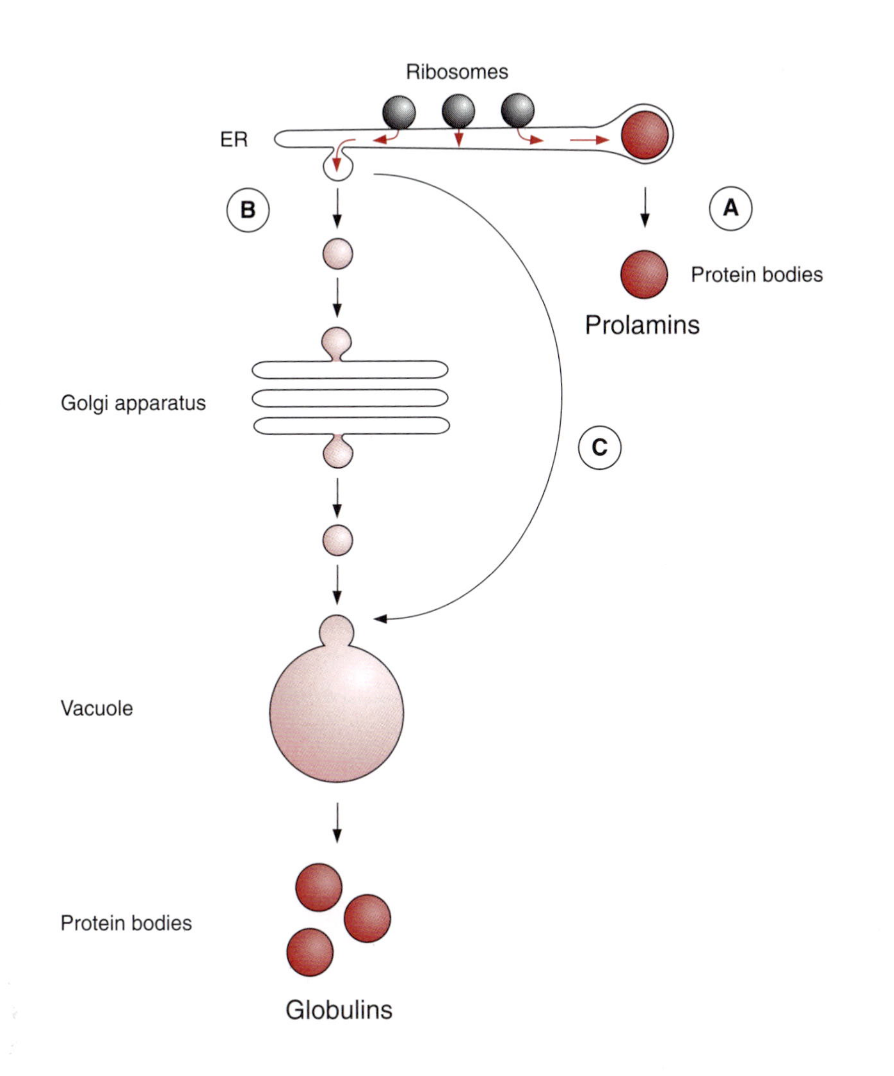

Figure 14.1 There are three ways of depositing storage proteins in protein bodies.
A. In the formation of prolamins in cereal grains, the prolamin aggregates in the lumen of the ER and the protein bodies are formed by budding off from the ER membrane.
B. The proteins appearing in the lumen of the ER are transferred via the Golgi apparatus to the vacuole. The protein bodies are formed by fragmentation of the vacuole. This is probably the most common pathway.
C. The proteins appearing in the lumen of the ER are directly transferred to the vacuole circumventing the Golgi apparatus.

normally cause large losses during storage of peas, the peas from the genetically engineered plants were protected against these losses.

14.5 Synthesis of the storage proteins occurs at the rough endoplasmic reticulum

Seed storage proteins are formed by ribosomes at the rough endoplasmic reticulum (ER) (Fig. 14.1). The newly synthesized proteins occur in the lumen of the ER, and the storage proteins are finally deposited in the **protein bodies**. In the case of 2S-proteins and prolamins, the protein bodies are formed by budding from the ER membrane. The globulins are mostly transferred from the ER by vesicle transfer via the Golgi apparatus (section 1.6), first to the vacuole, from which protein bodies are formed by fragmentation. There also exists a pathway by which certain proteins (e.g., globulins in wheat endosperm) are transported directly by vesicle transfer from the ER membrane to the vacuole without passing the Golgi apparatus.

Figure 14.2 shows the formation of legumin in detail. The protein formed by the ribosome contains at the N-terminus of the polypeptide chain a hydrophobic section called a **signal sequence**. After the synthesis of this signal sequence, translation comes to a halt, and the signal sequence forms a complex with three other components:

1. A **signal recognition particle**,
2. A **binding protein** located on the ER membrane, and
3. A **pore protein** present in the ER membrane.

The formation of this complex results in opening a pore in the ER membrane; translation continues and the newly formed protein chain (e.g., pre-pro-legumin) reaches the lumen of the ER and anchors the ribosome on the ER membrane for the duration of protein synthesis. Immediately after the peptide chain enters the lumen, the signal sequence is removed by a **signal peptidase** located on the inside of the ER membrane. The remaining polypeptide, termed a **pro-legumin**, contains the future α- and β-chains of the legumin. An S-S linkage within the pro-legumin is formed in the ER lumen. Three pro-legumin molecules form a **trimer**, facilitated by chaperones (section 21.2). During this association, a quality control occurs: Trimers without the correct conformation are degraded. The trimers are transferred via the Golgi apparatus to the vacuoles, where the α- and β-chains are separated by a peptidase. The subunits of the legumins assemble to **hexamers** and are deposited in this form. The protein bodies, the final storage site of the legumins, are derived from fragmentation of the vacuole. The carbohydrate chains of glycosylated vicilins (e.g., of the phaesolins from the bean *Phaseolus vulgaris*) are processed in the Golgi apparatus.

The pre-pro-forms of newly synthesized **2S-proteins** and **prolamins**, which occur in the lumen of the ER, also contain a signal sequence. Completion and aggregation of these proteins takes place in the lumen of the ER, from which the protein bodies are formed by budding.

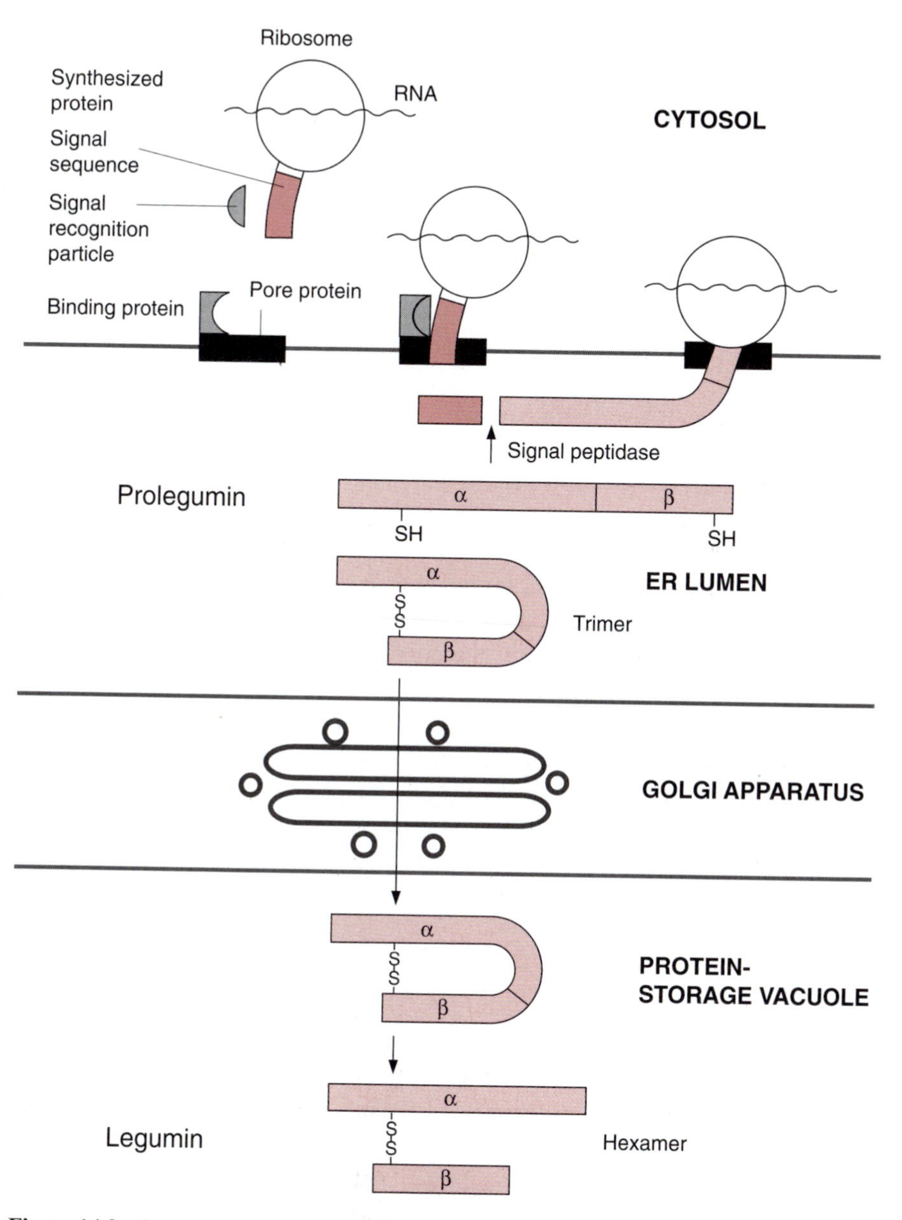

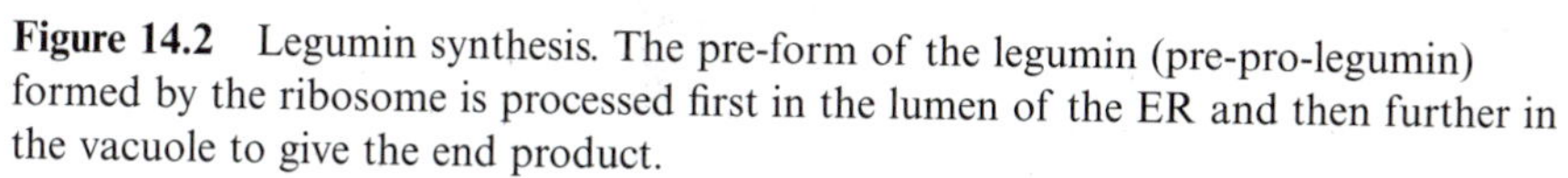
Figure 14.2 Legumin synthesis. The pre-form of the legumin (pre-pro-legumin) formed by the ribosome is processed first in the lumen of the ER and then further in the vacuole to give the end product.

14.6 Proteinases mobilize the amino acids deposited in storage proteins

Our knowledge about the mobilization of the amino acids from the storage proteins derives primarily from investigations of processes during seed germination. In most cases, germination is induced by the uptake of water, causing the protein bodies to form a vacuole. The hydrolysis of the storage proteins is catalyzed by proteinases, which are in part deposited as inactive pro-forms together with the storage proteins in the protein bodies. Other proteinases are synthesized anew and transferred via the lumen of the ER and the Golgi apparatus to the vacuoles (Fig. 14.2). These enzymes are synthesized initially as inactive pro-forms. Activation of these pro-proteinases proceeds by limited proteolysis, in which a section of the sequence is removed by a specific peptidase. The remainder of the polypeptide represents the active proteinase.

The degradation of the storage proteins is also initiated by limited proteolysis. A specific proteinase first removes small sections of the protein sequence, resulting in a change in the conformation of the storage protein. In cereal grains, S-S bridges of storage proteins are cleaved by reduced thioredoxin (section 6.6). The unfolded protein is then susceptible to hydrolysis by various proteinases: for example, exopeptidases, which split off amino acids one after the other from the end of the protein molecule, and endopeptidases, which cleave within the molecule. In this way storage proteins are completely degraded in the vacuole and the liberated amino acids are provided as building material to the germinating plant.

Further reading

Adachi, M., Takenaka, Y., Gidamis, A. B., Mikami, B., Utsumi, S. Crystal structure of soybean proglycinin A1aB1b homotrimer. J Mol Biol 305, 291–305 (2001).

Bethke, P. C., Jones, R. L. Vacuoles and prevacuolar compartments. CurrOpin Plant Biol 3, 469–475 (2000).

Hadlington, J. L., Denecke, J. Sorting of soluble proteins in the secretory pathway of plants. Curr Opin Plant Biol 3, 461–468 (2000).

Hillmer, S., Movafeghi, A., Robinson, D. G., Hinz, G. Vacuolar storage proteins are sorted in the *cis*-Cisternae of the pea cotyledon Golgi apparatus. J Cell Biol 152, 41–50 (2001).

Matsuoka, K., Neuhaus, J.-M. *Cis*-elements of protein transport to the plant vacuoles. J Exp Bot 50, 165–174 (1999).

Morton, R. L., Schroeder, H. E., Bateman, K. S., Chrispeels, M. J., Armstrong, E. Bean α-amylase inhibitor 1 in transgenic peas (*Pisum sativum*) provides complete protection from pea weevil (*Bruchus pisorum*) under field conditions. PNAS 97, 3820–3825 (2000).

Müntz, K. Deposition of storage proteins. Plant Mol Biol 38, 77–99 (1998).

Muentz, K., Shutov, A. D. Legumains and their functions in plants. Trends Plant Sci 7, 340–344 (2002).

Peumans, W. J., van Damme, E. J. M. Lectins as plant defense proteins. Plant Physiol. 109, 247–352 (1995).

Robinson, D. G., Hinz, G. Golgi-mediated transport of seed storage proteins. Seed Sci Res 9, 267–283 (1999).

Shewry, P. R., Napier, J. A., Tatham, A. S. Seed storage proteins: Structures and biosynthesis. Plant Cell 7, 945–956 (1995).

Somerville, C. R., Bonetta, D. Plants as factories for technical materials. Plant Physiol 125, 168–171 (2001).

Vitale, A., Galili, G. The endomembrane system and the problem of protein sorting. Plant Physiol 125, 115–118 (2001).

15

Glycerolipids are membrane constituents and function as carbon stores

Glycerolipids are fatty acid esters of glycerol (Fig. 15.1). **Triacylglycerols** (also called triglycerides) consist of a glycerol molecule that is esterified with three fatty acids. In contrast to animals, in plants triacylglycerols do not serve as an energy store but mainly as a carbon store in seeds, and they are used as vegetable oils. In **polar glycerolipids**, the glycerol is esterified with only two fatty acids and a hydrophilic group is linked to the third—OH group. These polar lipids are the main constituent of membranes.

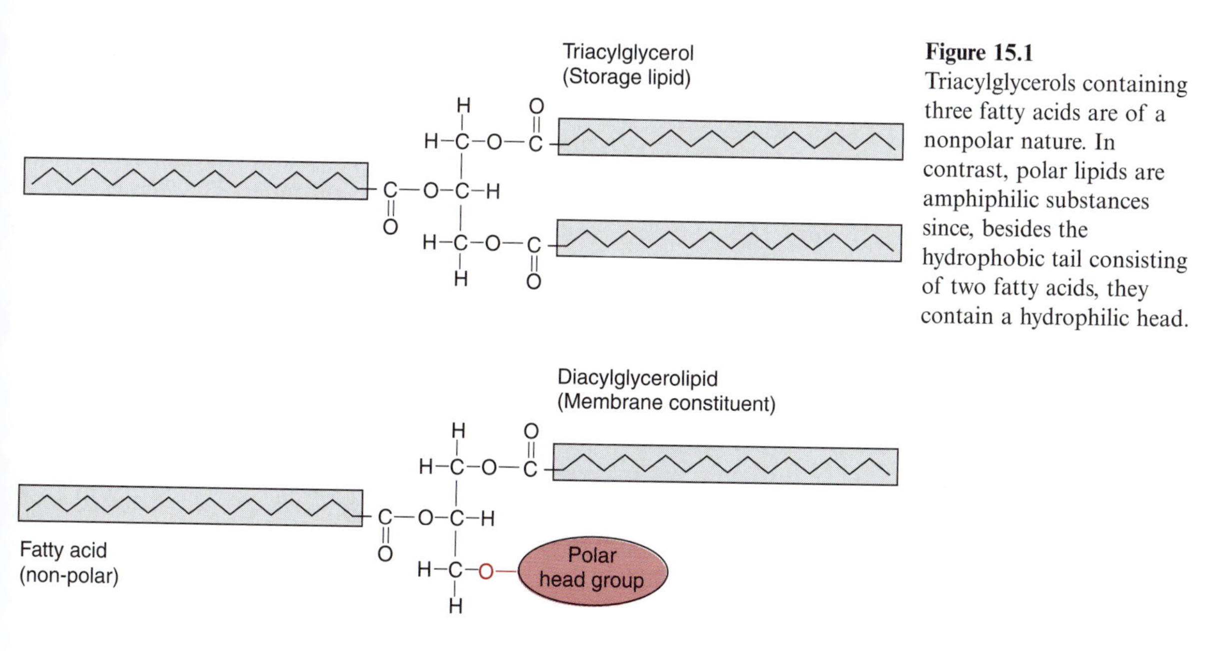

Figure 15.1
Triacylglycerols containing three fatty acids are of a nonpolar nature. In contrast, polar lipids are amphiphilic substances since, besides the hydrophobic tail consisting of two fatty acids, they contain a hydrophilic head.

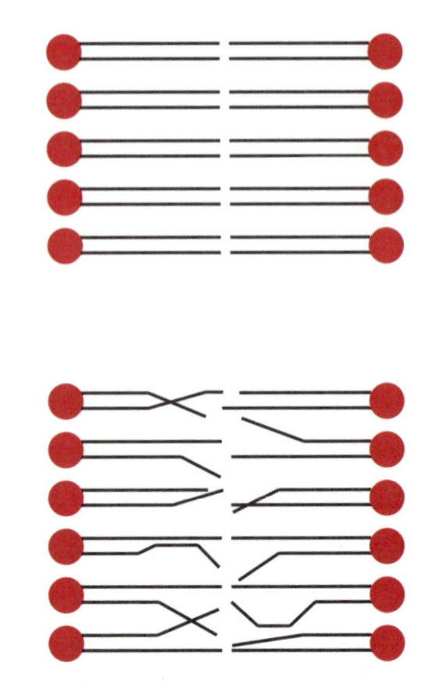

Figure 15.2 Membrane lipids with saturated fatty acids form a very regular lipid bilayer. The kinks caused in the hydrocarbon chain by *cis*-carbon-carbon double bonds in unsaturated fatty acids result in disturbances in the lipid bilayer and lead to an increase in its fluidity.

15.1 Polar glycerolipids are important membrane constituents

The polar glycerolipids are **amphiphilic molecules**, consisting of a hydrophilic head and a hydrophobic tail. This property enables them to form **lipid bilayers**, in which the hydrocarbon tails are held together by hydrophobic interactions and the hydrophilic heads protrude into the aqueous phase, thus forming the basic structure of a membrane (Fig. 15.2). Since the middle C atom of the glycerol in a polar glycerolipid is asymmetric, a distinction can be made between the two esterified groups of glycerol at the C-1 and C-3 positions.

But there are also other membrane lipids in plants. **Sphingolipids** (Fig. 15.5C) are important constituents of plasma membranes. The **sterols** shown in Figure 15.3 are also amphiphilic, the hydroxyl groups form the hydrophilic head and the sterane skeleton with the side chain serves as the hydrophobic tail. In addition to the sterols shown here, plants contain a large variety of other sterols as membrane constituents. They are contained primarily in the outer membrane of mitochondria, in the membranes of the endoplasmic reticulum, and in the plasma membrane. Sterols determine to a large extent the properties of these membranes (see below).

The polar glycerolipids mainly contain fatty acids with 16 or 18 carbon atoms (Fig. 15.4). The majority of these fatty acids are unsaturated and contain one to three carbon-carbon double bonds. These double bonds are present almost exclusively in the ***cis***-configuration and are present only rarely in the ***trans***-configuration. The double bonds normally are not conjugated.

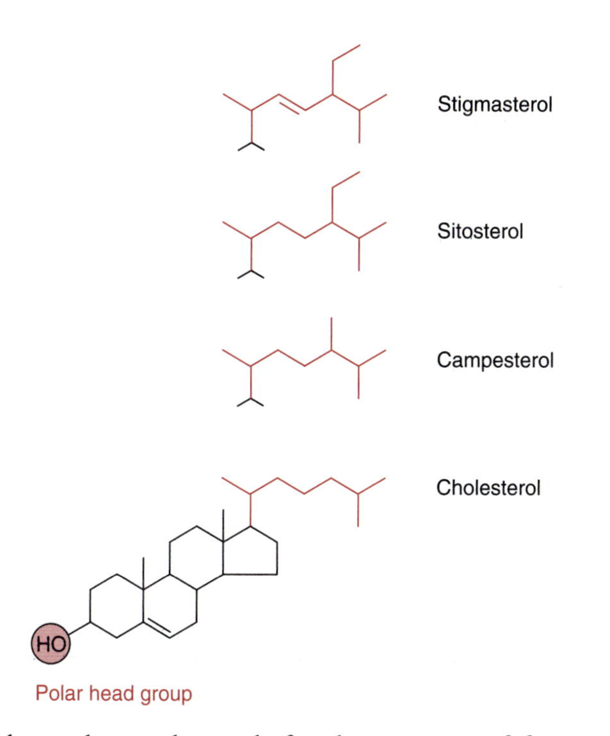

Figure 15.3 Cholesterol and the related sterols (only side chains shown) are membrane constituents.

Figure 15.4 shows the number code for the structure of fatty acids in the following format: number of C-atoms: number of double bonds, Δ position of the first C-atom of the double bond, c = *cis*-configuration (t = *trans*).

Plants often contain unusual fatty acids in their storage lipids (e.g., with conjugated double bonds, carbon-carbon triple bonds, or hydroxyl-, keto- or epoxy groups). By now more than 500 of these unusual fatty acids are known. There are also glycerolipids, in which the carbon chain is connected with the glycerol via an ether linkage.

The fluidity of the membrane is governed by the proportion of unsaturated fatty acids and the content of sterols

The hydrocarbon chains in saturated fatty acids are packed in a regular layer (Fig. 15.2), whereas in unsaturated fatty acids the packing is disturbed due to kinks in the hydrocarbon chain, caused by the *cis*-carbon-carbon double bonds, resulting in a more fluid layer. This is obvious when comparing the melting points of various fatty acids (Table 15.1). The melting point increases with increasing chain length as the packing becomes tighter, whereas the melting point decreases with an increasing number of double bonds. This also applies to the corresponding fats. Coco fat, for instance, consisting only of saturated fatty acids, is solid, whereas plant oils, with a very high natural content of unsaturated fatty acids, are liquid.

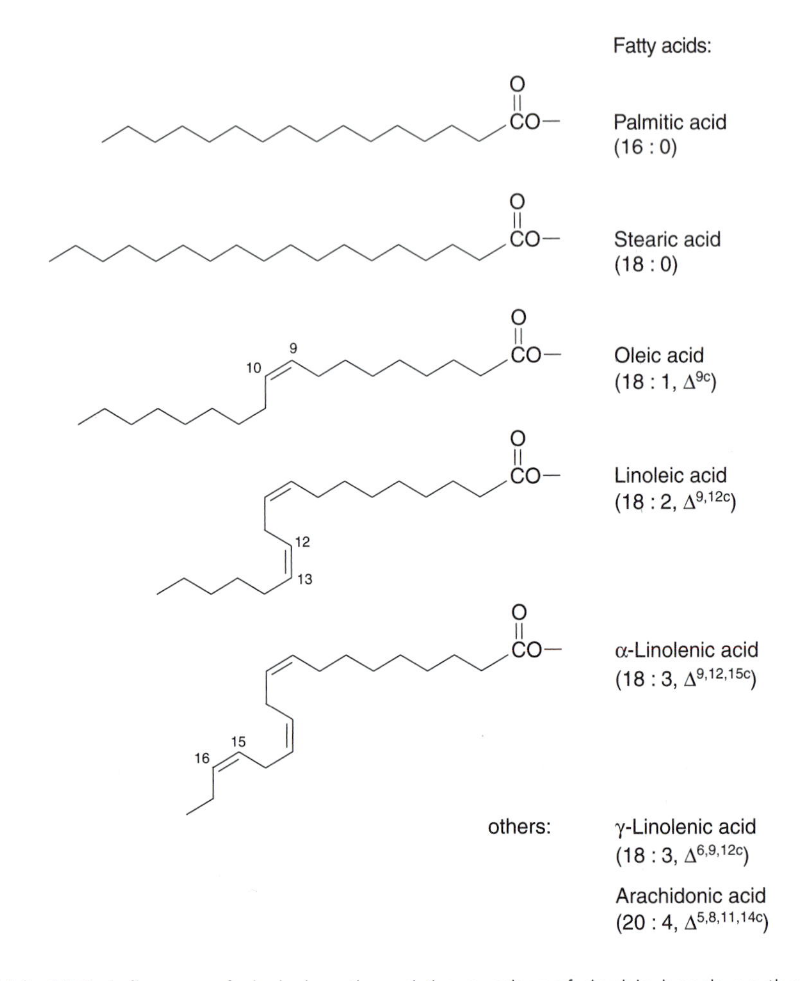

Figure 15.4 Fatty acids as hydrophobic constituents of membrane lipids.

Table 15.1: Influence of chain length and the number of double bonds on the melting point of fatty acids

Fatty acid	Chain length: double bonds	Melting point
Lauric acid	12:0	40°C
Stearic acid	18:0	70°C
Oleic acid	18:1	13°C
Linoleic acid	18:2	–5°C
Linolenic acid	18:3	–11°C

Likewise, the **fluidity of membranes** is governed by the proportion of **unsaturated fatty acids** in the membrane lipids. This is why in some plants, during growth at a low temperature, more highly unsaturated fatty acids are incorporated into the membrane to compensate for the decrease in membrane fluidity caused by the low temperature. Recently, the cold tolerance of tobacco has been enhanced by increasing the proportion of unsaturated fatty acids in the membrane lipids by genetic engineering. Sterols (Fig. 15.3) decrease the fluidity of membranes and probably also play a role in the adaptation of membranes to temperature. On the other hand, a decrease of unsaturated fatty acids in the lipids of thylakoid membranes, as achieved by genetic engineering, made tobacco plants more tolerant to heat.

Membrane lipids contain a variety of hydrophilic head groups

The head groups of the polar glycerolipids in plants are formed by a variety of compounds. However, these all fulfill the same function of providing the lipid molecule with a polar group (Fig. 15.5). In the **phospholipids**, the head group consists of a phosphate residue that is esterified with a second alcoholic compound such as ethanolamine, choline, serine, glycerol, or inositol. Phosphatidic acid is only a minor membrane constituent, but it plays a role as a signal substance (section 19.1).

The phospholipids mentioned previously are found as membrane constituents in bacteria as well as in animals and plants. A speciality of plants and cyanobacteria is that they contain, in addition to phospholipids, the **galactolipids** monogalactosyldiacylglycerol (MGDG) and digalactosyldiacylglycerol (DGDG), and the sulfolipid sulfoquinovosyldiacylglycerol (SL) as membrane constituents. The latter contains a glucose moiety to which a sulfonic acid residue is linked at the C-6 position as the polar head group (Fig. 15.5B).

There are great differences in the lipid composition of the various membranes in a plant (Table 15.2). Chloroplasts contain galactolipids, which are the main constituents of the thylakoid membrane and also occur in the envelope membranes. The membranes of the mitochondria and the plasma membrane contain no galactolipids, but have phospholipids as the main membrane constituents. **Cardiolipin** is a specific component of the inner mitochondrial membrane in animals and plants.

In a green plant cell, about 70% to 80% of the total membrane lipids are constituents of the thylakoid membranes. Plants represent the largest part of the biosphere, and this is why the galactolipids MGDG and DGDG are the most abundant membrane lipids on earth. Since plant growth is limited in many habitats by the phosphate content in the soil, it was probably advantageous for plants during evolution to be independent of the phosphate supply in the soil for the synthesis of the predominant part of their membrane lipids.

Figure 15.5A. Hydrophilic constituents of membrane lipids: phosphate and phosphate esters.

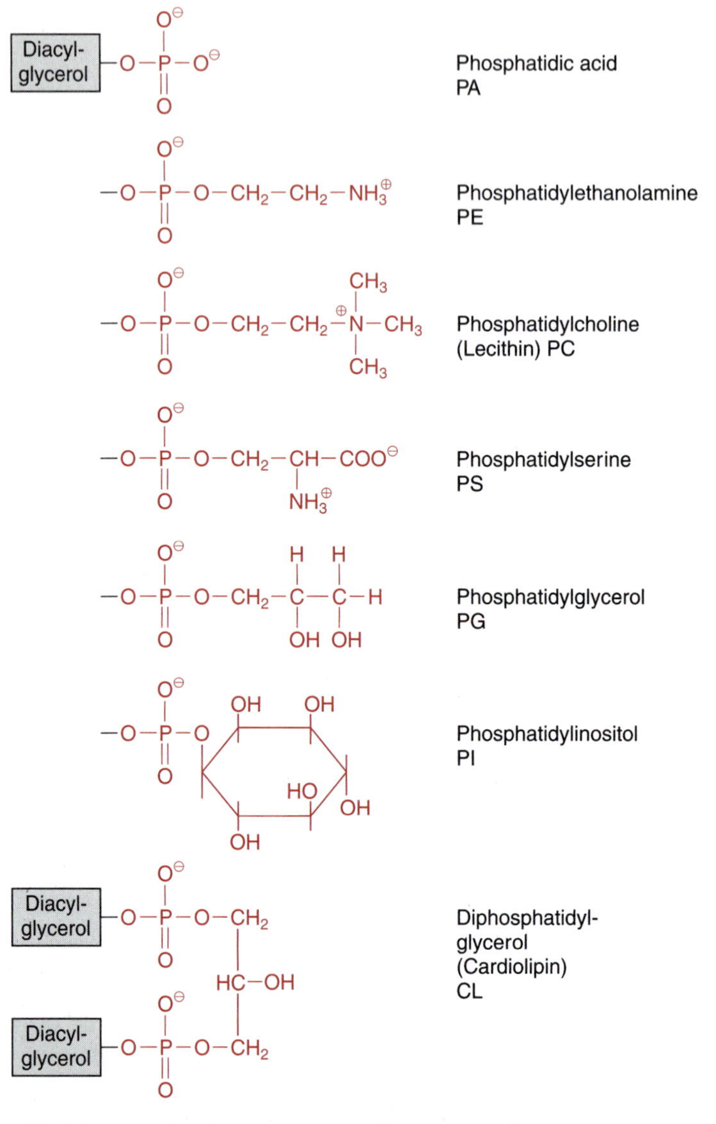

Sphingolipids are important constituents of the plasma membrane

Sphingolipids (Fig. 15.5C) occur in the plasma and ER membranes. The sphingolipids consist of a so-called **sphingo base**. This is a hydrocarbon chain containing double bonds, an amino group in position 2, and two to three hydroxyl groups in positions 1, 3, and 4. The sphingo base is connected by an amide link to a fatty acid (C16–24, with up to two double bonds). As an example of one of the many sphingolipids, **ceramide**, with sphinganine as base, is shown in Figure 15.5C. In **glucosylsphingolipids**, the terminal

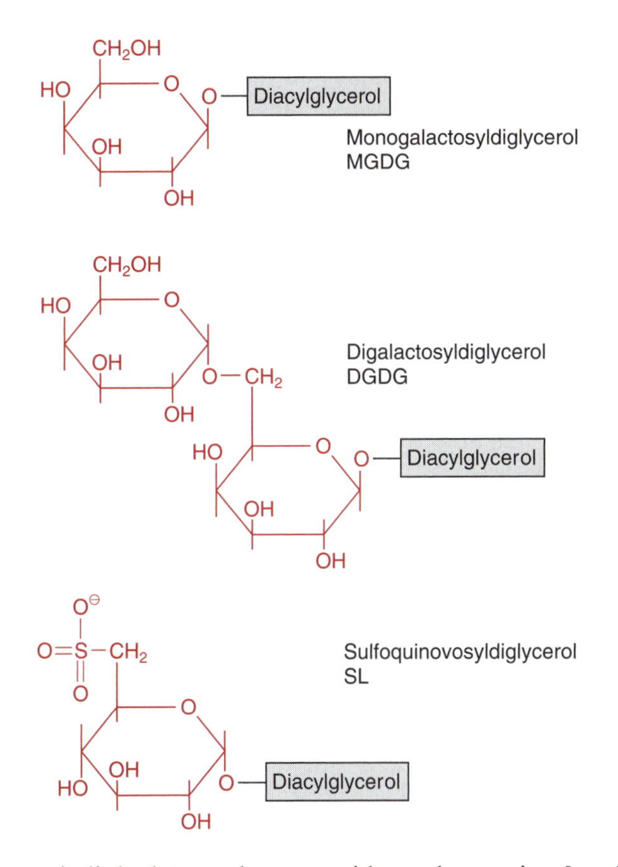

Figure 15.5B. Hydrophilic constituents of membrane lipids: hexoses.

hydroxyl group is linked to a glucose residue, whereas in **phosphorylsphingolipids** (not shown in figure 15C), these are esterified with phosphate or choline phosphate.

Sphingolipids are known to have an important signaling function in animals and yeast. Research on the function of sphingolipids in plants is still in its infancy. Recent results indicate that sphinganine 1-phosphate acts in guard cells as a **Ca^{++}-mobilizing messenger**, which is released upon the action of abscisic acid (**ABA**) (see also sections 8.2 and 19.6). Plant sphingolipid metabolites may also be involved as signals in **programmed cell death**.

15.2 Triacylglycerols are storage substances

Triacylglycerols are contained primarily in seeds but also in some fruits such as olives or avocados. The purpose of triacylglycerols in fruits is to attract animals to consume these fruits for distributing the seeds. The triacylglycerols in seeds function as a carbon store to supply the carbon required for biosynthetic processes during seed germination. Triacylglycerols have

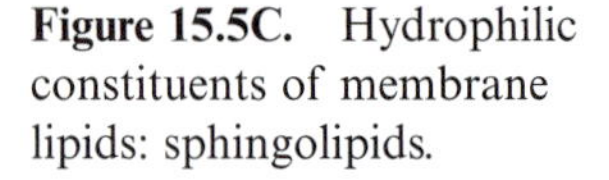

Figure 15.5C. Hydrophilic constituents of membrane lipids: sphingolipids.

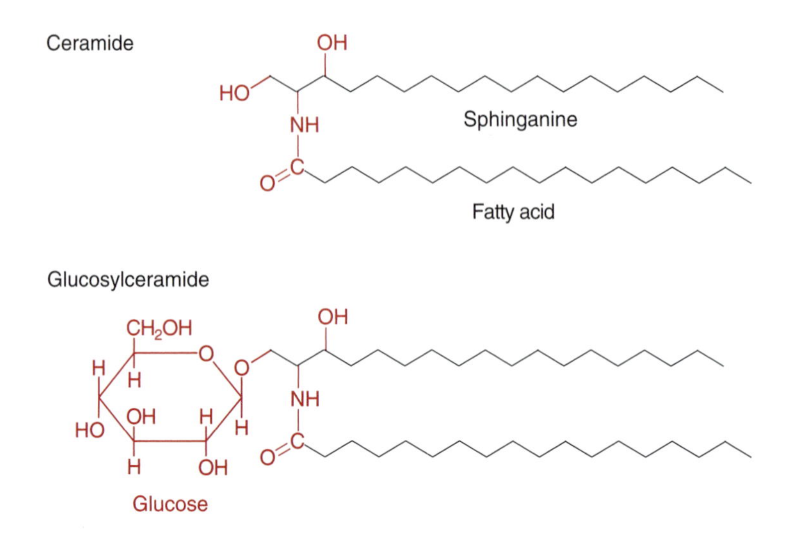

Table 15.2: The composition of membrane lipids in various organelle membranes

	Chloroplast thylakoid membrane	ER membrane	Plasma membrane
Membrane lipids*	Mol%		
Monogalactosyldiglyceride	42	1	2
Digalactosyldiglyceride	33	2	3
Sulfolipide	5	0	0
Phosphatidylcholine	5	45	19
Phosphatidylethanolamine	1	15	17
Phosphatidylserine	0	1	3
Phosphatidylglycerol	11	6	12
Phosphatidylinositol	0	8	2
Sphingolipids	0	10	7
Sterols	0	5	31

* Leaves from Rye, after Lochnit et al. (2001).

an advantage over carbohydrates as storage compounds, because their weight/carbon content ratio is much lower. A calculation illustrates this: In starch the glucose residue, containing six C atoms, has a molecular mass of 162 Da. The mass of one stored carbon atom thus amounts to 27 Da. In reality, this value is higher, since starch contains water molecules because it is hydrated. A triacylglycerol with three palmitate residues contains 51 C atoms and has a molecular mass of 807 Da. The mass of one stored carbon atom thus amounts to only 16 Da. Since triacylglycerols, in contrast to starch, are not hydrated, carbon stored in the seed as fat requires less than half the weight as when it is stored as starch. Low weight is advantageous for seed dispersal.

Triacylglycerols are deposited in **oil bodies**, also termed **oleosomes** or **lipid bodies** (Fig. 15.6). These consist of oil droplets, which are surrounded by a lipid monolayer. Proteins collectively termed **oleosins** are anchored to the lipid monolayer of the oil bodies and catalyze the mobilization of fatty acids from the triacylglycerol store during seed germination (section 15.6). These oleosins are present only in oil bodies contained in the endosperm and embryonic tissue of seeds. The oil bodies in the pericarp of olives or avocados, where the triacylglycerols are not used as a store but only to lure animals, contain no oleosins and, with a diameter of 10 to 20 μm, are very

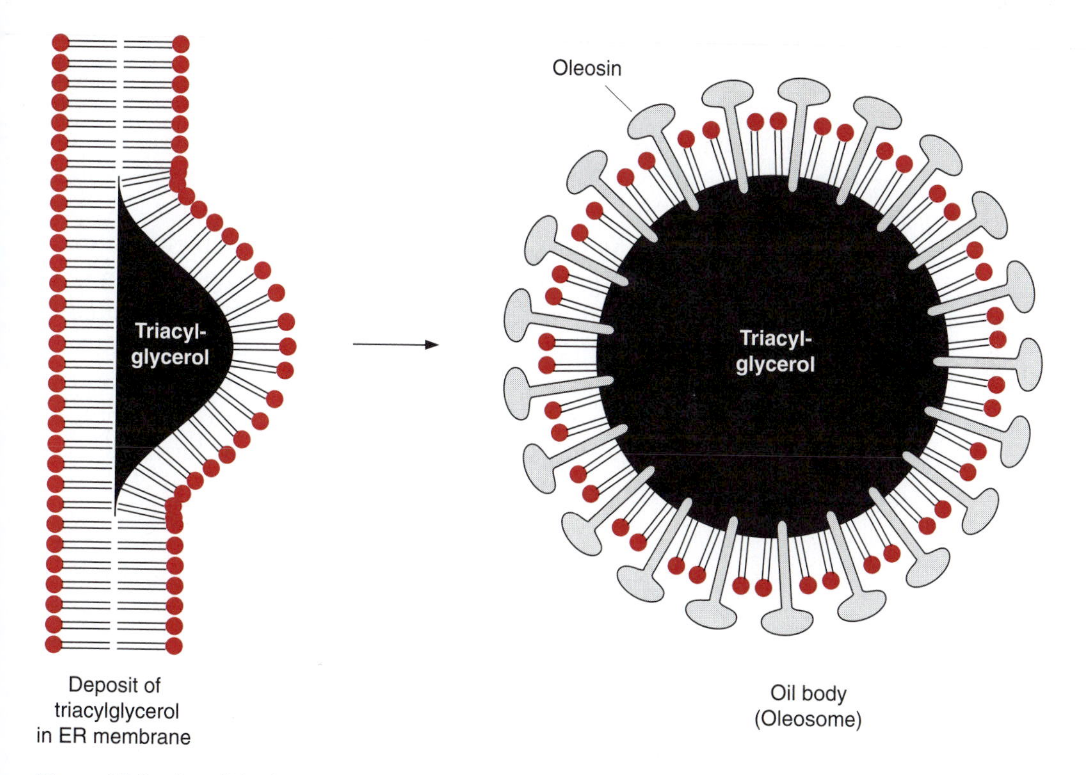

Figure 15.6 An oil body enclosed by a lipid monolayer is formed by the incorporation of triacylglycerols in an ER membrane. Oleosins are anchored to the lipid monolayer.

much larger than the oil bodies of storage tissues (diameter 0.5–2 μm). As discussed in section 15.5, synthesis of triacylglycerols occurs in the ER membrane. It is assumed that the newly synthesized triacylglycerol accumulates between the lipid double layer of the ER membrane, until the full size of the oil body is reached (Fig. 15.6). When the oil body buds off from the ER membrane, it is surrounded by a single phospholipid layer.

15.3 The de novo synthesis of fatty acids takes place in the plastids

The carbon fixed by CO_2 assimilation in the chloroplasts is the precursor not only for the synthesis of carbohydrates and amino acids, but also for the synthesis of fatty acids and various secondary metabolites discussed in Chapters 16 to 18. Whereas the production of carbohydrates and amino acids by the mesophyll cells is primarily destined for export to other parts of the plants, the synthesis of fatty acids occurs only for the cell's own requirements, except in seeds and fruits. Plants are not capable of long-distance fatty acid transport. Since fatty acids are present as constituents of membrane lipids in every cell, each cell must contain the enzymes for the synthesis of membrane lipids and thus also for the synthesis of fatty acids.

In plants the *de novo* synthesis of fatty acids always occurs in the plastids: in the chloroplasts of green cells and the leucoplasts and chromoplasts of non-green cells. Although in plant cells enzymes of fatty acid synthesis are also found in the membrane of the ER, these enzymes appear to be involved only in the **modification** of fatty acids, which have been synthesized earlier in the plastids. These modifications include a chain elongation of fatty acids, as catalyzed by **elongases**, and the introduction of further double bonds by **desaturases** (Fig.15.15B).

Acetyl CoA is the precursor for the synthesis of fatty acids

Like mitochondria (see Fig. 5.4), plastids contain a **pyruvate dehydrogenase**, by which pyruvate is oxidized to acetyl CoA, accompanied by the reduction of NAD (Fig. 15.7).

In chloroplasts, photosynthesis provides the **NADPH** required for the synthesis of fatty acids. But it is still not clear how **acetyl CoA** is formed from the products of CO_2 fixation. In chloroplasts, depending on the developmental state of the cells, the activity of pyruvate dehydrogenase is often

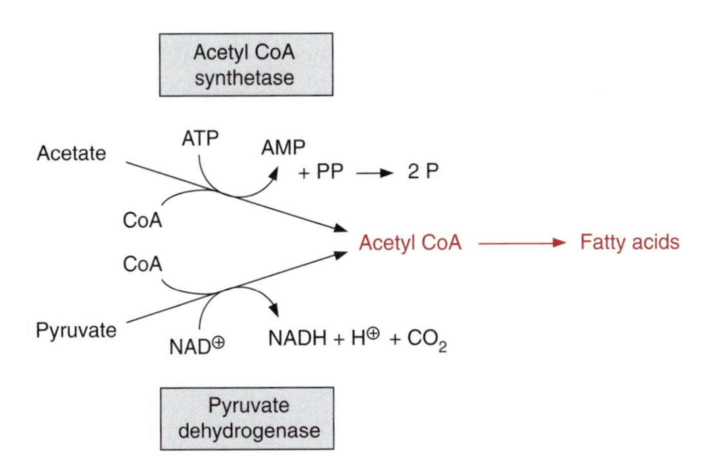

Figure 15.7 Acetyl CoA can be formed in two ways.

low. On the other hand, chloroplasts contain a high activity of **acetyl CoA synthetase**, by which acetate, with consumption of ATP, can be converted to acetyl CoA. When chloroplasts are supplied with radioactively labeled acetate, the radioactivity is very rapidly incorporated into fatty acids. In many plants, **acetate** is often a major precursor for the formation of acetyl CoA in the chloroplasts and leucoplasts. The origin of acetate is still not clear. It could be formed in the mitochondria by hydrolysis of acetyl CoA, derived from the oxidation of pyruvate by mitochondrial pyruvate dehydrogenase.

In leucoplasts, the reductive equivalents required for fatty acid synthesis are provided by the oxidation of glucose 6-phosphate. The latter is transported via a **glucose phosphate-phosphate translocator** to the plastids (see Fig. 13.5). The oxidation of glucose 6-phosphate by the **oxidative pentose phosphate pathway** (Fig. 6.21) provides the NADPH required for fatty acid synthesis.

Fatty acid synthesis starts with the carboxylation of acetyl CoA to malonyl CoA by **acetyl CoA carboxylase**, with the consumption of ATP (Fig. 15.8). In a subsequent reaction, CoA is exchanged for **acyl carrier protein (ACP)** (Fig. 15.9). ACP contains a serine residue to which a pantetheine residue is linked via a phosphate group. Since the pantetheine residue is also a functional constituent of CoA, ACP can be regarded as a CoA, which is covalently bound to a protein. The enzyme **β-ketoacyl-ACP synthase III** catalyzes the condensation of acetyl CoA with malonyl-ACP. The liberation of CO_2 makes this reaction irreversible. The acetoacetate thus formed remains bound as a thioester to ACP and is reduced by NADPH to β-D-hydroxyacyl-ACP. Following the release of water, the carbon-carbon double bond formed is reduced by NADPH to acyl-ACP. The product is a **fatty acid** that has been elongated by two carbon atoms (Fig. 15.8).

Figure 15.8 Reaction sequence for the synthesis of fatty acids: activation, condensation, reduction, release of water, and further reduction elongate a fatty acid by two carbon atoms.

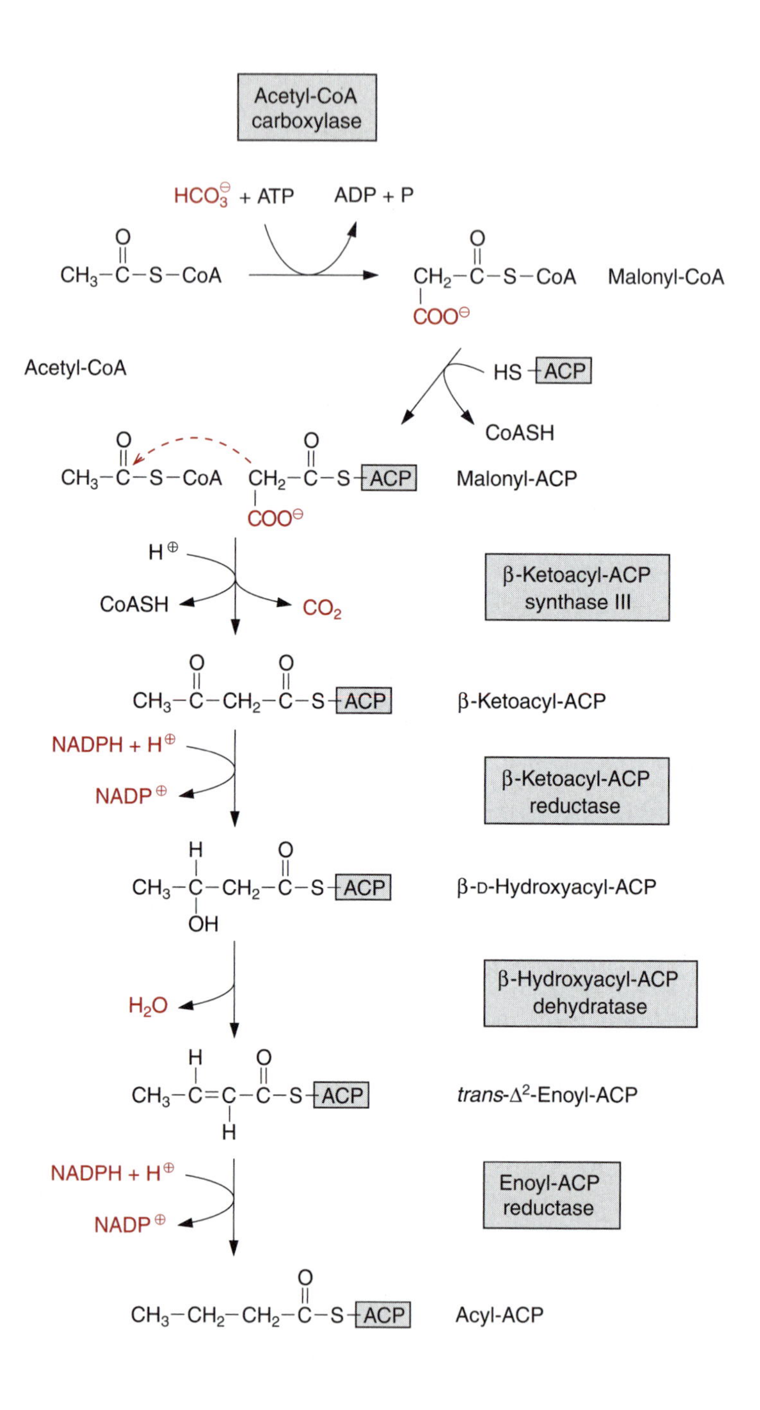

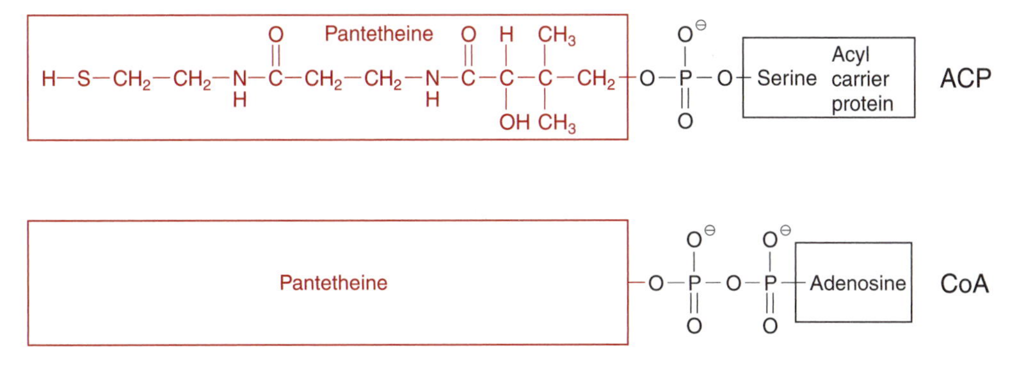

Figure 15.9 The acyl carrier protein (ACP) contains pantetheine, the same functional group as coenzyme-A.

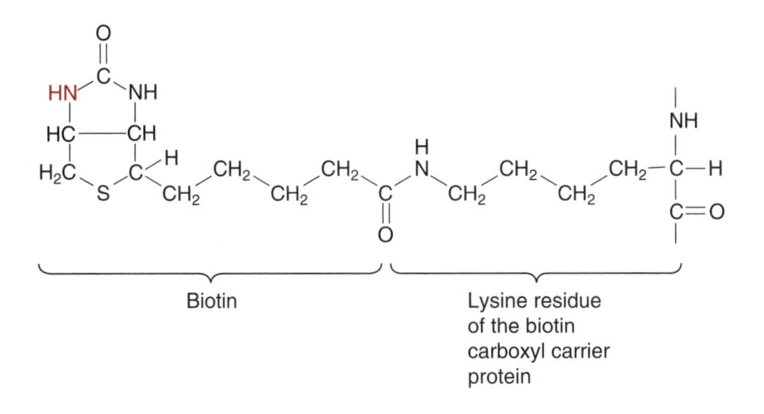

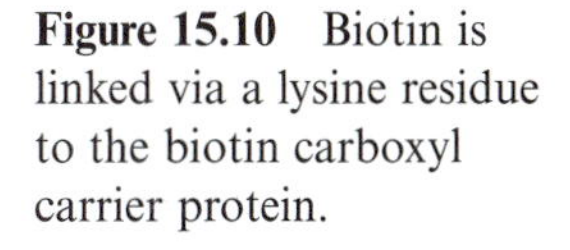

Figure 15.10 Biotin is linked via a lysine residue to the biotin carboxyl carrier protein.

Acetyl CoA carboxylase is the first enzyme of fatty acid synthesis

In the carboxylation of acetyl CoA, **biotin** acts as a carrier for "**activated CO_2**" (Fig. 15.10). Biotin is covalently linked by its carboxyl group to the ε-amino group of a lysine residue of the **biotin carboxyl carrier protein**, and it contains an -NH-group that forms a carbamate with HCO_3^- (Fig. 15.11). This reaction is driven by the hydrolysis of ATP. Acetyl CoA carboxylation requires two steps:

1. Biotin is carboxylated at the expense of ATP by **biotin carboxylase**.
2. Bicarbonate is transferred to acetyl CoA by **carboxyl transferase**.

All three enzymes—the biotin carboxyl carrier protein, biotin carboxylase, and carboxyl transferase—form a single multienzyme complex. Since the

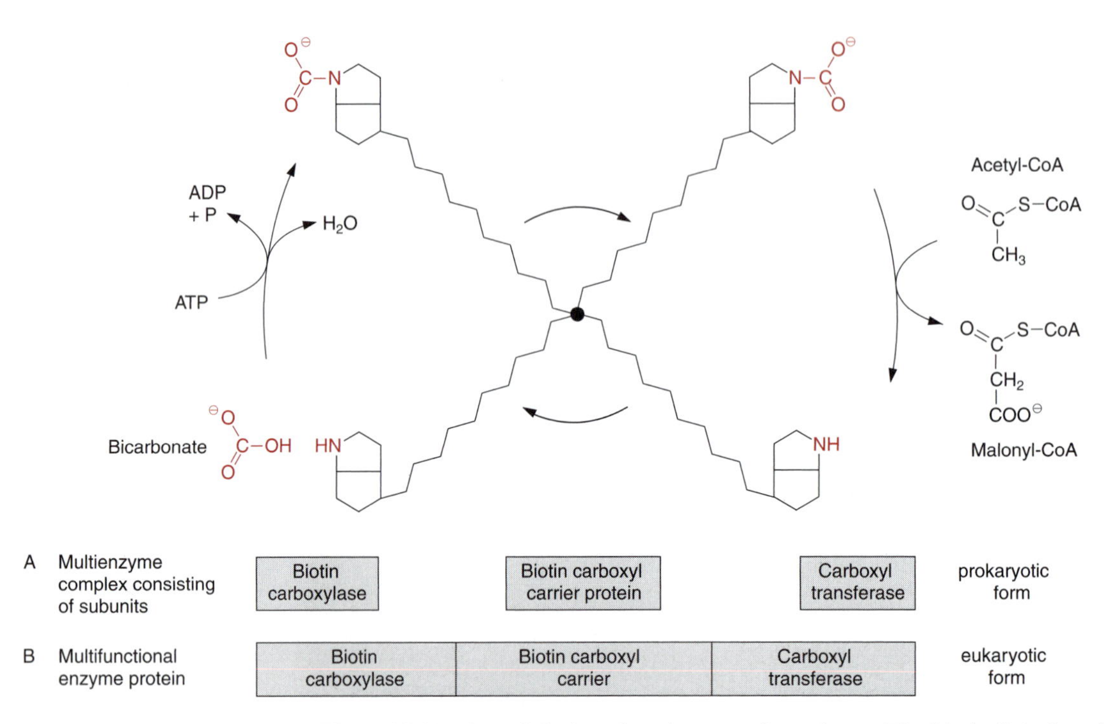

Figure 15.11 Acetyl CoA carboxylase reaction scheme. The biotin linked to the biotin carboxyl carrier protein reacts in turn with biotin carboxylase and carboxyl transferase. The circular motion has been chosen for the sake of clarity; in reality, there probably is a pendulum-like movement. The eukaryotic form of acetyl CoA carboxylase is present as a multifunctional protein.

biotin is attached to the carrier protein by a long flexible hydrocarbon chain, it reacts alternately with the carboxylase and carboxyl transferase in the multienzyme complex (Fig. 15.11).

The acetyl CoA carboxylase multienzyme complex in the stroma of plastids consists of several subunits, resembling the acetyl CoA carboxylase in cyanobacteria and other bacteria, and is referred to as the **prokaryotic form** of the acetyl CoA carboxylase. Acetyl CoA carboxylase is also present outside the plastids, probably in the cytosol. The malonyl CoA formed outside the plastids is used for chain elongation of fatty acids and is the precursor for the formation of flavonoids (see section 18.5). The extraplastid acetyl CoA carboxylase, in contrast to the prokaryotic type, consists of a single large **multifunctional protein** in which the biotin carboxyl carrier, the biotin carboxylase, and the carboxyl transferase are located on different sections of the same polypeptide chain. Since this multifunctional protein also occurs in a very similar form in the cytosol of yeast and animals, it is referred to as the **eukaryotic form**. It should be emphasized, however, that the eukary-

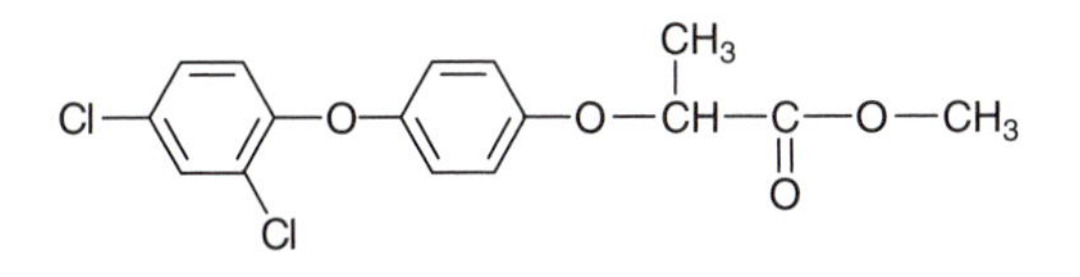

Figure 15.12 Diclofop methyl, an herbicide (Hoe-Grass, Aventis), inhibits the eukaryotic multifunctional acetyl CoA carboxylase.

otic form as well as the prokaryotic form of acetyl CoA carboxylase are encoded in the nucleus. Possibly only one protein of the prokaryotic enzyme is encoded in the plastid genome.

In *Gramineae* (grasses), including the various species of cereals, the prokaryotic form is not present. In these plants, the multifunctional eukaryotic acetyl CoA carboxylase is located in the cytosol as well as in the chloroplasts. The eukaryotic acetyl CoA carboxylase is inhibited by various arylphenoxypropionic acid derivatives, such as, for example, **diclofop methyl** (Fig. 15.12). Since eukaryotic acetyl CoA carboxylase in *Gramineae* is involved in the *de novo* fatty acid synthesis of the plastids, this inhibitor severely impairs lipid biosynthesis in this group of plants. Diclofop methyl (trade name Hoe-Grass, Aventis) and similar substances are therefore used as **selective herbicides** (section 3.6) to control grass weeds.

Acetyl CoA carboxylase, the first enzyme of fatty acid synthesis, is an important regulatory enzyme and its reaction is regarded as a rate-limiting step in fatty acid synthesis. In chloroplasts, the enzyme is fully active only during illumination and is inhibited during darkness. In this way fatty acid synthesis proceeds mainly during the day, when photosynthesis provides the necessary NADPH. The mechanism of light regulation is similar to the light activation of the enzymes of the Calvin cycle (section 6.6): The acetyl CoA carboxylase is reductively activated by thioredoxin and the activity is further enhanced by the increase of the pH and the Mg^{++} concentration in the stroma.

Further steps of fatty acid synthesis are also catalyzed by a multienzyme complex

β-Ketoacyl-ACP formed by the condensation of acetyl CoA and malonyl-ACP (Fig. 15.8) is reduced by NADPH to β-D-hydroxyacyl-ACP, and after the release of water the carbon-carbon double bond of the resulting enoyl-ACP is reduced again by NADPH to acyl-ACP. This reaction sequence resembles the reversal of the formation of oxaloacetate from succinate in the citrate cycle (Fig. 5.3). Fatty acid synthesis is catalyzed by a multienzyme complex. Figure 15.13 shows a scheme of the interplay of the various reactions. The ACP, containing the acyl residue bound as a thioester, is located

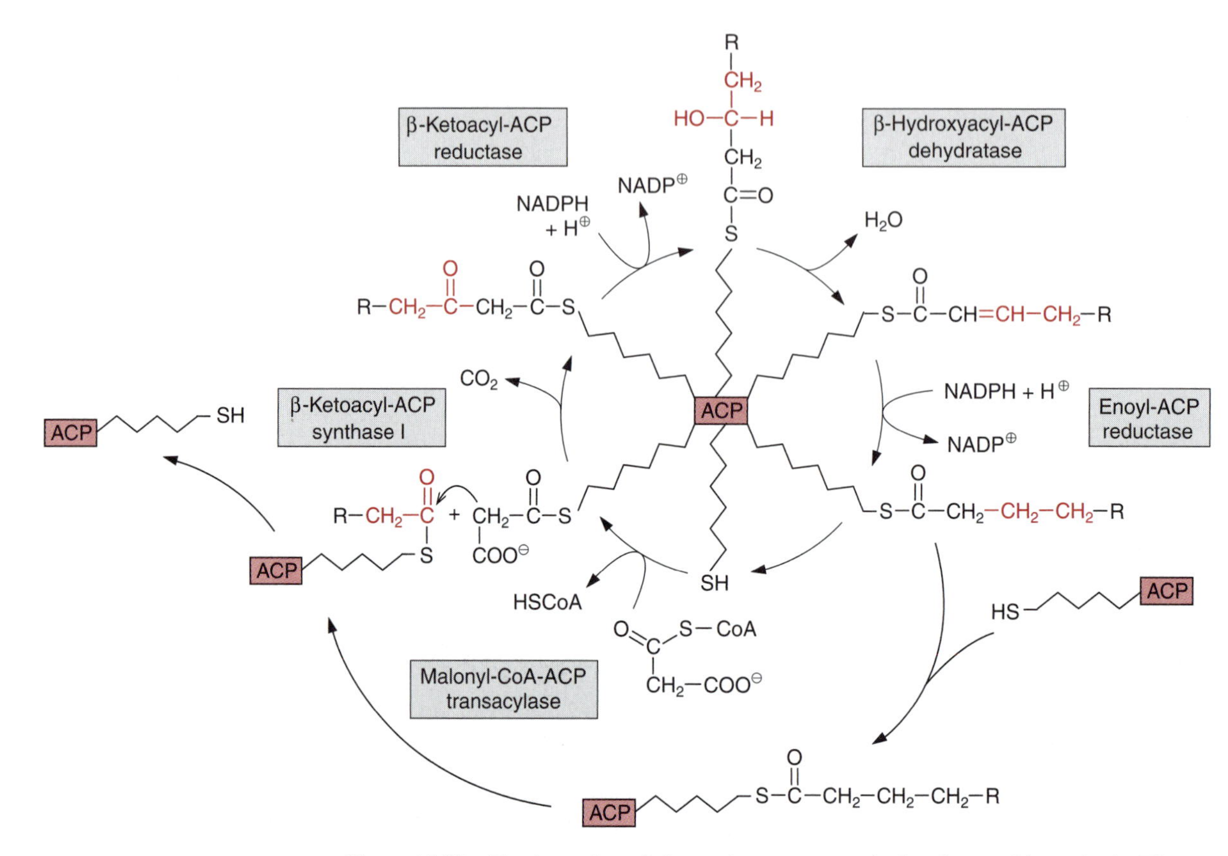

Figure 15.13 The interplay of the various enzymes during fatty acid synthesis. The acyl carrier protein (ACP), located in the center, carries the fatty acid residue, bound as thioester, from enzyme to enzyme. The circular movement has been chosen for the sake of clarity, but it does not represent reality.

in the center of the complex. Thus the acyl residue is attached to a flexible chain, to be transferred from enzyme to enzyme during the reaction cycle.

The fatty acid is elongated by first transferring it to another ACP and then condensing it with malonyl-ACP. The enzyme **β-ketoacyl-ACP synthase I**, catalyzing this reaction, enables the formation of fatty acids with a chain length of up to C-16. A further chain elongation to C-18 is catalyzed by **β-ketoacyl-ACP synthase II** (Fig. 15.14).

It should be mentioned that in animals and fungi, the enzymes of fatty acid synthesis shown in Figure 15.13 are contained in only one or two multifunctional proteins, which form a complex (eukaryotic fatty acid synthase complex). Since the fatty acid synthase complex of the plastids, consisting of several proteins, is similar to those of many bacteria, it is called **prokaryotic fatty acid synthase complex**.

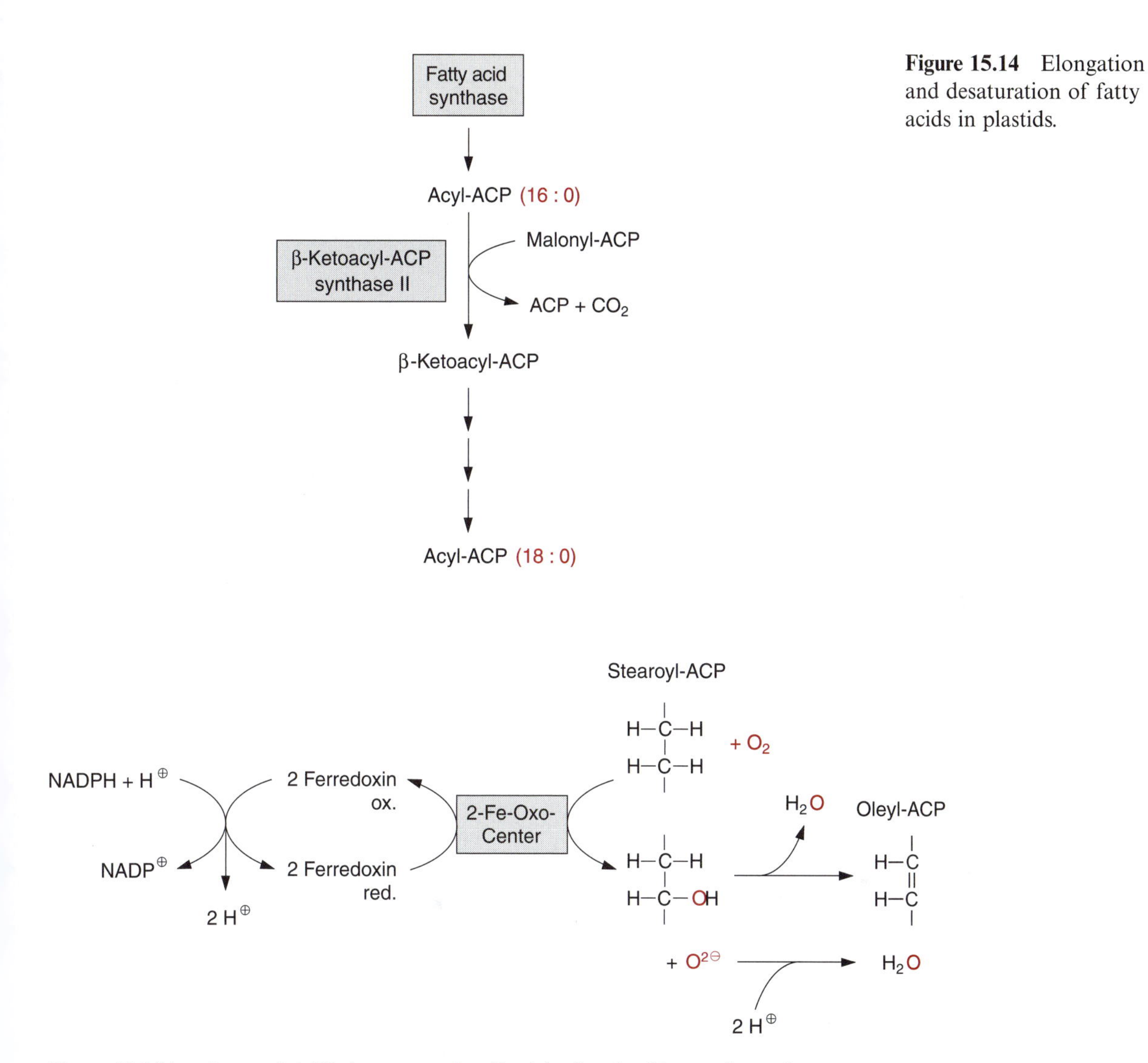

Figure 15.14 Elongation and desaturation of fatty acids in plastids.

Figure 15.15A. Stearoyl-ACP desaturase, localized in the plastids, catalyzes the desaturation of stearoyl-ACP to oleyl-ACP. The reaction can be regarded as a monooxygenation with the subsequent release of water.

The first double bond in a newly formed fatty acid is formed by a soluble desaturase

The stearoyl-ACP (18:0) formed in the plastid stroma is desaturated there to oleoyl-ACP (18:1) (Fig. 15.15A). This reaction can be regarded as a **monooxygenation** (section 18.2), in which one O-atom from an O_2 molecule is reduced to water and the other is incorporated into the hydrocarbon chain

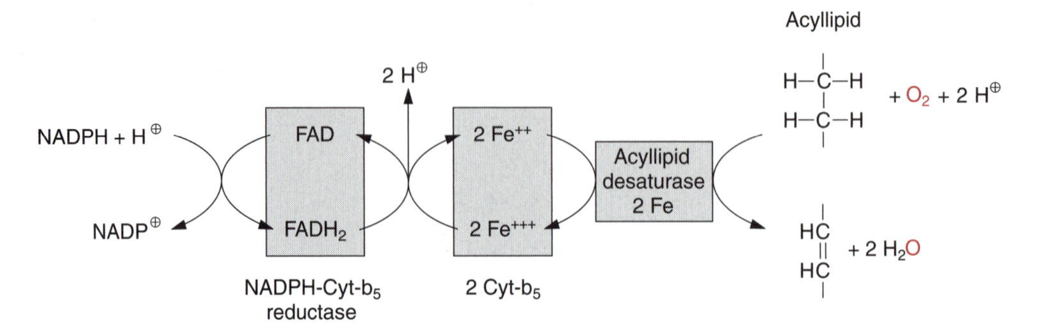

Figure 15.15B. Acyl lipid desaturases, integral proteins of the ER membrane, catalyze the desaturation of fatty acids, which are parts of phospholipids. Here, too, the reaction can be understood as monooxygenation followed by water cleavage (not shown in the figure). Electron transport from NADPH to the desaturase requires two more proteins, one NADPH-Cyt-*b* and cytochrom-b_5.

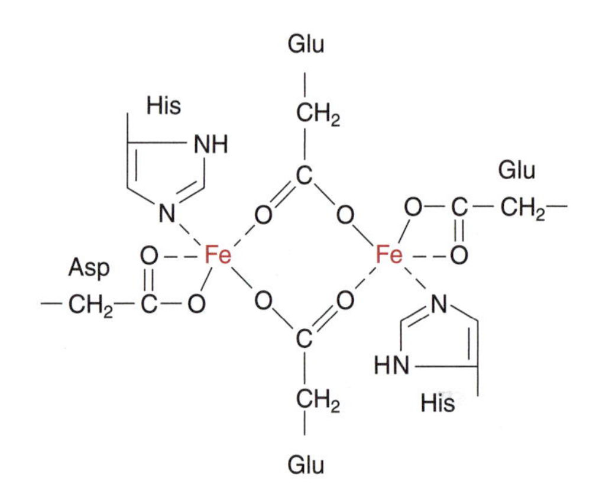

Figure 15.16 In a di-iron-oxo cluster, two Fe atoms are bound to glutamate, aspartate, and histidine side chains of the protein (After Karlin 1993).

of the fatty acid as hydroxyl group. A carbon-carbon double bond is formed by subsequent liberation of H_2O (analogous with the β-hydroxyacyl-ACP dehydratase reaction, Fig. 15.8), which, in contrast to fatty acid synthesis, has a *cis*-configuration. The monooxygenation requires two electrons, which are provided by NADPH via reduced ferredoxin. Monooxygenases are widespread in bacteria, plants, and animals. In most cases, O_2 is activated by a special cytochrome, cytochrome P_{450}. However, in the **stearoyl-ACP desaturase**, the O_2 molecule reacts with a **di-iron-oxo cluster** (Fig. 15.16). In previous sections, we have discussed iron-sulfur clusters as redox carriers, in which the Fe-atoms are bound to the protein via cysteine residues (Fig. 3.26). In the di-iron-oxo cluster of the desaturase, two iron atoms are bound to the

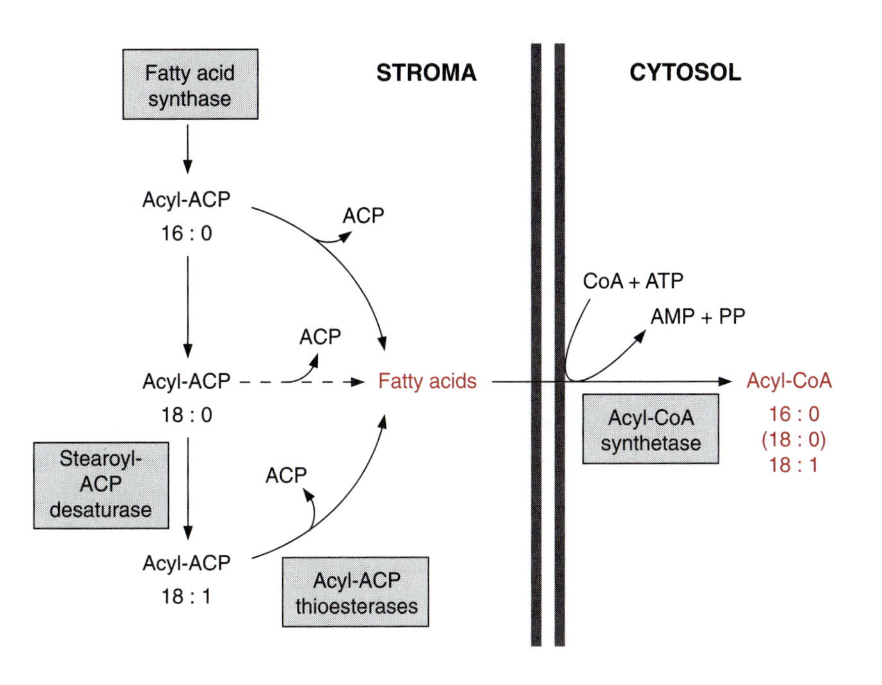

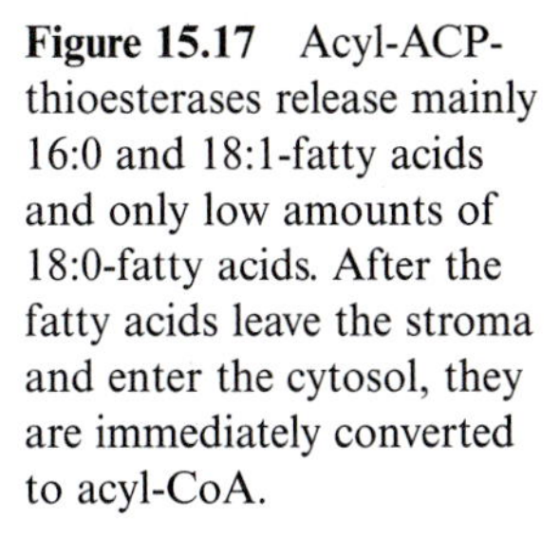
Figure 15.17 Acyl-ACP-thioesterases release mainly 16:0 and 18:1-fatty acids and only low amounts of 18:0-fatty acids. After the fatty acids leave the stroma and enter the cytosol, they are immediately converted to acyl-CoA.

enzyme via the carboxyl groups of glutamate and aspartate. The two Fe-atoms alternate between oxidation state +IV +III and +II. An O_2 molecule is activated by the binding of the two Fe-atoms.

Stearoyl-ACP desaturase is a soluble protein that is localized in chloroplasts and other plastids. The enzyme is so active that normally the newly formed stearoyl-ACP is almost completely converted to oleyl-ACP (18:1) (Fig. 15.17). This soluble desaturase is capable of introducing only one double bond into fatty acids. The introduction of further double bonds is catalyzed by other desaturases, which are integral membrane proteins of the ER and react only with fatty acids that are constituents of membrane lipids. For this reason, they are termed **acyl lipid desaturases**. These membrane-bound desaturases also require O_2 and reduced ferredoxin, similar to the aforementioned ACP-desaturases, but have a different electron transport chain (Fig. 15.15B). The required reducing equivalents are transferred from NADPH via an FAD containing **NADPH-cytochrome b_5-reductase** to cytochrome b_5 and from there further to the actual desaturase, which contains **two Fe atoms** probably bound to histidine residues of the protein. The acyl lipid desaturases belong to a large family of enzymes. Members of this family catalyze the introduction of hydroxyl groups (hydroxylases), epoxy groups (epoxygenases), conjugated double bonds (conjugases), and carbon triple bonds (acetylenases) into fatty acids of acyl lipids.

Acyl-ACP formed as product of fatty acid synthesis in the plastids serves two purposes

Acyl-ACP formed in the plastids has two important functions:

1. It acts as an acyl-donor for the synthesis of plastid membrane lipids. The enzymes of glycerolipid synthesis are in part located in both the inner and outer envelope membranes, and in lipid biosynthesis there is a division of labor between these two membranes. To avoid going into too much detail, no distinction will be made between the lipid biosynthesis of the inner and outer envelope membranes in the following text.
2. For biosynthesis outside the plastids, acyl-ACP is hydrolyzed by **acyl-ACP thioesterases** to **free fatty acids**, which then leave the plastids (Fig. 15.17). It is not known whether this export proceeds via nonspecific diffusion or by specific transport. These free fatty acids are immediately captured outside the outer envelope membrane by conversion to acyl-CoA, as catalyzed by an **acyl-CoA synthetase** with consumption of ATP. Since the thioesterases in the plastids hydrolyze primarily 16:0- and 18:1-acyl-ACP, and to a small extent 18:0-acyl-ACP, the plastids mainly provide CoA-esters with the acyl residues of 18:1 and 16:0 (also a low amount of 18:0) for lipid metabolism outside the plastids.

15.4 Glycerol 3-phosphate is a precursor for the synthesis of glycerolipids

Glycerol 3-phosphate is formed by reduction of dihydroxyacetone phosphate with NADH as reductant (Fig. 15.18). Dihydroxyacetone phosphate reductases are present in the plastid stroma as well as in the cytosol. In plastid lipid biosynthesis, the acyl residues are transferred directly from acyl-ACP to glycerol 3-phosphate. For the first acylation step, mostly an 18:1-, less frequently a 16:0-, and more rarely an 18:0-acyl residue is esterified to position 1. The C-2-position, however, is always esterified with a 16:0-acyl residue. Since this specificity is also observed in cyanobacteria, the glycerolipid biosynthesis pathway of the plastids is called the **prokaryotic pathway**.

For glycerolipid synthesis in the ER membrane, the acyl residues are transferred from acyl-CoA. Here again, the hydroxyl group in C-1 position is esterified with an 18:1-, 16:0-, or 18:0-acyl residue, but in position C-2 always with a somewhat desaturated 18:n-acyl residue. The glycerolipid pathway of the ER membrane is called the **eukaryotic pathway**.

Linkage of the polar head group to diacylglycerol proceeds mostly via an activation of the head group, but in some cases also by an activation of

Plastidal compartment: prokaryotic pathway

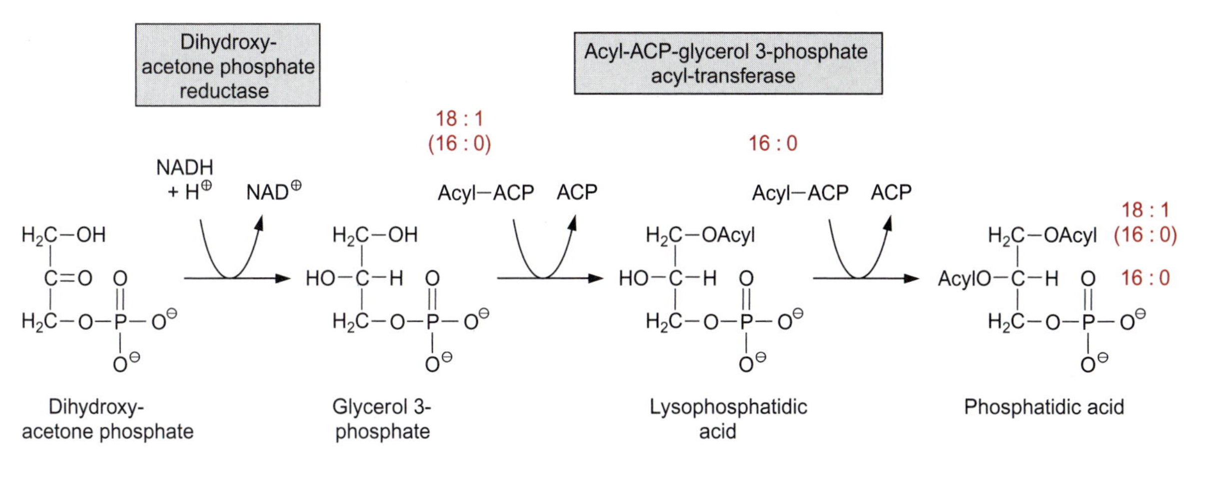

Endoplasmic reticulum: eukaryotic pathway

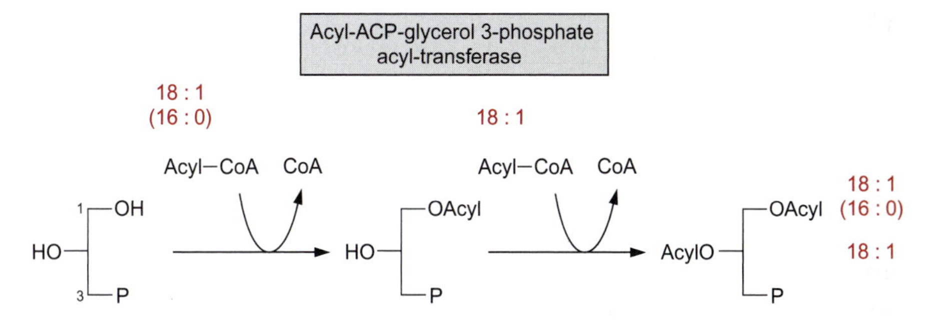

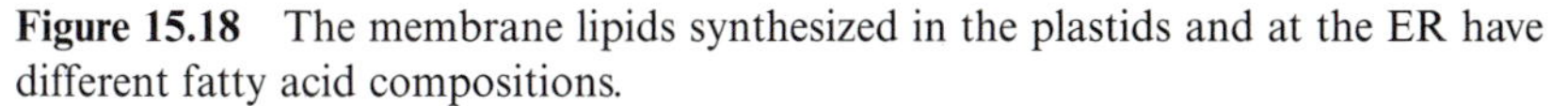

Figure 15.18 The membrane lipids synthesized in the plastids and at the ER have different fatty acid compositions.

diacylglycerol. Choline and ethanolamine are activated by phosphorylation via specific kinases and are then converted via cytidyl transferases by reaction with CTP to **CDP-choline** and **CDP-ethanolamine** (Fig. 15.19). A galactose head group is activated as **UDP-galactose** (Fig. 15.20). The latter is formed from glucose 1-phosphate and UTP via the UDP-glucose pyrophosphorylase (section 9.2) and UDP-glucose epimerase (Fig. 9.21). For the synthesis of digalactosyl diacylglycerol (DGDG) from monogalactosyl diacylglycerol (MGDG), a galactose residue is transferred again from UDP galactose. Also, sulfoquinovose is activated as a UDP derivative, but details of the synthesis of this moiety will not be discussed here. The acceptor for the activated head group is diacylglycerol, which is formed from phosphatidic acid by the hydrolytic release of the phosphate residue.

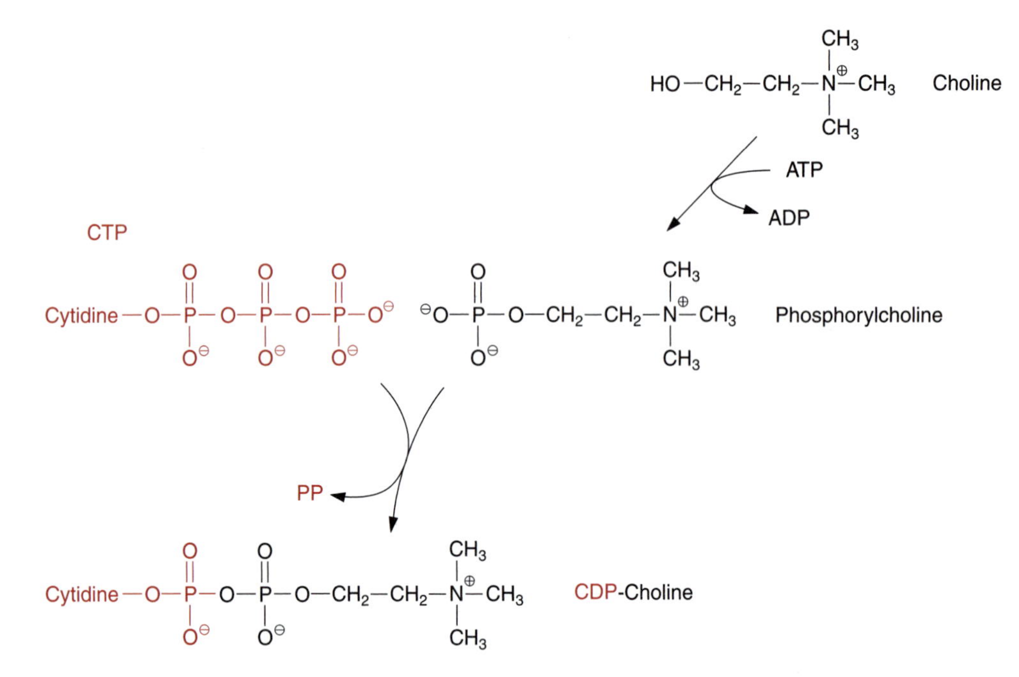

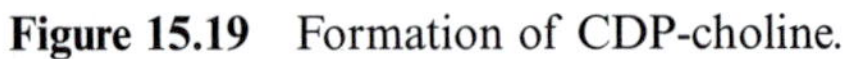

Figure 15.19 Formation of CDP-choline.

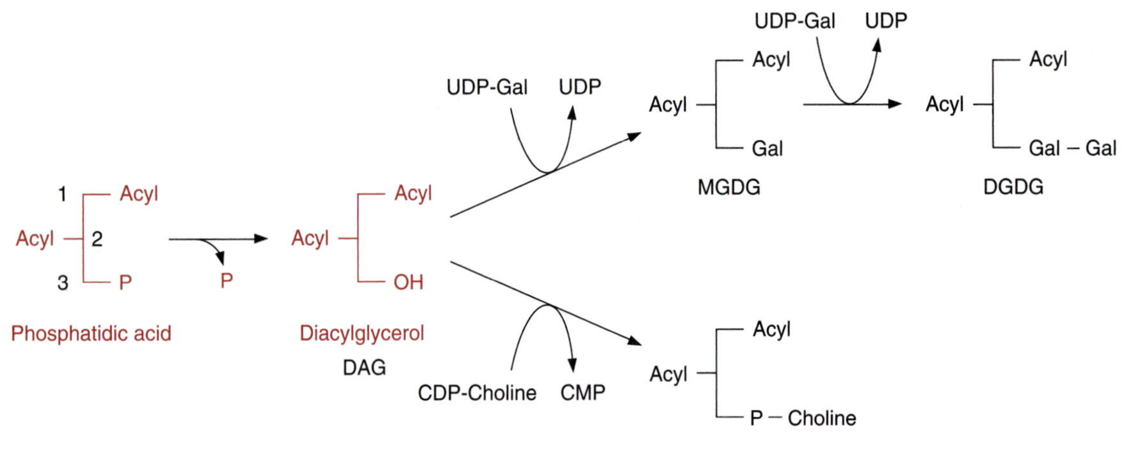

Figure 15.20 Overview of the synthesis of membrane lipids.

The ER membrane is the site of fatty acid elongation and desaturation

As shown in Figure 15.17, the plastids produce 16:0-, 18:1-, and to a lesser extent 18:0-acyl residues. However, some storage lipids contain fatty acids with a greater chain length. This also applies to waxes, which are esters of a long chain fatty acid (C_{20}–C_{24}) with very long chain acyl alcohols (C_{24}–C_{32}). The elongation of fatty acids greater than C_{18} is catalyzed by **elongases**, which are located in the membranes of the ER (Fig. 15.21). Elongation proceeds in the same way as fatty acid synthesis, with the only differences being

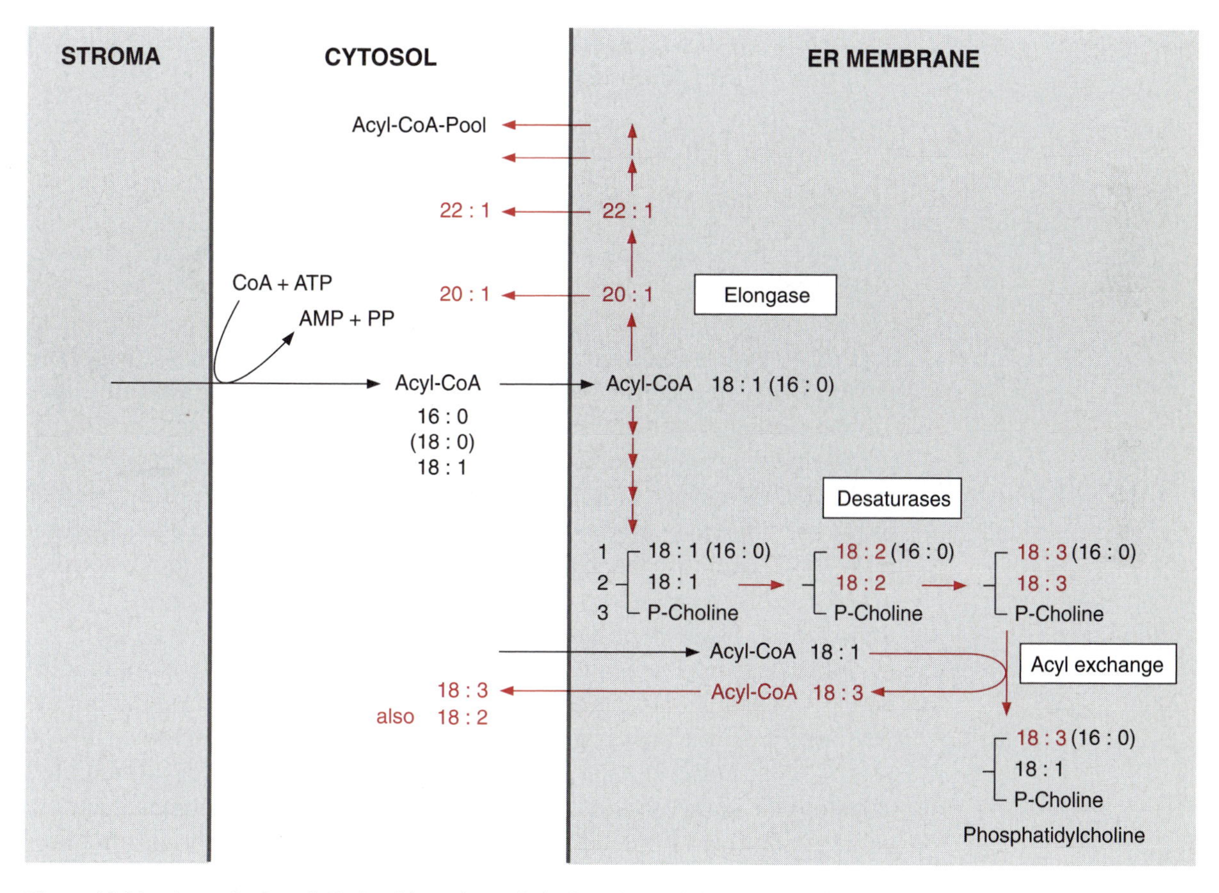

Figure 15.21 A pool of acyl-CoA with various chain lengths and desaturation is present in the cytosol. Acyl residues delivered from the plastids as acyl CoA, are elongated by elongases located at the ER. After incorporation in phosphatidylcholine, 18:1 acyl residues are desaturated to 18:2 and 18:3 by desaturases present in the ER membrane. The more highly desaturated fatty acids in position-2 can be exchanged for 18:1-acyl CoA. In this way the cytosolic pool is provided with 18:2 and 18:3-acyl CoA.

that different enzymes are involved and that the acyl- and malonyl residues are activated as acyl CoA thioesters.

The ER membrane is also the site for further desaturations of the acyl residue. For desaturation, the acyl groups are first incorporated into phospholipids, such as phosphatidylcholine (Fig. 15.21). **Desaturases** bound to the membrane of the ER convert oleate (18:1) to linoleate (18:2) and then to linolenate (18:3). The 18:2- or 18:3-acyl residues in the C-2- position of glycerol can be exchanged for an 18:1-acyl residue, and the latter can then be further desaturated.

The interplay of the desaturases in the plastids and the ER provides the cell with an **acyl CoA pool** to cover the various needs of the cytosol. The 16:0-, 18:0-, and 18:1-acyl residues for this pool are delivered by the plastids, and the longer chain and more highly unsaturated acyl residues are provided through modifications by the ER membrane.

Some of the plastid membrane lipids are formed via the eukaryotic pathway

The synthesis of glycerolipids destined for the plastid membranes occurs in the envelope membranes of the plastids. Besides the prokaryotic pathway of glycerolipid synthesis, in which the acyl residues are directly transferred from acyl-ACP, glycerolipids are also synthesized via the eukaryotic pathway. The desaturases of the ER membrane can provide double unsaturated fatty acids for plastid membrane lipids. A precursor for this can be, for instance, phosphatidylcholine with double unsaturated fatty acids, which has been formed in the ER membrane (Fig. 15.21). This phosphatidylcholine is transferred to the envelope membrane of the plastids and hydrolyzed to diacylglycerol (Fig. 15.22). In most cases, the latter is provided with a head group consisting of one or two galactose residues. The acyl residues can be further desaturated to 18:3 by a desaturase present in the envelope membrane.

Some desaturases in the plastid envelope are able to desaturate lipid-bound 18:1- and 16:0-acyl residues. A comparison of the acyl residues in the C2-position (in the prokaryotic pathway 16:0 and in the eukaryotic pathway 18:*n*) shows, however, that a large proportion of the highly unsaturated galactolipids in the plastids are formed via a detour through the eukaryotic pathway. The membrane lipids present in the envelope membrane are probably transferred by a special transfer protein to the thylakoid membrane.

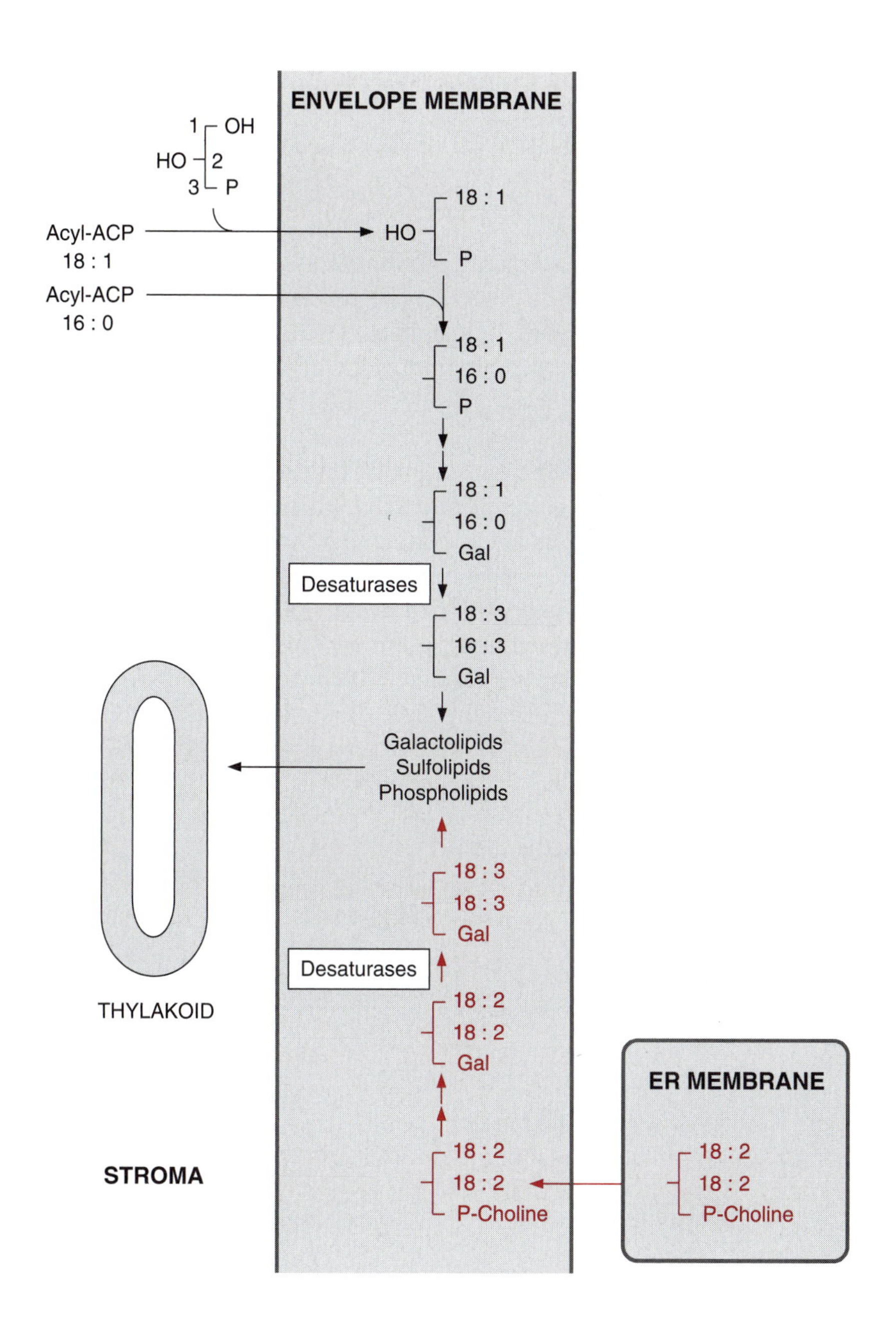

Figure 15.22 Part of the membrane lipids of the thylakoid membranes are synthesized via the eukaryotic pathway (red) and other parts via the prokaryotic pathway (black).

15.5 Triacylglycerols are formed in the membranes of the endoplasmic reticulum

The lipid content of mature seeds can amount to 45% of the dry weight, but in some cases (e.g., ricinus) up to 85%. The precursor for the synthesis of triacylglycerol is again glycerol 3-phosphate. There are at least four pathways for its synthesis, the two most important of which will be discussed here (Fig. 15.23):

1. **Phosphatidic acid** is formed by acylation of the hydroxyl groups in the C-1 and C-2 positions of glycerol, and after hydrolysis of the phosphate residue, the C-3 position is also acylated. The total cytosolic acyl CoA pool is available for these acylations, but because of the eukaryotic pathway, position-2 is mostly esterified by a C_{18} acyl residue (Fig. 15.21).
2. **Phosphatidylcholine** is formed first, and its acyl residues are further desaturated. The choline phosphate residue is then cleaved off by transfer to CMP, and the corresponding diacylglycerol is esterified with a third acyl

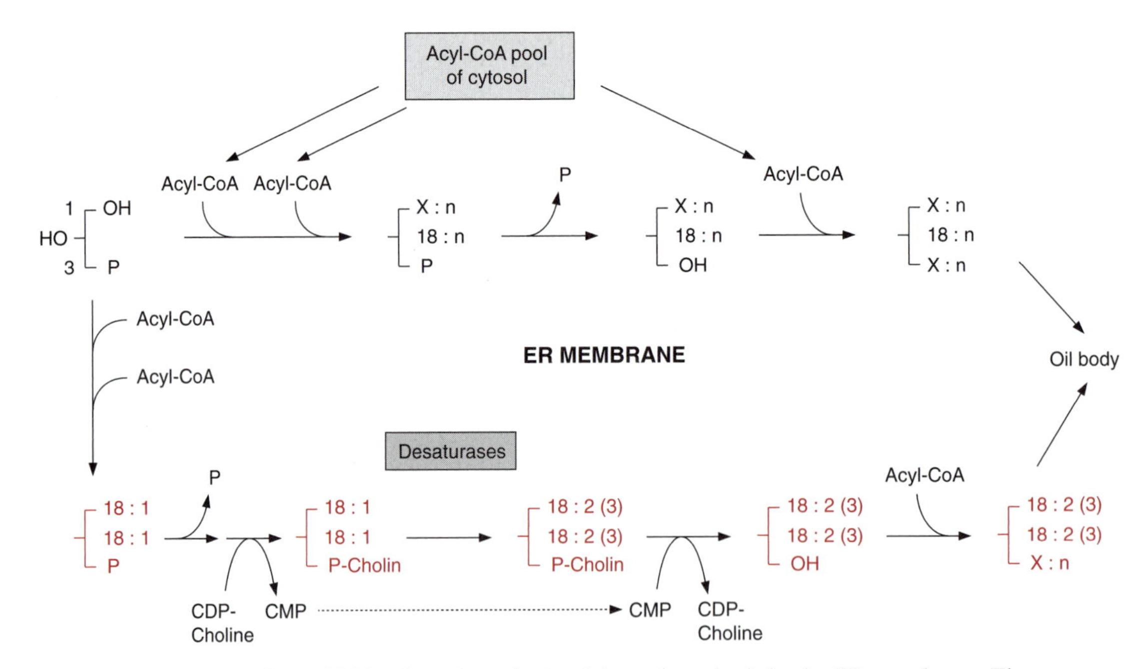

Figure 15.23 Overview of triacylglycerol synthesis in the ER membrane. The synthesis occurs either by acylation of glycerolphosphate (black) or via the intermediary synthesis of a phospholipid, of which the fatty acids are desaturated by desaturases in the ER membrane (red).

residue. This pathway operates frequently in the synthesis of highly unsaturated triacylglycerols.

Cross-connections exist between the two pathways, but for the sake of simplicity these are not discussed in Figure 15.23.

Plant fat is used not only for nutrition but also as a raw material in industry

About 20% of the human caloric nutritional uptake in industrialized countries is due to the consumption of plant fats. Plant fats have a higher content of unsaturated fatty acids than animal fats. Human metabolism requires unsaturated fatty acids with two or more carbon-carbon double bonds, but is not able to introduce double bonds in the 12 and 15 positions. This is why **linoleic acid** and **linolenic acid** (Fig. 15.4) are essential fatty acids, which are absolutely necessary in the human diet.

About 2 % of the world production of plant fats is utilized as a raw material for industrial purposes. Moreover, fatty acid methylesters synthesized from rapeseed oil are used in some countries as automobile fuel (biodiesel).

Table 15.3 shows that the fatty acid composition of various plant oils is variable. Triacylglycerols of some plants contain large quantities of rare fatty acids, which are used for industrial purposes (Table 15.4). Oil from palm kernels has a high content of the short-chained **lauric acid** (12:0), which

Table 15.3: Fatty acid composition of the most important vegetable oils

	Soy oil	Palm oil	Rapeseed oil*	Sunflower oil	Peanut oil
		% of total fatty acids			
16:0	11	42	4	5	10
18:0	3	5	1	1	3
18:1	22	41	60	15	50
18:2	55	10	20	79	30
18:3	8	0	9	0	0
20:1	0	0	2	0	3
others	1	0	2	0	0
World production 1994 (106 metric tons per year)**	18,6	14,7	10,3	7,9	4,2

* Erucic acid free varieties, after Baumann et al. 1988.
** Töpfer International, statistical information.

Table 15.4: Industrial utilization of fatty acids from vegetable oils

	Main source	Utilization for
Lauric acid (12:0)	Palm kernel, coconut	Soap, detergents, cosmetics
Linolenic acid (18:3)	Linseed	Paints, lacquers
Ricinoleic acid (18:1, Δ9, 12-OH)	Castor bean	Surface protectors, lubricant
Erucic acid (22:1, Δ13)	Rapeseed	Tensides
		Foam control for detergents, lubricant, synthesis of artificial fibers

is used as a raw material for the production of detergents and cosmetics. Large plantations of oil palms in Southeast Asia ensure its supply to the oleochemical industry. **Linolenic acid** from European linseed is used for the production of paints. **Ricinoleic acid**, a rare fatty acid that contains a hydroxyl group in the C-12-position and makes up about 90% of the fatty acids in castor oil, is used in industry as a lubricant and as a means of surface protection. **Erucic acid** is used for similar purposes. Earlier varieties of rapeseed contained erucic acid in their triacylglycerols, and for this reason rapeseed oil was then of inferior value in terms of nutrition. About 40 years ago, successful crossings led to a breakthrough in the breeding of rapeseed free of erucic acid, making the rapeseed oil suitable for human consumption. The values in Table 15.3 are for this type, which is cultivated worldwide. However, rapeseed varieties containing a high percentage of erucic acid are again being cultivated for industrial demands.

Plant fats are customized by genetic engineering

The progress in gene technology now makes it possible to alter the quality of plant fats in a defined way by changing the enzymatic profile of the cell. The procedures for the introduction of a new enzyme into a cell, or for eliminating the activity of an enzyme present in the cell, will be described in detail in Chapter 22. Three cases of the alteration of oil crops by genetic engineering will be illustrated here.

1. The lauric acid present in palm kernel and coconut oil (12:0) is an important raw material for the production of soaps, detergents, and cosmetics. Recently, rapeseed plants that contained oil with a lauric acid content of 66 % were generated by genetic transformation. The synthesis of fatty acids is terminated by the hydrolysis of acyl-ACP (Fig. 15.17). An **acyl-**

ACP thioesterase, which specifically hydrolyzes lauroyl-ACP, was isolated and cloned from the seeds of the California bay tree (*Umbellularia californica*), which contains a very high proportion of lauric acid in its storage lipids. The introduction of the gene for this acyl-ACP thioesterase into the developing rapeseed terminates its fatty acid synthesis at acyl (12:0), and the lauric acid released is incorporated into the seed oil. Field tests have shown that these plants grow normally and produce normal yields.

2. A relatively high content of stearic acid (18:0) improves the heat stability of fats for deep frying and for the production of margarine. The stearic acid content in rapeseed oil has been increased from 1%–2% to 40% by decreasing the activity of stearoyl-ACP desaturase using antisense technique (see section 22.5). On the other hand, genetic engineering has been employed to increase highly unsaturated fatty acids (e.g., in rapeseed) in order to give the oil a higher nutritional quality. In soybeans, the proportion of twice unsaturated fatty acids in the oil could be increased to 30%. Nutritional physiologists have found out that highly unsaturated fatty acids (20:4, 20:5, 22:6), which are present only in fish oil, are very beneficial for human health. It is tried by genetic engineering to produce rape lines with highly unsaturated fatty acids of C20 and C22. This would be an attempt to use gene technology for the production of "health food."

Erucic acid (Fig. 15.24) could be an important industrial raw material, for instance for the synthesis of synthetic fibers and foam control in detergents. At the moment, however, its utilization is limited, since conventional breeding has not succeeded in increasing the erucic acid content to more than 50% of the fatty acids in the seed oil. The cost of separating erucic acid and disposing of the other fatty acids is so high that for many purposes the industrial use of rapeseed oil as a source of erucic acid from the cultivars available at present is not economically viable. Attempts are being made to increase further the erucic acid content of rapeseed oil by overexpression of genes for elongases and by transferring the genes encoding enzymes that catalyze the specific incorporation of erucic acid to the C-2-position of triacylglycerols. If this were to be successful, present petrochemical-based industrial processes could be replaced with processes using rapeseed oil as a renewable raw material.

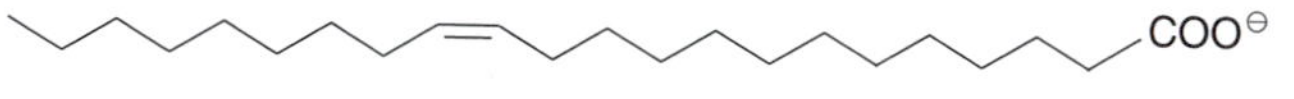

$COO^{\ominus}$

Erucic acid (22 : 1, Δ^{13}-*cis*)

Figure 15.24 Erucic acid, an industrial raw material.

15.6 During seed germination, storage lipids are mobilized for the production of carbohydrates in the glyoxysomes

At the beginning of germination, storage proteins (Chapter 14) are degraded to amino acids, from which the enzymes required for the mobilization of the storage lipids are synthesized. These enzymes include **lipases**, which catalyze the hydrolysis of triacylglycerols to glycerol and fatty acids. Lipases bind to the oleosins of the oil bodies (section 15.2). The **glycerol** formed by the hydrolysis of triacylglycerol, after phosphorylation to glycerol 3-phosphate and its subsequent oxidation to dihydroxyacetone phosphate, can be fed into the **gluconeogenesis** pathway (Chapter 9). The released **free fatty acids** are first activated as CoA-thioesters and then degraded to acetyl CoA by **β-oxidation** (Fig. 15.25). This process occurs in specialized peroxisomes called **glyoxysomes**.

Although in principle β-oxidation represents the reversal of the fatty acid synthesis shown in Figure 15.8, there are distinct differences that enable high metabolic fluxes through these two metabolic pathways operating in opposite directions. The differences between β-oxidation and fatty acid synthesis are:

1. In acyl CoA dehydrogenation, hydrogen is transferred via an **FAD-dependent oxidase** to form H_2O_2. A catalase irreversibly eliminates the toxic H_2O_2 at the site of its production by conversion to water and oxygen (section 7.1).
2. **β-L Hydroxyacyl CoA** is formed during the hydration of enoyl CoA, in contrast to the corresponding D-enantiomer during synthesis.
3. Hydrogen is transferred to **NAD** during the second dehydrogenation step. Normally, the NAD system in the cell is highly oxidized (section 7.3), driving the reaction in the direction of hydroxyacyl CoA oxidation. It is not known which reactions utilize the NADH formed in the peroxisomes.
4. In an irreversible reaction, **CoASH-mediated thiolysis** cleaves β-ketoacyl CoA to form one molecule of acetyl CoA and one of acyl CoA shortened by two C atoms.

During the degradation of unsaturated fatty acids, intermediate products are formed that cannot be metabolized by the reactions of β-oxidation. **Δ3-*cis*-enoyl CoA** (Fig. 15.26), which is formed during the degradation of oleic acid, is converted by an **isomerase** shifting the double bond to **Δ2-*trans*-enoyl CoA**, an intermediate of β-oxidation. In the β-oxidation of linoleic or linolenic acid, the second double bond in the corresponding intermediate is

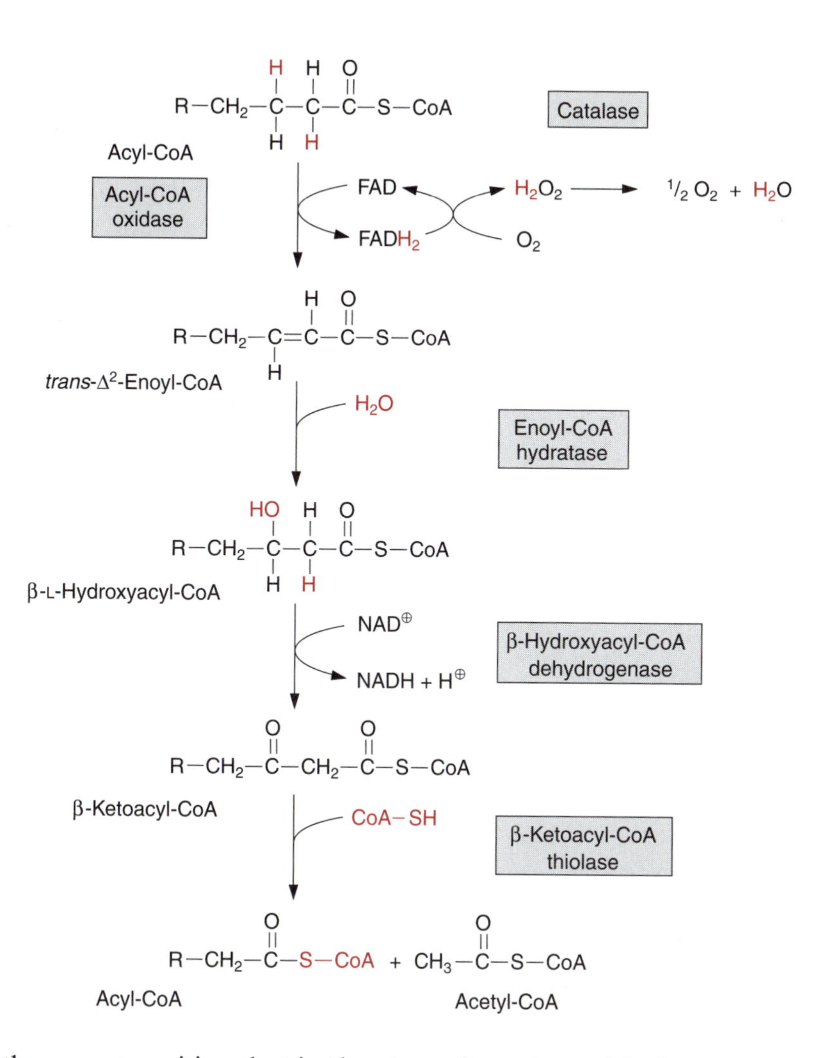

Figure 15.25 β-Oxidation of fatty acids in the glyoxysomes. The fatty acids are first activated as CoA-thioesters and then converted to acetyl CoA and a fatty acid shortened by two carbon atoms, involving dehydrogenation via an FAD-dependent oxidase, addition of water, a second dehydrogenation (by NAD), and thiolysis by CoASH.

in the correct position, but in the *cis*-configuration, with the consequence that its hydration by enoyl CoA hydratase results in the formation of β-D-hydroxyacyl CoA. The latter is converted by a dehydratase to Δ2-*trans*-enoyl-CoA, an intermediate of β-oxidation.

The glyoxylate cycle enables plants to synthesize hexoses from acetyl CoA

In contrast to animals, which are unable to synthesize glucose from acetyl CoA, plants are capable of gluconeogenesis from lipids, as they possess the enzymes of the glyoxylate cycle (Fig. 15.27). Like β-oxidation, this cycle is located in the glyoxysomes, which are named after the cycle. The two starting reactions of the cycle are identical to those of the citrate cycle (see

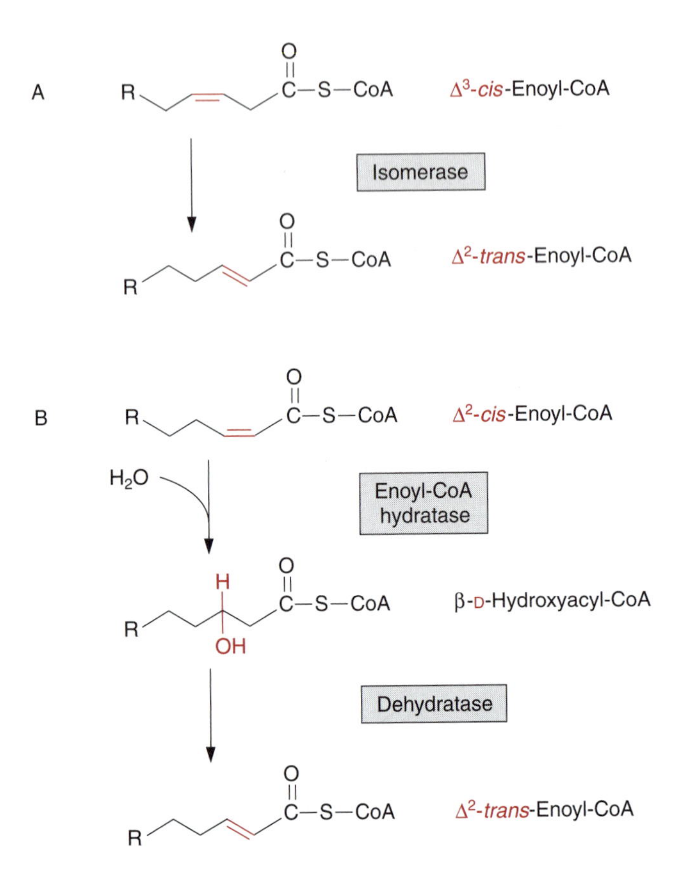

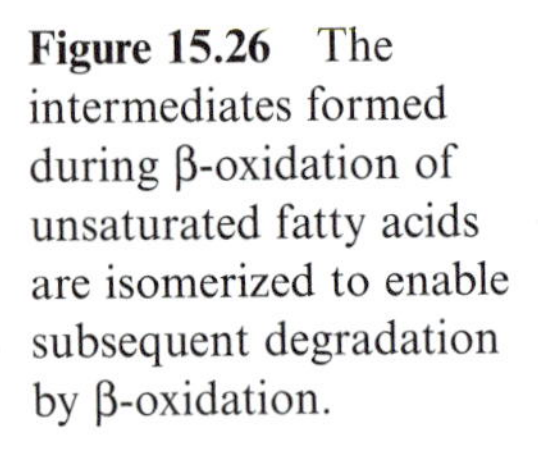
Figure 15.26 The intermediates formed during β-oxidation of unsaturated fatty acids are isomerized to enable subsequent degradation by β-oxidation.

Fig. 5.3). Acetyl CoA condenses with oxaloacetate, as catalyzed by **citrate synthase**, to form citrate, and the latter is converted in the cytosol by aconitase to isocitrate. The further reaction of isocitrate, however, is a speciality of the glyoxylate cycle: isocitrate is split by **isocitrate lyase** into succinate and glyoxylate (Fig. 15.28). **Malate synthase**, the second special enzyme of the glyoxylate cycle, catalyzes the instantaneous condensation of glyoxylate with acetyl CoA to form malate. The hydrolysis of the CoA-thioester makes this reaction irreversible. As in the citrate cycle, malate is oxidized by **malate dehydrogenase** to oxaloacetate, thus completing the glyoxylate cycle. In this way one molecule of succinate is generated from two molecules of acetyl CoA. The succinate is transferred to the mitochondria and converted there to oxaloacetate by a partial reaction of the citrate cycle. The oxaloacetate is released from the mitochondria by the oxaloacetate translocator and converted in the cytosol by phosphoenolpyruvate carboxykinase to phosphoenolpyruvate (Fig. 8.10). Phosphoenolpyruvate is a precursor for the synthesis of hexoses by the **gluconeogenesis pathway** and for other biosynthetic processes.

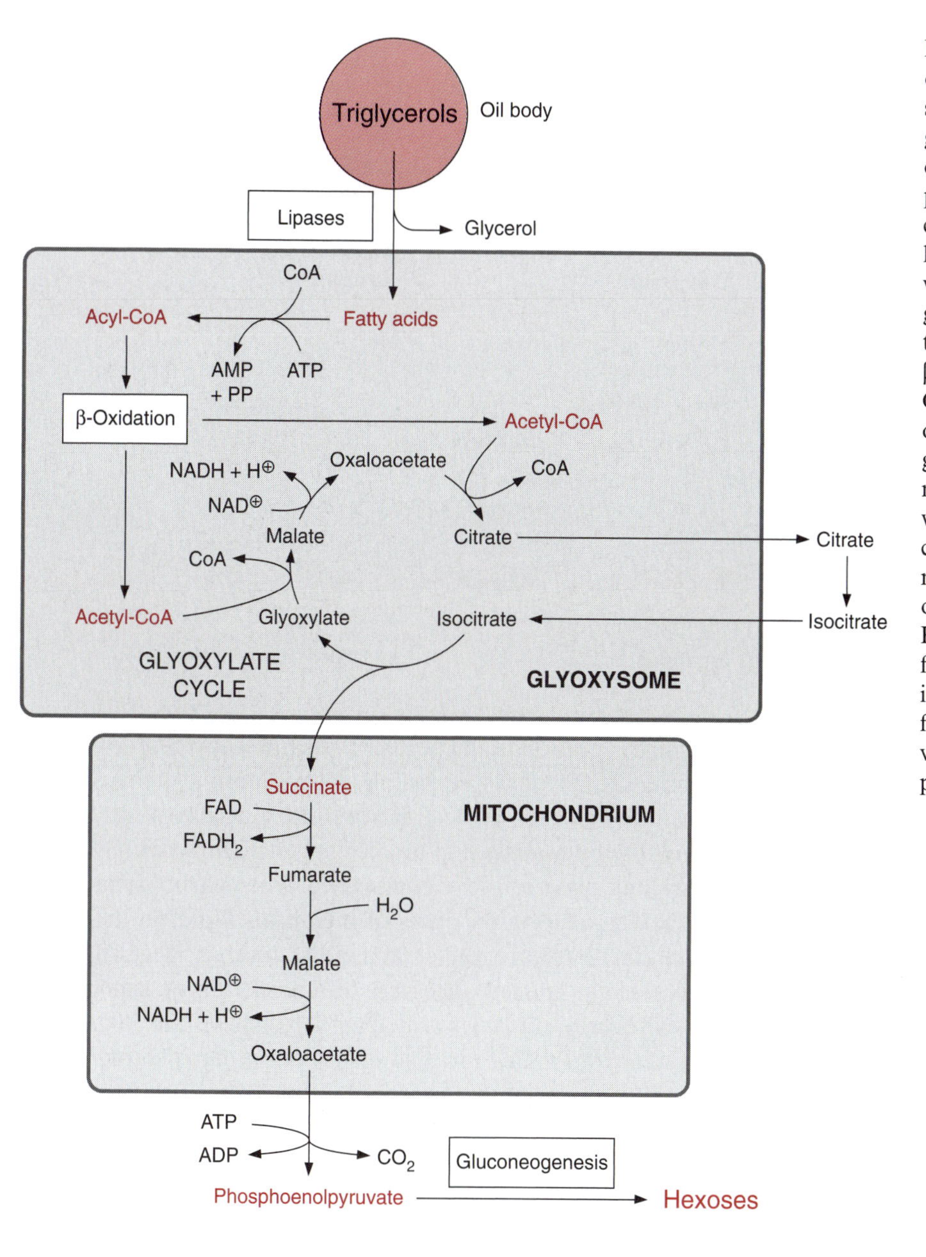

Figure 15.27 Mobilization of storage lipids for the synthesis of hexoses during germination. The hydrolysis of the triacylglycerols present in the oil bodies, as catalyzed by oleosin-bound lipases, yields fatty acids, which are activated in the glyoxysomes as CoA-thioesters and degraded by β-oxidation into acetyl CoA. From two molecules of acetyl CoA, the glyoxylate cycle forms one molecule of succinate, which is converted by the citrate cycle in the mitochondria to oxaloacetate. Phosphoenolpyruvate formed from oxaloacetate in the cytosol is a precursor for the synthesis of hexoses via the gluconeogenesis pathway.

Reactions with toxic intermediates take place in peroxisomes

There is a simple explanation for the fact that β-oxidation and the closely related glyoxylate cycle proceed in peroxisomes. As in photorespiration, in which peroxisomes participate (section 7.1), the toxic substances $\mathbf{H_2O_2}$ and **glyoxylate** are also formed as intermediates in the conversion of fatty acids

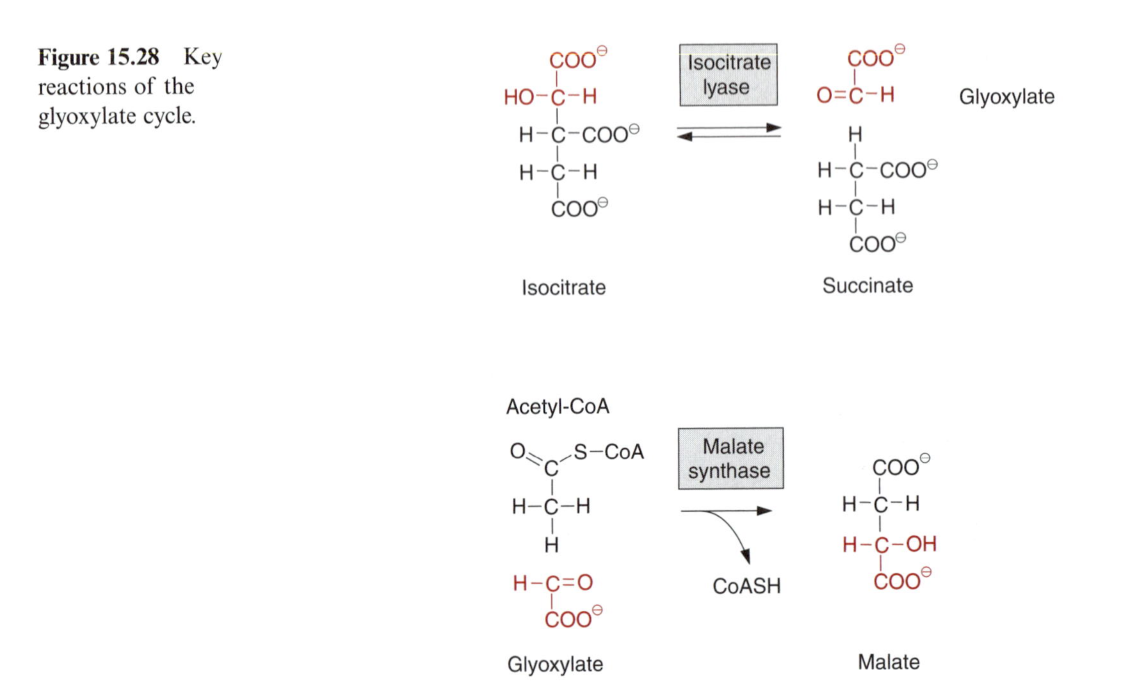

Figure 15.28 Key reactions of the glyoxylate cycle.

to succinate. Compartmentation in the peroxisomes prevents these toxic substances from reaching the cytosol (section 7.4). β-Oxidation also takes place, although at a much lower rate, in leaf peroxisomes, where it serves the purpose of recycling the fatty acids that are no longer required or have been damaged. This recycling plays a role in **senescence**, when carbohydrates are synthesized from the degradation products of membrane lipids in the senescent leaves and are transferred to the stem via the phloem (section 19.7). During senescence, one can indeed observe a differentiation of leaf peroxisomes into **glyoxysomes** (alternatively termed **gerontosomes**). The hydrophobic amino acids leucine and valine are also degraded in peroxisomes.

15.7 Lipoxygenase is involved in the synthesis of oxylipins, which are acting as defense and signal substances

Oxylipins, which derive from the oxygenation of unsaturated fatty acids, comprise a multiplicity of various signal substances in the animal and plant kingdoms. In animals, oxylipins include prostaglandins, leucotriens, and thromboxans, of which the specific roles in the regulation of physiological processes are known to a large extent. Likewise, plant oxylipins comprise a very large number of substances mostly different from those in animals. They are involved in defense reactions (e.g., as signal substances to regulate defense cascades), but also as fungicides, bactericides, and insecticides, or as volatile signals to attract predators, insects that feed on herbivores. Moreover, they participate in wound healing, regulate vegetative growth, and induce senescence. Our knowledge about these important substances is still fragmentary.

In plants, oxylipins are synthesized primarily via **lipoxygenases**. These are **dioxygenases**, mediating the incorporation of both atoms of the oxygen molecule into the fatty acid molecule, in contrast to monooxygenases (section 15.3), which catalyze the incorporation of only one O atom of an O_2 molecule. The lipoxygenases are a family of enzymes catalyzing the dioxygenation of multiple-unsaturated fatty acids, such as linoleic acid and linolenic acid, which contain a ***cis,cis*-1,4-pentadiene** sequence (colored red in Fig. 15.29). At the end of this sequence, a hydroperoxide group is introduced by reaction with O_2 and the neighboring double bond is shifted by one C-position in the direction of the other double bond, thereby attaining a *trans*-configuration. The **hydroperoxide lyase** catalyzes the cleavage of hydroperoxylinolenic acid into a 12-oxo-acid and a 3-*cis*-hexenal. More hexenals are formed by shifting of the double bond, and their reduction leads to the formation of hexenols (Fig. 15.29). In an analogous way, hydroperoxylinoleic acid yields hexanal and its reduction leads to hexanol.

Hexanals, **hexenals**, **hexanols**, and **hexenols** are volatile **aromatic substances** that are important components of the characteristic odor and taste of many fruits and vegetables. The wide range of aromas includes fruity, sweet, spicy, and grasslike. Work is in progress to improve the taste of tomatoes by increasing their hexenol content using genetic engineering. The quality of olive oil, for instance, depends on its content of hexenals and hexenols. Hexenals are responsible for the aroma of black tea. Green tea is processed to black tea by heat, whereby hexenals condense to aromatic compounds, which give black tea its typical taste. Large amounts of hexenals and hexenols are produced industrially as aromatic substances in the food industry or as ingredients in the preparation of perfumes.

The characteristic smell of freshly cut grass is caused primarily by the release of hexenals and hexenols, indicating that the activity of lipoxygenase and hydroperoxide lyase is greatly increased by tissue wounding. This is part of a defense reaction. When a leaf is damaged by feeding larvae, it attracts natural enemies of the herbivores by the emission of the volatiles mentioned previously. To give an example: After the wounding of corn or cotton,

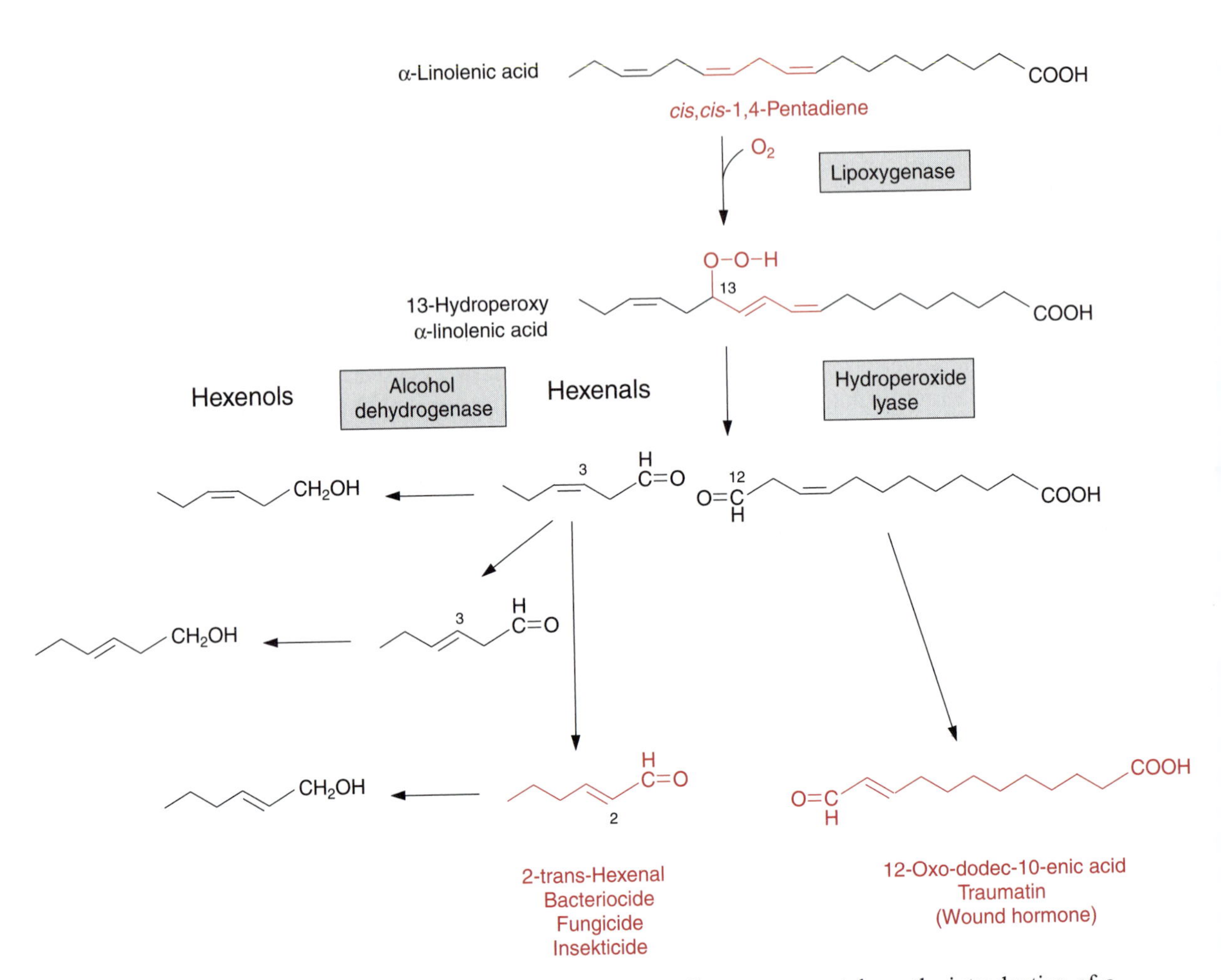

Figure 15.29 By reaction with O_2, lipoxygenase catalyzes the introduction of a peroxide group at the end of a *cis,cis*-1,4-pentadiene sequence (red). Hydroperoxide lyase cleaves the C-C bond between C atoms 12 and 13. The hexenal thus formed can be isomerized by shifting of the double bond, probably due to enzymatic catalysis. The hexenals are reduced to the corresponding hexenols by an alcohol dehydrogenase. The 12-oxo-acid formed as a second product is isomerized to traumatin. There are also lipoxygenases, which insert the peroxygroup in position 9.

parasitic wasps are attracted, which inject their eggs into the feeding larvae. Moreover, 2-*trans* hexenal (colored red in Fig. 15.29) is a strong bactericide, fungicide, and insecticide. But hexenals also interact with transcription factors in defense reactions. 12-Oxo-dodec-10-enic acid, which is formed from the cleavage product of hydroperoxylinolenic acid by the shifting of a double bond, has the properties of a wound hormone and has therefore been named **traumatin**. Traumatin induces cell division in neighboring cells. This

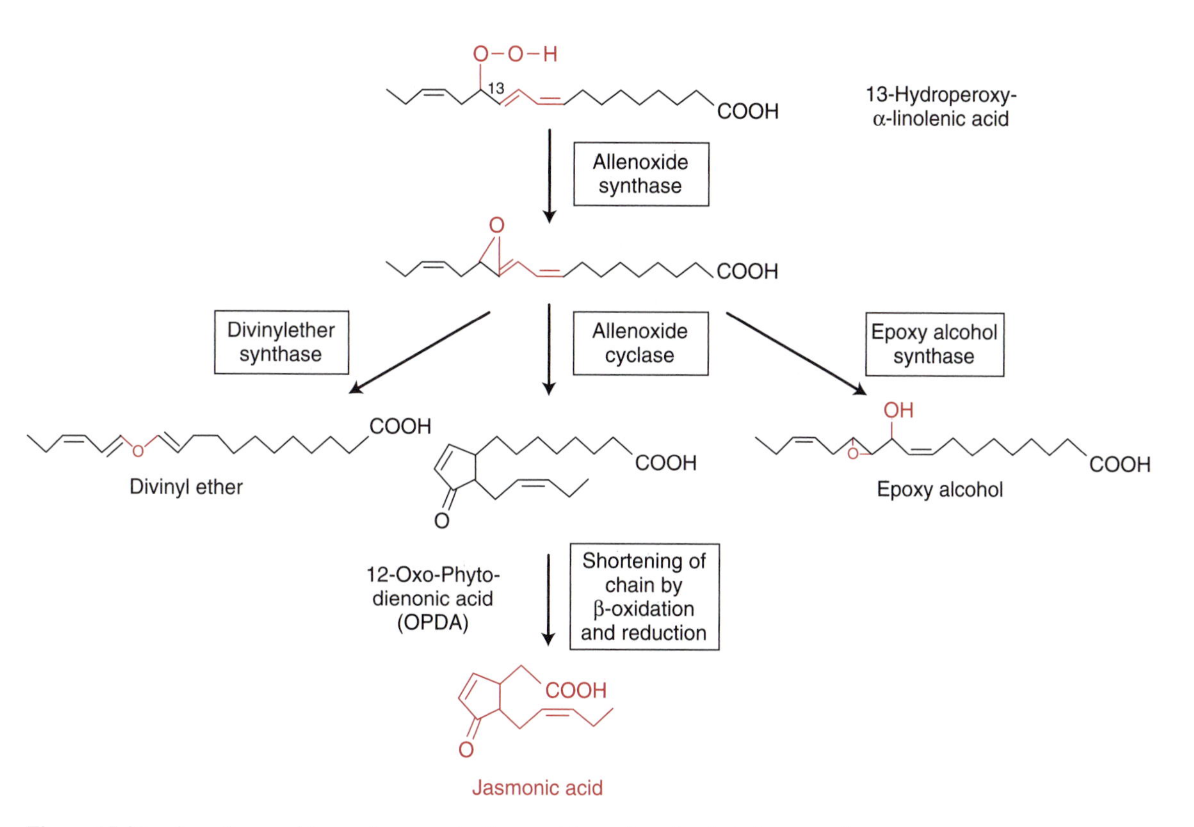

Figure 15.30 An allene oxide synthase and allene oxide cyclase (both belong to the P_{450} family of enzymes; see section 18.2) catalyze the cyclization of the hydroperoxyl linolenic acid by shifting the oxygen. These reactions take place in the chloroplasts. The shortening of the fatty acid chain by six C-atoms via β-oxidation leads to the formation of jasmonic acid, a phytohormone and signal substance. The peroxisomes are the site of the β-oxidation. Divinyl ether synthase catalyzes the conversion to divinyl ethers and epoxy alcohol synthase to epoxy alcohols. Both substances are formed as fungicides in response to fungal infection.

results in the formation of calli and the wound is sealed. However, our knowledge of these defense processes is still rather fragmentary.

Hydroperoxy α-linonelic acid is converted by divinylether synthase into a **divinylether** (Fig. 15.30). Such divinylethers are formed as fungicide in very high amounts in potato after infection with the noxious fungus *Phytophtera infestans*. Allene oxide synthase and cyclase catalyze the cyclization of 13-hydroperoxy-α-linolenic acid (Fig. 15.30). Shortening of the hydrocarbon chain by β-oxidation of the product (Fig. 15.25) results in the formation of **jasmonic acid**. Plants contain many derivatives of jasmonic acid, including sulfatated compounds and methyl esters, which are collectively termed

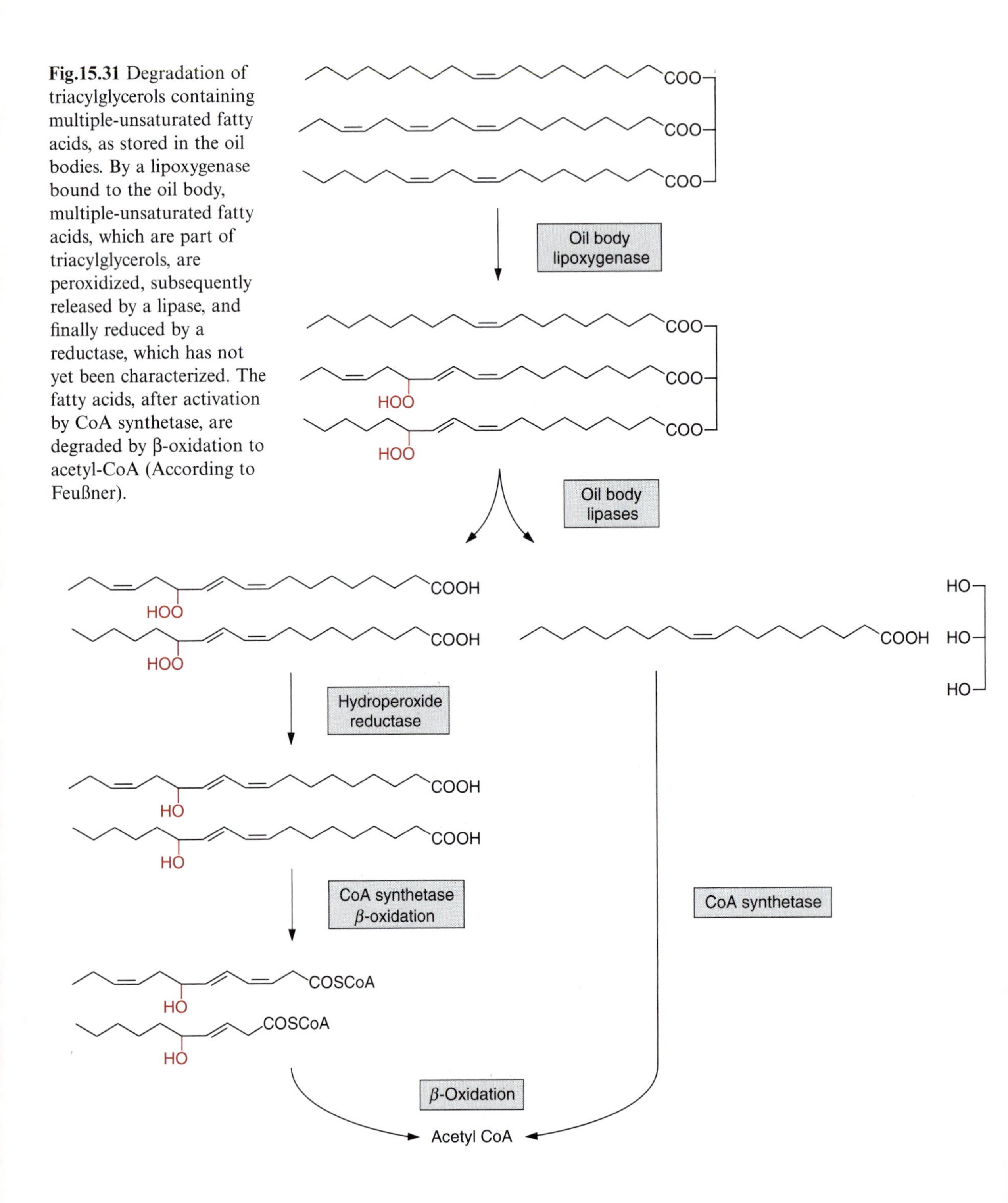

Fig.15.31 Degradation of triacylglycerols containing multiple-unsaturated fatty acids, as stored in the oil bodies. By a lipoxygenase bound to the oil body, multiple-unsaturated fatty acids, which are part of triacylglycerols, are peroxidized, subsequently released by a lipase, and finally reduced by a reductase, which has not yet been characterized. The fatty acids, after activation by CoA synthetase, are degraded by β-oxidation to acetyl-CoA (According to Feußner).

jasmonates. They represent a family of substances with distinct hormone-like functions. It has been estimated that the jasmonates in total regulate the expression of several hundred genes. They play, for instance, an important role in plant resistance to insects and disease; the formation of flowers, fruits, and seeds; and the initiation of senescence (section 19.9).

It was recently shown that lipoxygenases are also involved in the mobilization of storage lipids contained in oil bodies. The lipid monolayer enclosing the oil bodies contains, among other proteins, lipoxygenases. These catalyze the introduction of peroxide groups into multiple-unsaturated fatty acids, as long as they are constituents of triacylglycerols (Fig.15.31). Only after this peroxidation are these triacylglycerols hydrolyzed by a lipase, which is also bound to the oil body. After the reduction of the released peroxy fatty acids to hydroxy fatty acids, the latter are degraded in the glyoxysomes by β-oxidation to acetyl-CoA (Fig.15.25). This pathway for the mobilization of storage lipids exists in parallel to the "classic" pathway of triacylglycerol mobilization initiated by the activity of lipases, as discussed previously. The contribution of both pathways in lipid mobilization seems to be dependent on the species. The acetyl-CoA thus generated is substrate for gluconeogenesis via the glyoxylate cycle (Fig.15.27).

Further reading

Blee, E. Impact of phyto-oxylipins in plant defense. Trends Plant Sci 7, 315–321 (2002).

Caboon, E. B., Ripp, K. G., Hall, S. E., Kinney, A. J. Formation of conjugated Δ8, Δ10 double bonds by delta12-oleic acid desaturase related enzymes. J Biol Chem 276, 2083–2087 (2001).

Chasan, R. Engineering of fatty acids: The long and short of it. Plant Cell 7, 235–237 (1995).

Dörmann, P., Benning, C. Galactolipids rule in seed plants. Trends Plant Sci 7, 112–117 (2002).

Eastmond P. J., Graham, I. A. Re-examining the role of the glyoxylate cycle in oilseeds. Trends Plant Sci 6, 72–77 (2001).

Feussner, I., Kühn, H., Wasternack, C. Lipoxygenase-dependent degradation of storage lipids. Trends Plant Sci 6, 1360–1385 (2001).

Feussner, I., Wasternack, C. The lipoxygenase pathway. Annu Rev Plant Physiol Plant Mol Biol 53, 275–297 (2002).

Hatanaka, A. The biogeneration of green odour by green leaves. Phytochemistry 34, 1201–1218 (1993).

Huang, A. H. C. Oleosin and oil bodies in seeds and other organs. Plant Physiol 110, 1055–1062 (1996).

Kader, J.-C. Lipid-transfer proteins: A puzzling family of plant proteins. Trends Plant Sci 2, 66–70 (1997).

Murakami, Y., Tsuyama, M., Kobayashi, Y., Kodama, H., Ida, K. Trienoic fatty acids and plant tolerance of high temperature. Science 287, 476–479, (2000).

Ohlrogge, J. B., Browse, J. Lipid biosynthesis. Plant Cell, 7, 957–970 (1995).

Paré, P. W., Tumlinson, J. H. Plant volatiles as a defense against insect herbivores. Plant Physiol 121, 325–333 (1999).

Porta, H., Rocha-Sosa, M. Plant lipoxygenases. Physiological and molecular features. Plant Physiol 130, 15–21 (2002).

Reumann, S., Bettermann, M., Benz, R., Heldt, H. W. Evidence for the presence of a porin in the membrane of glyoxysomes of castor bean. Plant Physiol 115, 891–899 (1997).

Ross; J. H. E., Sanchez, J., Millan, F., Murphy, D. J. Differential presence of oleosins in oleogenic seed and mesocarp tissue in olive (*Olea europea*) and avocado (*Persea americana*). Plant Sci 93, 203–210 (1993).

Schaller, H. The role of sterols in plant growth. Prog Lipid Res 42, 163–175 (2003).

Shanklin, J., Cahoon, E. B. Desaturation and related modifications of fatty acids. Annu Rev Plant Physiol Plant Mol Biol 49, 611–641 (1998).

Spassieva, S., Hille, J. Plant Sphingolipids today—are they still enigmatic? Plant Biol 5, 125–136 (2003).

Sperling, P., Heinz, E. Plant sphingolipids: Structural diversity, biosynthesis, first genes and functions. Biochim Biophys Acta 1632, 1–5 (2003).

Sperling, P., Ternes, P., Zank, T. K., Heinz, E. The evolution of desaturases. Prostaglandins, Leukotriens, Essential Fatty Acids 68, 73–95 (2003).

Stumpf, P. K. (ed.) Lipids: Structure and Function. The Biochemistry of Plants, Vol. 9, Academic Press, Orlando, San Diego, New York, (1987).

Voelker, T., Kinney, A. J. Variations in the biosynthesis of seed-storage lipids. Annu Rev Plant Physiol Plant Mol Biol 52, 335–361 (2001).

Weber, H. Fatty acid derived signals in plants. Trends Plant Sci 7, 217–224 (2002).

Weber, S., Zarhloul, K., Friedl, W. Modification of oilseed quality by genetic transformation. Prog Bot 62, 140–174 (2001).

Worrall, D., Ng, C. K.-J., Hetherington, A. M. Sphingolipids, new players in plant signalling. Trends Plant Sci 8, 317–320 (2003).

Yalovsky, S., Rodriguez-Concepción, M., Gruissem, W. Lipid modifications of proteins—slipping in and out of membranes. Trends Plant Sci 4, 439–445 (1999).

16

Secondary metabolites fulfill specific ecological functions in plants

In addition to **primary** metabolites, such as carbohydrates, amino acids, fatty acids, cytochromes, chlorophylls, and metabolic intermediates of the anabolic and catabolic pathways, which occur in all plants and where they all have the same metabolic functions, plants also contain a large variety of substances, called **secondary** metabolites, with no apparent direct metabolic function. Certain secondary metabolites are restricted to a few plant species, where they fulfill specific ecological functions, such as attracting insects to transfer pollen, or animals to consume fruits and in this way to distribute seed, and last, but not least, to act as **natural pesticides**.

16.1 Secondary metabolites often protect plants from pathogenic microorganisms and herbivores

Plants, because of their protein and carbohydrate content, are an important source of food for many animals, such as insects, snails, and many vertebrates. Since plants cannot run away, they have had to evolve strategies that make them indigestible or poisonous to protect themselves from being eaten. Many plants protect themselves by producing toxic proteins (e.g., amylase– or proteinase inhibitors, or lectins), which impair the digestion of herbivores (section 14.4). In response to caterpillar feeding, maize plants mobilize a protease that destroys the caterpillar intestine. In the following, groups of secondary metabolites will now be discussed, comprising alkaloids (this chapter), isoprenoids (Chapter 17), and phenylpropanoids (Chapter 18), all of which include natural pesticides that protect plants against herbivores and pathogenic microorganisms. In some plants, these natural pesticides amount to 10% of the dry matter.

Defense substances against herbivores are mostly part of the permanent outfit of plants, they are **constitutive**. There are, however, also cases in which the plant forms a defense substance after browsing damage. Section 18.7 describes how acacias, after feeding damage, produce more tannins, thus making the leaves inedible. Another example, as described in section 15.7, is when plants damaged by caterpillars use the synthesis of scents to attract parasitic wasps, which lay their eggs in the caterpillars, thus killing them.

Microbes can be pathogens

Certain fungi and bacteria infect plants in order to utilize their resources for their own nutritional requirements. As this often leads to plant diseases, these infectants are called **pathogens**. In order to infect the plants effectively, the pathogenic microbes produce aggressive substances, such as enzymes, which attack the cell walls, or toxins, which damage the plant. An example is **fuscicoccin**, (section 10.3), which is produced by the fungus *Fusicoccum amygdalis*. The production of substances for the attack requires the presence of specific **avirulence genes**, which have developed during evolution. Plants defend themselves against pathogens by producing defense substances that are encoded by specific **resistance genes**. The interaction of the avirulence genes and resistance genes decides the success of attack and defense.

When a plant is susceptible and the pathogen is aggressive, it leads to disease, and the disease is **virulent**. Such is termed a **compatible reaction.** If, on the other hand, the infecting pathogen is killed or at least its growth is very much retarded, this is called an **incompatible reaction,** and the plant is regarded as **resistant**. Often just a single gene decides on compatibility and resistance between pathogen or host.

Plants form phytoalexins in response to microbial infection

Defense substances against microorganisms, especially fungi, are synthesized mostly in response to an infection. These inducible defense substances, which are formed within hours, are called **phytoalexins** (*alekein*, Greek, to defend). Phytoalexins comprise a large number of substances with very different structures, such as isoprenoids, flavonoids, and stilbenes, many of which act as antibiotics against a broad spectrum of pathogenic fungi and bacteria. Plants also produce as defense substances aggressive oxygen compounds, such as superoxide radicals ($\bullet O_2^-$) and H_2O_2, as well as nitrogen monoxide (NO) (section 19.9), and enzymes, such as β-glucanases, chitinases, and proteinases, which damage the cell walls of bacteria and fungi. The synthesis of these various defense substances is induced by so-called **elicitors**. Elicitors are often proteins excreted by the pathogens to attack plant cells (e.g., cell-degrading enzymes). In some cases, polysaccharide segments of the cell's own wall, produced by degradative enzymes of the pathogen, function as elicitors. But elicitors also can be fragments from the cell wall of the pathogen, released by defense enzymes of the plant. These various elicitors are bound to specific receptors on the outer surface of the plasma membrane of the plant cell. The binding of the elicitor releases signal cascades, in which protein kinases (section 19.1) and signal substances such as salicylic acid (section 18.2) and jasmonic acid (section 15.7) participate, and which finally induce the **transcription of genes** for the synthesis of phytoalexins, reactive oxygen compounds, and defense enzymes (section 19.9).

Elicitors may also cause an infected cell to die and the surrounding cells to die with it. In other words, the infected cells and those surrounding them commit suicide. This can be caused, for instance, by the infected cells producing phenols, with which they poison not only themselves but also their surroundings. This programmed cell death, called a **hypersensitive response**, serves to protect the plant. The cell walls around the necrotic tissue are strengthened by increased biosynthesis of lignin, and in this way the plant barricades itself against further spreading of the infection.

Plant defense substances can also be a risk for humans

Substances toxic for animals are, in many cases, also toxic for humans. In crop plants, toxic or inedible secondary metabolites have been eliminated or at least decreased by breeding. This is why cultivated plants usually are more sensitive than wild plants to pests, thus necessitating the use of pest control, which is predominantly achieved by the use of chemicals. Attempts to breed more resistant culture plants by crossing them with wild plants can lead to problems: A newly introduced variety of insect-resistant potato had to be taken off the market, because the high toxic solanine content (an alkaloid, see following section) made these potatoes unsuitable for human consumption. In a new variety of insect-resistant celery cultivated in the United States, the 10-fold increase in the content of psoralines (section 18.2) caused severe skin damage to people harvesting the plants. This illustrates that natural pest control also is not without risk.

A number of plant constituents that are harmful to humans [e.g., proteins such as lectins, amylase inhibitors, proteinase inhibitors, and cyanogenic glycosides or glucosinolates (dealt with in this chapter)] decompose when cooked. But most secondary metabolites are not destroyed in this way. In higher concentrations, many plant secondary metabolites are carcinogenic. It has been estimated that in industrialized countries more than 99% of all carcinogenic substances that humans normally consume with their diet are plant secondary metabolites that are natural constituents of the food. However, experience has shown that the human metabolism usually provides sufficient protection against these harmful natural substances.

16.2 Alkaloids comprise a variety of heterocyclic secondary metabolites

Alkaloids belong to a group of secondary metabolites that are synthesized from **amino acids** and contain one or several N atoms as constituents of **heterocycles**. Many of these alkaloids act as defense substances against animals and microorganisms. Since alkaloids usually are bases, they are stored in the protonated form, mostly in the vacuole, which is acidic. Since ancient times humans have used alkaloids in the form of plant extracts as poisons, stimulants, and narcotics, and, last but not least, as medicine. In 1806 the pharmacy assistant Friedrich Wilhelm Sertürner isolated morphine from poppy seeds. Another 146 years had to pass before the structure of morphine was finally resolved in 1952. More than 10,000 alkaloids of very different structures are now known. Their synthesis pathways are very diverse, to a large extent still not known, and will not be discussed here.

Figure 16.1 shows a small selection of important alkaloids. Alkaloids are classified according to their heterocycles. **Coniine**, a piperidine alkaloid, is a very potent poison in hemlock. Socrates died when he was forced to drink this poison. **Nicotine**, which also is very toxic, contains a pyridine and a pyrrolidine ring. It is formed in the roots of tobacco plants and is carried along with the xylem sap into the stems and leaves. Nicotine sulfate, a byproduct of the tobacco industry, is used as a very potent insecticide (e.g., for fumigating greenhouses). No insect is known to be resistant to nicotine. **Cocaine**, the well-known narcotic, contains **tropane** as a heterocycle, in which the N atom is a constituent of two rings. A further well-known tropane alkaloid is **atropine** (formula not shown), a poison contained in deadly nightshade (*Atropa belladonna*). In low doses, it dilates the pupils of the eye and is therefore used in medicine for eye examination. Cleopatra allegedly used extracts containing atropine to dilate her pupils to appear more attractive. **Quinine**, a quinoline alkaloid from the bark of *Chinchuna officinalis* growing

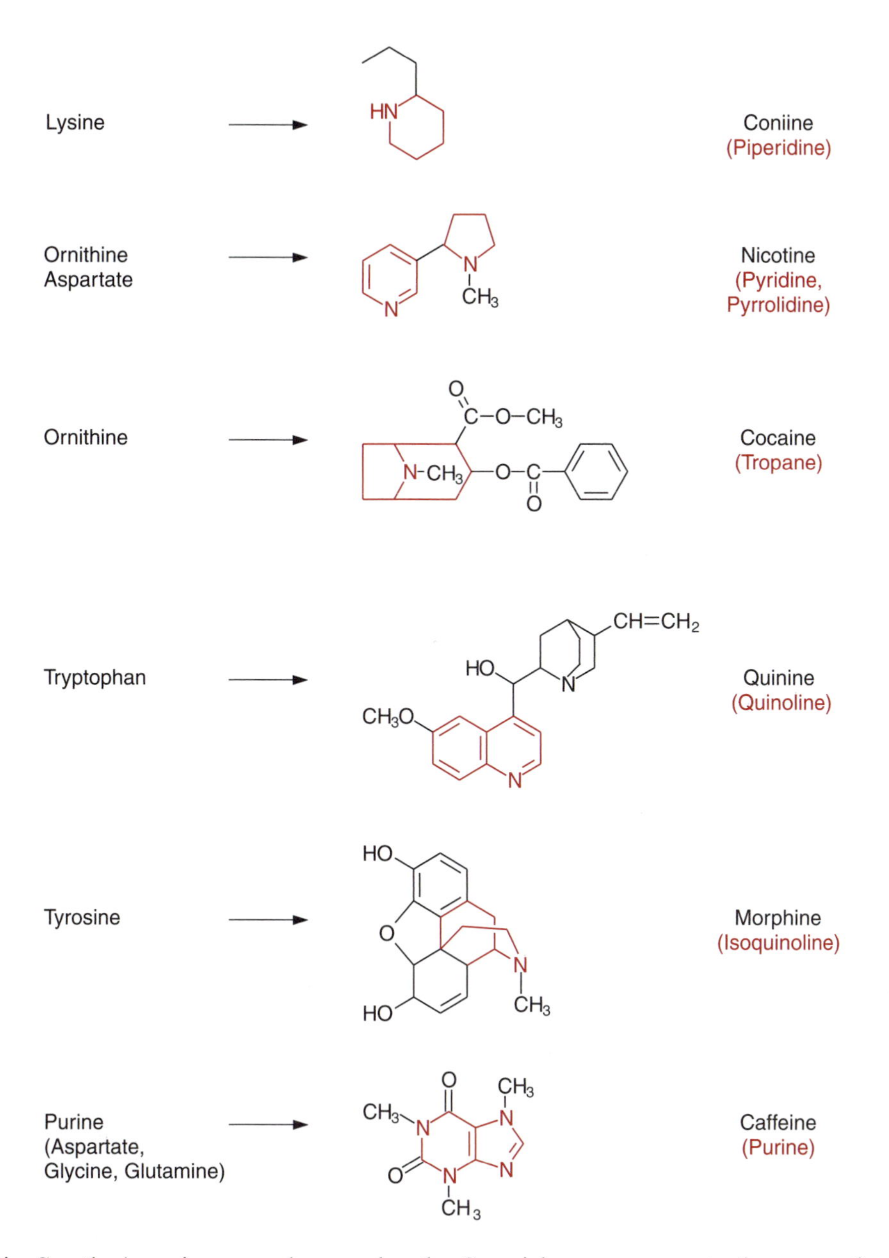

Figure 16.1 Some alkaloids and the amino acids from which they have been formed. The heterocycles, after which the alkaloids are classified, are colored red with their names in brackets. For coniine, a synthesis from acetyl CoA has also been described. Purine is synthesized from aspartate, glycine, and glutamine.

in South America, was known by the Spanish conquerors to be an antimalarial drug. The isoquinoline alkaloid **morphine** is an important painkiller and is also a precursor for the synthesis of heroin. **Caffeine**, the stimulant of coffee, contains a purine as the heterocycle.

In the search for new medicines, large numbers of plants are being analyzed for their content of secondary metabolites. As a result of a systematic search, the alkaloid **taxol**, isolated from the yew tree *Taxus brevifolia*, is now used for cancer treatment. Derivatives of the alkaloid camptothezine from the Chinese "happy tree" *Camptotheca acuminata* are also being clinically tested as cancer therapeutics. The search for new medicines against malaria and viral infections continues. Since large quantities of pharmacologically interesting substances often cannot be gained from plants, attempts are being made with the aid of genetic engineering either to increase production by the corresponding plants or to transfer the plant genes into microorganisms in order to use the latter for production.

16.3 Some plants emit prussic acid when wounded by animals

Since **prussic acid (HCN)** inhibits cytochrome oxidase, which is the final step of the respiratory chain, it is a very potent poison (section 5.5). Ten percent of all plants are estimated to use this poison as a defense against being eaten by animals. The consumption of peach kernels, for instance, or bitter almonds can have fatal results for humans. As also plants possess a mitochondrial respiratory chain and, in order not to poison themselves, they contain prussic acid in the bound form as **cyanogenic glycoside**. The amygdalin in the kernels and roots of peaches is an example of this (Fig. 16.2).

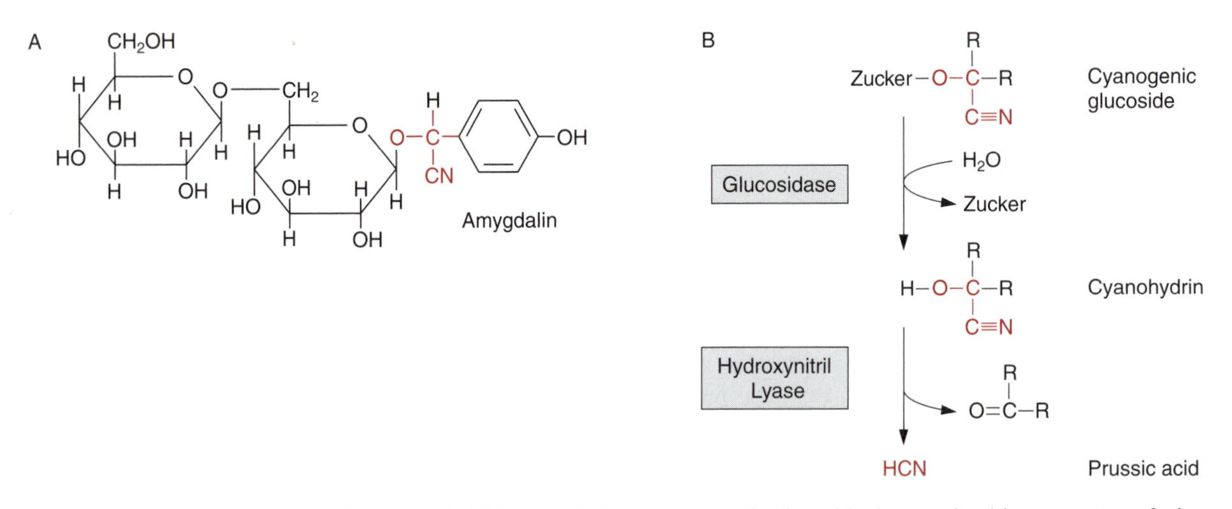

Figure 16.2 (A). Amygdalin, a cyanogenic glycoside, is contained in some stone fruit kernels. (B). After the sugar residue has been released by hydrolysis, cyanhydrin is formed from cyanogenic glycosides and decomposes spontaneously to prussic acid and a carbonyl compound.

The cyanogenic glycosides are stored as stable compounds in the vacuole. The **glycosidase** catalyzing the hydrolysis of the glycoside is present in another compartment. If the cell is wounded by feeding animals, the compartmentation is disrupted and the glycosidase comes into contact with the cyanogenic glycoside. After the hydrolysis of the glucose residue, the remaining cyanhydrin is very unstable and decomposes spontaneously to prussic acid and an aldehyde. A **hydroxynitrile lyase** enzyme accelerates this reaction. The aldehydes formed from cyanogenic glycosides are often very toxic. For a feeding animal, the detoxification of these aldehydes can be even more difficult than that of prussic acid. The formation of two different toxic substances makes the cyanogenic glycosides a very effective defense system.

16.4 Some wounded plants emit volatile mustard oils

Glucosinolates, also called mustard oil glycosides, have a similar protective function against herbivores. Glucosinolates can be found, for instance, in radish, cabbage, and mustard plants. Cabbage contains the glycoside glucobrassicin (Fig. 16.3), which is formed from tryptophan. The hydrolysis of the glycoside by a **thioglucosidase** results in a very unstable product from which, after the liberation and rearrangement of the sulfate residue, an

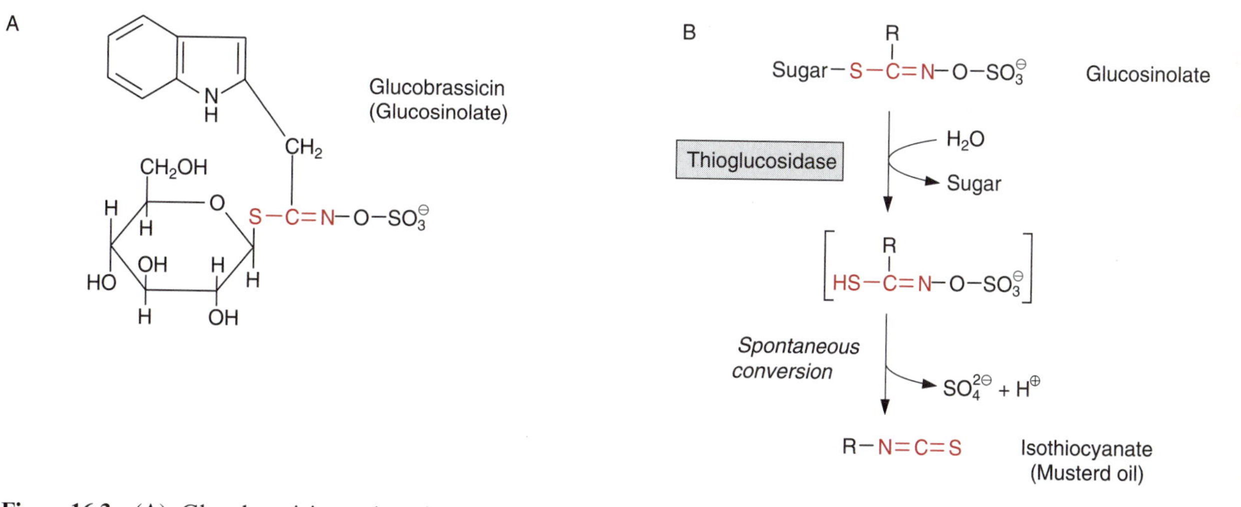

Figure 16.3 (A). Glucobrassicin, a glucosinolate from cabbage. (B). The hydrolysis of the glycoside by thioglucosidase results in an unstable product, which decomposes spontaneously into sulfate and isothiocyanate.

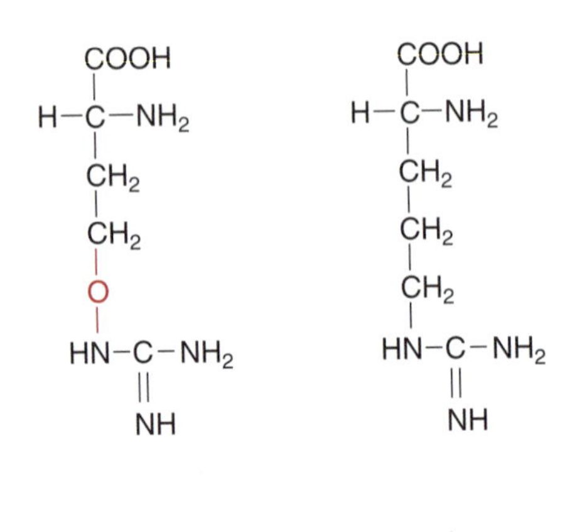

Figure 16.4 Canavanine is a structural analogue of arginine.

isothiocyanate, also termed **mustard oil**, is formed spontaneously. Mustard oils are toxic in higher concentrations. Like the cyanogenic glycosides (see preceding), glucosinolates and the hydrolyzing enzyme thioglucosidase also are located in separate compartments of the plant tissues. The enzyme comes into contact with its substrate only after wounding. When cells of these plants have been damaged, the pungent smell of mustard oil can easily be detected (e.g., in freshly cut radish). The high glucosinolate content in early varieties of rapeseed made the pressed seed unsuitable for fodder. Nowadays, as a result of successful breeding, rapeseed varieties that contain no glucosinolate in the seeds are cultivated.

16.5 Plants protect themselves by tricking herbivores with false amino acids

Many plants contain unusual amino acids with a structure very similar to that of protein building amino acids [e.g., **canavanine** from Jack bean (*Canavalia ensiformis*), a structural analogue of **arginine** (Fig. 16.4)]. Herbivores take up canavanine with their food. During protein biosynthesis, the arginine-transfer RNAs of animals cannot distinguish between arginine and canavanine and incorporate canavanine instead of arginine into proteins. This exchange can alter the three-dimensional structure of proteins, which then lose their biological function partially or even completely. Therefore canavanine is toxic for herbivores. In those plants synthesizing canavanine, the arginine transfer RNA does not react with canavanine, so it is not toxic for these plants. This same protective mechanism is used by some insects, which are specialized in eating leaves containing canavanine. Plants contain a large variety of false amino acids, which are toxic for herbivores in an analogous way to canavanine.

Further reading

Ames, B. N., Gold, L. S. Pesticides, risk and applesauce. Science 244, 755–757 (1989).

Anurag, A. A. Overcompensation of plants in response to herbivory and the by-product benefits of mutualism. Trends Plant Sci 5, 309–313 (2000).

Asai, T., Tena, G., Plotnikova, J., Willmann, M. R., Chiu, W.-L., Gomez-Gomez, L., Boller, T., Ausubel, F. M., Sheen, J. MAP kinase signaling cascade in *Arabidopsis* innate immunity. Nature 415, 977–983 (2002).

Beers, E. P., McDowell, J. M. Regulation and execution of programmed cell death in response to pathogens, stress and developmental cues. Curr Opin Plant Biol 4, 561–567 (2001).

Dangi, J. L., Jones, J. D. G. Plant pathogens and integrated defense responses to infection. Nature 411, 826–833 (2001).

De Luca, V., St Pierre, B. The cell and developmental biology of alkaloid biosynthesis. Trends Plant Sci 5, 168–173 (2000).

Facchini, P. J. Alkaloid biosynthesis in plants: Biochemistry, cell biology, molecular regulation, and metabolic engineering applications. Annu Rev Plant Physiol Plant Mol Biol 52, 29–66 (2001).

Grant, J. J., Loake, G. J. Role of reactive oxygen intermediates and cognate redox signaling in disease resistance. Plant Physiol 124, 21–29 (2000).

Halkier, B. A., Du, L. The biosynthesis of glucosinolates. Trends Plant Sci 2, 425–431 (1997).

Jones, A. M. Programmed cell death in development and defense. Plant Physiol 125, 94–97 (2001).

Kessler, A., Baldwin, I. T. Plant responses to insect herbivory: The emerging molecular analysis. Annu Rev Plant Biol 53, 299–328 (2002)

Kutchan, T. M. Alkaloid biosynthesis. The basis for metabolic engineering of medicinal plants. Plant Cell 7, 1059–1070 (1995)

Maleck, K., Dietrich, R. A. Defense on multiple fronts: How do plants cope with diverse enemies? Trends Plant Sci 4, 215–219 (1999).

Nürnberger, T., Scheel, D. Signal transmission in the plant immune response. Trends Plant Sci 6, 372–379 (2001).

Paul, N. D., Hatcher, P. E. Taylor, J. E. Coping with multiple enemies: An integration of molecular and ecological perspectives. Trends Plant Sci 5, 220–225 (2000).

Pechan, T., Cohen, A., Williams, W. P., Luthe, D. S. Insect feeding mobilizes a unique plant protease that disrupts the peritrophic matrix of caterpillars. Proc Natl Acad Sci USA 99,13319–13323 (2002).

Romeis, T. Protein kinases in the plant defense response. Curr Opin Plant Biol 4, 407–414 (2001).

Somssich, I. E., Hahlbrock, K. Pathogen defense in plants—a paradigm of biological complexity. Trends Plant Sci 3, 86–90 (1998).

Southon, I. W., Buckingham, J. (eds.). Dictionary of Alkaloids. London, Chapman & Hall (1989).

Stahl, E. A. Bishop, J. G. Plant-pathogen arms races at the molecular level. Curr Opin Plant Biol 3, 299–304 (2000).

Veronese, P., Ruiz, M. T., Coco, M. A., Hernandez-Lopez, A., Lee, H., Ibeas, J. I., Damsz, B., Pardo, J. M., Hasegawa, P. M., Bressan, R. A., Narasimhan, K. L. In defense against pathogens. Both plant sentinels and foot soldiers need to know the enemy. Plant Physiol 131, 1580–1590, (2003).

Zhnag, S., Klessig, D. F. MAPK cascades in plant defense signaling. Trends Plant Sci 6, 520–527 (2001).

17

A large diversity of isoprenoids has multiple functions in plant metabolism

Isoprenoids are present in all living organisms, but with an unusual diversity in plants. By 1997, more than 23,000 different plant isoprenoids had been listed and new substances are being constantly identified. These isoprenoids have many different functions (Table 17.1). In primary metabolism, they function as membrane constituents, photosynthetic pigments, electron transport carriers, growth substances, and plant hormones. They act as glucosyl carriers in glucosylation reactions and are involved in the regulation of cell growth. In addition, they have ecological functions as secondary metabolites: The majority of the different plant isoprenoids are to be found in resins, latex, waxes, and oils, and they make plants toxic or indigestible as a defense measure against herbivores. They act as antibiotics to protect the plant from pathogenic microorganisms. Many isoprenoids are formed only in response to infection by bacteria or fungi. Some plants synthesize isoprenoids, which inhibit the germination and development of competing plants. Other isoprenoids, in the form of pigments as well as scent in flowers or fruit, attract insects to distribute pollen or seed.

Plant isoprenoids are important commercially, for example, as aroma substances for food, beverages, and cosmetics, vitamins (A, D, and E), natural insecticides (e.g., pyrethrin), solvents (e.g., turpentine), and as rubber and gutta-percha. The plant isoprenoids also comprise important natural substances, which are utilized as pharmaceuticals or their precursors. Investigations are in progress to increase the ability of plants to synthesize isoprenoids by genetic engineering.

Isoprenoids are also called **terpenoids**. Plant ethereal oils have long been of interest to chemists. A number of mainly cyclic compounds containing

Table 17.1: Isoprenoids of higher plants

Precursor	Class	Example	Function
C_5: Dimethylallyl-PP	Hemiterpene	Isoprene	Protection of the photosynthetic apparatus against heat
Isopentenyl-PP		Side chain of cytokinin	Growth regulator
C_{10}: Geranyl-PP	Monoterpene	Pinene Linalool	Defense substance attractant
C_{15}: Farnesyl-PP	Sesquiterpene	Capsidiol	Phytoalexin
C_{20}: Geranylgeranyl-PP	Diterpene	Gibberellin Phorbol Casbene	Plant hormone Defense substance Phytoalexin
C_{30}: 2 Farnesyl-PP	Triterpene	Cholesterol Sitosterol	Membrane constituents
C_{40}: 2 Geranylgeranyl-PP	Tetraterpene	Carotenoids	Photosynthesis pigments
Geranylgeranyl-PP or Farnesyl-PP	Polyprenols	prenylated proteins	Regulation of cell growth
		Prenylation of plastoquinone, ubiquinone, chlorophyll, Cyt *a*	Membrane solubility of photosynthesis pigments and electron transport carriers
		Dolichols	Glucosyl carrier
		Rubber	

10, 15, 20, or correspondingly more C atoms have been isolated from turpentine oil. Such substances have been found in many plants and were given the collective name **terpenes**. Figure 17.1 shows some examples of terpenes. Limonene, an aromatic substance from lemon oil, is a terpene with 10 C atoms and is called a monoterpene. Carotene, with 40 C atoms, is accordingly a tetraterpene. Rubber is a polyterpene with about 1,500 C atoms. It is obtained from the latex of the rubber tree *Hevea brasiliensis*.

Otto Wallach (Bonn, Göttingen), who in 1910 was awarded the Nobel Prize in Chemistry for his basic studies on terpenes, recognized that isoprene is the basic constituent of terpenes (Fig. 17.1). Continuing these studies, Leopold Ruzicka (Zurich) found that isoprene is the universal basic element for the synthesis of many natural substances, including steroids, and for this he was awarded the Nobel Prize in Chemistry in 1939. He postulated the biogenic isoprene rule, according to which all terpenoids (derivatives of terpenes) are synthesized via a hypothetical precursor, which he named **active isoprene**. This speculation was verified by Feodor Lynen in Munich (1964 Nobel Prize in Medicine), when he identified **isopentenyl pyrophosphate** as the much sought after "active isoprene."

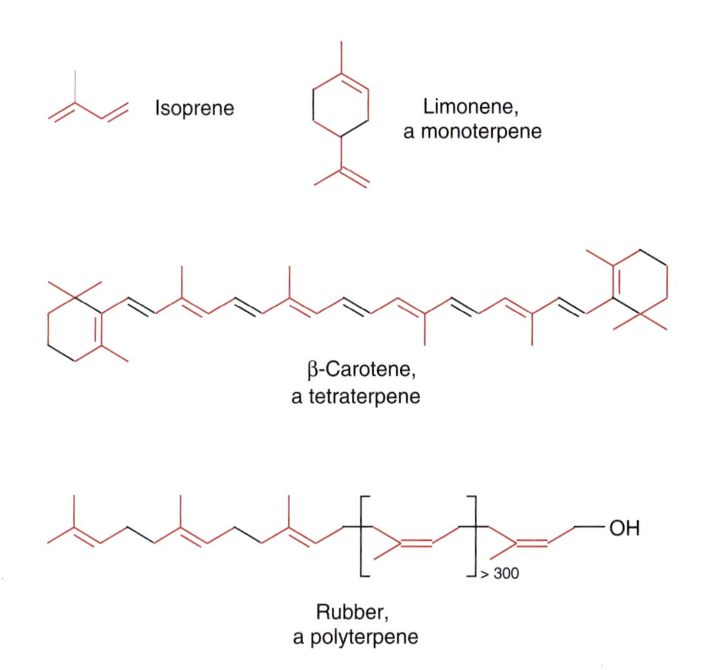

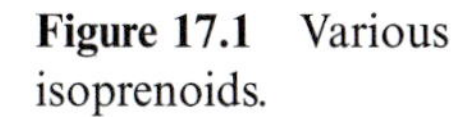
Figure 17.1 Various isoprenoids.

17.1 Higher plants have two different synthesis pathways for isoprenoids

Higher plants and some groups of algae contain the enzymes for synthesizing isoprenoids not only in the cytosol, but also in the plastids. The synthesis of the precursor isopentenyl pyrophosphate proceeds in the cytosol and in the plastids in different ways.

Acetyl-CoA is the precursor for the synthesis of isoprenoids in the cytosol

The basis for the elucidation of this isoprenoid biosynthesis pathway was the discovery by Konrad Bloch (United States, likewise a joint winner of the Nobel Prize in Medicine in 1964) that **acetyl-CoA** is a precursor for the biosynthesis of steroids. Figure 17.2 shows the synthesis of the intermediary product isopentenyl pyrophosphate: Two molecules of acetyl CoA react to produce **acetoacetyl-CoA** and then with a further acetyl-CoA to give **β-hydroxy-β-methylglutaryl-CoA** (HMG-CoA). In yeast and animals, these reactions are catalyzed by two different enzymes, whereas in plants a single enzyme, **HMG-CoA synthase**, catalyzes both reactions. The esterified carboxyl group of HMG-CoA is reduced by two molecules of NADPH to a

Figure 17.2
Isopentenyl pyrophosphate synthesis in the cytosol proceeds via the acetate-mevalonate pathway.

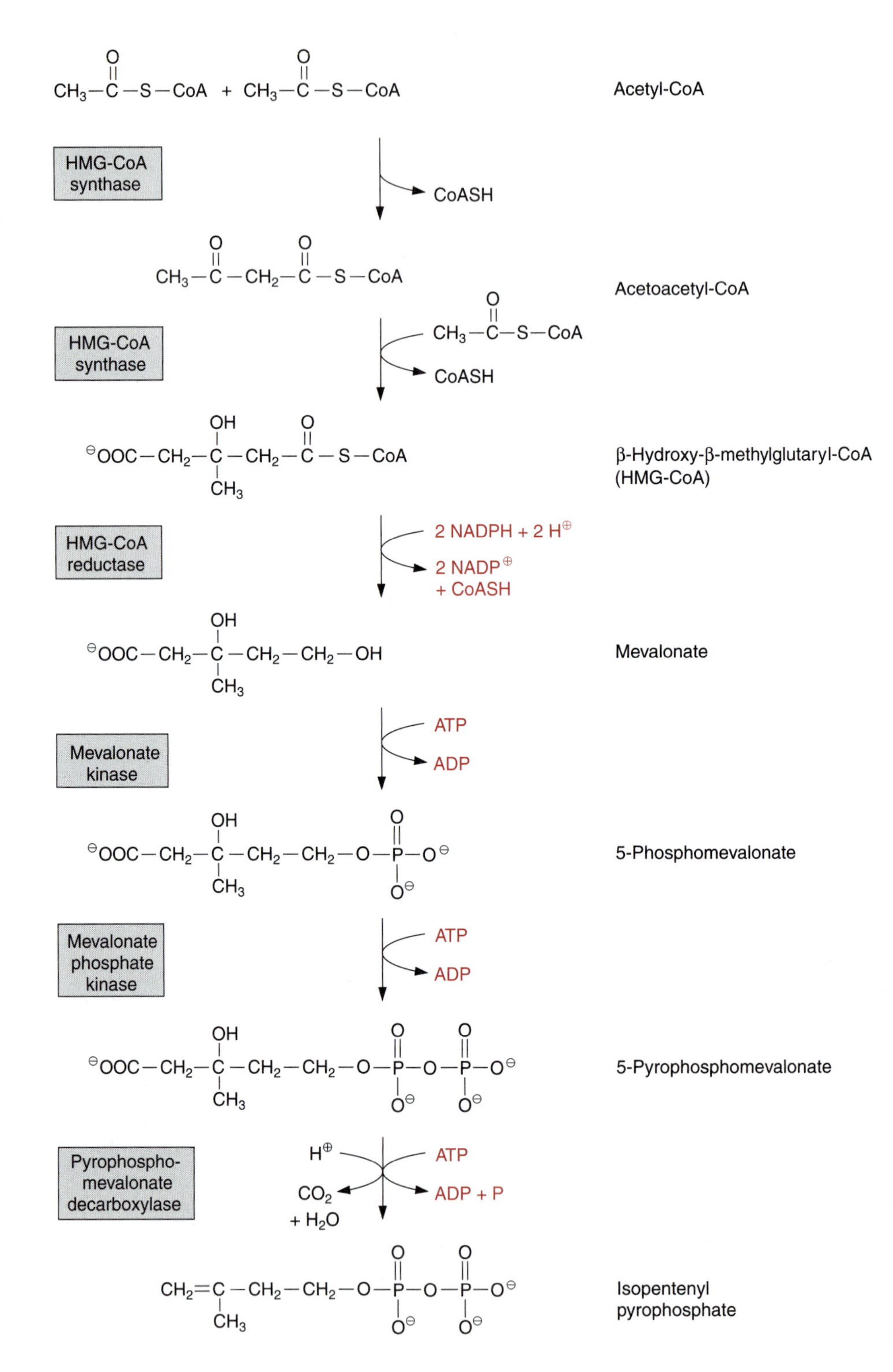

hydroxyl group, accompanied by hydrolysis of the energy-rich thioester bond. Thus **mevalonate** is formed in an irreversible reaction. The formation of mevalonate from HMG-CoA is an important regulatory step of isoprenoid synthesis in animals. It has not yet been resolved whether this also applies to plants. A pyrophosphate ester is formed in two successive phosphorylation steps, catalyzed by two different kinases. With consumption of a third molecule of ATP, involving the transitory formation of a phosphate ester, a carbon-carbon double bond is generated and the remaining carboxyl group is removed. Isopentenyl pyrophosphate thus formed is the basic element for the formation of an isoprenoid chain. The synthesis of isopentenyl pyrophosphate, termed the **acetate-mevalonate pathway**, is located in the cytosol. It is responsible for the synthesis of sterols, certain sesquiterpenes, and the side chain of ubiquinone.

Pyruvate and D-glyceraldehyde-3-phosphate are the precursors for the synthesis of isopentyl pyrophosphate in plastids

The acetate-mevalonate pathway is blocked by **mevilonin**, a very specific inhibitor of HMG-CoA reductase. Experiments with plants showed that mevilonin inhibits the isoprenoid synthesis in the cytosol, but not in the plastids. These findings led to the discovery that the synthesis of isopentenyl pyrophosphate follows a different pathway in the plastids from that in the cytosol (Fig. 17.3). For the plastidal synthesis pathway, **pyruvate** and **D-glyceraldehyde-3-phosphate** are the precursors. As in the pyruvate dehydrogenase reaction (Fig. 5.4), pyruvate is decarboxylated via thiamine pyrophosphate (TPP), and then, as in the transketolase reaction (Fig. 6.17), is transferred to D-glyceraldehyde-3-phosphate to yield **1-deoxy-D-xylulose-5-phosphate (DOXP)**. Via a series of reactions that have not all been resolved in detail, by rearrangement and reduction to 2-C-methyl-D-erythritol-4-phosphate, followed by two reduction steps, dehydration and phosphorylation, **isopentenyl pyrophosphate** is formed. The **DOXP-synthase pathway** for isoprenoids is present in bacteria, algae, and plants, but not in animal organisms. A large part of plant isoprenoids, including the hermiterpene isoprene, monoterpenes like pinene and limonene, diterpenes (e.g., phytol chains, gibberellin, abietic acid as oleoresin constituent) as well as tetraterpenes (carotenoids), are synthesized via the DOXP synthase pathway located in the plastids. Also, the side chains of chlorophyll and plastoquinone are synthesized by this pathway. It is still not quite clear to what extent there can be an exchange between the cytosolic and plastidal isopentenyl pyrophosphate pools, or at the level of other prenyl pyrophosphates (Fig. 17.4).

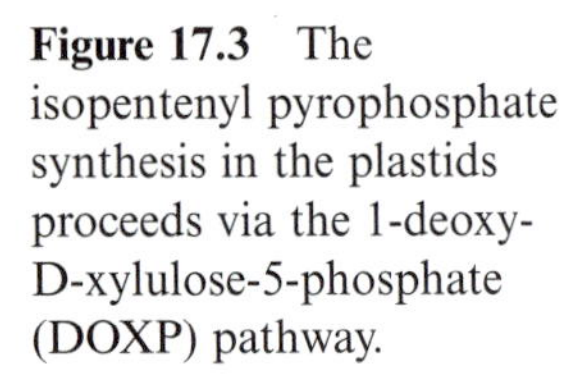

Figure 17.3 The isopentenyl pyrophosphate synthesis in the plastids proceeds via the 1-deoxy-D-xylulose-5-phosphate (DOXP) pathway.

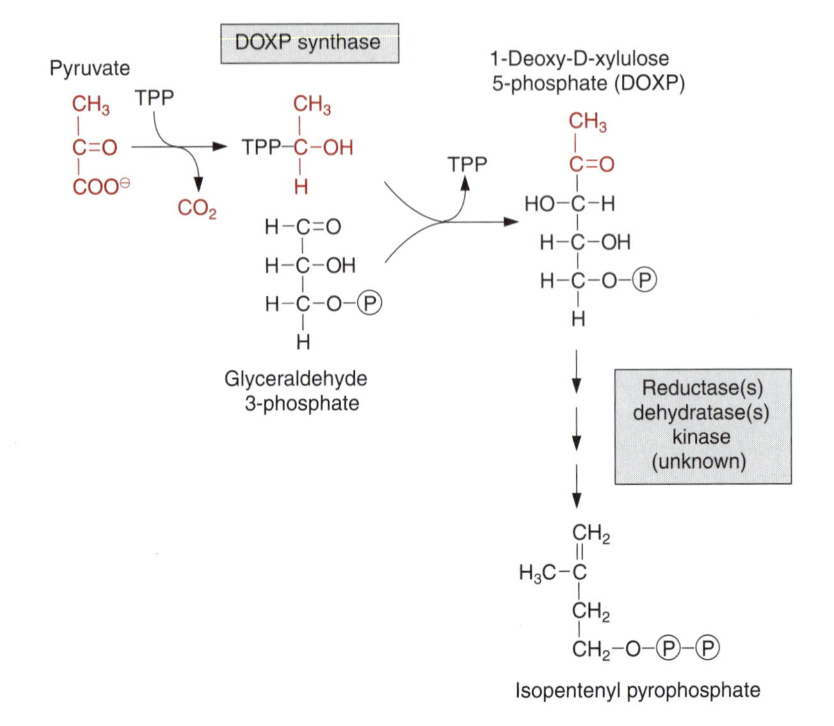

17.2 Prenyl transferases catalyze the association of isoprene units

Dimethylallyl pyrophosphate, which is formed by isomerization of isopentenyl pyrophosphate, is the acceptor for successive transfers of isopentenyl residues (Fig. 17.4). With the liberation of the pyrophosphate residue, dimethylallyl-PP condenses with isopentenyl-PP to produce geranyl-PP. In an analogous way, chain elongation is attained by further head-to-tail condensations with isopentenyl-PP, and so farnesyl-PP and geranylgeranyl-PP are formed one after the other.

The transfer of the isopentenyl residue is catalyzed by **prenyl transferases**. Prenyl residues are a collective term for isoprene or polyisoprene residues. Recent results have shown that a special prenyl transferase is required for the production of each of the prenyl pyrophosphates mentioned. For example, the prenyl transferase termed **geranyl-PP synthase** catalyzes only the synthesis of geranyl-PP. However, **farnesyl-PP synthase** synthesizes farnesyl-PP in two discrete steps: from dimethylallyl-PP and isopentenyl-PP, first geranyl-PP is formed, but this intermediate remains bound to the enzyme and reacts further with another isopentenyl-PP to give farnesyl-PP.

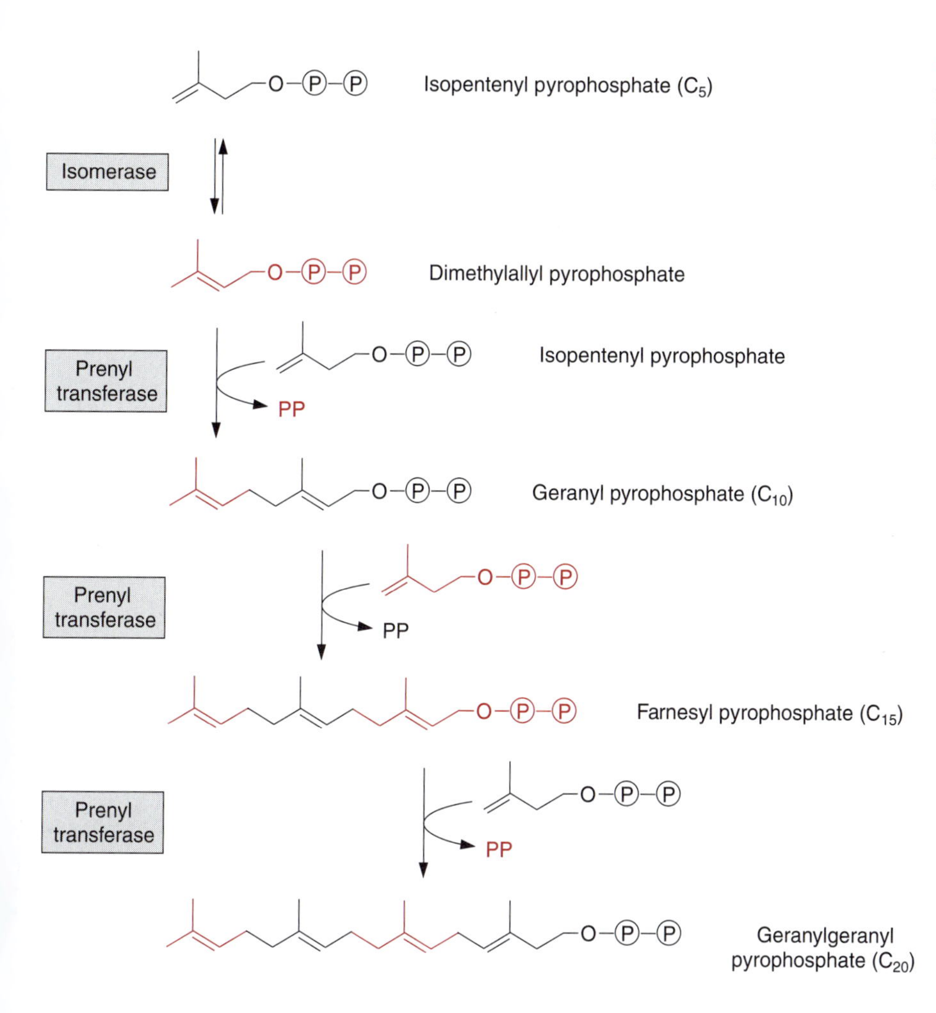

Figure 17.4 Higher molecular prenyl phosphates are formed by head-to-tail addition of active isoprene units.

Analogously, **geranylgeranyl-PP synthase** catalyzes all three steps of the formation of geranylgeranyl-PP. Table 17.1 shows that each of these prenyl pyrophosphates is the precursor for the synthesis of certain isoprenoids. As these prenyl pyrophosphates are synthesized by different enzymes, the synthesis of a certain prenyl pyrophosphate can be regulated by induction or repression of the corresponding enzyme. It appears that there is a synthesis pathway from isopentenyl pyrophosphate to geranylgeranyl pyrophosphate, not only in the cytosol but also in the plastids. The differences between these two pathways have not yet been resolved in detail.

The formation of a C-C linkage between two isoprenes proceeds by nucleophilic substitution (Fig. 17.5): An Mg^{++} ion, bound to the prenyl transferase, facilitates the release of the negatively charged pyrophosphate residue from the acceptor molecule, whereby a positive charge remains at

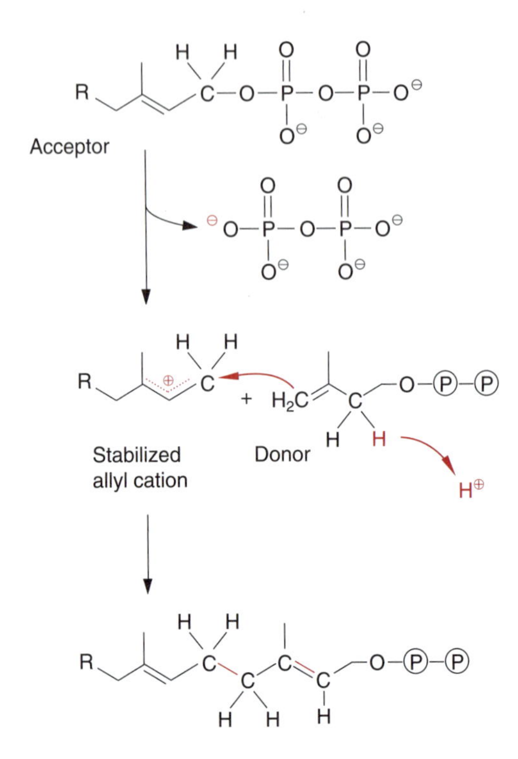

Figure 17.5 The head-to-tail addition of two prenyl phosphates by prenyl transferase is a nucleophilic substitution according to the S_N1 mechanism. First, pyrophosphate is released from the acceptor molecule. An allyl cation is formed, which reacts with the double bond of the donor molecule and forms a new C-C bond. The double bond is restored by release of a proton, but it is shifted by one C-atom. The reaction scheme is simplified.

the terminal C atom (C-1), which is stabilized by the neighboring double bond. The allyl cation thus formed reacts with the terminal C-C double bond of the donor molecule and a new C-C bond is formed with the release of a proton. According to the same reaction mechanism, not only isoprene chains, but also rings are formed, producing the unusual diversity of isoprenoids.

As shown in Table 17.1, the various prenyl phosphates are the precursors for isoprenoids with very different structures and functions, which are classified as hemiterpenes, monoterpenes, sesquiterpenes, and so on.

17.3 Some plants emit isoprenes into the air

The **hemiterpene** isoprene is formed from dimethyallyl-PP upon the release of pyrophosphate by an **isoprene synthase**, which is present in many plants (Fig. 17.6). Isoprene is volatile (boiling point 33°C) and leaks from the plant in gaseous form. Certain trees, such as oak, willow, planes, and poplar, emit isoprene during the day at temperatures of 30°C to 40°C . At such high temperatures, as much as 5% of the photosynthetically fixed carbon in oak

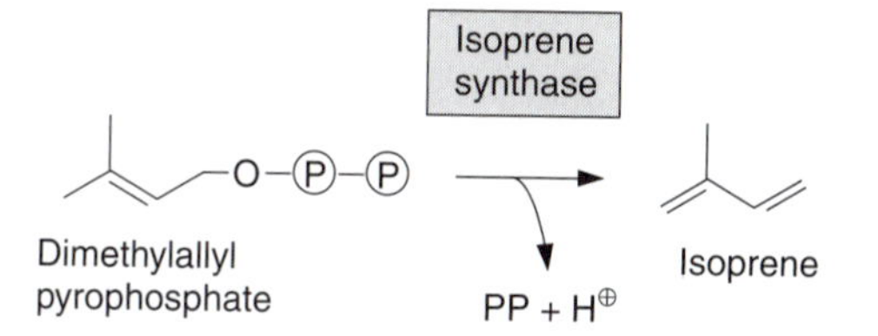

Figure 17.6 Via isoprene synthase some leaves form isoprene, which escapes as gas.

leaves can be emitted as isoprenes. Isoprene emissions of up to 20% of the total photoassimilate have been observed for the kudzu vine (*Pueraria lobota*), a climbing plant that is grown in Asia as fodder. Together with monoterpenes and other substances, isoprene emission is responsible for the blue haze that can be observed over woods during hot weather.

Isoprene is produced in the chloroplasts from dimethylallyl pyrophosphate, which is formed via the DOXP synthase pathway (Fig. 17.3). Isoprene synthase is induced when leaves are exposed to high temperatures. The physiological function of isoprene formation is not clear; there are indications that low amounts of isoprenes stabilize photosynthetic membranes against high temperature damage, but this is disputed. The global isoprene emission by plants is considerable. It is estimated to be about as high as the global methane emission. But, in contrast to methane, isoprene decomposes in the atmosphere rather rapidly.

17.4 Many aromatic substances are derived from geranyl pyrophosphate

The monoterpenes comprise a large number of open chain and cyclic isoprenoids, many of which, due to their high volatility and their lipid character, are classed as essential oils. Mostly, they have a distinctive, often pleasant scent and are, for example, responsible for the typical scents of pine needles, thyme, lavender, roses, and lily of the valley. As flower scents, they attract insects for the distribution of pollen, but their main function is to repel insects and other animals and thus protect the plants from herbivores.

The hydrolysis of geranyl-PP results in the formation of the alcohol geraniol (Fig. 17.7), the main constituent of rose oil. Geraniol has the typical scent of freshly cut geraniums. Accompanied by a rearrangement, the hydrolysis of geranyl-PP can also yield linalool (Fig. 17.7), an aromatic substance in many flowers. In several members of the *Compositae,* linalool is synthesized in the petals, where its biosynthesis commences simultaneously with the beginning of flowering.

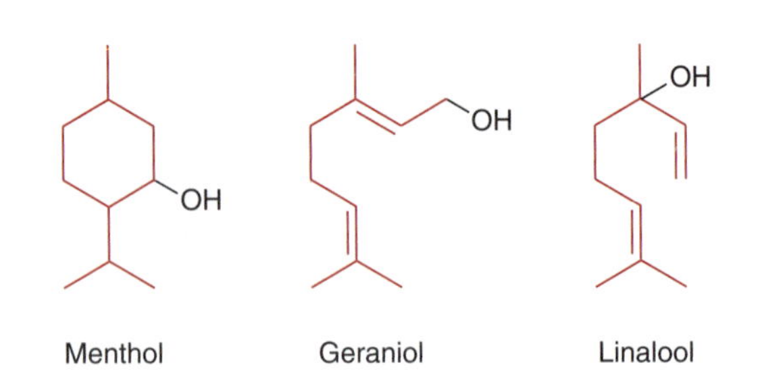

Figure 17.7 Menthol, a constituent of peppermint oil; geraniol, a constituent of rose oil, an aromatic substance in geraniums; and linalool, an aromatic substance of the *Compositae.*

The **cyclization** of geranyl-PP proceeds basically according to the same mechanism as the prenyl transferase reaction mentioned previously, with the only difference being that the reaction proceeds within the same molecule. An intramolecular prenylation takes place, often accompanied by an isomerization. Figure 17.8 shows as an example the synthesis of limonene by limonene synthase. Limonene is formed with a yield of 94% by this enzyme-catalyzed reaction, but also, as by-products, the very different compounds myrcene, α-pinene and β-pinene are formed with yields of 2% each. Thus a variety of substances can be synthesized by a single enzyme. These four monoterpenes are all major constituents of the resin (termed **olioresin**) of conifers. They are toxic for many insects and thus act as a protection against herbivores. Conifers respond to an attack by bark beetles with a strong

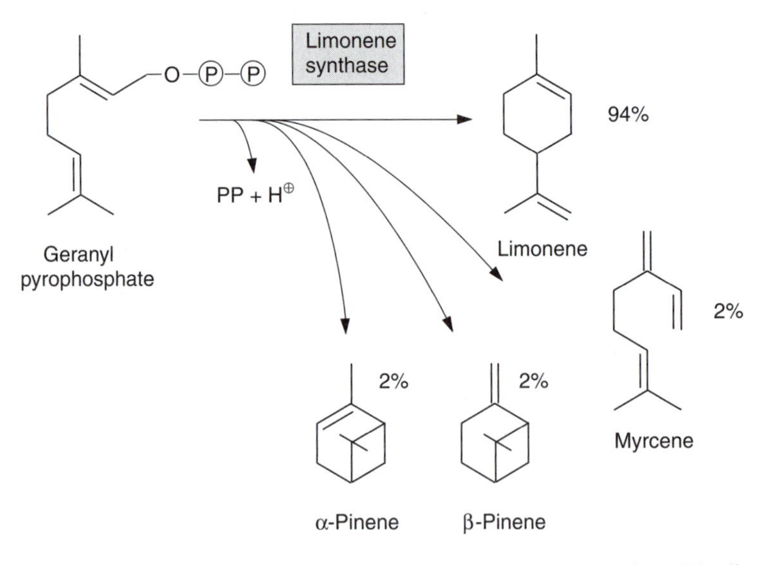

Figure 17.8 Examples of reactions catalyzed by a monoterpene cyclase. The limonene synthase forms limonene with a yield of 94%, and as by-products myrcene and α- and β-pinene, with yields of 2% each.

increase of cyclase activity, which results in enhanced resin formation. Limonene is also found in the leaves and peel of lemons. Another example of a monoterpene is menthol (Fig. 17.7), the main constituent of peppermint oil. It serves the plant as an insect repellent. There are many other monoterpenes containing carbonyl and carboxyl groups that are not discussed here.

17.5 Farnesyl pyrophosphate is the precursor for the formation of sesquiterpenes

The number of possible products is even greater for the cyclization of farnesyl-PP, proceeding according to the same mechanism as the cyclization of geranyl-PP described previously. This is illustrated in Figure 17.9. The reaction of the intermediary carbonium ion with the two double bonds of the molecule alone can lead to four different products. The number of pos-

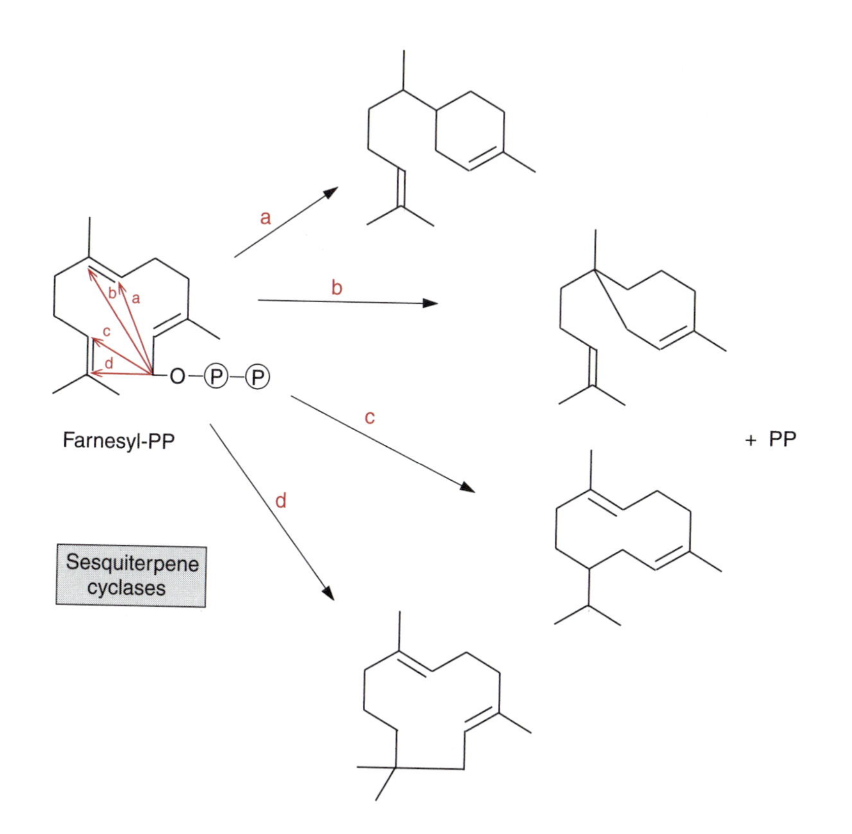

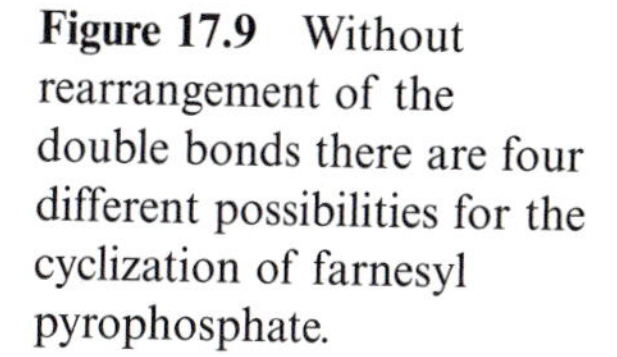
Figure 17.9 Without rearrangement of the double bonds there are four different possibilities for the cyclization of farnesyl pyrophosphate.

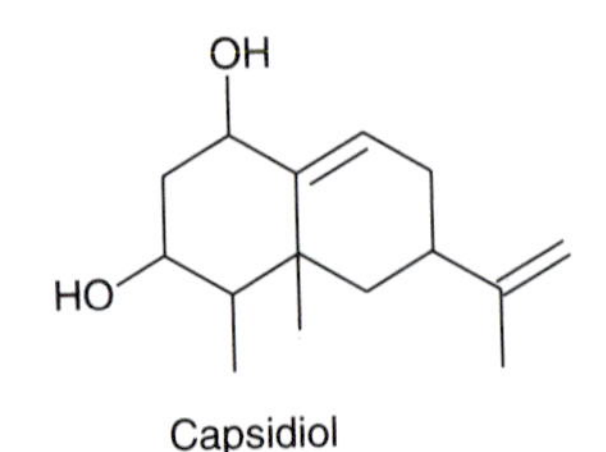

Figure 17.10 Capsidiol, a phytoalexin from pepper and tobacco.

sible products is multiplied by simultaneous rearrangements. Sesquiterpenes form the largest group of isoprenoids; they comprise more than 200 different ring structures. The sesquiterpenes include many aromatic substances such as eucalyptus oil and several constituents of hops. Capsidiol, (Fig. 17.10), a phytoalexin (section 16.1) formed in pepper and tobacco, is a sesquiterpene.

Steroids are synthesized from farnesyl pyrophosphate

The triterpene squalene is formed from two molecules of farnesyl-PP by an NADPH-dependent **reductive head to head condensation** (Fig. 17.11). Squalene is the precursor for membrane constituents such as cholesterol and sitosterol, the functions of which have been discussed in section 15.1, and also for **brassinosteroids**, which function as phytohormones (section 19.8).

A class of glucosylated steroids, named **saponins** because of their soap-like properties (Fig. 17.12), function in plants as toxins against herbivores and fungi. The glucosyl moiety of the saponins consists of a branched oligosaccharide formed from glucose, galactose, xylose, and other hexoses. The hydrophilic polysaccharide chain and the hydrophobic steroid give the saponins the properties of a **detergent**. Saponins are toxic, as they dissolve the plasma membranes of fungi and cause haemolysis of the red blood cells in animals. Some grasses contain saponins and are therefore a hazard for grazing cattle. **Yamonin**, a saponin from the yam plant (*Dioscorea*), is used in the pharmaceutical industry as a raw material for the synthesis of progesterones, a component of contraceptive pills. A number of very toxic glucosylated steroids called **cardenolides**, which inhibit the **Na^+/K^+ pump** present in animals, also belong to the saponins. A well-known member of this class of substances is **digitoxigenin**, a poison from foxglove (formula not shown). Larvae of certain butterflies can ingest cardenolides without being harmed. They store these substances, which then make them poisonous for birds. In low doses, cardenolides are widely used as a medicine against heart disease. Other defense substances are the **phytoecdysones**, a group of steroids with a structure similar to that of the insect hormone ecdysone. Ecdysone controls the pupation of larvae. When insects eat plants containing phytoecdysone, the pupation process is disturbed and the larvae die.

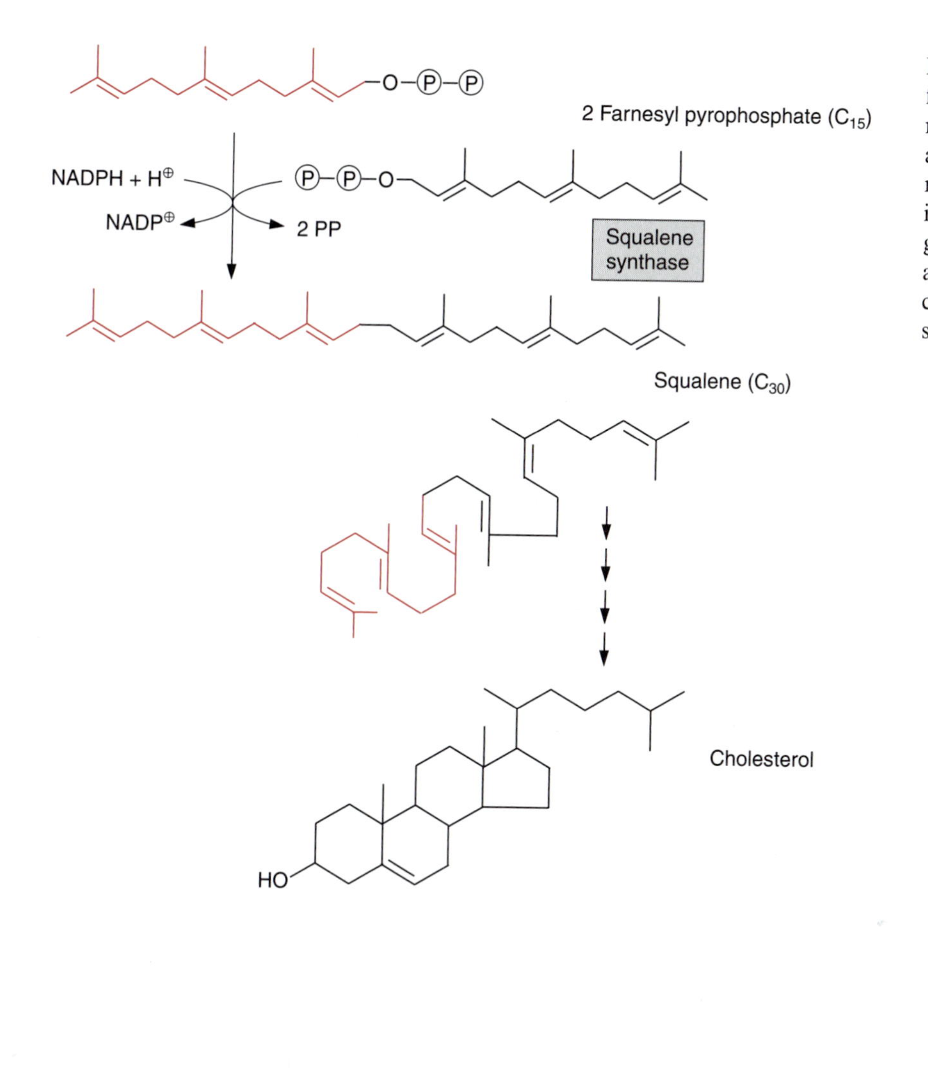

Figure 17.11 Squalene is formed from 2 farnesyl-PP molecules by head-to-head addition, accompanied by a reduction. After the introduction of an -OH group by a monooxygenase and a cyclization, cholesterol is formed in several reaction steps.

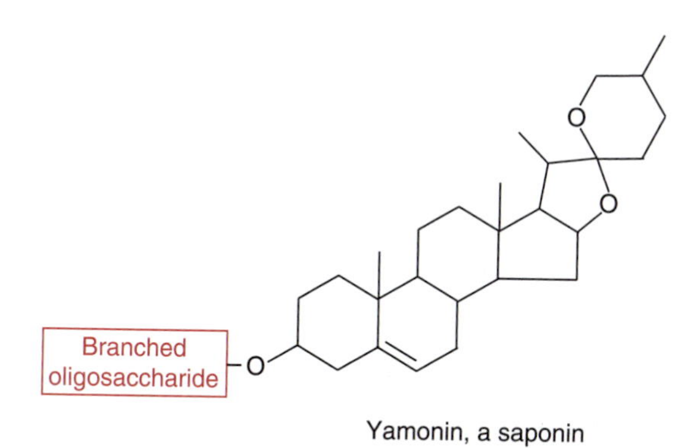

Figure 17.12 Yamonin, a saponin.

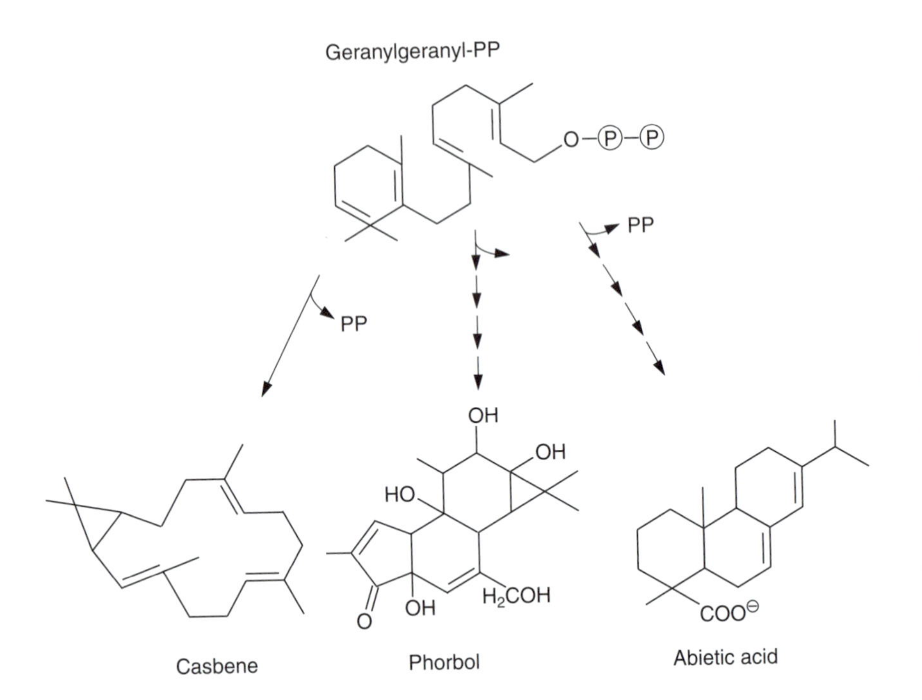

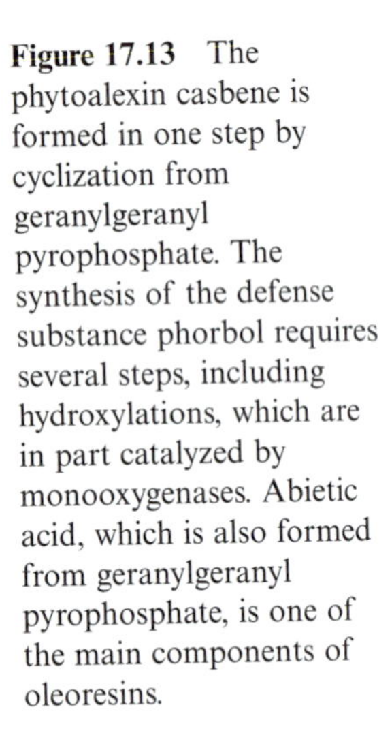
Figure 17.13 The phytoalexin casbene is formed in one step by cyclization from geranylgeranyl pyrophosphate. The synthesis of the defense substance phorbol requires several steps, including hydroxylations, which are in part catalyzed by monooxygenases. Abietic acid, which is also formed from geranylgeranyl pyrophosphate, is one of the main components of oleoresins.

17.6 Geranylgeranyl pyrophosphate is the precursor for defense substances, phytohormones, and carotenoids

The cyclization of geranylgeranyl-PP leads to the formation of the diterpene casbene (Fig. 17.13). Casbene is formed as a phytoalexin (section 16.1) in castor bean. The diterpene phorbol is present as an ester in the latex of plants of the spurge family (*Euphorbiae*). Phorbol acts as a toxin against herbivores; even skin contact causes severe inflammation. Since phorbol esters induce the formation of tumors, they are widely used in medical research. Geranylgeranyl-PP is also the precursor for the synthesis of gibberellins, a group of phytohormones (section 19.4).

Oleoresins protect trees from parasites

In temperate zone forests, conifers are widely spread and often reach an old age, some species being far over 1,000 years old. This demonstrates that conifers have been very successful in protecting themselves from browsing enemies. The greatest threat is the bark beetle, which not only causes damage itself, but also the destroyed bark is open to fungal infections. To protect themselves, the trees secrete **oleoresins** (tree resins), which seal the wound site and kill insects and fungi. The conifer oleoresins are a complex mixture of terpenoids, about half of which consist of a volatile **turpentine fraction** (monoterpenes and partly also sesquiterpenes) and the other half of a non-volatile **rosin fraction** (diterpenes). The turpentine fraction contains a series of substances that are toxic for insects and fungi [e.g., **limonene** (Fig. 17.5)]. The rosin fraction contains resin acids, the main component of which is **abietic acid** (Fig. 17.13). When the tree is wounded, the oleoresin stored in the resin channels flows out or is synthesized directly at the infected sites. It is presently being investigated how the toxic properties of the different components of the oleoresins affect different insects and fungi. Scientist are hopeful that such knowledge will make it possible to employ genetic engineering to enhance the parasite resistance of trees growing in large forests.

Carotene synthesis delivers pigments to plants and provides an important vitamin for humans

The function of **carotenoids** in photosynthesis has been discussed in detail in Chapters 2 and 3. Additionally, carotenoids function as pigments (e.g., in flowers and paprika fruits). The synthesis of carotenoids requires two molecules of geranylgeranyl-PP, which, as in the formation of squalene, are linked by head-to-head condensation (Fig. 17.14). Upon release of the first pyrophosphate, the intermediate pre-phytoene pyrophosphate is formed, and the subsequent release of the second pyrophosphate results in the formation of **phytoene**. Here the two prenyl residues are linked to each other by a carbon-carbon double bond. Catalyzed by two different desaturases, phytoene is converted to **lycopene**. According to recent results, these desaturations proceed via dehydrogenation reactions, in which hydrogen is transferred via FAD to O_2. Cyclization of lycopene then results in the formation of **β-carotene**. Another cyclase generates α-carotene. The hydroxylation of β-carotene leads to the xanthophyll **zeaxanthin**. The formation of the xanthophyll violaxanthin from zeaxanthin is described in Figure 3.41.

β-Carotene is the precursor for the synthesis of the visual pigment **rhodopsin**. Since β-carotene cannot be synthesized by humans, it is as **provitaminA** an essential part of human nutrition. Hundreds of millions of people, especially in Asia, who live mainly on rice that contains no β-carotene, suffer from severe provitaminA deficiency. Because of this, many children become blind. A recent success was the introduction of all the enzymes of the synthesis pathway from geranylgeranyl pyrophosphate to β-carotene into the endosperm of rice grains by genetic engineering. These transgenic rice lines produce β-carotene containing grains, with a yellowish color, and have therefore been called **"golden rice."** Nonprofit organizations have placed these transgenic rice lines at the disposal of many breeding

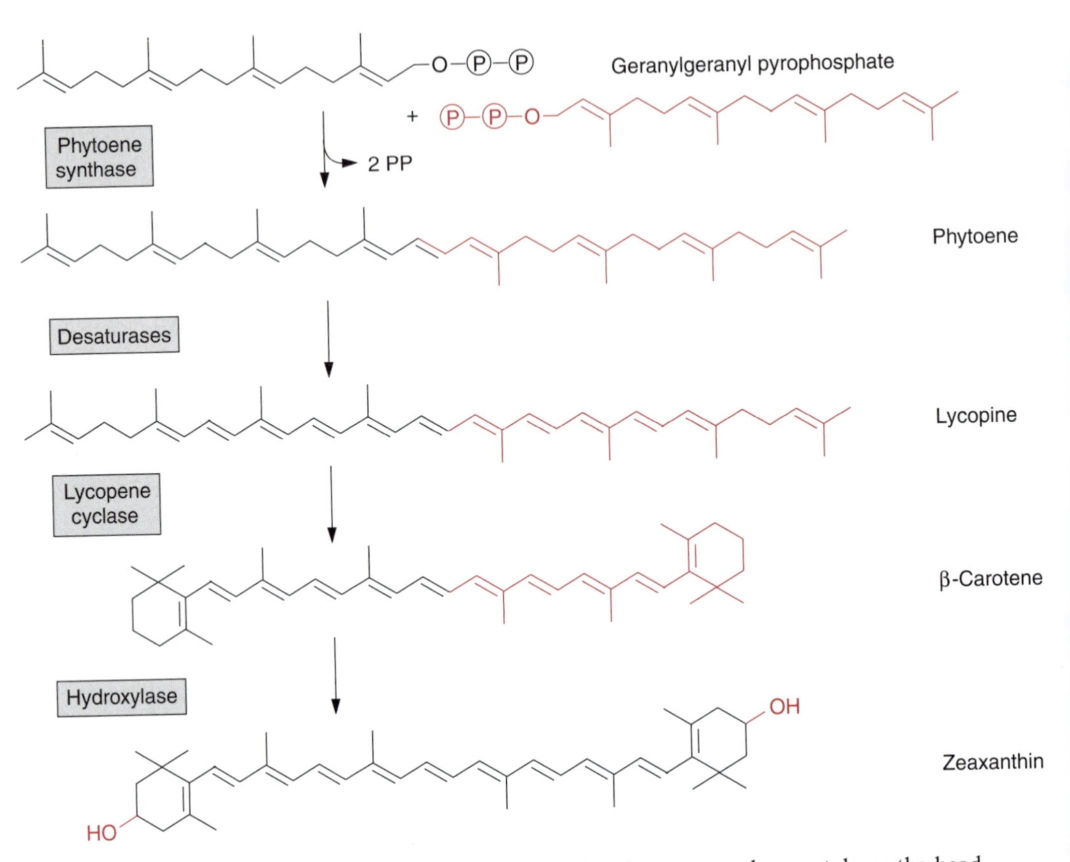

Figure 17.14 Carotenoid biosynthesis. The phytoene synthase catalyzes the head-to-head addition of two molecules of geranylgeranyl-PP to phytoene. The latter is converted by desaturases with neurosporene as the intermediate (not shown) to lycopene. β-Carotene is formed by cyclization and zeaxanthin is formed by hydroxylation.

stations in Asian countries, where they are at present crossed with local rice varieties. It is hoped that the serious provitaminA deficiency in wide parts of the world populations can be overcome through the cultivation of "golden rice".

17.7 A prenyl chain renders substances lipid-soluble

Ubiquinone (Fig. 3.5), plastoquinone (Fig. 3.19), and cytochrome-*a* (Fig. 3.24) are anchored in membranes by isoprenoid chains of various sizes. In

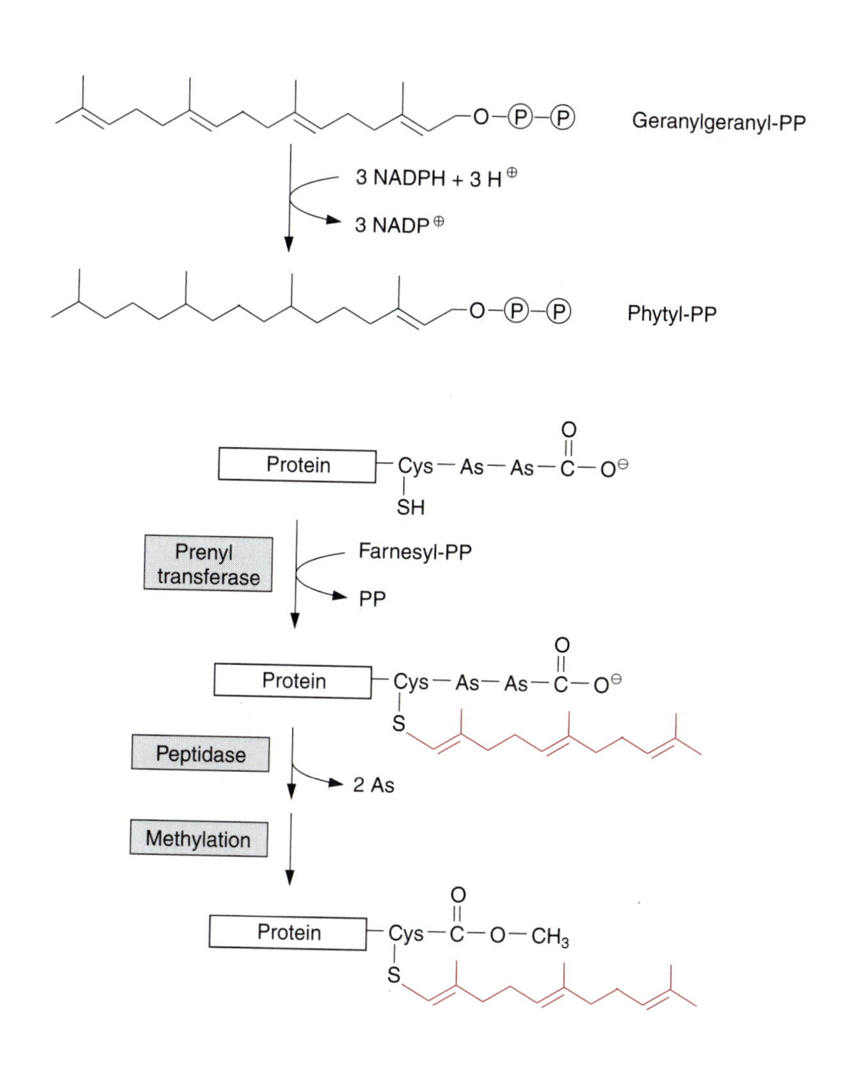

Figure 17.15 Synthesis of phytyl-PP from geranylgeranyl-PP.

Figure 17.16 Prenylation of a protein. A farnesyl residue is transferred to the SH group of a cysteine residue of the protein by a prenyl transferase. After hydrolytic release of the terminal amino acids (AS), the carboxyl group of the cysteine is methylated. The prenyl residue provides the protein with a membrane anchor.

the biosynthesis of these electron carriers, the **prenyl chains** are introduced from prenyl phosphates by reactions similar to those catalyzed by **prenyl transferases**. Chlorophyll (Fig. 2.4), tocopherols, and phylloquinone (Fig. 3.32), on the other hand, contain phytol side chains. These are formed from geranylgeranyl-PP by reduction with NADPH and are incorporated correspondingly (Fig. 17.15).

Proteins can be anchored in a membrane by prenylation

Recently, a large number of proteins that are linked at a cysteine residue near the *C*-terminus by a thioether bond to a farnesyl or geranyl residue have been found in yeast and animals (Fig. 17.16). The linking of these molecules is catalyzed by a specific prenyl transferase. In many cases, the

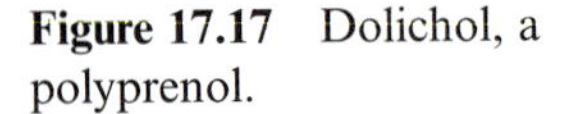

Figure 17.17 Dolichol, a polyprenol.

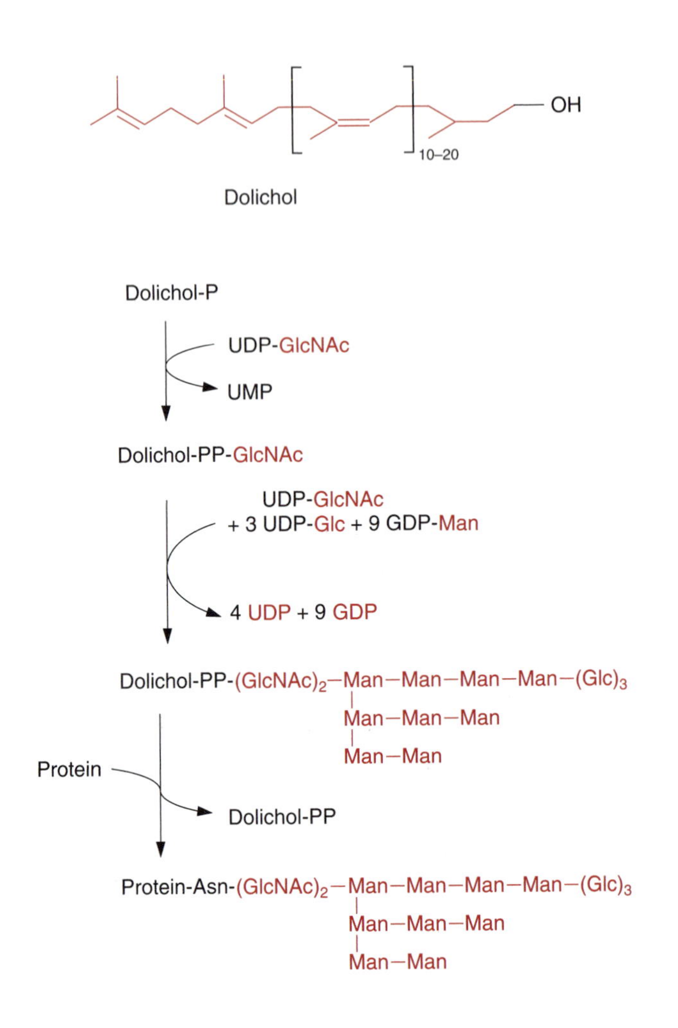

Figure 17.18 Dolichol as glucosyl carrier. For the synthesis of a branched oligosaccharide structure, from the corresponding UDP- and GDP-compounds, N-acetylglucosamine (GlcNAc), mannose (Man), and glucose (Glc) are transferred to dolichol. The first *N*-acetylglucose residue is attached to the -OH group of the dolichol via a pyrophosphate group. The complete oligosaccharide is then transferred to an asparagine residue of the protein. Asn = asparagine.

terminal amino acids linked to the cysteine residue are cut off after prenylation by a peptidase, and the carboxylic group of the cysteine is methylated. In this way the protein becomes lipid-soluble and can be anchored in a membrane. Recent results indicate that the prenylation of proteins also plays an important role in plants.

Dolichols mediate the glucosylation of proteins

Dolichols (Fig. 17.17) are isoprenoids with very long chains, occurring in the membranes of the endoplasmic reticulum and the Golgi network. They have an important function in the **transfer of oligosaccharides**. Many membrane proteins and secretory proteins are *N*-glucosylated by branched oligosaccharide chains. This glucosylation proceeds in the endoplasmic reticulum utilizing membrane-bound dolichol (Fig. 17.18). The oligosaccharide structure is first synthesized when attached to dolichol, and only after completion is transferred to an asparagine residue of the protein to be glucosylated. By subsequent modification in the Golgi network, in which certain carbohydrate residues are split off and others are added, a large variety of oligosaccharide structures are generated.

17.8 The regulation of isoprenoid synthesis

In plants, isoprenoids are synthesized at different sites according to demand. Large amounts of hydrophobic isoprenoids are synthesized mostly in specialized tissues, such as the glandular hairs of leaves (menthol) or the petals (linalool). The enzymes for synthesis of isoprenoids are usually present in the plastids, the cytosol, and the mitochondria. Each of these cellular compartments is essentially self-sufficient with respect to its isoprenoid content. Some isoprenoids, such as the phytohormone gibberellic acid, are synthesized in the plastids and then supplied to the cytosol of the cell. As mentioned in section 17.2, the various prenyl pyrophosphates, from which all the other isoprenoids are derived, are synthesized by different enzymes.

This spatial distribution of the synthetic pathways makes it possible that, despite their very large diversity, the different isoprenoids formed by basically similar processes can be efficiently controlled in their rate of synthesis via regulation of the corresponding enzyme activities in the various compartments. Results so far indicate that the synthesis of the different isoprenoids is regulated mainly by gene expression. This is especially obvious when, after infections or wounding, the isoprenoid metabolism is changed very rapidly by elicitor-controlled gene expression (section 16.1). This can lead to competition between the pathways. In tobacco, for instance, phytoalexin synthesis, induced by a fungal elicitor, blocks steroid synthesis. In such a case, the cell focuses its capacity for isoprenoid synthesis on defense.

17.9 Isoprenoids are very stable and persistent substances

Little is known about the catabolism of isoprenoids in plants. Biologically active derivatives, such as phytohormones, are converted by the introduction of further hydroxyl groups and by glucosylation into inactive substances, which are often finally deposited in the vacuole. It is questionable whether, after degradation, isoprenoids can be recycled in a plant. Some isoprenoids are remarkably stable. Large amounts of isoprenoids are found as relics of early life in practically all sedimentary rocks as well as in crude oil. In archaebacteria, the plasma membranes contain glycerol ethers with isoprenoid chains instead of fatty acid glycerol esters. Isoprenoids probably are constituents of very early forms of life.

Further reading

Bartley, G. E., Scolnik, P. A. Plant carotenoids: Pigments for photoprotection, visual attraction and human health. Plant Cell 7, 1027–1038 (1995).

Canae, D. E. (Ed.) Comprehensive natural products. Chemistry 2, Isoprenoids including carotenoids and steroids. Pergamon/Elsevier, Amsterdam (1999).

Chappel, J. Biochemistry and molecular biology of the isoprenoid biosynthetic pathway in plants. Annu Rev Plant Physiol Plant Mol Biol 45, 521–547 (1995).

Connolly, J. D., Hill R. A. Dictionary of Terpenoids. Chapman & Hall, London (1991).

Dudareva, N., Pichersky, E. Biochemical and molecular genetic aspects of floral scents. Plant Physiol 122, 627–633 (2000).

Eisenreich, W., Rohdich, F., Bacher, A. Deoxyxylulose phosphate pathway. Trends Plant Sci 6, 78–84 (2001).

Giuliano, G., Aquilani, R., Dharmapuri, S. Metabolic engineering of plant carotenoids. Trends Plant Sci 5, 406–409 (2000).

Hirschberg, J. Carotenoid biosynthesis in flowering plants. Curr Opin Plant Biol 4, 210–218 (2001).

Laule, O., Fuerholz, A., Chang, H.-S., Zhu, T., Wang, X., Heifetz, P. B., Gruissem, W., Lange, B. M. Crosstalk between cytosolic and plastidial pathways of isoprenoid biosynthesis in *Arabidopsis thaliana*. Proc Natl Acad Sci USA 100, 6866–6871 (2003).

Lichtenthaler, H. K. The 1-deoxy-D-xylulose-5-phosphate pathway of isoprenoid biosynthesis in plants. Annu Rev Plant Physiol Plant Mol Biol 50, 47–65 (1999).

Logan, B. A., Anchordoquy, T. J., Monson, R. K., Pan, R. S. The effects of isoprene on the properties of spinach thylakoids and phosphatidylcholine liposomes. Plant Biol 1, 602–606 (1999).

Logan, B. A., Monson, R. K., Potosnak, M. J. Biochemistry and physiology of foliar isoprene production. Trends Plant Sci 5, 477–481 (2000).

Mahmoud, S. S., Croteau, R. B. Strategies for transgenic manipulation of monoterpene biosynthesis in plants. Trends Plant Sci 7, 366–373 (2002).

McGarvey, D. J., Croteau, R. Terpenoid metabolism. Plant Cell 7, 1015–1026 (1995).

Osbourn, A. Saponins and plant defense—a soap story. Trends Plant Sci 1, 4–9 (1996).

Sacchettini, J. C., Poulter, C. D. Creating isoprenoid diversity. Science 277, 1788–1789 (1997).

Sharkey, T. T., Yeh, S. Isoprene emission from plants. Annu Rev Plant Physiol Plant Mol Biol 52, 408–436 (2001).

Trapp, S., Croteau, R. Defensive resin biosynthesis in conifers. Annu Rev Plant Physiol Plant Mol Biol 52, 689–724 (2001).

Ye, X., Al-Babili, S., Kloeti, A., Zhang, J., Lucca, P., Beyer, P., Potrykus, I. Engineering the provitamin A (β-carotene) biosynthetic pathway into (carotenoid-free) rice endosperm. Science 287, 303–305 (2000).

18

Phenylpropanoids comprise a multitude of plant secondary metabolites and cell wall components

Plants contain a large variety of phenolic derivatives. As well as simple phenols, these comprise flavonoids, stilbenes, tannins, lignans, and lignin (Fig. 18.1). Together with long chain carboxylic acids, phenolic compounds are also components of suberin and cutin. These rather varied substances have important functions as antibiotics, natural pesticides, signal substances for the establishment of symbiosis with *Rhizobia*, attractants for pollinators, protective agents against ultraviolet (UV) light, insulating materials to make cell walls impermeable to gas and water, and structural material to give plants stability (Table 18.1). All these substances are derived from phenylalanine, and in some plants also from tyrosine. Phenylalanine and tyrosine are formed by the **shikimate pathway**, described in section 10.4. Since the phenolic compounds derived from the two amino acids contain a phenyl ring with a C_3 side chain, they are collectively termed **phenylpropanoids**. The

Table 18.1: Some functions of phenylpropanoids

Coumarins	Antibiotics, toxins against browsing animals
Lignan	Antibiotics, toxins against browsing animals
Lignin	Cell wall constituent
Suberin and Cutin	Formation of impermeable layers
Stilbenes	Antibiotics, especially fungicides
Flavonoids	Antibiotics, signal for interaction with symbionts, flower pigments, light protection substances
Tannin	Tannins, fungicides, protection against herbivores

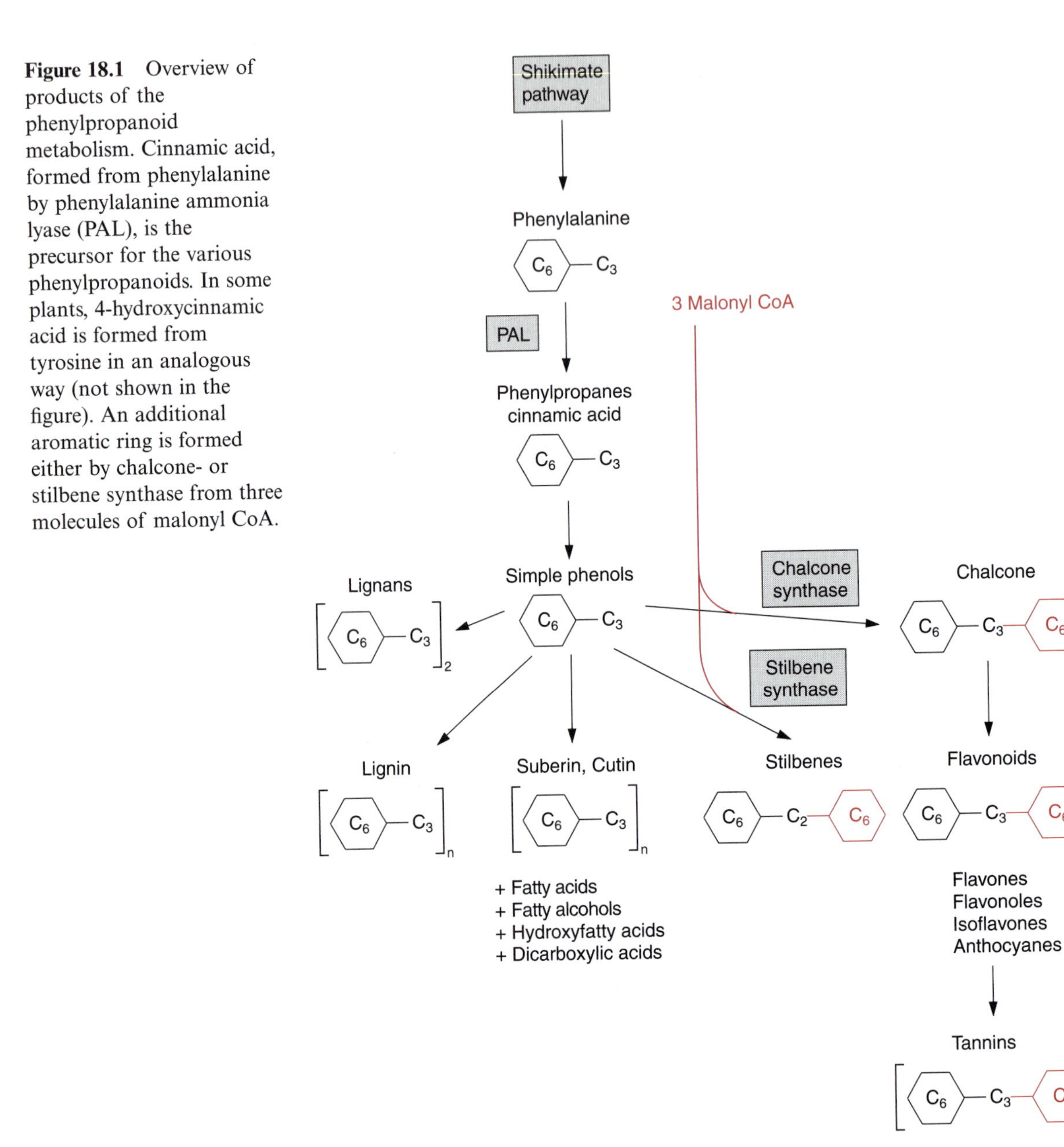

Figure 18.1 Overview of products of the phenylpropanoid metabolism. Cinnamic acid, formed from phenylalanine by phenylalanine ammonia lyase (PAL), is the precursor for the various phenylpropanoids. In some plants, 4-hydroxycinnamic acid is formed from tyrosine in an analogous way (not shown in the figure). An additional aromatic ring is formed either by chalcone- or stilbene synthase from three molecules of malonyl CoA.

flavonoids, including flavones, isoflavones, and anthocyanidins, contain, as well as the phenylpropane structure, a second aromatic ring that is formed from three molecules of malonyl CoA (Fig. 18.1). This also applies to the stilbenes, but here, after the introduction of the second aromatic ring, one C atom of the phenylpropane is split off.

18.1 Phenylalanine ammonia lyase catalyzes the initial reaction of phenylpropanoid metabolism

Phenylalanine ammonia lyase, abbreviated **PAL**, catalyzes a deamination of phenylalanine (Fig. 18.2): A carbon-carbon double bond is formed with the release of NH_3, yielding *trans*-cinnamic acid. In some grasses, tyrosine is converted to 4-hydroxycinnamic acid in an analogous way by **tyrosine ammonia lyase**. The released NH_3 probably is refixed by the glutamine synthetase reaction (section 10.1).

PAL is one of the most intensively studied enzymes of plant secondary metabolism. The enzyme consists of a tetramer with subunits of 77 to 83 kDa. The formation of phenylpropanoid phytolalexins after fungal infection involves a very rapid induction of PAL. PAL is inhibited by its product *trans*-cinnamic acid. The phenylalanine analogue aminoxyphenylpropionic acid (Fig. 18.3) is also a very potent inhibitor of PAL.

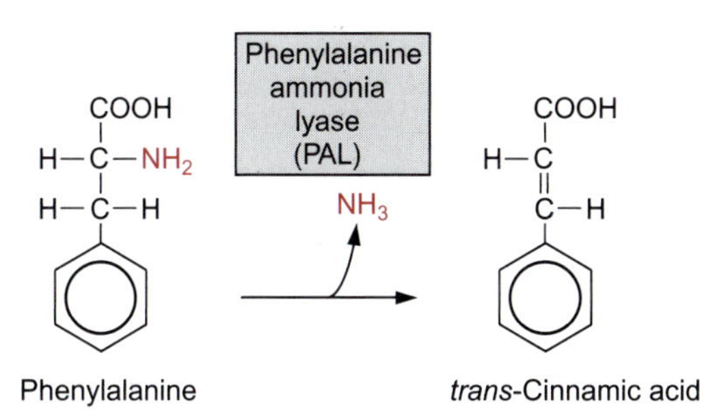

Figure 18.2 Formation of *trans*-cinnamic acid.

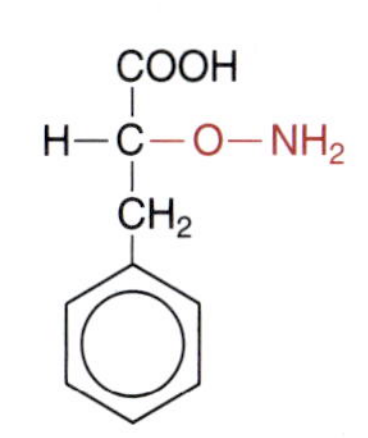

Figure 18.3 Aminooxyphenylpropionic acid, a structural analogue of phenylalanine, inhibits PAL.

18.2 Monooxygenases are involved in the synthesis of phenols

The introduction of the hydroxyl group into the phenyl ring of cinnamic acid (hydroxylation, Fig. 18.4) proceeds via a **monooxygenase catalyzed reaction** utilizing cytochrome P_{450} as the O_2 binding site according to:

$$NADPH + H^+ + R\text{-}CH_3 + \mathbf{O_2} \longrightarrow NADP^+ + RCH_2\text{-}\mathbf{OH} + H_2\mathbf{O}$$

In this reaction, electrons are transferred from NADPH via FAD (Fig. 5.16) to cytochrome-P_{450} (pigment with absorption maximum at 450 nm) and from there to O_2. From the O_2 molecule, only one O atom is incorporated in the formed hydroxyl group; the remaining O atom is reduced to yield H_2O. For this reason, the reaction is termed **monooxygenase**. Like cyt-a_3 (section 5.5), cyt-P_{450} can bind CO instead of O_2. Therefore, P_{450}-monooxygenases are inhibited by CO.

P_{450} monooxygenases are widely distributed in the animal and plant kingdoms. Genomic analyses have shown that the model plant *Arabidopsis*

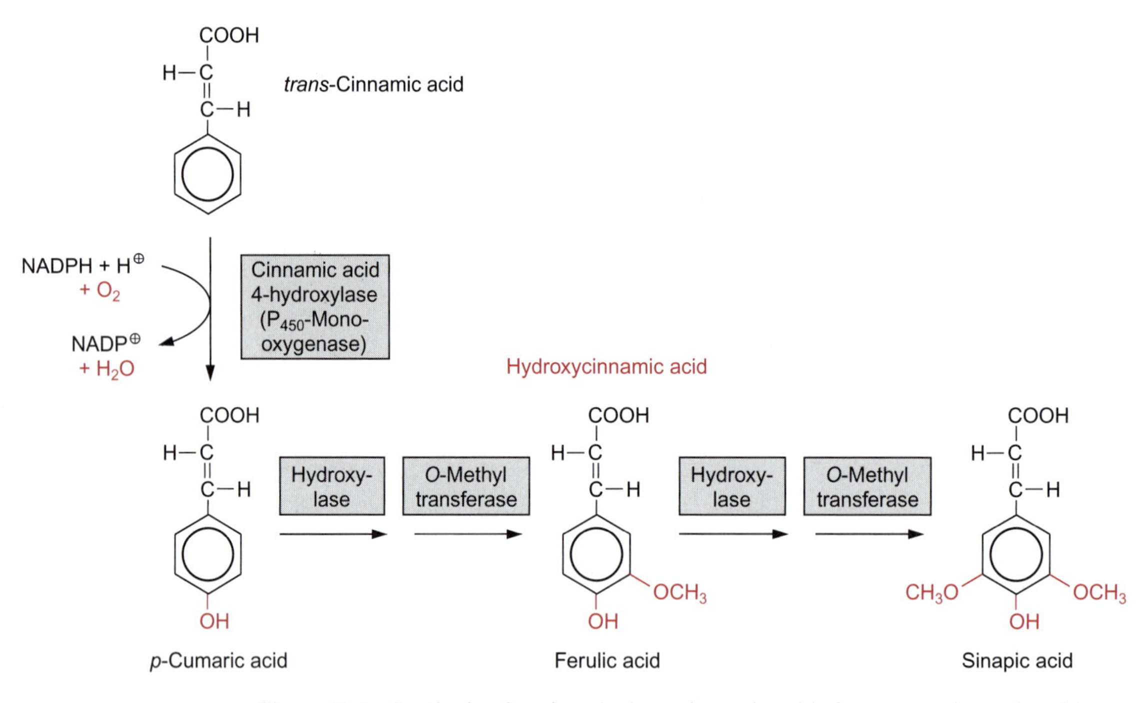

Figure 18.4 Synthesis of various hydroxycinnamic acids from *trans*-cinnamic acid.

thaliana contains about 300 different genes encoding P_{450} proteins. It seems to be the largest gene family in plants. The majority of these proteins probably are involved in the generation of hydroxyl groups for the synthesis of plant hormones and secondary metabolites, but they also play an important role in detoxification processes (e.g., the detoxification of herbicides) (section 3.6).

Like all P_{450}-monooxygenases, the cinnamic acid hydroxylase is bound at the membranes of the endoplasmic reticulum. *p*-Coumaric acid formed can be hydroxylated further in positions 3 and 5 by hydroxylases, again of the P_{450} monooxygenase type. The -OH groups thus generated are mostly methylated via *O*-methyl transferases with *S*-**adenosylmethionine** as the methyl donor (Fig. 12.10). In this way ferulic acid and sinapic acid are formed, which, together with *p*-coumaric acid, are the precursors for the synthesis of lignin (section 18.3).

Benzoic acid derivatives, including **salicylic acid** as well as a derivative of benzaldehyde, **vanillin**, the aroma substance of vanilla, are formed by cleavage of a C_2 fragment from the phenylpropanes (Fig. 18.5). Under the trade name aspirin, the acetyl ester of salicylic acid is widely used as a remedy against pain, fever, and other illnesses and probably is the most frequently used pharmaceutical worldwide. The name salicylic acid is derived from *Salix*, the Latin name for willow, since it was first isolated from the bark of the willow tree, where it occurs in particularly high amounts. Since ancient times, the salicylic acid content of willow bark has led to its being used as medicine in the Old and New Worlds. In the fourth century B.C. Hippocrates gave women willow bark to chew to relieve pain during childbirth. Native Americans also used extracts from willow bark as painkillers.

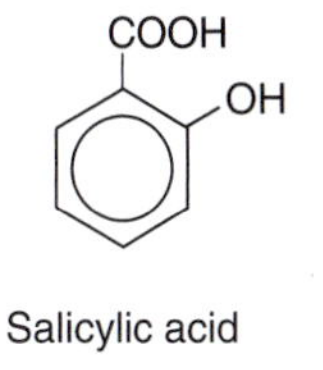

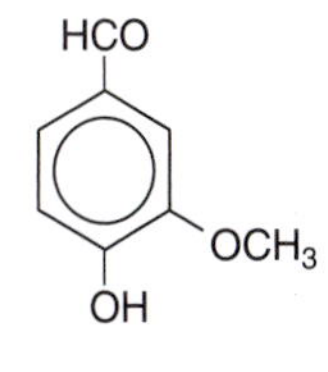

Figure 18.5 Salicylic acid and vanillin are formed from phenylpropanoids.

Salicylic acid also affects plants. It has been observed that tobacco plants treated with aspirin or salicylic acid have enhanced resistance to pathogens, such as the *Tobacco mosaic virus*. Many plants show an increase in their salicylic acid content after being infected by viruses or fungi, but also after being exposed to UV radiation or ozone stress. Salicylic acid functions as an important signal substance by activating signal chains for the expression of enzymes that are part of **defense reactions** against viruses, bacteria, and fungi, (sections 16.1 and 19.9). *Arabidopsis* mutants, which have lost the ability to produce salicylic acid, are much more prone to infection. As discussed previously, a dose of salicylic acid can give a plant better protection against pathogens. This principle is now used commercially. A salicylic acid analogue with the trade name **Bion** (Syngenta) is being sprayed on wheat to prevent infection by mildew.

However, salicylic acid not only triggers defense reactions, but also can induce blooming in some plants. By stimulating the mitochondrial alterna-

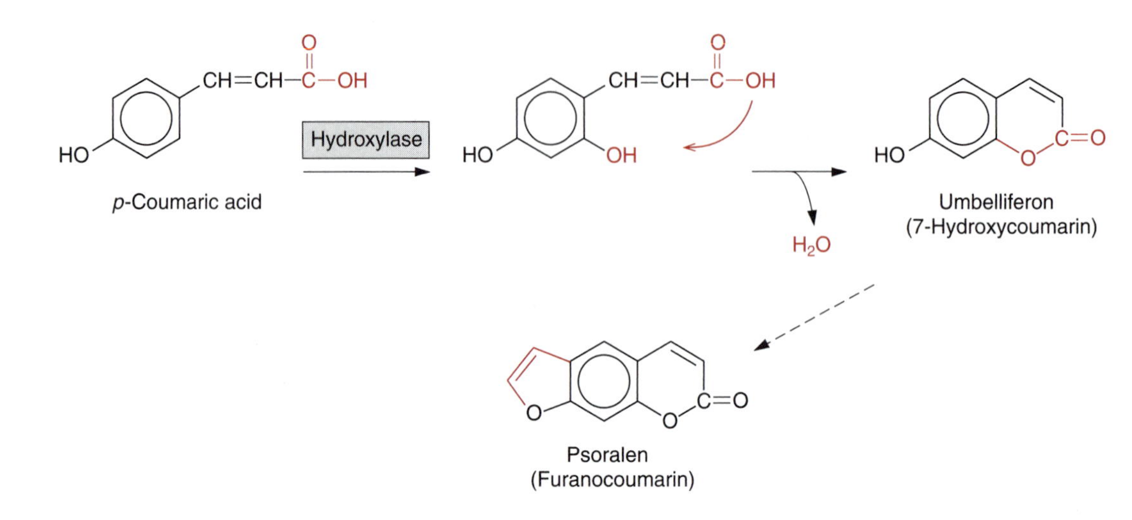

Figure 18.6 Umbelliferone, which is the precursor for the synthesis of the defense substance psoralen, is formed by hydroxylation of *p*-coumaric acid and the formation of a ring.

tive oxidase (section 5.7), it activates the production of heat in the spadix of the voodoo lily, emitting a carrion-like stench.

7-Hydroxycoumarin, also called **umbelliferone**, is synthesized from *p*-coumaric acid by hydroxylation and the formation of an intermolecular ester, a lactone (Fig. 18.6). The introduction of a C_2 group into umbelliferone yields psoralen, a **furanocoumarin**. Illumination with UV light turns psoralen into a toxic substance. The illuminated psoralen reacts with the pyrimidin bases of the DNA, causing blockage of transcription and DNA repair mechanisms and finally resulting in the death of the cell. As mentioned in section 16.1, some celery varieties contain very high concentrations of psoralen and caused severe skin inflammation in workers involved in harvesting. Many furanocoumarins have antibiotic properties. In some cases, they are constitutive components of the plants, whereas in other cases, they are formed only after infection or wounding as phytoalexins.

18.3 Phenylpropanoid compounds polymerize to macromolecules

As mentioned in section 1.1, after cellulose, **lignin** is the second most abundant natural substance on earth. The basic components for lignin synthesis are *p*-coumaryl, sinapyl, and coniferyl alcohols, which are collectively termed **monolignols** (Fig. 18.7). Synthesis of the monolignols requires

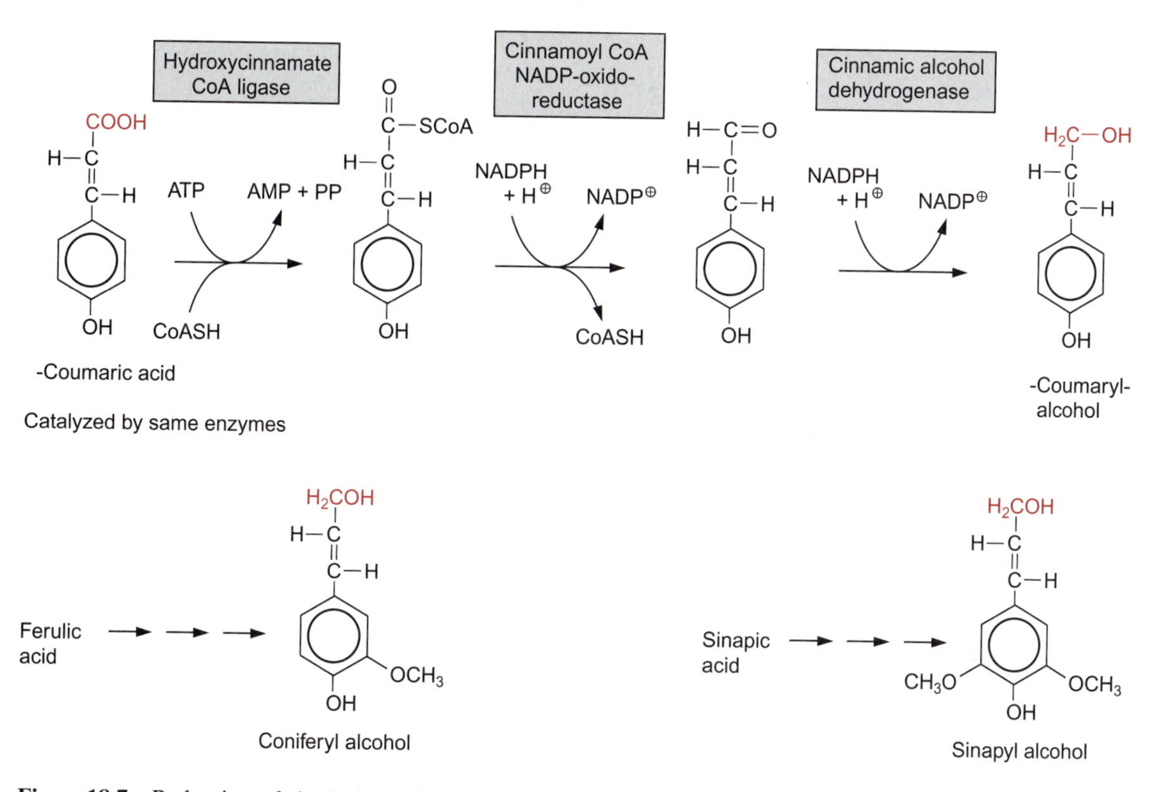

Figure 18.7 Reduction of the hydroxycinammic acids to the corresponding alcohols (monolignols).

reduction of the carboxylic group of the corresponding acids to an alcohol. When discussing the glyceraldehyde phosphate dehydrogenase reaction in section 6.3, it was shown that a carboxyl group can be reduced by NADPH to an aldehyde only if it is activated beforehand via the formation of a thioester. For the reduction of *p*-coumaric acid by NADPH (Fig. 18.7), a similar activation occurs. The required thioester is formed with CoA at the expense of ATP in a reaction analogous to fatty acid activation described in section 15.6. The cleavage of the energy-rich thioester bond drives the reduction of the carboxylate to the aldehyde. In the subsequent reduction to an alcohol, NADPH is again the reductant. The same three enzymes that catalyze the conversion of *p*-coumaric acid to *p*-coumaryl alcohol (Fig. 18.7) may also catalyze the formation of sinapyl and coniferyl alcohols from sinapic and ferulic acids. Alternatively, coniferyl and sinapyl alcohols can also be formed from *p*-cumaryl alcohol by hydroxylation followed by methylation (Fig. 18.4).

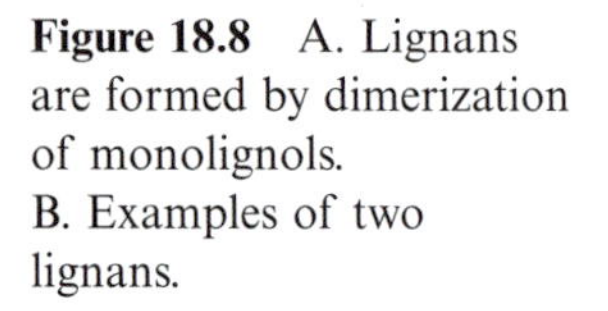

Figure 18.8 A. Lignans are formed by dimerization of monolignols.
B. Examples of two lignans.

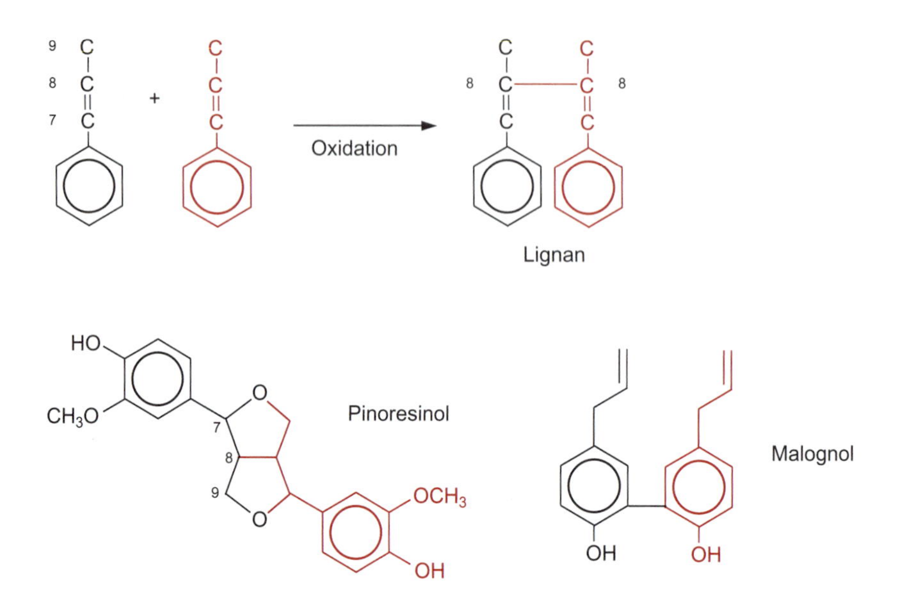

Lignans act as defense substances

The dimerization of monolignols leads to the formation of **lignans** (Fig. 18.8). This takes place mostly by a reductive linkage of the side chains in the 8,8-position, but sometimes also by a condensation of the two phenol rings. The mechanism of lignan formation is still not clear. Probably free radicals are involved (see next section). Plant lignans also occur as higher oligomers. In the plant world, lignans are widely distributed as defense substances. The lignan pinoresinol is a constituent of the resin of forsythia and is formed when the plant is wounded. Its toxicity to microorganisms is caused by an inhibition of cAMP phosphodiesterase. Pinoresinol thus averts the regulatory action of cAMP, which acts as a messenger substance in most organisms, possibly including plants (section 19.1). Malognol inhibits the growth of bacteria and fungi.

Some lignans have interesting pharmacological effects. Two examples may be used to illustrate this: Podophyllotoxin, from *Podophyllum*, a member of the *Berberidacea* family occurring in America, is a mitosis toxin. Derivatives of podophyllotoxin are used to combat cancer. Arctigenin and tracheologin (from tropical climbing plants) have antiviral properties. Investigations are under way to try to utilize this property to cure AIDS.

Lignin is formed by radical polymerization of phenylpropanoid derivatives

Lignin is formed by polymerization of a mixture of the three monolignols *p*-coumaryl, sinapyl, and coniferyl alcohol. The synthesis of lignin takes place outside the cell, but the mechanism by which the monolignols are exported from the cell for lignin synthesis is still not known. There are indications that the monolignols are exported as glucosides that are hydrolyzed outside the cell by glucosidases, but this is still a matter of controversy. The mechanism of lignin formation also remains unclear. Both **laccase** and **peroxidases** probably play a role in the linkage of the monolignols. Laccase is a monophenol oxidase that oxidizes a phenol group to a radical and transfers hydrogen via an enzyme-bound Cu^{++} ion to molecular oxygen. The enzyme was given this name as it was first isolated from the lac tree (*Rhus vermicifera*) growing in Japan. In the case of peroxidases, H_2O_2 functions as an oxidant, but the origin of the required H_2O_2 is not certain. As shown in Figure 18.9A, the oxidation of a phenol by H_2O_2 presumably results in the formation of a resonance-stabilized phenol radical. These phenol radicals can dimerize nonenzymatically and finally polymerize nonenzymatically (Fig. 18.9B). Due to the various resonance structures, many combinations are possible in the polymer. Mostly monolignols react to form several C-C or C-O-C linkages, building a highly branched phenylpropanoid polymer. Free hydroxyl groups are present in only a few side chains of lignin and sometimes are oxidized to aldehyde and carboxyl groups.

Earlier it was assumed that the polymeric structures are formed through random reactions of monomeric, dimeric, and higher oligomeric lignol radicals. Although until now the primary structure of lignins has not been fully established in all its details, it is recognized that even in a single cell, lignins of **different structure** are deposited in **discrete sections of the cell wall**. It has been postulated that certain extracellular glycoproteins, termed **dirigent proteins**, control the polymerization of monolignol radicals in such a way that defined lignin structures are formed. It is still a matter of debate to what extent lignin is formed by chance and formed specifically through the action of dirigent proteins.

The composition of lignin varies greatly in different plants. Lignin of conifers, for instance, has a high coniferyl content, whereas the coumaryl moiety prevails in the straw of cereals. Lignin is covalently bound to cellulose in the cell walls. **Lignified cell walls** have been compared with **reinforced concrete**, in which the cellulose fibers are the steel and lignin is the concrete. In addition to giving mechanical strength to plant parts such as stems or twigs, or providing stability for the vascular tissues of the xylem, lignin has

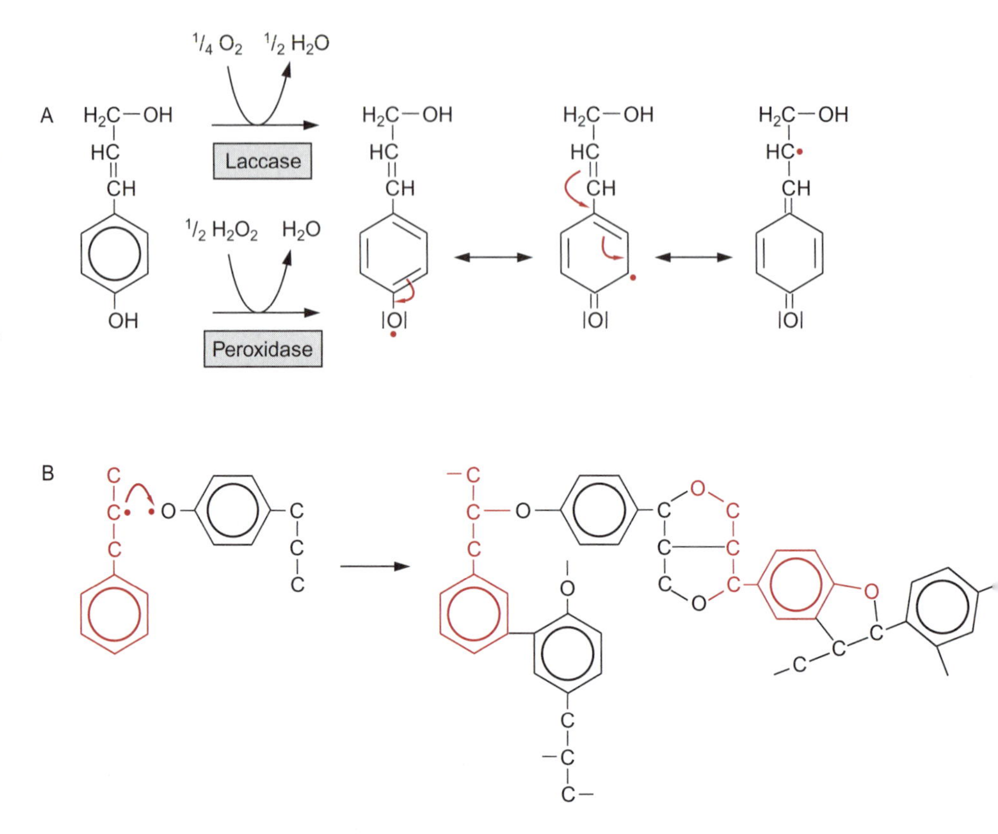

Figure 18.9
A. Oxidation of a monolignol by laccase or a peroxidase results in the formation of a phenol radical. The unpaired electron is delocalized and can react with various resonance structures of the monolignol. B. Two monolignols can form a dimer and polymerize further. Finally, highly branched lignan polymers are formed.

a function in defense. Its mechanical strength and chemical composition make plant tissues difficult for herbivores to digest. In addition, lignin inhibits the growth of pathogenic microorganisms. Lignin is formed in many plants in response to wounding. Only a few bacteria and fungi are able to cleave lignin. A special role in the **degradation of lignin** is played by **woodrot fungi**, which are involved in the rotting of tree trunks.

Often one-third of dry wood consists of lignin. For the production of cellulose and paper, this lignin has to be removed, which is very costly and also pollutes rivers. Attempts are being made to reduce the formation of lignin in wood by means of genetic engineering.

Suberins form gas- and water-impermeable layers between cells

Suberin is a polymeric compound of phenylpropanoids, long chain fatty acids, and fatty alcohols (C_{18}–C_{30}), as well as hydroxyfatty acids and dicarboxylic acids (C_{14}–C_{20}) (Fig. 18.10). In suberin, the phenylpropanoids are partly linked with each other as in lignin. However, the 9′-OH groups are

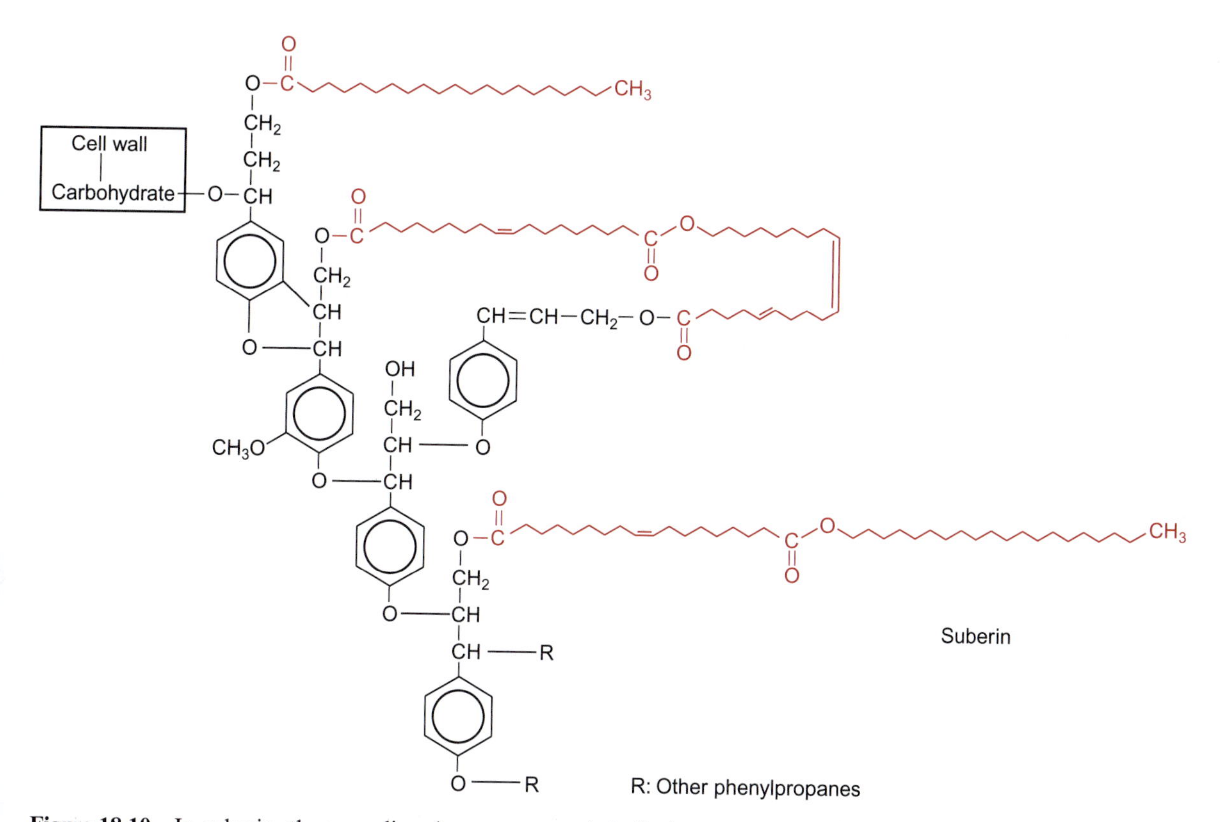

Figure 18.10 In suberin, the monolignols are connected similarly to those in lignin, but the 9′–OH groups usually do not participate. Instead they form esters with long-chain fatty acids and hydroxyfatty acids. Carboxylic acid esters provide a link between two monolignols.

mostly not involved in these linkages and instead form esters with the fatty acids mentioned previously. Often two phenylpropanoids are connected by dicarboxylic acids via ester linkages, and fatty acids and hydroxyfatty acids also can form esters with each other. Although the mechanism of suberin synthesis is to a large extent still not known, it appears that peroxidases are also involved in this process.

Suberin is a cell wall constituent that forms gas- and watertight layers. It probably is contained in the **Casparian strip** of the root endodermis, where it acts as a diffusion barrier between the apoplast of the root cortex and the central cylinder. Suberin is present in many C_4 plants as an impermeable layer between the bundle sheath and mesophyll cells. Cork tissue, consisting of dead cells surrounded by alternating layers of suberin and wax, has a particularly high suberin content. **Cork cells** are found in a secondary protective layer called the periderm and in the bark of trees. Cork layers containing suberin protect plants against loss of water, infection by microorganisms, and heat exposure. Due to this, some plants even survive short fires and are able to continue growing afterward.

Cutin is a gas- and water-impermeable constituent of the cuticle

The epidermis of leaves and other shoot organs is surrounded by a gas- and water-impermeable cuticle (Chapter 8). It consists of a cell wall that is impregnated with cutin and in addition is covered by a wax layer. **Cutin** is a polymer similar to suberin, but with a relatively small proportion of phenylpropanoids and dicarboxylic acids, consisting mainly of esterified hydroxyfatty acids (C_{16}-C_{18}).

18.4 For the synthesis of flavonoids and stilbenes, a second aromatic ring is formed from acetate residues

Probably the largest group of phenylpropanoids is that of the flavonoids, in which a second aromatic ring is linked to the 9′-C atom of the phenylpropanoid moiety. A precursor for the synthesis of flavonoids is chalcone (Fig. 18.11), synthesized by **chalcone synthase (CHS)** from *p*-coumaryl-CoA and three molecules of malonyl CoA. This reaction is also called the **malonate pathway**. The release of three CO_2 molecules and four CoA molecules makes chalcone synthesis an irreversible process. In the overall reaction, the new aromatic ring is formed from three acetate residues. Since CHS represents the first step of flavonoid biosynthesis, this enzyme has been thoroughly investigated. In some plants, one or two different isoforms of the enzyme have been found, while in others there are up to nine. CHS is the most abundant enzyme protein of phenylpropanoid metabolism in plant cells, probably due to the fact that this enzyme has only a low catalytic activity. As in the case of phenylalanine ammonia lyase (section 18.1), the *de novo* synthesis of CHS is subject to multiple controls of gene expression by internal and external factors, including elicitors.

The stilbenes include very potent natural fungicides

Some plants, including pine, grapevine, and peanuts, possess a **stilbene synthase** activity, by which *p*-coumaroyl CoA reacts with three molecules of

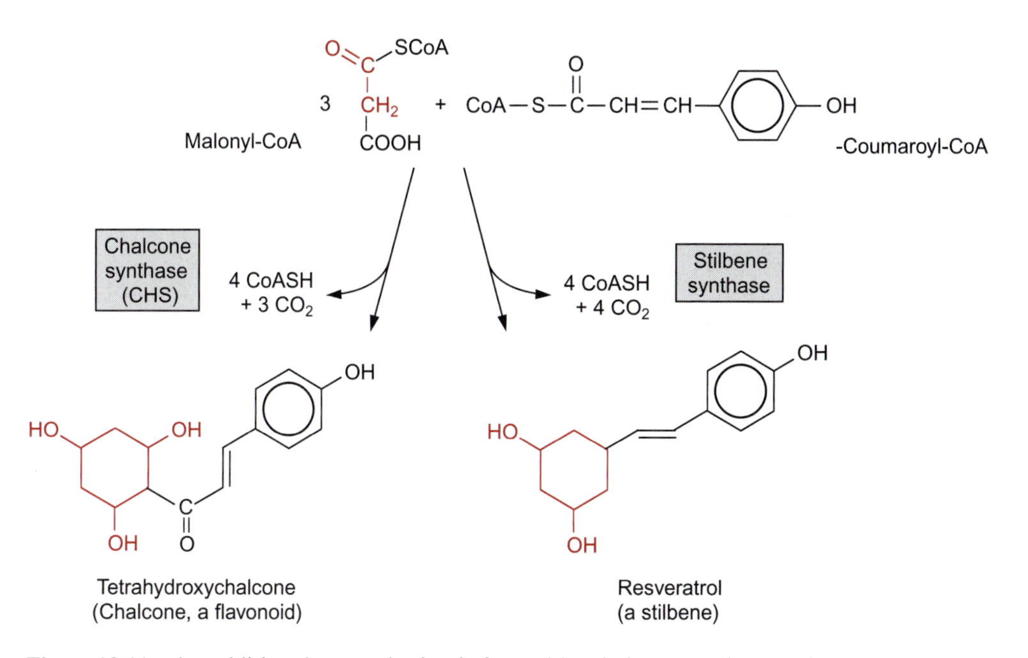

Figure 18.11 An additional aromatic ring is formed by chalcone synthase and stilbene synthase.

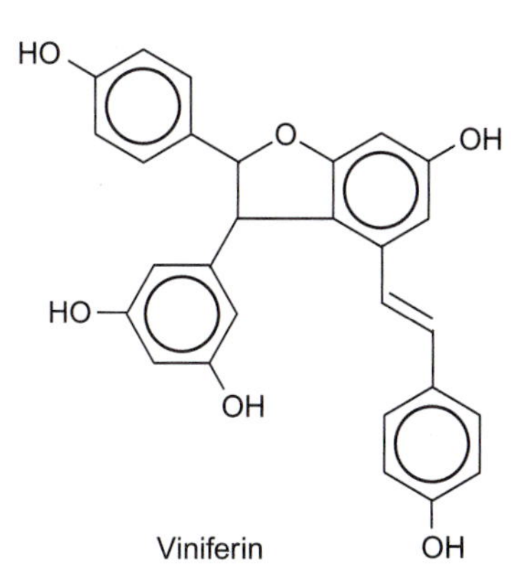

Figure 18.12 A natural fungicide from grapevine.

malonyl CoA. In contrast to CHS, the 9′-C atom of the phenylpropane is released as CO_2 (Fig. 18.11). **Resveratrol**, synthesized by this process, is a phytoalexin belonging to the stilbene group. A number of very potent plant fungicides derive from the stilbenes, including **viniferin** (Fig. 18.12), which is formed in grapevine. The elucidation of stilbene synthesis has opened new possibilities to combat fungal infections. Recently, a gene from grapevine for the formation of resveratrol has been expressed by genetic engineering in tobacco, and the resultant transgenic tobacco plants were resistant to the pathogenic fungus *Botrytis cinerea.*

18.5 Flavonoids have multiple functions in plants

Chalcone is converted to flavanone by **chalcone isomerase** (Fig. 18.13). The ring structure is formed by the addition of a phenolic hydroxyl group to the

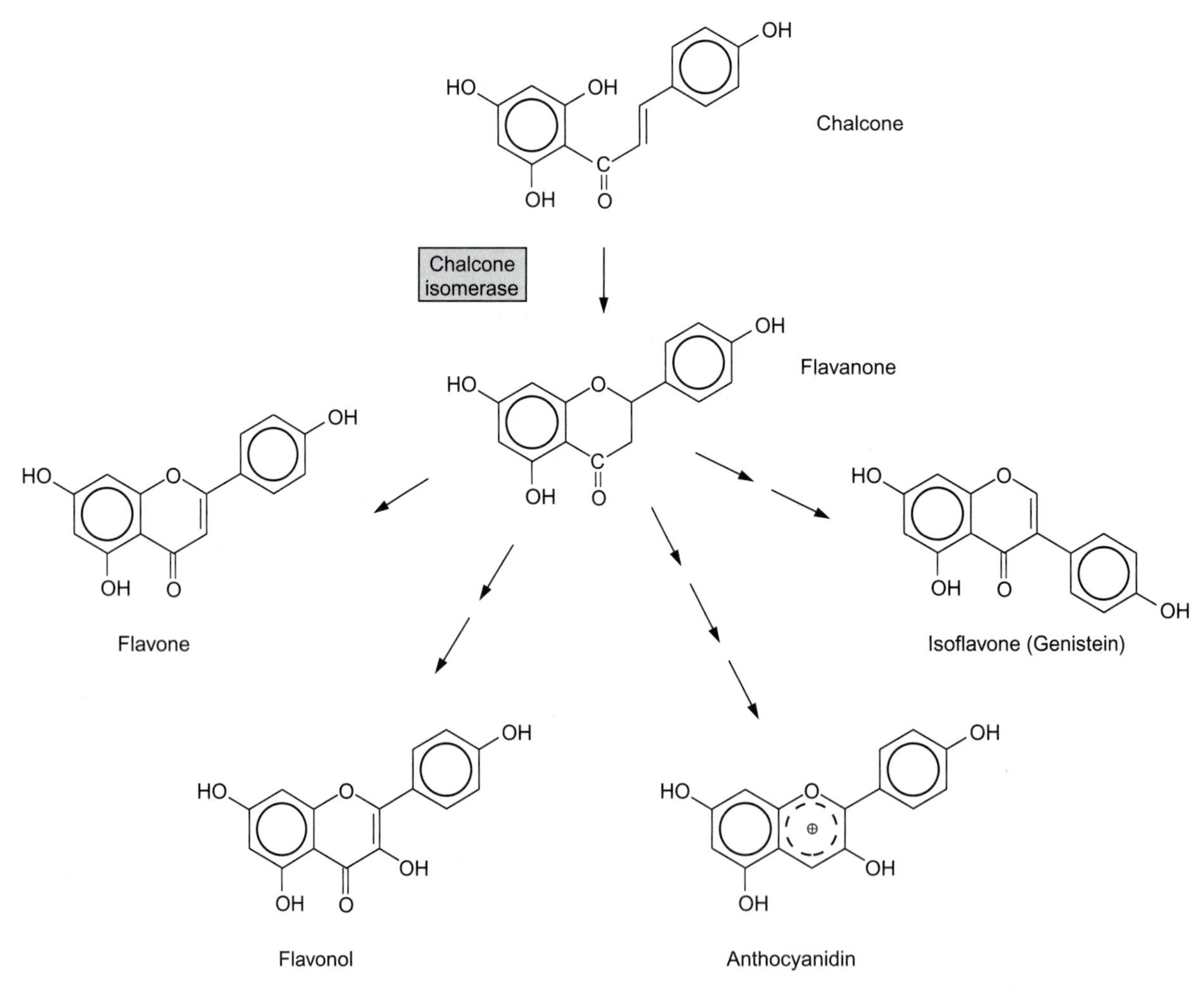

Figure 18.13 Chalcone is the precursor for the synthesis of various flavonoids.

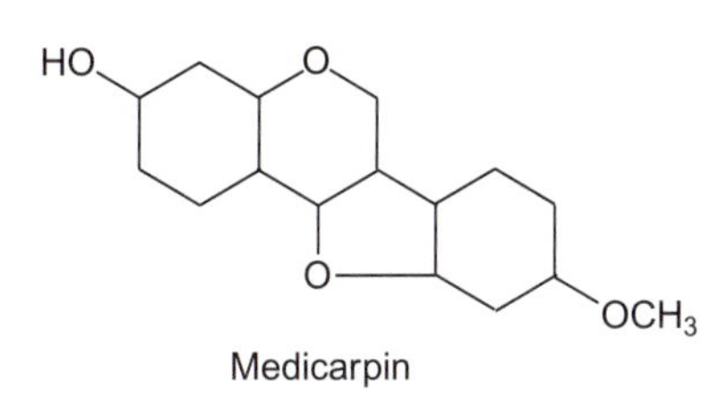

Figure 18.14 A phytoalexin from lucerne.

double bond of the carbon chain connecting the two phenolic rings. Flavanone is the precursor for a variety of flavonoids, the synthesis of which will not be described here. As a key enzyme of flavonoid synthesis, the synthesis of the enzyme protein of chalcone isomerase is subject to strict control. It is induced like PAL and CHS by elicitors.

The flavonoids include protectants against herbivores and many phytoalexins. An example of this is the poisonous isoflavone dimer **rotenone**, an inhibitor of the respiratory chain (section 5.5), which is contained in the leaves of a tropical legume. Aboriginals in South America used to kill fish by flinging the leaves of these plants into the water. The isoflavone **medicarpin** from lucerne (alfalfa) is a phytoalexin (Fig. 18.14). Flavonoids also serve as signals for interactions of the plant with symbionts. Flavones and flavonols are emitted as signal substances from leguminous roots in order to induce in *Rhizobia* the expression of the genes required for nodulation (section 11.1).

Flavones and flavonols have an absorption maximum in the UV region. As **protective pigments**, they shield plants from the damaging effect of UV light. The irradiation of leaves with UV light induces a strong increase in flavonoid biosynthesis. Mutants of *Arabidopsis thaliana*, which, because of a defect in either chalcone synthase or chalcone isomerase, are not able to synthesize flavones, are extremely sensitive to the damaging effects of UV light. In some plants, fatty acid esters of sinapic acid (section 18.2) can also act as protective pigments against UV light.

Many flavonoids act as **antioxidants**. As constituents of nutrients, they are assumed to be protectants against cardiovascular disease and cancer. For this reason, nutrients containing flavonoids (e.g., green tea, soy sauce, and red wine) have been regarded as beneficial for health.

Recently, particular attention has been focused on certain isoflavones that are found primarily in legumes. It had been observed earlier that sheep became infertile after grazing on certain legumes. It turned out that these forage plants contained isoflavones, which in animals (and in humans) have an effect similar to that of estrogens. For this reason, they have been named **phytoestrogens**. A strong estrogen effect has **genistein**, shown in Figure 18.13. Investigations are in progress to use phytoestrogens for medical purposes.

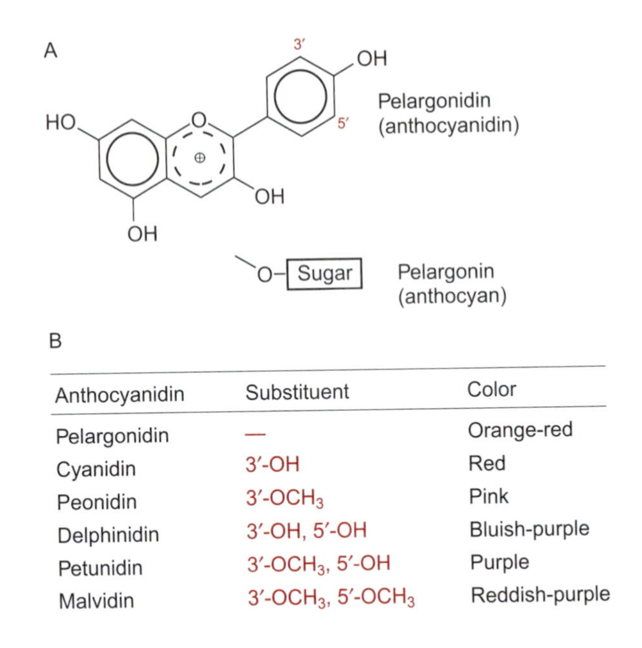

Anthocyanidin	Substituent	Color
Pelargonidin	—	Orange-red
Cyanidin	3′-OH	Red
Peonidin	3′-OCH_3	Pink
Delphinidin	3′-OH, 5′-OH	Bluish-purple
Petunidin	3′-OCH_3, 5′-OH	Purple
Malvidin	3′-OCH_3, 5′-OCH_3	Reddish-purple

Figure 18.15 A. Pelargonidin, an anthocyanidin, is a flower pigment. It is present in the petals as a glucoside, named pelargonin. B. More plant pigments are formed by additional -OH groups in 3′ and 5′ positions and their methylation.

18.6 Anthocyanins are flower pigments and protect plants against excessive light

We have already discussed carotenoids as yellow and orange flower pigments (section 17.6). Other widely distributed flower pigments are the yellow **chalcones**, light yellow **flavones**, and red and blue **anthocyanins**. Anthocyanins are glucosides of anthocyanidins (Fig. 18.15) in which the sugar component, consisting of one or more hexoses, is usually linked to the -OH group of the pyrylium ring. Anthocyanins are deposited in the vacuole. It has been shown that anthocyanins are transported via the glutathione translocator to the vacuole as glutathione conjugates (section 12.2). The anthocyanin pelargonin, shown in Figure 18.15, contains **pelargonidin** as chromophore. The introduction of two -OH groups in 3′ and 5′ positions of the phenyl residue by P_{450} dependent monooxygenases (section 18.2) and their successive methylation yields five additional flower pigments, each with a different color. Substitutions at other positions result in even more pigments. A change in the pH leads to a change of color. This in part explains the change of color when plants fade. Moreover, the color of the pigment is altered by the formation of complexes with metal ions. Thus, upon complexation with Al^{+++} or Fe^{+++}, the color of pelargonin changes from orange red to blue. These various pigments and their mixtures lead to the many color nuances of flowers. With the exception of pelargonidin, all the pigments listed in

Figure 18.15 are found in the flowers of petunia. To date, 35 genes that are involved in the coloring of flowers have been isolated from petunia.

Anthocyanins not only comprise flower pigments to attract pollen-transferring insects, but also function as protective pigments for shading leaf mesophyll cells. Plants in which growth is limited by environmental stress factors, for instance phosphate deficiency, chilling, or high salt content of the soil, often have red leaves, due mainly to the accumulation of anthocyanins. Stress conditions, in general, reduce the utilization of NADPH and ATP, which are provided by the light reactions of photosynthesis. Shading the mesophyll cells by anthocyanins decreases the light reactions and thus prevents overenergization and overreduction of the photosynthetic electron transport chain (see section 3.10).

18.7 Tannins bind tightly to proteins and therefore have defense functions

Tannins are a collective term for a variety of plant polyphenols used in the tanning of rawhides to produce leather. Tannins are widely distributed in plants and occur in especially high amounts in the bark of certain trees (e.g., oak) and in galls. One differentiates between condensed or hydrolyzable tannins (Fig. 18.16A). The **condensed tannins** are flavonoid polymers and thus are products of phenylpropanoid metabolism. Radical reactions probably are involved in their synthesis, but few details of the biosynthesis pathways are known. The **hydrolyzable tannins** consist of **gallic acids** (Fig. 18.16 B). Many of these gallic acids are linked to hexose molecules. Gallic acid in plants is formed mainly from shikimate (Fig. 10.19).

The phenolic groups of the tannins bind so tightly to proteins by forming hydrogen bonds with the -NH groups of peptides that these bonds cannot be cleaved by digestive enzymes. In the tanning process, tannin binds to the collagen of the animal hides and thus produces leather that is able to withstand the attack of degrading microorganisms. Tannins have a sharp unpleasant taste; binding of tannins to the proteins of the mucous membranes and saliva draws the mouth together. In this way animals are discouraged from eating plants that contain tannin. In addition, when an animal eats plant material, destruction of plant cells results in binding of tannins to plant proteins, making them less digestible and thus unsuitable for fodder. Tannins also react with enzymes of the herbivore digestive tract. For these reasons, tannins are very effective in protecting leaves from being eaten by animals. As an example of this, in the South African savannah, the

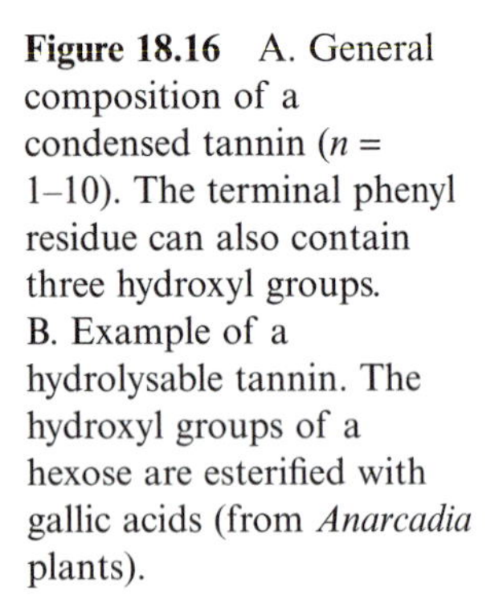

Figure 18.16 A. General composition of a condensed tannin (n = 1–10). The terminal phenyl residue can also contain three hydroxyl groups. B. Example of a hydrolysable tannin. The hydroxyl groups of a hexose are esterified with gallic acids (from *Anarcadia* plants).

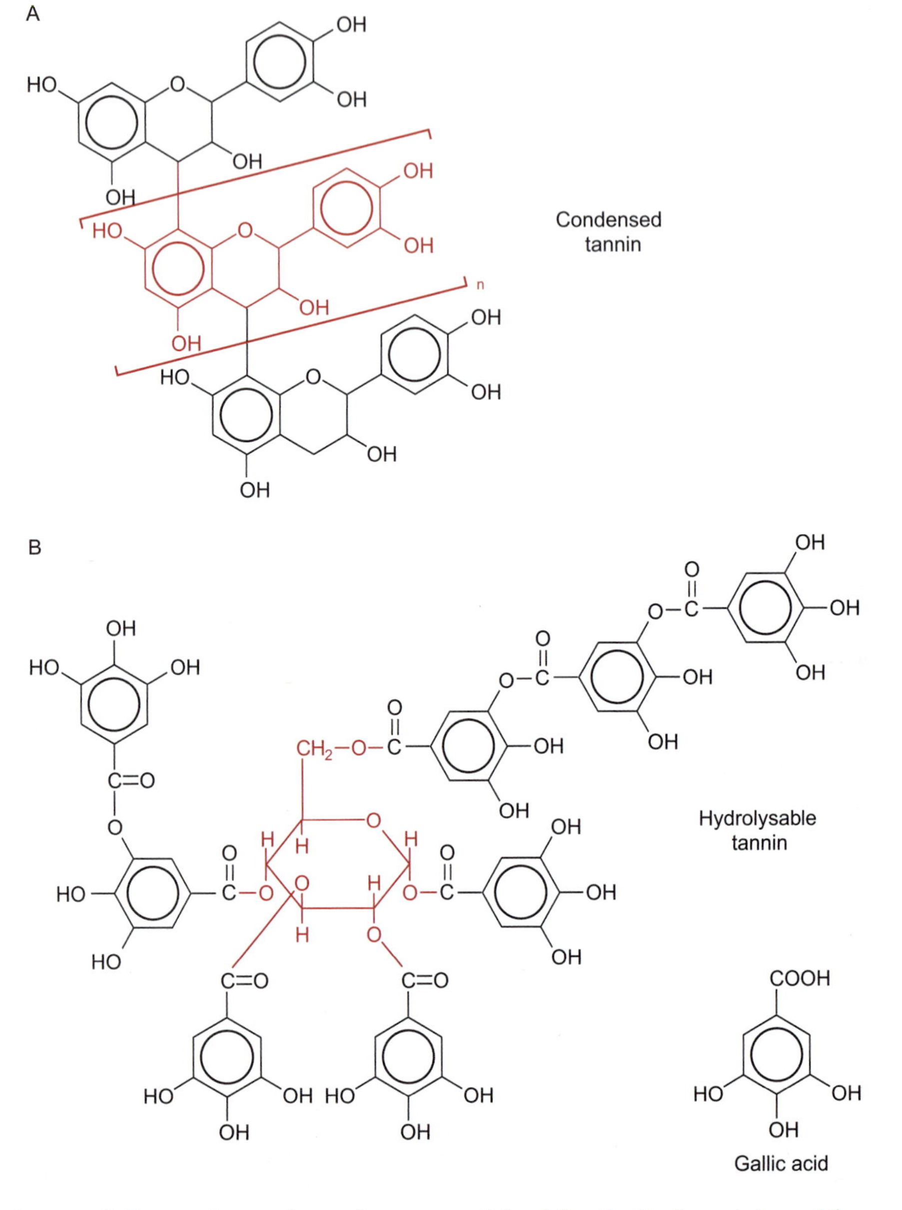

leaves of the acacia are the main source of food for the kudu antelope. These leaves contain tannin, but in such low amounts that it does not affect the nutritional quality. Trees injured by feeding animals emit gaseous ethylene (section 19.5), and within 30 minutes the synthesis of tannin is induced in the leaves of neighboring acacias. If too many acacia leaves are eaten, the tannin content can increase to such a high level that the kudu could die when feeding from these leaves. Thus the acacias protect themselves from complete defoliation by a collective warning system. Investigations are in progress to decrease the tannin content of forage plants by genetic engineering.

Tannins also protect plants against attack by microorganisms. Infection of plant cells by microorganisms is often initiated by the secretion of enzymes for lytic digestion of plant cell walls. These aggressive enzymes are inactivated when tannins are bound to them.

Further reading

Boerjan, W., Ralph, J., Baucher, M. Lignin biosynthesis. Annu Rev Plant Biol 54, 519–546 (2003).

Boudet, A.-M. A new view of lignification. Trends Plant Sci 3, 67–71 (1998).

Davin, L. B., Lewis, N. G. Phenylpropanoid metabolism: Biosynthesis of monolignols, lignans and neolignans, lignins and suberins. In Phenolic Metabolism in Plants, H. A. Stafford, R. K. Ibrahim (eds.). Plenum Press, New York, pp. 325–375 (1992).

Dixon, R. A., Paiva, N. L. Stress-induced phenylpropanoid metabolism. Plant Cell 7, 1085–1097 (1995).

Dixon, R. A., Steele, C. L. Flavonoids and isoflavonoids—a gold mine for metabolic engineering. Trends Plant Sci 4, 394–400 (1999).

Douglas, C. J. Phenylpropanoid metabolism and lignin biosynthesis: From weeds to trees. Trends Plant Sci 1, 171–178 (1996).

Durner, J., Shah, J., Klessig, D. F. Salicylic acid and disease resistance in plants. Trends Plant Sci 2, 266–274 (1997).

Hatfield, R., Vermeris, W. Lignin formation in plants. The dilemma of linkage specifity. Plant Physiol 126, 1351–1357 (2001).

Holton, T. A., Cornish, E. C. Genetics and biochemistry of anthocyanin biosynthesis. Plant Cell 7, 1071–1083 (1995).

Humphreys, J. M., Chappie, C. Molecular "pharming" with plant P450s. Trends Plant Scie 5, 271–273 (2000).

Lewis, N. G. A 20th century roller coaster ride: A short account of lignification. Curr Opin Plant Biol 2, 153–162 (1999).

Lois, R., Buchanan, B. B. Severe sensitivity to ultraviolet radiation in *Arabidopsis* mutants deficient in flavonoid accumulation: Planta 194, 504–509 (1994).

Lynn, D. G., Chang, M. Phenolic signals in cohabitation: Implications for plant development. Annu Rev Plant Physiol Plant MolBiol 41, 497–526 (1990).

Murphy, A. M., Chivasa, S., Singh, D. P., Carr, J. P. Salicylic acid-induced resistance to viruses and other pathogens: A parting of the ways? Trends Plant Sci 4, 155–160 (1999).

Schuler, M. A., Werck-Reichart, D. Functional genomics of B450. Annu Rev Plant Biol 54, 629–667 (2003).

Shirley, B. W. Flavonoid biosynthesis: New functions for an old pathway. Trends Plant Sci 1, 377–382 (1996).

Strid, A., Chow, W.S., Anderson, J. M. UV-B damage and protection at the molecular level in plants. Photosynth Res 39, 475–489 (1994).

19

Multiple signals regulate the growth and development of plant organs and enable their adaptation to environmental conditions

In complex multicellular organisms, such as higher plants and animals, metabolism, growth, and development of the various organs are coordinated by the emission of signal substances. In animals, such signals can be hormones, which are secreted by glandular cells. One differentiates between paracrine hormones, which function as signals to neighboring cells, and endocrine hormones, which are emitted to distant cells (e.g., via the blood circulation). Also in plants, signal substances are formed in certain organs, often signaling to neighboring cells, but also being distributed as signals to distant cells via the xylem or the phloem. All these plant signal substances are termed **phytohormones**. Some of the phytohormones (e.g., brassinosteroids) resemble in their structure animal hormones, whereas others are totally different. Like animal hormones, phytohormones also have many different signal functions. They control the adjustment of plant metabolism to environmental conditions, such as water supply, temperature, and day length, and regulate plant development. In addition, the quality of the light, in particular the intensity of the red, blue, and ultraviolet (UV) light, controls the growth and the differentiation of plants. Major light sensors are **phytochromes**, which recognize red and far-red light and **cryptochromes** and **phototropin** monitoring blue light. Cryptochromes also function as circadian photoreceptors in animals and plants for controlling the day-night cycle.

The signal transduction chain between the binding of a certain hormone to the corresponding receptor and its effect on specific parameters, such as the transcription of genes or the activity of enzymes, is now known for many animal hormones. In contrast, signal transduction chains have not been fully resolved for any of the phytohormones or light sensors. However, partial results indicate that certain components of the signal transduction chain in plants may be similar to those in animals. The phytohormone receptors and light sensors apparently act as a **multicomponent system**, in which the signal transduction chains are interwoven to a **network**.

19.1 Signal transduction chains known from animal metabolism also function in plants

G-proteins act as molecular switches

A family of proteins, which by binding of GTP or GDP can alternate between two conformational states, is widely distributed in the animal and plant kingdoms. These proteins are called **GTP-binding proteins**, or simply **G-proteins**. The heterotrimeric G-proteins, discussed in the following, interact with receptor proteins located in the plasma membrane. The receptors have a binding site for signal substances at the outside and the binding site for G-proteins at the cytoplasmic site of the plasma membrane, and are therefore well suited to pass external signals into the cell. Our basic knowledge about G-proteins derives primarily from investigations in animals. **Heterotrimeric-G-proteins** are composed of three different subunits: $\mathbf{G_\alpha}$ (molecular mass 45–55 kDa), $\mathbf{G_\beta}$ (molecular mass 35 kDa), and $\mathbf{G_\gamma}$ (molecular mass 8 kDa) (Fig. 19.1). Subunit G_α contains a binding site that can be occupied by either GDP or GTP. In animals, binding of the heterotrimer to a receptor (e.g., an adrenaline receptor occupied by adrenaline) enables the exchange of the GDP for GTP bound at G_α.

The binding of GTP results in a **conformational change** of the G_α subunit and thereby in its dissociation from the trimer. The liberated G_α unit, loaded with GTP, functions as an activator of various enzymes forming messengers for signal transduction. For instance, G_α-GTP stimulates a **GMP-cyclase** that forms the messenger substance guanosine-3′-5′-monophosphate (**cGMP**) from GTP (Fig. 19.2), as has been found in plants and animals. G_α-GTP also stimulates **phospholipase C** (see Fig. 19.4). The function of this reaction in the liberation of Ca^{++} as a messenger will be discussed in the following section. In fungi and animals, G_α-GTP stimulates the synthesis of the messenger substance adenosine-3′-5′-monophosphate (**cAMP**) from ATP via an activation of AMP-cyclase. It was believed that cAMP played

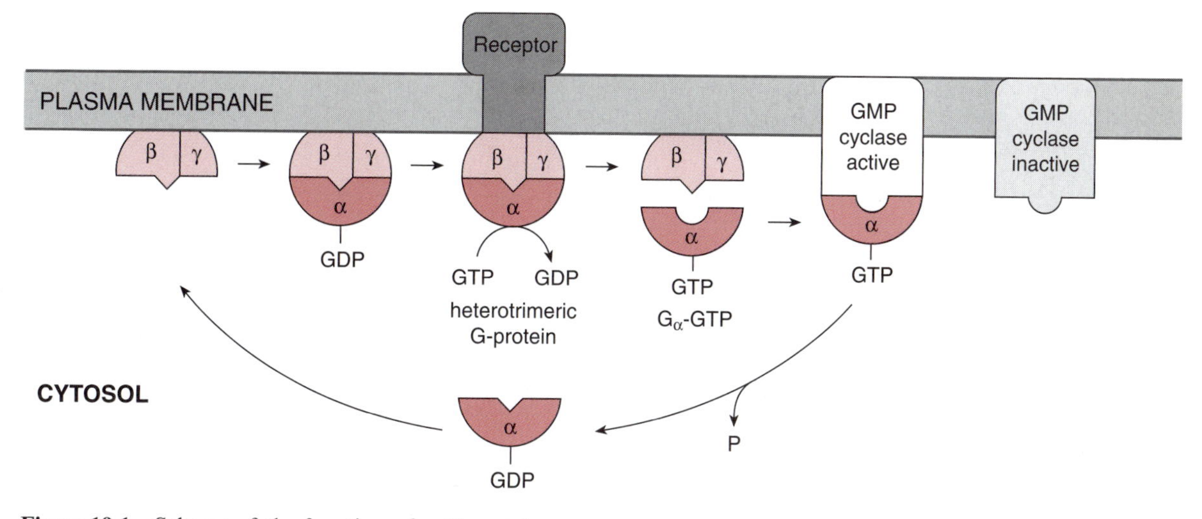

Figure 19.1 Scheme of the function of a G-protein.

no role in plant metabolism. However, there now are indications that cAMP might also take part in the regulation of plant metabolism, but this is still a matter of debate.

G_α-GTP has a half-life of only a few minutes. Bound GTP is hydrolyzed to GDP by an intrinsic GTPase activity, and the resulting conformational change causes G_α to lose its activator function. It binds once more to the dimer to form a trimer and a new cycle can begin again. The short life of G_α-GTP makes the signal transduction very efficient.

Small G-proteins have diverse regulatory functions

All eukaryotes also contain **small G-proteins**, which have only one subunit and are related to the α-subunit of heterotrimeric G-proteins discussed in the preceding. All small G-proteins belong to a superfamily termed the **Ras superfamily**. These small G-proteins, located in the cytosol, have binding domains for GDP/GTP and an effector domain. When stimulated by a signal, the small G-protein interacts with an exchange factor, which converts the GDP-bound inactive form to a GTP-active form by GTP/GDP replacement. Through its effector domain, the active GTP-form interacts with effector proteins in analogy to the effect of G_α-GTP of the heterotrimeric G-protein. It has been predicted from genomic analyses that the *Arabidopsis* genome encodes 93 small G-proteins. Small G-proteins were found to have various regulatory functions, such as the regulation of defense reactions, ABA responses, vesicle transport, cell polarity, and the growth of

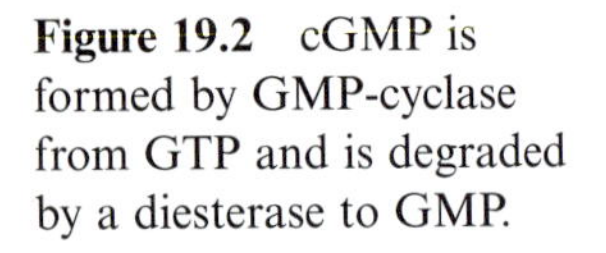
Figure 19.2 cGMP is formed by GMP-cyclase from GTP and is degraded by a diesterase to GMP.

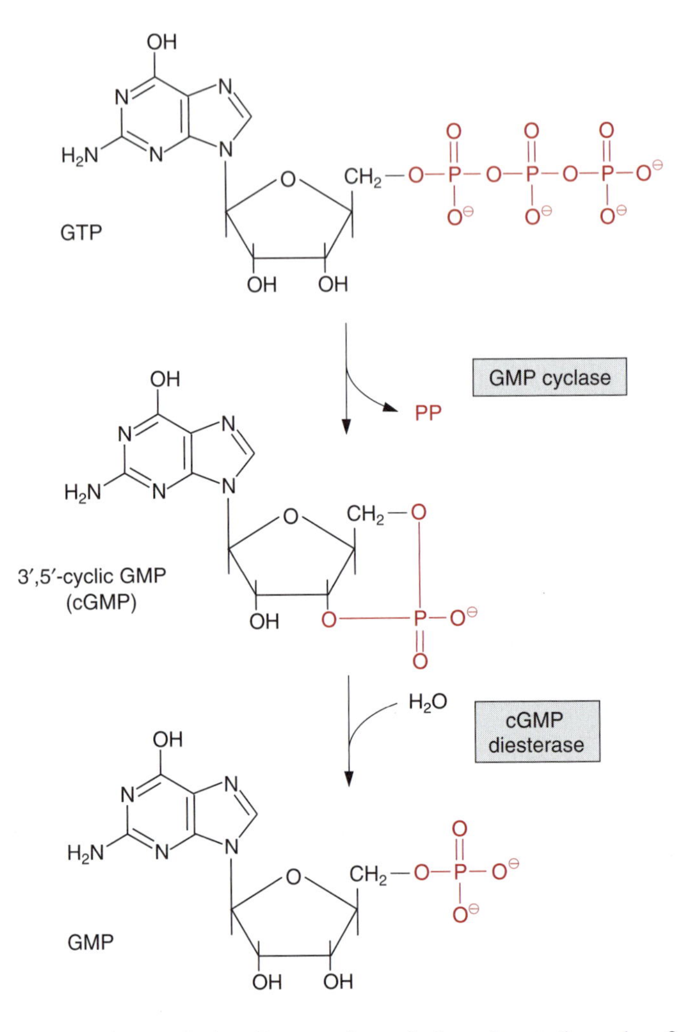

pollen tubers and root hairs. Present knowledge about the role of small G-proteins in plants is still at an early stage.

Ca^{++} acts as a messenger in signal transduction

In animal cells as well as in plant cells, the cytosolic concentration of free Ca^{++} is normally lower than about 10^{-7} mol/L. These very low Ca^{++} concentrations are maintained by ATP-dependent pumps (Ca^{++}-P-ATPases, section 8.2), which accumulate Ca^{++} in the lumen of the endoplasmic reticulum and the vacuole (in plants) or transport Ca^{++} via the plasma membrane to the extracellular compartment (Fig. 19.3). Alternatively, Ca^{++} can be taken up into mitochondria by H^+/Ca^{++} antiporters. Signals induce Ca^{++} channels on endomembranes of intracellular stores to open for a short time, resulting in

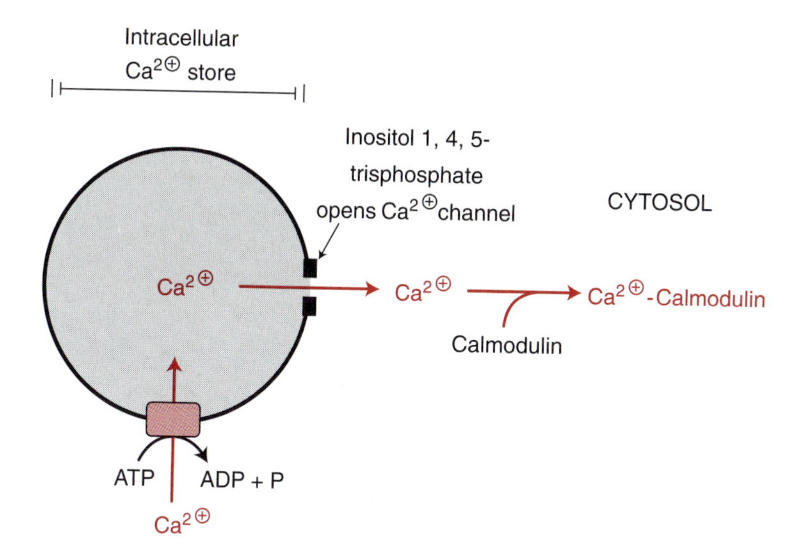

Figure 19.3 The endoplasmic reticulum of animals and plants and the plant vacuole (designated here as intracellular Ca^{++} store) contain in their membrane a Ca^{++}-P-ATPase (section 8.2), which pumps Ca^{++} into the lumen or into the vacuole. Ca^{++} can be released by an IP_3-dependent Ca^{++} channel to the cytosol.

a rapid increase in the cytosolic concentration of free Ca^{++}, which stimulates in almost all cells regulatory enzymes including protein kinases.

The phosphoinositol pathway controls the opening of Ca^{++} channels

Ca^{++} channels can be controlled by the phosphoinositol signal transduction cascade (Fig. 19.4), which has initially been resolved in animal metabolism, but has also been shown to exist in plants. **Phosphatidyl inositol** is present, although in relatively low amounts, as a constituent of cell membranes. In animal cells, the two fatty acids of phosphatidyl inositol are usually stearic acid and arachidonic acid. The inositol residue is phosphorylated at the hydroxyl groups in 4′ and 5′ position by a kinase. **Phospholipase C**, stimulated by a G-protein, cleaves the lipid to **inositol-1,4,5-trisphosphate (IP_3)** and **diacylglycerol** (DAG). IP_3 causes a rise in the cytosolic Ca^{++} concentration, whereas diacylglycerol stimulates in animals a Ca^{++}-dependent protein kinase. In plants, diacylglycerol as such does not seem to play a role in metabolism. However, it has an indirect effect, since **phosphatidic acid** (Fig.15.5) deriving from the phosphorylation of diacylglycerol affects protein kinases and ion channels.

Patch-clamp studies (see section 1.10) demonstrated that in plant vacuoles and other Ca^{++} stores, such as the endoplasmic reticulum, IP_3 causes Ca^{++} channels to open. The rapid influx of Ca^{++} into the cytosol is limited by the very short life of IP_3 (often less than 1 s). The rapid elimination of IP_3 proceeds via either further phosphorylation of IP_3 or a hydrolytic

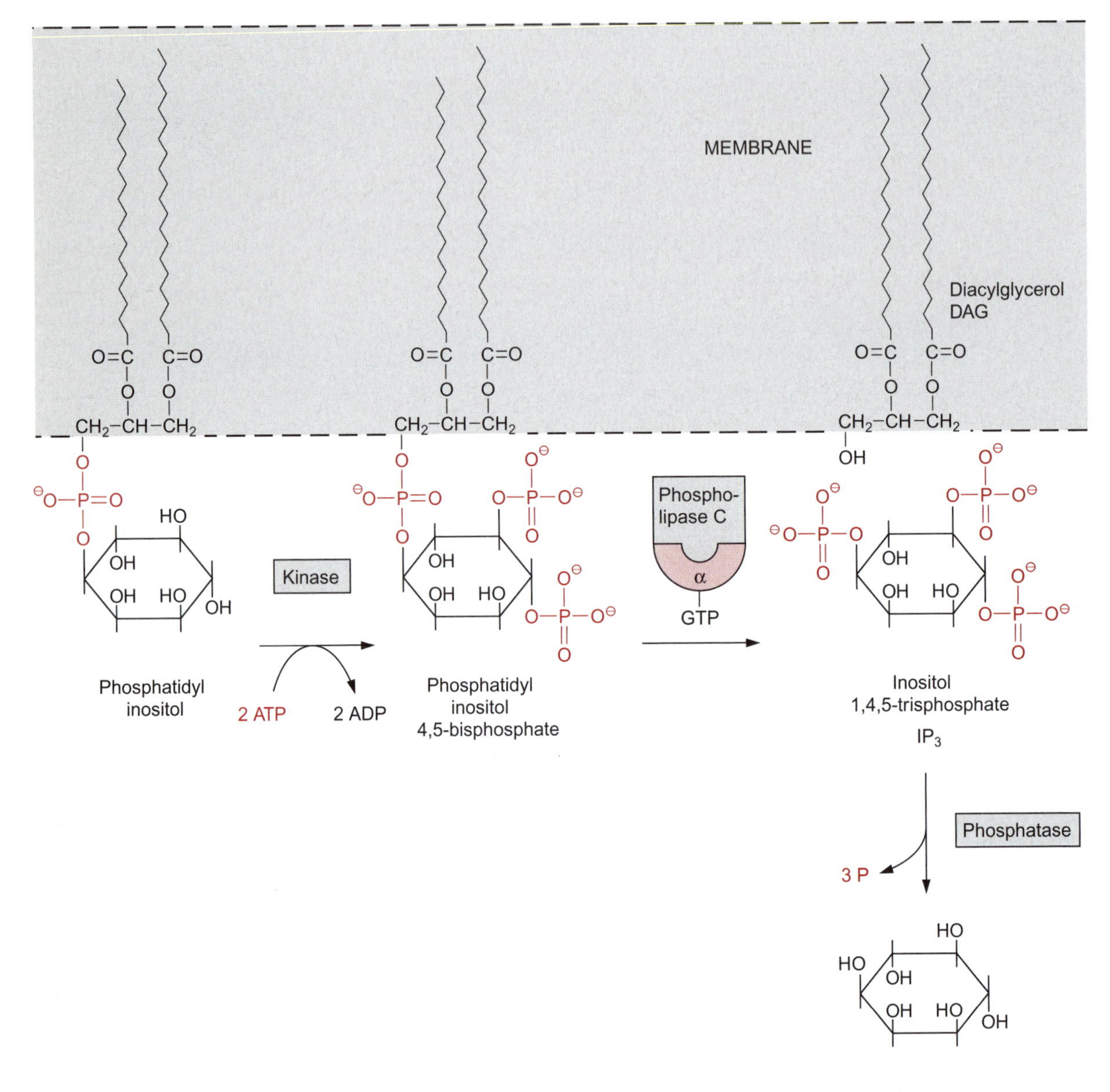

Figure 19.4 Inositol-1,4,5-trisphosphate (IP_3) as part of a signal transduction chain. Two hydroxyl groups of the inositol residue of a membrane phospholipid are phosphorylated by a kinase, and the resultant IP_3 is liberated by a G-protein ($G_\alpha GTP$)-dependent phospholipase C. The messenger IP_3 formed in this way is degraded by phosphatases.

liberation of the phosphate groups by a phosphatase. The short life of IP_3 enables a very efficient signal transduction.

The phosphoinositol cascade has in plants an important role in passing signals from the environment to cellular functions (e.g., in adjusting the stomata opening to the water supply). A specific kinase has been identified in plants, which catalyzes the phosphorylation of phosphatidyl inositol to **phosphatidyl inositol-3-phosphate**. This modified membrane lipid functions as a signal for vesicle transfer (Fig. 1.16) (e.g., in the transfer of hydrolytic enzymes from the ER to the vacuole).

Calmodulin mediates the messenger function of Ca^{++} ions

Often Ca^{++} acts as a messenger not directly but by binding to calmodulin. **Calmodulin** is a soluble protein (molecular mass 17 kDa) that occurs in animals as well as in plants. It is a highly conserved protein; the identity of the amino acid sequences between the calmodulin from wheat and cattle is as high as 91%. Calmodulin is present mainly in the cytosol. It consists of a flexible helix containing at each end two loops, of which each loop possesses a binding site for a Ca^{++} ion. Since these loops contain glutamate (E) and phenylalanine (F), they have been called **EF hands** (Fig. 19.5). The occupation of all four EF hands by Ca^{++} results in a conformational change of calmodulin by which its hydrophobic domain is exposed. This domain interacts with certain protein kinases (**calmodulin-binding kinases (CBK)**), by which the latter are activated. In the case of the protein kinase-CBK II, after activation, the kinase first catalyzes its own phosphorylation (autophosphorylation), and only then reaches its full activity, and even retains it after the dissociation of calmodulin, until the phosphate residue is released by hydrolysis. Ca^{++}-calmodulin also binds to other proteins, thus changing their activity, and is therefore an important member of signal transduction chains.

Moreover, plants contain a family of protein kinases in which Ca^{++}-binding EF hands are part of their protein. These are termed Ca^{++}-dependent

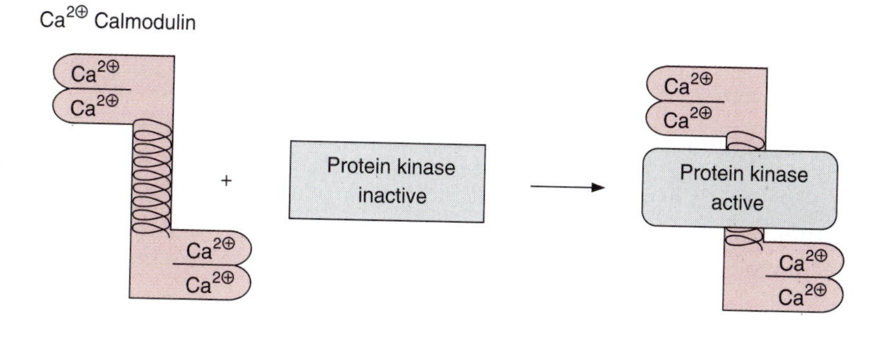

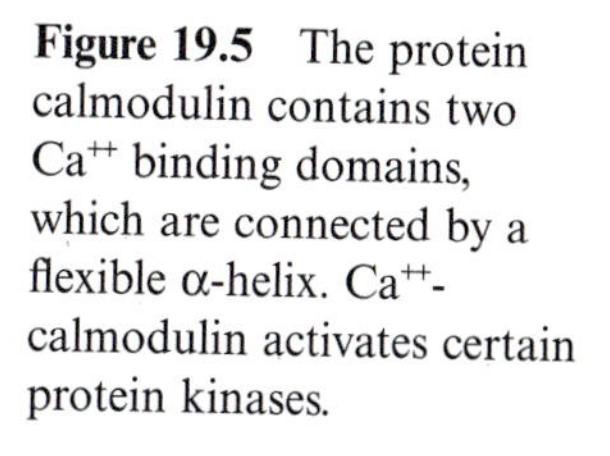

Figure 19.5 The protein calmodulin contains two Ca^{++} binding domains, which are connected by a flexible α-helix. Ca^{++}-calmodulin activates certain protein kinases.

protein kinases (**CDPK**). Numerous protein kinases of this type are present in plants. Thus, in *Arabidopsis,* by now 34 genes of the CDPK-family have been detected, although the function of part of them is still not known. CDP-kinases are involved in the phosphorylation of sucrose phosphate synthase (Fig. 9.18) and nitrate reductase (Fig. 10.9), pathogen defense reaction, and the response to various abiotic stress.

There are also other proteins with calmodulin domains, the so-called **calmodulin-related proteins (CRK)**, the functions of which are to a large extent not known.

Phosphorylated proteins form elements of signal transduction

Protein kinases, several of which have been discussed previously, and **protein phosphatases** are important elements in the regulation of intracellular processes. Since proteins change by phosphorylation and dephosphorylation between two states of different effectiveness, and many protein kinases are switched on or off by phosphorylation, protein kinases represent a **network of on-off switches in the cell**, comparable to those of computer chips, and they control differentiation, metabolism, defense against pests, and many other cell processes. It is estimated that in a eukaryotic cell 1% to 3% of the functional genes encode protein kinases. Initially protein kinases have been investigated mainly in yeast and animals, but in the meantime several hundred genes encoding protein kinases have been identified in plants, although the physiological functions of only a small part of them is known. Research about the function of protein kinases is at present a dynamic field in plant biochemistry.

Most protein kinases in eukaryotes, such as fungi, animals, or plants, contain 12 conserved regions. These protein kinases phosphorylate mainly the -OH group of **serine** and/or **threonin**e and in some cases also of **tyrosine**. Since all these protein kinases are homologous and thus descend from a common ancestor, they are grouped in a **superfamily of eukaryotic protein kinases**. Table 19.1 shows some members of this family. There is a family of protein kinases, which is regulated by cGMP (**protein kinases G**). The existence in plants of **protein kinases A**, regulated by cAMP, is still a matter of dispute. The protein kinases regulated by Ca^{++}-calmodulin (**CBK**) were already mentioned, as have the Ca^{++}-dependent protein kinases (**CDPK**). Further members of the superfamily are the **receptor-like protein kinases (RLK)**. These protein kinases generally are located in plasma membranes. They contain an extra cytoplasmatic domain with a receptor (e.g., for a phytohormone). The occupation of this receptor results in the activation of a protein kinase at the cytoplasmatic side of the membrane, reacting with cel-

Table 19.1: Some members of the eukaryotic protein kinase super family

	Modulator
Protein kinase-A	cAMP
Protein kinase-G	cGMP
Ca^{++} dependent protein kinase (CDPK)	Ca^{++}
Calmodulin-binding kinase (CBK)	Ca^{++}-Calmodulin
Receptor-like protein kinase (RLK)	z. B. Phytohormone
Cyclin-dependent protein kinase (CDK)	Cyclin
Mitogen-activated protein kinase (MAPK)	Mitogen
MAPK-activated protein kinase (MAPKK)	MAPK
MAPKK-activated protein kinase (MAPKKK)	MAPKK

lular proteins. Genome sequencing revealed that the *Arabidopsis* genome contains more than 400 genes encoding RLKs.

The superfamily of eukaryotic protein kinases also encompasses the **cyclin-dependent-protein kinases (CDK)**. **Cyclin** is a protein that is present in all eukaryotic cells, as it has an essential function in the cell cycle. CDkinases activate a number of proteins that are involved in mitosis. Further members of the superfamily are the **mitogen-activated-protein kinases (MAPK)**. **Mitogen** is a collective term for a variety of substances, many of them of unknown nature, which stimulate mitosis, and thus the cell cycle, but also other reactions. G-proteins and phytohormones may act as mitogens. MAPK plays an important role in **protein kinase cascades**, where protein kinases are regulated through phosphorylation by other protein kinases. In such a cascade, a G-protein, for example, activates a MAP-kinase-kinase-kinase (**MAPKKK**), which activates by phosphorylation a MAP-kinase-kinase (**MAPKK**), which activates a MAP-kinase (**MAPK**). The MAP-kinase in turn phosphorylates various substances. By phosphorylation of a series of transcription factors (section 20.2), it regulates the expression of different genes. The **MAP-kinase-cascade** thus has an important regulatory function in the process of cell development and differentiation. Moreover, the MAP-kinase system is also involved in the signal cascades of **pathogen defense systems**, which are triggered by elicitors (section 16.1), and in the response to abiotic stress. Genome sequencing revealed that 20 MAPKs, 10 MAPKKs, and 60 MAPKKKs exist in *Arabidopsis*.

Recently, protein kinases have been identified that phosphorylate **histidine** and **aspartate** residues of proteins and which do not belong to the superfamily mentioned previously. As will be discussed in section 19.7, histidine protein kinases are involved in the function of the receptors for ethylene and cytokinin.

Also, **protein phosphatases** exist in eukaryotes as a superfamily, with **serine-threonine-phosphatases** and **tyrosine-phosphatases** as different groups. Many of these phosphatases are regulated similarly to protein kinases (e.g., by binding of Ca^{++} plus calmodulin, or by phosphorylation). In this way the protein phosphatases also play an active role in signal transduction cascades. Research in this field is still at the beginning.

19.2 Phytohormones comprise a variety of very different compounds

Phytohormones (Fig. 19.6) have very diverse structures and functions. As examples, only few of these functions may be summarized here. Indole acetic

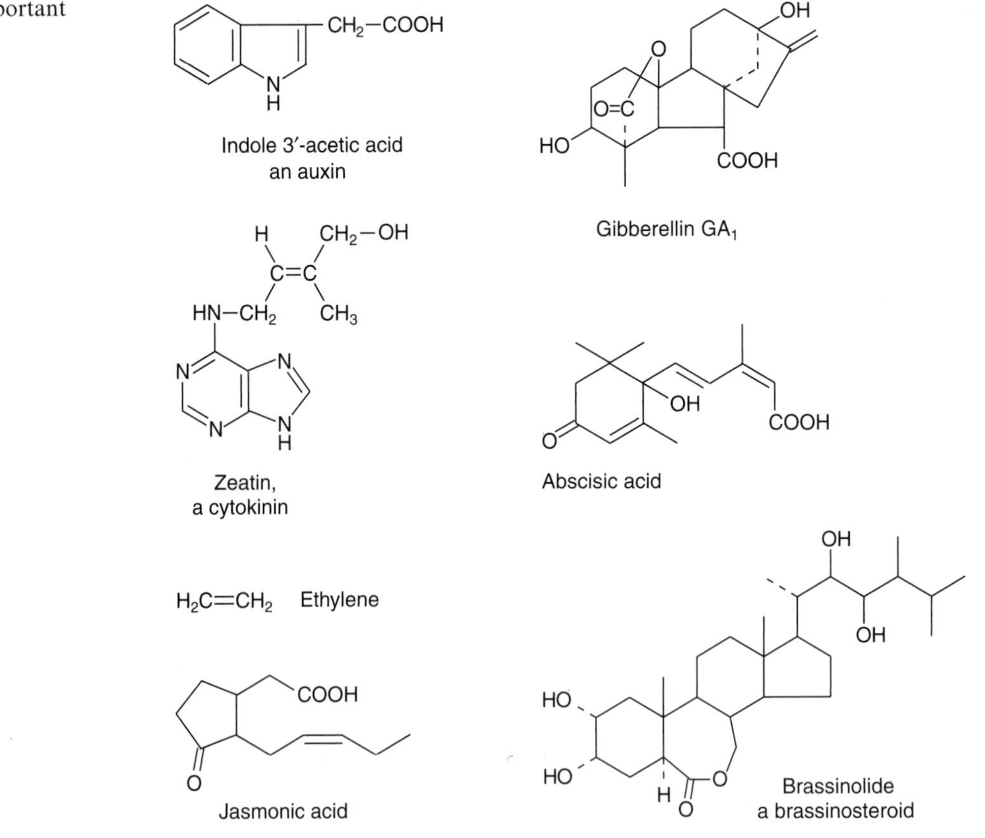

Figure 19.6 Important phytohormones.

acid, an **auxin** derived from indole, stimulates cell elongation. **Gibberellins**, derivatives of gibberellane, induce elongation growth of internodes. Zeatin, a **cytokinin**, is a prenylated adenine and stimulates cell division. **Abscisic acid**, which is formed from carotenoids, regulates the water balance. **Ethylene** and **jasmonic acid** (a derivative of fatty acids, section 15.7) enhance senescence. **Brassinosteroids** have a key function in the regulation of cell development. **Peptide hormones** regulate plant development, and, in addition to **salicylic acid** and jasmonic acid, play a role in pathogen defense. In many cases, phytohormone function is caused by a pair of antagonistic phytohormones. Thus abscisic acid induces seed dormancy, and gibberelic acid terminates it. Let us now look at the phytohormones in detail.

19.3 Auxin stimulates shoot elongation growth

Charles Darwin and his son Francis had already noted in 1880 that growing plant seedlings bend toward sunlight. They found that illumination of the tip initiated the bending of seedlings of canary grass (*Phalaris canariensis*). Since the growth zone is a few millimeters distant from the tip, they assumed that a signal is transmitted from the tip to the growth zone. In 1926 the Dutch researcher Frits Went isolated from the tip of oat seedlings a growth-stimulating substance, which he named **auxin** and which was later identified as **indoleacetic acid (IAA)**. The synthesis of IAA proceeds in different ways in different plants. Figure 19.7 shows two synthetic pathways starting from tryptophan. But precursors of tryptophan synthesis (Fig. 10.19) may also act as precursors of IAA synthesis. Besides IAA, some other substances are known with auxin properties (e.g., phenylacetic acid) (Fig. 19.8). The synthetic auxin **2,4-dichlorophenoxyacetic acid (2,4-D**, Rohm & Haas) is used as an **herbicide**. It kills plants by acting as an especially powerful auxin, resulting in disordered morphogenesis and an increased synthesis of ethylene, thus leading to a premature senescence of leaves. As **agent orange**, it was used in the Vietnam War to defoliate forests. 2,4-D is a selective herbicide that destroys dicot plants. Monocots are insensitive to it, because they eliminate the herbicide by degradation. For this reason, 2,4-D is used for combating weeds in cereal crops.

Auxin functions in many ways. During **early embryogenesis**, auxin governs the formation of the main **axis of polarity**, with the shoot meristem at the top and the root meristem at the opposite pole. Auxin generally influences cell division and differentiation. One effect of IAA is to enhance the

Figure 19.7 Two biosynthetic pathways for the formation of indole-3-acetic acid from tryptophan.

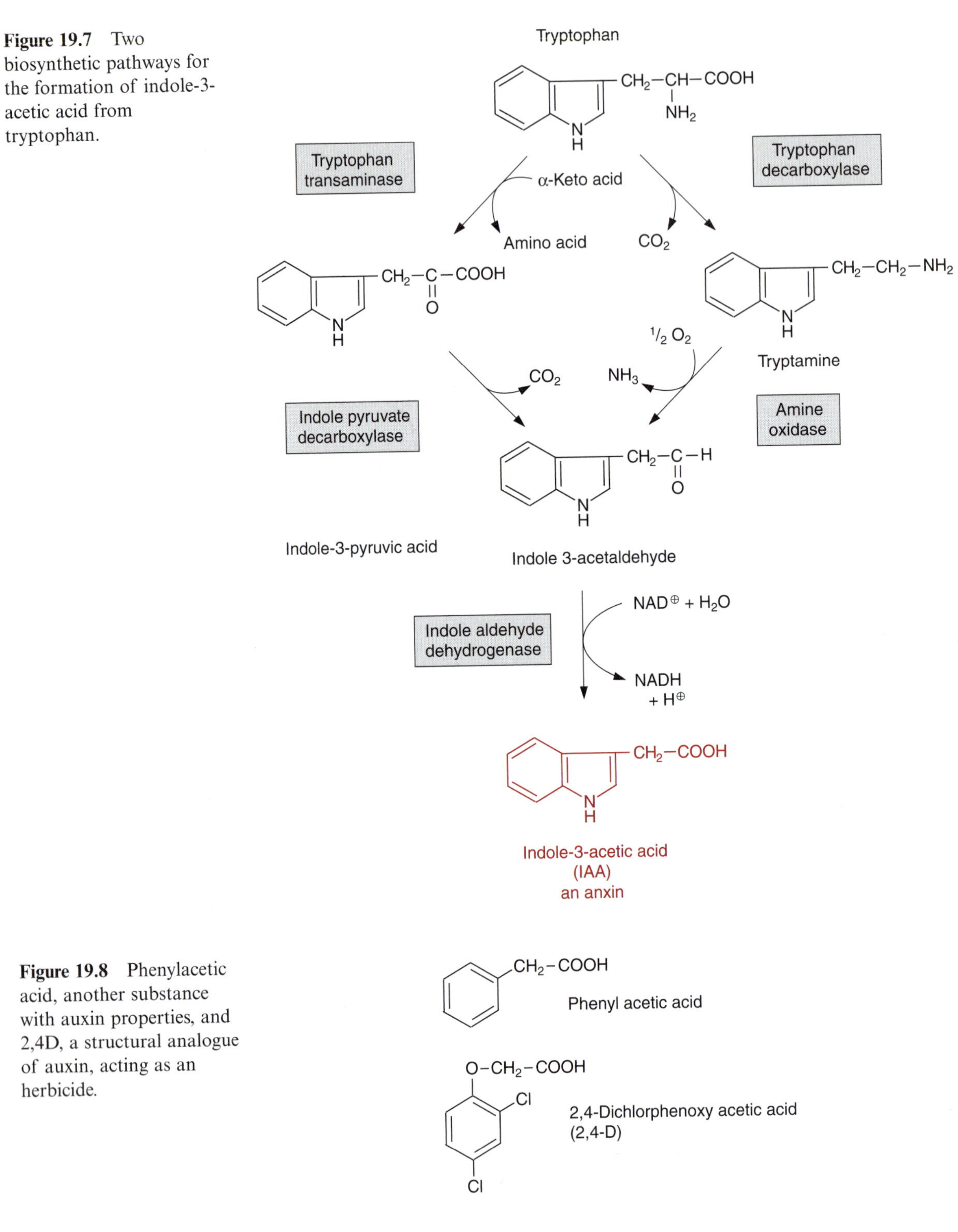

Figure 19.8 Phenylacetic acid, another substance with auxin properties, and 2,4D, a structural analogue of auxin, acting as an herbicide.

elongation growth of cells (see below). Therefore the highest IAA concentrations are found in the main growth zones of the shoot. However, IAA is formed primarily at the tip of the shoot. From there it is transported from cell to cell by an energy-dependent **polar transport**. The transport of auxin from cell to cell proceeds via specific **efflux** and **influx carriers** of the plasma membrane. The polar transport is caused by an asymmetric distribution of these carriers. The membrane-bound efflux carrier proteins are transferred in a reversible fashion between membrane regions by vesicle transport via the Golgi apparatus. In this way the efflux carriers can be rapidly moved from one area of the plasma membrane to another to facilitate a polar transport. During the curvature of the coleoptile, IAA is transported laterally to one side. The resulting differential stimulation of cell elongation at only one side of the shoot leads to the bending. IAA is also transported via the phloem from the leaves to distant parts of the plant.

The action of IAA on cell growth in the shoot can be traced back to two effects:

1. A few minutes after IAA is added to a cut tissue, a loosening of the normally rigid cell wall can be observed. How this happens has not been resolved unequivocally. One hypothesis is that auxin activates an H^+-P-ATPase of the plasma membrane via a signal transduction chain in which a 14.3.3 protein (section 10.3) and probably a protein kinase are involved, resulting in the acidification of the cell wall region, which in turn activates enzymes that loosen the rigid cell wall and in this way enable a turgor-driven cell enlargement.
2. About one hour after the addition of IAA, proteins required for growth are found to be synthesized. This is attributed to the effect of auxin on gene expression. Auxin is known to induce or repress specific sets of genes.

IAA shows rather diverse effects in different tissues. IAA stimulates cell division in the cambium, enhances **apical dominance** by suppression of lateral bud growth, and controls embryo development. Moreover, IAA prevents the formation of an abscission layer for leaves and fruits and is thus an antagonist to ethylene (section 19.7). On the other hand, increased IAA concentrations can induce the synthesis of ethylene.

Moreover, auxin induces the **formation of fruits**. Normally, seeds produce IAA only after fertilization. Transformants of eggplants that expressed a bacterial enzyme of IAA synthesis in the unfertilized seed were generated by genetic engineering. This IAA prevents the formation of seeds, resulting in seedless eggplants of normal consistency but being four times larger than usual. This is an impressive example of the importance of auxin for fruit growth and shows the possibilities of generating genetically altered vegetables.

Despite intensive research in this field, so far relatively little is known about the signal transduction chains of the various effects of auxin. A recently identified **auxin-binding protein** may function as an auxin receptor, although it is not certain where it is localized in a cell or how it is connected to signal transduction chains. At the other end of the unknown auxin signal transduction chain are probably so-called **Aux/IAA-proteins**, of which 25 different genes have been found in *Arabidopsis.* The Aux/IAA-proteins are repressor proteins, which bind to further regulator proteins, the so-called **auxin responsive factors (ARF)**, of which 20 genes are known in *Arabidopsis.* ARF enhances the transcription of many genes. Auxin causes, via a signal transduction chain, the conjugation of Aux/IAA-repressor proteins with **ubiquitin**, whereby the repressors are labeled for degradation through the **proteasome** machinery (section 21.4). In this way auxin enhances the expression of genes by **relief of a restraint**.

19.4 Gibberellins regulate stem elongation

The discovery of gibberellins goes back to a plant disease. The infection of rice by the fungus *Gibberella fujikuroi* results in the formation of extremely tall plants that fall over and bear no seed. In Japan this disease was called "foolish seedling disease." In 1926 Eiichi Kurozawa and collaborators in Japan isolated from this fungus a substance that induces unnatural growth. It was named **gibberellin**. These results were known in the West only after World War II. Structural analysis revealed that gibberellin is a mixture of various substances with similar structures, which also occur in plants and act there as phytohormones.

Gibberellins are derived from the hydrocarbon ***ent*-gibberellane** (Fig. 19.9). More than 100 gibberellins are now known in plants, which are numbered in the order of their identification. Therefore the numbering gives no

Figure 19.9 Hydrocarbon from which gibberellins are derived.

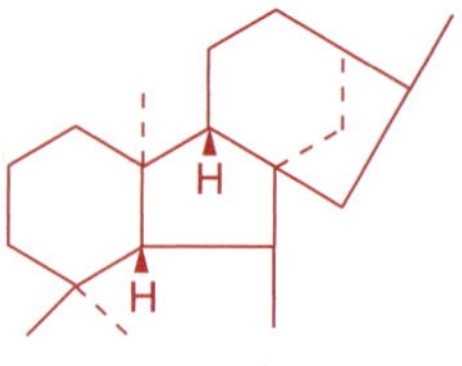

H
H
ent-Gibberellane

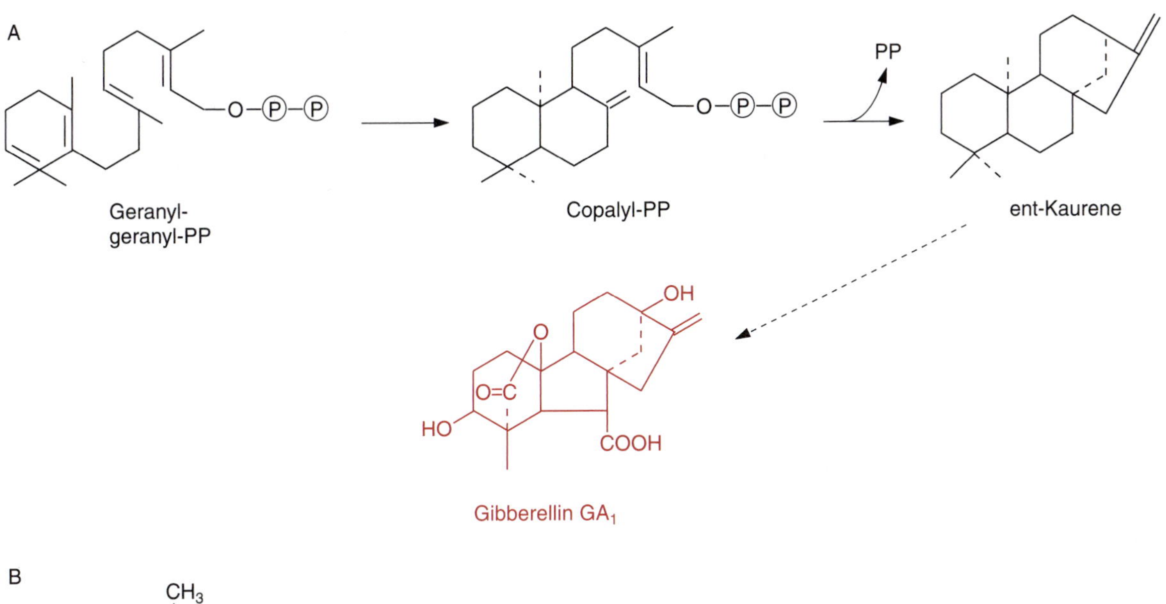

B

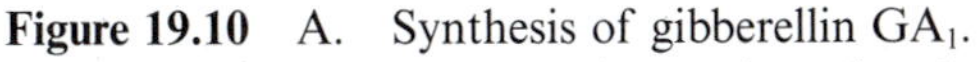

Figure 19.10 A. Synthesis of gibberellin GA_1.
B. Cycocel (BASF), a growth retardant that decreases the growth of stalks in wheat and other cereals, inhibits kaurene synthesis and thus also the synthesis of gibberellins.

information about structural relationships or functions. Many of these gibberellins are intermediates or by-products of the biosynthetic pathway. Only a few of them have been shown to act as phytohormones. Whether other gibberellins have a physiological function is not known. The most important gibberellins are GA_1 (Fig. 19.10A) and GA_4 (not shown). The synthesis of GA_1 occurs from the isoprenoid geranylgeranyl pyrophosphate in 13 steps, with *ent*-kaurene as an intermediate. In some cases, gibberellin synthesis is controlled by light via phytochromes (section 19.10).

Similarly to IAA, gibberellins **stimulate shoot elongation**, especially in the internodes of the stems. A pronounced gibberellin effect is that it induces rosette plants (e.g., spinach or lettuce) to shoot up for the formation of flowers, and they also regulate flowering. Additionally, gibberellins have a number of other functions, such as the preformation of fruits and the stimulation of their growth. Gibberellins terminate seed dormancy, probably by softening of the seed coat, and facilitate seed germination by the expression of genes for the necessary enzymes (e.g., amylases).

The use of gibberellins is of economic importance for the production of long, seedless grapes. In these grapes, GA_1 causes not only extension of the cells, but also parthenocarpy (the generation of the fruit as a result of parthogenesis). Moreover, in the malting of barley for beer brewing, gibberellin is added to induce the formation of α-amylase in the barley grains. The gibberellin GA_3, produced by the fungus *Gibberella fujikuroi* mentioned previously, is generally used for these purposes. Inhibitors of gibberellin biosynthesis are commercially used as **retardants** (growth inhibitors). A number of substances that inhibit the synthesis of the gibberellin precursor *ent*-kaurene, such as **chloroethyltrimethyl ammonia chloride**, (trade name, Cycocel, BASF) (Fig. 19.10B) are sprayed on cereal fields to decrease the **growth of the stalks**. This enhances the strength of the cereal stalks and at the same time increases the proportion of total biomass in seeds. Slowly degradable gibberellin synthesis inhibitors are used in horticulture to keep household plants small.

Gibberellins influence gene expression. Present knowledge about the signal transduction chain of gibberellin action is still rather fragmentary. There appear to be gibberellin receptors of unknown nature present in the plasma membrane. The signal transduction from the receptor to gene function probably involves a **heterotrimeric G-protein** (Fig. 19.1), **cGMP**, **Ca^{++} ions**, calmodulin (Fig. 19.5) and a cascade of protein kinases, possibly of the **MAPK** and **CDPK** type (section 19.1), resulting in the modulation of transcription factors for various genes. One effect of GA is to reduce the action of repressor proteins, named **DELLA proteins**, which suppress growth. It has been observed recently that due to a GA-dependent protein kinase cascade, DELLA proteins are phosphorylated and thus labeled for proteolytic degradation by the **proteasome** machinery via conjugation with **ubiquitin** (section 21.4). In this way GA enhances growth by **relief of restraint** (see also section 19.3).

Mutants in which the synthesis of GA or the function of GA on growth was impaired turned out to be important for agriculture. The dramatic increase in the yield of cereal crops achieved after 1950, often named the **green revolution**, is in part due to the introduction of **dwarf wheat lines**. At that time, attempts to increase the crop yield of traditional wheat varieties by an increased application of nitrogen fertilizer failed, since it rather produced more straw biomass than enhanced grain yield. This was averted by breeding wheat varieties with **reduced stalk growth**, where the portion of grains in the total biomass (harvest index) was increased considerably. It turned out that the decreased stalk growth in these varieties was due to the mutation of genes encoding transcription factors of the gibberellin signal transduction chain. In wheat, the mutated genes have been termed **Rht**

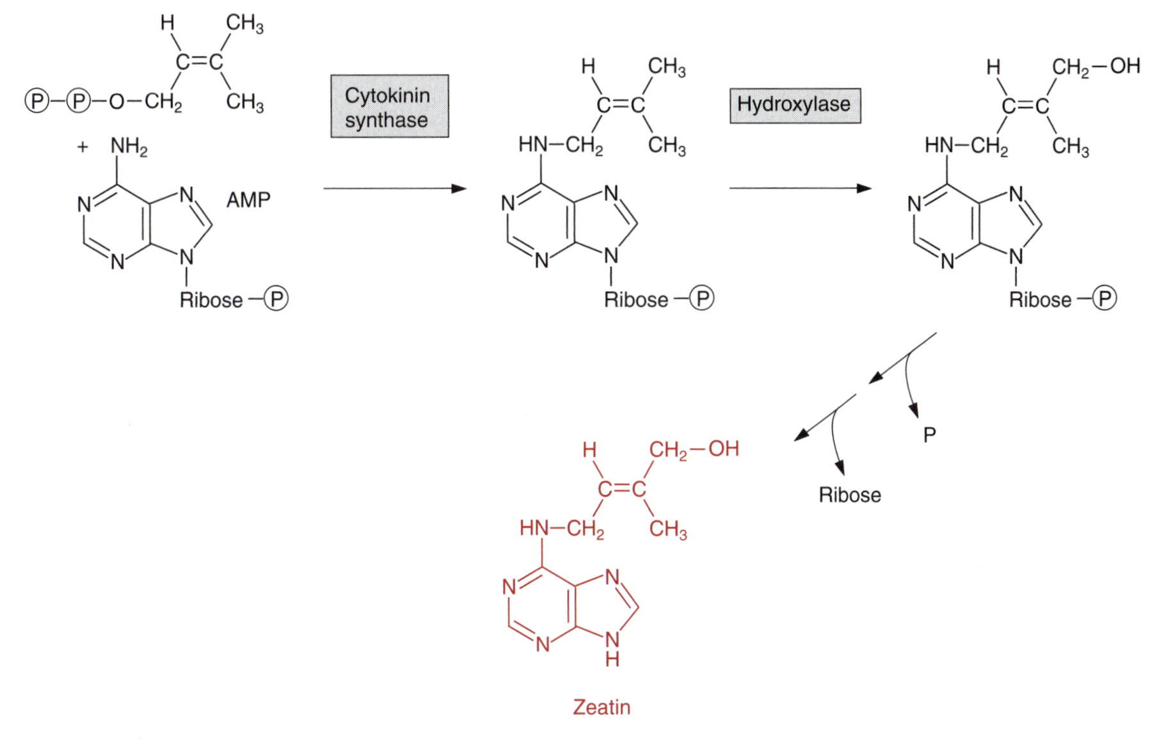

Figure 19.11 Synthesis of zeatin, a cytokinin.

(reduced height). In one mutant, the ubiquitin-dependent degradation of repressor proteins (see preceding) is inhibited and therefore the stimulation of shoot elongation by gibberellin is impaired. Analogous genes have been identified in other plants, such as maize and rice, and corresponding mutants with reduced shoot elongation have been generated, yielding cultivars with an increased harvest index.

19.5 Cytokinins stimulate cell division

Cytokinins are prenylated derivatives of adenine. In **zeatin**, which is the most common cytokinin, the amino group of adenine is linked with the hydroxylated isoprene residue in the *trans*-position (Fig. 19.11). Cytokinins enhance plant growth by stimulating **cell division** and increase the sprouting of lateral buds. As cytokinins override apical dominance, they are antagonists of the auxin IAA. Cytokinins **retard senescence** and thus counteract the phytohormone ethylene (section 19.7). The larvae of some butterflies (e.g., *Stigmella,* which invade beech trees) use this principle for their nutrition. They excrete cytokinin with their saliva and thus prevent senescence of the leaves on which they are feeding. As a result, **green islands** of intact leaf material remain in yellowing autumn leaves, which provide, beyond the actual vegetation period, the caterpillars with the forage they need to form pupae.

Mature (i.e., differentiated) plant cells normally stop dividing. By adding cytokinin and auxin, differentiated cells can be induced to start dividing again. When a leaf piece is placed on a solid culture medium containing auxin and cytokinin, leaf cells start unrestricted growth, resulting in the formation of a callus that can be propagated in tissue culture. Upon the application of a certain cytokinin/auxin ratio, a new complete shoot can be regenerated from single cells of this callus. The use of tissue culture for the generation of transgenic plants will be described in section 22.3.

Some bacteria and fungi produce auxin and cytokinin to induce unrestricted cell division, which results in tumor growth in a plant. The formation of the **crown gall** induced by ***Agrobacterium tumefaciens*** (see section 22.2) is caused by a stimulation of the production of cytokinin and auxin. The bacterium does not produce these phytohormones itself, but transfers the genes for the biosynthesis of cytokinin and auxin from the T_i plasmid of the bacterium to the plant genome.

Zeatin is formed from AMP and dimethyallylpyrophosphate (Fig. 19.11). The isoprene residue is transferred by **cytokinin synthase** (a prenyl transferase, see section 17.2) to the amino group of the AMP and is then hydroxylated. Cytokinin synthesis takes place primarily in the meristematic tissues. Transgenic tobacco plants in which the activity of cytokinin synthase in the leaves is increased have been generated. The leaves of these tobacco plants have a much longer life span than normal, since their senescence is suppressed by the enhanced production of cytokinin.

Cytokinin receptors, like ethylene receptors, are **dimeric histidine kinases**. They are located in the plasma membrane, where the receptor site is directed to the extracellular compartment and the kinase is directed to the cytoplasm. The kinase moiety of the dimer contains two histidine residues and two aspartyl residues. Upon binding of cytokinin, the two histidine kinases phosphorylate their histidine residues reciprocally in a process called **autophosphorylation**. Subsequently, the phosphate groups are transferred to histidine residues or aspartyl residues of **transmitter proteins**. The transmitter proteins pass into the nucleus, where they function as transcription factors and thus regulate the expression of many genes.

19.6 Abscisic acid controls the water balance of the plant

When searching for substances that cause the abscission of leaves and fruits, **abscisic acid (ABA)** (Fig. 19.12) was found to be an inducing factor and was named accordingly. Later it turned out that the formation of the abscission layer for leaves and fruit is induced primarily by ethylene (described in the following section). An important function of the phytohormone ABA, however, is the induction of **dormancy** (endogenic rest) of seeds and buds. Moreover, ABA has a major function in maintaining the **water balance of plants**, since it induces with nitric oxide (**NO**) the closure of the stomata during water shortage (section 8.2). In addition, ABA prevents vivipary (i.e., seed embryos from germinating before the seeds are mature). Mutants deficient in ABA have been found in tomatoes, which, due to the resulting disturbance of the water balance, bear wilting leaves and fruits. In these wilting mutants, the immature seeds germinate in the tomato fruits while they are still attached to the mother plant.

ABA is a product of isoprenoid metabolism. The synthesis of ABA proceeds via oxidation of **violaxanthin** (Fig. 19.12, see also Fig. 3.41). ABA synthesis occurs in leaves and also in roots, where a water shortage is directly monitored. ABA can be transported by the transpiration stream via the xylem vessels from the roots to the leaves, where it induces closure of the stomata (section 8.2).

ABA causes alterations in metabolism by influencing gene expression. The signal transduction chain of the ABA-action involves G-proteins, protein kinases, protein phosphatases, and the synthesis of the messenger

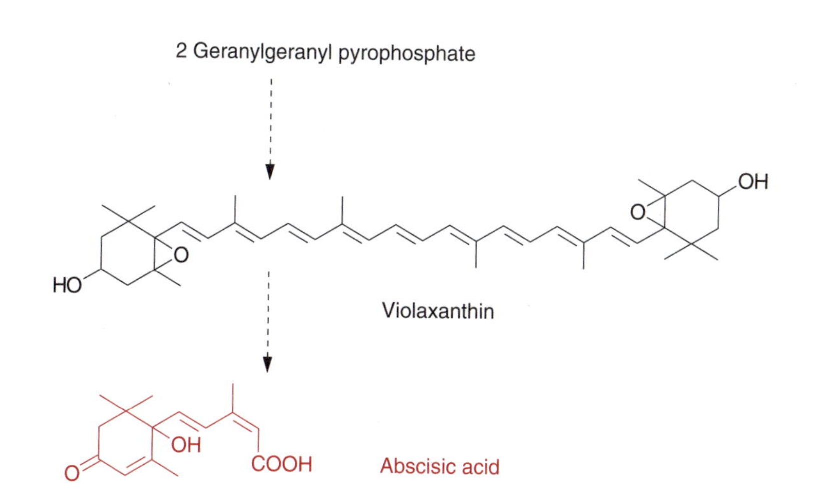

Figure 19.12 Abscisic acid is formed in several steps by the oxidative degradation of violaxanthin.

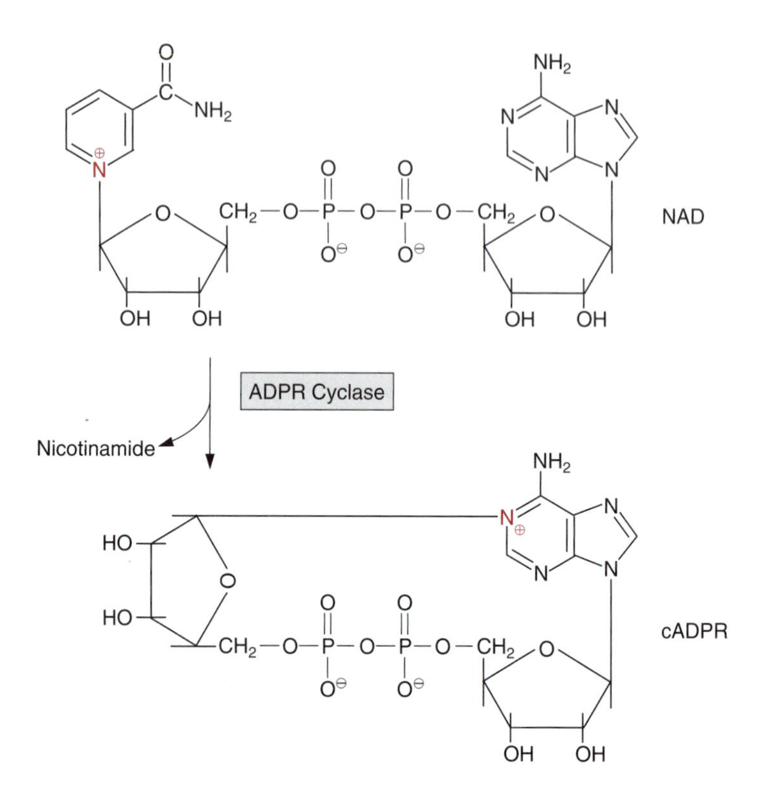

Figure 19.13 Cyclic ADP-ribose, a messenger substance recently discovered in plants, causes Ca^{++} ions to be released into the cytosol. This messenger substance seems ubiquitous in the plant and animal kingdoms. In the formation of cyclic ADP-ribose from NAD by an ADPR-cyclase, the ribose moiety on the left side of the figure is transferred from the positively charged N-atom of the pyridine ring to the likewise positively charged N-atom in the adenine ring.

substance **cyclic ADP-ribose (cADPR)** (Fig.19.13). Earlier investigations of the animal system suggested that cADPR, similar to inositol trisphosphate, acts as a messenger substance causing the discharge of Ca^{++} ions. This function was clearly confirmed for the plant *Arabidopsis*. There is still another signal transduction chain of ABA action, where the release of Ca^{++} ions occurs via the phosphoinositol pathway, with mediation by a phospholipase C (Fig. 19.4). The rise in the cytosolic Ca^{++} level, caused either by cADPR or by phosphoinositol trisphosphate, results in the activation of ion channels involved in the opening of stomata and also the modulation of transcription factors, via a cascade of protein kinases (e.g., of the MAPK-type).

19.7 Ethylene makes fruit ripen

As mentioned in the previous section, **ethylene** is involved in the induction of **senescence**. During senescence, the degradation of leaf material is initiated. Proteins are degraded to amino acids, which, together with certain ions

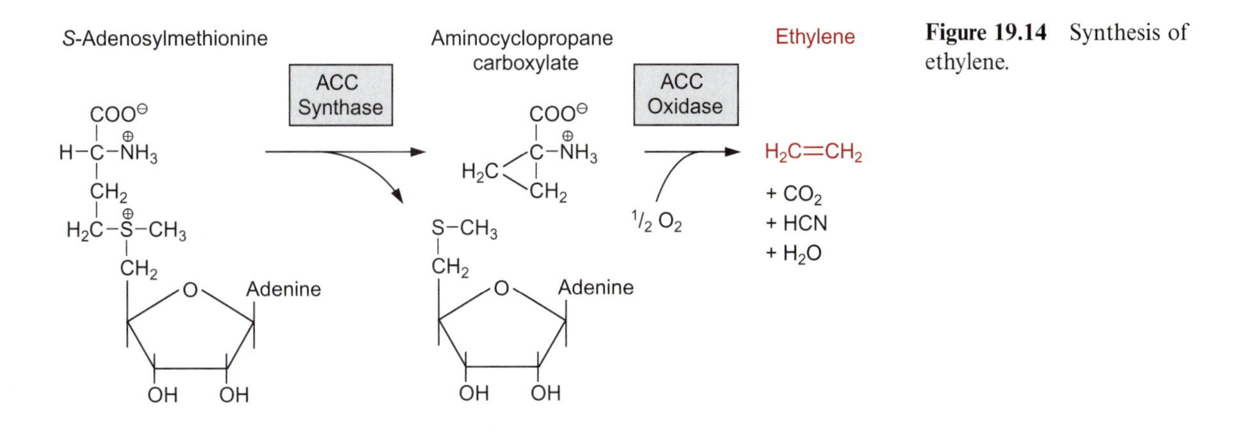

Figure 19.14 Synthesis of ethylene.

(e.g., Mg^{++}), are withdrawn from the senescing leaves via the phloem system for reutilization. In perennial plants, these substances are stored in the stem or in the roots, and in annual plants they are utilized to enhance the formation of seeds. Ethylene induces **defense reactions** after infection by fungi or when plants are wounded by feeding animals. As an example, the induction of the synthesis of tannins by ethylene in acacia as a response to feeding antelopes has been discussed in section 18.7.

In addition to stimulating the **abscission of fruit**, ethylene has a general function in **fruit ripening**. The ripening of fruit is to be regarded as a special form of senescence. The effect of gaseous ethylene can be demonstrated by placing a ripe apple and a green tomato together in a plastic bag; ethylene produced by the apple accelerates the ripening of the tomato. Bananas are harvested green and transported halfway around the world under conditions that suppress ethylene synthesis (low temperature, CO_2 atmosphere). Before being sold, these bananas are ripened by gassing them with ethylene. Also, tomatoes are often ripened only prior to sale by exposure to ethylene.

***S*-adenosylmethionine** (Fig. 12.10), is the precursor for the synthesis of ethylene (Fig. 19.14). The positive charge of the sulfur atom in *S*-adenosylmethionine enables its cleavage to form a cyclopropane, in a reaction catalyzed by aminocyclopropane carboxylate synthase, abbreviated **ACC synthase**. Subsequently, **ACC oxidase** catalyzes the oxidation of the cyclopropane to ethylene, CO_2, HCN, and water. HCN is immediately detoxified by conversion to β-cyanoalanine (reaction not shown).

Genetic engineering has been employed to suppress ethylene synthesis in tomato fruits in two different ways:

1. The activity of ACC synthase and ACC oxidase was decreased by antisense technique (section 22.5).

2. By introducing a bacterial gene into the plants, an enzyme was expressed that degraded the ACC formed in the tomato fruits so rapidly that it could no longer be converted to ethylene by the ACC oxidase.

The aim of this genetic engineering is to produce tomatoes that keep better during transport. It may be noted that transgenic tomato plants have also been generated, in which the durability of the harvested fruits is prolonged by an antisense repression of the polygalacturonidase enzyme, which lyses the cell wall.

The effect of ethylene is caused by an alteration of gene expression. Since ethylene, like other phytohormones, exerts its effect at very low concentrations (~10^{-9} mol/liter), the **ethylene receptor** is expected to have a very high affinity. Like the cytokinin receptor, it consists of a dimer of **histidine receptor kinases**, each containing a histidine residue, which, after autophosphorylation, transfers the phosphate group to histidine residues or aspartyl residues of target proteins. By binding of ethylene to the receptor dimer, in which a copper-cofactor is involved, the kinase is inactivated and autophosphorylation is prevented. Depending on the phosphorylation state of the ethylene receptor, a signal is transmitted via protein kinase cascades, in which MAPKK and MAPK (section 19.1) participate, to transcription factors, which control the expression of certain genes. It may be noted that histidine kinases occur in plants, yeast, and bacteria, but not in animals.

19.8 Plants also contain steroid and peptide hormones

Earlier it was believed that steroid and peptide hormones, which have many different functions in animals, do not play any major role in plants. Recent results have clearly shown that this is not true. Steroids as well as peptides in plants indeed have essential functions as phytohormones.

Brassinosteroids control plant development

Steroid phytohormones are the so-called **brassinosteroids**. The most effective member of the brassinosteroids is **brassinolide** (Fig. 19.15). Brassinosteroids are synthesized from campesterol (a membrane lipid, Fig. 15.3). Brassinosteroids regulate **plant development**. Among other things, they stimulate shoot growth, folding of leaves, and differentiation of the xylem. They retard root growth and the formation of anthocyan. Since this effect also can be

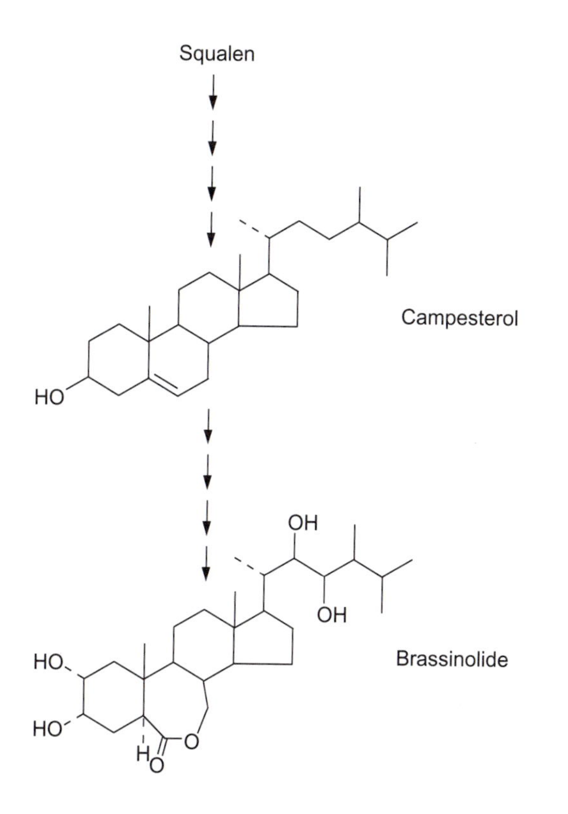

Figure 19.15 Brassinolide is formed from the membrane lipid campesterol via a synthesis chain involving cyt P_{450}-dependent hydroxylases, a reductase, and others. The synthesis pathway is very similar to corresponding steroid synthesis pathways in animals.

produced by an overlapping of the functions of the "classic" hormones (auxin, cytokinin, gibberellin, etc.), there had been earlier great doubts that the brassinosteroids acted as phytohormones.

Brassinosteroids were first isolated from pollen. It was known that pollen contain a growth factor. In 1979 scientists from the U.S. Department of Agriculture isolated from 40 kg of rape pollen, collected by bees, 4 mg of a substance that they identified as brassinolide. Later, using very sensitive analysis techniques, it was shown that plants in general contained brassinolide and other brassinosteroids. The function of brassinosteroids as essential phytohormones was clearly established from the study of *Arabidopsis* mutants with developmental defects, such as **dwarf growth**, **reduced apical dominance**, and **lowered fertility**. The search for the responsible defective gene revealed that such mutations affected enzymes of the brassinolide synthesis pathway, which turned out to have very great similarities with the synthesis pathway of animal steroid hormones. These developmental defects could not be prevented by the addition of "classic" phytohormones, but only by an injection of a small amount of brassinolide, as contained in plants. These results clearly demonstrated the essential function of brassinosteroids for the growth and development of plants.

In **animal cells**, steroid hormones are perceived by binding to defined **steroid receptors**, which are present in the cytoplasm. Once activated, the receptor complex is transferred to the nucleus to promote or repress the expression of certain genes. It seems that plant steroids do not function in this way, as plants lack homologues of animal steroid receptors. There are indications that a **brassinolide receptor** in plants interacts with or is identical to a **receptor-like kinase** (**RLK**, section 19.1), which is possibly connected to a **MAP kinase cascade**, regulating transcription factors. Research about brassinosteroids is at an early stage, and many questions are still unanswered.

Various phytohormones are polypeptides

Following the discovery or the polypeptide hormone insulin, a large variety of polypeptide hormones with very diverse functions have been identified in animals. It appears now that plants also contain a multiplicity of polypeptide hormones. Research on this presumably very important field is still in its infancy. Until now, owing to the small size of the molecules, only a few functional peptides have been identified in plants. The following paragraphs deal with three examples from the large number of polypeptide phytohormones.

Systemin induces defense against herbivore attack

Many plants respond to insect attacks by accumulating **proteinase inhibitors**, which are toxic to insects because they impair their digestion of proteins. It was shown in tomato plants that the polypeptide **systemin**, made up of 18 amino acids, is involved in defense reactions. In response to insect attack, a systemin precursor protein is synthesized, which is subsequently processed by endoproteases to the active polypeptide. Systemin is perceived by a membrane-bound receptor of very high affinity. A systemin concentration of as low as 10^{-10} moles/liter is sufficient for a half saturation of the receptor. The receptor was identified as a receptor-like kinase (**RLK**) (section 19.1). It is part of a signal cascade that generates **jasmonic acid** (section 15.7), a key signal in the transcription activation of defense-related genes (e.g., for the synthesis of proteinase inhibitor) (see also following section). It may be noted that the effect of systemin is species-specific. Systemin from tomato also affects potato and pepper, but it has no effect on the closely related tobacco or on other plants. Tobacco produces a systemin-like polypeptide, with a structure similar to the tomato systemin, and has analogous effects. It remains to be elucidated whether systemin-like polypeptide hormones may be involved in defense reactions of other plants.

The addition of systemin-like polypeptides to tobacco cell suspension cultures was seen to cause an alkalization of the culture medium, reflecting its hormone action. A polypeptide of 49 amino acids led to a more rapid alkalization and therefore has been termed **RALF**. Homologues of RALF are found in many plant species. In *Arabidopsis,* nine different RALF genes that are expressed in different organs of the plant have been identified. The ubiquity of RALF polypeptides suggests that they play a general role in plants, one that is still to be elucidated. Recent observations indicate that RALF inhibits cell division in root tissues.

Phytosulfokines regulate cell proliferation

In media of cell suspension cultures, a factor that enhanced the proliferation of the cells was found. This factor was identified as a mixture of two small polypeptides, named phytosulfokines (PSK_α and PSK_β) containing two tyrosine residues each, of which both hydroxyl groups are esterified with phosphate.

PSK_α: Tyr (SO_3H)-Ile-Tyr (SO_3H)-Thr-Gln

PSK_β: Tyr (SO_3H)-Ile-Tyr (SO_3H)-Thr

Phytosulfokines with identical structures occur in many plants and have, in addition to auxin and cytokinins, an important regulatory effect on the **dedifferentiation** of cells. Plant cells can retain the ability of totipotency, which means that they can be dedifferentiated in such a way that they can reenter the cell cycle to form all organs of a new plant. As a receptor for phytosulfokines, a receptor-like protein kinase (**RLK**) has been identified, which probably is connected via signal cascades to transcription factors regulating genes of dedifferentiation and proliferation.

19.9 Defense reactions are triggered by the interplay of several signals

As discussed in section 16.1, plants defend themselves against pathogenic bacteria and fungi by producing **phytoalexins**, and in some cases, in order to contain an infection, also by programmed cell death (**hypersensitive reaction**). Animals feeding on plants may stimulate the plants to produce defense substances, which make the herbivore's meal poisonous or indigestible.

These various defense reactions are initiated by the interplay of several signal substances in a network. After an attack by pathogens or as a response to abiotic stress, signal cascades, including the phosphoinositol cascade (section 19.1), are initiated, which lead to an increase of the **Ca^{++}** concentration in the cytosol, by which Ca^{++} dependent protein kinases (**CDK**) are activated. This in turn activates **protein kinase cascades**, which modulate gene expression via transcription factors (section 19.1). Moreover, in an early response, superoxide ($\bullet O_2^-$) and/or H_2O_2 [reactive oxygen species (**ROS**)] are formed via an NADPH oxidase located in the plasma membrane. The ROS represent chemical weapons for a direct attack on the pathogens, but they also are messengers for inducing signal cascades to initiate the production of other defense substances. H_2O_2 is involved in the lignification process (section 18.3) and thus plays a role in the solidification of the cell wall as a defense against pathogens. Another early response to pathogen attack is the formation of nitric oxide (**NO**), a radical. NO, known as a messenger in animals and plants, is formed by the oxidation of arginine, as catalyzed by nitric oxide-synthase.

$$\text{Arginine} + O_2 + \text{NADPH} \longrightarrow \text{Citrulline} + \text{NADP} + \text{NO}$$

Alternatively, NO can be formed from nitrite, as catalyzed in a by-reaction by nitrate reductase (section 10.1).

$$NO_2^- + \text{NAD(P)H} + H^+ \longrightarrow NO + H_2O$$

At present it is not known which of the two pathways plays a major role in NO synthesis or how NO synthesis is regulated. In plants, NO is an important messenger in hormonal and defense responses. It promotes the Ca^{++} release from intracellular stores to raise the cytoplasmic Ca^{++} concentration and activates signal cascades in this way. In addition to abscisic acid, it induces the opening of stomata (section 8.2). NO is involved in the initiation of programmed cell death and the formation of phytoalexins. It also induces the synthesis of **salicylic acid (SA)** (Fig. 18.5).

Salicylic acid has a central function in pathogen defense. Transgenic tobacco plants, in which the synthesis of salicylic acid was intercepted, proved to be very vulnerable to bacterial or fungal infections. Enzymes induced by salicylic acid include β1.3 glucanase, which digests the cell wall of fungi, and lipoxygenase, a crucial enzyme in the pathway of the synthesis of **jasmonic acid (JA)** (Figs. 15.29 and 15.30).

The formation of jasmonic acid (e.g., induced by systemin) is regulated by a signal cascade involving **Ca^{++} ions** and **MAP kinases**. Also, in the perception of jasmonic acid, a MAP kinase cascade regulating transcription factors appears to be involved. Jasmonic acid and its methylester, as well as its precursor 12-oxo-phytodienoic acid (**OPDA**) play a central role in defense reactions. As a response to fungal infection, jasmonic acid induces the synthesis of **phenylammonium lyase (PAL)** (Fig. 18.2), the entrance enzyme of phenyl propanoid biosynthesis, as well as of **chalcone synthase (CHS)** (Fig. 18.11), the key enzyme of flavonoid biosynthesis. As discussed in Chapter 18, many defense substances are derived from these compounds. For example, as a response to wounding by herbivores, jasmonic acid induces plants to produce proteinase inhibitors. As a response to mechanical stress (e.g., by wind), jasmonic acid induces increased growth in the thickness of stems or tendrils to give the plants higher stability. Moreover, jasmonic acid regulates the **development of pollen** in some plants. *Arabidopsis* mutants, which are unable to synthesize jasmonic acid, cannot produce functioning pollen and therefore are **male-sterile**. Jasmonic acid, like auxin and gibberellin (sections 19.3 and 19.4), uses the ubiquitin-proteasome pathway (section 21.4) to control gene expression through protein degradation.

An attack by herbivores in many cases initiates defense responses not only in the wounded leaves, but also in more distant parts of a plant (**systemic response**). This requires a long-distance transport of signal substances within the plant. Recent results indicate that **jasmonic acid** may function as such a **systemic wound signal**.

19.10 Light sensors regulate growth and development of plants

Light controls plant development from germination to the formation of flowers in many different ways. Important light sensors in this control are **phytochromes**, which sense red light. Phytochromes are involved when light initiates the germination and greening of the seedling and in the adaptation of the photosynthetic apparatus of the leaves to full sunlight or shade. For their adaptation to the full spectrum of the sunlight, plants also have photoreceptors for blue and UV light, So far, three proteins have been identified as blue light receptors; these are **cryptochrome** 1 and 2, each containing a flavin (Fig. 5.16) and a pterin (Fig. 10.3); and **phototropin**, containing one flavin as a blue light–absorbing pigment. There also seem to be sensors for UV light that have not yet been identified. The signal transduction chains for cryptochromes and phototropin are to a large extent not known. Phototropin is a receptor kinase. Upon exposure to blue light, it is autophosphorylated by its protein kinase activity. Also, cryptochrome undergoes blue

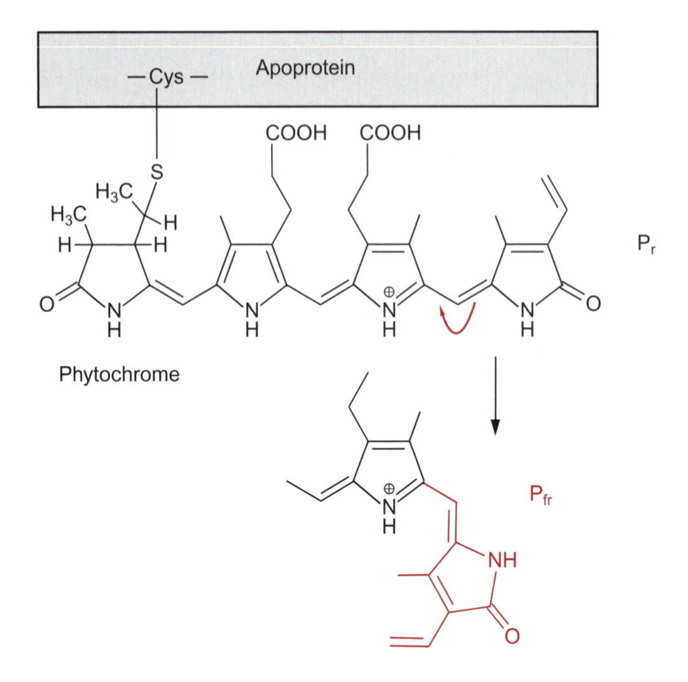

Figure 19.16 The chromophore of phytochrome consists of an open-chain tetrapyrrole, which is linked via a thioether bond to the apoprotein. The absorption of red light results in a *cis-trans*-isomerization of a double bond, causing a change in the position of one pyrrole ring (colored red).

light– dependent phosphorylation. All these photoreceptors probably affect the activity of transcription factors via signal transduction chains, involving the change of cytoplasmic Ca^{++} concentration and protein kinases, and in this way regulate the expression of genes.

Since the structure and function of **phytochromes** have been studied extensively in the past, they offer a good example for a more detailed discussion of the problems of signal transduction in plants. Phytochromes are soluble **dimeric proteins** that probably function without being bound to a membrane. The monomer consists of an apoprotein (molecular mass 120–130 kDa), in which the sulfydryl group of a cysteine residue is linked to an open-chain tetrapyrrole, functioning as a chromophore (Fig. 19.16). The autocatalytic binding of the tetrapyrrole to the apoprotein results in the formation of a phytochrome $\mathbf{P_r}$ (r = red) with an absorption maximum at about 660 nm (**light red light**) (Fig. 19.17). The absorption of this light results in a change in the chromophore; a double bond between the two pyrrole rings changes from the ***trans-*** to the ***cis-***configuration (colored red in Fig. 19.16), resulting in a change in the conformation of the protein. The phytochrome in this new conformation has an absorption maximum at about 730 nm (**far-red light**) and in this state is named $\mathbf{P_{fr}}$. P_{fr} represents the active form of the phytochrome. It signals the state of illumination. P_{fr} is reconverted to P_r by the absorption of far-red light. Since the light absorption of P_r and P_{fr} overlaps (Fig. 19.17), depending on the color of the irra-

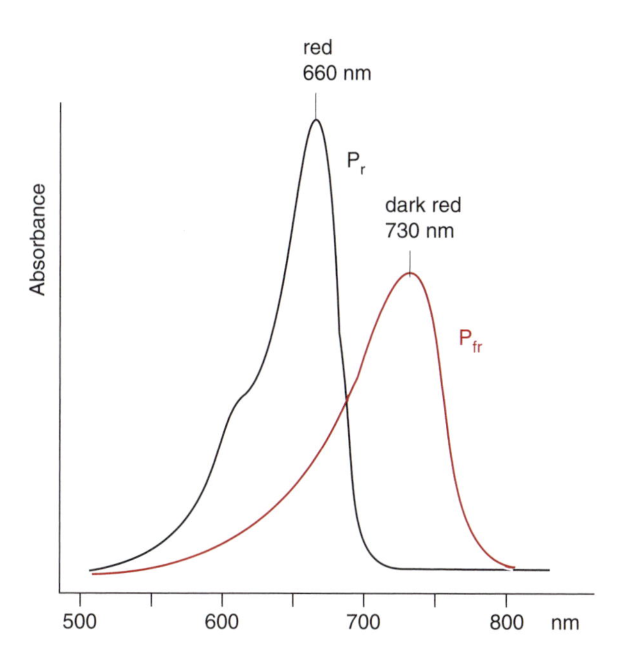

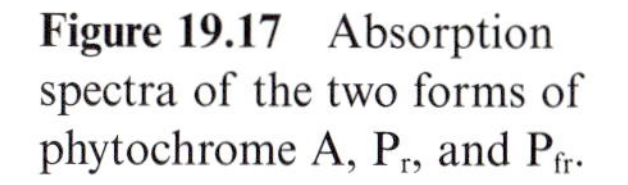
Figure 19.17 Absorption spectra of the two forms of phytochrome A, P_r, and P_{fr}.

diated light, a reversible equilibrium between P_r and P_{fr} is attained. Thus, with light of 660 nm, 88% of the total phytochrome is present as P_{fr} and at 720 nm, only 3% is present. In bright sunlight, where the red component is stronger than the far-red component, the phytochrome is present primarily as P_{fr} and signals the state of illumination to the plant. Altogether five different phytochromes (A–E) have been identified in the well-characterized model plant *Arabidopsis thaliana* (section 20.1). **Phytochrome A** monitors light in the far-red region, whereas **phytochrome B** monitors red light. Depending on the light conditions, both phytochromes are involved in the morphogenesis of plants. The phytochromes C, D, and E fulfill special functions, which have not yet been resolved in all details.

Whereas the inactive form of phytochrome (P_r) has a rather long lifetime (~100 h), the lifetime of the active form (P_{fr}) is reduced to 30 to 60 minutes. P_{fr} can be recovered by a reversal of its formation (Fig. 19.18). In the case of phytochrome A, however, the signal function can be terminated by conjugation of P_{fr} with **ubiquitin**, which marks it for proteolytic degradation by the **proteasome pathway** (section 21.4).

It has been shown that phytochrome A affects the transcription of **10% of all genes** in *Arabidopsis*. This effect of phytochromes is modulated by phytohormones, such as cytokinins and brassinosteroids, and the signal cascades of pathogen defense. Light sensors, phytohormones, and defense reactions appear to be interwoven in a **broad network**. The active phytochrome P_{fr} influences gene expression via **transcription factors**. Since

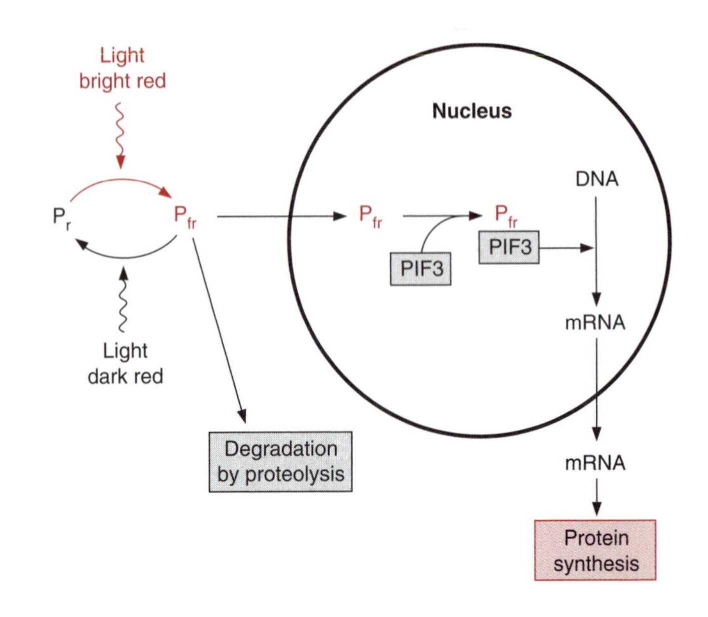

Figure 19.18 Direct effect of phytochromes A and B on gene expression. Phytochrome is converted by irradiation with red light to the active form P_{fr} and is reconverted by far-red light to the inactive P_r. P_{fr} enters the nucleus, where it binds to a transcription factor PIF3. After loaded with P_{fr}, PIF3 regulates gene expression by binding to promoter regions of the DNA. In the case of phytochrome A, P_{fr} is irreversibly degraded by proteolysis. At a recent conference (2004), the function of PIF3 has been a matter of debate.

phytochromes control the expression of so many different genes, it is to be expected that different signal transduction chains are involved in this. Investigations carried out so far support this notion.

1. It was recently demonstrated that phytochrome A, as well as phytochrome B, which are present in the cytosol, after their conversion into the active form P_{fr} enter the nucleus, where each of them can associate with a phytochrome interacting factor 3 (**PIF3**) (Fig. 19.18). It is postulated that the complex of PiF3 and P_{fr} binds to the promoter region of genes and thus regulates their expression. Although further studies are required to establish details of this, these findings may for the first time characterize in plants a signal transduction chain from a sensor to a gene. It may be noted that this very short signal transduction chain does not involve second messengers or protein kinases
2. **Protein kinase** activity was detected in phytochrome A of certain plants. This suggests that phytochromes are linked to protein kinase cascades regulating transcription factors.
3. There also are indications for another signal transduction chain. Important objects for the research on this topic have been the cells of tomato mutants, which, due to the lack of phytochrome, are unable to green. The micro-injection of Ca^{++} ions and calmodulin into single cells of these mutants simulated the action of phytochromes in inducing the synthesis of light harvesting complexes. These results indicated that Ca^{++} is involved as a messenger in signal transduction from phytochromes

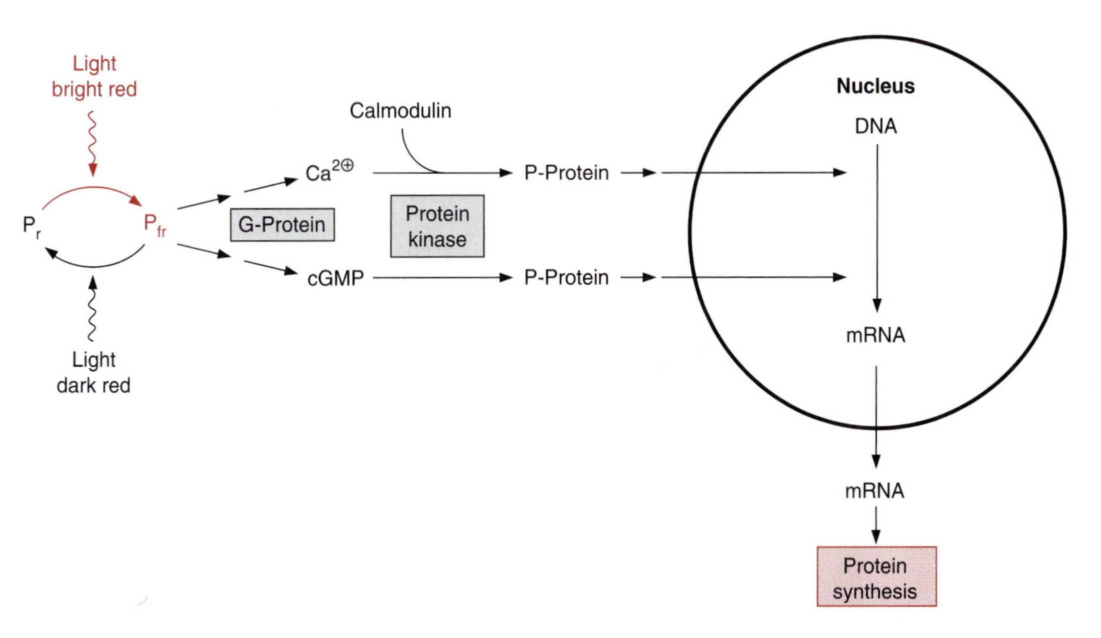

Figure 19.19 Postulated signal transduction chains for the effect of phytochromes on gene expression.

(Fig. 19.19). Phytochromes also induce the enzyme chalcone synthase (Fig. 18.11). In the mutant cells already mentioned, such an induction of chalcone synthase also could be achieved by the injection of cGMP (Fig. 19.2). This led to the conclusion that cGMP participates as another messenger to Ca^{++} in P_{fr}-induced signal transduction chains. Further investigations have indicated that a **G-protein** is involved in both transduction chains. More studies are required to verify these postulated signal transduction chains.

From the present intensive research activities on the signal transduction chains of light sensors it is to be expected that the coming years will bring more certainty about the corresponding signal networks.

Further reading

Bishop, G. J., Koncz, C. Brassinosteroids and plant steroid hormone signaling. Plant Cell, Supplement, 97–110 (2002).

Chang, C. Ethylene signaling: The MAPK module has finally landed. Trends Plant Sci 8, 365–368 (2003).

Cheng, S.-H., Willmann, M. R., Chen, H.-C., Sheen, J. Calcium signaling through protein kinases. The *Arabidopsis* calcium-dependent protein kinase gene family. Plant Physiol 129, 469–485 (2002).

Cohen, J. D., Slovin, J. P., Hendrickson, A. M. Two genetically discrete pathways convert tryptophan to auxin: More redundancy in auxin biosynthesis. Trends Plant Sci 8, 197–199 (2003).

Finkelstein, R. R., Gampala S. S. L., Rock, C. D. Abscisic acid signaling in seeds and seedlings. Plant Cell, Supplement, 15–45 (2002).

Friml, J., Vieten, A., Sauer, M., Weijers, D., Schwarz, H., Hamann, T., Offringa, R., Juergens, G. Efflux-dependent auxin gradients establish the apical-basal axis of *Arabidopsis.* Nature 426, 147–153 (2003).

Frohnmeyer, H., Staiger, D. Ultraviolet-B radiation-mediated responses in plants. Balancing damage and protection. Plant Physiol 133, 1420–1428 (2003).

Fujioka, S., Yokota, T. Biosynthesis and metabolism of brassinosteroids. Annu Rev Plant Biol 54, 137–164 (2003).

García-Martínez, J. L., Huq, E., Quail, H. Direct targeting of light signals to a promoter element-bound transcription factor. Science 288, 859–862 (2000).

Geldner, N., Frimi, J., Stierhof, Y. D., Juergens, G., Palme, K. Auxin transport inhibitors block PIN1 cycling and vesicle trafficking. Nature 413, 425–428 (2001).

Grossmann, K. Mode of action of auxin herbicides: A new ending to a long, drawn out story. Trends Plant Sci 5, 506–508 (2000).

Guillaume, T., Asai, T., Chiu, W.-L., Sheen, J. Plant mitogen-activated protein kinase signaling cascades. Curr Opin Plant Biol 4, 329–400 (2001).

Guo, F.-Q., Okamoto, M., Crawford, N. M. Identification of a plant nitric oxide synthase gene involved in hormonal signaling. Science 302, 100–103 (2003).

Hardie, D. G. Plant protein serine/threonine kinases: Classification and functions. Annu Rev Plant Physiol Plant Mol Biol 50, 97–131 (1999).

Harmon, A. C., Gribskov, M., Harper, J. F. CDPKs—a kinase for every Ca^{2+} signal? Trends Plant Sci 5, 154–159 (2000).

Itoh, H., Matsuoka, M., Steber, M. A role for the ubiquitin-26S-proteasome pathway in gibberellin signaling. Trends Plant Sci 8, 492–497 (2003).

Kakimoto, T. Perception and signal transduction of cytokinins. Annu Rev Plant Biol 54, 605–627 (2003).

King, R. W., Evans, L. T. Gibberellins and flowering of grasses and cereals: Prizing open the lid of the "florigen" black box. Annu Rev Plant Biol 54, 307–328 (2003).

Lamattina, L., García-Mata, C., Graziano, M., Pagnussat, G. Nitric oxide: The versatility of an extensive signal molecule. Annu Rev Plant Biol 54, 109–135 (2003).

Lee, J., Rudd, J. J. Calcium-dependent protein kinases: Versatile plant signalling components necessary for pathogen defence. Trends Plant Sci 7, 97–99 (2002).

Leyser, O. Auxin signalling: The beginning, the middle and the end. Curr Opin Plant Biol 4, 382–386 (2001).

Li, J., Nam, K. H. Regulation of brassinosteroid signaling by a GSK3/SHAGGY-like kinase. Science 295, 1299–1301 (2002).

Lin, C., Shalitin D. Cryptochrome structure and signal transduction. Annu Rev Plant Biol 54, 469–496 (2003).

Liscum, E., Hodgson, D. W., Campbell, T. J. Blue light signaling through the cryptochromes and phototropins. So that's what the blues is all about. Plant Physiol 133, 1429–1436 (2003).

Lovegrove, A., Hooley, R. Gibberellin and abscisic acid signaling in aleurone. Trends Plant Sci 5, 102–110 (2000).

MAPK Group (Ichimura, K. et al.). Mitogen-activated protein kinase cascades in plants: A new nomenclature. Trends Plant Sci 7, 301–308 (2002).

Matsubayashi, Y., Ogawa, M., Morita, A., Sakagami, Y. An LRR receptor kinase involved in perception of a peptide plant hormone, phytosulfokine. Science 296, 1470–1472 (2002).

Meijer, H. J. G., Munnik, T. Phospholipid-based signaling in plants. Annu Rev Plant Biol 54, 265–306 (2003).

Mok, D. W. S., Mok, M. C. Cytokinin metabolism and action. Annu Rev Plant Physiol Plant Mol Biol 52, 89–118 (2001).

Møller, S. G., Kim, Y.-S., Kunkel, T., Chua, N.-H. PP7 is a positive regulator of blue light signaling in *Arabidopsis*. Plant Cell 15, 1111–1119 (2003).

Muday, G. K., Peer, W. A., Murphy, A. S. Vesicular cycling mechanisms that control auxin transport polarity. Trends Plant Sci 8, 301–304 (2003).

Munnik, T. Phosphatidic acid: An emerging plant lipid second messenger. Trends Plant Sci 6, 227–233 (2001).

Nagy, F., Schäfer, E. Control of nuclear import and phytochromes. Curr Opin Plant Biol 3, 450–454 (2000).

Olszewski, N., Sun, T.-p., Gubler, F. Gibberellin signaling. Biosynthesis, catabolism, and response pathways. Plant Cell, Supplement, 61–80 (2002).

Parks, B. M. The red side of photomorphogenesis. Plant Physiol 133, 1437–1444 (2003).

Reed, J. W. Roles and activities of Aux/IAA proteins in *Arabidopsis*. Trends Plant Sci 6, 420–425 (2001).

Richards, D. E., King, K. E., Ait-ali, T., Harberd, N. P. How Gibberellin regulates plant growth: A molecular genetic analysis of gibberellin signaling. Annu Rev Plant Physiol Plant Mol Biol 52, 67–88 (2001).

Ross, J., O'Neill, D. New interactions between classical plant hormones. Trends Plant Sci 6, 2–4 (2001).

Ryan, C. A., Pearce, G. Polypeptide hormones. Plant Physiol 125, 65–68 (2001).

Ryan, C. A., Pearce, G., Scheer, J., Moura, D. S. Polypeptide hormones. Plant Cell, Supplement 251–264 (2002).

Sanders, D., Pelloux, J., Brownlee, C., Harper, J. F. Calcium at the crossroads of signaling. Plant Cell, Supplement, 401–417 (2002).

Sasaki, A., Itoh, H., Gomi, K., Ueguchi-Tanaka, M., Ishiyama, K., Kobayashi, M., Jeong, D.-H., An, G., Kitano, H., Ashikari, M., Matsuoka, M. Accumulation of phosphorylated repressor for gibberellin signaling in an F-box mutant. Science 299, 1896–1898 (2003).

Scheel, D., Wasternack, C. (eds.). Plant signal transduction. Oxford University Press, Oxford, UK (2002).

Seo, M., Koshiba, T. Complex regulation of ABA biosynthesis in plants. Trends Plant Sci 1, 41–48 (2002).

Silverstone, A. L., Sun, T-P. Gibberellins and the green revolution. Trends Plant Sci 5, 1–2 (2000).

Stepanova, A. N., Ecker, J. R. Ethylene signaling: From mutants to molecules. Curr Opin Plant Biol 3, 353–360 (2000).

Stevenson, J. M., Perera, I. Y., Heilmann, I., Person, S., Boss, W. F. Inositol signaling and plant growth. Trends Plant Sci 5, 252–258 (2000).

Stoelzle, S., Kagawa, T., Wada, M., Hedrich, R., Dietrich, P. Blue light activates calcium-permeable channels in *Arabidopsis* mesophyll cells via the phototropin signaling pathway. Proc Natl Acad Sci USA 100, 1456–1461 (2003).

Stratmann, J. W. Long distance run in the wound response—jasmonic acid is pulling ahead. Trends Plant Sci 8, 247–250 (2003).

Tichtinsky, G., Vanoosthuyse, V., Cock, J. M., Gaude, T. Making inroads into plant receptor kinase signalling pathways. Trends Plant Sci 8, 231–237 (2003).

Timpte, C. Auxin binding protein: Curiouser and curiouser. Trends Plant Sci 6, 586–590 (2001).

Torii, K. U. Receptor kinase activation and signal transduction in plants: An emerging picture. Curr Opin Plant Biol 3, 361–367 (2000).

Turner, J. G., Ellis, C., Devoto, A. The jasmonate signal pathway. Plant Cell, Supplement, 153–164 (2002).

Wang, H., Deng, X. W. Dissecting the phytochrome A-dependent signaling network in higher plants. Trends Plant Sci 8, 172–178 (2003).

Wu, Y., Kuzma, J., Maréchal, E., Graeff, R., Lee, H. C., Foster, R., Chua N.-H. Abscisic acid signaling through cyclic ADP-ribose in plants. Science 278, 2126–2130 (1997).

Xiong, L., Zhu, J.-K. Regulation of abscisic acid biosynthesis. Plant Physiol 133, 29–36 (2003).

Yang, Z. Small GTPases: Versatile signaling switches in plants. Plant Cell, Supplement, 375–388 (2002).

Zhao, Y., Dai, X., Blackwell, H. E., Schreiber, S. L., Chory, J. SIR1, an upstream component in auxin signalling identified by chemical genetics. Science 301, 1107–1110 (2003).

Zhang, L., Lu, Y.-T. Calmodulin-binding protein kinases in plants. Trends Plant Sci 8, 123–127 (2003).

Zhang, S., Klessig, D. F. MAPK cascades in plant defense signaling. Trends Plant Sci 6, 520–527 (2001).

20

A plant cell has three different genomes

A plant cell contains three genomes: in the **nucleus**, the **mitochondria**, and the **plastids**. Table 20.1 lists the size of the three genomes in three plant species and, in comparison, the two genomes of man. The size of the genomes is given in base pairs (bp). As discussed in Chapter 1, the genetic information of the mitochondria and plastids is located on one circular DNA double strand (although sometimes on several), with many copies present in each organelle. During the multiplication of the organelles by division, these copies are distributed randomly between the daughter organelles. Each cell contains a large number of plastids and mitochondria, which also are randomly distributed during cell division between both daughter cells. In this way the genetic information of mitochondria and plastids is inherited predominantly maternally from generation to generation, where the large number of inherited gene copies protects against mutations. The structures and functions of the plastid and mitochondrial genomes are discussed in sections 20.6 and 20.7.

Table 20.1: Size of the genome in plants and in humans

	Arabidopsis thaliana	*Zea mays* (maize)	*Vicia faba* (broad bean)	*Homo sapiens* (human)
	Number of base pairs in a single genome			
Nucleus (haploid chromosome set)	7×10^7	390×10^7	1450×10^7	280×10^7
Plastid	156×10^3	136×10^3	120×10^3	
Mitochondrium	370×10^3	570×10^3	290×10^3	17×10^3

20.1 In the nucleus the genetic information is divided among several chromosomes

For almost the whole of their developmental cycle, eukaryotic cells, as they are **diploid**, normally contain two chromosome sets, one set from the mother and the other set from the father. Only the generative cells (e.g., egg and pollen) are **haploid**, that is, they possess just one set of chromosomes. The DNA of the genome is replicated during the interphase of mitosis and generates a quadruple chromosome set. During the anaphase, this set is distributed by the spindle apparatus to two opposite poles of the cell. Thus, after cell division, each daughter cell contains a double chromosome set and is diploid once again. **Colchicine**, an alkaloid from the autumn crocus, inhibits the spindle apparatus and thus interrupts the distribution of the chromosomes during the anaphase. In such a case, all the chromosomes of the mother cell can end up in only one of the daughter cells, which then possesses four sets of chromosomes, making it **tetraploid**. Infrequently this tetraploidy occurs spontaneously due to mitosis malfunction.

Tetraploidy is often stable and is then inherited by the following generations via somatic cells. Hexaploid plants can be generated by crossing tetraploid plants with diploid plants, although this is seldom successful. Tetraploid and hexaploid (polyploid) plants often show a higher growth rate, and this is why many crop plants are polyploid. Polyploid plants can also be generated by protoplast fusion, a method that can produce hybrids between two different breeding lines. When very different species are crossed, the resulting diploid hybrids often are sterile due to incompatibilities in their chromosomes. In contrast, polyploid hybrids generally are fertile.

The chromosome content of various plants is listed in Table 20.2. The crucifer ***Arabidopsis thaliana*** (Fig. 20.1), an inconspicuous weed growing at the roadside in central Europe, has only 2×5 chromosomes with altogether just 7×10^7 base pairs (Table 20.1). *Arabidopsis* corresponds in all details to a typical dicot plant and, when grown in a growth chamber, has a life cycle of only six weeks. The breeding of a defined line of *Arabidopsis thaliana* by the botanist Friedrich Laibach (Frankfurt) in 1943 marked the beginning of the worldwide use of *Arabidopsis* as a model plant for the investigation of plant functions. In 1965 an international symposium on *Arabidopsis* research took place in Goettingen, Germany, under Gerhard Roebbelen. By now the genomes of *Arabidopsis thaliana* and also of *Oriza sativa* (rice) have been completely sequenced.

The often much higher number of chromosomes that can be found in other plants is in part the result of a gene combination. Hence, rapeseed (*Brassica napus*) with 2×19 chromosomes is a crossing of *Brassica rapa*

Table 20.2: Number of chromosomes

n: Ploidy m: Number of chromosomes in the haploid genome	n × m
A Dicot plants	
Arabidopsis thaliana	2 × 5
Vicia faba (broad bean)	2 × 6
Glycine maximum (soybean)	2 × 20
Brassica napus (rapeseed)	2 × 19
Beta vulgaris (sugar beet)	6 × 19
Solanum tuberosum (potato)	4 × 12
Nicotiana tabacum (tobacco)	4 × 12
B Monocot plants	
Zea mays (maize)	2 × 10
Hordeum vulgare (barley)	2 × 7
Triticum aestivum (wheat)	6 × 7
Oryza sativa (rice)	2 × 12

Figure 20.1 *Arabidopsis thaliana*, an inconspicuous small weed growing in central Europe at the roadsides, from the family *Brassicaceae* (crucifers), has become the most important model plant worldwide, because of its small genome. In the growth chamber with continuous illumination, the time from germination until the formation of mature seed is only six weeks. The plant reaches a height of 30 to 40 cm. Each fruit contains about 20 seeds. (By permission of M. A. Estelle and C. R. Somerville.)

(2×10 chromosomes) and *Brassica oleracea* (2×9 chromosomes) (Table 20.2). In this case, a diploid genome is the result of crossing. In wheat, on the other hand, the successive crossing of three wild forms resulted in hexaploidy: Crossing of einkorn wheat and goat grass (2×7 chromosomes each) resulted in wild emmer wheat (4×7 chromosomes), and this crossed with another wild wheat, resulting in the wheat (*Triticum aestivum*) with 6×7 chromosomes cultivated nowadays. Tobacco (*Nicotiana tabacum*) is also a cross between two species (*Nicotiana tomentosiformis* and *N. sylvestris*), each with 2×12 chromosomes.

The nuclear genome of broad bean with 14.5×10^9 base pairs has a 200-fold higher DNA content than *Arabidopsis*. However, this does not mean that the number of protein-encoding genes (structural genes) in the broad bean genome is 200-fold higher than in *Arabidopsis*. Presumably the number of structural genes in both plants does not differ by more than a factor of two to three. The difference in the size of the genome is due to a different number of identical DNA sequences of various sizes arranged in sequence, termed **repetitive DNA**, of which a very large part may contain no encoding function at all. In broad bean, for example, 85% of the DNA represents repetitive sequences. This includes the **tandem repeats**, a large number (sometimes thousands) of identical repeated DNA sequences (of a unit size of 170–180 bp, sometimes also 350 bp). The tandem repeats are spread over the entire chromosome, often arranged as blocks, especially at the beginning and the end of the chromosome, and sometimes also in the interior. This highly repetitive DNA is called **satellite DNA**. This also includes microsatellite DNA, which is discussed in section 20.3. Its sequence is genus- or even species-specific. In some plants, more than 15% of the total nuclear genome consists of satellite DNA. So far its function is not known. Perhaps it plays a role in the segregation of species. The sequence of the satellite DNA can be used as a species-specific marker in generating hybrids by protoplast fusion in order to check the outcome of the fusion in cell culture.

The genes for ribosomal RNAs also occur as repetitive sequences and, together with the genes for some transfer RNAs, are present in the nuclear genome in several thousand copies.

In contrast, structural genes are present in only a few copies, sometimes just one (**single-copy gene**). Structural genes encoding for structurally and functionally related proteins often form a **gene family**. Such a gene family, for instance, is formed by the genes for the small subunit of the ribulose bisphosphate carboxylase, which exist several times in a slightly modified form in the nuclear genome (e.g., five times in tomato). So far 14 members of the light harvesting complex (LHC) gene family (section 2.4) have been identified in tomato. Zein, which is present in maize kernels as a storage protein

(Chapter 14), is encoded by a gene family of about 100 genes. In *Arabidopsis* almost 40% of the proteins predicted from the genome sequence belong to gene families with more than five members, and 300 genes have been identified that encode P_{450} proteins (see section 18.2).

The DNA sequences of plant nuclear genomes have been analyzed in a dicot and a monocot plant

It was a breakthrough in the year 2000 when the entire nuclear genome of *Arabidopsis thaliana* was completely sequenced. The sequence data revealed that the nuclear genome contains about **25,000 structural genes**, twice as many as in the insect *Drosophila*. A comparison with known DNA sequences from animals showed that about one-third of the *Arabidopsis* genes are plant-specific. By now the rice genome also has been sequenced. It is estimated to contain about **50,000 structural genes**.

The function of many of these genes is not yet known. The elucidation of these functions is a great challenge. One approach to solve this is a comparison of sequence data with identified sequences from microorganisms, animals, and plants, available in data banks, by means of **bioinformatics**. Another way is to eliminate the function of a certain gene by mutation and then investigate its effect on metabolism. A gene function can be eliminated by random mutations (e.g., by Ti-plasmids) (section 22.5) (**T-DNA insertion mutant**). In this case, the mutated gene has to be identified. Alternatively, defined genes can be mutated by the **RNAi technique**, as discussed in section 22.5. All these investigations require an automated evaluation with a very high technical expenditure. The project *Arabidopsis* 2010 (United States) and the German partner program AFGN (*Arabidopsis* functional genomics network) have set their goal at fully elucidating the function of the *Arabidopsis* genome by the year 2010.

20.2 The DNA of the nuclear genome is transcribed by three specialized RNA polymerases

Of the two DNA strands, only the **template strand** is transcribed (Fig. 20.2). The DNA strand complementary to the template strand is called the **encoding strand**. It has the same sequence as the transcription product RNA, with the exception that it contains thymine instead of uracil. The DNA of the

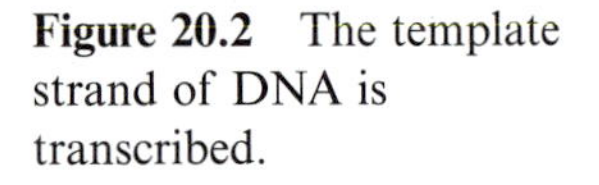
Figure 20.2 The template strand of DNA is transcribed.

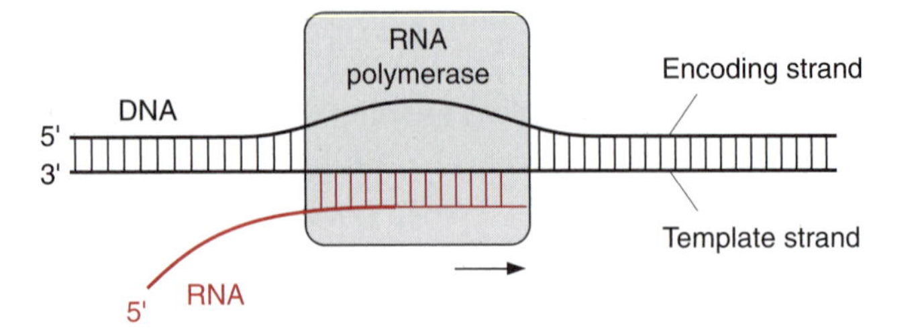

Table 20.3: Three RNA polymerases

RNA-Polymerase	Transcript	Inhibition by α-Amanitin
Type I	Ribosomal RNA (5,8S-, 18S-, 25S-rRNA)	None
Type II	Messenger-RNA-precursors, small RNA (snRNA)	In concentrations of ca. 10^{-8} mol/L
Type III	Transfer-RNA, ribosomal RNA (5S-rRNA)	Only at higher concentrations (10^{-6} mol/L)

nuclear genome is transcribed by **three specialized RNA polymerases (I, II, and III)** (Table 20.3). The division of labor between the three RNA polymerases, along with many details of the gene structure and principles of gene regulation, are valid for all eukaryotic cells. RNA polymerase II catalyzes the transcription of the structural genes and is strongly inhibited by **α-amanitin** at a concentration as low as 10^{-8} mol/L. α-Amanitin is a deadly poison of the toadstool *Amanita phalloides* (also called death cap). People frequently die from eating this toadstool.

The transcription of structural genes is regulated

In a plant containing about 25,000 to 50,000 structural genes, most of these genes are switched off and are activated only in certain organs, and then often only in certain cells. Moreover, many genes are switched on only at specific times (e.g., the genes for the synthesis of phytoalexins after pathogenic infection) (section 16.1). Therefore the transcription of most structural genes is subject to very complex and specific regulation. The genes for enzymes of metabolism or protein biosynthesis, which proceed in all cells, are transcribed more often. The genes that every cell needs for such basic functions, independent of its specialization, are called **housekeeping genes**.

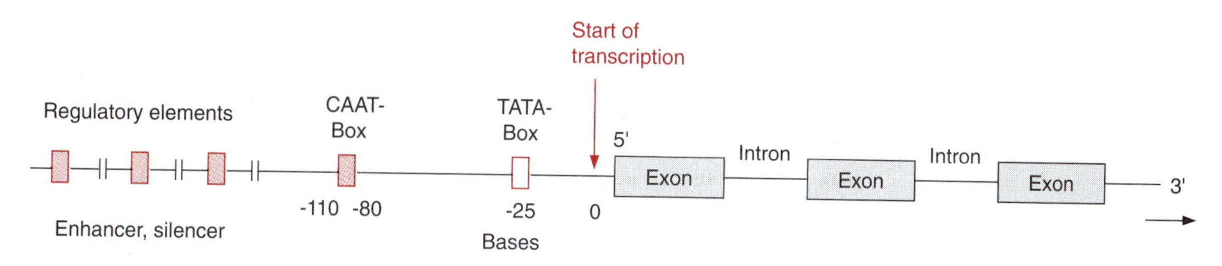

Figure 20.3 Sequence elements of a eukaryotic gene. The numbers mark the distance of the bases from the transcription start.

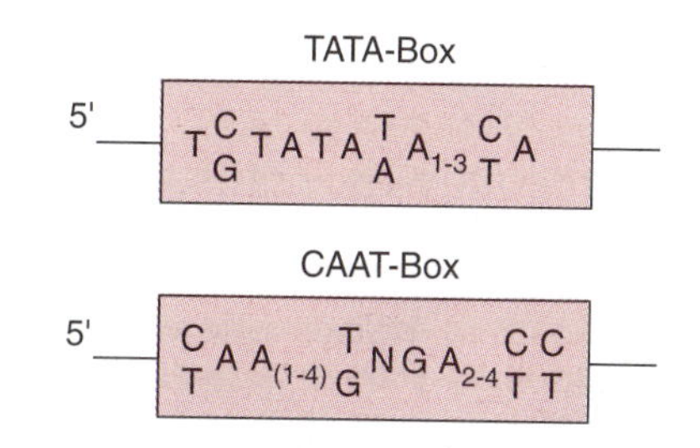

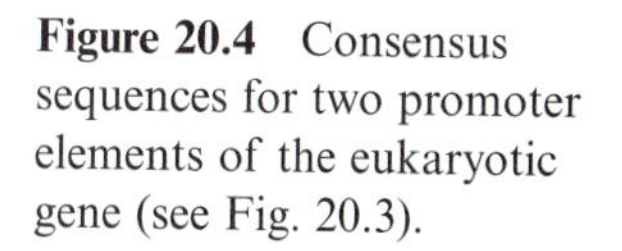

Figure 20.4 Consensus sequences for two promoter elements of the eukaryotic gene (see Fig. 20.3).

Promoter and regulatory sequences regulate the transcription of genes

Figure 20.3 shows the basic design of a structural gene. The section of the DNA on the left of the transcription starting point is termed 5′ or upstream and that to the right is referred to as 3′ or downstream. The encoding region of the gene is distributed mostly among several exons, which are interrupted by introns.

About 25 bp upstream from the transcription start site is situated a **promoter element**, which is the position where RNA polymerase II binds. The sequence of this promoter element can vary greatly between genes and between species, but it can be depicted as a **consensus sequence**. (A consensus sequence is an idealized sequence in which each base is found in the majority of the promoters. Most of the promoter elements differ in their DNA sequences only by one or two bases from this consensus sequence.) This consensus sequence is named the **TATA box** (Fig. 20.4). About 80 to 110 bp upstream another consensus sequence is often found, the **CAAT box** (Fig. 20.4), which influences the rate of transcription. The housekeeping genes mentioned earlier often contain a C-rich region instead of the CAAT box. Additionally, sometimes more than 1,000 bp upstream, several sequences can be present, which function as **enhancer** or as **silencer** regulatory elements.

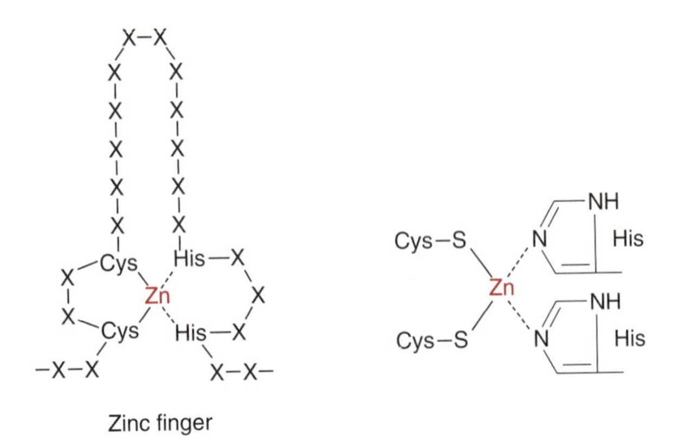

Figure 20.5 Many transcription factors have the structure of a zinc finger. An amino acid sequence (X = amino acid) contains 2 cysteine residues separated by 2 to 4 other amino acids and, after a further 12 amino acids, 2 histidine residues, which are separated from each other by 3 to 4 amino acids. A zinc ion is bound between the cysteine and the histidine residues. A transcription factor contains 3 to 9 such zinc fingers, with each finger binding to a sequence of 3 bases on the DNA.

Transcription factors regulate the transcription of a gene

The regulatory elements contain binding sites for **transcription factors**, which are proteins modifying the rate of transcription. It has been estimated that the *Arabidopsis* genome encodes 1,500 factors for the regulation of gene expression. Different transcription factors often have certain structures in common. One type of transcription factor consists of a peptide chain containing two cysteine residues and, separated from these by 12 amino acids, two histidine residues (Fig. 20.5). The two cysteine residues bind covalently to a zinc atom, which is also coordinatively bound to the imidazole rings of two histidine residues, thus forming a so-called **zinc finger**. Such a finger binds to a base triplet of a DNA sequence. Zinc finger transcription factors usually contain several (up to 9) fingers and so are able to cling tightly to certain DNA sections.

Another type of transcription factor consists of a dimer of DNA binding proteins, where each monomer contains a DNA binding domain and an α-helix with three to nine leucine residues (Fig. 20.6). The hydrophobic leucine residues of the two α-helices are arranged in such a way that they are exactly opposite each other and are held together by hydrophobic interaction like a zipper. This typical structure of a transcription factor has been named the **leucine zipper**. The activity of transcription factors is frequently regulated by signal chains linked to the perception of phytohormones or other stimuli (section 19).

Micro-RNAs inhibit gene expression by inactivating messeger RNAs

There is another way to regulate gene expression in plants. As recently as 2002 it was reported that plants contain so-called **micro-RNAs (miRNAs)**.

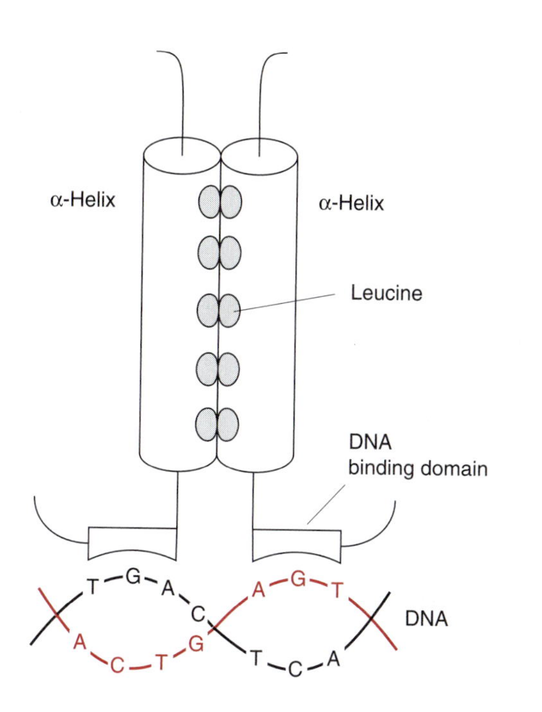

Figure 20.6 A frequent structural motif of transcription factors is the leucine zipper. The factor consists of two polypeptide chains each with an α-helix, which contains a leucine residue at about every seventh position. The leucine residues, which are all located at one side of the α-helix, hold the two α-helices together by hydrophobic interactions in the manner of a zipper. The two DNA binding domains contain basic amino acids, which enable binding to the DNA. Simplified representation.

These are noncoding RNAs of 21 to 22 nucleotides, which are highly conserved and also occur in animal cells. miRNAs inhibit gene expression by binding to complementary regions of **mRNAs** (e.g., of those encoding key proteins for plant development). This binding inactivates the mRNAs and labels them for a subsequent degradation by ribonuclease. In this way miRNAs appear to function as negative regulators to control processes such as meristem cell identity, organ polarity, and other developmental processes. Many miRNAs have until now been identified in plants, and their number is expected to greatly increase in the course of further research.

The transcription of structural genes requires a complex transcription apparatus

RNA polymerase II consists of 8 to 14 subunits, but, on its own, it is unable to start transcription. Transcription factors are required to direct the enzyme to the start position of the gene. The **TATA binding protein**, which recognizes the TATA box and binds to it, has a central function in transcription (Fig. 20.7). The interaction of the TATA binding protein with RNA polymerase requires a number of additional transcription factors (designated as A, B, F, E, and H in the figure). They are all essential for transcription and are termed **basal factors**.

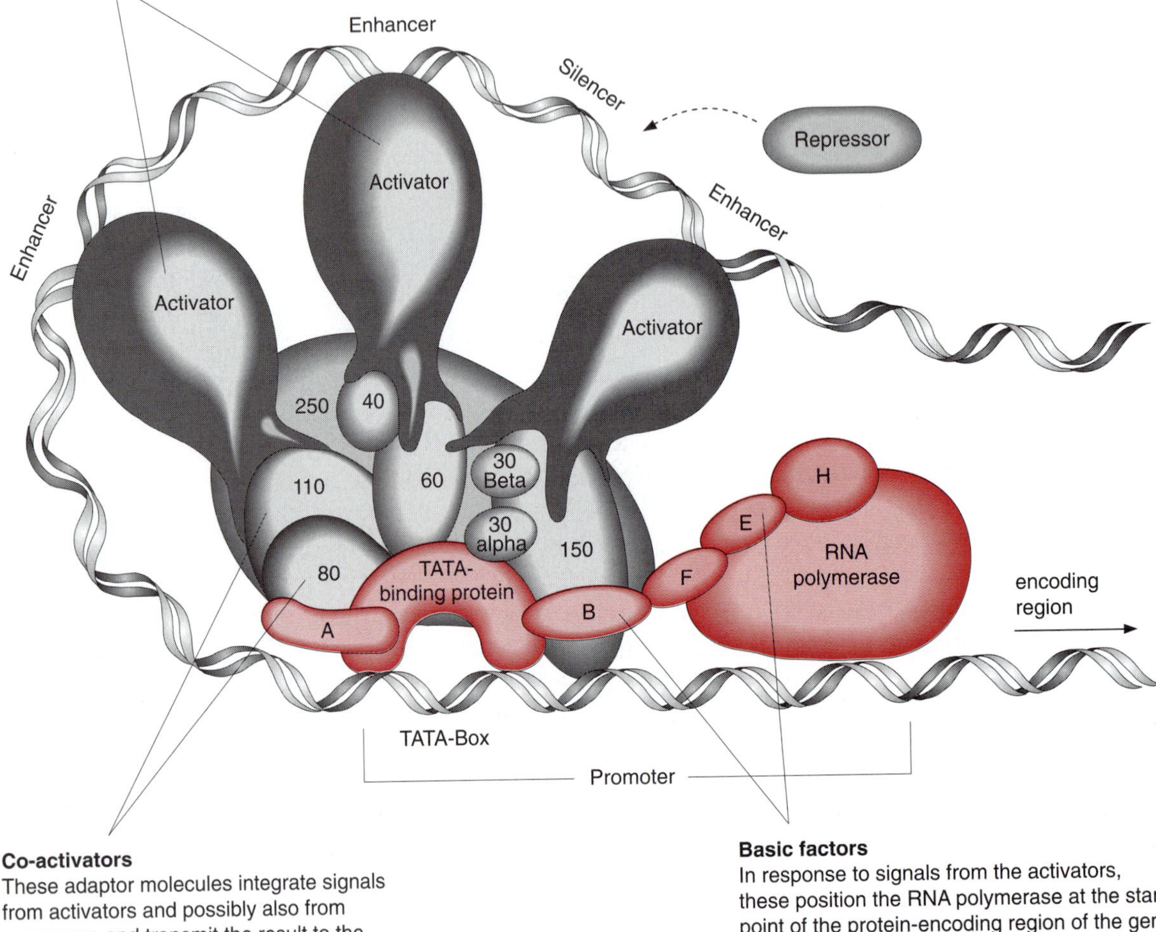
Activators
These proteins bind to enhancer elements.
They control which gene is turned on and enhance the rate of transcription.
Repressors
These proteins bind to silencer elements, interfere with the function of activators, and thus lower the rate of transcription.
Enhancer
Silencer
Repressor
Enhancer
Enhancer
Activator
Activator
Activator
250
40
110
60
30
Beta
30
alpha
150
80
TATA-
binding protein
H
E
RNA
polymerase
F
B
A
encoding
region
TATA-Box
Promoter
Co-activators
These adaptor molecules integrate signals from activators and possibly also from repressors and transmit the result to the basic factors.
Basic factors
In response to signals from the activators, these position the RNA polymerase at the start point of the protein-encoding region of the gene and thus enable transcription to start.

Figure 20.7 Eukaryotic transcription apparatus of mammals, which is thought to be similar to the transcription apparatus of plants. The basal factors (TATA binding protein, peptides A-H, colored red) are indispensable for transcription, but they can neither enhance nor slow down the process. This is brought about by regulatory molecules: activators and repressors, the combination of which is different for each gene. They bind to regulatory sequences of the DNA, termed enhancer or silencer, which are located far upstream from the transcription start. Activators (and possibly repressors) communicate with the basal factors via co-activators, which form tight complexes with the TATA binding protein. This complex docks first to the nuclear promoter, a control region close to the protein gene. The co-activators are designated according to their molecular weight (given in kDa). (From Tijan, R., Spektrum der Wissenschaft, 4, 1995, with permission.)

The transcription apparatus is a complex of many protein components, around which the DNA is wrapped in a loop. In this way regulatory elements positioned far upstream or downstream from the encoding gene are able to influence the activity of RNA polymerase. The rate of transcription is determined by transcription factors, either **activators** or **repressors**, which bind to the upstream regulatory elements (**enhancer**, **silencer**; Fig. 20.3).

These transcription factors interact through a number of **co-activators** with the TATA binding protein and modulate its function on RNA polymerase. Various combinations of activators and repressors thus lead to activation or inactivation of gene transcription. The scheme of the transcription apparatus shown in Figure 20.7 was derived from investigations with animals. Results obtained so far in plants indicate that the regulation of transcription occurs there in essentially the same way.

Knowledge of the promoter and enhancer/silencer sequences is very important for genetic engineering of plants (Chapter 22). For this it is not always essential to know all these boxes and regulatory elements in detail. It can be sufficient for practical purposes if the DNA region that is positioned upstream of the structural gene and influences its transcription in a specific way is identified. In eukaryotic cells, this entire regulatory section is often simply called a promoter. For example, promoter sequences have been identified, which determine that a gene is to be transcribed in a leaf, and there only in the mesophyll cells or the stomata, or in potato tubers, and there only in the storage cells. In such cases, the specificity of gene expression is explained by the effect of cell-specific transcription factors on the corresponding promoters.

The formation of the messenger RNA requires processing

The transcription of DNA in the nucleus by RNA polymerase II yields a **primary transcript** (pre-mRNA, Fig. 20.8), which is processed in the nucleus to mature mRNA. During transcription, a special GTP molecule is linked to the 5′-OH group of the RNA terminus by a 5′-5′ pyrophosphate bridge (Fig. 20.9). Moreover, guanosine and the second ribose (sometimes also the third, not shown in the figure) are methylated using S-adenosylmethionine as a donor (Fig. 12.10). This modified GTP at the beginning of RNA is called the **cap** and is contained only in mRNA. It functions as a binding site in the formation of the initiation complex during the start of protein biosynthesis at the ribosomes (section 21.1) and probably also provides protection against degradation by exo-ribonucleases.

The introns have to be removed for further processing of the pre-mRNA. Their size can vary from 50 to over 10,000 bases. The beginning and end of an intron are defined by certain base sequences. The last two bases on the

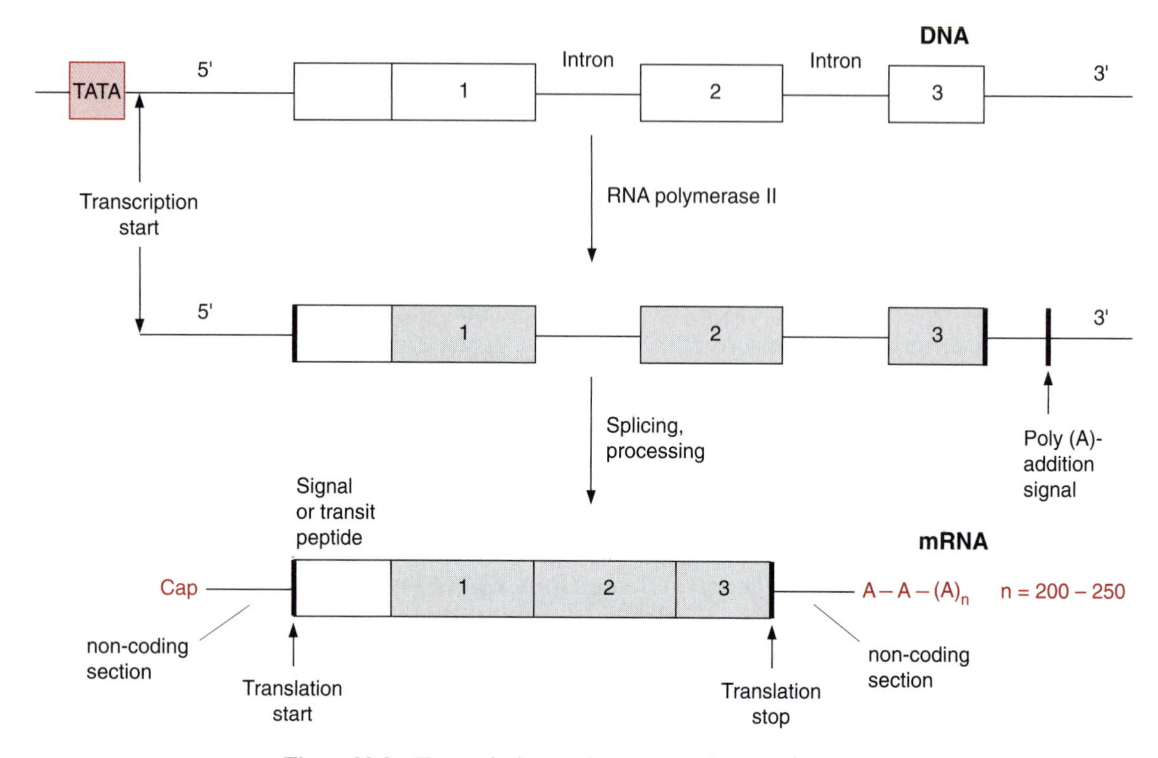

Figure 20.8 Transcription and posttranscriptional processing of a eukaryotic structural gene. The introns are removed from the primary transcript formed by RNA polymerase II by splicing (Figs. 20.10 and 20.11). The mature mRNA is formed by the addition of a cap sequence (Fig. 20.9) to the 5′ end and a poly(A) sequence to the 3′ end of a cleavage site marked by a poly(A) addition signal.

Figure 20.9 The cap sequence consists of a 7′ methyl guanosine triphosphate, which is linked to the 5′ terminal of the mRNA. The ribose residues of the last two nucleotides of the mRNA are often methylated at the 2′ position.

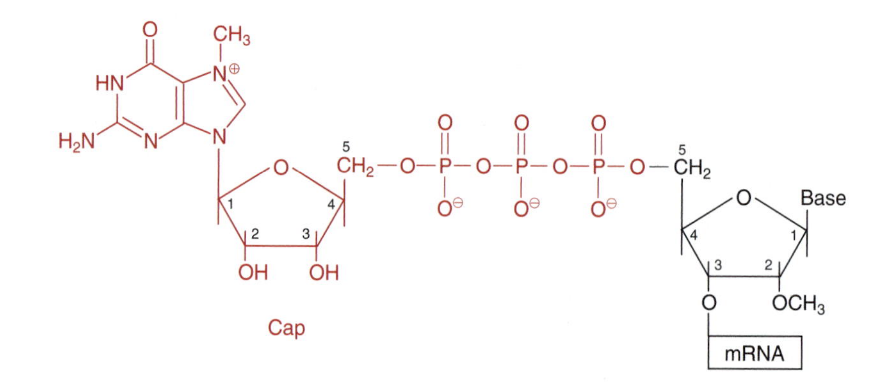

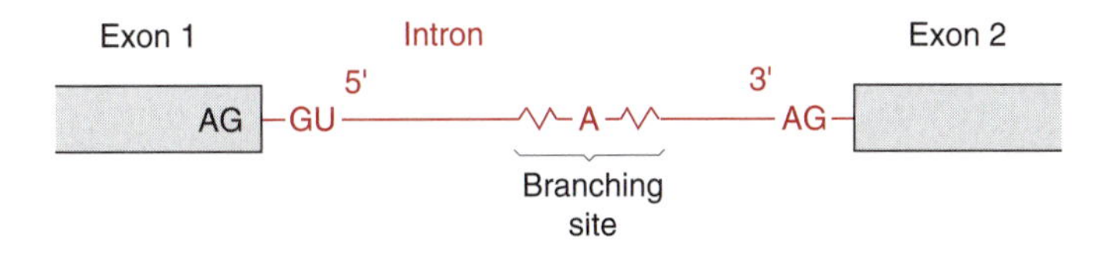

Figure 20.10 The beginning of an intron is often marked by the sequence AG/GU and the end by the sequence AG. At about 20 to 50 bases before the end of the intron, there is a consensus sequence of about seven nucleotides containing an A, which forms the branching site during splicing (see Fig. 20.11).

exon are mostly AG and the first bases of the intron are GU (Fig. 20.10). The intron ends with AG. About 20 to 50 bases upstream from the 3′ end of the intron an adenyl residue known as the branching site is in the sequence.

The excision of introns (**splicing**) is catalyzed by **riboprotein complexes**, composed of RNA and proteins. Five different RNAs of 100 to 190 bases, named **snRNA** (sn = small nuclear), are involved in the splicing procedure. Together with proteins and the RNA to be spliced, these snRNA form the **spliceosome** particle (Fig. 20.11). The first step is that the 2′-OH group of the ribose at the adenyl residue of the branching site forms a phosphate ester with the phosphate residue linking the end of exon 1 with the start of the intron, cleaving the ester bond between exon 1 and the intron. It is followed by a second ester formed between the 3′-OH group of exon 1 and the phosphate residue at the 5′-OH of exon 2, accompanied by a cleavage of the phosphate ester with the intron, thus completing the splicing. The intron remains in the form of a lasso and later is degraded by ribonucleases.

As a further step in RNA processing, the 3′ end of the pre-mRNA is cleaved behind a poly(A) addition signal (Fig. 20.12) by an endonuclease, and a **poly(A)-sequence** of up to 250 bp is added at the cleaving site. The resulting mature mRNA is bound to special proteins and leaves the nucleus as a DNA-protein complex.

rRNA and tRNA are synthesized by RNA polymerase I and III

Eukaryotic ribosomes of plants contain four different rRNA molecules named 5S-, 5.8S-, 18S-, and 25S-rRNA according to their sedimentation coefficients. The genes for 5S-rRNA are present in many copies, arranged in tandem on certain sections of the chromosomes. The transcription of these genes and also of tRNA genes is catalyzed by **RNA polymerase III**. The three remaining ribosomal RNAs are encoded by a continuous gene sequence, again in tandem and in many copies. These genes are transcribed by **RNA polymerase I**. The primary transcript is subsequently processed by **methylation**, especially of -OH groups of ribose residues, and the **cleavage of RNA** to produce mature 18S-, 5.8S-, and 25S-rRNA (Fig. 20.13). The excised RNA spacers between these rRNAs are degraded. Because of their rapid

Figure 20.11 In the splicing procedure, several RNAs and proteins assemble at the splicing site of the RNA to form a spliceosome. The terminal phosphate at the 5′ end of the introns forms an ester linkage with the 2′-OH group of the A in the branching site and the RNA chain after the exon 1/intron junction is cleaved. The 3′ end of exon 1 forms a new ester linkage with the phosphate residue at the 5′ end of exon 2, releasing the intron in the form of a lasso.

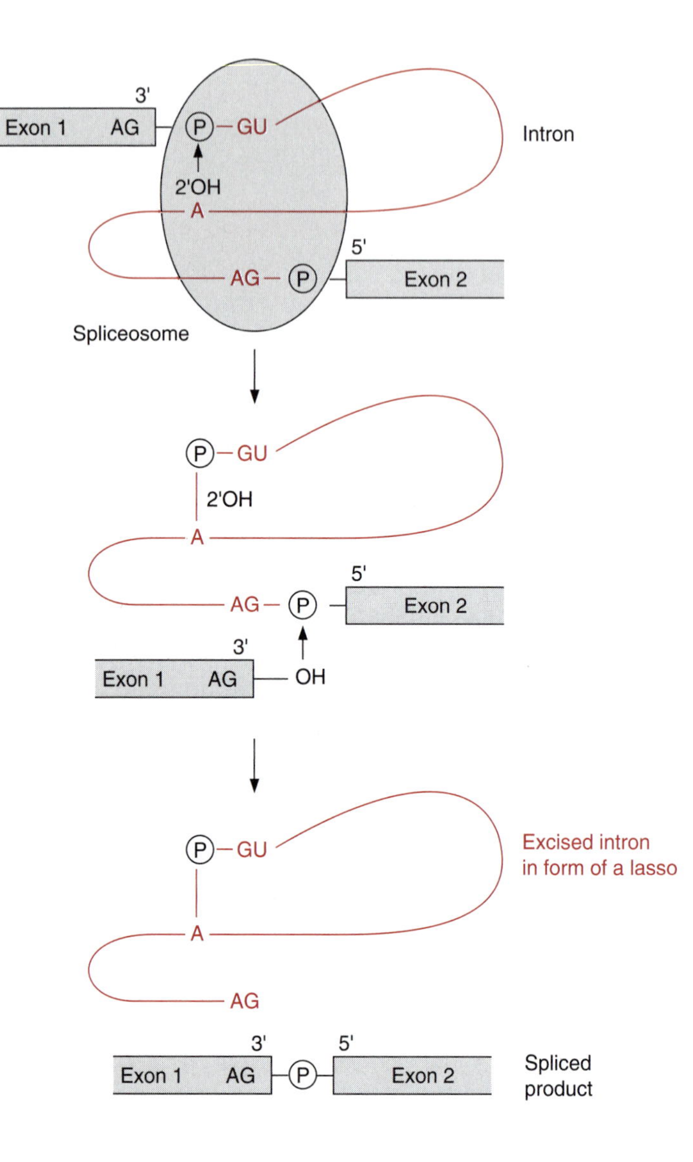

Figure 20.12 Consensus sequence for the poly(A) addition signal.

Poly (A) addition signal

5' $^{G}_{A}$ATAA$_{(1-3)}$

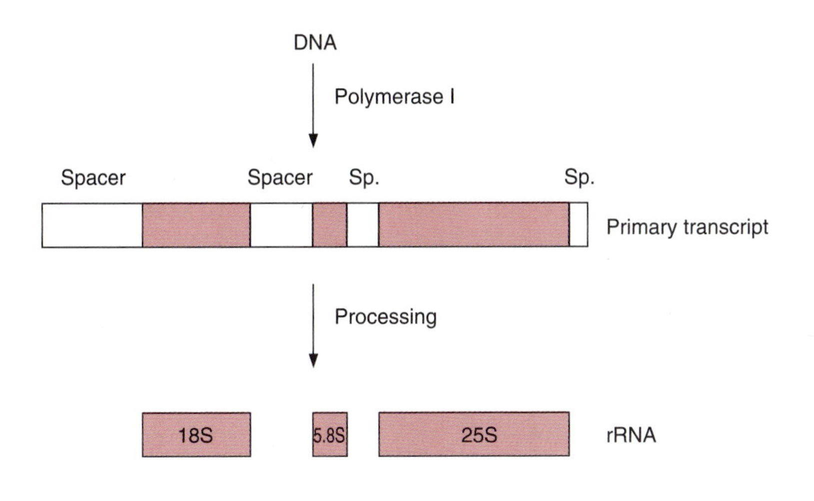

Figure 20.13 Three of the four rRNA molecules are transcribed polycistronically. The spacers are removed during processing.

evolution, comparative sequence analyses of these spacer regions can be used to establish a phylogenetic classification of various plant species within a genus.

20.3 DNA polymorphism yields genetic markers for plant breeding

An organism is defined by the DNA base sequences of its genome. Differences between the DNA sequences (DNA polymorphisms) exist not only between different species, but also to some extent between individuals of the same species. Between two varieties of a cultivar, in a structural gene often 0.1% to 1% of the bases are altered, mainly in the introns.

Plants usually are selected for breeding purposes by external features (e.g., for their yield or resistance to certain pests). In the end, these characteristics are all due to differences in the base sequence of structural genes. Selecting plants for breeding would be much easier if it were not necessary to wait until the phenotypes of the next generation were evident, but if instead the corresponding DNA sequences could be analyzed directly. A complete comparative analysis of these genes is not practical, since most of the genes involved in the expression of the phenotypes are not known, and DNA sequencing is expensive and time-consuming. Other techniques that are easier and less expensive to carry out are available.

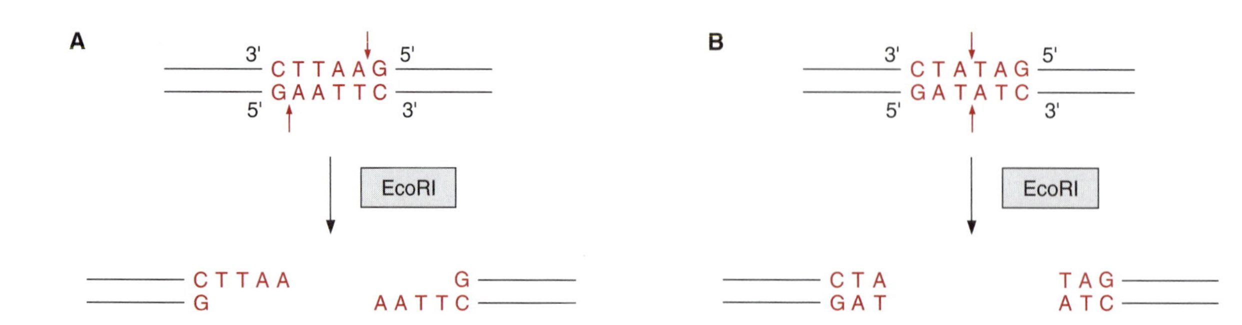

Figure 20.14 Restriction endonucleases of the type II cleave the DNA at restriction sites, which consist of a palindromic recognition sequence. As an example, the restriction sites for two enzymes from *Escherichia coli* strain are shown. A. The restriction endonuclease Eco RI (Fig. 20.14A) causes staggered cuts of the two DNA strands, leaving four nucleotides of one strand unpaired at each resulting end. These unpaired ends are called sticky ends because they can form a base pair with each other, or with complementary sticky ends of other DNA fragments. B. In contrast, the restriction endonuclease Eco RV produces blunt ends.

Individuals of the same species can be differentiated by restriction fragment length polymorphism

It is possible to detect differences in the genes of individuals within a species even without a detailed sequence comparison and to relate these differences empirically to analyzed properties. One of the methods for this is the analysis of restriction fragment length polymorphism (**RFLP**). This is based on the use of bacterial **restriction endonucleases**, which cleave a DNA at a palindromic recognition sequence, known as a **restriction site** (Fig. 20.14). The various restriction endonucleases have specific recognition sites of four to eight bp. Since these restriction sites appear at random in DNA, those recognition sequences with four bp appear more frequently than those with eight bp. Therefore it is possible to cleave the genomic DNA of a plant into thousands of defined DNA fragments by using a particular restriction endonuclease (usually enzymes with 6 bp restriction sites). The exchange of a single base in a DNA may eliminate or newly form a restriction site, resulting in a polymorphism of the restriction fragment length.

The digestion of genomic DNA by restriction enzymes results in an immense number of fragments. In order to mark fragments of defined sections of the genome, labeled **DNA probes** are required. Such probes are prepared by first identifying a certain DNA section of a chromosome of about 10 to 20 kbp, located as near as possible to the gene responsible for the trait of interest, or which is even part of that gene. This DNA section is introduced into bacteria (mostly *Escherichia coli*) using plasmids or bacteriophages as a vector and is propagated there (see Chapter 22). The plasmids or bacteriophages are isolated from the bacterial suspension, the multiplied DNA sections are cut out again, isolated, and afterward radioactively labeled or provided with a fluorescence label. Such probes used for this purpose are called **RFLP markers**.

The analysis of DNA restriction fragments by the labeled probes described previously is carried out by the **Southern blot** method, developed

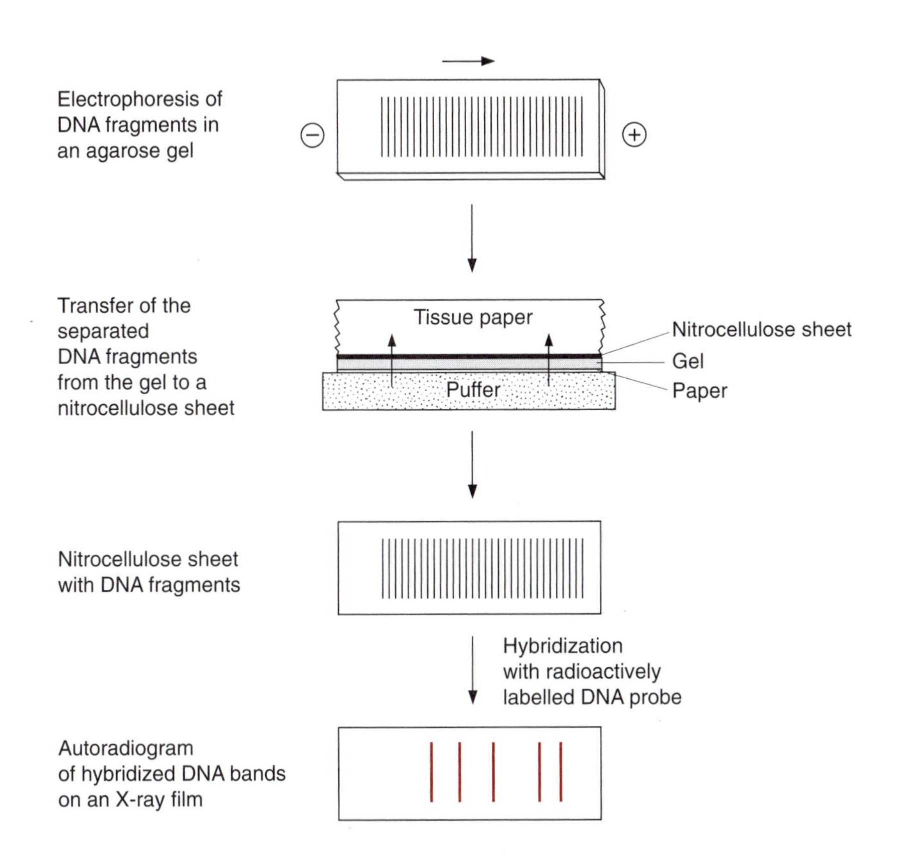

Figure 20.15 The Southern blot procedure.

by Edward Southern in 1975. The restriction fragments are first separated according to their length by electrophoresis in an agarose gel (the shortest fragment moves the farthest). The separated DNA fragments in the gel are transferred to a nitrocellulose or nylon membrane by placing the membrane on the gel. By covering them with a stack of tissue paper, a buffer solution is drawn through the gel and the membrane, and the washed out DNA fragments are bound to the membrane. The buffer also causes the dissociation of the DNA fragments into single strands (Fig. 20.15). When a labeled DNA probe is added, it hybridizes to complementary DNA sequences on the membrane. Only those DNA fragments that are complementary to the probe are labeled, and after removal of the nonbound DNA probe molecules by washing, are subsequently identified by autoradiography (in the case of a radioactive probe) or by fluorescence measurement. The position of the band on the blot is then related to its migration and hence its size.

Figure 20.16 explains the principles of RFLP. Figure 20.16A shows in (a) a section of a gene with three restriction sites (R_1, R_2, and R_3). Since the probe binds only to the DNA region between R_1 and R_3, just two noticeable restriction fragments result from the treatment with the restriction endonu-

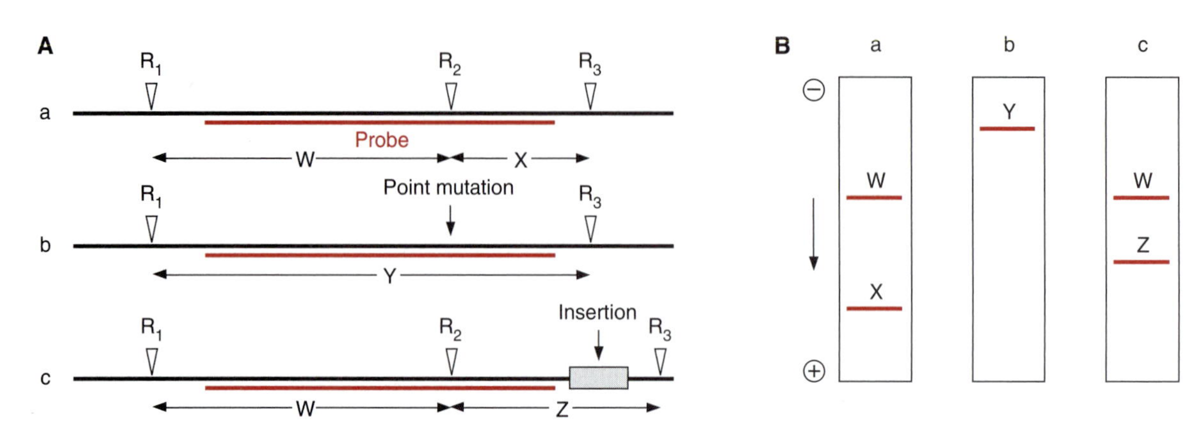

Figure 20.16 The molecular formation of a restriction fragment length polymorphism. A. The restriction sites for the restriction endonuclease in the genotypes a, b, and c are numbered R_1, R_2, and R_3. The probe by which the fragments are identified is marked red. B. (a). The electrophoretic movement of the fragments (W, X) that are labeled by the probe. (b). By point mutation one restriction site is eliminated and only one fragment is formed (Y), which, because of its larger size, migrates less during electrophoresis. (c). After insertion fragment X is enlarged (Z).

clease. Due to their different lengths, they are separated by gel electrophoresis and detected by hybridization with a probe (Fig. 20.16B(a)). Upon the exchange of one base (point mutation) (b), the restriction site R_2 is eliminated and therefore only one labeled restriction fragment is detected, which, because of its larger size, migrates in gel electrophoresis for a lesser distance than the fragments of (a). When a DNA section is inserted between the restriction sites R_2 and R_3 (c), the corresponding fragment is longer.

The RFLP represent **genetic markers**, which are inherited according to Mendelian laws and can be employed to characterize a certain variety. Normally several probes are used in parallel measurements. RFLP is also used in plant systematics to establish phylogenetic trees. Moreover, defined restriction fragments can be used as labeled probes to localize certain genes on the chromosomes. In this way chromosome maps have been established for several plants (e.g., *Arabidopsis*, potato, tomato, and maize).

The RAPD technique is a simple method for investigating DNA polymorphism

An alternative method for analyzing the differences between DNA sequences of individuals or varieties of a species is the amplification of randomly obtained DNA fragments (random amplified polymorphic DNA, **RAPD**). This method, which has been in use only since 1990, is much easier to work with, compared to the RFLP technique mentioned previously, and its application has become widespread in a very short time.

The basis for the RAPD technique is the **polymerase chain reaction (PCR)**. The method enables selected DNA fragments of a length of up to two to three kbp to be amplified by DNA polymerase. This requires an **oligonucleotide primer**, which binds to a complementary sequence of the DNA to be amplified and forms the starting point for the synthesis of a DNA daughter strand at the template of the DNA mother strand. In the polymerase chain reaction, two primers (A, B) are needed to define the beginning and end of the DNA strand that is to be amplified. Figure 20.17 shows the principle of the reaction. In the first step, by heating to about 95°C, the DNA double strands are separated into single strands. During a subsequent cooling period, the primers hybridize with the DNA single strands and thus enable, in a third step at a medium temperature, the synthesis of DNA from deoxynucleoside triphosphates. Mostly a DNA polymerase originally isolated from the thermophilic bacterium *Thermus aquaticus* living in hot sprigs (**Taq polymerase**) is used, since this enzyme is not affected by the preceding heat treatment. Subsequently, the DNA double strands thus formed are separated again by being heated at 95°C, the primer binds during an ensuing cooling period, and this is followed by another cycle of DNA synthesis by Taq polymerase. By continuous alternating heating and cooling, this reaction can be continued for 30 to 40 cycles, with the amount of DNA being doubled in each cycle. In the first cycle, the length of the newly formed DNA is not yet restricted at one end. By the primer binding to the complementary base sequence of the newly formed DNA strand, in the next DNA synthesis cycle a product that is restricted in its length by both primers is formed. With the increasing number of cycles, DNA fragments of uniform length are amplified. Since in the polymerase chain reaction the number of the DNA molecules formed is multiplied exponentially by the number of cycles (e.g., after 25 cycles by the factor 34×10^6) very small DNA samples (in the extreme case a single molecule) can be multiplied *ad libitum*.

In the RAPD technique mentioned previously, genomic DNA and only one oligonucleotide primer typically consisting of 10 bases are required for the polymerase chain reaction. Since the probability of the exact match of 10 complementary bases on the genomic DNA is low, the primer binds at only a few sites of the genomic DNA. The characteristics of the DNA polymerases used for the amplification require the distance between the two bound primers to be no larger than two to three kbp. Therefore, only a few sections of the genome are amplified by the polymerase chain reaction and a further selection of the fragments by a probe is not necessary. The amplification produces such high amounts of single DNA fragments that, after

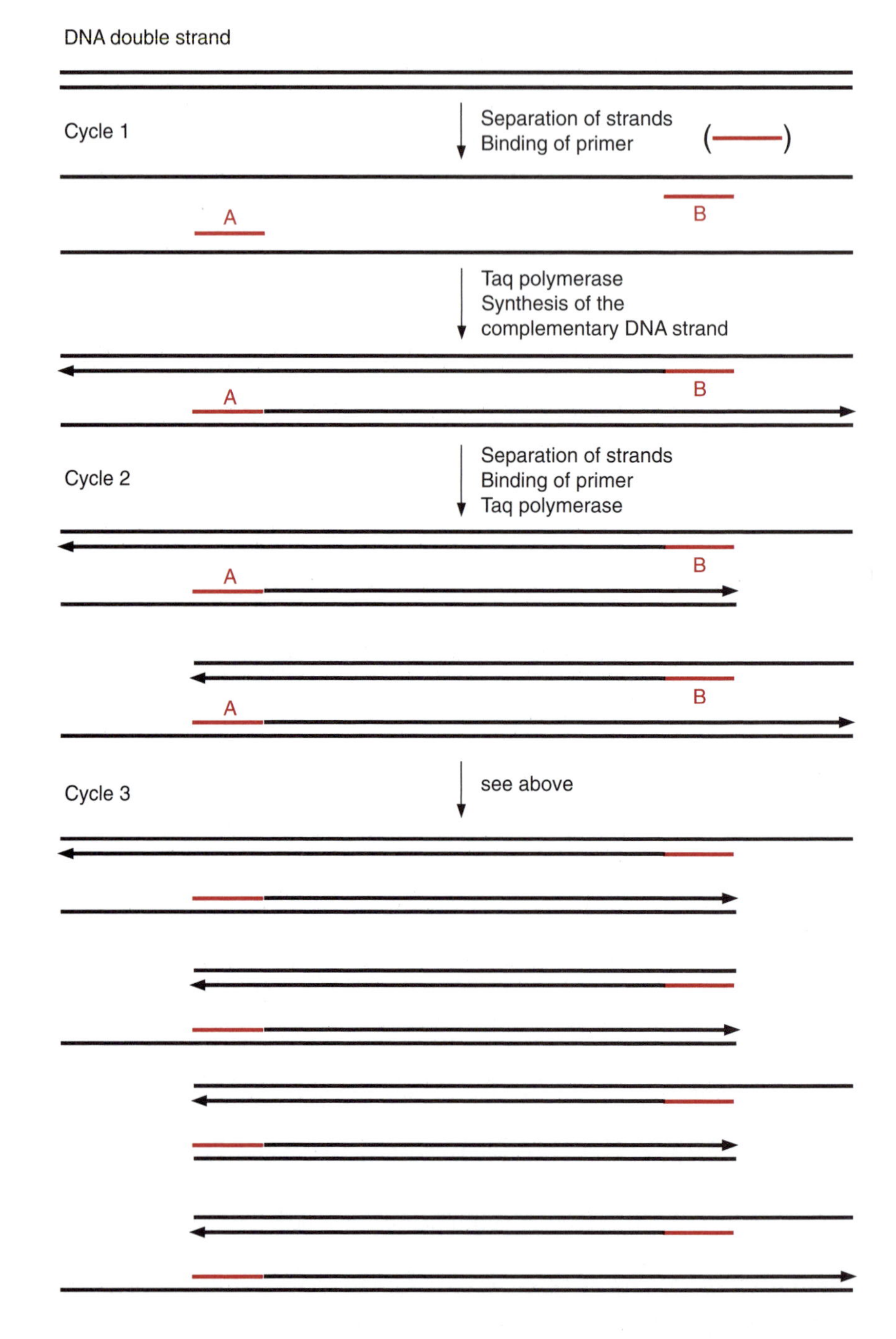

Figure 20.17 Principle of the polymerase chain reaction.

being separated by gel electrophoresis and stained with ethidium bromide, the fragments can be detected as fluorescent bands under ultraviolet (UV) light. Point mutations, which eliminate primer binding sites or form new ones, and deletions or insertions, all of which affect the size and number of the PCR products, can change the pattern of the DNA fragments in an analogous way to that of the RFLP technique (Fig. 20.16). Changing the primer sequence can generate different DNA fragments. Defined primers of 10 bases are commercially available in many variations. In the RAPD technique, different primers are tried until, by chance, bands of DNA fragments that correlate with a certain trait are obtained. The RAPD technique takes less work than the RFLP technique because it requires neither the preparation of probes nor the time-consuming procedure of a Southern blot. It has the additional advantage that only very small amounts of DNA (e.g., the amount that can be isolated from the embryo of a plant) are required for analysis. However, it is not possible to define from which gene these fragments derive. Since the RAPD technique allows differentiation between varieties of a species, it has become an important tool in breeding.

The polymorphism of micro-satellite DNA is used as a genetic marker

Recently **micro-satellite DNA** (section 20.1) has become an important tool for identifying certain plant lines. Micro-satellite DNAs contain a sequence of one to two, sometimes also three to six base pairs, which are located in 10 to 50 repetitions at certain sites of the genome, in the region of the intron, or directly before or behind a gene, whereby the number of the repetitions is highly polymorphic. Also, in this method PCR is utilized for detection. Micro-satellite polymorphism not only is used to identify individual humans (e.g., in criminal cases), but also is employed as a genetic marker for plant breeding.

20.4 Transposable DNA elements roam through the genome

In certain maize varieties, a cob may contain some kernels with different pigmentation from the others, indicating that a mutation has changed the pigment formation. Snapdragons normally have red flowers, but occasionally have mutated progeny in which parts of the flower no longer form red pigment, resulting in white stripes in the flowers. Sometimes the descendants of these defective cells regain the ability to synthesize the red pigment, forming flowers with not only white stripes but also red dots.

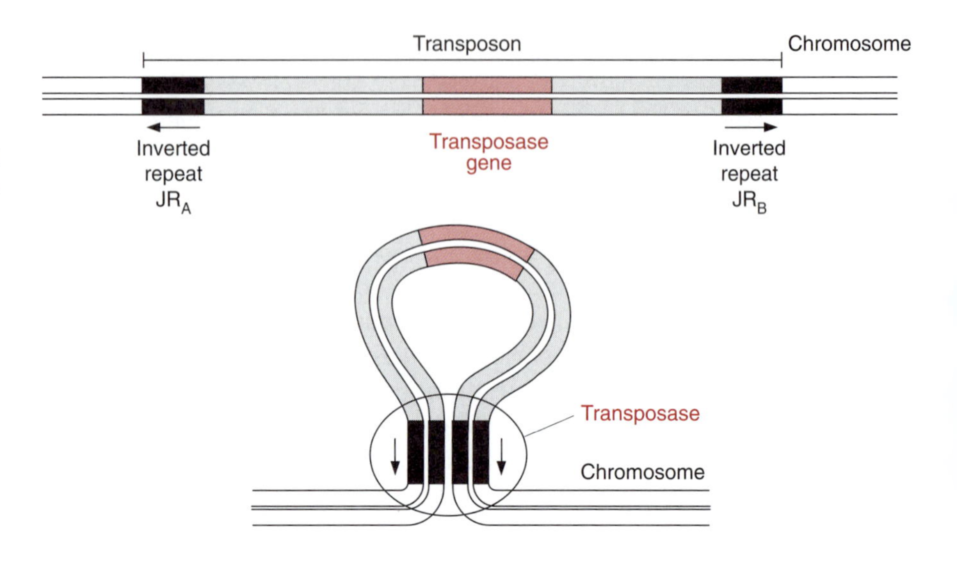

Figure 20.18 A transposon is defined by inverted repeats at both ends. The structural gene for transposase is included in the transposon. When a transposon leaves a chromosome, the two inverted repeats bind to each other and the remaining gap in the chromosome is closed. In an analogous way the transposon enters the chromosome at another site.

Barbara McClintock (United States) studied these phenomena in maize for many years, using the methods of classic genetics. In the genome of maize, she found **mobile DNA elements**, which jump into a structural gene and thus inactivate it. Generally, this mobile element does not stay there permanently, but sooner or later jumps into another gene, whereby, in most cases, the function of the first structural gene is restored. For these important findings Barbara McClintock was awarded in 1983 the Nobel Prize in Medicine. Later it became apparent that these transposable elements, which were to be named **transposons**, are not unique to plants, but also occur in bacteria, fungi, and animals.

Figure 20.18 shows the structure of the transposon Ac (activator) from maize, consisting of double-stranded DNA with 4,600 bp. Both ends contain a 15 bp long inverted repeat sequence (IR_A, IR_B). Inside the transposon is a structural gene that encodes **transposase**, an enzyme catalyzing the transposition of the gene. This enzyme binds to the flanking inverted repeats and catalyzes the transfer of the transposon to another location. It can happen that the excision of the transposon is sometimes imprecise, so that after its exit the remaining gene may have a slightly modified sequence, which could result in a lasting mutation.

In maize, besides the transposon AC, a transposon has been found, named DS, in which the structural gene for the transposase is defective. Therefore, the transposon DS is mobile only in the presence of the transposon AC, which encodes the transposase required for the transposition of the DS.

A transposon can thus be regarded as an autonomous unit encoding the proteins required for jumping. There are controversial opinions about the origin and function of the transposons. Many see the transposons as a kind of parasitic DNA, with features comparable to the viruses, which exploit the cell to multiply themselves. But it is also possible that the transposons offer the cells a selection advantage by increasing the mutation rate in order to enhance adaptation to changed environmental conditions.

As the transposons can be used for **tagging genes**, they have become an interesting tool in biotechnology. It has already been discussed that the insertion of a transposon in a structural gene results in the loss of the encoding function. For example, when a transposon jumps into a gene for anthocyanin synthesis in snapdragons, the red flower pigment can no longer be formed. The transposon inserted in this inactivated gene can be used as a DNA probe (marker) to isolate and characterize a gene of the anthocyanin biosynthesis pathway. The relevant procedures will be discussed in section 22.1.

20.5 Most plant cells contain viruses

With the exception of meristematic cells, all other plant cells are generally infected by viruses. In many cases, viruses do not kill their host since they depend on the host's metabolism to reproduce themselves. The viruses encode only a few special proteins and use the energy metabolism and the biosynthetic capacity of the host cell to multiply. This often weakens the host plant and lowers the yield of virus-infected cultivars. Infection by some viruses can lead to the destruction of the entire crop. Courgettes and melons are extremely susceptible to the *Cucumber mosaic virus*. In some provinces of Brazil, 75% of the orange trees were destroyed by the *Tristeza virus* within 12 years.

The similarity between viruses and transposons was mentioned in the previous section. Both are elements that can insert themselves into the genome of the host and can also leave it again, and both contain the enzymes necessary for this process. The main difference between a transposon and a virus, however, is that the latter at certain stages of its life cycle is enclosed by a protein coat, a **capsid**, which enables the virus to survive outside the host.

The virus genome consists of **RNA** or **DNA** surrounded by a protein coat. In the majority of plant viruses, the genome consists of a single-stranded RNA, called the plus RNA strand. In some viruses (e.g., *Brome mosaic virus*, which infects certain cereals), the plus RNA strand shows the characteristics of an mRNA and is translated by the host. In other viruses [e.g., the ***Tobacco mosaic virus*** **(TMV)**], the plus RNA strand is first transcribed to a complementary minus RNA strand and the latter then serves as a template

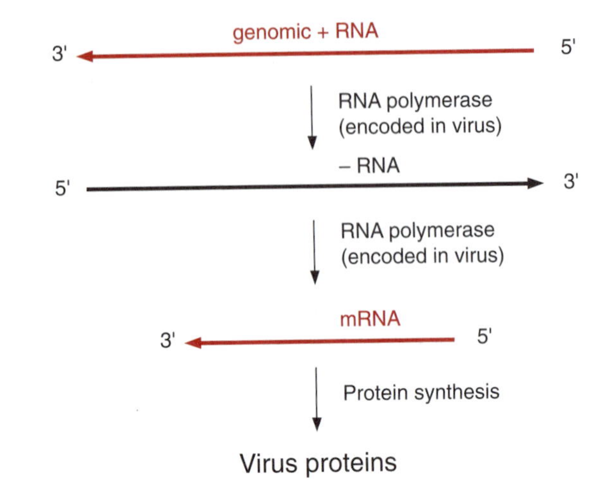

Figure 20.19 The single-stranded genomic RNA of many viruses is first transcribed to a minus strand-RNA, and the latter then to an mRNA for the synthesis of proteins.

for the formation of mRNAs (Fig. 20.19). The translation products of these mRNAs encompass **replicases**, which catalyze the replication of the plus and minus RNAs, **movement proteins** that enable the spreading of the viruses from cell to cell (section 1.1), and **coat proteins** for packing the viruses. Normally, a virus reaches a cell through wounds, which, for instance, have been caused by insects, such as aphids feeding on the plant (section 13.2). Once viruses have entered the cell, their movement proteins widen the plasmodesmata between the single cells enough to let them pass through and spread over the entire symplast.

In the retroviruses, the genome consists also of a single-stranded RNA, but in this case, when the cell has been infected, the RNA is transcribed by a **reverse transcriptase** into DNA, which is in part integrated in the nuclear genome. So far, infections by retroviruses are known only in animals.

The ***Cauliflower mosaic virus*** **(CaMV)**, which causes pathogenic changes in leaves of cauliflower and related plants, is somewhat similar to a retrovirus. The genome of the CaMV consists of a double-stranded DNA of about 8 kbp, with gaps in it (Fig. 20.20). When a plant cell is infected, the virus loses its protein coat and the gapped DNA strands are filled by repair enzymes of the host. The virus genome acquires a double helical structure and forms in the nucleus a chromatin-like aggregate with the histones. This permits the viral genome to stay in the nucleus as a **mini-chromosome**. The viral genome contains promoter sequences that are similar to those of nuclear genes. They possess a TATA box and a CAAT box as well as enhancer elements. The virus promoter is recognized by the RNA polymerase II of the host cell and is transcribed at a high rate. The transcript is subsequently processed into individual mRNAs, which encode the synthesis of six virus proteins, including the coat protein and the reverse transcrip-

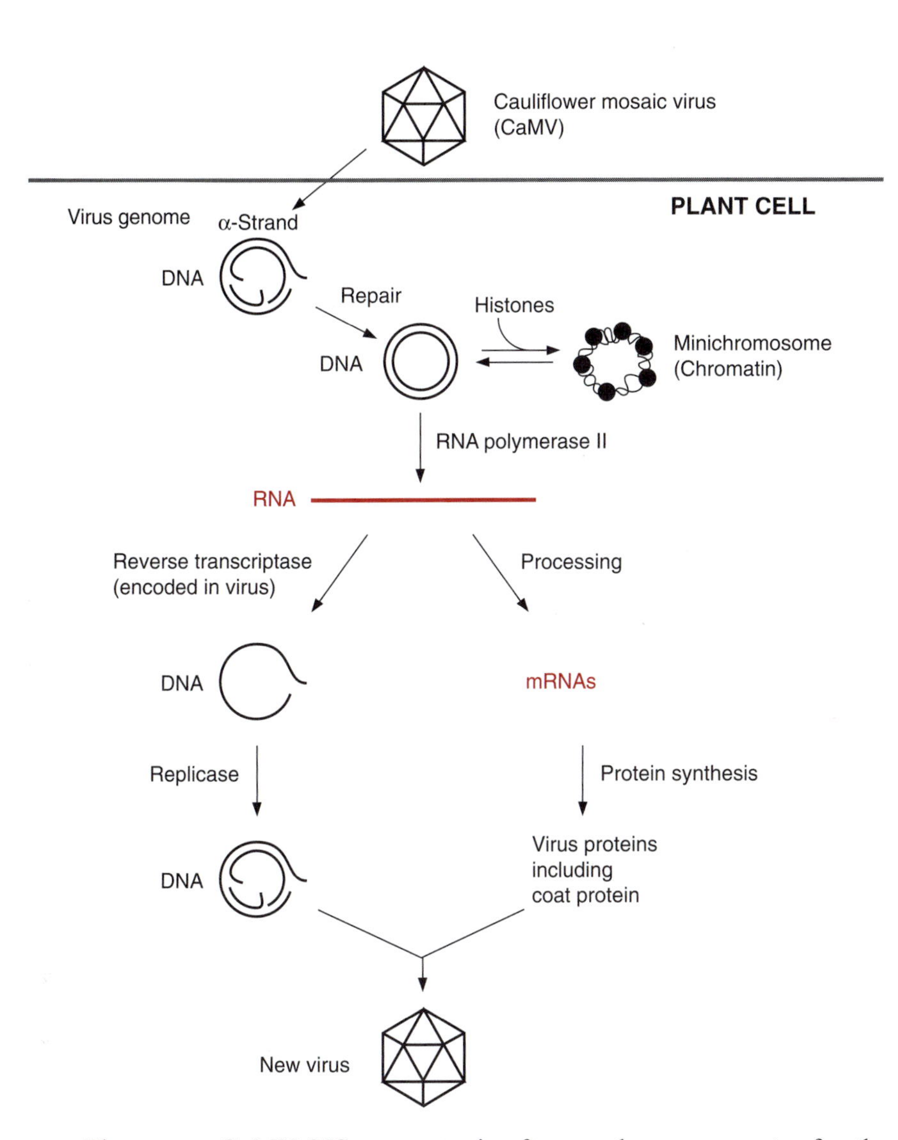

Figure 20.20 Infection of a plant cell by the *Cauliflower mosaic virus* (CaMV).

tase. The strong CaMV 35S promoter is often used as a promoter for the expression of foreign genes in transgenic plants (Chapter 22).

The transcript formed by RNA polymerase II is also transcribed in reverse by the virus-encoded reverse transcriptase into DNA and, after synthesis of the complementary strand, packed as double-stranded genome into a protein coat. This completed virus is now ready to infect other cells.

Retrotransposons are degenerated retroviruses

Besides the transposons described in the previous section, there is another class of mobile elements, which are derived from the **retroviruses**. They do

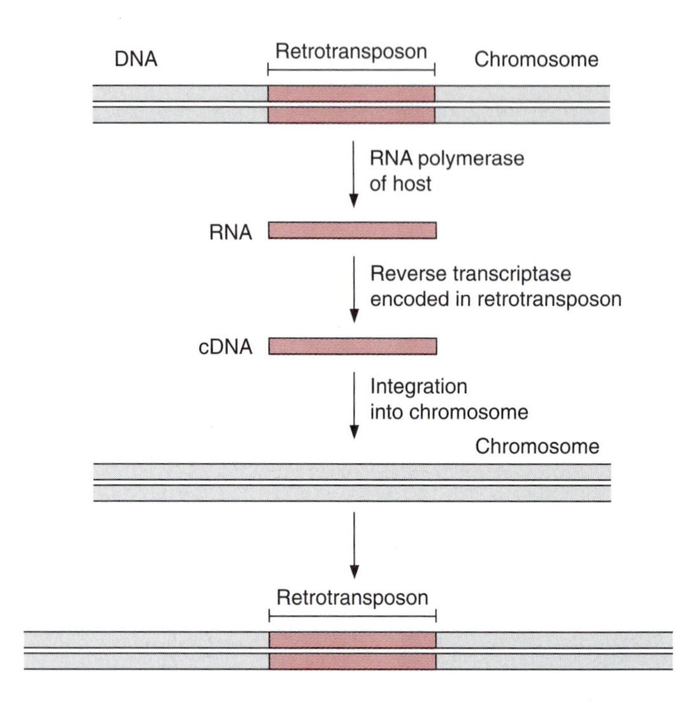

Figure 20.21
Retrotransposons consist of a DNA sequence that is integrated into a chromosome. Flanking sequences contain the recognition signal for transcription by an RNA polymerase of the host cell. The RNA is transcribed into cDNA by a reverse transcriptase encoded in the retrotransposon and integrated into another section of the genome.

not jump out of a gene like the transposons but just multiply. **Retrotransposons** contain at both ends sequences, which carry the signals for the transcription of the retrotransposon DNA by the RNA polymerase of the host cell. The retrotransposon RNA encodes some proteins, but no coat protein. It encodes a reverse transcriptase, which is homologous to the retrovirus enzyme and transcribes the retrotransposon RNA into DNA (Fig. 20.21). This DNA is then inserted into another site of the genome. It is assumed that these retrotransposons are retroviruses that have lost the ability to form a protein coat. Several different retrotransposons containing all the constituents for their multiplication have been found in *Arabidopsis*, but so far an insertion of a retrotransposon into a gene of *Arabidopsis* has never been monitored. About 0.1% of the genome of *Arabidopsis* consists of these retrotransposons, suggesting that these actually multiply, albeit at a slow rate.

20.6 Plastids possess a circular genome

Many arguments support the hypothesis that plastids have evolved from prokaryotic endosymbionts (see section 1.3). The circular genome of the plastids is similar to the genome of the prokaryotic cyanobacteria, although much smaller. The DNA of the plastid genome is named **ctDNA** (chloroplast) or **ptDNA** (plastid). In the majority of the plants investigated so far, the **circular plastid genome** has the size of 120 to 160 kbp. Depending on the plant, this is only 0.001% to 0.1% of the size of the nuclear genome (Table 20.1). But the cell contains many copies of the plastid genome for two reasons. First, each plastid contains many genome copies. In young leaves, the number of ctDNA molecules per chloroplast is about 100, whereas in older leaves, it is between 15 and 20. Second, a cell contains a large number of plastids, a mesophyll cell for instance 20 to 50. Thus, despite the small size of the genome, the plastid DNA can amount to 5% to 10% of the total cell DNA.

The first complete analysis of the base sequence of a plastid genome was carried out in 1986 by the group of Katzuo Shinozaki in Nagoya with chloroplasts from tobacco and by Kanji Ohyama in Kyoto with chloroplasts from the liverwort *Marchantia polymorpha*. Although the two investigated plants are very distantly related, their plastid genomes are rather similar in gene composition and arrangements. Obviously, the plastid genome has changed little during recent evolution. Analysis of the DNA sequence of plastid genes from many plants supports this notion.

Figure 20.22A shows a complete gene map of the chloroplast genome of tobacco and Figures 20.22B and C show schematic representations of the plastid genomes of other plants. The plastid genome of most plants contain so-called **inverted repeats** (IR), which divide the remaining genome into large or small single copy regions. The repeat IR_A and IR_B each contain the genes for the four ribosomal RNAs as well as the genes for some transfer RNAs, and can vary in size from 20 to 50 kbp. These inverted repeats are not found in the plastid genomes of pea, broad bean, and other legumes (Fig. 20.22C), where the inverted repeats probably have been lost during the course of evolution. On the remainder of the genome (single copy region), genes are present usually only in a single copy.

Analysis of the ctDNA sequence of tobacco revealed that the genome contains 122 genes (146 if the genes of each of the two inverted repeats are counted) (Table 20.4). The gene for the large subunit of ribulose bisphosphate carboxylase/oxygenase (RubisCO, section 6.2) is located in the large single copy region, whereas the gene for the small subunit is present in the nuclear genome. The single copy region of the plastid genome also encodes six subunits of F-ATP synthase, whereas the remaining genes of F-ATP synthase are encoded in the nucleus. Also encoded in the plastid genome are part of the subunits of photosystem I and II, of the cytochrome b_6/f complex, and of an NADH dehydrogenase (which also occurs in mitochondria, see sections 3.8 and 5.5), and furthermore, proteins of plastid protein synthesis and gene transcription. Some of these plastid structural

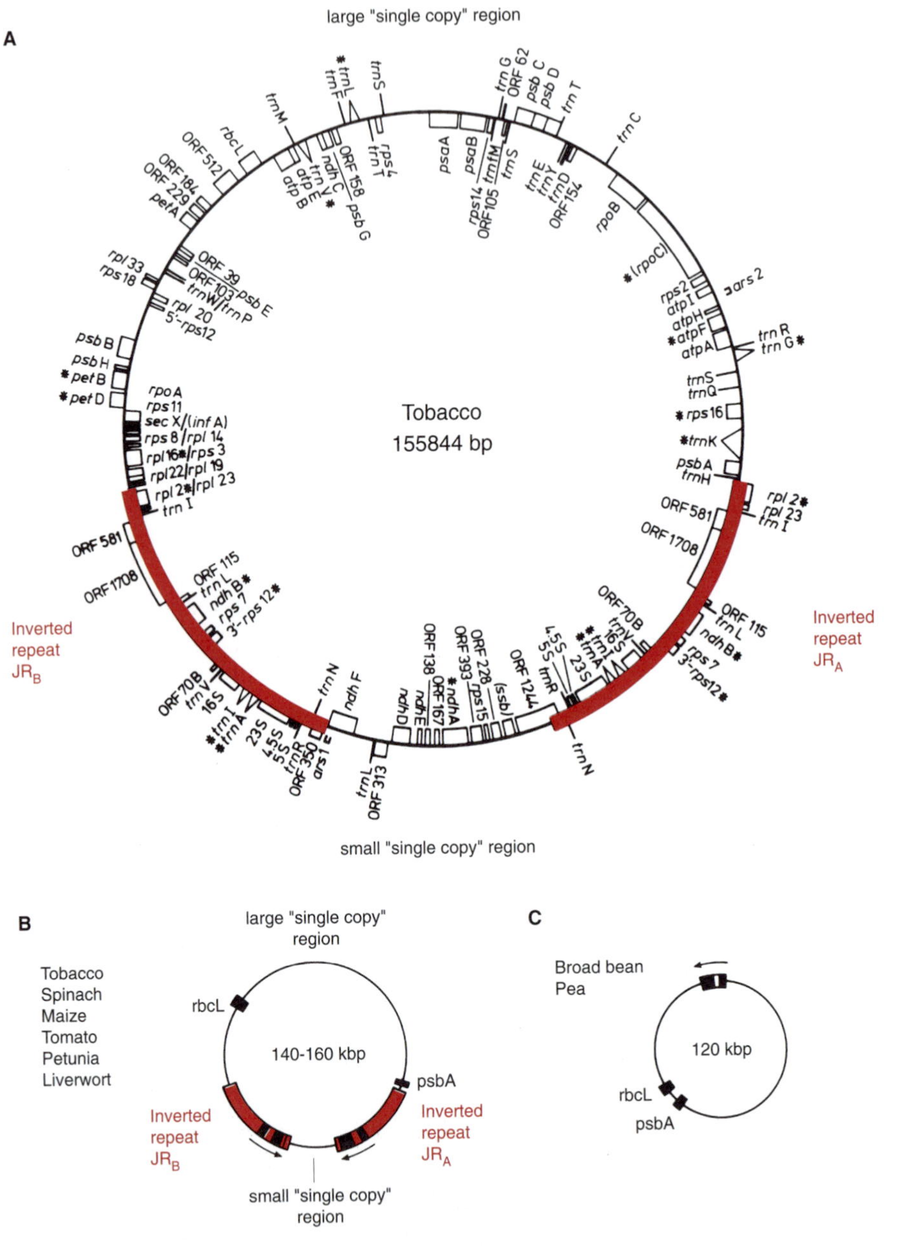

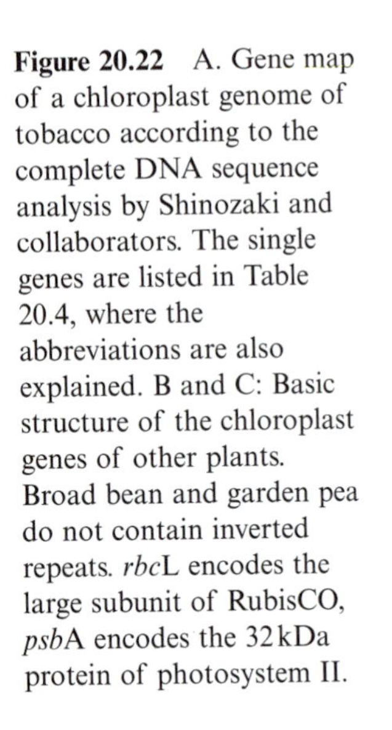
Figure 20.22 A. Gene map of a chloroplast genome of tobacco according to the complete DNA sequence analysis by Shinozaki and collaborators. The single genes are listed in Table 20.4, where the abbreviations are also explained. B and C: Basic structure of the chloroplast genes of other plants. Broad bean and garden pea do not contain inverted repeats. *rbc*L encodes the large subunit of RubisCO, *psb*A encodes the 32 kDa protein of photosystem II.

genes contain introns. In addition, there are putative genes on the genome with so-called open reading frames (ORF), which, like the other genes, are bordered by a start and a stop codon, but where the encoded proteins are not yet known. The plastid genome encodes only a fraction of plastid

Table 20 4: Identified genes in the genome of maize chloroplasts (Shinozaki et al.)

Name of the gene	Gene product (protein or RNA)
Photosynthesis Apparatus	
rbcL	RubisCO: large subunit
atpA, -B, -E	F-ATP-SYNTHASE: subunits α, β, ε,
atpF, -H, -I	F-ATP-SYNTHASE: subunits I, III, IV
psaA, -B, -C	PHOTOSYSTEM I: subunit A1, A2, 9-kDa protein
*psb*A, -B, -C, -D	PHOTOSYSTEM II: subunit D1, 51 kDa, 44 kDa, D2
psb-E, -F, -G, -H, -I	PHOTOSYSTEM II: subunit Cytb_{559}-9 kDa, -4 kDa, G, 10Pi, I-protein
*pet*A, -B, -D	CYT-b_6/*f*-COMPLEX: Cyt-*f*, Cyt-b_6, subunit IV
*ndh*A, -B, -C, -D,	NADH-DEHYDROGENASE (ND) subunits 1, 2, 3, 4
ndh-E, -F	NADH-DEHYDROGENASE (ND) subunits ND4L, 5
Protein synthesis	
rDNA	RIBOSOMAL RNAs (16S, 23S, 4,5S, 5S)
trn	TRANSFER RNAs (30 species)
*rps*2, -3, -4, -7, -8, -11	30S-RIBOSOMAL PROTEINS (CS) 2, 3, 4, 7, 8, 11
rps-12, -14, -15, -16, -18, -19	30S-RIBOSOMAL PROTEINS (CS) 12, 14, 15, 16, 18, 19
*rpl*2, -14, -16, -20, -22	50S-RIBOSOMAL PROTEINS (CL) 2, 14, 16, 20, 22
Rpl-23, -33, -36	50S-RIBOSOMAL PROTEINS CL 23, 33, 36
infA	Initiation factor 1
Gene Transcription	
rpoA, -B, -C	RNA Polymerase-α, -β -β'
ssb	ssDNA Binding protein

proteins, as the majority are encoded in the nucleus. It is assumed that many genes of the original endosymbiont have been transferred during evolution to the nucleus, but there also are indications for gene transfer between the plastids and the mitochondria (section 20.7).

All four rRNAs, which are constituents of the plastid ribosome (4.5S-, 5S-, 16S-, and 23S-rRNA) are encoded in the plastid genome. The plastid ribosomes (sedimentation constants 70S) are smaller than the eukaryotic ribosomes (80S) contained in the cytosol, but are similar in size to the ribosomes of bacteria. As in bacteria, these four rRNAs are contained in the plastid genome in one transcription unit (Fig. 20.23). Between the 16S- and 23S-rRNA is situated a large spacer, containing the sequence for one or two tRNAs. In total, about 30 tRNAs are encoded in the plastid genome. It is assumed that additional tRNAs needed in the plastids are encoded in the nucleus.

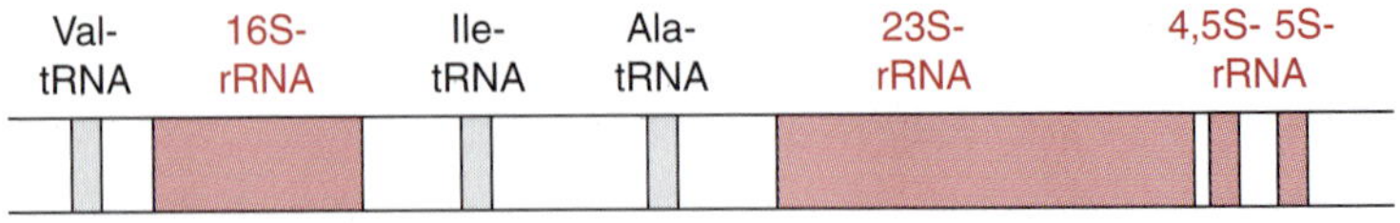

Figure 20.23 In the plastids of maize, all four ribosomal rRNAs and two tRNAs are transcribed as one transcription unit.

The transcription apparatus of the plastids resembles that of bacteria

In the plastids **two types of RNA polymerases** are active, of which only one is encoded in the plastid genome and the other in the nucleus.

1. The RNA polymerase encoded in the **plastids** enables the transcription of plastid genes for subunits of the photosynthesis complex. This RNA polymerase is a multienzyme complex resembling that of **bacteria**. But in contrast to the RNA polymerase of bacteria, the plastid enzyme is insensitive to **rifampicin**, a synthetic derivative of an antibiotic from *Streptomyces*.
2. The plastid RNA polymerase, which is encoded in the **nucleus**, is derived from the duplication of mitochondrial RNA polymerase. This "imported" RNA polymerase is homologous to RNA polymerases from **bacteriophages**. The nucleus-encoded RNA polymerase transcribes the so-called **housekeeping genes** in the plastids. These are the genes that have general functions in metabolism, such as the synthesis of rRNA or tRNA.

As in bacteria, many plastid genes contain a box 10 bp upstream from the transcription start with the consensus sequence TATAAT and at 35 bp upstream a further promoter site with the consensus sequence TTGACA. Some structural genes are polycistronic, which means that several are contained in one transcription unit and are transcribed together in a large primary transcript, as also often happens with bacterial genes. In some cases, the primary transcript is subsequently processed by ribonucleases of which many details are still not known.

20.7 The mitochondrial genome of plants varies largely in its size

In contrast to animals, plants possess a very large mitochondrial (mt) genome. In *Arabidopsis* it is 20 times, and in melon 140 times larger than in

Table 20.5: Size of the mitochondrial DNA (mtDNA) in plants in comparison to the mtDNA of humans

Organism	Size of the mtDNA (kbp)
Arabidopsis thaliana	367
Vicia faba (broad bean)	290
Zea mays (maize)	570
Citrullus lanatus (water melon)	330
Curcubita pepo (pumpkin)	850
Cucumis melo (honey melon)	2400
Marchantia polymorpha (liverwort)	170
Homo sapiens (human)	17

humans (Table 20.5). The plant mitochondrial genome also contains more genetic information: The number of encoding genes in a plant mt-genome is about seven times higher than in humans.

The size of the mt-genome varies largely in higher plants, even within a family. *Citrullus lanatus* (330 kbp), *Curcubita pepo* (850 kbp), and *Cucumis melo* (2400 kbp), listed in Table 20.5, all belong to the family of the *Curcubitaceae* (squash plants). In contrast, the size of the plastid genome is in most plants relatively constant at 120 to 160 kbp.

The mitochondrial genome in plants often consists of one large DNA molecule and several smaller ones. In some mitochondrial genomes, this partitioning may be permanent, but in many cases, the fragmentation of the mt-genome seems to be derived from homologous recombination of repetitive elements (e.g., maize contains six such repeats). Figure 20.24 shows how an interaction of two repeats can lead by **homologous recombination** to a fragmentation of a DNA molecule into two parts. The 570 kbp mt-genome of maize is present in the form of a **master circle** as well as up to four **subcircles** (Fig. 20.25). The recombination of DNA molecules can also form larger units. This may explain the large variability in the size of the mitochondrial genome in plants.

The number of mitochondria in a plant cell can range between 50 and 2,000, with each mitochondrion containing 1 to 100 genomes. Therefore, at each cell division, the mitochondrial genome is inherited in many copies.

In animals and in yeast, the mtDNA is normally circular, like the bacterial DNA. Plant mtDNA also is thought to be circular. This is undisputed for the small mtDNA molecules (subcircles, Fig. 20.25), but it remains unclear whether this circular structure also generally applies to the master mtDNA. There are indications that the master genome can also occur as open strands.

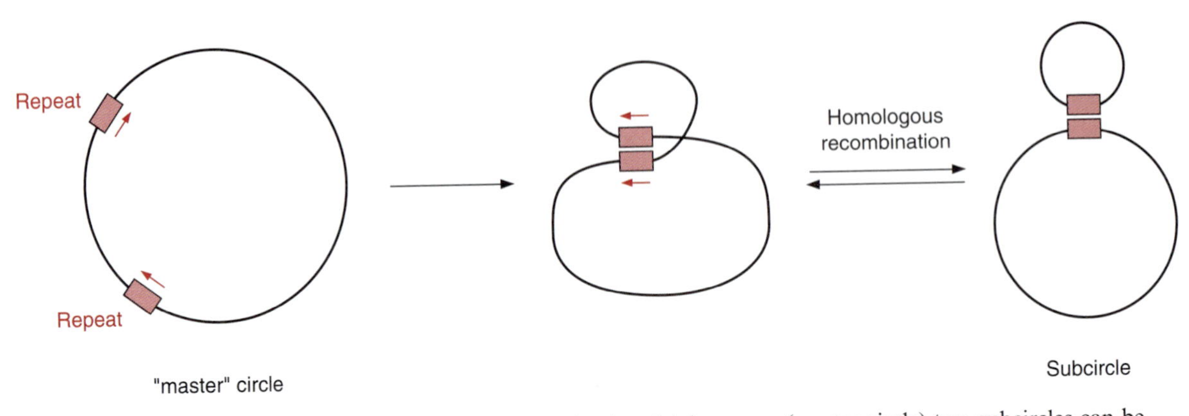

Figure 20.24 From a mitochondrial genome (master circle) two subcircles can be formed by recombination of two repetitive sequences by reversible homologous recombination.

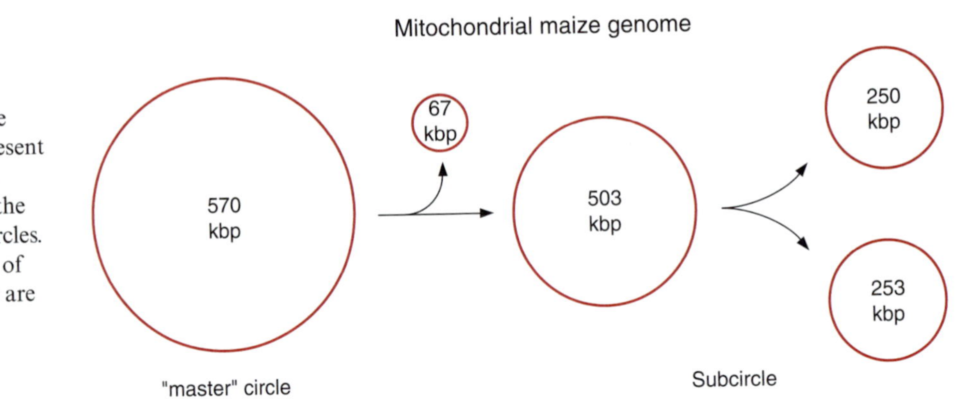

Figure 20.25 Due to homologous recombination, the mitochondrial genome from maize can be present as a continuous large genome as well as in the form of several subcircles. In many mt-genomes of plants only subcircles are observed.

Figure 20.26 shows the complete gene map of the mt-genome of *Arabidopsis thaliana*, and Table 20.6 shows the average content of genes in the mt-genome of a higher plant. A comparison with the number of genes in the plastid genome (Table 20.4) shows that the mt-genome of the plant, although mostly much larger than the plastid genome, contains much less genetic information. The relatively small information content of the mt-genome in relation to its size is due to a high content of repetitive sequences, which probably are derived from gene duplication. The mt-genome contains much DNA with no recognizable function. Some researchers regard this as **junk DNA**, which has accumulated in the mt-genome during evolution. Part of this junk DNA has its origin in plastid DNA and another part in nuclear DNA. The mitochondrial genome, like the nuclear genome, apparently

Table 20.6: Genes identified in the genome of plant mitochondria

Translation apparatus:	5S-, 18S-, 26S-rRNA 10 ribosomal proteins 16 transfer-RNA
NADH-dehydrogenase	9 subunits
Succinate dehydrogenase	1–3 subunits
Cytochrome-b/c_1 complex	1 subunits
Cytochrome-a/a_3 complex	3 subunits
F-ATP synthase	4 subunits
Cytochrome-*c*-biogenesis	>3 genes
Conserved open reading frame of Unknown coding	>10 genes

After Schuster and Brennicke.

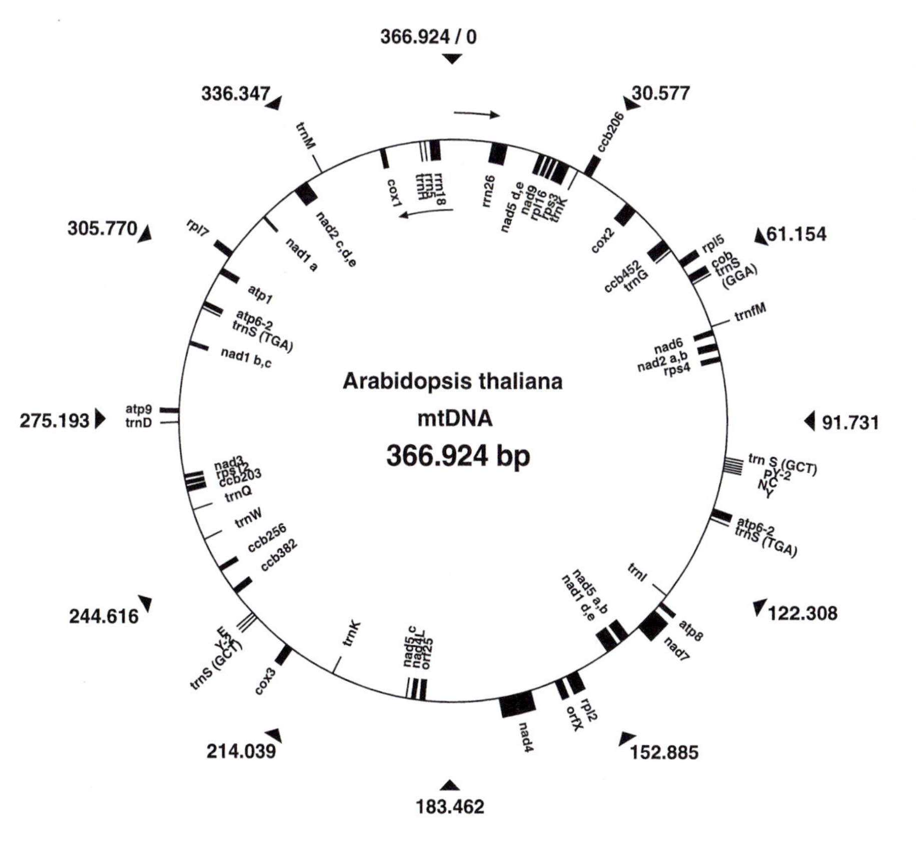

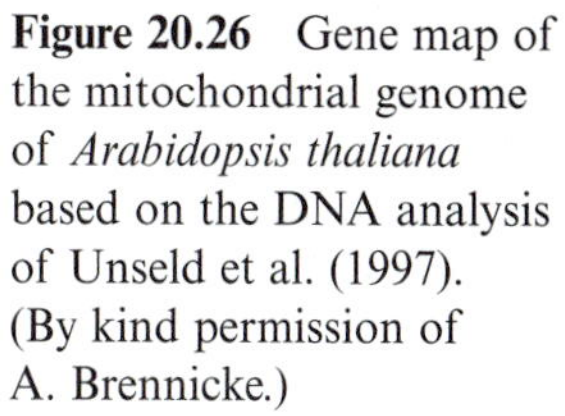

Figure 20.26 Gene map of the mitochondrial genome of *Arabidopsis thaliana* based on the DNA analysis of Unseld et al. (1997). (By kind permission of A. Brennicke.)

tolerates a large portion of senseless sequences and passes these on to following generations. It is interesting that a large part of mtDNA is transcribed. About 30% of the mt-genome of *Brassica rapa* (218 kbp) is transcribed, as is that of the six times larger genome of *Cucumis melo* (2,400 kbp). Why so many transcripts are formed is obscure, considering that in the two abovementioned mt-genomes the total number of the encoded proteins, tRNAs, and rRNAs amounts only to about 60.

The mitochondrial genome encodes parts of the translation machinery, including 3 rRNAs, about 16 tRNAs, and about 10 ribosomal proteins. These are involved in the formation of various hydrophobic membrane proteins, which also are encoded in the mt-genome, e.g. some subunits of the respiratory chain (section 5.5), of F-ATP synthase (section 4.3), and also at least three enzymes of cytochrome-*c* synthesis (section 10.5). About 95% of the mitochondrial proteins, including most subunits for the respiratory chain and F-ATP synthase, as well as several tRNAs, are encoded in the nucleus. Considering that mitochondria have derived from endosymbionts, it must be assumed that the greatest part of the genetic information of the endosymbiont genome has been transferred to the nucleus. Such gene transfers occur quite often in plants; the gene content in the mitochondria can vary between the different species. But a gene transfer apparently has also occurred from the plastids to the mitochondria. From their base sequence, several tRNA genes of the mt-genome seem to originate in the plastid genome.

The promoters of plant mitochondrial genes are heterogeneous. The sequences signaling the start and end of transcription are quite variable, even for the genes of the same mitochondrion. Most likely plant mitochondria contain several mtRNA polymerases, as in plastids. The corresponding transcription factors have not yet been unequivocally characterized. Most of the mitochondrial genes are transcribed monocistronically.

Mitochondrial DNA contains incorrect information that is corrected after transcription

A comparison of the amino acid sequences of proteins encoded in mitochondria, with the corresponding nucleotide sequences of the encoding genes, revealed strange discrepancies: The amino acid sequences did not correspond to the DNA sequences of the genome according to the rules of the genetic code. At sites of the DNA sequence where, according to the protein sequence, a T was to be expected, a C was found, and sometimes vice versa. More detailed studies showed that the transcription of mtDNA yielded an mRNA that did not contain the correct information for the protein to be synthesized. It was all the more astonishing to discover that this incorrect mRNA subsequently is processed in the mitochondria by several replacements of C by U, but sometimes also of U by C, until the correct mRNA is reconstructed as a template for synthesizing the proper protein. This process is called **RNA editing**.

Subsequent editing of the initially incorrect mRNA to the correct, translatable mRNA is not an exception taking place only in some exotic genes, but is the rule for the mitochondrial genes of higher plants. In some mRNAs produced in the mitochondria, 40% of the C are replaced by U in the editing process. Mitochondrial tRNAs are also edited in this way. Such RNA editing has been observed in the mitochondria of all higher plants investigated so far. The question arises whether, by differences in the editing, a structural gene can be translated into different proteins. Now and again mitochondrial proteins were found which had been translated from only partially edited mRNA. Since these proteins normally are nonfunctional, they are probably rapidly degraded.

RNA editing was first shown in the mitochondria of trypanosomes, the unicellular pathogen of sleeping sickness. It also occurs in mitochondria of animals and, in a few cases, has also been found in plastids.

Only since 1989 has RNA editing been known to exist in plant mitochondria, leaving many questions still open. What mechanism is used for RNA editing? How are the bases exchanged? Also, the question of the physiological meaning of RNA editing is still unanswered. Is a higher mutation rate of the maternally inherited mt-genome corrected by the editing? From where does the information for the proper base sequence of the mRNA come? Is this information provided by the nucleus or is it also contained in the mitochondrial genome? It is feasible that the very large mitochondrial genome contains, in addition to the structural genes, single fragments, the transcripts of which are utilized for the correction of the mRNA, as has actually been observed in the mitochondria from trypanosomes.

Male sterility of plants caused by the mitochondria is an important tool in hybrid breeding

When two selected inbreeding lines are crossed, the resulting **F_1 hybrids** are normally larger, are more robust, and produce higher yields of harvest products than the parent plants. This effect, called **hybrid vigor**, was observed before the rediscovery of the Mendelian rules of inheritance and was first utilized in 1906 for breeding hybrid maize by George Schull in the laboratory at Cold Spring Harbor in the United States. The success of these studies brought about a revolution in agriculture. Based on the results of

Schull's research, private seed companies bred maize F_1 hybrids that provided much higher yields than the customary varieties. In 1965, 95% of the maize grown in the Corn Belt of the United States was hybrid. The use of F_1 hybrids was, to a large extent, responsible for the increase in maize yields per acre by a factor of 3.5 between 1940 and 1980 in the United States.

F_1 hybrids cannot be further propagated, since according to the Mendelian laws the offspring of the F_2 generation is heterogeneous. Most of the second-generation (F_2) plants have some homozygosity, resulting in yield depression. Each year, therefore, farmers have to purchase new hybrid seed from the seed companies, whereby the seed companies gained a large economical importance. Hybrid breeding has been put to use for the production of many varieties of cultivated species. Taking maize as an example, the following describes the principles and problems of hybrid breeding.

For the production of F_1 hybrid seed (Fig. 20.27), the pollen of a paternal line A is transferred to the pistil of a maternal line B, and only the cobs of B are harvested for seed. These crossings are carried out mostly in the field. Plants of lines A and B are planted in separate, but neighboring rows, so that the pollen is transferred from A to B by the wind. To prevent the pistils of line B from being fertilized by the pollen of the same line, the plants of line B are emasculated. Since in maize the pollen producing male flowers are separated from the female flowers in a panicle, it is possible to remove only the male flowers by cutting them off. To produce hybrid seed in this way on a commercial scale, however, requires a great expenditure in manual labor. This method is totally unworkable on any practical scale in plants such as rye, where the male and the female parts of the flower are combined.

It was a great step forward in the production of hybrids when maize mutants that produced sterile pollen were found. In these male-sterile plants, the fertility of the pistil was not affected as long as it was fertilized by pollen of other lines. This male sterility is inherited maternally by the genome of the mitochondria. Several male-sterile mutants of maize and other plants are now known which are the result of the mutation of mitochondrial genes.

The relationship between the mutation of a mitochondrial gene and the male sterility of a plant has been thoroughly investigated in the maize mutant T (Texas). The mitochondria of this mutant contain a gene designated as **T-*urf*13**, which encodes a **13 kDa protein**. This gene is probably the product of a complex recombination. The 13 kDa protein has no apparent effect on the metabolism of the mitochondria under conditions of vegetative growth, and the mutants are of normal phenotypes. Only the formation of pollen is disturbed by this protein, for reasons not known so far. A possible explanation might be that the *tapetum cells* of the pollen sac, which are involved in pollen production, have an unusual abundance of mitochondria and apparently depend very much on mitochondrial metabolism. Therefore,

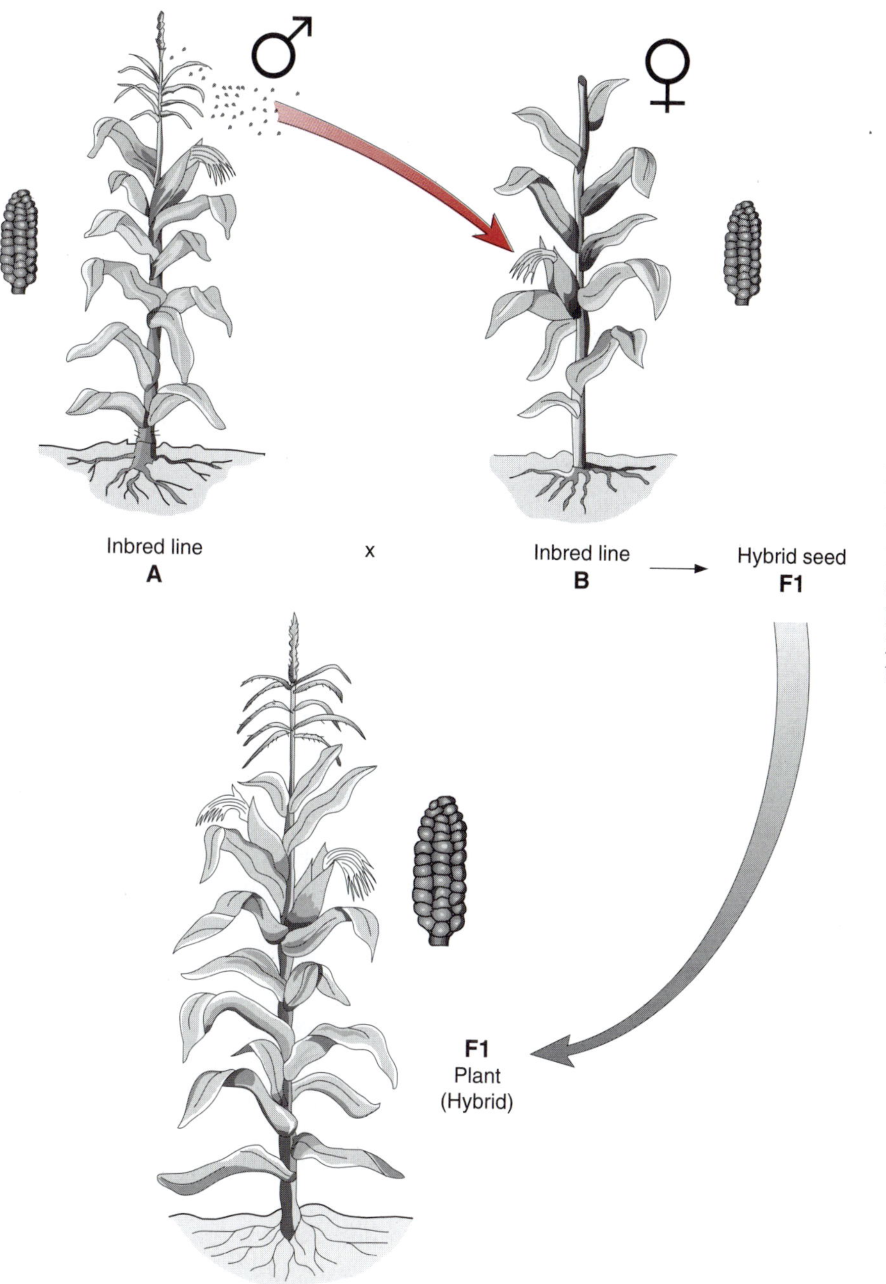

Figure 20.27 Principle of hybrid breeding shown with maize as an example. The lines A and B are inbred lines, which produce a relatively low yield of harvest products. When these lines are crossed, the resulting F_1 progeny is much more robust and produces high yields. The seeds are obtained from line B, which is pollinated by line A. To prevent B being fertilized by its own pollen, the male flowers of B are removed by cutting them off. (From Patricia Nevers, Pflanzenzüchtung aus der Nähe gesehen, Max Planck Institut für Züchtungsforschung, Köln, by permission.)

in these cells a mitochondrial defect, which normally does not affect metabolism, might interfere with pollen production.

The successful use of these male-sterile mutants for breeding is based on a second discovery: Maize lines that contain so-called **restorer genes** in their

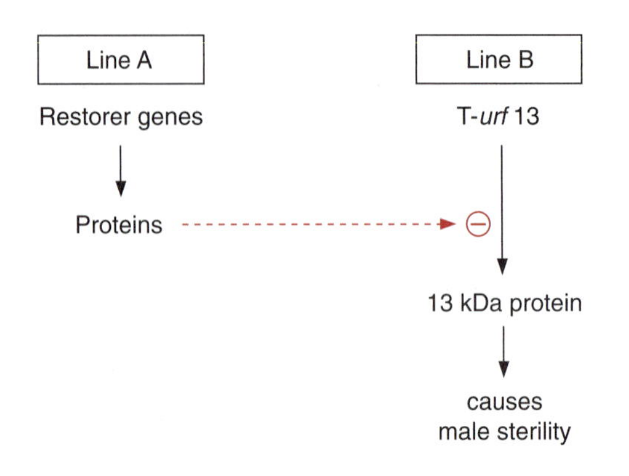

Figure 20.28 A maize mutant T (here designated as line B) contains in the mitochondrial genome a gene named T-*urf*13. The product of this gene, a 13 kDa protein, prevents the formation of sterile pollen and thus causes male sterility in this mutant. Another maize line (A) contains in its nucleus one or several so-called restorer genes, encoding proteins, which suppress the expression of the T-*urf*13 gene in the mitochondria. Thus, after a crossing of A with B, the pollen is fertile again and the male sterility is abolished.

nucleus were found. These encode proteins that repress the expression of the T-*urf*13 gene in the mitochondria (Fig. 20.28). The crossing of a paternal plant A containing these restorer genes, with a male-sterile maternal plant B results in an F_1 generation, in which the fertility of the pollen is restored and corn cobs are produced normally.

The crossing of male-sterile T maize lines with lines containing restorer genes enables F_1 hybrids to be produced very efficiently. Unfortunately, however, the 13 kDa protein encoded by the T-*urf*13 makes a maize plant more sensitive to the toxin of the fungus *Bipolaris maydis T*, the pathogen of the much dreaded fungal disease "southern corn blight," which destroyed a large part of the American maize crop in 1971. The 13 kDa protein reacts with the fungal toxin to form a pore in the inner mitochondrial membrane and thus eliminates mitochondrial ATP production. In order to continue hybrid seed production, it was then necessary to return to the manual removal of the male flowers. Male-sterile lines are now known not only in maize, but also in many other plants, in which the sterility is caused by proteins encoded in the mt-genome, and also other lines that suppress the formation of the inhibiting protein by nuclear encoded proteins. Nowadays these lines are used for the production of fertile F_1 hybrid seed. Presumably the reaction of nuclearly encoded proteins on the expression of mitochondrial genes such as T-*urf*13 is a normal reaction of mitochondrial metabolism in plants and therefore, after the production of corresponding mutants, can be used in many ways to generate male sterility.

Today intensive research is being carried out all over the world to find ways of generating male sterility in plants by genetic engineering. Success can be noted. Using a specific promoter, it is possible to express a ribonuclease from the bacterium *Bacillus amyloliquefaciens* exclusively in the tapetum cells of the pollen sac in tobacco and rapeseed. This ribonuclease degrades the mRNA formed in tapetum cells, thus preventing the development of pollen. Other parts of the plants are not affected and the plants grow normally. For the generation of a restorer line, the gene of a ribonuclease inhibitor (from the same bacterium) was transferred to the tapetum cells. The great advantage of such a synthetic system is its potential for general application. In this way male sterility can be introduced into species in which this cannot be achieved by manual removal of the stamen, and where male sterility due to mutants is not available. Genetically engineered rapeseed hybrids are nowadays grown to a large extent in the United States and Canada. It is to be expected that the generation of male-sterile plants by genetic engineering and the resultant use of hybrid seed might lead to increased harvests of many crops.

Further reading

Arabidopsis Genome: A milestone in plant biology. Special Issue. Plant Physiol 124, 1449–1865 (2000).

Arabidopsis functional genomics. Special Issue. Plant Physiol 129, 389–925 (2002).

Bachmann, K. Molecular markers in plant ecology. New Phytologist 126, 403–418 (1994).

Backert, S., Nielsen, B. L., Börner, T. The mystery of the rings: Structure and replication of mitochondrial genomes from higher plants. Trends Plant Sci 2, 477–483 (1997).

Bartel, B., Bartel, D. P. MicroRNAs: At the root of plant development? Plant Physiol 132, 709–717 (2003).

Brennicke, A., Kück, U. (eds.). Plant mitochondria with emphasis on RNA editing and cytoplasmic male sterility. Verlag Chemie, Weinheim (1993).

Chrispeels, M. J., Sadava, D. E. Plants, Genes and Agriculture. Jones and Bartlett, Boston, London (1994).

Fosket, D. E. Plant Growth and Development. A Molecular Approach. Academic Press, San Diego, New York (1994).

Goff, S. A., *et al.* A draft sequence of the rice genome (*Oryza sativa L. ssp. indica*) Science 296, 92–100 (2002).

Marienfeld, J., Unseld, M., Brennicke, A. The mitochondrial genome of *Arabidopsis* is composed of both native and immigrant information. Trends Plant Sci 4, 495–502 (1999).

Mayfield, S. P., Yohn, C. B., Cohen, A., Danon, A. Regulation of chloroplast gene expression. Annu Rev Plant Physiol Plant Mol Biol 46, 147–166 (1995).

Meyerowitz, E. M. Prehistory and history of *Arabidopsis* research. Plant Physiol 125, 15–19 (2001).

Moore, G. Cereal chromosome structure, evolution, and pairing. Annu Rev Plant Physiol Plant Mol Biol 51, 195–222 (2000).
Ohyama, K., Fukuzuwa, H., Kochi, T., *et al.* Chloroplast gene organization deduced from the complete sequence of liverwort *Marchantia polymorpha* chloroplast DNA. Nature 322, 572–574 (1986).
Rochaix, J.-D. Chloroplast reverse genetics: New insights into the function of plastid genes. Trends Plant Sci 2, 419–425 (1997).
Saedler, H., Gierl, A. Transposable Elements. Current Topics in Microbiology and Immunology 204 Springer-Verlag, Heidelberg (1996).
Schnable, P. S., Wise, R. P. The molecular basis of cytoplasmic male sterility and fertility restoration. Trends Plant Sci 3, 175–180 (1998).
Shimamoto, K., Kyozuka, J. Rice as model for comparative genetics in plants. Annu Rev Plant Biol 53, 399–420 (2002).
Shinozyki, K., Ohme, M., Wakasugi, T., *et al.* The complete nucleotide sequence of the chloroplast genome: Its gene organization and expression. EMBO J 9, 2043–2049 (1986).
Stern, D. B., Higgs, D. C., Yang, J. Transcription and translation in chloroplasts. Trends Plant Sci 2, 308–315 (1997).
Sundaresan, V. Horizontal spread of transposon mutagenesis: New uses of old elements. Trends Plant Sci 1, 184–190 (1996).
The *Arabidopsis* genome initiative. Analysis of the genome sequence of the flowering plant *Arabidopsis thaliana*. Nature 408, 796–815 (2000).
The rice full-length cDNA consortium, Kikuchi, S. *et al.* Collection, mapping and annotation of over 28,000 cDNA clones from *japonica* rice. Science 301, 376–379 (2003).
Unseld, M., Marienfeld, J. R., Brandt, P., Brennicke, A. The mitochondrial genome of *Arabidopsis thaliana* contains 57 genes in 366,924 nucleotides. Nat Genet 15, 57–61 (1997).
Westhoff, P. Molecular Plant Development.
Oxford University Press. Oxford, UK (1998).
Zdravko, J. L., Wieczorek Kirk, D. A., Lambernon, M. H. L., Flipowicz, W. Pre-mRNA splicing in higher plants. Trends Plant Sci 5, 160–167 (2000).
Yamada, K., *et al* (69). Empirical analysis of transcriptional activity in the *Arabidopsis* genome. Science 302, 842–846 (2003).
Yu, J., *et al.* A draft sequence of the rice genome (*Oryza sativa L. ssp. indica*) Science 296, 79–92 (2002).

21

Protein biosynthesis occurs at different sites of a cell

During protein biosynthesis, the base sequence of mRNA is translated into an amino acid sequence. The "interpreters" are **transfer ribonucleic acids** (tRNAs), small RNAs of 75 to 85 ribonucleotides, which have a defined structure with three hairpin loops. The middle loop contains the **anticodon**, which is complementary to the mRNA **codon**. For each amino acid, there is at least one and sometimes even several tRNAs. The covalent binding of the amino acid to the corresponding tRNA is catalyzed by its specific **aminoacyl tRNA synthetase** with the consumption of ATP, and the mixed anhydride aminoacyl-AMP is formed as an intermediate (Fig. 21.1).

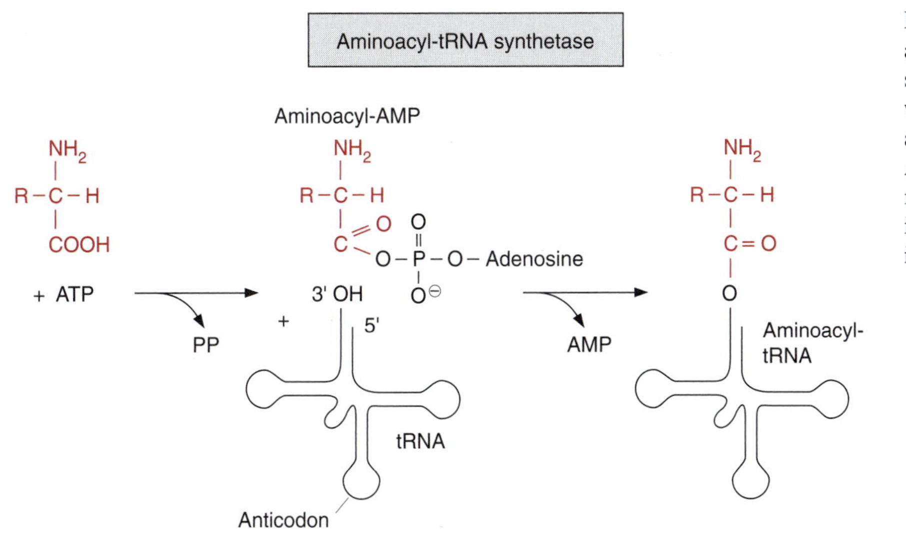

Figure 21.1 By amionoacyl-tRNA synthetase, tRNA is loaded with its corresponding amino acid, during which ATP is consumed. In this reaction, aminoacyl-AMP is formed as an intermediate.

In a plant cell, protein biosynthesis takes place at three different sites. The translation of the nuclearly encoded mRNA proceeds in the cytosol, and that of the mRNAs encoded in the plastidal or mitochondrial genome takes place in the plastid stroma and mitochondrial matrix, respectively.

21.1 Protein synthesis is catalyzed by ribosomes

Ribosomes are large riboprotein complexes that contain three to four different rRNA molecules and a large number of proteins. In the intervals between the end of the translation of one mRNA and the start of the translation of another mRNA, the ribosomes dissociate into two subunits. The ribosomes of the cytosol, plastids, and mitochondria are different in size and composition (Table 21.1). The **cytosolic** ribosomes (termed **eukaryotic ribosomes**), with a sedimentation constant of 80S, dissociate into a small subunit of 40S and a large subunit of 60S. In contrast, the **mitochondrial ribosomes**, with a size of about 78S, varying from species to species, and the **plastidic ribosomes** (70S) are smaller. Due to their relationship to the bacterial ribosomes, the mitochondrial and plastidic ribosomes are classified as **prokaryotic ribosomes.** Ribosomes of the bacterium *Escherichia coli* have a sedimentation constant of 70S.

Table 21.1: Composition of the ribosomes in the cytosol, chloroplast stroma, and mitochondrial matrix in plants

	Complete ribosome	Ribosomal subunits	rRNA-components	Proteins
Cytosol (eukaryotic ribosome in plants)	80S	small. UE 40S	18S-rRNA	ca. 30
		large UE 60S	5S-rRNA 5,8S-rRNA 25S-rRNA	ca. 50
Chloroplast (prokaryotic ribosome)	70S	small. UE 30S	16S-rRNA	ca. 24
		large UE 50S	4,5S-rRNA 5S-rRNA 23S-rRNA	ca. 35
mitochondrion (prokaryotic ribosome)	78S	small UE ≈30S	18S-rRNA	ca. 33
		large UE ≈50S	5S-rRNA 26S-rRNA	ca. 35

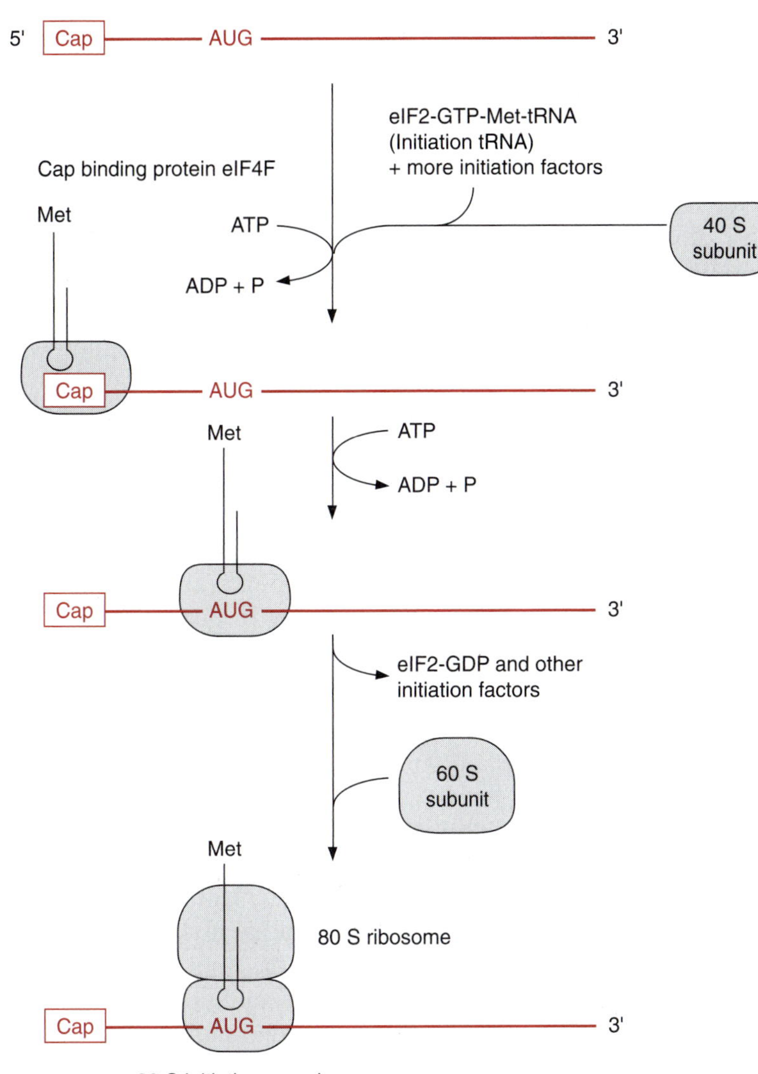

Figure 21.2 Formation of the initiation complex of eukaryotic (80S) ribosomes.

At the beginning of translation, mRNA forms an **initiation complex** with a ribosome. A number of **initiation factors** participate in this process. Let us first look at the eukaryotic translation occurring in the cytosol. First, the eukaryotic initiation factor2 (eIF2), together with GTP and a transfer RNA loaded with methionine, form an **initiation-transfer-RNA.** With the participation of other initiation factors, it is then bound to the small 40S subunit. The initiation factor eIF4F, which consists of several protein components (also known as Cap-binding protein), and under consumption of ATP, mediates the binding of the small 40S subunit, loaded with the initiation transfer-RNA, to the **Cap sequence** present at the 5′ end of the mRNA (see Fig. 20.9). Driven by the hydrolysis of ATP, the 40S subunit migrates downstream (5′ → 3′) until it finds an AUG start codon. Usually, but not always, the first AUG triplet on the mRNA is the one recognized. In some mRNAs, the translation starts at a later AUG triplet. The neighboring sequences on the mRNA decide which AUG triplet is recognized as the start codon. The large 60S subunit is then bound to the 40S subunit, accompanied by the dissociation of several initiation factors and of GDP. The formation of the initiation complex is now completed and the resulting ribosome is able to translate.

The mitochondrial and plastidal mRNA have no cap sequence. Plastidal mRNA has a special ribosome binding site for the initial binding to the small subunit of the ribosome, consisting of a purine-rich sequence of about 10 bases. This sequence, called the **Shine-Dalgarno sequence**, binds to the 16S-rRNA of the small ribosome subunit. A Shine-Dalgarno sequence is also found in bacterial mRNA, but it is not known whether it also plays this role in the mitochondria. In mitochondria, plastids, and bacteria, the initiation tRNA is loaded with N-formyl methionine (instead of methionine as in the cytosol). After peptide formation the formyl residue is cleaved from the methionine.

A peptide chain is synthesized

A ribosome, completed by the initiation process, contains two sites where the tRNAs can bind to the mRNA. The **peptidyl site (P)** allows the binding of the initiation tRNA to the AUG start codon (Fig. 21.3). The **aminoacyl (A) site** covers the second codon of the gene and at first is unoccupied. On the other side of site P is the exit **(E) site** where the empty tRNA is released. The **elongation** begins after the corresponding aminoacyl-tRNA occupies the A site by forming base pairs with the second codon. Two **elongation factors** participate in this. The eukaryotic elongation factor (eEF1α) binds GTP and guides the corresponding aminoacyl-tRNA to the A site, during which the GTP is hydrolyzed to GDP and P. The cleavage of the energy-rich anhydride bond in GTP enables the aminoacyl-tRNA to bind to the codon at the A site. Afterward the GDP, still bound to eEF1α, is exchanged for GTP, as mediated by the elongation factor eEF1βγ. The eEF1α-GTP is now ready for the next cycle.

Subsequently a peptide linkage is formed between the carboxyl group of methionine and the amino group of the amino acid of the tRNA bound to the A site. The **peptidyl transferase**, which, as part of the ribosome, catalyzes this reaction, is a complex enzyme consisting of several ribosomal proteins. The 25S-rRNA has a decisive function in the catalysis. The enzyme facili-

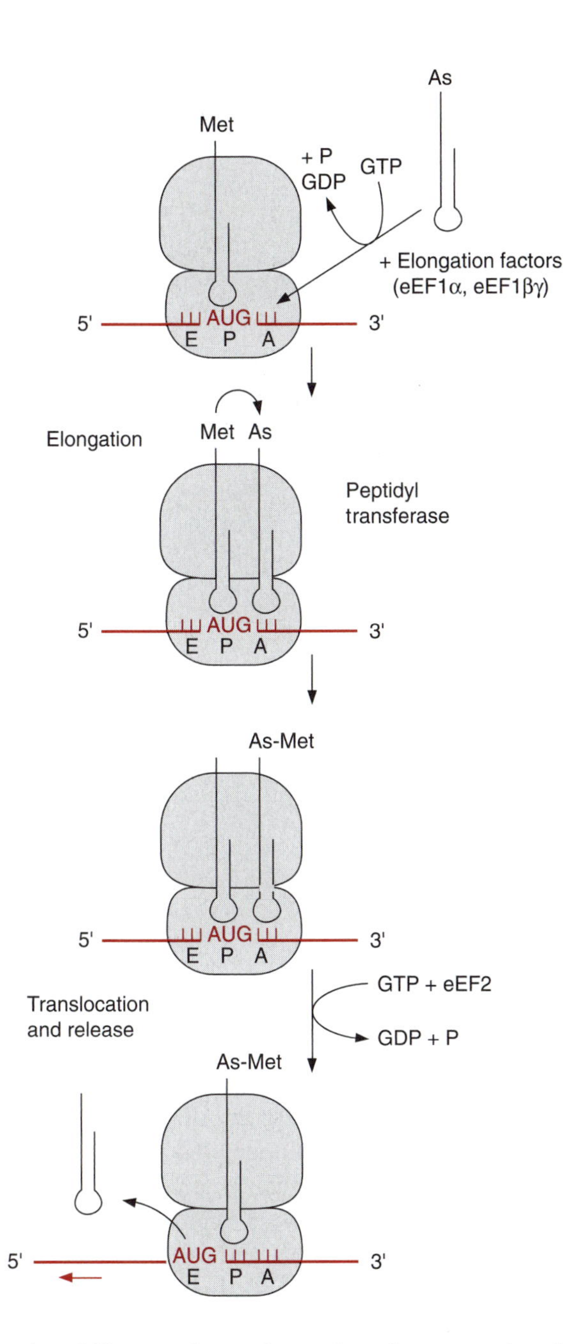

Figure 21.3 Elongation cycle of protein biosynthesis. After binding of the corresponding aminoacyl-tRNA to the A site, a peptide bond is formed by peptidyl transferase. By subsequent translocation, the remaining empty tRNA is moved to the E site and released, while the tRNA loaded with the peptide chain now occupies the P site. The binding of a new tRNA to site A starts another elongation cycle. As = amino acid.

tates the *N*-nucleophilic attack on the carboxyl group, whereby the peptide bond is formed. This results in the formation of a dipeptide bound to the tRNA at the A site (Fig. 21.4A).

Accompanied by the hydrolysis of one molecule of GTP into GDP and P, the elongation factor eEF2 facilitates the **translocation** of the ribosome

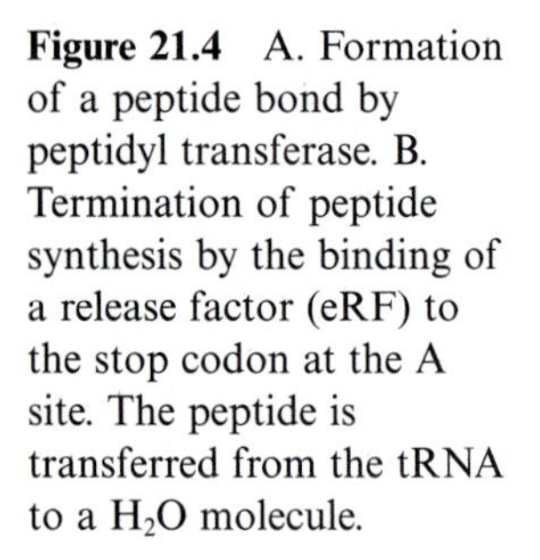

Figure 21.4 A. Formation of a peptide bond by peptidyl transferase. B. Termination of peptide synthesis by the binding of a release factor (eRF) to the stop codon at the A site. The peptide is transferred from the tRNA to a H_2O molecule.

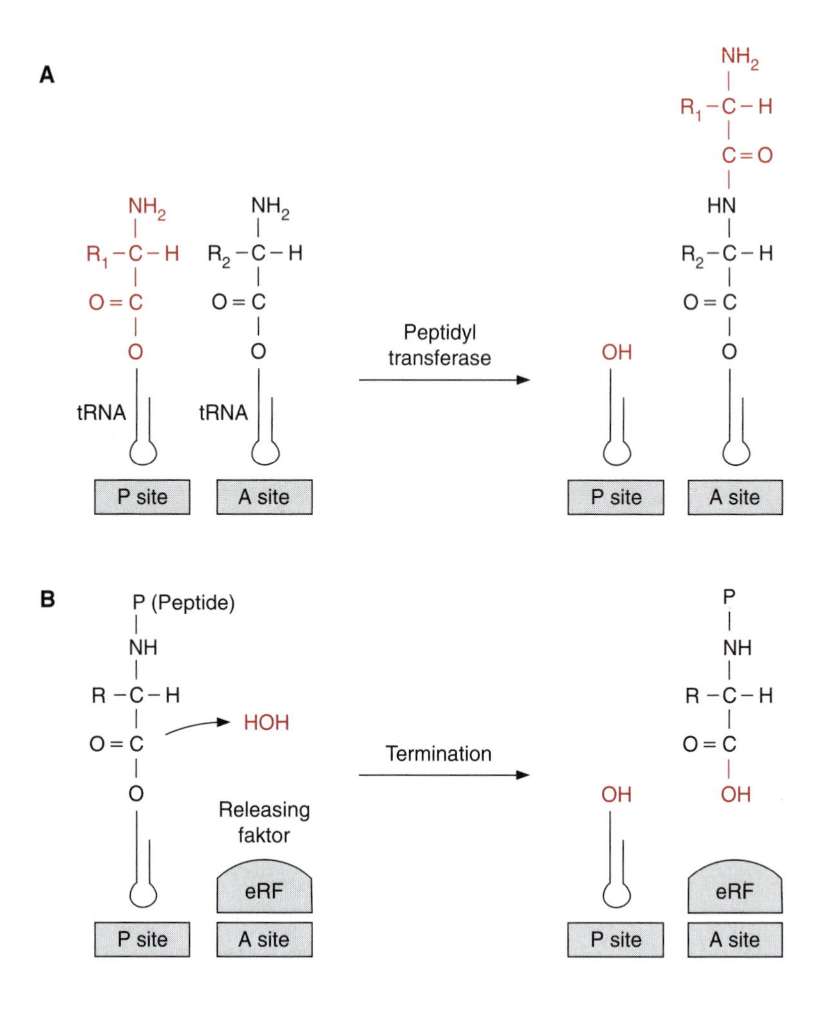

along the mRNA to three bases downstream (Fig. 21.3). In this way the free tRNA arrives at site E, is released, and the tRNA loaded with the peptide now occupies the P site. The third aminoacyl-tRNA binds to the now vacant site A and a further elongation cycle can begin.

After several elongation cycles, the 5′ end of the mRNA is no longer bound to the ribosome and can start a new initiation complex. An mRNA that is translated simultaneously by several ribosomes is called a **polysome** (Fig. 21.5).

Translation is terminated when the A site finally binds to a **stop codon** (UGA, UAG, or UAA) (Fig. 21.4B). These stop codons bind the **release factor** (eRF) accompanied by hydrolysis of GTP to form GDP and P. Binding of eRF to the stop codon alters the specificity of the peptidyl transferase: A water molecule instead of an amino acid is now the

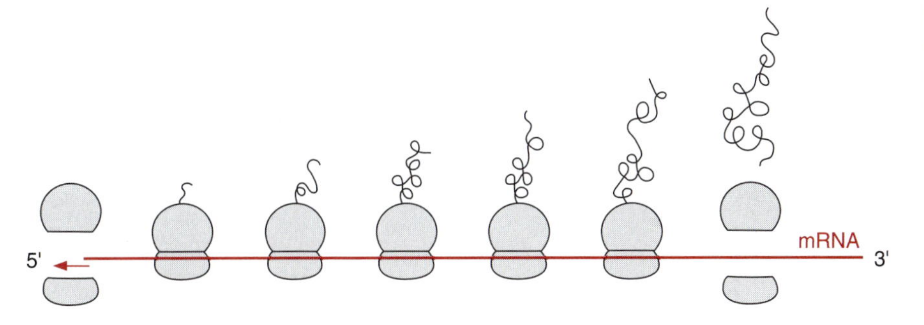

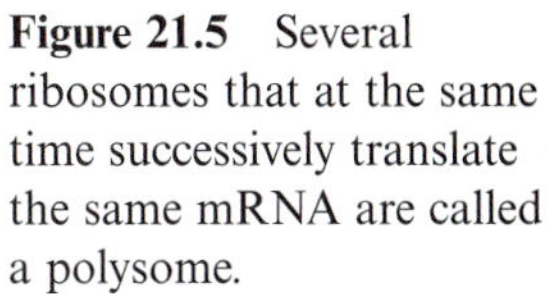
Figure 21.5 Several ribosomes that at the same time successively translate the same mRNA are called a polysome.

acceptor for the peptide chain. In this way the formed protein is released from the tRNA.

Specific inhibitors of the translation can be used to decide whether a protein is encoded in the nucleus or the genome of plastids or mitochondria

Elongation, translocation, and termination occur in prokaryotic ribosomes in an analogous way as described previously with only minor alterations, but in this case the termination requires no GTP. However, eukaryotic and prokaryotic translation can react differently to certain antibiotics (Fig. 21.6, Table 21.2). **Puromycin**, an analogue of tRNA, is a general inhibitor of protein synthesis, whereas **cycloheximide** inhibits only protein synthesis by **eukaryotic** ribosomes. **Chloramphenicol**, **tetracycline**, and **streptomycin** primarily inhibit protein synthesis by **prokaryotic** ribosomes. These inhibitors can be used to determine whether a certain protein is encoded in the nucleus or in the genome of plastids or mitochondria. A relatively simple method for monitoring protein synthesis is to measure the incorporation of a radioactively labeled amino acid (e.g., ^{35}S-labeled methionine) into proteins. If the incorporation of this amino acid into a particular protein is inhibited by cycloheximide, this indicates that the protein is encoded in the nucleus. Likewise, inhibition by chloramphenicol shows that the corresponding protein is encoded in the genome of plastids or mitochondria.

The translation is regulated

The synthesis of many proteins is specifically regulated at the level of translation. This regulation may involve protein kinases by which proteins participating in translation are phosphorylated. Since the rate-limiting step of translation is mostly the initiation, this step is especially suited for a regulation of translation. In animals, the initiation factor eIF2 is inactivated by

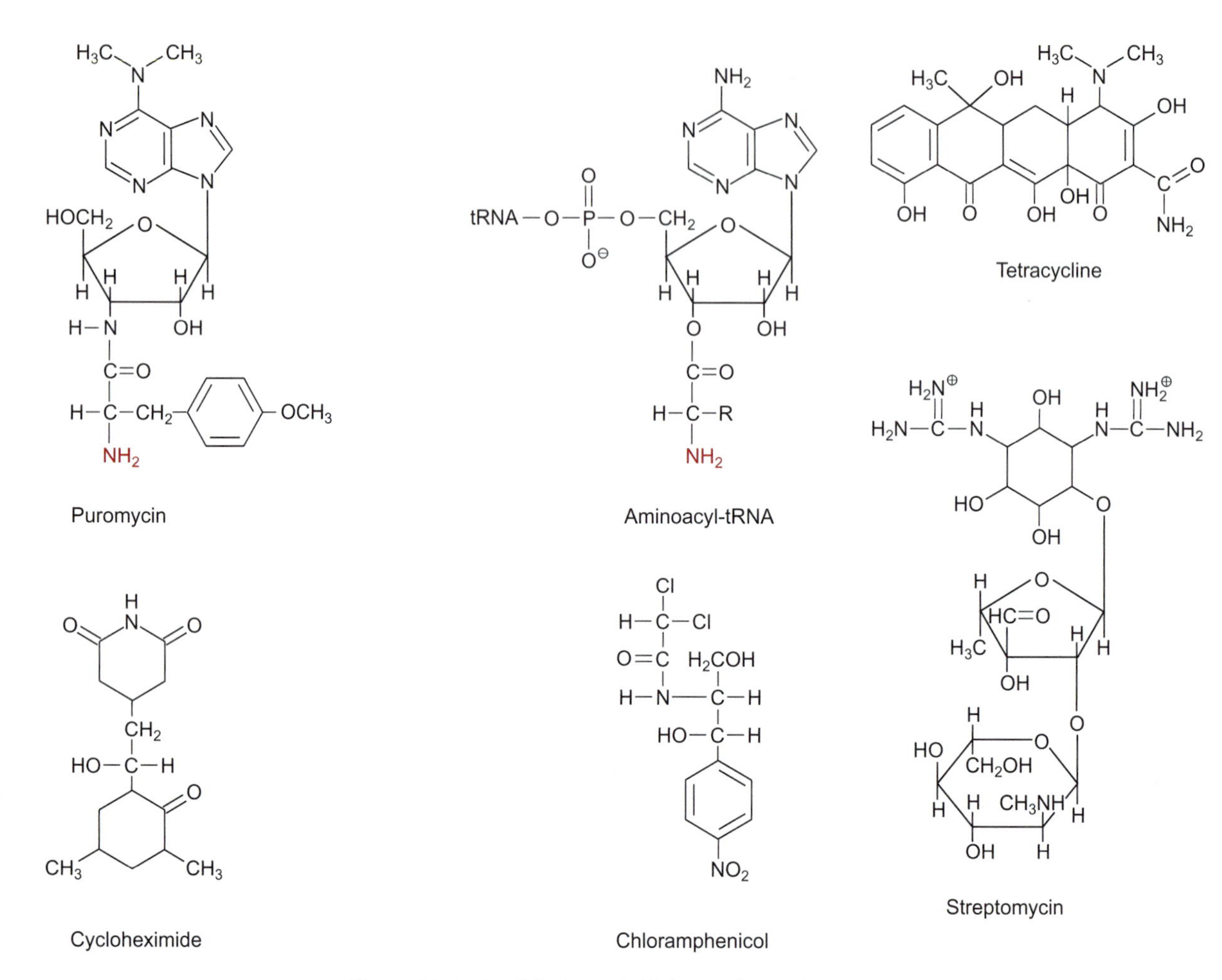

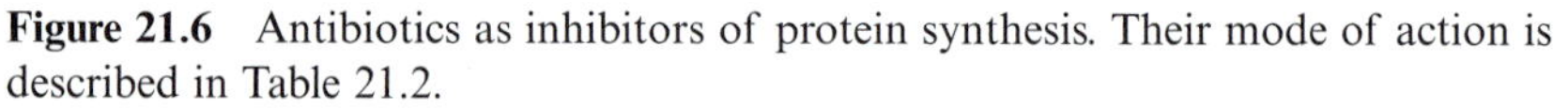

Figure 21.6 Antibiotics as inhibitors of protein synthesis. Their mode of action is described in Table 21.2.

phosphorylation and initiation is therefore inhibited. Little is known about the regulation of translation in plants.

21.2 Proteins attain their three-dimensional structure by controlled folding

Protein biosynthesis by ribosomes first yields an unfinished polypeptide. This is converted (if necessary after sequence segments being cut off) by specific folding to the biological active form, the **native protein**. The three-

Table 21.2: Antibiotics as inhibitors of protein synthesis. The listed antibiotics are all derived from *Streptomycetae*

Antibiotic	Inhibitor action
Puromycin	Binds as an analogue of an aminoacyl-tRNA to the A site and participates in all elongation steps, but prevents the formation of a peptide bond, thus terminating protein synthesis in prokaryotic and eukaryotic ribosomes.
Cycloheximide	Inhibits peptidyl transferase in eukaryotic ribosomes.
Chloramphenicol	Inhibits peptidyl transferase in prokaryotic ribosomes.
Tetracycline	Binds to the 30S subunit and inhibits the binding of aminoacyl-tRNA to prokaryotic ribosomes much more than to eukaryotic ones.
Streptomycin	The interaction with 70S-ribosomes results in an incorrect recognition of mRNA sequences and thus inhibits initiation in prokaryotic ribosomes.

dimensional structure of the native protein normally represents the lowest energy state of the molecule and is determined to a large extent by the amino acid sequence of the molecule.

The folding of a protein is a multistep process

Theoretically there are about 10^{100} possible conformations for a peptide with 100 amino acids, which is a rather small protein of about 11 kDa. Since the reorientation of a single bond requires about 10^{-13} s, it would take the incredibly long time of 10^{87} s to try out all possible folding states one after the other. By comparison, the age of the earth is about 1.6 10^{17} s. In reality, a protein attains its native form within seconds or minutes. Apparently the folding of the molecule proceeds in a multistep process. It begins by forming secondary structures such as **α-helices** or **β sheets**. They consist of 8 to 15 amino acid residues, which are formed or dissolved again within milliseconds (Fig. 21.7). The secondary structures then associate stepwise with increasingly larger associates and in this way also stabilize the regions of the molecule that do not form secondary structures. The hydrophobic interaction between the secondary structures is the driving force in these folding processes. After further conformational changes, the correct three-dimensional structure of the molecule is attained rapidly by this coopera-

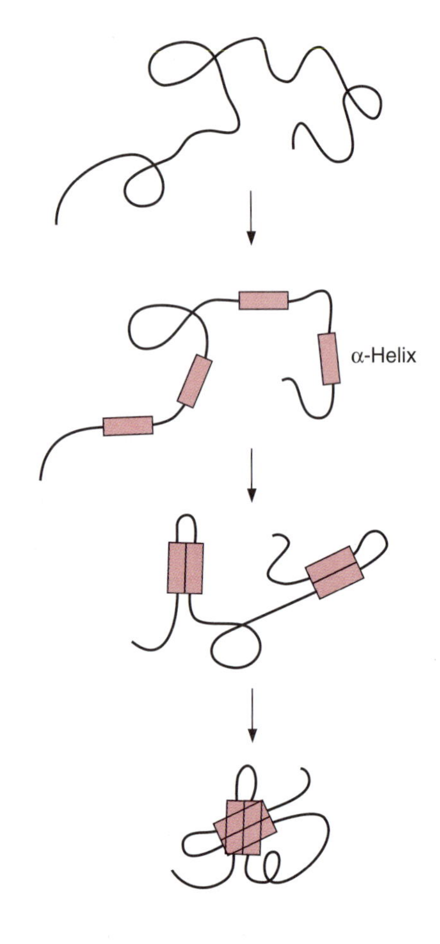

Figure 21.7 Protein folding is a stepwise hierarchic process. First, secondary structures are formed, which then aggregate successively, until finally, after slight corrections to the folding, the tertiary conformation of the native protein is attained.

tive folding procedure. In proteins with several subunits, the subunits associate to form a quarternary structure.

Proteins are protected during the folding process

The folding process can be severely disturbed when the secondary structures in the molecule associate incorrectly, or particularly when secondary structures of different molecules associate resulting in an undesirable aggregation of proteins (hydrophobic collapse). This danger is especially high during protein synthesis, when the incomplete protein is still attached to the ribosome (Fig. 21.5), or during the transport of an unfolded protein through a membrane, when only part of the peptide chain has reached the other side. Moreover, incorrect intermolecular associations are likely to occur when the concentration of a newly synthesized protein is very high, as can be the case in the lumen of the rough endoplasmic reticulum (Chapter 14).

To prevent such incorrect folding, a family of proteins present in the various cell compartments helps newly formed protein molecules to attain their correct conformation by avoiding incorrect associations. These proteins have been named **chaperones**.

Heat shock proteins protect against heat damage

Chaperones not only have a function in protein biosynthesis but also protect cell proteins that have been denatured by exposure to high temperatures against aggregation, thus assisting their reconversion to the native conformation. Bacteria, animal, and plant cells react to a temperature increase of about 10% above the temperature optimum with a very rapid synthesis of so-called **heat shock proteins**, most of which are chaperones. Many plants can survive otherwise lethal high temperatures if they have been previously exposed to a smaller temperature increase, which induces the synthesis of heat shock proteins. This phenomenon is called **acquired thermal tolerance**. Investigations with soybean seedlings showed that such tolerance coincides with an increase in the content of heat shock proteins. However, most of these heat shock proteins are constitutive, which means that they are also present in cells, and under normal conditions have important functions in the folding of proteins.

Chaperones bind to unfolded proteins

Since chaperones were initially characterized as heat shock proteins, they, as well as protein factors modulating the chaperone function, are commonly designated by the abbreviation Hsp followed by the molecular mass in kDa.

Chaperones of the **Hsp70 family** have been found in bacteria, mitochondria, chloroplasts, and the cytosol of eukaryotes, as well as in the endoplasmic reticulum. These are highly conserved proteins. Hsp70 has a binding site for adenine nucleotides, which can be occupied either by ATP or ADP. When occupied by ADP, Hsp70 forms with the chaperone **Hsp40** a tight complex with unfolded segments of a protein, but not with native proteins (Fig. 21.8). The ADP bound to Hsp70 is replaced by ATP. The resultant ATP-Hsp70 complex has only a low binding affinity and therefore dissociates from the protein segment. Due to the subsequent hydrolysis of the bound ATP to ADP, Hsp70 is ready to bind once more to an unfolded peptide segment. In this way Hsp70 binds to a protein only for a short time, dissociates from it, and, if necessary, binds to the protein again. This stabilizes an unfolded protein without restricting its folding capacity. The mechanism of the ADP-dependent binding of Hsp70 to unfolded peptides as mediated by Hsp40 has been conserved during evolution. Fifty percent of

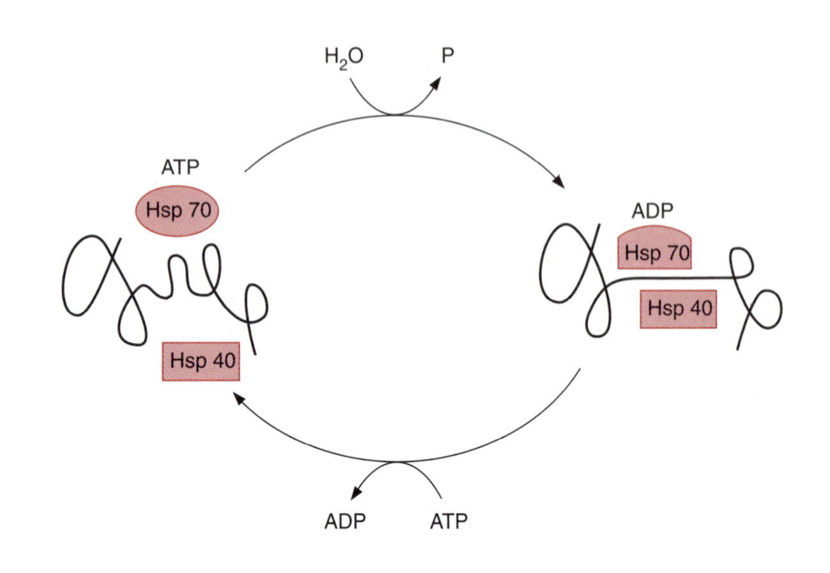

Figure 21.8 The Hsp70 chaperone contains a binding site for ATP and hydrolyzes ATP to ADP. The Hsp70 paired with ADP binds tightly with Hsp40 to an unfolded segment of a protein. The ADP bound to Hsp70 is exchanged for ATP. The Hsp70 paired with ATP has only a low binding affinity for the protein, which is therefore released. A protein can be bound again only after ATP hydrolysis. This simplified scheme does not deal with intermediates involved and also does not represent the real structure of the binding complex.

the amino acids in the sequences of the Hsp70 protein in *E. coli* and in humans are identical.

The proteins of the **Hsp60 family**, contained in bacteria, plastids, and mitochondria, also bind to unfolded proteins. They were first identified as the **GroEL factor** in *E. coli* and as RubisCO binding protein in chloroplasts, until it was realized that both proteins are homologous and act as chaperones. A **GroES factor**, called **Hsp10** in mitochondria and in chloroplasts, is involved in binding the Hsp60 chaperones. In bacteria, 14 GroEL and 7 GroES molecules are assembled to a **superchaperone complex**, forming a large cavity into which an unfolded protein fits (Figs. 21.9, 21.10). The protein is temporarily bound to Hsp60 molecules of the cavity analogously to the binding to Hsp70 in Figure 21.8. Correct folding to the native protein

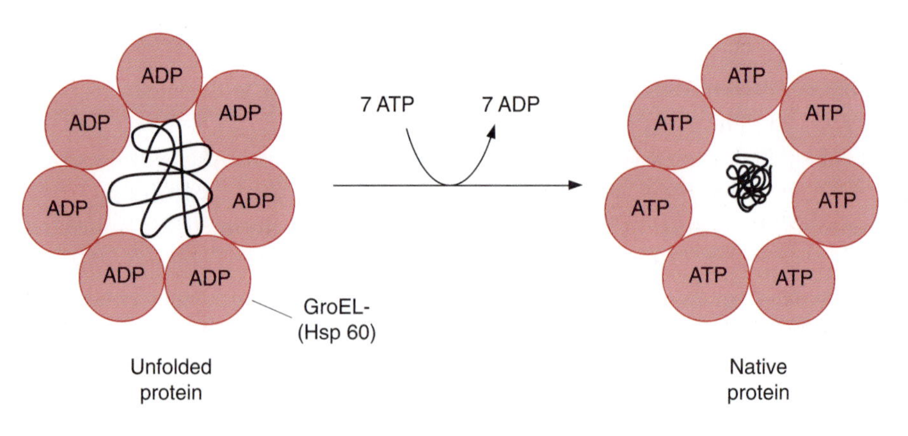

Figure 21.9 Section through the superchaperone complex of prokaryotes consisting of 14 molecules of GroEL (Hsp60) and 7 molecules of GroES (Hsp10). The chaperone molecules, paired with ADP, bind unfolded segments of the newly formed protein. Repeated release of the protein after exchange with ATP, and consecutive binding after ATP hydrolysis (see previous figure), enables the protein to fold. The native protein is released because in the end it no longer binds to chaperones.

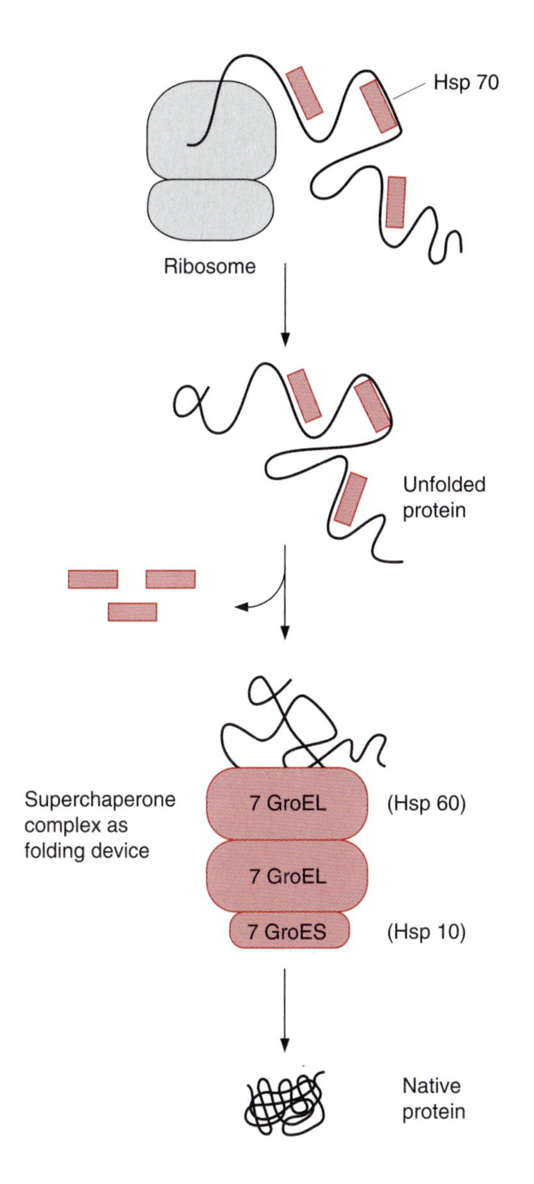

Figure 21.10 The folding of proteins in prokaryotes, plastids, and mitochondria. The unfolded protein is protected by being bound to an Hsp70 chaperone paired with ADP and is then folded to the native protein in the cavity of the superchaperone complex, which consists of GroEL(Hsp60) and GroES(Hsp10). ATP is consumed in this reaction.

is aided by several ATP hydrolysis cycles, involving dissociation and rebinding of the protein segments. In this way the unfolded protein can reach the native conformation by avoiding association with other proteins.

Hsp70 as well as Hsp60 and Hsp10 participate in the protein folding in plastids and mitochondria (Fig. 21.10). Hsp70 protects single segments of the growing peptide chain during protein synthesis, and the super chaperone complex from Hsp60 and Hsp10 finally enables the undisturbed folding of the total protein. The chaperone **Hsp90**, which is regarded as the most abundant cytosolic protein, is found in very high concentrations in the cytosol of eukaryotes. Hsp90 is believed to play a central role in the folding and assembling of cytosolic proteins. Moreover, chaperones named CCT (cytosolic complex T), somewhat resembling the prokaryotic Hsp60, have been identified in the cytosol of eukaryotes. They act as a folding device by forming oligomeric chaperone complexes, which are probably similar to the Hsp60-Hsp10 superchaperone complex (Fig. 21.10).

Furthermore, there are proteins facilitating other processes, which limit protein folding. Such limiting processes include the formation of disulfide bridges and the *cis-trans* isomerization of the normally non-rotatable prolyl peptide bonds. Since thorough investigation of chaperones began only a few years ago, many questions about their structure and function are still unanswered.

21.3 Nuclearly encoded proteins are distributed throughout various cell compartments

The ribosomes contained in the cytosol also synthesize proteins destined for cell organelles, such as plastids, mitochondria, peroxisomes, and vacuoles, as well as proteins to be secreted from the cell. To reach their correct location, these proteins must be specifically transported across various membranes.

Proteins destined for the vacuole are transferred during their synthesis to the lumen of the ER (section 14.5). This is aided by a signal sequence at the terminus of the synthesized protein, which binds with a signal recognition particle to a pore protein present in the ER membrane and thus directs the protein to the ER lumen. In such a case the ribosome is attached to the ER membrane during protein synthesis and the synthesized protein appears immediately in the ER lumen (Fig. 14.2). This process is called **co-translational protein transport**. These proteins are then transferred from the ER lumen by vesicle transfer across the Golgi apparatus to the vacuole or are exported by secretory vesicles from the cell.

In contrast, protein uptake into plastids, mitochondria, and peroxisomes occurs mainly, if not exclusively, by **post-translational transport**, which means the proteins are transported across the membrane only after completion of protein synthesis and their release from the ribosomes. So far, transport into mitochondria has been investigated most thoroughly and therefore will be described first.

Most of the proteins imported into the mitochondria have to cross two membranes

More than 95% of the mitochondrial proteins in a plant are encoded in the nucleus and translated in the cytosol. Our present knowledge about the import of proteins from the cytosol into the mitochondria derives primarily from studies with yeast. In order to direct proteins from the cytosol to the mitochondria, they have to be provided with a **targeting signal**. Some proteins destined for the mitochondrial inner membrane or the intermembrane compartment, as well as all the proteins for the mitochondrial outer membrane, contain internal targeting signals that have not yet been identified. Other proteins of the mitochondrial inner membrane and most of the proteins of the mitochondrial matrix are synthesized in the cytosol as **precursor proteins**, which contain as a targeting signal a **signal sequence** of 12 to 70 amino acids at their amino terminus. When these signal sequences are removed after import, as is usually the case, they are called **transit peptides**. These targeting presequences have a high content of positively charged amino acids and are able to form α-helices in which one side is positively charged and the other side is hydrophobic. The three-dimensional structure of the amphiphilic α-helices, rather than a certain amino acid sequence, forms the targeting signal. The directing function of this presequence can be demonstrated in an experiment. When a foreign protein, such as the dihydrofolate reductase from mouse, is provided with a targeting signal sequence for the mitochondrial matrix, this protein is actually taken up into the mitochondrial matrix.

For the import of proteins into the mitochondrial matrix, both the outer and inner membranes have to be traversed (see Figure 1.12). This protein import occurs primarily at so-called **translocation sites** where the inner and outer membranes are closely attached to each other (Fig. 21.11). Each membrane contains its own translocation apparatus, which transfers the proteins in the **unfolded state** through the membranes.

The precursor proteins formed by the ribosomes associate in the cytosol with chaperones (e.g., ctHsp70, ct = cytosol) in order to prevent premature folding or aggregation of the often hydrophobic precursor proteins. The association with ctHsp70 is accompanied by the hydrolysis of ATP (Fig. 21.8). The transport across the outer membrane is catalyzed by a so-called **TOM complex** (translocase of the outer mitochondrial membrane) consisting of at least eight different proteins. The TOM20 and TOM22 subunits function as receptors for the targeting signal sequence. An electrostatic interaction between the positively charged side of the α-helix of the presequence and the negative charge on the surface of TOM22 is probably involved in the specific recognition of the targeting signal. TOM22 and TOM20 then

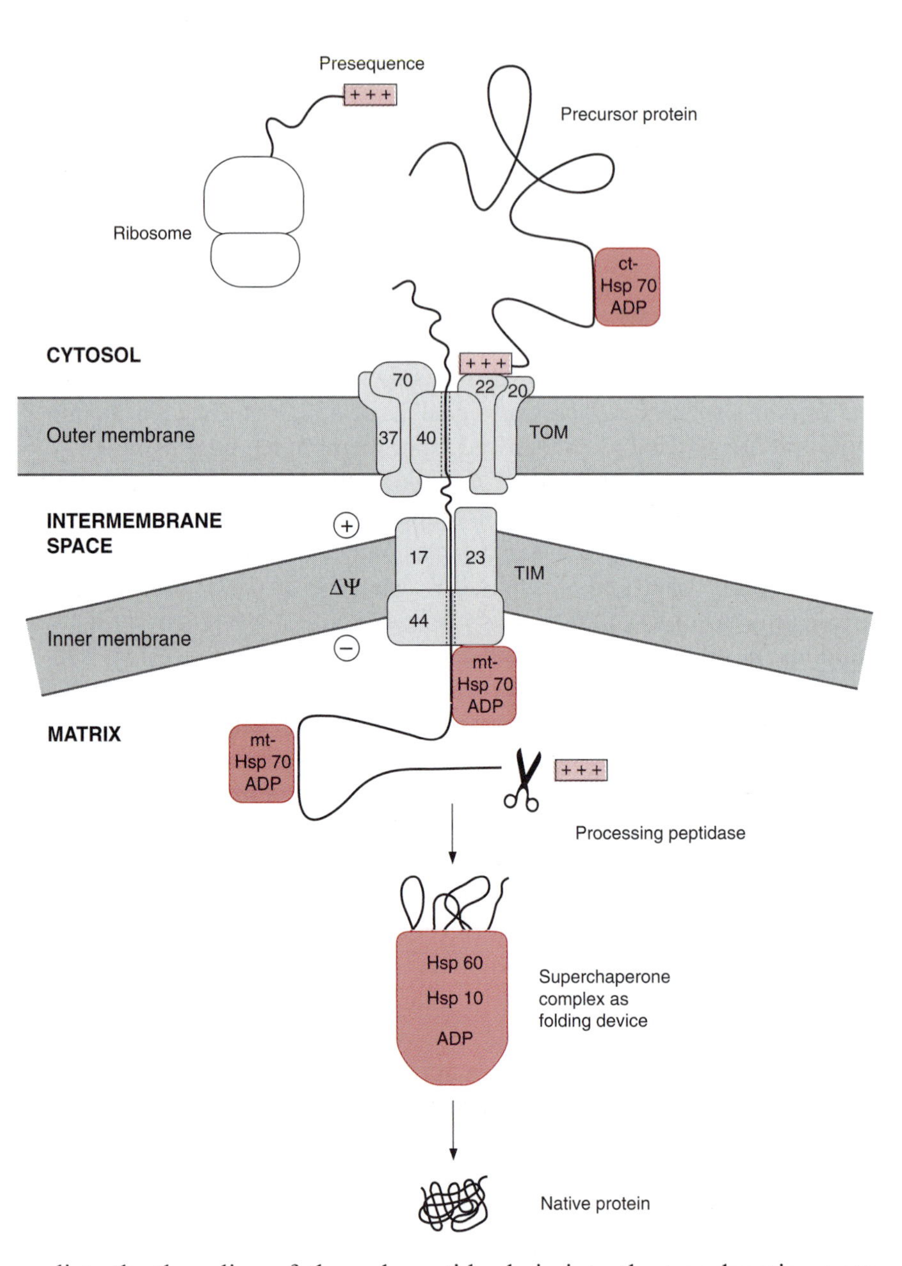

Figure 21.11 Protein import into mitochondria (after Lill and Neupert). The precursor protein formed at the ribosomes present in the cytosol is stabilized in its unfolded conformation by the cytosolic Hsp70 chaperone. A positively charged presequence binds to the receptors TOM20 and TOM22. The presequence threads the precursor protein into the translocation pore of the outer and the inner membranes. Mitochondrial Hsp70 chaperones bind to the peptide chain appearing in the matrix, and thus enable the chain to slide through the translocation pore. The presequence is cut off by a matrix processing peptidase and afterward the protein attains its native conformation in a superchaperone complex.

mediate the threading of the polypeptide chain into the translocation pore. Another receptor for the transport of proteins is TOM70. This receptor, together with TOM37, mediates the uptake of the ATP-ADP translocator protein and other translocators of the inner membrane, which contain an internal targeting signal instead of a presequence. Probably TOM40 as well as the small subunits TOM5, 6, 7 (not shown in Fig. 21.11) participate in the formation of the translocation pore.

The subsequent transport across the inner membrane is catalyzed by the **TIM complex**, consisting of the proteins TIM 17, 23, 44 (translocase of the inner mitochondrial membrane) and several others not yet identified. A precondition for protein transport across the inner membrane is the presence of **a membrane potential** $\Delta\Psi$ (section 5.6). Presumably the positively charged presequence is driven through the translocation pore by the negative charge at the matrix side of the inner membrane. The peptide chain appearing in the matrix is first bound to TIM44 and is then bound with hydrolysis of ATP (Fig. 21.8) to an mtHsp70 chaperone and also to other chaperones not dealt with here. It is assumed that Brownian movement causes a section of the peptide chain to slip through the translocation pore, which is then immediately bound to the mtHsp70 inside, thus preventing the protein from slipping back. It is postulated that repetitive binding of Hsp70 converts a random movement of the protein chain in the translocation channel into a unidirectional motion. According to this model of a **molecular ratchet**, the ATP required for the reversible binding of mtHsp70 probably is not required for pulling the polypeptide chain through the pore, but, instead, to change its free diffusion across the two translocation pores into unidirectional transport. But an alternative hypothesis is also under discussion, according to which the protein entering the pore is pulled into the matrix by ATP-dependent conformational changes of the mtHsp70 bound to the peptide.

When the peptide chain arrives in the matrix, the presequence is immediately cleaved from the protein by a **processing peptidase** (Fig. 21.11). The folding of the matrix protein probably occurs via a superchaperone folding apparatus consisting of the chaperones Hsp60 and Hsp10 (see Figs. 21.9 and 21.10). Proteins destined for the **mitochondrial outer membrane** can, after being bound to the receptors of the TOM complex, be inserted directly into the membrane.

In most cases, proteins destined for the **mitochondrial inner membrane**, after transport through the outer membrane, are inserted directly from the intermembrane space into the inner membrane. In some cases, proteins destined for the inner membrane contain a presequence, which first directs them to the matrix space. After this presequence has been removed by processing peptidases, they are then integrated from the matrix side into the inner membrane via a second targeting sequence.

The import of proteins into chloroplasts requires several translocation complexes

Proteins of the chloroplasts are also encoded mainly in the nucleus. Although much less is known about protein transport into chloroplasts than into mitochondria, parallels as well as differences between the two transport processes are apparent. Transport into the chloroplasts also proceeds **post-translationally**. The precursor proteins formed in the cytosol contain a targeting signal sequence (so-called transit peptides) with 30 to 100 amino acid residues at the N-terminus. As in the mitochondria, the targeting signal probably does not consist of a specific amino acid sequence, but its function is due to the secondary structure of the signal presequence. The precursor proteins of the chloroplasts are stabilized by Hsp70 chaperones during their passage through the cytosol.

In order to be imported into the stroma, the protein must cross two membranes (Fig. 21.12). The translocation apparatus of the outer chloroplast envelope membrane contains four functional proteins, which, according to their molecular mass (in kDa), are named **TOC** (translocase of the outer chloroplast membrane) 34, 64, 70, 75, 160 and together represent about 30% of the total membrane proteins of the outer envelope membrane. TOC75 seems to form the translocation pore for the passage of the unfolded peptide chain. TOC160 acts as receptor for the precursor protein. At least five proteins are involved in the subsequent transport via the inner envelope membrane.

In contrast to mitochondrial protein transport, protein transport into the chloroplast stroma does not require a membrane potential $\Delta\Psi$, but ATP, which is supplied from the stroma. The protein transport appears to be regulated by the binding of GTP to TOC160 and to TOC34. In this the translocation apparatus of the chloroplasts differs quite clearly from that of mitochondria. But also in the chloroplasts a unidirectional motion of the unfolded peptide chain through the translocation pore is caused by a repetitive binding of Hsp70 chaperones according to the model of a molecular ratchet, accompanied by the hydrolysis of ATP.

After it has delivered the protein chain to the stroma, the presequence is removed by a processing peptidase of the stroma. The resulting protein is folded to the native conformation, probably with the aid of an Hsp60-Hsp10 superchaperone complex, and is then released. In this way also the small subunit of RubisCO (section 6.2) is delivered to the stroma, where it is assembled with the large subunit encoded in the chloroplasts.

Those proteins destined for the thylakoid membrane are first delivered to the stroma and then directed by internal targeting signals into the thylakoid membrane. Some proteins of the thylakoid membrane, including proteins of the water-splitting apparatus, are transported first into the thylakoid space before being incorporated into the membrane. Three different transport systems are now known to carry this out. The precursor protein of plastocyanin, destined for the thylakoid space, contains two targeting signal sequences. The first targeting signal sequence directs the protein into the stroma and is cut off by the stroma processing endopeptidase, and the

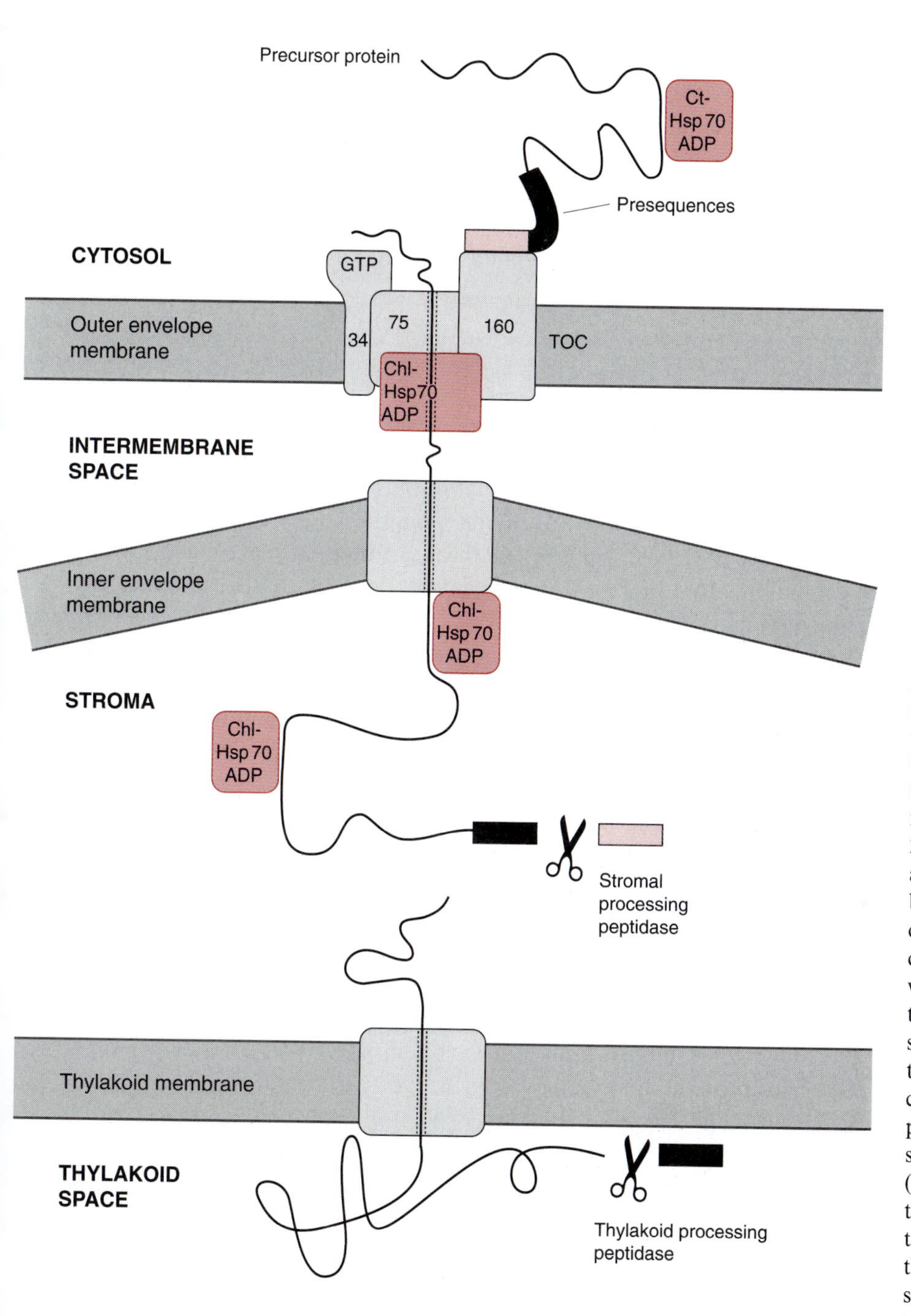

Figure 21.12 Simplified diagram of the protein import into chloroplasts (after Soll 1995). A protein formed in the cytosol and destined for the thylakoid lumen contains two presequences as a targeting signal. The first presequence (colored red) binds to the receptor TOC160 of the translocation apparatus of the outer envelope membrane. It is presumed that transport is regulated by GTP. The membrane protein TOC75 probably represents the translocation pore. The inner translocation pore consists of at least five proteins with the molecular mass 20, 22, 40, 55, and 110 kDa. The import requires ATP for the release of the protein from Hsp 70 (see Fig. 21.8). The peptide chain appearing in the stroma is bound to several chloroplastic Hsp70 chaperones and in this way makes it easier for the unfolded chain to slide through the translocation pore. After cleavage of the first presequence (red), the second presequence (black) serves as a targeting signal for transport across the thylakoid membrane. The second sequence is removed by a membrane-bound thylakoid processing peptidase.

second is for transport across the thylakoid membrane, which is afterward cut off by a thylakoid membrane-bound processing peptidase. The transport of plastocyanin and also of some proteins of the water-splitting apparatus is driven by the hydrolysis of ATP. Since this mode of transport also occurs in cyanobacteria, it is called the prokaryotic pathway. Other proteins of the water-splitting apparatus are transported into the thylakoid space first, as driven by a pH gradient, before being inserted into the thylakoid membrane.

Proteins are imported into peroxisomes in the folded state

The peroxisomes, in contrast to mitochondria and chloroplasts, contain no individual genome. All the peroxisomal proteins are **nuclear-encoded**. It is still not clear how the peroxisomes multiply. Peroxisomes, like mitochondria and chloroplasts, can **multiply by division** and thus can be inherited from mother cells. But there are also observations that a ***de novo* synthesis** of peroxisomal membranes can take place at the endoplasmatic reticulum. Signal sequences cause some peroxisomal membrane proteins to be incorporated into certain sections of the ER membrane, which are subsequently detached as vesicles regarded as **pre-peroxisomes**. It is postulated that these vesicles fuse to the peroxisomes already present or that they can form new peroxisomes by fusion.

Independent of how the peroxisomes are formed, whether by division or by *de novo* synthesis, it is necessary to import the peroxisomal proteins, which are encoded in the nucleus and synthesized in the cytosol (Fig. 21.13). Two different signal sequences are known as targeting sequences (**peroxisomal targeting signals**) PTS1 and PTS2. **PTS1** contains at the C terminus the consensus sequence serine-lysine-leucine (**SKL**) which is not detached after the corresponding protein has been transported into the peroxisomes. **PTS2** consists of a sequence of about nine amino acids near the N terminus of certain proteins and is removed after the import of the protein via proteolysis. The proteins targeted by one of the two signals, bind to the corresponding **soluble receptor proteins** [Pex5 and Pex7 (peroxisomal) biogenesis factor], which bind them to the translocation apparatus (**docking complex**), consisting of several membrane proteins. After dissociation from the receptor proteins, the proteins are transferred upon the consumption of ATP across the membrane into the peroxisomal matrix, in a process not yet fully elucidated. According to present knowledge, the import of proteins into the peroxisomes proceeds in the **folded state**, which is in contrast to mitochondrial and chloroplastic protein transport, proceeding in the unfolded conformation. It seems that protein import into the peroxisomes is entirely different from protein transport into the ER, mitochondria, and plastids.

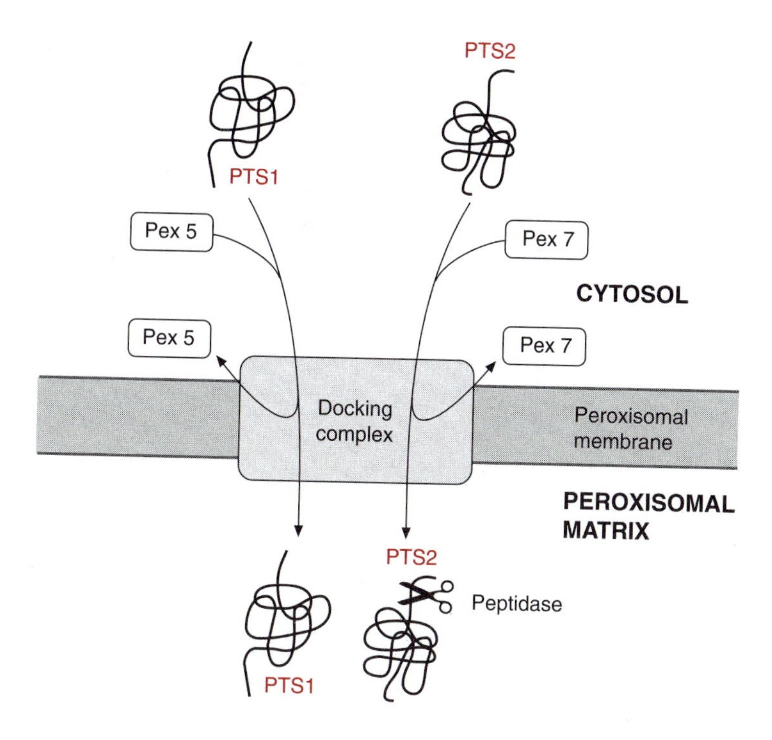

Figure 21.13 Protein import in peroxisomes. The proteins synthesized in the cytosol contain as target for the import into the peroxisomes either the signal sequence PTS1 or PTS2. They associate in the cytosol with the relevant soluble receptor proteins Pex5 or Pex7, which conduct them to the docking complex, where the folded proteins are carried through the peroxisomal membrane. After import, a peptidase detaches the signal sequence PTS2, and the signal sequence PTS1 remains as part of the mature protein.

21.4 Proteins are degraded in a strictly controlled manner by proteasomes

In a eukaryotic cell the protein outfit is regulated not only by synthesis but also by degradation. Eukaryotes possess a very highly conserved machinery for controlled protein degradation, consisting of a multienzyme complex termed **proteasome**. The outstanding role of this pathway in plants may be illustrated by the fact that more than 5% of all structural genes in *Arabidopsis* participate in this degradation device. In order to be degraded, the corresponding proteins are labeled by covalent attachment of **ubiquitin** molecules. Ubiquitin occurs as a highly conserved protein in all eukaryotes. It has an identical sequence of 76 amino acids in all plants. The carboxyl end of the molecule contains a glycine residue, with the terminal carboxyl group exposed to the outside. Proteins destined for degradation are conjugated to ubiquitin by forming a so-called iso-peptide link between the carboxyl group already referred to and the amino group of a lysine residue of the target protein.

The attachment of ubiquitin to a target protein requires the interplay of three different enzymes (Fig. 21.14A). The **ubiquitin-activating enzyme (E1)** activates ubiquitin upon the consumption of ATP to form a thioester with

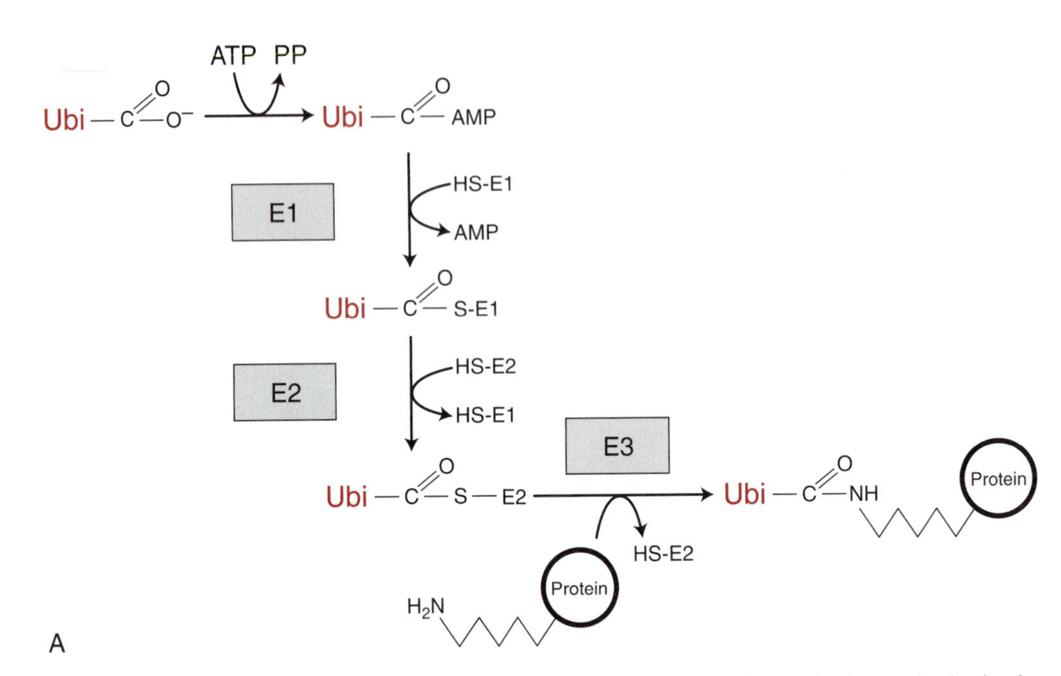

Figure 21.14 Protein degradation by the proteasome pathway. A. A protein destined for degradation (target protein) is tagged by conjugation with ubiquitin (Ubi). Ubiquitin is activated through reaction with ATP by the ubiquitin-activating enzyme (E1) to form via acyl-AMP a thioester link with a cysteine residue of the enzyme. Ubiquitin is then transferred to a cysteine residue of the ubiquitin-conjugating enzyme (E2). Ubiquitin-protein ligase (E3) mediates the transfer of ubiquitin to a lysine residue of a target protein to form an isopeptide. By repetition of the process, further ubiquitin molecules can be linked to lysine residues of the existing ubiquitin tag to form a ubiquitin chain or to other lysine residues of the target protein. In this way a target protein is often tagged with several ubiquitin molecules.

an SH-group of the enzyme. The ubiquitin is then transferred to a **ubiquitin-conjugating enzyme (E2)**. Subsequently the target protein and the ubiquitin attached to E2 react with a specific **ubiquitin-protein ligase (E3)** to form the ε-isopeptide linkage bond. More ubiquitin molecules can be conjugated, either to lysine residues of the ubiquitin already attached to the target protein or by linkage to other lysine residues of the target protein. In this way target proteins can be labeled with a chain of ubiquitin molecules or by several ubiquitin molecules at various sites. Genome analyses indicated that *Arabidopsis* contains 2 genes for E1, 24 for E2, and **1,200 genes for E3**. Apparently, the specificity of protein degradation is governed by the various E_3 proteins.

The proteolysis of the labeled target protein is catalyzed by the **proteasome**. This multienzyme complex can be divided into two different particles,

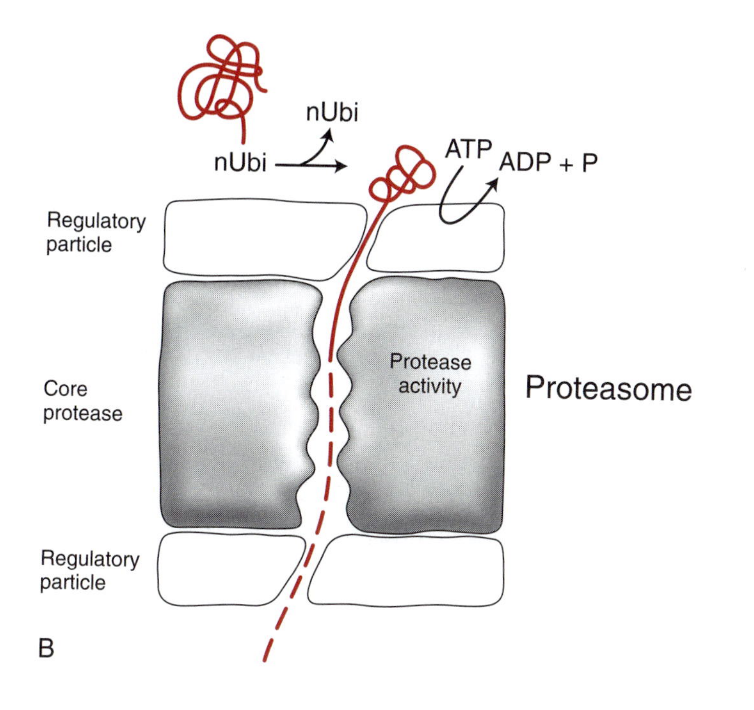

B. Scheme of how a proteasome functions. The multienzyme complex consists of a core protease (CP, 14 subunits) with the shape of a barrel, which is sealed from both sides by regulatory particles (RP, 20 subunits) like lids. The ubiquitin-tagged protein is bound to RP and the attached ubiquitin molecules are released by hydrolysis. The target protein is unfolded by an ATP-consuming reaction and the resultant polypeptide chain passes the interior of CP, where it is degraded by the proteolytic enzymes present.

a **core protease (CP)** consisting of 14 subunits and a **regulatory particle (RP)** consisting of 20 subunits. The core protease has a **barrel-like structure**, with the catalytic sites for proteolysis inside. The openings on both sides are sealed by a regulatory particle. (Fig. 21.14B). The regulatory particle recognizes the ubiquitin-labeled target proteins, and catalyzes the hydrolytic cleavage of the ubiquitin molecules, which are thus available for further ubiquitination of proteins. The target protein bound to the RP subunit is unfolded at the expense of ATP, and the peptide chain is allowed to pass through the interior of the barrel, where it is split by the proteolytic activity inside into peptides of 7 to 9 amino acid residues, which are released from the barrel and further digested by cytosolic peptidases.

The controlled protein degradation by the proteasome plays a role in the regulatory functions of the phytohormones **gibberellin** and **auxin**. These hormones induce via signal chains the degradation of transcription **repressors** and in this way enhance the expression of genes (sections 19.3 and 19.4). Proteasomes are also involved in the degradation of **activated phytochrome A** (section 19.10).

The ubiquitin-dependent proteasome pathway is not the only way to degrade cellular proteins. During senescence, proteins of the cytoplasm or organelles are surrounded by membranes to form **autophagic vesicles**, which fuse to **lytic vacuoles**, where the proteins are subsequently degraded by proteolysis.

Further reading

Ban, N., Nissen, P., Hansen, J., Moore, P. B., Steitz, T. A. The complete atomic structure of the large ribosomal subunit at 2.4 resolution. Science 289, 905–930 (2000).

Fuks, B., Schnell, D. J. Mechanism of protein transport across the chloroplast envelope. Plant Physiol 114, 405–410 (1997).

Cech, T., R. The ribosome is a ribozyme. Science 289, 878–879, (2000).

Glaser, E., Sjöling, S., Tanudji, M., Whelan, J. Mitochondrial protein import in plants: Signals, sorting, targeting, processing and regulation. Plant Mol Biol 38, 311–338 (1998).

Gutierrez, R. A., MacIntosh, G. C., Green, P. Current perspectives on mRNA stability in plants: Multiple levels and mechanisms of control. Trends Plant Sci 4, 429–438 (1999).

Johnson, T. L., Olsen, L. J. Building new models for peroxisome biogenesis. Plant Physiol 127, 731–739 (2001).

Keegstra, K., Cline, K. Protein import and routing systems of chloroplasts. Plant Cell 11,557–570 (1999).

Mogk, A., Mayer, M. P., Deuerling, E. Mechanismus der Proteinfaltung. Molekulare Chaperone und ihr biotechnologisches Potential. Biologie in unserer Zeit 3, 182–192 (2001).

Mullen, R. T., Flynn, C. R., Trelease, R. N. How are peroxisomes formed? The role of the endoplasmic reticulum and peroxins. Trends Plant Sci 23, 68–73 (1998).

Netzer, W. J., Hartl, F. U. Protein folding in the cytosol: Chaperonin-dependent and-independent mechanisms. Trends Plant Sci 6, 256–261 (2001).

Pfanner, N., Douglas, M. G., Endo, T., Hogenraad, N. J., Jensen, R. E., Meijer, M., Neupert, W., Schatz, G., Schmitz, U. K., Shore, G. C. Uniform nomenclature for the protein transport machinery of the mitochondrial membranes. Trends Biochem Sci 21, 51–52 (1996).

Robinson, C., Hynds, P. J., Robinson, D., Mant, A. Multiple pathways for the targeting of thylakoid proteins in chloroplasts. Plant Mol Biol 38, 209–221 (1998).

Schatz, G., Dobberstein, B. Common principles of protein translocation across membranes. Science 271, 1519–1526 (1996).

Schleiff, E. Signals and receptors—The translocation machinery on the mitochondrial surface. J Bioenergetics and Biomembranes 32, 55–66 (2000).

Vierstra, R. D. The ubiquitin/26S proteasome pathway, the complex last chapter in in the life of many plant proteins. Trends Plant Sci 8. 136–142 (2003).

Vothknecht, U. C., Soll, J. Protein import: The hitchhikers guide into chloroplasts. Biol Chem 381, 887–897 (2000).

Wiedemann, N., Kozjak, V., Chacinska, A., Schoenfisch, B., Rospert, S., Ryan, M. T., Pfanner, N., Meisinger, C. Machinery for protein sorting and assembly in the mitochondrial outer membrane. Nature 424, 565–571 (2003).

22

Gene technology makes it possible to alter plants to meet requirements of agriculture, nutrition, and industry

Recent years have witnessed spectacular developments in plant gene technology. In 1984 the group of Marc van Montagu and Jeff Schell in Gent and Cologne, and the group of Robert Horsch and collaborators of the Monsanto Company in St. Louis, Missouri (United States) simultaneously published procedures for the transfer of foreign DNA into the genome of plants utilizing the Ti plasmids of ***Agrobacterium tumefaciens*** (new nomenclature: ***Rhizobium radiobacter***). This method has made it possible to alter the protein complement of a plant specifically to meet special requirements: for example, to render plants resistant to pests or herbicides, to achieve a qualitative or quantitative improvement in the productivity of crop plants, and to adapt plants so that they can produce defined sustainable raw materials for chemical industry.

Hardly any other discovery in botany has had such far- reaching consequences in such a short time, when one considers that in 2003 in the United States about 80% of the planted soy beans, 70% of cotton, and 38% of maize were varieties altered by genetic engineering using *Agrobacterium tumefaciens*. Here we have the case in which the results of basic research on an exotic theme, namely, the gall formation in a plant, has led to a technique that brought about a revolution in agriculture.

The following sections will describe how a plant can be altered by genetic engineering. From the abundance of established procedures, only the principles of some major methods can be outlined here. For the sake of brevity, details or complications in methods will be omitted. Some practical examples will show how genetic engineering can be used to alter crop plants.

22.1 A gene is isolated

Let us consider the case where a transgenic plant A is to be generated, which synthesizes a foreign protein (e.g., a protein from another plant B). For this, the gene encoding the corresponding protein first has to be isolated from plant B. Since a plant probably contains between 25,000 and 50,000 structural genes, it will be difficult to isolate a single gene from this very large number.

A gene library is required for the isolation of a gene

To isolate a particular gene from the great number of genes existing in the plant genome, it is advantageous to make these genes available in the form of a gene DNA library. Two different kinds of gene libraries can be prepared.

To prepare a **genomic DNA library**, the total genome of the organism is cleaved by restriction endonucleases (section 20.3) into fragments of about 15 to several 100 kbp. Digestion of the genome in this way results in a very large number of DNA sequences, which frequently contain only parts of genes. These fragments are inserted into a vector (e.g., a plasmid or a bacteriophage) and then each fragment is amplified by cloning, usually in bacteria.

To prepare a **cDNA library**, the mRNA molecules present in a specific tissue are first isolated and then transcribed into corresponding cDNAs by **reverse transcriptase** (see section 20.5). The cDNAs are inserted into a vector and amplified by cloning. The mRNA is isolated from a tissue in which the corresponding gene is expressed to a high extent. In contrast to the fragments of the genomic library, the resulting cDNAs contain entire genes without introns and can therefore, after transformation, be expressed in prokaryotes to synthesize proteins. Since a cDNA contains no promoter regions, such an expression requires a prokaryotic promoter to be added to the cDNA.

To prepare a cDNA library from leaf tissue, for example, the total RNA is isolated from the leaves, of which the mRNA may amount to only 2%. To separate the mRNA from the bulk of the other RNA species, one makes use of the fact that eukaryotic mRNA contains a **poly(A) tail** at the 3′ terminus (see Fig. 20.8). This allows mRNA to be separated from the other RNAs by affinity chromatography. The column material consists of solid particles of cellulose or other material to which a poly-deoxythymidine oligonucleotide [poly-(dT)] is linked. When an RNA mixture extracted from leaves is applied to the column, the mRNA molecules bind to the column by hybridization of their poly-(A) tail to the poly-(dT) of the column material, whereas

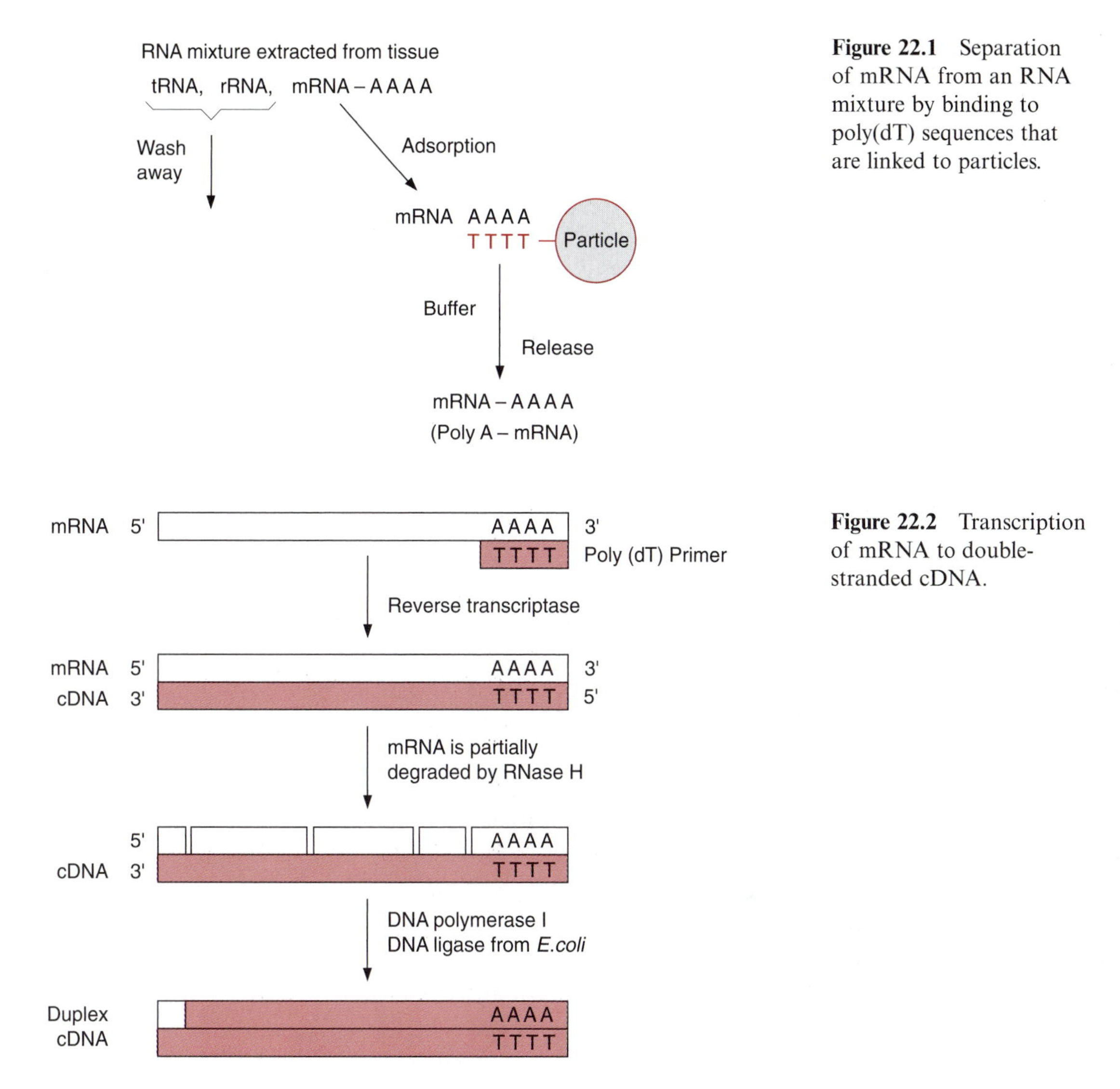

Figure 22.1 Separation of mRNA from an RNA mixture by binding to poly(dT) sequences that are linked to particles.

Figure 22.2 Transcription of mRNA to double-stranded cDNA.

the other RNAs run through (Fig. 22.1). With a suitable buffer, the bound mRNA is eluted from the column.

To synthesize by reverse transcriptase a cDNA strand complementary to the mRNA, a poly-(dT) is used as a primer (Fig. 22.2). Subsequently, the mRNA is hydrolyzed by a ribonuclease either completely or, as shown in the figure, only partly. The latter way has the advantage that the mRNA fragments can serve as primers for the synthesis of the second cDNA strand by DNA polymerase. By using DNA polymerase I, these mRNA fragments are successively replaced by DNA fragments and these are linked to each other

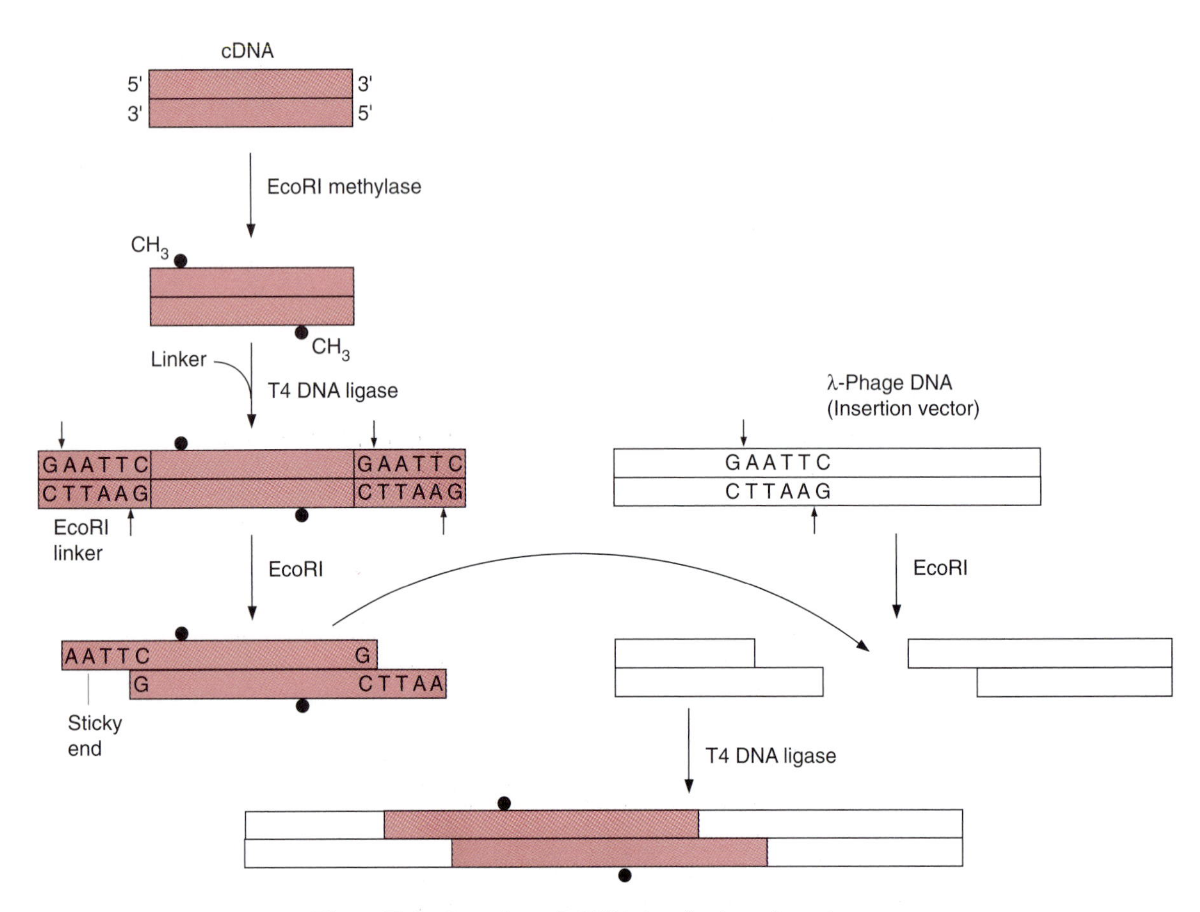

Figure 22.3 Insertion of cDNA in a λ-phage insertion vector.

by DNA ligase. A short RNA section remains, which is not replaced at the end of the second cDNA strand, but this is of minor importance, since in most cases the mRNA at the 5′ terminus contains a non-encoding region (see Fig. 20.8).

The double-stranded cDNA molecules thus formed from the mRNA molecules are amplified by cloning. **Plasmids or bacteriophages** can be used as **cloning vectors**. Nowadays a large variety of made-to-measure phages and plasmids are commercially available for many special purposes. A distinction is made between vectors that only amplify DNA and **expression vectors** by which the proteins encoded by the amplified genes can also be synthesized.

A gene library can be kept in phages

Figure 22.3 shows the insertion of cDNA into the DNA of a **λ phage**. In the example shown here, the phage DNA possesses a cleavage site for the

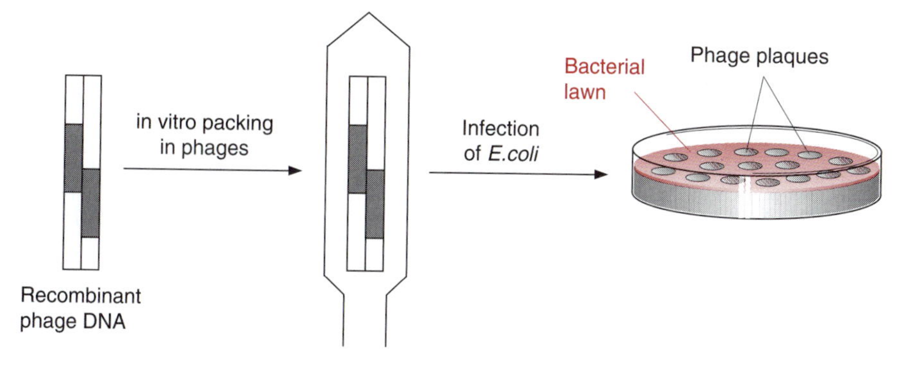

Figure 22.4 A recombinant phage DNA is packed into a virus particle. *E. coli* cells are infected with the formed phage and plated on agar plates. The cells of the infected colonies are lysed by the multiplying phages and show as transparent spots (plaques) in the bacterial lawn.

restriction endonuclease *Eco*RI (section 20.3). To protect the restriction sites within the cDNA, the cDNA double strand is first methylated by an *Eco*RI methylase at the *Eco*RI restriction sites. DNA ligase is then used to link chemically synthesized double-stranded oligonucleotides with an inbuilt restriction site, (in this case for *Eco*RI) to both ends of the double-stranded cDNA. These oligonucleotides are called **linkers**. The restriction endonuclease *Eco*RI cleaves this linker as well as the λ phage DNA and thus generates **sticky ends** at which the complementary bases of the cDNA and the phage DNA can anneal by base pairing. The DNA strands are then linked by DNA ligase, and in this way the cDNA is inserted into the vector.

The phage DNA with the inserted cDNA is packed *in vitro* into a **phage protein coat** (Fig. 22.4), using a packing extract from phage-infected bacteria. In this way one obtains a gene library, in which the cDNA formed from the many different mRNAs of the leaf tissue are packed in phages, which, after infecting bacteria, can be amplified *ad libitum,* whereby each packed cDNA forms a clone.

The bacteria are infected by mixing them with the phages and they are then plated on agar plates containing cultivation medium. At first the infected bacteria grow on the agar plates to produce a bacterial lawn, but then are lysed by the phages, which have been multiplied within the bacteria. The lysed bacterial colonies appear on the agar plate as clear spots called **plaques**. These plaques contain newly formed phages, which can be multiplied further. It is customary to plate a typical cDNA gene library on about 10 to 20 agar plates. Ideally, each of these plaques contains only one clone. From these plaques, the clone containing the cDNA of the desired gene is selected, using specific probes as described later.

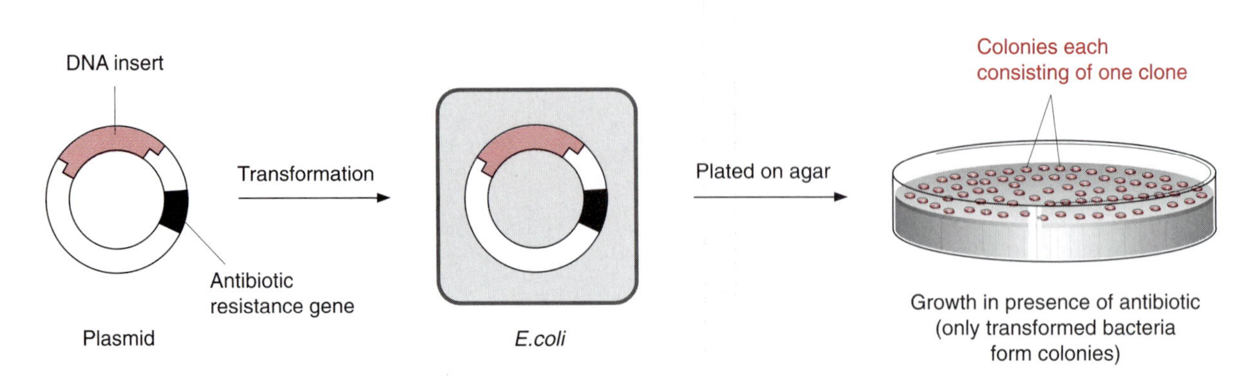

Figure 22.5 cDNA can be propagated via a plasmid vector in *E. coli.* An antibiotic resistance gene on the plasmid enables the selection of the transformed cells.

A gene library can also be kept in plasmids

To clone a gene library in plasmids, cDNA is inserted into plasmids via a restriction cleavage site in more or less the same way as in the insertion into phage DNA (Fig. 22.5). The plasmids are then transferred to *E. coli* cells. The transfer is brought about by treating the cells with $CaCl_2$ to make their membrane more permeable to the plasmid. The cells are then mixed with plasmid DNA and exposed to a short heat shock. In order to select the transformed bacterial cells from the large majority of untransformed cells, the transformed cells are provided with a marker. The plasmid vector contains an **antibiotic resistance gene,** which makes bacteria resistant to a certain antibiotic, such as ampicillin or tetracycline. When the corresponding antibiotic is added to the culture medium, cells containing the plasmid survive and grow, whereas the other non-transformed cells die. After plating on an agar culture medium, bacterial colonies develop, which can be recognized as spots.

In order to check whether a plasmid actually contains an inserted DNA sequence (**insert**), plasmid vectors have been constructed in which the restriction cleavage site for insertion of the foreign DNA is located inside a gene, which encodes the enzyme β-galactosidase (Fig. 22.6). This enzyme hydrolyzes the colorless compound X-Gal into an insoluble blue product. When X-Gal is added to the agar plate culture medium, all the clones that do not contain a DNA insert, and therefore contain an intact β-galactosidase gene, form blue colonies. If a DNA segment is inserted into the cleavage site of the β-galactosidase gene, this gene is interrupted and is no longer able to encode a functional β-galactosidase. Therefore

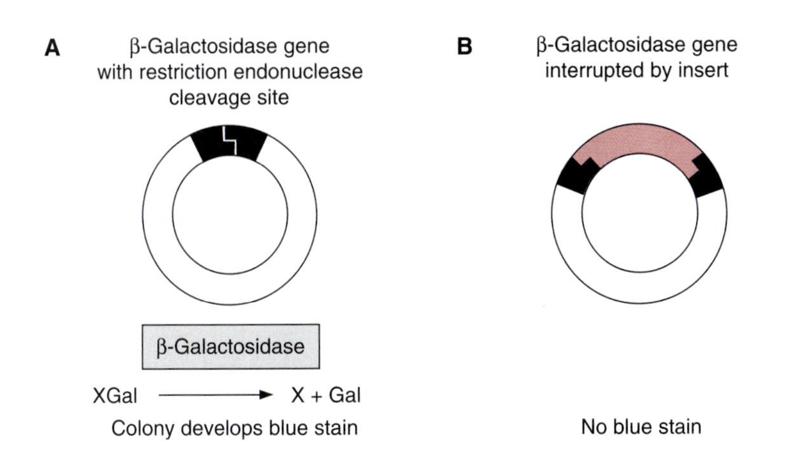

Figure 22.6 To check whether the plasmid of a bacterial colony carries a DNA insert, the cleavage site of the plasmid vector is contained within a β-galactosidase gene. In colonies with the intact gene (i.e., no insert), the colorless chemical X-Gal (5-bromo-4-chloro-3-indolyl-β-D-galactopyranoside) is hydrolyzed by the corresponding gene product into galactose and an indoxyl derivative, which oxidizes to form a blue colored dimer. This results in blue staining of the colony. When the function of the β-galactosidase gene is disrupted by the DNA insert, the gene product can no longer be formed and the corresponding colonies are not stained blue. X = 5-Brom-4-chlor 3-indol.

the corresponding colonies are not stained blue but remain white (blue/white selection).

A gene library is screened for a certain gene

Specific probes are employed to screen the bacterial colonies or phage plaques for the desired gene. A blot is made of the various agar plates by placing a nylon or nitrocellulose membrane on top of them (Fig. 22.7). Some of the phages contained in the plaques, or the bacteria contained in the colonies, bind to the blotting membrane, although most of the contents of the plaques and the colonies remain on the agar plate. Two kinds of probes can be used to screen the phage or bacterial clones bound to the blotting membrane:

1. Specific antibodies to identify the protein formed as gene product of the desired clone (Western blot); and
2. Specific DNA probes to label the cDNA of the desired clone by hybridization.

A clone is identified by antibodies against the gene product

Antibodies against a certain protein are often used to identify the corresponding gene by **Western blot**. This method requires sufficient amounts of the corresponding protein to be purified beforehand in order to obtain **polyclonal antibodies** by immunization of animals. If antibodies are to be used to identify a gene product, the cDNA library must be inserted in an appropriate expression vector. This is a cloning vector, which facilitates transcription of the inserted DNA molecule, and the resultant mRNA is then

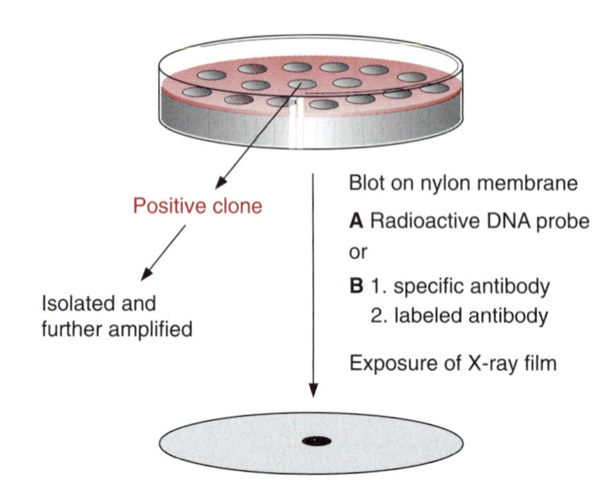

Figure 22.7 For screening the gene library, a blot is prepared by placing a nylon or nitrocellulose membrane on the agar plate. By means of a labeled probe (A, DNA probe or antibody), the desired DNA or the corresponding gene product is detected on the blot as a dark spot on the X-ray film, caused by radioactivity or by chemoluminescent light. The corresponding clone is identified on the agar plate by comparing the positions and can be picked up with a toothpick for further propagation.

translated into the corresponding protein, which might be recognized by the antibody. The vector contains a promoter sequence, which controls initiation of transcription of the inserted gene, and often also a sequence for termination of the transcription at its end.

In practice, bacteria bound to the blotting membrane are disrupted and the released bacterial proteins are fixed to the membrane. When phages are used as vectors, cell disruption is not required, since the phages themselves lyse the cell and thus liberate the cell proteins, which are then fixed to the blotting membrane. Afterward antibodies are added, which bind specifically to the corresponding protein, but are washed off from all other parts of the membrane. Usually a second antibody, which recognizes the first antibody, is used to detect the bound antibody (Fig. 22.8). Formerly the second antibody was labeled with radioactive ^{125}iodine. Nowadays mostly the **ECL-technique** (*e*nhan*c*ed *l*uminescence) is employed. This entails a peroxidase from radish being attached to the second antibody. This peroxidase catalyzes the oxidation of added luminol (3-aminophtalic acid hydrazide) by H_2O_2, accompanied by the emission of blue luminescent light. Using a luminescence enhancer, the intensity of this chemoluminescence can be increased by a factor of 1,000. After the incubation with the second antibody, the blotting membrane is washed; H_2O_2, luminol, and luminescence enhancer are added to it; and it is exposed for about 1 hour to an X-ray film. A positive colony can be recognized as a dark spot in the autoradiography (Fig. 22.7). The position of the positive clone on the agar plate can be identified from the position of the spot on the blotting membrane. After the first screening, an apparently positive clone may actually contain several clones, due to a high density of bacteria. For this reason, the colony, picked up from the positive region with a toothpick, is diluted and plated again on an agar plate.

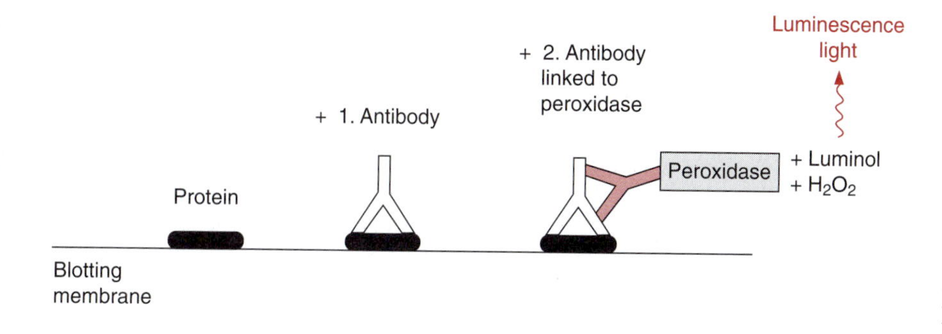

Figure 22.8 An antibody bound to a protein is recognized by a second antibody, which is linked to a peroxidase. The peroxidase catalyzes the oxidation of luminol by H_2O_2 upon emitting luminescent light, which is detected by exposure to an X-ray film.

By repeating the screening procedure described above (**rescreening**), single pure clones finally are obtained. Positive phage plaques are also regrown in bacteria and plated again in order to obtain pure clones.

A clone can also be identified by DNA probes

In this procedure, the phages or bacteria present on the blotting membrane are first lysed and the proteins are removed. The remaining DNA is then dissolved into single strands, which are tightly bound to the membrane. Complementary DNA sequences, which are radioactively labeled by ^{32}P-labeled deoxynucleotides, are used as probes. These probes bind by hybridization to the desired cDNA present on the blotting membrane. The identification of the positive clones proceeds via autoradiography, similar to the detection procedures described previously.

Chemically synthesized oligonucleotides of about 20 bases are also employed as DNA probes. These probes are radioactively labeled at the 5′ end by ^{32}P-phosphate and are used particularly when only low quantities of the purified proteins are available, which are not sufficient for the generation of antibodies. However, with very low amounts of protein, it is possible to determine by micro-sequencing part of the N-terminal amino acid sequence of the protein. From such a partial amino acid sequence, the corresponding DNA sequence can be deduced according to the genetic code, in order to produce oligonucleotide probes by chemical synthesis using automatic synthesizers. However, the degeneracy of the genetic code means that amino acids are often encoded by more than one nucleotide triplet (Fig. 22.9). This is taken into account during the design of oligonucleotides. When, for example, the third base of the triplet encoding lysine can be either an A or a G, a mixture of precursors for both is added to the synthesizer during the linking of the third base of this triplet to the oligomer. In order to introduce the third base of the alanine triplet, a mixture of all four

Trp	Lys	Ala	Met	Asn	Ile
UGG .	AAA .	GCU .	AUG .	AAU .	AUU
	G	C		C	C
		A			A
		G			

Figure 22.9 A degenerate oligonucleotide contains all the possible sequences for encoding the given amino acid sequence.

nucleotide precursors is added to the synthesizer. The "degenerate" oligonucleotide shown in Figure 22.9 is thus, in fact, a mixture of 48 different oligonucleotides, only one of which contains the correct sequence of the desired gene.

When the protein encoded by the desired gene has not been purified, a corresponding gene section from related organisms sometimes can be used as a probe. Domains with specific amino acid sequences are often conserved in enzymes or other proteins (e.g., translocators), even from distantly related plants, and are encoded by correspondingly similar gene sequences.

After a cDNA has been successfully isolated, the next step usually is to determine its base sequence. To an increasing extent, DNA sequencing is carried out by automatic analyzers, which will not be described here.

Genes encoding unknown proteins can be isolated by complementation

In some cases, it has been possible to isolate the cDNA encoding an unknown protein, which is defined only by its function. One method for identifying the gene encoding such an unknown protein is by the complementation of **deficiency mutants** of bacteria or yeast after transformation with plasmids from a cDNA library (Fig. 22.10). Plasmids that can be amplified in *E. coli* as well as in yeast are available as cloning vectors and can be used to express the encoded protein. Several plant translocators, including the sucrose translocator involved in phloem loading (section 13.1), have been identified by complementation. To identify the gene of the sucrose translocator, a yeast deficiency mutant that had lost the ability to take up sucrose and therefore could no longer use sucrose as a nutritional source was employed. This mutant was transformed with plasmids from a plant cDNA library, where the plasmid DNA was provided with a yeast promoter in order to express the plasmid DNA within the yeast cell. After plating the transformed yeast cells on a culture medium with sucrose as the only carbon source, a yeast clone that grew on the sucrose cultivation medium was found. This indicated that transformation with the corresponding plasmid from the

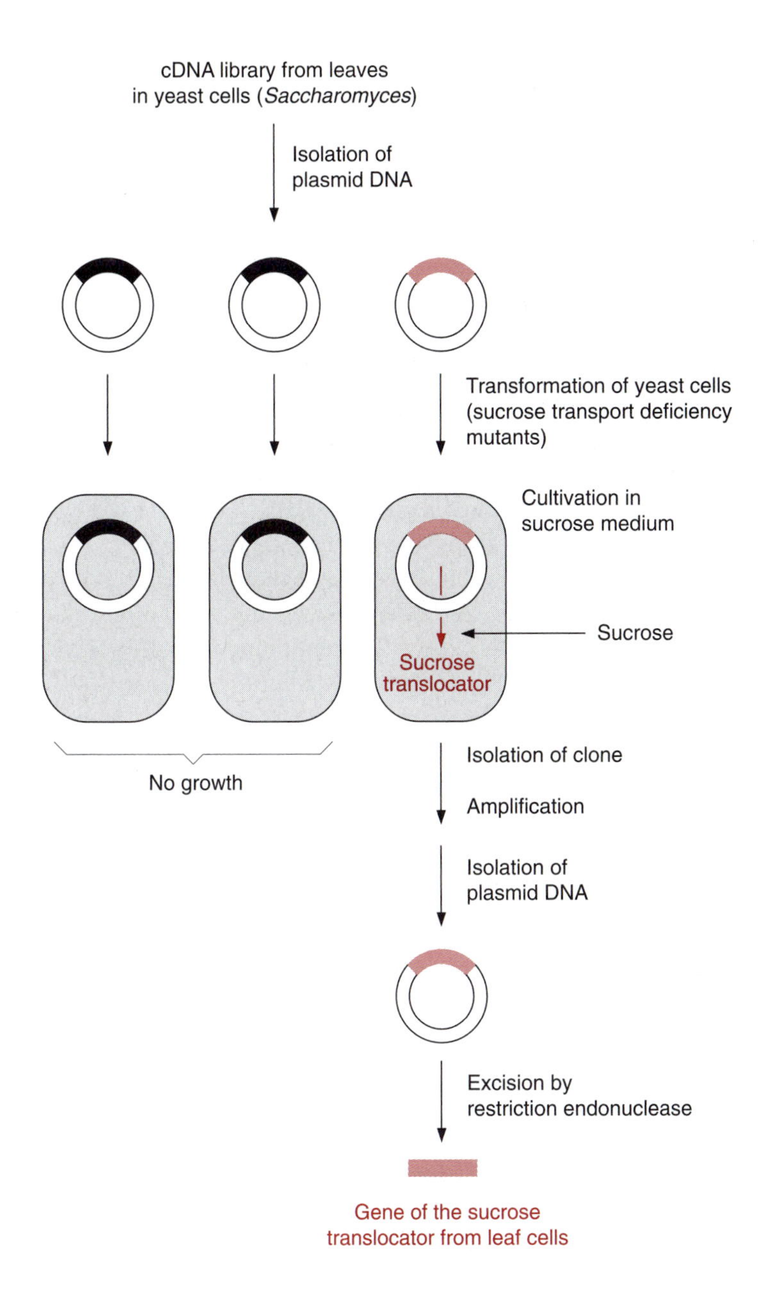

Figure 22. 10
Identification of a plant gene by complementation of a yeast mutant deficient in sucrose uptake.

plant cDNA library had generated yeast cells that produced the plant sucrose translocator and incorporated it into their cell membrane. After this positive yeast clone had been amplified, plasmid cDNA was isolated and sequenced. The cDNA sequence yielded the amino acid sequence of the previously unknown sucrose translocator.

Genes can be tracked down with the help of transposons or T-DNA

Another possibility to identify genes, aside from the function of their gene products, is to use transposons. Transposons are DNA sequences that can "jump" within the genome of a plant, which sometimes results in a recognizable elimination of a gene function (section 20.4). This could, for example, be the loss of the ability to synthesize a flower pigment. In such a case, a gene probe based on the known transposon sequence is used to identify by DNA hybridization the region of the genome in which the transposon has been inserted. By using the cloning and screening procedures already mentioned, it is possible to identify a gene, in our example a gene for the enzyme of flower pigment synthesis, and to determine the amino acid sequence of the corresponding enzyme by DNA sequence analysis. Labeling a gene with an inserted transposon is called **gene tagging**.

An alternative method, which has superseded the transposon technique of tracking genes, is the T-DNA technique, which will be described in the next section. Inserting T-DNA into the genome causes random mutations (**T-DNA-insertion mutants**). The gene locus into which the T-DNA has "jumped" is identified by using a gene probe and subsequently is sequenced. The complete sequence of the mutated gene is then identified, usually by comparing the relevant gene section with a database of the known sequence of T-DNA insertion mutants. For instance, the firm Syngenta makes a data bank available with the sequences of about 100,000 T-DNA insertion mutants of *Arabidopsis*.

22.2 Agrobacteria have the ability to transform plant cells

Gram-negative soil bacteria of the species ***Agrobacterium tumefaciens*** (new nomenclature: *Rhizobium radiobacter*) induce a tumor growth at wounding sites in various plants, often on the stem, which can lead to the formation of **crown galls**. The tumor tissue from these galls continues to grow as a callus in cell culture. As described in sections 19.3 and 19.5, mature differentiated plant cells, which normally no longer divide, can be stimulated to unrestricted growth by the addition of the phytohormones **auxin** and **cytokinin**. In this way a tumor, in the form of a callus, can be formed from a differentiated plant cell. The gall is formed by agrobacteria in basically the same way. The stricken plants are forced to produce high concentrations

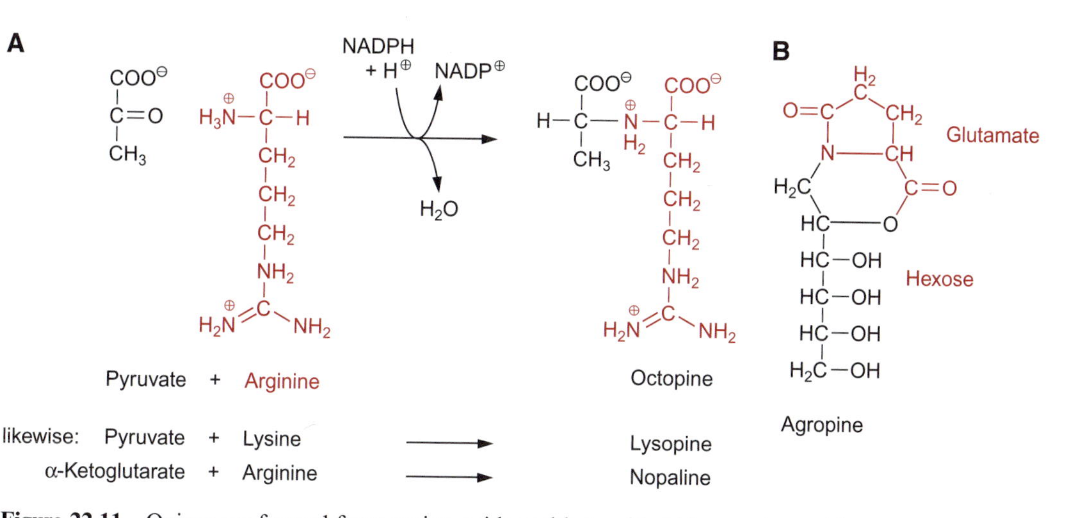

Figure 22.11 Opines are formed from amino acids and ketoacids (A) or amino acid and a hexose (B).

of auxin and cytokinin, resulting in a proliferation of the plant tissue leading to tumor growth. Since the capacity for increased cytokinin and auxin synthesis is inherited after cell division by all the succeeding cells, a callus culture of crown gall tissue can be multiplied without adding phytohormones.

The crown gall tumor cells have acquired yet another ability: They can produce a variety of products named **opines** by condensating amino acids and α-ketoacids or amino acids and sugars. These opines are formed in such high amounts that they are excreted from the crown galls. Figure 22.11 shows **octopine**, **lysopine**, **nopaline,** and **agropine** as examples of opines. Each *Agrobacterium* strain induces the synthesis of only a single opine. The synthesis of the first three opines mentioned proceeds via condensation to form a Schiff base and a subsequent reduction by NADPH. The opines are so stable that they cannot be metabolized by most soil bacteria. *Agrobacterium tumefaciens* strains have specialized in utilizing these opines. Normally, a particular opine can be catabolized only by that strain which has induced its synthesis in the plant. In such a way the stricken plants are forced to produce a special nutrient that can be consumed only by the corresponding *Agrobacterium.*

The conversion of differentiated plant cells to opine-producing tumor cells occurs without the agrobacteria entering those cells. This reprogramming of plant cells is caused by the transfer of functional genes from the bacteria to the genome of the stricken plant cells. *The agrobacteria have acquired the ability to transform plants in order to use them as a*

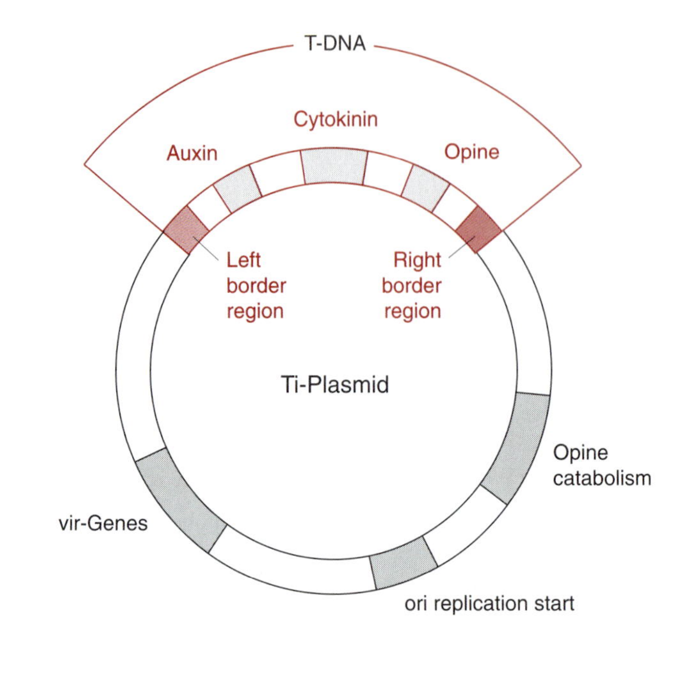

Figure 22.12 Diagram of a Ti plasmid, (not to scale). The T-DNA that is transferred to the genome of the plant, representing about 7% to 13% of the Ti plasmid, is defined by its left and right border sequences. The T-DNA contains genes encoding enzymes for the synthesis of the phytohormones cytokinin and auxin, and of a specific opine. The T-DNA is transcribed and translated only in the plant cell. The remaining part of the plasmid is transcribed and translated in the bacterium and contains several *vir* genes as well as a single or several genes for opine catabolism. The *ori* region represents the replication start. (After Glick and Pasternak.)

production site for their nutrient. Jeff Schell named this novel parasitism **genetic colonization**.

The Ti plasmid contains the genetic information for tumor formation

The phytopathogenic function of *A. tumefaciens* is encoded in a tumor-inducing **Ti plasmid**, with a size of about 200 kbp (Fig. 22.12). Strains of *A. tumefaciens* containing no Ti plasmids are unable to induce the formation of crown galls.

Plants are infected by bacteria at wounds, frequently occurring at the stem base. After being wounded, plants excrete phenolic substances as a defense against pathogens (Chapter 18). These phenols are used by the agrobacteria as a signal to initiate the attack. Phenols stimulate the expression of about 11 virulence genes (**vir genes**) located within a region of the Ti plasmid. These vir genes encode virulence proteins, which enable the transfer of the bacterial tumor-inducing genes to the plant genome. From a 12 to 25 kbp long section of the Ti plasmid, named **T-DNA** (T, transfer), a single strand is excised by a **vir nuclease**. The cleavage sites are defined by **border sequences** present at both ends of the T-DNA. The transfer of the T-DNA single strand from the bacterium to the plant cell nucleus (Fig. 22.13)

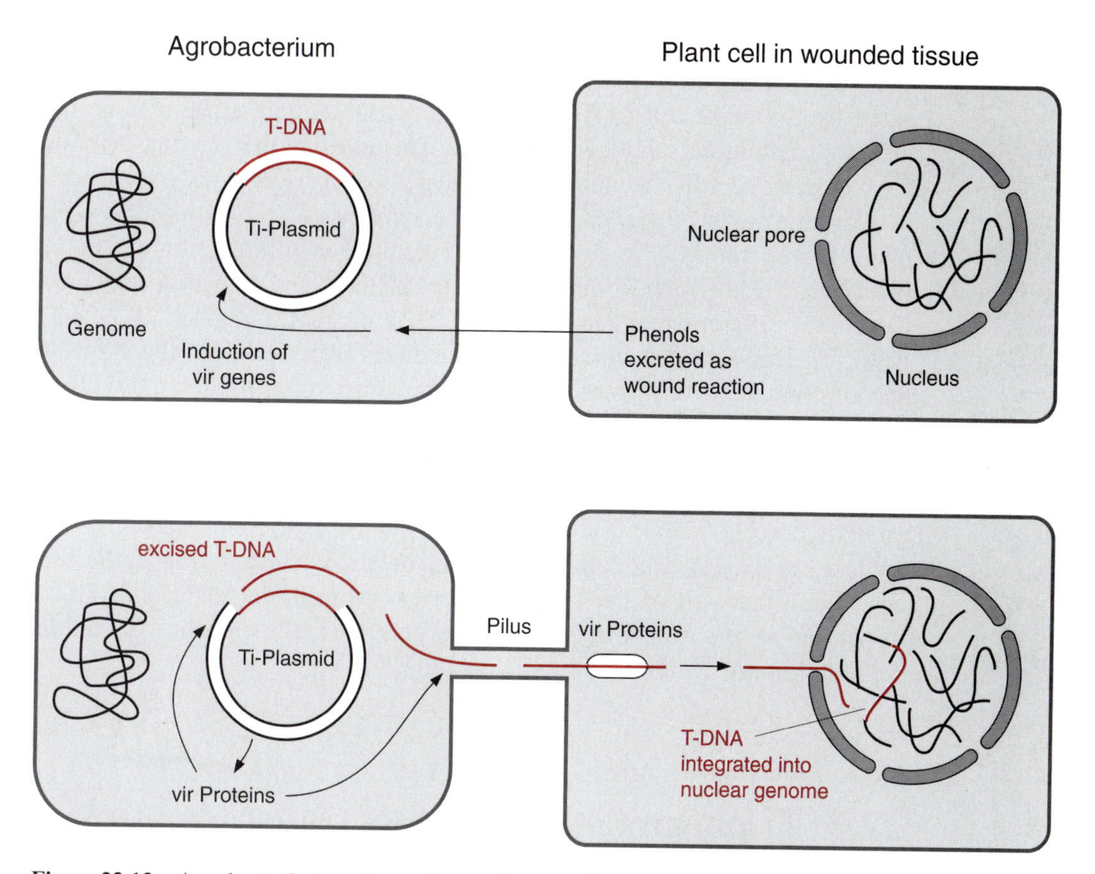

Figure 22.13 Agrobacteria transform plant cells to make them produce opines as their nutrient. Wounded plant tissues induce the expression of virulence genes on the Ti plasmid of the agrobacterium, resulting in the synthesis of virulence proteins, which cause a single-stranded DNA segment, known as T-DNA, to be excised from the plasmid, transferred to the plant cell, and integrated in the nuclear genome.

proceeds in analogy to the conjugation, which is a form of bacterial sexual propagation. To connect the bacterium to the plant cell, vir proteins form a **pilus**, a threadlike structure through which the T-DNA is conducted. After being transferred into the plant cell, the T-DNA strand migrates further into the nucleus. Vir-encoded proteins protect the T-DNA from being attacked en route by DNA-degrading plant enzymes and also facilitate the transport through the nuclear pores to the nucleus. The right border region of the T-DNA has an important function in its integration into the plant nuclear genome. The T-DNA is integrated randomly into chromosomes and, when inserted in a gene, can eliminate the function of this gene. This can be used to identify a gene by gene tagging in an analogous way as the gene tagging by the transposons described at the end of section 22.1.

The T-DNA integrated in this way in the nuclear genome has the properties of a eukaryotic gene and is inherited in a Mendelian fashion. It is replicated by the plant cell as if it were its own DNA and, since it contains eukaryotic promoters, is also transcribed. The resultant mRNA corresponds to a eukaryotic mRNA and is translated as such. The T-DNA encodes cytokinin synthase, a key enzyme for the synthesis of the cytokinin **zeatin** (Fig. 19.11), as well as two enzymes for the synthesis of the auxin **indoleacetic acid** (IAA) from tryptophan. This bacterial IAA synthesis proceeds in a different manner than plant IAA biosynthesis (Fig. 19.7), but details of this pathway will not be considered here. Moreover, the T-DNA encodes one or two enzymes, different from strain to strain, for the synthesis of a special opine. The T-DNA integrated in the plant genome thus carries the genetic information for the synthesis of cytokinin and auxin to induce tumor growth, as well as the information for the plant to synthesize an opine.

The enzymes required for catabolism of the corresponding opine are encoded in that part of the Ti plasmid that remains in the bacterium. Thus, in parallel to the transformation of the plant cell, the enzymes for opine degradation are synthesized by the bacteria.

22.3 Ti plasmids are used as transformation vectors

Its ability to transform plants has made *A. tumefaciens* an excellent tool for integrating foreign genes in their functional state in a plant genome. It was necessary, however, to modify Ti plasmids before using them as vectors (Fig. 22.14). The genes for auxin and cytokinin synthesis were removed to prevent tumor growth in the transformed plants. Since synthesis of an opine is unnecessary for a transgenic plant and would be a burden on its metabolism, the genes for the opine synthesis also were removed. Thus the T-DNA is defined only by the two border sequences. In order to insert a foreign gene between these two border sequences, it was necessary to incorporate a DNA sequence containing cleavage sites for several restriction endonucleases (known as a **polylinker sequence** or a multicloning site) within the T-DNA region of the Ti plasmid. The Ti plasmid could then be cleaved in the T-DNA region by a certain restriction endonuclease. A foreign DNA sequence, excised by the same restriction endonuclease, can be inserted in this cleavage site (see Fig. 22.3). Since the polylinker sequence contains cleavage sites for several restriction endonucleases, several DNA molecules can be inserted

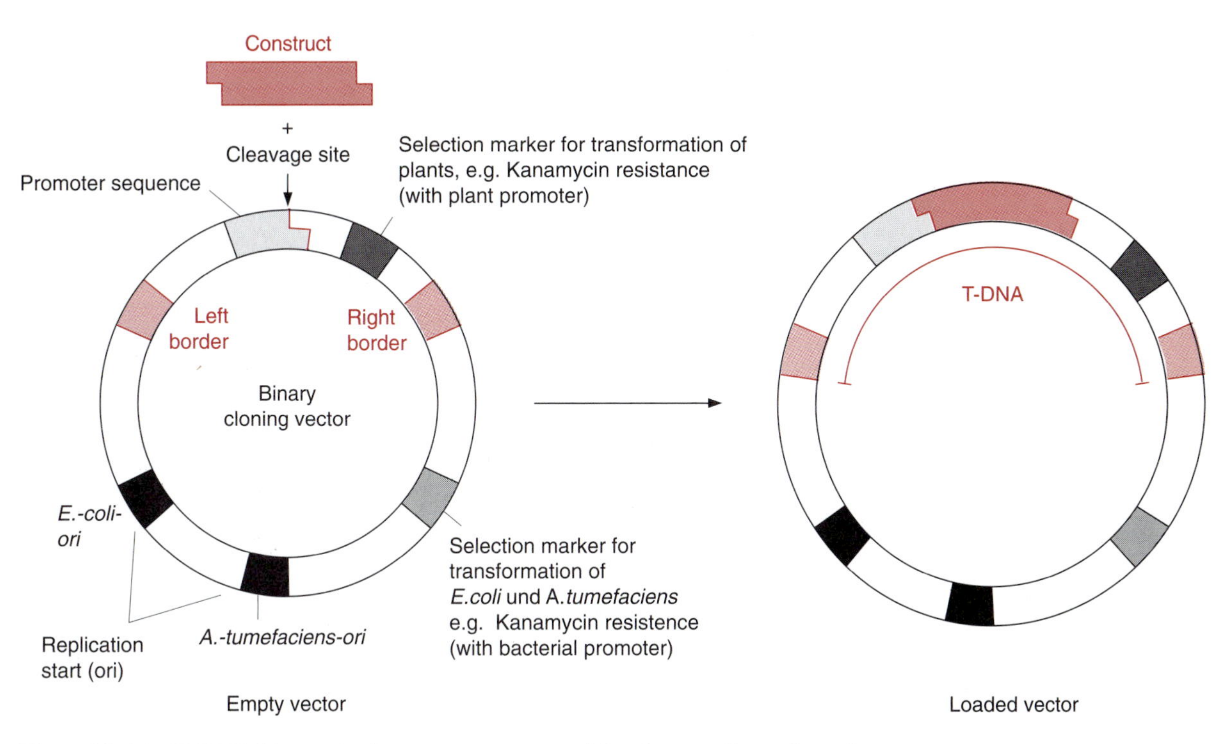

Figure 22. 14 T-DNA binary cloning vector constructed for the transformation of plant cells.

sequentially in the T-DNA. In Figure 22.14, a promoter, enabling the expression of the inserted DNA in the host cell, is located to the left of the cleavage site.

In modern transformation systems, vectors derived from the Ti plasmid no longer contain any vir genes and are therefore unable to transform a plant cell on their own. For transformation they require the assistance of a second so-called **helper plasmid**, which contains the vir genes, but no T-DNA and is therefore also unable to transform a plant on its own. Since both of these vectors are required together for a transformation, they are called **binary vectors**. A large variety of such vectors have been designed for special applications and are available commercially.

In order to transform a plant, sufficient amounts of vectors containing the gene to be transferred must be present. It proved to be advantageous to first amplify the vectors by cloning them in *E. coli* (Fig. 22.15). Since *E. coli* does not recognize the replication start site of the natural Ti plasmid (***A. tumefaciens*-ori**), a second replication start (***E. coli-ori***) is introduced into the plasmid.

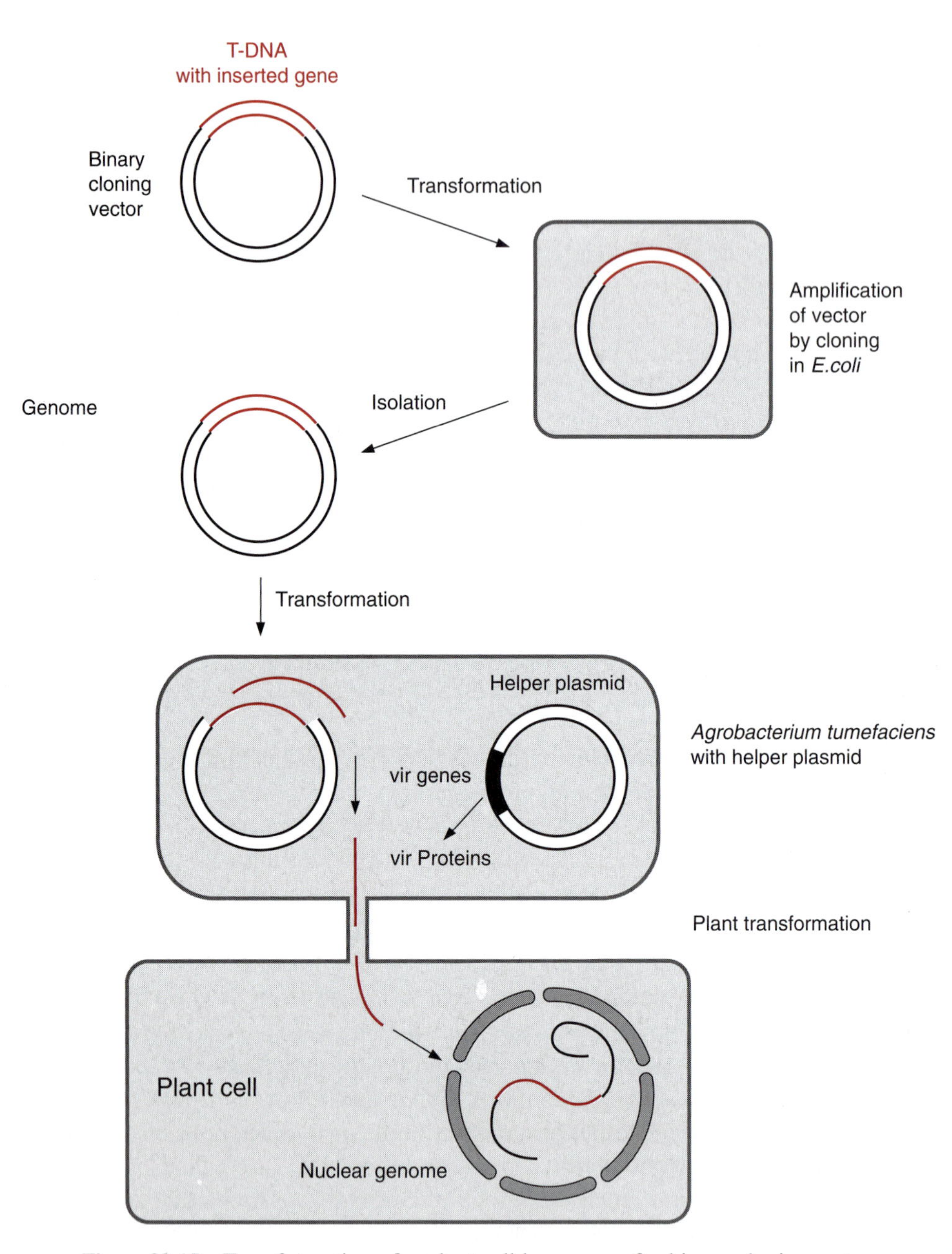

Figure 22.15 Transformation of a plant cell by means of a binary cloning system. After bacterial transformation, the T-DNA vector containing the DNA insert is propagated in *E. coli*, isolated, and transferred into *Agrobacterium*. A helper plasmid, already present in *Agrobacterium*, encodes the virulence proteins required for the transformation of plant cells.

The plasmid is provided with a selection marker in order to select those *E. coli* cells that have been transformed by the Ti plasmid. For this a gene encoding **neomycin phosphotransferase** is frequently used. This enzyme degrades the antibiotic **kanamycin**, thus rendering the cell resistant to this antibiotic. Kanamycin is very seldomly used in medicine, which is an important aspect in regard to transgenic plants containing this antibiotic resistance gene. The kanamycin resistance gene is provided with a bacterial promoter and, after transformation of the vector, is therefore expressed in *E. coli* cells. When kanamycin is added to the culture medium of the bacteria, only the transformed bacteria survive, which are protected from the antibiotic by the resistance gene on the vector. Thus the Ti plasmid can be propagated efficiently by cloning in *E. coli.*

As hosts for the binary vectors, *A. tumefaciens* strains are used that do not contain complete Ti plasmids, but only the helper plasmids mentioned previously. These encode the vir proteins required for the transfer of the T-DNA from the binary vector into the plant genome (Fig. 22.15).

A new plant is regenerated following transformation of a leaf cell

After transformation, the few transformed plant cells are selected from the large number of nontransformed cells by another selection marker contained in the T-DNA vector. Mostly the kanamycin resistance gene described previously is also used for this purpose, but in this case provided with a plant promoter.

It was mentioned in the introduction to this chapter that *A. tumefaciens* attacks plants at wounds. Leaf discs therefore, with their cut edges, are a good target for performing a transformation (Fig. 22.16). The leaf discs are immersed in a suspension of *A. tumefaciens* cells transformed by the vector. After a short time, the discs are transferred to a culture medium containing agarose, which, besides nutrients, contains the phytohormones **cytokinin** and **auxin** to induce the cells of the leaf disc to grow a callus. The addition of the antibiotic kanamycin kills all plant cells except the transformed cells, which are protected from the antibiotic by the resistance gene. The remaining agrobacteria are killed by another antibiotic specific for bacteria. The cut edges of the leaf discs are the site where the calli of the transformed cells develop. When the concentrations of cytokinin and auxin are appropriate, these calli can be propagated almost without limit in tissue culture on agarose culture media. In this way transformed plant cells can be kept and propagated in tissue culture for very long periods of time. If required, new plants can be regenerated from these tissue cultures.

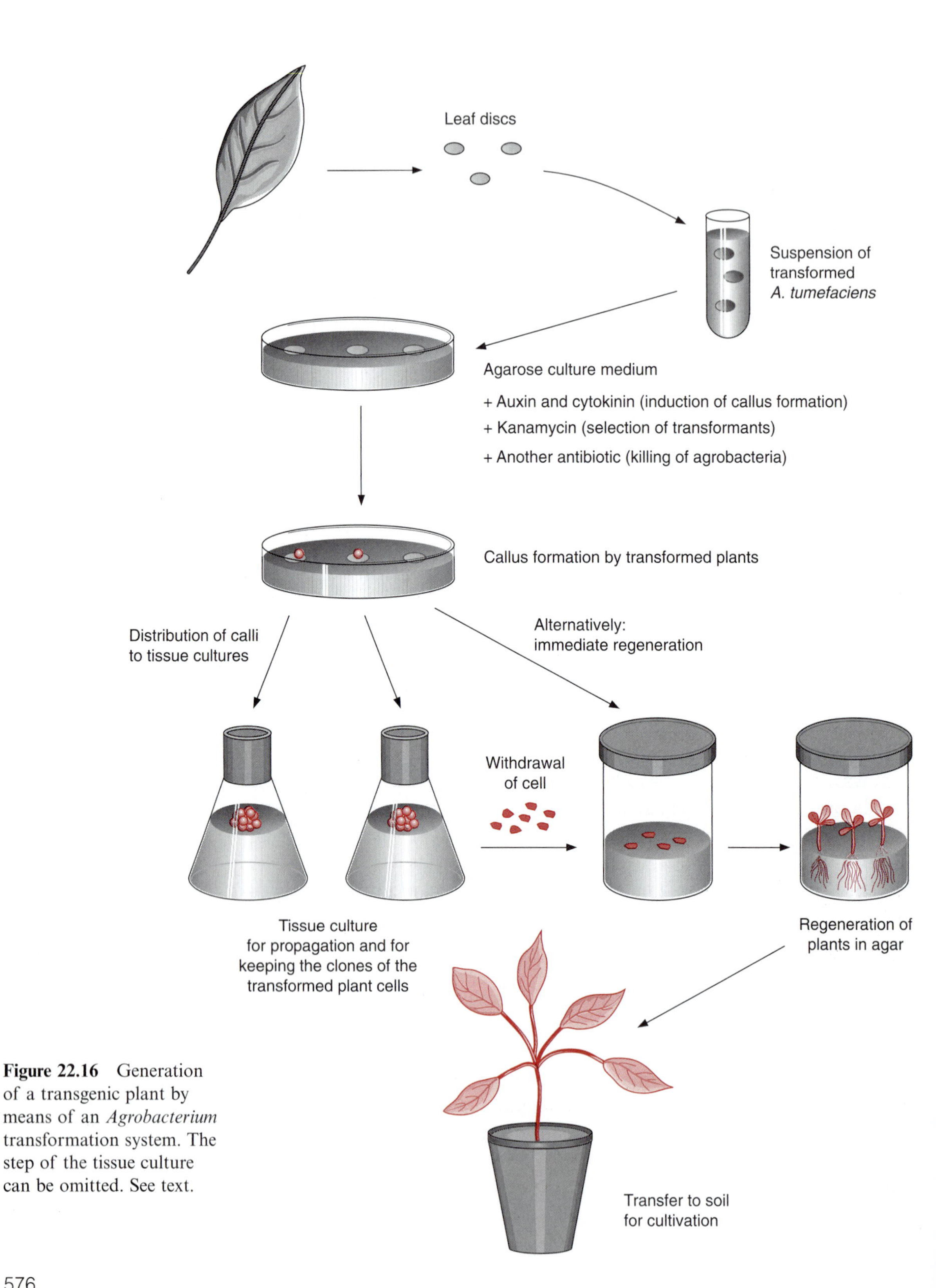

Figure 22.16 Generation of a transgenic plant by means of an *Agrobacterium* transformation system. The step of the tissue culture can be omitted. See text.

To regenerate new plants, cells of the callus culture are transferred to a culture medium containing more cytokinin than auxin, and this induces the callus to form shoots. Root growth is then stimulated by transferring the shoots to a culture medium containing more auxin than cytokinin. After plantlets with roots have developed and grown somewhat, they can be transplanted to soil, where in most cases they develop to normal plants, capable of being multiplied by flowering and seed production.

The pioneering work of Jeff Schell, Marc van Montagu, Patricia Zambryski, Robert Horsch, and several others has developed the *A. tumefaciens* transformation system to a very easy method for transferring foreign genes to cells of higher plants. Nowadays it is often possible even for students to produce several hundred different transgenic tobacco plants with no great difficulty.

Using this method, more than 100 different plant species have been transformed successfully. Initially, it was very difficult or even impossible to transform monocot plants with the *Agrobacterium* system. Recently, this transformation method has been improved to such an extent, that it also can now be successfully applied to transform several monocots, such as rice. An alternative way to transform plant cells is a physical gene transfer, the most successful being the bombardment of plant cells by **microprojectiles**.

Plants can be transformed by a modified shotgun

Transformation by bombardment of plant cells with microprojectiles was developed in 1985. The microprojectiles are small spheres of tungsten or gold with a diameter of 1 to 4 μm, which are coated with DNA. A **gene gun** (similar to a shotgun) is used to shoot the pellets into plant cells (Fig. 22.17). Initially, gunpowder was used as propellant, but nowadays the microprojectiles are often accelerated by compressed air, helium, or other gases. The target materials include calli, embryonic tissues, and leaves. In order to penetrate the cell wall of the epidermis and mesophyll cells, the velocity of the projectiles must be very high and can reach about 1,500 km/h in bombardments with a gene gun in a vacuum chamber. The cells in the center of the line of fire may be destroyed during such a bombardment but, because the projectiles are so small, the cells nearer the periphery survive. The DNA transferred to the cells by these projectiles can be integrated not only in the nuclear genome, but also in the genome of mitochondria and chloroplasts. This makes it possible to transform mitochondria and chloroplasts. In some plants, the gene gun works especially well. Thus, by bombardment of embryonic callus cells of sugarcane, routinely 10 to 20 different transformed plant lines are obtained by one shot.

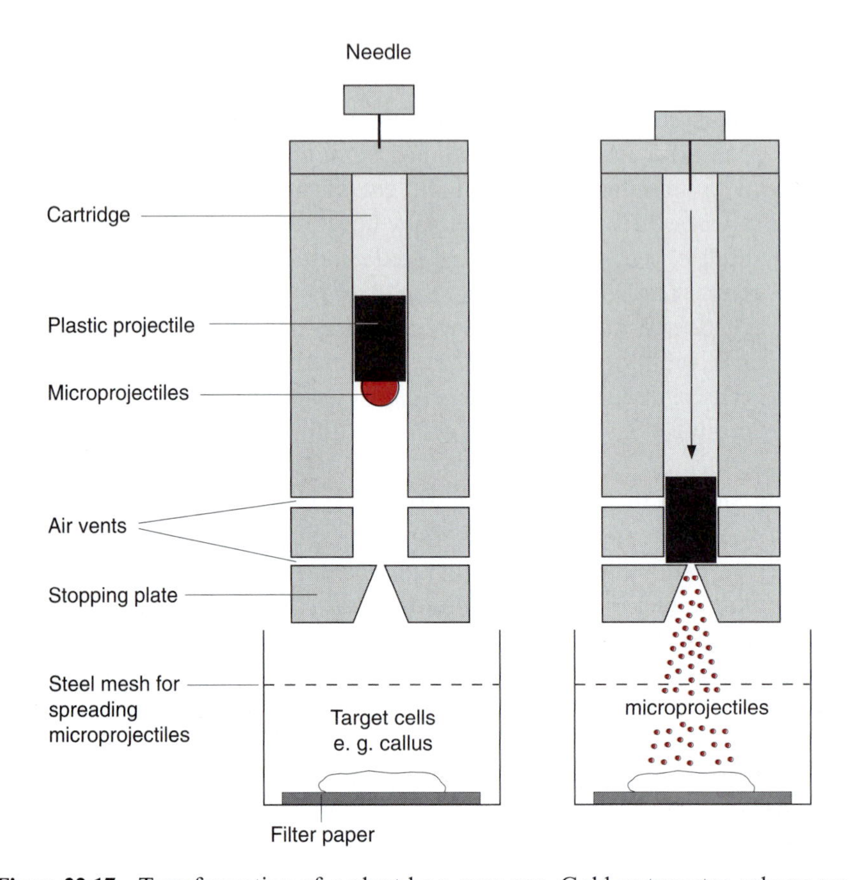

Figure 22.17 Transformation of a plant by a gene gun. Gold or tungsten spheres are coated with a thin DNA layer by a deposit of $CaCl_2$. The spheres are inserted in front of a plastic projectile into the barrel of the gun. The gun and the target are in an evacuated chamber. When the gun is fired, the plastic projectile is driven to the stopping plate, by which the microprojectiles are driven through holes and shot at a high velocity into the cells of the plant tissue. The DNA carried in this way into the plant cells contains an antibiotic resistance gene (e.g., for kanamycin resistance) as well as the transferred gene. Transformed cells can be selected by this marker in the same way as for cells transformed with *A. tumefaciens*. The transformed cells are propagated by callus tissue culture and from these plants are regenerated. (After Hess, Biotechnologie der Pflanze, Verlag Ulmer, Stuttgart, 1992.)

Protoplasts can be transformed by the uptake of DNA

The transformation of protoplasts is another way to transfer foreign genetic information to a plant cell. Protoplasts can be obtained from plant tissues by digestion of the cell walls (section 1.1). Protoplasts are able to take up foreign DNA in the presence of $CaCl_2$ and polyethylene glycol, and often integrate it in their genome. This transformation resembles that of bacteria by plasmids. In protoplast transformation, the gene to be transferred is linked with a selection marker encoding resistance to an antibiotic. After the antibiotic has been added, only the transformed protoplasts survive. In principle, the protoplasts of all plants can be transformed in this way, since there is no host specificity involved, as in the case of transformation by *A. tumefaciens.* However, the use of this method is restricted to a few plant species where it has been possible to regenerate intact fertile plants from protoplasts. Protoplast transformation has been employed successfully for the transformation of maize and rice.

The use of plastid transformation for genetic engineering of plants is of advantage for the environment

Recently, the **transformation of chloroplasts** has gained importance. Transgenic plants obtained by transformation of the nuclear genome (e.g., by the Ti-plasmid) could pass their genetic information via pollen to cultivars in neighboring fields or in some cases also to related wild plants. This can lead to undesirable cross-breeding (e.g., causing the generation of herbicide-resistant weeds in the neighborhood of herbicide-resistant cultivars). This problem can be avoided when the genetic transformations are made in the plastid genome of the plant. Since in most cases the plastid genome is **inherited maternally**, the genetic alteration will not spread to other plants by pollen.

The **gene gun** is usually used for the transformation of chloroplasts. The foreign DNA that is to be integrated into the plastid genome is at both ends provided with sequence sections, which are identical to sequences in the plastid genome (Fig. 22.18). After the foreign DNA, with the aid of the gene gun, has entered the plastids, it can be integrated into the plastid genome by **homologous recombination** at a site defined by the sequences at both its ends. Whereas in the plant nuclear genome, homologous recombinations are rare events, these occur frequently in the plastid genome. In this way random mutations are avoided, as in the case when Ti-plasmids are used for transformation of the nuclear genome. It is a problem that plant cells contain many plastids, each with 10 to 100 genomes. By repetitive selection and regeneration, it is possible to achieve transgenic lines in which practically all the plastid genomes have integrated the foreign DNA. Plants in which each cell contains many hundred copies of a foreign gene can be cultivated. This has the advantage that these transformed plants can produce large amounts of foreign proteins (up to 46% of the soluble protein). This is interesting when the plants are to be used for the synthesis of defense substances or pharmaceuticals. Another advantage of the plastid transformation is that

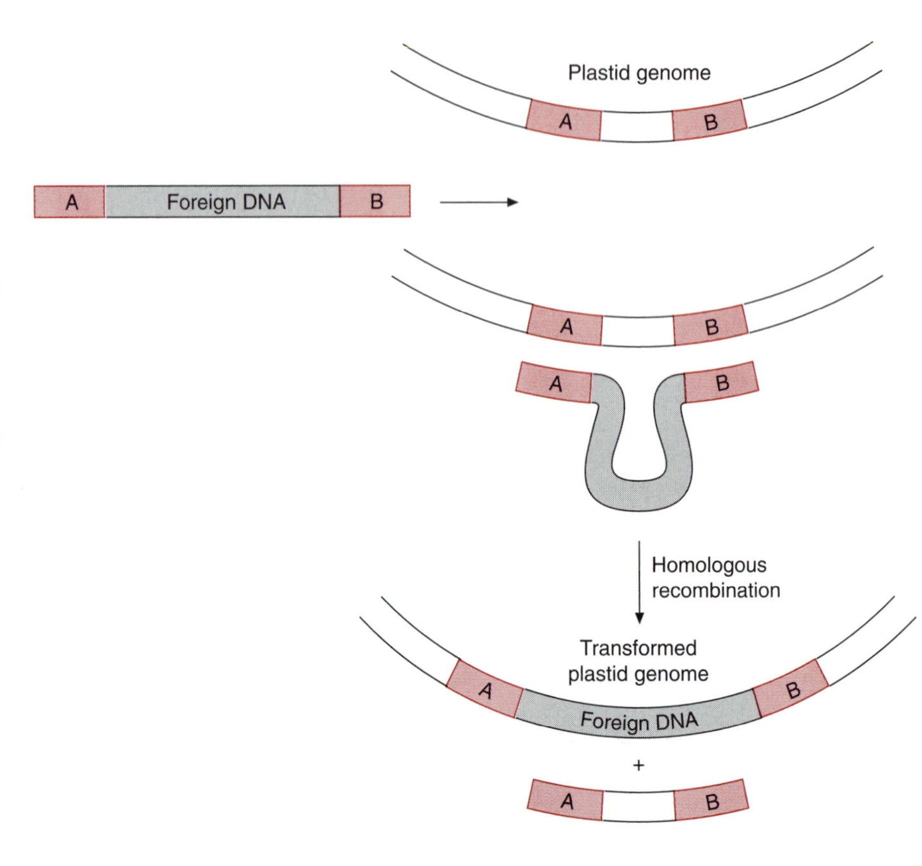

Figure 22.18
Transformation of the plastid genome. With the aid of a gene gun, foreign DNA is transported into the plastids. Beforehand both sides of the DNA were given sequence sections analogous to those of the plastid genome. The foreign DNA is integrated in the plastid genome by being exchanged by homologous recombination for the sections of the plastid DNA bordered by the A and B sequences.

plastid genes can be transcribed **polycistronically**. While the DNA present in the nucleus usually is transcribed to monocistronic mRNAs, which are then translated into only one protein, most of the plastid genes are transcribed to polycistronic mRNAs, which encode several proteins and afterward are processed to single translatable mRNAs. This property of the plastid genome makes it relatively easy to integrate several foreign genes (e.g., for a synthesis pathway) in one step into the plastid genome. It should be noted, however, that in certain cases such can also be achieved by the transformation of the nuclear genome. Another advantage of the plastid transformation is that proteins with a **disulfide bridge** can be formed in the plastids, which is not possible in the cytosol. As described earlier, chloroplast enzymes are regulated by the oxidation of adjacent -SH groups (Fig. 6.25). Because of this ability, the plastid compartment is well suited to produce, after genetic transformation, animal proteins, such as antibodies or oral vaccines, with disulfide bridges in the correct folding.

Plastid transformation as a method of plant biotechnology, at present still at its beginning, may have a great future, because the transformants are compatible with the environment, they cannot spread their foreign genes to other plants by pollen, and very high amounts of foreign proteins can be produced in this way in a plant.

22.4 Selection of appropriate promoters enables the defined expression of an inserted gene

Any foreign gene transferred to a plant can be expressed only when it has been provided with a suitable promoter sequence (section 20.2). The selection of the promoter determines where, when, how much, and under what conditions gene expression takes place. These promoters usually are already included in the commercially available vectors (see Figure 22.14).

To have a high expression of an inserted foreign gene in all parts of the plants, the **CaMV-35S promoter** from the *Cauliflower mosaic virus* (section 20.5) is often used. This promoter contains transcription enhancers that enable a particularly high transcription rate in different tissues of very many plants. The ***nos* promoter** from the Ti plasmid of *A. tumefaciens*, normally regulating the gene for nopaline synthesis, is also often used as a nonspecific promoter in transgenic plants.

The DNA sequences of specific plant promoters are determined by the analysis of plant genes obtained from a genomic gene library (section 22.1). A promoter for specific gene expression in the potato tuber, for instance, has been identified by analysis of the DNA sequence of the gene for patatin, the storage protein of the potato (Chapter 14).

A **reporter gene** can be used to determine whether an isolated promoter sequence is tissue-specific. Such reporter genes encode proteins that are easily detected in a plant, for instance, the **green fluorescent protein** (GFP) from the jellyfish *Aequorea victoria*, which can be directly seen by confocal microscopy in intact leaves. Frequently, the gene for the **β-glucuronidase (GUS)** enzyme from *E. coli* also is used as a reporter gene (Fig. 22.19). This enzyme resembles β-galactosidase (mentioned in section 22.1) and does not occur in plants. It hydrolyzes a synthetic X-glucuronide to give a blue hydrolysis product, which can be identified readily under the microscope. When a potato plant is transformed with the GUS reporter gene fused to the patatin promoter mentioned previously, after addition of X-glucuronide, a deep blue color develops only in cuts of the tubers, but not in other tissues of the plant. This indicates that the patatin promoter acts as a specific promoter for gene expression in the potato tubers. Many promoters have been isolated that are active only in certain organs or tissues, such as leaves, roots, phloem,

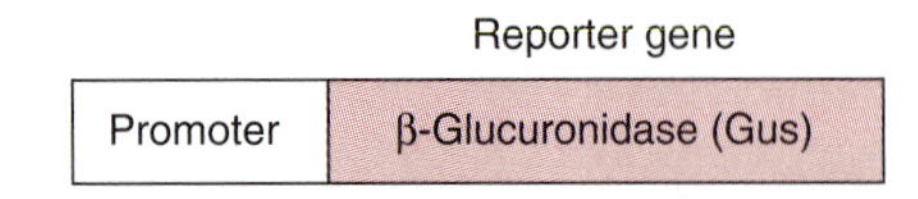

Figure 22 19 In order to check the function of a promoter, it is linked to a reporter gene. In the example shown, the reporter gene encodes the enzyme β-glucuronidase from *E. coli*. X-Glucuronide is hydrolyzed by this enzyme and a blue-stained product is formed analogously to the β-galactosidase reaction (see Fig. 22.6).

flowers, or seed. Moreover, promoters have been isolated that control the expression of gene products only under certain environmental conditions, such as light, high temperatures, water stress, or pathogenic infection. Frequently, promoters are also active in heterologous plant species, although the extent of expression by a certain promoter can vary between different plant species.

Gene products are directed to certain subcellular compartments by targeting sequences

In transgenic plants, the expression of a foreign gene can be restricted to certain tissue or cell types by the use of defined promoters. Furthermore, a protein encoded by a foreign gene can be directed to a particular subcellular destination (e.g., the chloroplast stroma or the vacuolar compartment) by the presence of additional amino acid **presequences,** which serve as **targeting signals** for transfer via the various subcellular membrane transport systems (sections 14.5, 21.3). Many targeting sequences are now available, making it possible to direct a protein encoded by a foreign gene to a defined subcellular compartment, such as the vacuolar compartment or the chloroplast stroma.

22.5 Genes can be turned off by transformation

In addition to producing transgenic plants with new properties by the transfer and expression of foreign genes, it is of great interest to decrease or even eliminate certain properties by inhibiting the expression of the corresponding gene, for instance the activity of a translocator or of an enzyme. Turning off a gene is also an important way to identify its function by investigating the consequences of the missing gene on plant metabolism.

In prokaryotic genomes (including the mitochondrial and plastid genomes) as well as in the nuclear genome of animals, **homologous recombinations** frequently occur. This property is utilized in animals to eliminate the function of a defined gene by generating a so-called **"knock out"** mutant.

To eliminate the function of a gene, a section of it is isolated and cloned, and after being altered by genetic techniques, it is reintroduced into the nucleus, where it can be incorporated by homologous recombinations into the gene in question and thus eliminate its function. In plants, however, this technique usually does not work, since in the nuclear genome of plants homologous recombinations are rare. Therefore alternative methods are required in plants. A decrease in the expression of a gene can be achieved by inactivating the encoding mRNA by the synthesis of a complementary RNA, called **antisense RNA** (Fig. 22.20). Normally, mRNA occurs as a single strand, but in the presence of a complementary RNA strand, it can form a double-stranded RNA, which is unstable and degraded rapidly by ribonucleases. This may be why mRNA loses the ability to encode protein synthesis when the corresponding antisense RNA is present.

To synthesize the antisense RNA, the corresponding gene is first isolated according to the methods already described and then inserted as cDNA in **reverse orientation** in a vector, which is utilized to transform the plant in question. As the orientation is reversed in this inserted gene (the promoter is positioned at the wrong side), its transcription results in the formation of the antisense RNA. This **antisense technique** has become a very important tool in plant genetic engineering for specifically decreasing the expression of a defined gene, but it does not always lead to the desired results. In many cases, it turned out to be impossible to decrease by this method the synthesis of a protein (e.g., an enzyme or a translocator) by more than 90%, when the activity of the remaining protein could be even enhanced by regulation.

An alternative way to reduce the activity of a gene is **co-suppression**, where an additional copy of an endogenous gene is introduced by transformation. The overexpression of the same gene or a similar gene frequently results in a decrease in activity not only of the inserted gene but also of the gene already present. The mechanism of this effect is not yet fully understood. Co-suppression can be more efficient than antisense inhibition, but it functions only in certain cases.

A very effective new technique for eliminating the expression of a defined gene is the **RNAi-technique** (RNA-mediated-interference). In this method, a double-stranded RNA (**dsRNA**) is prepared of a small RNA section (up to 30 bases) of the mRNA to be inactivated. It was found that the presence of such dsRNA very effectively inhibits the translation of the respective mRNA and causes its degradation by the cell's nucleases. Therefore it is possible to eliminate the expression of a certain gene by transforming a plant with DNA encoding both strands of the dsRNA. Since the RNAi-technique appears to have advantages over the antisense technique mentioned previously, it is widely used now.

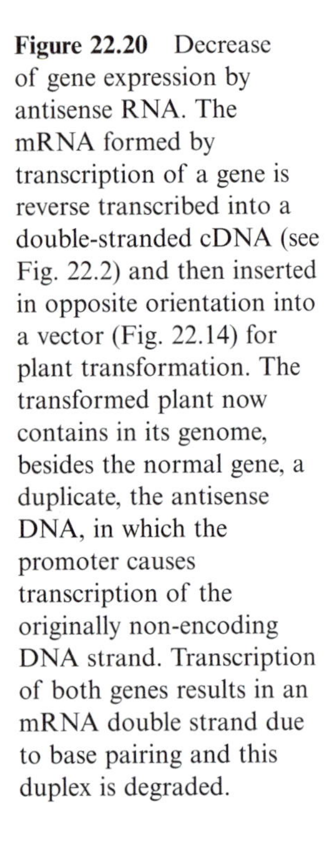

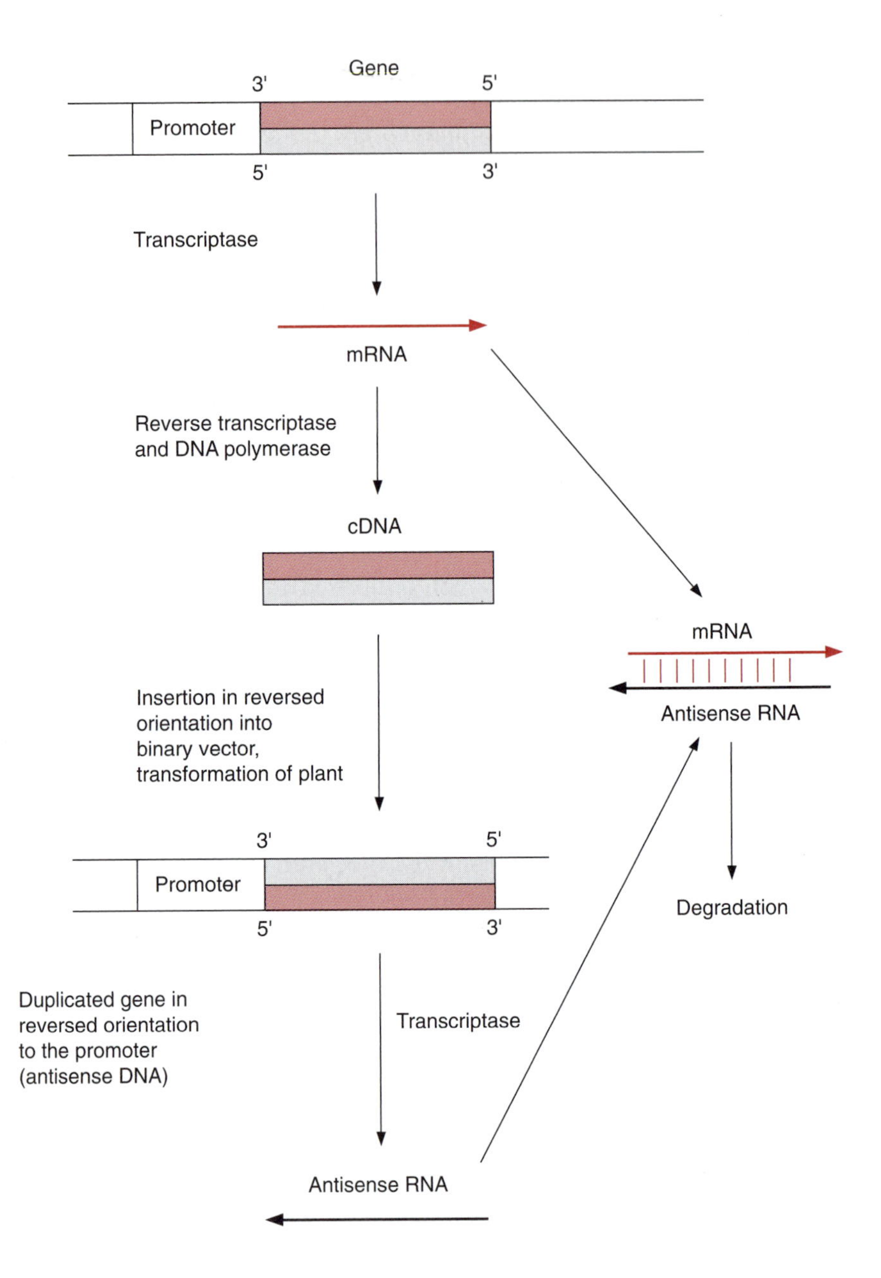

Figure 22.20 Decrease of gene expression by antisense RNA. The mRNA formed by transcription of a gene is reverse transcribed into a double-stranded cDNA (see Fig. 22.2) and then inserted in opposite orientation into a vector (Fig. 22.14) for plant transformation. The transformed plant now contains in its genome, besides the normal gene, a duplicate, the antisense DNA, in which the promoter causes transcription of the originally non-encoding DNA strand. Transcription of both genes results in an mRNA double strand due to base pairing and this duplex is degraded.

For many plants, the transformation by *A. tumefaciens* is a simple way to eliminate gene functions at random (see Chapter 20.1). Plants take up the T-DNA of the Ti plasmids and integrate them randomly into the genome. When the T-DNA is integrated into a gene, this gene is mutated (**T-DNA insertion mutant**) and therefore loses its function. The mutant can be subsequently identified by a probe for the employed T-DNA.

22.6 Plant genetic engineering can be used for many different purposes

The method of genetic engineering via the *Agrobacterium* system has produced revolutionary results in basic science as well as in applied plant science. In basic science, it has led in a very short time to the identification and characterization of many new proteins, such as enzymes and translocators. When investigating the function of a protein in a plant, it is now common to increase or decrease the expression of the protein by molecular genetic transformation (see previous sections). From the effects of these changes, conclusions can be drawn about the role of the corresponding protein in metabolism.

In agriculture, plant genetic engineering has been utilized in many ways to augment protection against pests, to increase the qualitative and quantitative yield of crop plants, and to produce sustainable raw materials for industrial purposes. By now many genetically altered (transgenic) cultivars are grown worldwide, especially in North and South America and in China. The exception is Europe, where until now only one single transgenic species of a cultivated plant (a BT maize, see below) has been grown, due to lack of public acceptance. By far the greatest part of the worldwide cultivated transgenic plants were developed with the aim of improving pest and weed management, with resistance to insects and herbicides (see section 10.4). In order to increase the harvest yield, genetically engineered male-sterile plants [e.g., of rapeseed (canola)] have been generated for producing hybrids. Transformants were also produced to improve the quality of the harvest [e.g., to improve storage properties of tomatoes (section 19.5)] or the production of customized fats in rape (section 20.7).

Plants are selectively protected against some insects by the BT protein

Field crops are in great danger of being attacked by insects. Some examples may illustrate this:

1. The Colorado beetle, originating in North America, can cause complete defoliation of potato fields.
2. The larvae of the corn borer, penetrating maize shoots, cause large crop damage by feeding inside the shoots.
3. In a similar way, the cotton borer prevents the formation of cotton flowers by feeding inside shoots.

It has been estimated that about one-sixth of the global plant food production is lost by insect pests. In order to avoid serious crop losses, the farmer often has no option but to use chemical pest protection. In former times, chlorinated hydrocarbons such as DDT or Aldrin were used as a very potent means of protection against insects. Since these substances are degraded only very slowly and therefore accumulate in the food chain, they cause damage to the environment and are now restricted in their use or even forbidden by law in many countries. Nowadays, organophosphorous compounds are mostly used as insecticides, which, as phosphoesterase inhibitors, impair the nerve function at the site of the synapses. These compounds are readily degraded, but unfortunately destroy not only pests, but also useful insects such as bees, and also are poisonous for humans. The threat to humans lies not so much in the pesticide residues in consumed plant material, but primarily to the people applying the insecticides.

For more than 30 years, preparations from *Bacillus thuringensis* have been used as alternative biological insecticides. These bacteria form toxic peptides (**BT proteins**) that bind to receptors in the intestine of certain insects, thus impairing the absorption of food. This inactivation of the intestinal function causes the insect to starve to death. More than 100 bacterial strains are known, which form different BT proteins with a relatively specific toxicity toward certain insects. Toxicological investigations have shown that BT proteins are not harmful to humans. For many years now, bacterial suspensions containing the BT protein have been used as a biological spray to protect crops from insects. Unfortunately, these preparations are relatively expensive and are easily washed off the leaves by rain. Spraying also has the disadvantage that it does not reach larvae inside plant shoots (e.g., the corn borer and cotton borer).

The genes for various BT proteins have been cloned and used to transform a number of plants. Although transgenic plants produce only very low amounts of the toxic BT protein (0.1% of total protein), it is more than enough to deter insects from eating the plant. The BT protein is decomposed in soil and is degraded in the human digestive tract just like all other proteins. On this basis, insect-resistant transformed varieties of maize, soybean, and cotton are widely grown in the United States. Their cultivation has resulted in a substantial reduction in the use of insecticides, easing threats to the entire fauna spectrum. The use of such transgenic plants thus may contribute to preserving the environment.

The insertion by genetic engineering of foreign genes encoding **proteinase inhibitors** is an alternative way to protect plants from insect pests (see section 14.4). After wounding (e.g., by insect attack or by fungal infection), the formation of proteinase inhibitors, which inhibit specific proteinases of animals and microorganisms, is induced in many plants. Insects feeding on these plants consume the inhibitor, whereby their digestive processes are disrupted with the result that the insect pest starves to death. The synthesis of the inhibitors is not restricted to the wound site, but often occurs in large parts of the plant and thereby protects them from further attacks. The introduction of suitable foreign genes in transgenic potato, lucerne (alfalfa), and tobacco plants enabled a high expression of proteinase inhibitors in these plants, efficiently protecting them from being eaten by insects. This strategy has the advantage that the proteinase inhibitors are not specific to certain insect groups. These proteinase inhibitors are contained naturally in many of our foods, sometimes in relatively high concentrations, but they are destroyed by cooking.

The expression of an **amylase inhibitor** in pea seeds, which prevents storage losses caused by the larvae of the pea beetle (section 14.4), is another example of how genetic engineering of plants offers protection from insect damage.

Plants can be protected against viruses by gene technology

Virus diseases can result in catastrophic harvest losses. Many crop plants are threatened by viruses. Infection with the ***Cucumber mosaic virus*** can lead to the total destruction of pumpkin, cucumber, melon, and courgette crops. In sugar beet, losses of up to 60% are caused by the viral disease **rhizomania**. In contrast to fungal or animal pests, viruses cannot be directly combated by the use of chemicals. Traditional procedures, such as decreasing the propagation of the viruses by crop rotation, are not always successful. Another way to control virus infections has been to attack the virus-transferring insects, especially aphids, with pesticides.

It has long been known that after infection with a weak pathogenic strain of a certain virus, a plant may be protected against infection by a more aggressive strain. This phenomenon has been applied successfully in the biological plant protection of squash plants. It was presumed that a single molecular constituent of the viruses caused this protective function, and this has been verified by molecular biology: The introduction of the **coat protein** gene of the *Tobacco mosaic virus* into the genome of tobacco plants makes them resistant to this virus. This has been confirmed for many other viruses: If a gene for a coat protein of a particular plant virus is expressed sufficiently in a plant, the plant usually becomes resistant to infection by this pathogen. This principle has already been used several times with success to generate virus-resistant plants by genetic engineering. In the United States, a virus-resistant squash variety generated in this way has been licensed for cultivation. In Germany, this method was utilized to produce **rhizomania-resistant sugar beet** (Planta, Einbeck).

The generation of fungus-resistant plants is still at an early stage

The use of gene technology to generate resistance to fungal infections in plants is still at an early stage. An attempt is being made to utilize the natural protective mechanisms of plants. Some plants protect themselves against fungi by attacking the cell wall of the fungi. The cell walls of most fungi contain chitin, an N-acetyl-D-glucosamine polymer, which does not occur in plants. Some plants contain **chitinases**, which lyse the cell wall of fungi, in their seeds. This protective function has been transferred to other plants. Plant cultivars have been generated by transformation that contain a chitinase gene from beans, thereby gaining an increased resistance against certain fungi. Another strategy lies in the expression of enzymes for the synthesis of fungicide phytoalexins (e.g., stilbenes) (see section 18.4). However, it may still take some time until fungus-resistant plants are ready for cultivation.

Nonselective herbicides can be used as selective herbicide following the generation of herbicide-resistant plants

The importance of herbicides for plant protection has been discussed in section 3.6, and examples of the effects of various herbicides have been dealt with in sections 3.6, 10.4, and 15.3. The most economically successful herbicide is **glyphosate** (trade name Round Up, Monsanto) (Fig. 10.18), which specifically inhibits the synthesis of aromatic amino acids via the shikimate pathway at the EPSP synthase step (Fig. 10.19). Since the shikimate pathway is not present in animals, animal metabolism is not markedly affected by glyphosate. Because of its simple structure, it is relatively rapidly degraded by soil bacteria. Therefore it can be applied only by spraying the leaves. As a nonselective herbicide, glyphosate even destroys many of the very persistent weeds. For example, it is widely used to clear vegetation from railway tracks, to control the weeds on the grounds of fruit and wine plantations, and to kill weeds before crops are planted. In order to apply this powerful herbicide as a selective **post-emergence herbicide**, glyphosate-resistant transformants have been generated for a number of crop plants by means of genetic engineering. To generate glyphosate resistance in plants, genes were isolated from bacteria encoding EPSP synthases, which are less sensitive to glyphosate than the plant enzymes. Transformant plants that express bacterial EPSP synthase activity acquired protection against the herbicide. Glyphosate-resistant cotton, rapeseed, and soybean are now available to the farmer. In a similar way, crop plants have been made resistant to the herbicide glufosinate (trade name Liberty, Aventis) (Figure 10.7) by the expression of bacterial detoxifying enzymes.

Plant genetic engineering is used for the improvement of the yield and quality of crop products

In the application of genetic engineering for generating resistance against pests or herbicides, usually only one additional gene is transferred into the plant. For altering the quality or the yield of harvest products, however, a transfer of several genes is often required and therefore is more difficult to achieve.

A promising way to increase crop yields is the generation of **hybrids** from genetically engineered male-sterile plants, as described in section 20.7. Another strategy for the improvement of crop yield is to alter the partitioning of biomass between the harvestable and nonharvestable organs of the plants. In transgenic potato plants, an improvement of the tuber yield has been observed (section 13.3), but these results have yet to be confirmed by field trials.

Genetic engineering is now being utilized in multiple ways to improve the quality and in particular the health value of food and fodder. Examples of this are the formation of highly unsaturated fatty acids in rapeseed oil (section 15.5), the generation of rice containing provitaminA (golden rice, section 17.6), or the increase of the methionine content in soy beans (section 14.3), as discussed earlier.

Genetic engineering is used to produce raw material for industry and pharmaceuticals

Genetic engineering shows great promise as a method for growing plants that can produce raw material for industry. Transgenic plants, which produce customized fats, with short chain fatty acids for the detergent and cosmetic industries, and with high erucic acid content for the production of synthetic materials, were discussed in section 15.5. Potato transformants, which produce starch consisting only of long-chain α-amylose (section 9.1) would be an interesting supplier of raw material for the production of plastics. Amylose ethers have polymer properties similar to those of polyethylenes, but with the advantage that they are biodegradable (i.e., can be degraded by microorganisms).

Transgenic plants, especially plastidal transformants, are well suited for the production of peptides and proteins, such as human serum albumin or interferon. Progress is being made in the attempts to use plants to produce human monoclonal antibodies (e.g., for curing intestinal cancer). Antibodies for bacteria causing caries have been produced in plants, and it is feasible that they could be added to toothpaste. For such uses, it would be necessary to produce very large amounts of antibody proteins at low costs. Plants would be suitable for this purpose. There also has been success in using transgenic plants for the production of oral vaccines. The fodder plant lucerne was transformed to produce an oral vaccine against foot and mouth disease. Oral vaccines for hepatitis B virus have been produced in potatoes and lupines, and a vaccine for rabies has been produced in tomatoes. Although these experiments are still at an initial phase, they open up the possibility of vaccinating by ingesting plant material (e.g. fodder for animals or fruit for humans).

Genetic engineering provides a chance for increasing the protection of crop plants against environmental stress

In the preceding chapters, various mechanisms have been described by which a plant protects itself against environmental stresses, such as heat (section 21.2), cold (sections 3.10 and 15.1), drought and soil salinity (Chapter 8 and section 10.4), xenobiotic and heavy metal pollution (section 12.2), and oxygen radical production (sections 3.9 and 3.10). Genetic engineering opens the prospect of increasing the resistance of cultivated plants to these stresses by overexpression of enzymes participating in the stress responses. Thus an increase of the number of double bonds in the fatty acids of membrane lipids through genetic engineering has improved the cold tolerance of tobacco (section 15.1). The generation of plants accumulating heavy metals, such as mercury and cadmium, in order to detoxify polluted soils (section 12.2), may have a great future.

With the growth of the world's population, the availability of sufficient arable land becomes an increasing problem. Because of high salt content, often caused by inadequate irrigation management, large areas of the world can no longer be utilized for agriculture. In 1990, 20% of the total area used worldwide for agriculture (including 50% of the artificially irrigated land) has been classified as salt-stressed. Investigations are in progress to determine how to make plants salt-tolerant by increasing the synthesis of osmotically compatible substances, such as mannitol, betaine, or proline (section 10.4), but also by increasing the expression of enzymes, which eliminate reactive oxygen species (**ROS**) (sections 3.9 and 3.10). ROS are a serious damage factor in drought and salt stress. However, it may still take considerable time until these efforts can be put into practice. If plant genetic engineering were to succeed in generating salt-resistant crop plants, this would be a very important contribution in securing the world's food supply.

The introduction of transgenic cultivars requires a risk analysis

At present, plant genetic engineering raises, on the one hand, sometimes exaggerated expectations and, on the other hand, induces fear in parts of the population. Responsible application of plant genetic engineering requires that for each plant licensed for cultivation, a risk analysis be made according to strict scientific criteria as to whether the corresponding plant represents a hazard to the environment or to human health. Among other criteria, it has to be examined whether crossing between the released transformants and wild plants is possible, and what the potential consequences of this are for the environment. For example, crossing can occur between transgenic rapeseed and other *Brassicaceae* such as wild mustard. This could be prevented by using chloroplast transformants (section 22.5). Moreover, transgenic plants themselves could grow in the wild. In this way herbicide-resistant weeds may develop from herbicide-resistant cultivars. However, it also should be noted that in the conventional application of herbicides, herbicide-resistant weeds have evolved, namely, by natural selection (see section 10.4). Experiments in the laboratory as well as controlled field tests are required for such risk analyses. It is beyond the scope of this book to deal with the criteria set by legislation to evaluate the risk entailed by the release of a transgenic cultivar. It is expected, when used responsibly, plant genetic engineering may contribute to the improvement of crop production, an increase of the health benefit of foods, and the provision of sustainable raw materials for industry. If plant protection based on plant genetic engineering were to simply have the result that "less chemicals are put on the field," then this would be an improvement for the environment.

Further reading

Allen, D. J., Ort, D. R. Impacts of chilling temperatures on photosynthesis in warm-climate-plants. Trends Plant Sci 6, 36–42 (2001).

Brandt, P. Zukunft der Gentechnik. Birkhäuser-Verlag, Basel (1997).

Chrispeels, M. J., Sadava, D. E. Plant Genes and Agriculture. Jones and Bartlett, Boston, London (1994).

Chilton, M.-D. Agrobacterium. A memoir. Plant Physiol 125, 9–14 (2001).

Daniell, H., Streatfield, S. J., Wycoff, K. Medical molecular farming: Production of antibodies, biopharmaceuticals and edible vaccines in plants. Trends Plant Sci 5, 219–226 (2001).

Daniell, H., Khan, M. S., Allison, L. Milestones in chloroplast genetic engineering: An environmentally friendly era in biotechnology. Trends Plant Sci 7, 84–91 (2002).

De Block, M., Herrera-Estrella, L., Van Montagu, M., Schell, J. Expression of foreign genes in regenerated plants and their progeny. EMBO J 3, 1681–1689 (1984).

Dale, M. F. B., Bradshaw, J. E. Progress in improving processing attributes in potato. Trends Plant Sci 8, 310–312 (2003).

Groot, A. T., Dicke, M. Insect-resistant transgenic plants in a multi-trophic context. Plant J 31, 387–406 (2002).

Grusak, M. A. Improving the nutrient composition of plants to enhance human nutrition and health. Annu Rev Plant Physiol Plant Mol Biol 50, 133–136 (1999).

Hansen, G., Wright, M. S. Recent advances in the transformation of plants. Trends Plant Sci 4, 226–231 (1999).

Hellens, R., Mullineaux, P., Klee, H. A guide to *Agrobacterium* binary Ti vectors. Trends Plant Sci 5, 446–451 (2000).

Iba, K. Acclimative response to temperature stress in higher plants: Approaches of gene engineering for temperature tolerance Annu Rev Plant Biol 53, 225–246 (2002).

Horsch, R. B., Fraley, R. T., Rogers, S. G., Sanders, P. R., Lloyd, A., Hoffman, N. Inheritance of functional foreign genes in plants. Science 223, 496–498 (1984).

Käppeli, O., Auberson, L. How safe enough is plant genetic engineering? Trends Plant Sci 3, 276–281 (1998).

Miflin, B. J. Crop biotechnology, where now? Plant Physiol 123, 17–27 (2000).

Moffat, A. S. Exploring transgenic plants as a new vaccine source. Science 268, 658–660 (1995).

Nap. J.-P., Metz, P. L. J., Escaler, M., Conner, A. J. The releaseof genetically modified crops into the environment Plant J 33, 1–18 (2003).

Pilon-Smits, E., Pilon M. Breeding mercury-breathing plants for environmental cleanup. Trends Plant Sci 5, 235–236 (2000).

Plant biotechnology: Food, feed. Special Issue. Science 285, 289–484 (1999).

Poppy, G. GM crops: Environmental risks and non-target effects. Trends Plant Sci 5, 4–6 (2000).

Pray, C. E., Huang, J., Hu, R., Rozelle, S. Five years of Bt cotton in China—the benefits continue. Plant J 31, 423–430 (2002).

Sarhan, F., Danyluk, J. Engineering cold-tolerant crops—throwing the master switch. Trends Plant Sci 3, 289–290 (1998).

Sommerville, C. R., Bonetta, D. Plants as factories for technical materials. Plant Physiol 125, 168–171 (2001).

Straughan R. Moral and ethical issues in plant biotechnology. Curr Opin Plant Biol 3, 163–165 (2000).

The Royal Society. An open review of the science relevant to GM crops and food based on interests and concerns of the public. The Royal Society, London (2004).

Thomashow M. E. So what's new in the field of plant cold acclimation? Lots! Plant Physiol 125, 89–93 (2001).

Tinland, B. The integration of T-DNA into plant genomes. Trends Plant Sci 1, 178–184 (1996).

Tzfira, T., Citowsky, V. The agrobacterium-plant cell interaction. Taking biology lessons from a bug. Plant Physiol 133, 943–947 (2003).

Ward, D. V., Zupan, J. R., Zambryski, P. C. *Agrobacterium* VirE2 gets the VIP1 treatment in plant nuclear import. Trends Plant Sci 7, 1–3 (2002).

Wilkinson, M. J., Sweet, J., Poppy, G. M. Risk assessment of GM plants: Avoiding gridlock? Trends Plant Sci 8, 208–212 (2003).

Willmitzer, L. Transgenic plants. In: H. J. Rehm, G. Reed, A. Pühler, P. Stadler (eds.): Biotechnology, A Multi-Volume Comprehensive Treatise. Band 2, 627–659 (1993). Verlag Chemie, Weinheim.

Ye, X., Al-Babili, S., Klöti, A., Zhang, J., Lucca, P., Beyer, P., Potrykus, I. Engineering the provitamin A (β-carotene) biosynthetic pathway into (carotenoid-free) rice endosperm. Science 287, 303–305 (2000).

Zhu, J.-K. Plant salt tolerance. Trends Plant Sci 6, 66–71 (2001).

Index

Note: Figures are indicated by *f* following the page on which they appear; tables are indicated by *t*.

C

O

Q

R